Fascinating Life Sciences

This interdisciplinary series brings together the most essential and captivating topics in the life sciences. They range from the plant sciences to zoology, from the microbiome to macrobiome, and from basic biology to biotechnology. The series not only highlights fascinating research; it also discusses major challenges associated with the life sciences and related disciplines and outlines future research directions. Individual volumes provide in-depth information, are richly illustrated with photographs, illustrations, and maps, and feature suggestions for further reading or glossaries where appropriate.

Interested researchers in all areas of the life sciences, as well as biology enthusiasts, will find the series' interdisciplinary focus and highly readable volumes especially appealing.

More information about this series at https://link.springer.com/bookseries/15408

Robert J. Marquis • Suzanne Koptur
Editors

Caterpillars in the Middle

Tritrophic Interactions in a Changing World

 Springer

Editors
Robert J. Marquis
Department of Biology
Whitney R. Harris World Ecology Center
University of Missouri–St. Louis
St Louis, MO, USA

Suzanne Koptur
Department of Biological Sciences
International Center for Tropical Botany
Florida International University
Miami, FL, USA

ISSN 2509-6745   ISSN 2509-6753 (electronic)
Fascinating Life Sciences
ISBN 978-3-030-86687-7   ISBN 978-3-030-86688-4 (eBook)
https://doi.org/10.1007/978-3-030-86688-4

© The Editor(s) (if applicable) and The Author(s), under exclusive license to Springer Nature Switzerland AG 2022, Corrected Publication 2022
Chapter "The Natural History of Caterpillar-Ant Associations" is licensed under the terms of the Creative Commons Attribution 4.0 International License (http://creativecommons.org/licenses/by/4.0/). For further details see license information in the chapter.
This work is subject to copyright. All rights are solely and exclusively licensed by the Publisher, whether the whole or part of the material is concerned, specifically the rights of translation, reprinting, reuse of illustrations, recitation, broadcasting, reproduction on microfilms or in any other physical way, and transmission or information storage and retrieval, electronic adaptation, computer software, or by similar or dissimilar methodology now known or hereafter developed.
The use of general descriptive names, registered names, trademarks, service marks, etc. in this publication does not imply, even in the absence of a specific statement, that such names are exempt from the relevant protective laws and regulations and therefore free for general use.
The publisher, the authors and the editors are safe to assume that the advice and information in this book are believed to be true and accurate at the date of publication. Neither the publisher nor the authors or the editors give a warranty, expressed or implied, with respect to the material contained herein or for any errors or omissions that may have been made. The publisher remains neutral with regard to jurisdictional claims in published maps and institutional affiliations.

Cover image: We thank Sam Jaffe of the Caterpillar Lab (jaffe@thecaterpillarlab.org) for his cover illustrations depicting the biology of *Synchlora aerata* (Geometridae), which cuts plant parts and attaches them with silk to its dorsal surface. Five *Synchlora* caterpillars, variously adorned with petals and leaf fragments, are shown across the first two rows. A fresh brown pupa, a developed and ready to eclose green pupa, and an adult *Synchlora* moth are pictured to the bottom left. In the third row, two unique parasitoid wasps that are closely associated with *Synchlora* appear: a specialist camplopegine ichneumonid, and a chalcid hyperparasitoid. The campoplegine's cocoon, constructed from its host caterpillar's cadaver, is above the wasps, at the end of the second row. Finally, to the bottom right a *Synchlora* caterpillar is shown actively feeding on a daisy fleabane blossom after decorating itself with the available petals.

This Springer imprint is published by the registered company Springer Nature Switzerland AG
The registered company address is: Gewerbestrasse 11, 6330 Cham, Switzerland

To our families
Carol, Emma, and Zoe
John and JJ
We love you even more than caterpillars

Foreword

The caterpillar is a poster child for insect holometaboly—a developmental style that involves a change from a sedentary feeding stage to a resting pupa in which massive internal changes occur, finally leading to the emergence of an adult. An important part of the story is that females must make the momentous decisions on where to place their eggs.

In a caterpillar, growth rate is maximized at the cost of food processing efficiency, and growth is largely independent from the tissue differentiation required to generate organs for dispersal and reproduction. Its function is to eat, its gut is the principal organ, and its stretchable cuticular protein is efficiently reused at each molt. Although the caterpillar life stage emphasizes feeding at the expense of finesse at fleeing from peril or ability to find scarce food, these predicaments are clearly offset in the evolution of the holometabolous insect life history. Holometabolous development is relatively more abundant than the ancestral hemimetabolous pattern.

But risks exist, and the study of caterpillars inevitably involves research on the multitude of inimitable traits that enhance survival. Caterpillars face threats from above from numerous species of natural enemies. The dangers are almost infinite, and their selection pressure has influenced every aspect of caterpillar biology. Lepidopteran larvae masquerade as any possible non-caterpillar object in the environment, and in many cases, their defenses against enemies depend on adaptations to specific host plants. Diverse behavioral traits provide endless ways to deceive parasites and predators, such as dropping down from the host plant on silk, flinging frass away from give-away feeding sites, or hiding in homemade houses. They must feed fast to avoid being noticed and consequently demonstrate mandibular adaptations uniquely suited to the particular food. Dietary specialists may become masters of chemical protection by sequestering chemicals that are deterrent or toxic to predators and parasites. Other specialist species excel at host-specific visual or chemical crypsis, making them harder to detect by their enemies. Generalists have found diverse mixtures of traits to contend with enemies—opportunistic sequestration of plant poisons, nimble behaviors, mimicking predators, and even making their own noxious chemicals. Different species have particular habits of food choice that enhance protection from particular groups of natural enemies that dominate in their microhabitats.

But caterpillars are "in the middle" in a changing tritrophic world. There are also hazards from below—the physical and chemical plant defenses. First are the jeopardies associated with physical structures that plants produce to make it difficult to maintain a purchase, or prevent easy walking, or make biting and chewing too challenging. There are secretions that are sticky or gluey, such as latex or resin, which can glue up their mouthparts. A remarkable evolutionary feat of the Lepidoptera has been the adoption of countermeasures to deal with each of the defenses plants have come up with.

The all-important plant chemistry has been studied in great depth, with the discovery of innumerable compounds that interfere with caterpillar growth and survival. These must be avoided or overcome physiologically one way or another. In addition, the details of sequestration of noxious plant chemicals have been elucidated in numerous species, different chemical structures being of particular value against different groups of natural enemies.

Plants present a food source that is typically low in protein and variable in nutrient quality. Plants also provide diverse physical challenges and an immense variety of potentially toxic metabolites that often increase with damage. Environmental changes involving climate and chemicals from human activity challenge not only the plants and their availability, but also the caterpillars that feed on them, and the mothers who must locate their foods.

Unlike many herbivores, caterpillars have the task of ingesting such vast amounts that the gut content may be up to twice the weight of its own tissues. An alimentary throughput time of just hours means there is virtually no role for symbiotic microorganisms to aid digestion. And the frenzy of feeding involves big doses of potential toxins. The first parts of this book address some of these problems and the remarkable ways that caterpillars have dealt with them.

Finally, there are gambles associated with availability of hosts; variation in quality of the host, genetically or as a result of the herbivore damage; the sensing of nearby herbivore damage; or the acquisition of a plant disease.

Presented here are updates about impact of the natural enemies and diseases on the ecology and evolution of caterpillars and the multiple interactive effects among the three trophic levels—plants, herbivores, and herbivore enemies. The exceptional diversity of anatomical, physiological, and behavioral traits to deal with the number and variety of problems provides a classic arena for the study of evolution.

Research on caterpillar behavior and ecology continues to be important in natural ecosystems for understanding the impact of climate change and loss of habitat on all three trophic levels, and the complexity of all those tangled webs. How will herbivorous Lepidoptera be affected by changing distribution of hostplants in relation to changes in climate? Can citizen scientists provide the quantity of information to monitor changes in each of the three trophic levels around the world?

Acknowledgments

We thank the following persons for reviewing chapters. These reviewers made substantial contributions to the quality of the material contained within this book through their careful reading and resulting constructive criticism and suggestions: Annette Aiello, Inge Armbrecht, Christina Baer, Alex Baranowski, Karina Boege, Deane Bowers, Laura Braga, Sue Carnahan, Phyllis Coley, Rex Cocroft, Tatiana Cornelissen, Jaret Daniels, Ek del-Val, David Dussourd, Lee Dyer, Marianne Espeland, James Fordyce, Jessica Forrest, Andres Freitas, Rieta Gols, David Heckel, Owen Lewis, John Lill, Eric LoPresti, Tanner Matson, Kailen Mooney, Xoaquin Moreira, Douglas Morse, Paul Ode, Paulo Oliveira, Steve Passoa, Moria Robinson, Jay Rosenheim, Carlo Seifert, Tim Schowalter, Mike Singer, Angela Smilanich, John Stireman, Doug Tallamy, Heiko Vogel, Martin Volf, David Webb, Martha Weiss, James Whitfield, and Meron Zalucki.

In addition, we would like to express our appreciation to the Entomological Society of America for the initial opportunity to bring many of the authors to St. Louis in November 2019 to contribute to the symposium that led to the initial proposal for this book. Finally, we appreciate the enthusiasm and support from our colleagues, employees, and students at Florida International University and the University of Missouri-St. Louis during the completion of this volume.

Contents

Part I   Introduction

Introduction – Caterpillars as Focal Study Organisms............... 3
Suzanne Koptur and Robert J. Marquis

On Being a Caterpillar: Structure, Function, Ecology, and Behavior.... 11
David L. Wagner and Alexela C. Hoyt

Part II   Impacts of the First Trophic Level on Caterpillar Ecology and Evolution

Surface Warfare: Plant Structural Defenses Challenge Caterpillar Feeding... 65
Ishveen Kaur, Sakshi Watts, Cristina Raya, Juan Raya, and Rupesh Kariyat

Impacts of Plant Defenses on Host Choice by Lepidoptera in Neotropical Rainforests 93
María-José Endara, Dale Forrister, James Nicholls, Graham N. Stone, Thomas Kursar, and Phyllis Coley

Ecology and Evolution of Secondary Compound Detoxification Systems in Caterpillars 115
Simon C. Groen and Noah K. Whiteman

Sequestered Caterpillar Chemical Defenses: From "Disgusting Morsels" to Model Systems....................... 165
M. Deane Bowers

Part III Impacts of the Third Trophic Level on Caterpillar Ecology and Evolution

Acoustic Defence Strategies in Caterpillars 195
Jayne E. Yack

Natural History and Ecology of Caterpillar Parasitoids 225
John O. Stireman III and Scott R. Shaw

**Predators and Caterpillar Diet Breadth:
Appraising the Enemy-Free Space Hypothesis** 273
Michael S. Singer, Riley M. Anderson, Andrew B. Hennessy,
Emily Leggat, Aditi Prasad, Sydnie Rathe, Benjamin Silverstone,
and Tyler J. Wyatt

Caterpillar Responses to Ant Protectors of Plants 297
Suzanne Koptur, Jaeson Clayborn, Brittany Harris, Ian Jones,
Maria Cleopatra Pimienta, Andrea Salas Primoli, and Paulo S. Oliveira

The Natural History of Caterpillar-Ant Associations 319
Naomi E. Pierce and Even Dankowicz

Part IV Multiple Interactive Effects Among All Three Trophic Levels

**Caterpillars, Plant Chemistry, and Parasitoids
in Natural vs. Agroecosystems** 395
Paul J. Ode

Host Plants as Mediators of Caterpillar-Natural Enemy Interactions ... 425
John T. Lill and Martha R. Weiss

Host Plant Effects on the Caterpillar Immune Response 449
Angela M. Smilanich and Nadya D. Muchoney

**Trophic Interactions of Caterpillars in the Seasonal Environment
of the Brazilian Cerrado and Their Importance in the Face
of Climate Change** 485
Laura Braga and Ivone R. Diniz

**The Impact of Construct Building by Caterpillars on Arthropod
Colonists in a World of Climate Change** 509
Robert J. Marquis, Christina S. Baer, John T. Lill, and H. George Wang

Part V Caterpillar Foodwebs in a World of Rapidly Changing Climate

**Caterpillar Patterns in Space and Time: Insights From and Contrasts
Between Two Citizen Science Datasets** 541
Grace J. Di Cecco and Allen H. Hurlbert

Impacts of Climatic Variability and Hurricanes on Caterpillar Diet Breadth and Plant-Herbivore Interaction Networks............. 557
Karina Boege, Ivonne P. Delgado, Jazmin Zetina, and Ek del-Val

Plant-Caterpillar-Parasitoid Natural History Studies Over Decades and Across Large Geographic Gradients Provide Insight Into Specialization, Interaction Diversity, and Global Change......... 583
Danielle M. Salcido, Chanchanok Sudta, and Lee A. Dyer

Part VI   Synthesis

Synopsis and the Future of Caterpillar Research.................... 609
Robert J. Marquis and Suzanne Koptur

Correction to: The Natural History of Caterpillar-Ant Associations...... C1

Subject Index... 623

Taxonomic Index.. 631

List of Contributors

Riley M. Anderson Department of Biology, Wesleyan University, Middletown, CT, USA

Christina S. Baer First-year Research Immersion Program, Binghamton University, Binghamton, NY, USA

Karina Boege Departamento de Ecología Evolutiva, Instituto de Ecología, Universidad Nacional Autónoma de México, Ciudad de México, Mexico

M. Deane Bowers Department of Ecology and Evolutionary Biology and Museum of Natural History, University of Colorado, Boulder, CO, USA

Laura Braga Departamento de Zoologia, Instituto de Ciências Biológicas, Universidade de Brasília, Brasília, Brazil

Grace J. Di Cecco Department of Biology, University of North Carolina, Chapel Hill, NC, USA

Jaeson Clayborn Department of Biological Sciences, International Center for Tropical Botany, Florida International University, Miami, FL, USA

Miami-Dade College, Padrón Campus, Miami, FL, USA

Phyllis Coley School of Biological Sciences, University of Utah, Salt Lake City, UT, USA

Even Dankowicz Harvard University, Museum of Comparative Zoology, Cambridge, MA, USA

Ivonne P. Delgado Instituto de Investigaciones en Ecosistemas y Sustentabilidad, Universidad Nacional Autónoma de México, Morelia, Michoacán, Mexico

Ek del-Val Instituto de Investigaciones en Ecosistemas y Sustentabilidad, Universidad Nacional Autónoma de México, Morelia, Michoacán, Mexico

Ivone R. Diniz Departamento de Zoologia, Instituto de Ciências Biológicas, Universidade de Brasília, Brasília, Brazil

Lee A. Dyer Ecology, Evolution, Conservation Biology, University of Nevada, Reno, NV, USA

María-José Endara Grupo de Investigación en Biodiversidad, Medio Ambiente y Salud-BIOMAS, Universidad de las Américas, Quito, Ecuador

School of Biological Sciences, University of Utah, Salt Lake City, UT, USA

Dale Forrister School of Biological Sciences, University of Utah, Salt Lake City, UT, USA

Simon C. Groen Department of Nematology, University of California, Riverside, CA, USA

Brittany Harris Department of Biological Sciences, International Center for Tropical Botany, Florida International University, Miami, FL, USA

Andrew B. Hennessy Department of Biology, Wesleyan University, Middletown, CT, USA

Alexela C. Hoyt Department of Ecology and Evolutionary Biology, University of Connecticut, Storrs, CT, USA

Allen H. Hurlbert Department of Biology, University of North Carolina, Chapel Hill, NC, USA

Environment, Ecology, and Energy Program, University of North Carolina, Chapel Hill, NC, USA

Ian Jones Department of Biological Sciences, International Center for Tropical Botany, Florida International University, Miami, FL, USA

Faculty of Forestry, University of Toronto, Toronto, ON, Canada

Rupesh Kariyat School of Earth Environmental and Marine Sciences, University of Texas Rio Grande Valley, Edinburg, TX, USA

Department of Biology, University of Texas Rio Grande Valley, Edinburg, TX, USA

Ishveen Kaur School of Earth Environmental and Marine Sciences, University of Texas Rio Grande Valley, Edinburg, TX, USA

Suzanne Koptur Department of Biological Sciences, International Center for Tropical Botany, Florida International University, Miami, FL, USA

Thomas Kursar School of Biological Sciences, University of Utah, Salt Lake City, UT, USA

Emily Leggat Department of Biology, Wesleyan University, Middletown, CT, USA

John T. Lill Department of Biological Sciences, George Washington University, Washington, DC, USA

Robert J. Marquis Department of Biology, Whitney R. Harris World Ecology Center, University of Missouri–St. Louis, St. Louis, MO, USA

Nadya D. Muchoney Program in Ecology, Evolution, and Conservation Biology, University of Nevada, Reno, NV, USA

Department of Biology, University of Nevada, Reno, NV, USA

James Nicholls Institute of Evolutionary Biology, University of Edinburgh, Edinburgh, UK

CSIRO, Australian National Insect Collection, Black Mountain Labs, Acton, ACT, Australia

Paul J. Ode Department of Agricultural Biology, Graduate Degree Program in Ecology, Colorado State University, Fort Collins, CO, USA

Paulo S. Oliveira Departamento de Biologia Animal, Universidade Estadual de Campinas, UNICAMP, Campinas, Brazil

Naomi E. Pierce Harvard University, Museum of Comparative Zoology, Cambridge, MA, USA

Maria Cleopatra Pimienta Department of Biological Sciences, International Center for Tropical Botany, Florida International University, Miami, FL, USA

Aditi Prasad Department of Biology, Wesleyan University, Middletown, CT, USA

Andrea Salas Primoli Department of Biological Sciences, International Center for Tropical Botany, Florida International University, Miami, FL, USA

Sydnie Rathe Department of Biology, Wesleyan University, Middletown, CT, USA

Cristina Raya Department of Biology, University of Texas Rio Grande Valley, Edinburg, TX, USA

Juan Raya Department of Biology, University of Texas Rio Grande Valley, Edinburg, TX, USA

Danielle M. Salcido Ecology, Evolution, Conservation Biology, University of Nevada, Reno, NV, USA

Scott R. Shaw UW Insect Museum, Department of Ecosystem Science and Management, University of Wyoming, Laramie, WY, USA

Benjamin Silverstone Department of Biology, Wesleyan University, Middletown, CT, USA

Michael S. Singer Department of Biology, Wesleyan University, Middletown, CT, USA

Angela M. Smilanich Program in Ecology, Evolution, and Conservation Biology, University of Nevada, Reno, NV, USA

Department of Biology, University of Nevada, Reno, NV, USA

John O. Stireman III Department of Biological Sciences, Wright State University, Dayton, OH, USA

Graham N. Stone Institute of Evolutionary Biology, University of Edinburgh, Edinburgh, UK

Chanchanok Sudta Ecology, Evolution, Conservation Biology, University of Nevada, Reno, NV, USA

David L. Wagner Department of Ecology and Evolutionary Biology, University of Connecticut, Storrs, CT, USA

H. George Wang Department of Biological and Environmental Sciences, East Central University, Ada, OK, USA

Sakshi Watts Department of Biology, University of Texas Rio Grande Valley, Edinburg, TX, USA

Martha R. Weiss Georgetown University, Washington, DC, USA

Noah K. Whiteman Department of Integrative Biology, University of California, Berkeley, CA, USA

Tyler J. Wyatt Department of Biology, Wesleyan University, Middletown, CT, USA

Jayne E. Yack Department of Biology, Nesbitt Biology Building, Carleton University, Ottawa, ON, Canada

Jazmin Zetina Departamento de Ecología Evolutiva, Instituto de Ecología, Universidad Nacional Autónoma de México, Ciudad de México, Mexico

Part I
Introduction

Introduction – Caterpillars as Focal Study Organisms

Suzanne Koptur and Robert J. Marquis

> *My cousin has great changes coming*
> *Someday he'll wake with wings...*
> *(Cousin Caterpillar (Mike Heron),*
> *The Incredible String Band: The Big Huge, 1968)*

Caterpillars are truly in the middle, ecologically speaking. They are both major consumers of plants and critical food for predators. Plants take the sun's energy and atmospheric carbon dioxide and use it to make sugars and oxygen, so they serve as the primary producers in most ecosystems (Jensen and Salisbury 1972). Herbivores (Crawley 1983), including caterpillars, eat plants, obtaining energy and building blocks for their bodies from the organic compounds composed of carbon, nitrogen, and other important molecules and essential elements from plant bodies; herbivores are therefore termed primary consumers. As the famously hungry (Carle 1969) larval stage of butterflies and moths, caterpillars make plant energy and nutrients available to predators by concentrating the essentials in their tissues as they feed and grow. But it is the consumed caterpillars that mediate the transfer of untold amounts of energy and nutrients from plants to carnivores. Many species of predators and parasitoids, termed secondary consumers, attack and eat caterpillars. The populations of these natural enemies depend on an abundance of caterpillars. Caterpillars are therefore at the center of food webs in terrestrial ecosystems, powering their life-supporting properties around the world.

S. Koptur (✉)
Department of Biological Sciences, International Center for Tropical Botany, Florida International University, Miami, FL, USA
e-mail: kopturs@fiu.edu

R. J. Marquis
Department of Biology, Whitney R. Harris World Ecology Center, University of Missouri–St. Louis, St. Louis, MO, USA

© The Author(s), under exclusive license to Springer Nature Switzerland AG 2022
R. J. Marquis, S. Koptur (eds.), *Caterpillars in the Middle*, Fascinating Life Sciences, https://doi.org/10.1007/978-3-030-86688-4_1

Evolution rewards the caterpillars that survive to become adult moths and butterflies and ultimately reproduce. Accordingly, caterpillars have evolved many fascinating behavioral and physiological traits to feed on plants despite the fluctuating availability of host plant material and the many defenses (chemical, physical, nutritional, phenological, and biotic) that plants have evolved in response to their herbivores. This coevolution between caterpillars and plants has been going on for millions of years. The diverse traits of plants are the bottom-up forces affecting caterpillar populations. At the same time, caterpillars have evolved myriad defenses in response to predation pressure from their many natural enemies, top-down forces that control caterpillar populations. Sandwiched as they are between the bottom-up and top-down forces that affect their survival, caterpillars have collectively evolved a staggering diversity of traits. Their lifestyle is a compromise wrought by the selective forces represented by the first and third trophic levels. The wide diversity in caterpillar form and behavior, which contributes in great part to their attraction for study, is the result of these evolutionary forces. This diversity is showcased in the 150,000 species of Lepidoptera known to science, with more still to be described. We are only beginning to catalogue this diversity, let alone understand the evolutionary forces driving the diversification of caterpillars. Within the pages of this volume, the reader will find our collective current understanding of caterpillars as components of food webs.

In this introduction we consider the discovery of connections between caterpillars and adult forms, the documentation of their natural history, the study of their development and physiology, and the significance of holometaboly. We will briefly recount and pay homage to the large body of work that laid the foundation for our understanding of the ecology and evolution of plant/insect/natural enemy interactions that have given rise to the contributions in this collection, *Caterpillars in the Middle*. In inviting the group of researchers represented in the pages of this book, we have sought to include insights from various levels of study, with the intention of providing a well-rounded look at advances in caterpillar biology, ecology, and evolution. Biologists may study organisms at any position of the food chain/web and often deal with the effects of one level on another; some even put all levels together and consider the effects of abiotic factors on their organisms of interest. In the chapters of this volume, you will find all these approaches, and we intend that this work will serve to inspire more research on caterpillars in all directions.

Who first recognized that caterpillars are immature stages of butterflies and moths? It is likely that early humans first noticed the beauty of flying adults, watching their movements with awe as the Lepidoptera visited flowers and landed on vegetation, some with their colors as beautiful and bright as flowers, others flying only at night and drawn to fire and lights as were the people who made those things. And while some ate caterpillars as an important source of protein (e.g., mopane worm, the larva of the emperor moth *Imbrasia belina*) (Baiyegunhi et al. 2016; Stack et al. 2003), did they connect the two life stages as parts of a single organism?

This realization may have taken place before written history, but the earliest recorded considerations of the phenomenon appear in the (384–322 BCE) writings of Aristotle (1942 translation; Ryan 2011). He described the caterpillar as a continuation of embryonic life, "a soft egg" that preceded the ultimate goal of adult butterfly. Aristotle was mistaken, however, that metamorphosis transpired despite lack of fertilization, seeing it as a process involved in the spontaneous generation of life (Reynolds 2019). He proposed that the eggs of holometabolous insects hatched "before their time," thus necessitating the extra stages of development outside the egg that preceded the perfect adult organism. Aristotle saw sperm as the agent that transformed the egg to another state, just as heat can cook an egg or curdle milk. These changes were necessary for the organism to achieve its "perfect" form, meaning the stage that could be used to determine its species or identity, the adult form. In the seventeenth century, scientists pursued their endeavors in the light of Aristotle's influence, and though his ideas have fallen out of favor in current understanding of the phenomena of fertilization and development, it is useful to consider the precedent to today's knowledge (Reynolds 2019).

We can find insight into the natural history of Lepidoptera in some of the artwork of early observers of nature. Maria Sibylla Merian's detailed illustrations of European insects and plants (Fig. 1) provided some of the earliest documentation of the life cycles of numerous Lepidoptera. In the seventeenth century, when women were not formally trained or educated and unusual interests led to suspicion and even accusations of witchcraft, this young German girl loved to draw insects (Sidman 2018). Merian, a craftsperson and the daughter of a tradesman (Todd 2007), lived from 1647 to 1717 and described and depicted what she observed around her, the first important step of scientific endeavor. She has been called one of

Fig. 1 Realistic depiction of moth life history including hostplant juxtaposed with high quality digital photographs of the twenty-first century. Left image – Maria Sibylla Merian's hawkmoth on morning glory – note all life history stages, even the shape of the frass produced by the large caterpillar. Upper right – *Agrius convolvuli* (L.) female dorsal view, photo by Didier Descouens (CC BY-SA 4.0); Lower right – *Agrius convolvuli korseby* caterpillar, photo by Kristian Peters (GNU Free Document License)

the first ecologists (Etheridge 2011), as unlike other artists of that time, she sought to show in her work the plants with which the insects were associated, putting the subjects in the context of their natural world. Using direct observation both in the wild and rearing insects in her home, focusing on the interactions between plants and animals, her work was foundational for modern-day ecology, the field of many authors in this book. In her quest to describe in her art the metamorphosis of butterflies and moths (as well as other animals), she raised many species from eggs and larvae, carefully recording all the stages of their life cycles. While people had long recognized that silk moths produced eggs that hatched into caterpillars that eventually made silken cocoons from which silk thread were obtained, not many had wondered about the origins of other beautiful moths and butterflies of all colors, sizes, and shapes. In her lifetime she illuminated the life cycles of European Lepidoptera and other insects in two major works of several volumes each (*Neues Blumenbuch*, New Book of Flowers, Merian 1675-1680; and *Der Raupen wunderbare Verwandelung, und sonderbare Blumen-nahrung*, The Wondrous Transformation of Caterpillars and Their Particular Nourishment from Flowers, Merian 1679, 1683, and 1718). Before Darwin, Humboldt, and Audubon, Merian traveled at the turn of the eighteenth century (1699) with her daughter on a voyage of discovery to Surinam to see and learn about neotropical Lepidoptera, after which she wrote *Metamorphosis insectorum Surinamensium*, The metamorphosis of the insects of Surinam, 1705, Amsterdam: G. Valck. Her work inspired many subsequent naturalists and artists, including Mark Catesby who pioneered depicting birds in their natural settings (Etheridge and Pieters 2015), the "Colonial Audubon" (Frick and Stearns 1961).

Scientific investigation continued with observation and illustration to investigations of physiology and development. In the mid-twentieth century, Wigglesworth (1934) discovered that hormones control transformation from larva to pupa to adult. Through his continued work (Wigglesworth 1954) and that of many others, we now know that transformation in all insects is regulated by the interplay of two hormones: ecdysone and juvenile hormone (Rolff et al. 2019). Some insects are hemimetabolous, developing through larval stages very similar in morphology to adults (e.g., Orthoptera) only lacking wings; others are holometabolous, with larvae of entirely different morphology than the adults (e.g., Diptera, Hymenoptera, and Lepidoptera). Why do these extreme changes in form exist within the life history of a single organism? Darwin postulated, in *The Origin of the Species by Means of Natural Selection* (1866), that the significance of holometaboly was that the different stages of development had different lifestyles and occupied different ecological niches; this idea was reiterated by Williams (1952) and reviewed by Wilbur (1980). Perhaps the most satisfying explanation is that complete metamorphosis is an adaptation permitting the decoupling of growth and differentiation (Rolff et al. 2019). This can be advantageous when food is sporadically available. Rapid growth is advantageous as the larva is a vulnerable stage, reducing the period of time it may be killed by predators, pathogens, and parasitoids. Plants can thwart herbivorous larvae through a variety of defenses: biotic, chemical, mechanical, and phenological. *It is this position of caterpillars in the middle of the top-down and bottom-up forces that provides the content of this volume.*

Many young ecologists of a generation ago were inspired by the influential volume *Coevolution of Animals and Plants* (Gilbert and Raven 1975) and by other volumes in which experts on different topics contributed chapters (Chapman and Bernays 1978; Futuyma and Slatkin 1983; Price et al. 1991). There have been many important books published in the last few decades about multitrophic interactions, many with a focus on arthropod/plant interactions. *Insects on Plants* (Strong et al. 1984) brought these interactions into more prominence in community ecology, and *Herbivory* (Crawley 1983) examined animal effects on plants at many levels. *Trophic Cascades* (Terborgh and Estes 2010) emphasized tri-trophic interactions involving vertebrate herbivores and vertebrate predators. An edited volume (Barbosa and Letourneau 1988) focused on mediation of complex interactions by plant allelochemicals, including their effects on higher trophic levels, while Rosenthal and Berenbaum (1991) focused on the effects of plant chemicals on herbivores. The multi-volume series published by CRC press, *Insect-Plant Interactions* (1979–1984), was made up of five volumes of contributed chapters, edited by Elizabeth Bernays. Two other edited volumes brought together the work of scientists examining plant-insect-enemy interactions, including *Multitrophic Interactions in Terrestrial Systems* (Gange and Brown 1997) and *Herbivores: Between Plants and Predators* (Olff et al. 1999). A volume entitled *Multitrophic Level Interactions* (Tscharntke and Hawkins 2002) included bottom-up and top-down effects in both above-ground and below-ground food webs. Some scholarly works have focused more on the first trophic level, plants (Fritz and Simms 1992); some on the second trophic level, herbivores (Tilmon 2008); and others on the third trophic level, predators and parasitoids (Hajek 2004; Hawkins 1994; Rico-Gray and Oliveira 2007; Wäckers et al. 2005). Still more were overviews of numerous kinds of interactions and how they have coevolved (Abrahamson 1989; Thompson 1982, 1994, 1997, 2005; Jolivet 1998; Schoonhoven et al. 2005; Herrera and Pellmyr 2002), some with a particular focus on global change (Kareiva et al. 1993; Post 2013; Tylianakis et al. 2008).

While caterpillars as herbivores were included in most of the abovementioned works and other books have focused on either identification (e.g., Wagner 2005) or the biology of particular groups (Tuskes et al. 1996; Tuttle 2007; Conner 2009), it was nearly 30 years ago when the first and only book on the ecology and evolution of caterpillars was published. That volume, edited by Nancy Stamp and Tim Casey (1993), became a treasured classic, bringing together the perspectives of a diverse and international group of researchers. Although the title suggested the focus was only caterpillar foraging ecology, its coverage included effects of abiotic and biotic forces on caterpillars, examination of interesting lifestyles, and how caterpillar feeding and associations varied in space and time. It is time for a new look at the ecological and evolutionary forces that affect larval Lepidoptera and to consider the effects of a changing planet on their continued existence. We are fortunate to have some contributors to the earlier caterpillar book (Stamp and Casey 1993) authoring chapters in this book. We have invited contributions from scientists whose interests and expertise range widely, from simple natural history to experiments and analyses that provide some insights that were not possible at an earlier time.

The idea for this book came from organizing a symposium for the Entomological Society of America meetings held in St. Louis in the Fall of 2019, and most of the participants in that symposium have contributed to this book. We were glad that other renowned scientists also agreed to contribute to this new compendium of research on caterpillars, their hostplants, and natural enemies, reflecting a variety of approaches and expertise. As this work is written in the early part of the twenty-first century, a time when we are well aware that human activity on the earth has changed the climate of our home planet, we attempted to include consideration of these forces in every contribution. We now present this book entitled *Caterpillars in the Middle: Tritrophic Interactions in a Changing World*, hoping it has something for everyone and may serve to inspire future research on and appreciation of caterpillars.

Acknowledgements We thank many of the contributors to this volume for their constructive comments on this introduction.

References

Abrahamson WG (1989) Plant-animal interactions. McGraw-Hill
Aristotle (1942 translation) Generation of animals. (Transl. by L. Peck; Loeb Classical Library, no. 366)
Baiyegunhi LJS, Oppong BB, Senyolo GM (2016) Mopane worm (*Imbrasia belina*) and rural household food security in Limpopo province, South Africa. Food Secur 8:153–165
Barbosa P, Letourneau DK (1988) Novel aspects of insect-plant interactions. Wiley
Bernays EA (ed) (1979–1984) Insect-plant relationships. Volumes I–V, CRC Press.
Carle E (1969) The very hungry caterpillar. The World Publishing Company
Chapman RF, Bernays EA (eds) (1978) Insects and host plants. Proc Int Symp Entomol Exp Appl 24:201–766
Conner WE (2009) Tiger moths and wooly bears: behavior, ecology and evolution of the Arctiidae. Oxford University Press
Crawley MJ (1983) Herbivory: the dynamics of animal-plant interactions (studies in ecology). Blackwell Science Ltd.
Etheridge K (2011) Maria Sibylla Merian and the metamorphosis of natural history. Endeavour 35:15–21
Etheridge K, Pieters FFJM (2015) Maria Sibylla Merian (1647-1717): pioneering naturalist, artist, and inspiration for Catesby. In: Nelson EC, Elliott DJ (eds) The Curious Mr. Catesby: a "truly ingenious" naturalist explores new worlds. University of Georgia Press, Athens, pp 205–218
Frick GF, Stearns RP (1961) Mark Catesby: the colonial Audubon. University of Illinois Press, Urbana
Fritz RS, Simms EL (eds) (1992) Plant resistance to herbivores and pathogens. University of Chicago Press
Futuyma DJ, Slatkin M (eds) (1983) Coevolution. Sinauer Associates Inc
Gange AC, Brown VK (eds) (1997) Multitrophic interactions in terrestrial systems. Blackwell Science Ltd
Gilbert LE, Raven PH (eds) (1975) Coevolution of animals and plants. University of Texas Press, Austin
Hajek A (2004) Natural enemies: an introduction to biological control. Cambridge University Press, Cambridge

Hawkins BA (1994) Pattern and process in host-parasitoid interactions. Cambridge University Press, Cambridge
Herrera CM, Pellmyr O (2002) Plant/animal interactions – an evolutionary approach. Blackwell Science Ltd
Jensen WA, Salisbury FB (1972) Botany: an ecological approach. Wadsworth Publishing Co, Belmont
Jolivet P (1998) Interrelationship between insects and plants. CRC press
Kareiva PM, Kingsolver JG, Huey RB (1993) Biotic interactions and global changes. Sinauer Associates Inc, Sunderland
Merian MS (1675-1680) Neues Blumenbuch – new book of flowers. Johann Andreas Graff, Nuremberg
Merian MS (1679, 1683, 1718) Der Raupen wunderbare Verwandelung, und sonderbare Blumennahrung – the wondrous transformation of caterpillars and their particular nourishment from flowers. Nuremberg and Frankfurt
Olff H, Brown VK, Drent RH (eds) (1999) Herbivores: Between plants and predators. Blackwell Science Ltd
Post E (2013) Ecology of climate change: the importance of biotic interactions. Princeton University Press, Princeton
Price PW, Lewinsohn TM, Fernandes GW, Benson WW (eds) (1991) Plant-animal interactions – evolutionary ecology in tropical and temperate regions. Wiley
Reynolds S (2019) Cooking up the perfect insect: Aristotle's transformational idea about the complete metamorphosis of insects. Phil Trans R Soc B 374:20190074. https://doi.org/10.1098/rstb.2019.0074
Rico-Gray V, Oliveira PS (2007) The ecology and evolution of ant-plant interactions. University Chicago Press, Chicago
Rolff J, Johnston PF, Reynolds S (2019) Complete metamorphosis of insects introduction. Philos Trans R Soc Lond Ser B Biol Sci 374:20190063
Rosenthal GA, Berenbaum, MR (eds) (1991) Herbivores: Their Interactions with Secondary Plant Metabolites. Academic Press, NY, 2 volumes, pp 468, 493
Ryan F (2011) The mysterious power of metamorphosis. New Scientist 211(2831):56–59. https://doi.org/10.1016/S0262-4079(11)62354-3
Schoonhoven LM, vanLoon JJA, Dicke M (2005) Insect-plant biology, 2nd edn. Chapman & Hall
Sidman J (2018) The girl who drew butterflies. How Maria Merian's art changed science. Houghton Mifflin Harcourt, New York
Stack J, Dorward A, Gondo T, Frost P, Taylor F, Kurebgaseka N (2003) Mopane worm utilisation and rural livelihoods in Southern Africa. In: International conference on rural livelihoods, forests and biodiversity, Bonn, Germany, pp 19–23
Stamp NE, Casey TM (eds) (1993) Caterpillars – ecological and evolutionary constraints on foraging. Chapman and Hall
Strong DR, Lawton JH, Southwood R (1984) Insects on plants – community patterns and mechanisms. Blackwell Scientific Publications
Terborgh J, Estes JA (eds) (2010) Trophic cascades. Island Press
Thompson JN (1982) Interaction and Coevolution. Wiley, New York
Thompson JN (1994) The coevolutionary process. University of Chicago Press, Chicago
Thompson JN (1997) Conserving interaction biodiversity. In: Pickett STA, Ostfeld RS, Shachak M, Likens GE (eds) The ecological basis of conservation: heterogeneity, ecosystems, and biodiversity. Chapman & Hall, New York, pp 285–293
Thompson JN (2005) The geographic mosaic of coevolution. Univ. of Chicago Press, Chicago
Tilmon KJ (2008) Specialization, speciation, and radiation – the evolutionary biology of herbivorous insects. University of California Press
Todd K (2007) Chrysalis: Maria Sibylla Merian and the secrets of metamorphosis. Harcourt, New York

Tscharntke T, Hawkins BA (eds) (2002) Mutitrophic level interactions. Cambridge University Press, Cambridge
Tuskes PM, Tuttle JP, Collins MM (1996) The wild silk moths of North America. Cornell University Press
Tuttle JP (2007) The hawk moths of North America. Wedge Entomological Research Foundation
Tylianakis JM, Didham RK, Bascompte J, Wardle DA (2008) Global change and species interactions in terrestrial ecosystems. Ecol Lett 11:1351–1363
Wäckers FL, van Rijn PCJ, Bruin J (eds) (2005) Plant-provided food for carnivorous insects: a protective mutualism and its applications. Cambridge University Press, Cambridge
Wagner DL (2005) Caterpillars of eastern North America: a guide to identification and natural history. Princeton Field Guides, Princeton University Press, Princeton
Wigglesworth VB (1934) The physiology of ecdysis in *Rhodnius prolixus* (Hemiptera) II. Factors controlling molting and 'metamorphosis'. J Cell Sci 77:191–222
Wigglesworth VB (1954) The physiology of insect metamorphosis. Cambridge University Press, Cambridge
Wilbur HM (1980) Complex life cycles. Annu Rev Ecol Syst 11:67–93. https://doi.org/10.1146/annurev.es.11.110180.000435
Williams CM (1952) Morphogenesis and the metamorphosis of insects. Harvey Lect 47:126–155

On Being a Caterpillar: Structure, Function, Ecology, and Behavior

David L. Wagner and Alexela C. Hoyt

Last instar *Eupackardia calleta* (Saturniidae). When alarmed the caterpillar secretes clear droplets rich in biogenic amines from its dorsal scoli, here from segments A3 and A7–A9; the droplets repel ants and other invertebrate enemies. Photo by Mike Thomas

D. L. Wagner (✉) · A. C. Hoyt
Department of Ecology and Evolutionary Biology, University of Connecticut, Storrs, CT, USA
e-mail: david.wagner@uconn.edu; alexela.hoyt@uconn.edu

Introduction

Whether measured in terms of species richness, abundance, or biomass, Lepidoptera are among the most successful lineages on this planet. More than 157,000 species of Lepidoptera have been described (van Nieukerken 2011; Mitter et al. 2017). The great majority of Lepidoptera are found in tropical forests, especially those of South America (Wardhaugh 2014). One 65-km elevational transect on the eastern slope of the Peruvian Andes has already yielded over 2500 butterfly species (Lamas Müller 2017; Lamas et al. 2021), which equates to 14% of the described global butterfly fauna. In well-studied faunas of the Northern Hemisphere, butterflies often comprise about 6% of the total lepidopteran species diversity. Assuming that ratio holds for Peru, the same transect should yield some 39,000 species of moths—approximately twice the number found in North America north of Mexico (Hodges et al. 1983; Pohl et al. 2016). Given the extraordinary richness of tropical insect faunas, it is our guess that global species richness for butterflies and moths will approach or exceed 350,000 species. Gaston (1991) and Kristensen et al. (2007) offered estimates as high as 500,000 species.

Caterpillars account for much of the above-ground insect biomass in many ecosystems: grasslands, deserts, chaparral associations, scrublands, savannas, and especially forest communities. In addition to their critical roles in natural systems, the order includes many of the most important defoliators of forests, cereals, and field crops. In some ecosystems caterpillars may transfer more energy from plants to other animals than all other herbivores combined (Janzen 1988). They are integral elements in terrestrial food webs; many lineages of birds are reliant on caterpillars, timing their nesting activities to the weeks of larval abundance, with both clutch size and fledging success tied to caterpillar availability (e.g., Lack 1950; Wesołowski and Rowiński 2014; Glądalski et al. 2015; Laney et al. 2015; Smith and Smith 2019, Fig. 1a). Many lizards and snakes are caterpillar hunters (Fig. 1b). Caterpillar- and

Fig. 1 Ecosystem function: caterpillars are a staple food of many terrestrial vertebrates: (**a**) white-eyed vireo (*Vireo griseus*), feeding a notodontid caterpillar to nestling and (**b**) broad-banded copperhead (*Agkistrodon laticinctus*) feeding on larva of white-lined sphinx (*Hyles lineata*) (Sphingidae), the most commonly identified prey from the gut of this diminutive viper endemic to west Texas. (Images (**a**) courtesy of Doug Tallamy; (**b**) courtesy of Gerry Salmon)

pupa-feeding mammals include mice, shrews, raccoons, skunks, foxes, and many human cultures. The early stages of Lepidoptera serve as the resource base for a multitude of insect parasitoids (Krombein et al. 1979) (see also Stireman and Shaw, Chapter "Natural History, Ecology, and Human Impacts on Caterpillar Parasitoids", this volume) and invertebrate predators (see below).

Caterpillars also play important roles in decomposition and nutrient cycling. They do so most obviously by consuming living plant tissues, defecating, and producing greenfall (Risley and Crossley 1988). Seasonal tropical forests can rain frass at the beginning of the wet season. Even unconsumed plant tissue can be shaped by caterpillars; in response to nearby feeding (Smith 1983; Kant et al. 2015; Chen and Mao 2020) or presence of frass (Ray et al. 2016), plants alter their leaf chemistry to deter further damage, with subsequent effects on leaf decomposition rates (Frost and Hunter 2007). Also important are those species that specialize on fallen leaves, wood, and other organic matter (Scoble 1992; Wagner 2013). Herminiine noctuids, for example, which commonly feed on senescent forest floor plant tissues, are especially diverse and numerically abundant in temperate oak woodlands (Hohn and Wagner 2000). Other caterpillars, especially erebids, tineids, and even members of one tribe of hairstreak butterflies, consume the decomposers themselves, feeding on mycelia and fruiting bodies of fungi (Powell 1980; Rawlins 1984; Nishida and Robbins 2020, and text below).

This chapter is a primer on caterpillars, intended to introduce those interested in caterpillar ecology to their life cycle, basic morphology, and natural history. Along the way, an effort is made to suggest aspects of caterpillar biology that are especially interesting, identify data gaps, introduce new phenomena, touch on emergent research frontiers, and share our passion for these creatures. Given the audience of this volume, our treatment is focused on the lineages that are easily studied by ecologists: macrolepidopterans, larger externally feeding microlepidoptera, leafminers, and gall makers; we deemphasize lineages and guilds comprised of shelter-forming and internal feeders, subterranean taxa, and microlepidopterans (but see Marquis et al., Chapter "The Impact of Construct Building by Caterpillars on Arthropod Colonists in a World of Climate Change", this volume), either because they are rarely encountered or present significant sampling and/or identification challenges. A shortcoming of this effort is its anchoring to temperate North American experiences and taxa, but we endeavored to be mindful of this bias. New life history observations, shared at various points in the chapter, are listed in Table 1 at the end of this chapter.

Basic Anatomy

Caterpillars are soft-bodied organisms, analogous to water balloons with internal sclerotized rods (apodemes) and plates (phragmata), external sclerotized plates (sclerites), and appendages that lend strength and promote specialized functions. They are distensible feeding machines whose muscles and hydrostatic skeleton (Lin and Trimmer 2010) allow them to minimally invest in their integument, grow rapidly, and still enjoy considerable mobility. Below we review morphological features, emphasizing those of ecological importance, especially those affecting their relationships with other species.

External Anatomy Caterpillars have a sclerotized head, ancestrally with six lateral image-forming eyes (stemmata) of limited acuity, short antennae, 3 pairs of feeding appendages (mandibles, maxilla, and labium), and a trunk composed of 13 serially homologous segments: 3 thoracic segments and 10 discernable abdominal segments (Figs. 2a and 3). The three pairs of thoracic legs, each bearing an apical claw, are homologous to those of the adult (Fig. 4b). Abdominal segments 3–6 and 10 often have fleshy, crochet-bearing prolegs (Fig. 4a), although many variations on this proleg complement are scattered across the order: e.g., geometrids are often missing the first three pairs of prolegs, and basal noctuid subfamilies (semi-loopers) the first two pairs (Wagner 2005). Below we refer to thoracic segments as T1–T3 and abdominal segments A1–A10 such that A2 would stand for the second abdominal segment.

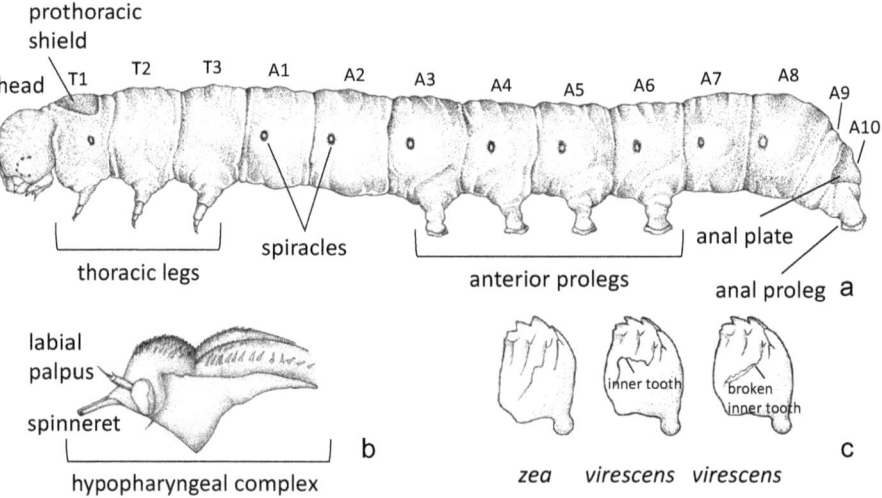

Fig. 2 External caterpillar anatomy: (**a**) habitus; (**b**) hypopharyngeal complex (spination patterns over upper surface often will separate closely related species); (**c**) right mandible of *Helicoverpa zea* (corn earworm) and *Chloridea virescens* (=*Heliothis virescens*) (cotton bollworm) (both Noctuidae). These two major crop pests are surprisingly difficult to differentiate from photographs, yet their mandibles immediately separate the two. Abdominal (A) and thoracic (T) segments; T2 = second thoracic segment. (Line art by Virginia R. Wagner (from Wagner et al. 2011); reproduced with permission from Princeton University Press)

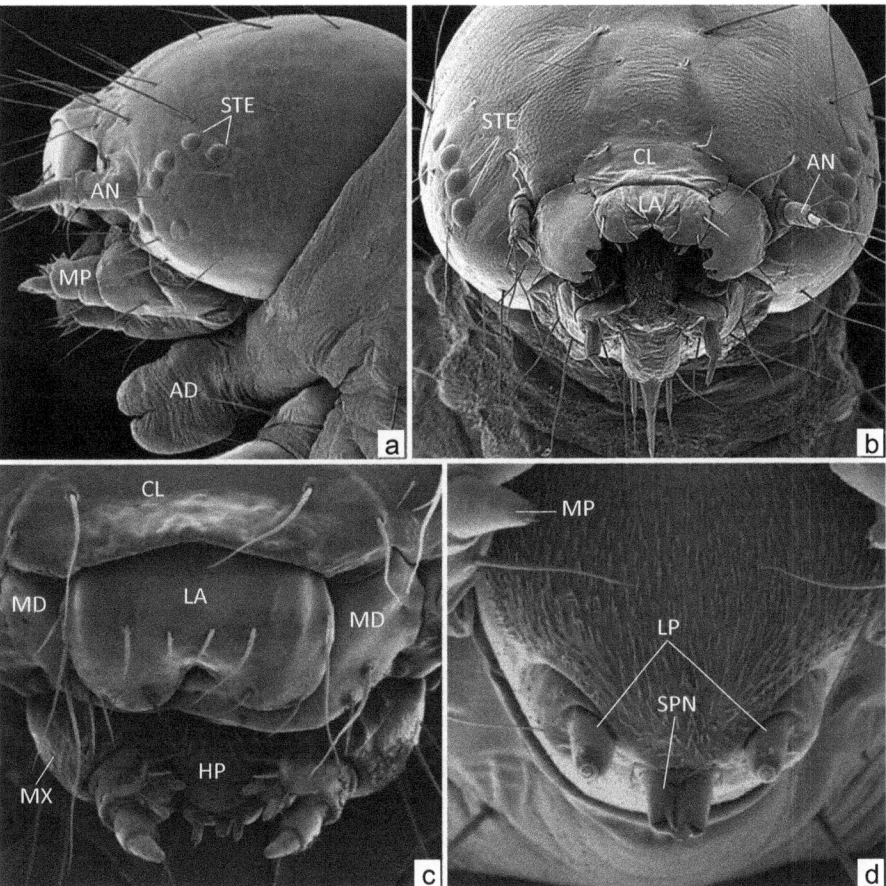

Fig. 3 Scanning electron micrographs of caterpillar external anatomy: (**a**) lateral view of head of *Spodoptera androgea* (Noctuidae); (**b**) frontal view of *Brenthia* sp. (Choreutidae); (**c**) frontal view of mouthparts of *Spodoptera androgea*; and (**d**) dorsal view of hypopharyngeal complex of *Spodoptera androgea*. Adenosma (AD), antenna (AN), clypeus (CL), hypopharynx (HP), labial palp (LP), labrum (LA), mandible (MD), maxillary palp (MP), maxilla (MX), spinneret (SPN), and stemmata (STE). The length, structure, and arrangement of setae on the upper side of the hypopharynx are often diagnostic for a given caterpillar species. (Image (**b**) courtesy of Jadranka Rota)

Color and patterning characters will allow the identification of many of externally feeding caterpillars in areas with modest species richness (Wagner 2005; Wagner et al. 2011). More detailed study of microscopic features and/or dissection may be necessary for others, especially for large genera, those that feed internally, or lineages in which coloration can be highly variable within a species. Chaetotaxy (the size and placement of the primary setae) can be used to distinguish closely

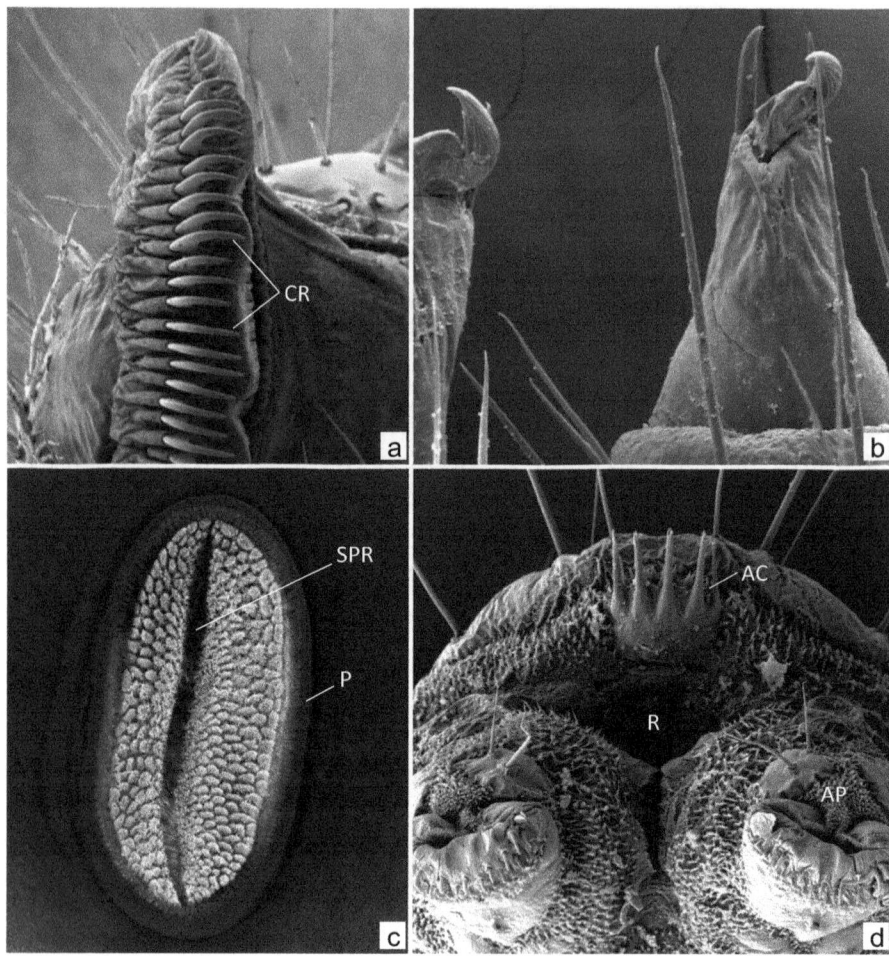

Fig. 4 Scanning electron micrographs of caterpillar external anatomy: (**a**) crochet series of *Hypercompe scribonia* (Erebidae: Arctiinae); (**b**) prothoracic legs of *Hyblaea* (Hyblaeidae), (**c**) spiracle of *Spodoptera androgea* (Noctuidae); and (**d**) posterior view of *Paralobesia viteana* (Tortricidae) abdomen. Anal comb (AC), anal proleg (AP), crochets (CR), peritreme (P), and spiracle (SPR). (Images (**b**) and (**d**) courtesy of Marc Epstein)

related species; in some, the length of setae relative to the closest spiracle can be used to differentiate congeners. The number and relative proportions of the teeth (incisors) and other mandibular details are sometimes used to distinguish caterpillars, e.g., those of *Helicoverpa zea* from the confusingly similar *Chloridea* (=*Heliothis*) *virescens* (both Noctuidae) (Fig. 2c), many plusiine loopers (Noctuidae), and others. The spination of the dorsal surface of the hypopharynx (Figs. 2b and 3b–d) will vary across sets of closely related species.

Sensilla (specialized setae) provide essential sensory information for caterpillars. Pre-oral and oral chemoreception, relevant to food consumption, are facilitated by chemosensilla located on the antenna, mandible, maxilla, labium, labrum, and epipharyngeal complex (Song et al. 2014; Men and Wu 2016). A taste sensillum typically has three to four individual taste cells, each of which responds most strongly to a single compound (Glendinning et al. 2000, 2001). These sensory setae allow caterpillars to differentiate among a sweep of plant compounds (Glendinning et al. 2002) that modulate feeding responses. For instance, myo-inositol (a sugar alcohol common in plants) incites feeding in *Manduca sexta*. Pyrrolizidine alkaloids act as feeding stimulants for both *Estigmene acrea* and *Grammia incorrupta* (both Erebidae: Arctiinae) (Bernays et al. 2002). In addition to food recognition, the sensilla of caterpillars function in mechanoreception, thermoreception, and hygroreception (Scoble 1992; Men and Wu 2016).

Caterpillars respond to both air- and substrate-borne vibrations (sounds). Specialized socketed thoracic setae play integral roles in sound detection, with many species responding to high frequencies, such as those produced by the wings of flying insects. Presumably a primary role is the detection of predatory wasps, as well as both dipteran and hymenopteran parasitoids (Tautz and Markel 1978; Taylor and Yak 2019). Such frequencies can trigger immediate stasis or a defense response, and even both reactions in the same individual, depending on the context and duration. Swallowtail caterpillars will sometimes evert the osmeterium, when spoken (or sung) to or at the sound of a clap (DLW and Sam Jaffe, unpubl. observations).

Internal Anatomy Insects have many of the same organ systems present in vertebrate animals, but given that the caterpillar is principally a feeding and growth stage, we emphasize the digestive system here and only briefly touch on silk glands and a few other systems. A caterpillar is essentially a walking digestive system, with a gut that volumetrically greatly exceeds that of all other systems combined. The remainder of a caterpillar's body (by decreasing volume) is given to fat bodies, Malpighian tubules, silk glands, and other organ systems common to animals (Fitzgerald 1995).

Insect digestive tracts are divided into three parts: the foregut, midgut, and hindgut (Figs. 5 and 6). The first and last of these are ectodermal in origin and consist of a lined cuticle that is shed at each molt. Caterpillars possess an enormous but surprisingly simple digestive tract: a comparatively short, unconvoluted tube, with an extended midgut adapted for rapid digestion of plant tissues (Dow 1986). The mouth is delimited anteriorly by a medially grooved labrum, laterally by the mandibles and maxillae, and posteriorly by the labium (Fig. 3b); the hypopharyngeal complex (Figs. 2b and 3b–d), principally derived from the labium, projects into the buccal cavity and acts as a tongue, moving food into the foregut and aiding in chemoreception (Traxler 1977).

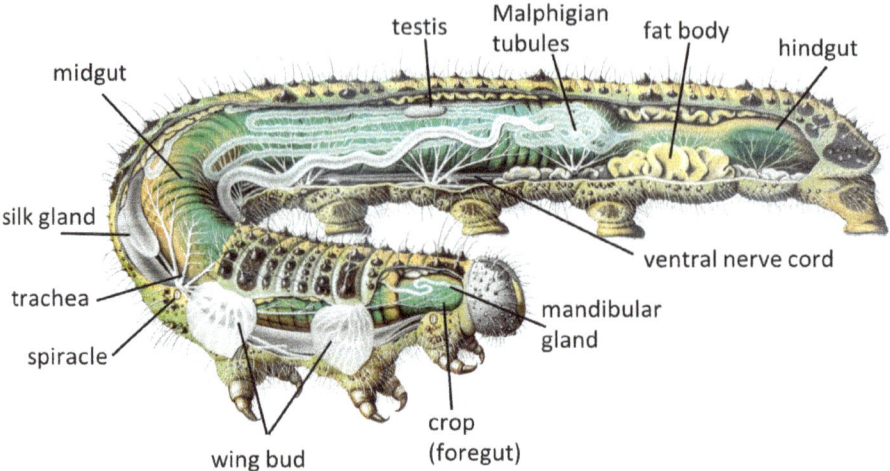

Fig. 5 Internal anatomy of *Pieris brassicae* (late last instar). (Original art by Paul Pfurtscheller circa 1908; this augmented reproduction used with permission from The Caterpillar Lab)

The esophagus is short and abruptly enlarges to a crop that serves for both food storage and defense (when emptied to discourage a would-be enemy) (Wang et al. 2018; Peterson et al. 1987). Exceptionally the foregut may have diverticula for the storage of defensive compounds, e.g., see Common and Bellas (1977). During a feeding bout, the crop can become greatly enlarged, pushing the midgut rearward, such that it occupies much of the anterior lumen (Fitzgerald 1995). Caterpillars that we have examined lack a muscularized and armored proventriculus as is found in grasshoppers, although in at least some caterpillars, there are small spines, plates, or denticles about the valve (not visible in figures) that separates the fore- and midgut, which contribute to mechanical digestion. Internally, the proventricular valve serves as the end of the foregut and controls entry of ingested material into the midgut, where most digestion and nutrient assimilation takes place (Fig. 6). As food is ingested and passed into the midgut, the foregut contracts, regaining its pre-meal dimensions in tent caterpillars (Lasiocampidae) (Fitzgerald 1995). External indications of the junction between the foregut and midgut of lepidopterans can be modest, in contrast to the guts of other insects.

The alkaline caterpillar midgut is lined with a chitinous sheath called the peritrophic membrane that acts as a protective, semi-permeable envelope around food boli (Peters 1992). The midgut serves as the primary site of chemical digestion, via an array of amylases, glycosidases, lipases, and proteases that pass through the membrane, although some enzymes remain embedded in the peritrophic membrane (Ferreira et al. 1994). The midgut has a characteristically high pH, typically falling between 9 and 11 (Ferreira et al. 1994; McMillan and Adamo 2020) with high titers

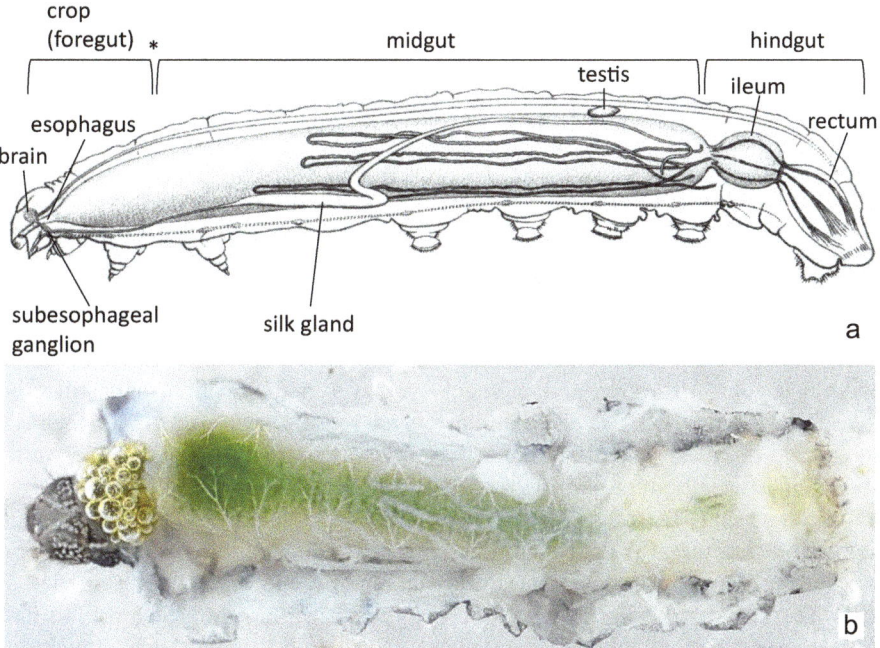

Fig. 6 Internal anatomy of caterpillars. (**a**) Monarch (*Danaus plexippus*) (Nymphalidae) lateral view from Scudder (1889). Note the volume occupied by the gut—a caterpillar is little more than a feeding machine. Unlike many insects, the foregut and midgut juncture is not externally differentiated in many Lepidoptera, and why we add the * to indicate that our demarcation is only an approximation. (**b**) Beet armyworm (*Spodoptera exigua*) dorsal view. The crop here is the yellow portion, filled with bubbles, which curiously appear to contain additional smaller bubbles. A (stomadeal) valve separates the foregut (crop) from the green midgut—the green being recently ingested leaf tissue

of phospholipases, which enhance digestion while discouraging the survival of bacteria and other microorganisms. Goblet cells, rich in mitochondria, line the midgut, serve in the uptake of salts and amino acids, and maintain the high alkalinity of the midgut (Levy et al. 2004). The large vacuoles of the goblet cells may also serve as storage areas for compounds that will be excreted.

The short hindgut consists of the pylorus, ileum, colon, and rectum and is the primary area for absorption of water and salt (Levy et al. 2004). Six Malpighian tubules originate near the junction of the midgut and hindgut; these extend anteriorly around the midgut, then switchback, and run to the posterior end of the caterpillar, where they form a structure known as the cryptonephridium around the rectal tissues. The Malpighian tubules are essentially the caterpillar's kidneys, regulating salt and water balance (Levy et al. 2004; Kolosov and O'Donnell 2019).

Powerful rectal muscles shape feculae or frass prior to release (Levy et al. 2009; Ramsay 1976). Both the size and shape of the feculae are diagnostic and consequently can be used to identify the caterpillars that produced a given pellet, at least to family but sometimes even to species (Haylett 2000). As such, feculae have the potential to be used in caterpillar monitoring and other ecological investigations (e.g., see Bernays and Janzen 1988). Because they contain volatiles, frass pellets are used by many predators and parasitoids to locate feeding caterpillars (Weiss 2003, 2006; Moraes et al. 2012, and discussion below). Even plants may respond to the presence of caterpillar frass e.g. by elevating inducible defenses (Ray et al. 2015).

The fat body appears as an amorphous, white, unconsolidated, mesodermal organ that functions in intermediary metabolism, fat storage and energy regulation, endocrine control, protein and pigment synthesis, detoxification, and still other roles (Hoshizaki 2012). Its cells (adipocytes) are loosely organized into trachea-rich sheets and nodules, linked by connective tissue, that are most conspicuous in the abdomen especially in proximity to portions of the gut (Figs. 5 and 6b).

Vegliante and Hasenfuss (2012) identify 21 lepidopteran exocrine glands and associated structures. We only introduce readers to about a third of these, with our treatment emphasizing those that are taxonomically widespread, large in size, or known to play significant roles in the ecology of caterpillars. Most caterpillars have both a mandibular and labial salivary gland. The thin mandibular gland, which may extend back into the thorax or more rarely into the abdomen, produces salivary fluids that are secreted from the base of the mandible. The secretion may contain proteins, lipids, sterols, and triglycerides (Felton 2008); the specific constituents, proportions, and functions of the gland appear to vary across taxa and are in need of more study. That of *Vanessa gonerilla* (Nymphalidae) contains digestive enzymes such as amylase lysozyme, α-amylase, as well as sericotropin, which is thought to play roles in defense and immunity (Celorio-Mancera et al. 2012). The 13-carbon alcohol mandibular gland secretions of *Cossus cossus* (Cossidae) appear to serve in defense (Eaton 1988). The labial salivary gland in most Lepidoptera is modified to produce silk. Much of our understanding of silk structure and function is anchored to the Oriental silk moth (*Bombyx mori*) (Bombycidae). In addition to silk, the labial gland may secrete enzymes—e.g., lysozyme, ascorbate peroxidase, and glucose oxidase—that serve in digestion, immunity, and the inhibition of induced defenses of their host plants (Celorio-Mancera et al. 2012). Glucose oxidase, in particular, appears to function principally in shutting down inducible host plant defenses (Musser et al. 2002). A cervical gland (adenosma) that opens through a medial pore on the venter of the prothorax and serves principally in defense, is present in many Hesperiidae, Noctuoidea, Notodontidae, Nymphalidae, Papilionoidea, Riodinidae, and Yponomeutidae (Vegliante and Hasenfuss 2012, and below).

Trail pheromones typically released from either the labial gland or the terminal abdominal segments are often co-mingled with silk deposition (Crump et al. 1987;

Vegliante and Hasenfuss 2012). They occur in Pieridae, processionary caterpillars (Notodontidae: Thaumetopoeinae), tent caterpillars (Lasiocampidae), and some Saturniidae (Vegliante and Hasenfuss 2012). We suspect that analogous trail signals are used by some solitary caterpillars that leave their feeding sites by day to shelter on bark, at the base of the host, or in soil or litter. These might be especially important for those species that rest well removed from the host (e.g., *Catocala illecta*), presumably to lower their predation risk. Such behaviors and the role of silk and other chemical markers in retaining site fidelity have received little attention.

Lycaenid and riodinid butterflies have highly specialized exocrine glands and organs that release sugary and amino acid secretions that encourage ant attendance (Malicky 1969, 1970; Fiedler 1991; Vegliante and Hasenfuss 2012), which in turn protects caterpillars from parasitoids and invertebrate predators as well as the attending ants (see Pierce and Dankowicz, Chapter "The Natural History of Caterpillar-Ant Associations", this volume). Most widespread are pore cupola cells—scattered across the body—that may secrete a substance that can be attractive or otherwise mediate caterpillar-ant interactions. A middorsal gland on A7 (Newcomber's gland), widespread among ant-attended lycaenids, secretes a sugary solution rich in amino acids that is eagerly fed upon by their retinue. Many lycaenids also have paired, eversible, mushroom-shaped, tentacular organs on A8. When everted, they release an air-borne signal that attracts nearby ants. Aphnaeine lycaenids have a series of middorsal dew patches (or dish organs) on A2–A5 that release a sugary reward (Vegliante and Hasenfuss 2012). Riodinids have comparable structures: paired tentacle nectary organs (TNOs) on A8 or a middorsal organ on T3 (metathorax) (ATO), both of which include a nutritive function. To learn more about the biology of the caterpillars of these butterflies, their exocrine structures, and galaxy of myrmecophilous interactions, consult the many works of Philip DeVries and Naomi Pierce.

Cellular Immune System Pathogens and other foreign entities detected within an insect's body are attacked by hemocytes, but arthropods lack the acquired immune response of vertebrates, i.e., they do not produce target-specific antibodies for different pathogens. Among the most important hemocytes, at least in lepidopterans, are the phagocytic granulocytes that envelop and destroy microbial parasites as well as foreign non-biological particles. Larger foreign entities are encapsulated by plasmocytes that adhere to the surface of the foreign threat in great number. This agglomeration triggers a chain reaction of melanin deposition, creating a physical barrier that prevents gas and nutrient exchange that can suffocate parasitoid eggs and larvae and other foreign bodies (Smilanich and Muchoney, Chapter "Host Plant Effects on the Caterpillar Immune Response", this volume).

Gut and Microbiome Many insects, e.g., aphids, bees, cicadas, termites, and others, have a microbiome associated with their digestive systems. Others, including many lepidopterans, may lack a functional gut microbiome (Hammer et al. 2017,

2019; Phalnikar et al. 2019). This may be due to their simplified, short digestive tract (lacking diverticula), the rapidity of food movement through the gut, and the high alkalinity of the caterpillar digestive tract, which collectively yield an unfavorable environment for symbionts (Appel 1994; Hammer et al. 2017). While the absence of beneficial gut microbes might appear evolutionarily disadvantageous, the sheer ecological abundance and diversity of caterpillars would seem to argue otherwise. Regardless, even in the absence of a microbiome, many lepidopterans are able to quickly process ingested tissues, detoxify myriad secondary plant compounds, and grow rapidly (Hammer et al. 2017; Phalnikar et al. 2019).

Caterpillars do have bacteria in their guts (Buchon et al. 2014); in many leaf feeders, these derive from the bacteriofauna of their host plants and, as far as is known, only infrequently contribute appreciably to digestion. *Manduca sexta* (Sphingidae) caterpillars treated with antibiotics have growth rates and development times comparable to those of untreated controls (Hammer et al. 2017). Many negative effects of ingested bacterial pathogens have been well documented (McMillan and Adamo 2020).

A novel exception to the above occurs in two pyralids that feed in honey bee nests: *Galleria mellonella* and *Achroia grisella*. With the help of gut symbionts, the caterpillars can digest and grow on polypropylene plastics (and, almost as surprisingly, excrete ethylene glycol antifreeze as an end product of plastic digestion) (Cressone et al. 2020). Acevedo et al. (2017) found that bacteria in the oral secretions of fall armyworm caterpillars (*Spodoptera frugiperda*) can trigger upregulation of some induced plant defenses and the down-regulation of others. We are unaware of instances where gut microbes play a positive role in detoxification of secondary plant compounds by Lepidoptera (but see Hammer and Bowers 2015). The degree to which microorganisms might alter a caterpillar's immune response is an area of active inquiry.

Ontogeny, Life Cycles, and Diapause

Lepidoptera have four life stages: egg, larva, pupa, and adult, with the larval stage emphasized in this volume. The number of larval instars varies from three to seemingly indeterminate in some wood feeders (Wagner 1985). While five instars are the median, both four and seven are common; some have greater numbers. Across most taxa, there is a fixed number of instars. Females may have an extra instar in some large-bodied species, e.g., many lymantriines (Erebidae) (Esperk and Tammaru 2006). Some arctiine erebids go through six to ten instars (Dyar 1890). Supernumerary instars are sometimes added when caterpillars are under nutritional stress (Grunert et al. 2015).

A caterpillar is a feeding machine whose charge is to eat and not get eaten. Some will increase their mass 1,000-fold or even 10,000-fold from first to final instar

(Reavey 1993; Lin et al. 2011). Most absolute growth (and plant consumption and, by extension, economic impact) occurs during the final instar—approximately 90% of the increase in mass in *Manduca sexta* (Sphingidae) occurs in the last (fifth) instar (Grunert et al. 2015). In contrast to these shifts in overall growth rate, an apparent constraint across lepidopterans is that strongly sclerotized structures (e.g. head capsule, mandibles, and prothoracic and anal plates) increase by a rather constant factor of 1.3 to 1.4 at each molt, resulting in geometric growth over the larval stage (Dyar 1890). While many exceptions are known (e.g. Cole 1980; Albert 1982), this relatively constant rate of increase has broad applications, e.g. can be used to infer the instar of shed capsule or a cadaver vanquished by a parasitoid.

Considerable changes in form, ecology, behavior, and associated selective pressures invariably transpire during the larval stage (see reviews by Reavey 1993; Boege et al. 2019). Morphologically, first instars are typically much differentiated from those seen across second to final instars. Additional changes, some striking, may occur across larval molts, a few of which are discussed below (see also Figs. 7 and 8). Rather modest attention has been focused on the ecology of early instars (Reavey 1993; Zarlucki et al. 2002)—a matter that likely will long remain a frontier for insect-plant ecologists. By way of example, trichomes on the underside of a host plant leaf might be readily consumed by a middle or late instar, yet represent an insurmountable threat to both egg and first instar (Zarlucki et al. 2002; Kaur et al., Chapter "Surface Warfare: Plant Structural Defenses Challenge Caterpillar Feeding", this volume).

Likewise, sweeping changes occur in a caterpillar's natural enemy complex over the course of its development (Reavey 1993; Hawkins et al. 1997; Frankfater et al. 2009; Boege et al. 2019). Many ecologists focus their attentions on late instars as they are more visible and confidently identifiable. Yet, given a typical invertebrate survivorship curve (Price 1997, Fig. 13), one might expect a cohort's numbers to have been halved 2–3 times before attaining a size likely to be tallied in many ecological studies (Zarlucki et al. 2002).

Important phenotypic and ecological changes can accrue in modest increments or in saltational steps across instars (Fig. 7), highlighting the mastery of gene regulation and metamorphosis in Lepidoptera. The phenotype, diet, and resting behaviors of *Stiria* (Noctuidae) caterpillars track the phenological changes of their Asteraceae hosts (Fig. 8). Some *Acronicta*, *Egira*, and *Lithophane* caterpillars (all Noctuidae) change from a green leaf-resting penultimate to a dark last instar that rests on bark—so different in phenotype and microhabitat that the two forms may be unrecognizable as a single species. In many swallowtails the early instars are bird-dropping mimics that rest on upper leaf surfaces by day, but after a single molt change into a strikingly different morph with new behavioral repertoires: *Papilio troilus* becomes a green shelter-former with false eyes, while *P. cresphontes* molts to a twig-resting viper imposter (Wagner 2005).

Dramatic morphological transitions occur in hypermetamorphic taxa, i.e., those with two or more distinct larval forms. The developmental changes in leafmining

Fig. 7 Ontogenetic, phenotypic, and behavioral changes across instars of *Arsenura batesii* (Saturniidae): (**a**) first instar, (**b**) third instar, and (**c**) last instar. Last instars rest on tree trunks. Images courtesy of Annette Aiello

Fig. 8 Ontogenetic, phenotypic, and behavioral changes across instars of *Stiria intermixta* (Noctuidae): (**a**) early instar, (**b**) penultimate instar, and (**c**) last instar. Early instar *Stiria* often have dark transverse bars that mimic the dark pollen-producing florets surrounding the disk flowers; in the right pane, the caterpillar has a more warted form, more complex pattern, and rests tightly curled in the cavity created while feeding on the disk flowers

Gracillariidae are especially noteworthy in this regard. The first two to three instars are prognathous (forward-directed jaws), silkless, legless, liquid feeders with reduced eyes (Needham et al. 1928; Wagner et al. 2000). The larvae at this stage tunnel through epidermal and parenchymatous plant tissues. By contrast, the later instars are often quite ordinary, being hypognathous (downward-directed jaws), and are silk-producing, with relatively unmodified thoracic and abdominal prolegs and normal eyes; these are feeding on whole cells (Needham et al. 1928). A functioning spinneret makes the differences possible: silk deposition within the mine allows the caterpillar to draw the mine into a bubble—a three-dimensional environment—with the two leaf surfaces well separated. Other Lepidoptera with hypermetamorphic development include a few gall formers (e.g., Bucculatricidae) (Needham 1948), some opostegids (Davis and Stonis 2007), and ectoparasitoids (e.g., Cyclotornidae and Epipyropidae) (Epstein et al. 1998). Non-feeding instars are rare, generally occurring in the first or last instars. The former occurs in "nettle group" limacodids (Zaspel et al. 2016); the latter is found in some gracillariids: e.g., *Cameraria* and *Marmara* have two non-feeding prepupal instars (Needham et al. 1928; Wagner et al. 2000).

Transitions in phenotype and behavior can occur within an instar, most commonly through the course of the final instar. *Adelpha serpa* (Nymphalidae), which normally is cryptically patterned, turns golden yellow the day before hanging for pupation (Aiello 1984). Acronictine noctuids that tunnel into bark or soft wood to pupate sometimes have a prepupal phenotype that is more cryptic on bark and thus helps the caterpillar to remain less apparent over the hours it takes to excavate its pupal crypt (Wagner et al. 2011). A dramatic example is *Polygrammate hebraeicum* (Noctuidae) and kin: larvae change from a green leaf-feeding form to a waxy-blue and black-dotted bark-tunneling form and then again into a red prepupa, all within the confines of a single instar (Wagner et al. 2006, Fig. 9). Virtually all caterpillars change as prepupae: the body contracts and thickens, patterning often dulls or is lost, and the integument frequently becomes shiny. Many lineages take on a rose flush or, in the extreme, become red (as in *Polygrammate*) (Fig. 9c). A famous example would be that of *Comadia redtenbacheri* (Cossidae), the gusano rojo, swirling at the bottom of many mescal bottles. We are unsure as to why the worms in bottles tend toward tan or brown—either their bright red coloration is lost in alcohol or they are collected and added before becoming prepupal—yet another case we are looking to crack. We have not seen literature explaining the basis of this transition. Our guess is that the phenomenon is quite taxonomically widespread but that it routinely happens inside the pupal crypt or cocoon, where it goes unnoticed (and exempt from appreciable selective pressure). To what degree this color change is linked to the bright red transition that happens in *Galleria mellonella* (Pyralidae) larvae (Fenton et al. 2011), when infected by the nematode *Heterorhabditis bacteriophora*, also remains unstudied.

The transitions described above are pre-programmed, i.e., the changes are part of every individual's development; but caterpillars also provide heralded instances of phenotypic plasticity, taking on different shapes and coloration in response to environmental stimuli. Common stimuli documented to trigger dramatic phenotypic changes in caterpillar morphology (and behavior) include light environment, crowding, diet, and temperature (Akino et al. 2004; Noor et al. 2008). That many caterpillars are given to phenotypic plasticity is well known. Poulton (1892) published papers on the phenomenon in the late nineteenth century, after observing that geometrid caterpillar phenotypes varied when reared in boxes lined with paper of different colors. Some ennomine geometrids are green when reared in green

Fig. 9 Phenotypic and behavioral changes sometimes occur within a single instar. *Polygrammate hebraeicum* (Noctuidae): (**a**) feeding last instar; (**b**) wood-boring prepupa excavating a chamber in which the winter will be passed; and (**c**) bright pink prepupal morph that normally occurs inside the pupal chamber, but this caterpillar was photographed while it was still on its search for pulpy wood of appropriate hardness for tunneling, long past the time it would normally have entombed itself

environments and brown when reared in environments dominated by earth tones (Noor et al. 2008). In larvae of the pepper moth (*Biston betularia*) (Geometridae), detection of the visual environment and subsequent color change are mediated by dermal receptors along the larval body—even with its eyes covered, the caterpillar's body can detect its light environment; across molts *Biston* caterpillars can change their color to better background match (Eacock et al. 2019). Perhaps most famously, the oak-feeding caterpillars *Nemoria arizonaria* (Geometridae) develop into catkin (flower) mimics when fed diets low in tannins (e.g., catkins) and twig mimics when fed older tissues with elevated titers of tannins (e.g., mature leaves) (McFarland 1988; Greene 1989) (Fig. 10).

A dramatic and not yet fully understood color change that happens across multiple lineages of Lepidoptera is the green-to-black and cryptic-to-"warningly" colored polyphenisms associated with high population densities, analogous to the transitions that transpire in the migratory locust (Wang and Kang 2014, Fig. 11). The caterpillars of many defoliators are green or otherwise cryptic at low densities but become increasingly blackened or "warningly" colored as densities rise and foliage becomes scarce or deteriorates. The phenomenon has evolved in dozens of lineages of Lepidoptera—Erebidae, Geometridae (both Ennominae and Larentiinae), Noctuidae (e.g., Bagisarinae, Noctuinae, and Plusiinae), Nymphalidae, Pieridae, Saturniidae, and Sphingidae (unpubl. data)—which is argument enough that the transition is adaptive, yet the raison d'être for its evolution remains poorly understood. Some suggest the polyphenism is merely an epiphenomenon of an upregulated immune response, due to the elevation of phenolic titers, to combat microbial and viral diseases (e.g., Lee and Wilson 2006). Surely going from a cryptic to a prominently non-cryptic phenotype has enormous ecological consequences, which are as yet unstudied. Moreover, the green-to-black change is sometimes reversible, again across instars; Wagner et al. (2011) were able to induce the appearance of black forms in *Orthosia alurina* (Noctuidae) by rearing larvae in sleeves at high density but reverse larvae back to green forms in later instars by re-sleeving the same cohort at lower densities.

Larval development can be as rapid as 2 weeks or drawn out over many years; the latter is common in polar and alpine regions and among large-bodied wood feeders.

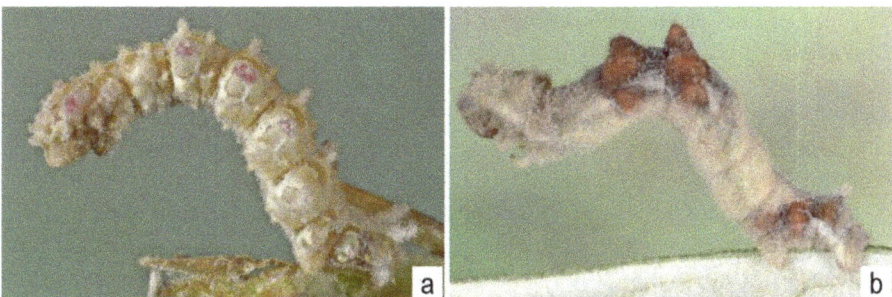

Fig. 10 *Nemoria arizonaria* (Geometridae): (**a**) spring-brood larvae feed on and mimic catkins; (**b**) summer- and fall-brood larvae feed on leaves and mimic twigs (McFarland 1988; Greene 1989). Both morphs can be generated from a single cohort if larvae are fed diets with low or high titers of tannin phenolics (Greene 1989)

Fig. 11 Green-to-black and cryptic-to-"warningly" colored polyphenism. *Ennomos subsignarius* (Geometridae) (top row) and *Hypocala andremona* (Erebidae) (bottom row). (Images (d) and (e) courtesy of Berry Nall)

The arctic woolly bear, *Gynaephora groenlandica* (Erebidae: Lymantriinae), takes 7 or more years to mature (Morewood and Ring 1998). In contrast, taxa in deserts and drylands can have exceeding rapid life cycles, e.g., some *Schinia* flower moths (Noctuidae: Heliothinae), which feed on highly nutritious, ripening seeds, will complete their larval development in as little as 13 days (Hardwick 1996). Development from egg to adult can be as short as 15–16 days in the snout butterfly (*Libytheana carinenta*) (Nymphalidae) (Nall 2020); two larger butterflies, *Danaus plexippus* and *Vanessa cardui* (both Nymphalidae), can complete their life cycle in 28 days.

In temperate and seasonal ecosystems, many species are univoltine, with the availability of suitable food and temperatures constraining the number of generations. Far more typical, however, and where resources and temperatures permit, additional generations are produced, with an overlay of facultative broods, i.e., where a fraction of a given population emerges and attempts to complete its development before conditions become unfavorable (Wagner et al., unpubl. observ.). Cohort failure is a common outcome for facultative broods. South of areas subjected to hard freezes, many Lepidoptera remain active year-round, with population pulses tied to availability of appropriate larval resources. In areas of challenging abiotic conditions, nearly all are bet hedgers, with facultative broods the norm with some individuals remaining in diapause.

Diapause occurs in all four life stages and all larval instars, most commonly in the egg and pupal stage in temperate Lepidoptera; exceptionally more than one stage may be programmed for diapause, especially in those taxa that take more than 1 year to develop. Many caterpillars diapause as first to third instars or as a pre-pupa in a protected cell or cocoon. The diapausing instar is nearly always taxon-specific: e.g., firsts in *Malacosoma americanum* (Lasiocampidae) and *Speyeria* (Nymphalidae); thirds or fourths in *Chlosyne* and *Euphydryas* (both Nymphalidae) and *Synchlora* (Geometridae); *Haploa* (Erebidae: Arctiinae) as fifth or sixth instars; and last instars in *Arachnis* and *Pyrrharctia* (Erebidae: Arctiinae) and as a prepupa, in a protected cell or cocoon, in many Notodontidae, Prodoxidae, many Zygaenoidea, and others. Diapause induction in Lepidoptera is most commonly triggered by photoperiod (day lengths) and less commonly temperature and host plant quality (Hunter and McNeil 1997). Across boreal and temperate areas, diapause is typically terminated by warmer temperatures that signal the return of favorable (growth) conditions. But in deserts, grasslands, and seasonal forests—where precipitation signals impend availability of new growth—rains also drive activity. Lepidoptera inhabiting drylands and deserts often have the capacity to remain in diapause for more than 1 year. Pupae of *Anthocharis* (orange tip) butterflies (Pieridae) can diapause up to 11 years (Todd 2018). Prepupae of the false yucca moth (*Prodoxus*) (Prodoxidae) yielded moths 19 years after their collection (Powell 1989).

Larval Diets

Perhaps 98% of Lepidoptera are plant feeders, with nearly all of these associated with vascular plants (Strong et al. 1984; Mitter et al. 1988; Pierce 1995; Powell et al. 1998). Lepidopteran species diversity is largely a reflection of the ecological and evolutionary successes of gymnosperms and angiosperms, but especially the latter (Ehrlich and Raven 1964; Wahlberg et al. 2013). To a large measure, the species richness on any given plant species is closely tied to the host plant's geographic range, architectural complexity, and abundance; other determinants include apparency, the degree of taxonomic isolation, and the plant's physical and chemical properties (Lawton 1976, 1978, 1983; Lawton and Schröder 1977; Strong and Levin 1979; Strong et al. 1984). Stated differently, large, ecologically dominant, widespread plants, with many congeners of similar ecological stature, have the richest lepidopteran faunas. In north temperate areas oaks (*Quercus*) are unrivaled: more than 900 species of Lepidoptera feed on oaks in America north of Mexico (Shropshire and Tallamy, unpubl. data). *Quercus, Prunus, Populus, Salix, Pinus,* and just ten other tree genera are thought to host about 80% of the lepidopteran species diversity of eastern deciduous forests (Tallamy and Shropshire 2009; Narango et al. 2020).

More than 85% of Lepidoptera are believed to specialize on one or just a set of closely related species (Jaenike 1990; Forister et al. 2015), i.e., plants in the same genus or taxonomically proximate genera. The remainder of the phytophagous species are either oligophagous or polyphagous, with some of the latter being diet mixers (Singer et al. 2002), feeding on more than one host over the course of the larval stage or even a single feeding bout. Few specialists are locked to a single host species, except in those cases where specialization is forced by the absence of congeneric and otherwise related species. Among the most dietarily specialized are those that feed internally within their host: e.g., leafminers, gallers, and fruit or seed borers. In arid regions of western North America, caterpillars that feed on ephemeral tissues, such as new leaves, flowers, fruits, and seeds, that demand phenological synchrony between larva and host plant, show great degrees of specialization (Posledovich et al. 2015, DLW pers. obs.).

Essentially all terrestrial plants and plant tissues, be these alive or dead, are consumed by some lepidopteran. While the lion's share are leaf feeding, reproductive tissues are frequently targeted, with leaf feeders often switching to flowers and fruits when available. Excluding defoliators that kill their hosts, the majority of lepidopteran lineages act ecologically as plant parasites. However, those that consume whole plants, pollen grains, ovules, seeds, or spores are functionally predators that remove individuals from a population. Early instars of some *Schinia*, *Spragueia*, and *Stiria* (each representing different Noctuidae subfamilies) will specialize on pollen if available (Wagner et al. 2011, unpubl. observ.). Fruits or seeds are consumed by many Lepidoptera. Greatly diversified seed-feeding lineages occur in Lycaenidae; Acontiinae, Chamaecleini, Heliothine, and Stiriine (all Noctuidae); Grapholitini (Tortricidae); and many other microlepidopteran lineages.

Stems and other woody tissues, as well as roots, are targeted by some lineages. Wood-feeding caterpillars include Cossidae, Hepialidae, and Sesiidae, but this niche largely belongs to the larvae of Coleoptera. However, stem boring of non-woody species is a niche occupied by several lineages of Lepidoptera whose numbers sometimes exceed those of beetles. Among leaf-feeders, caterpillars further specialize. Some leafminers target just spongy or palisade parenchyma; *Phyllocnistis* and some *Marmara* (both Gracillariidae) may feed only on epidermal cells.

Ferns, and even mosses, appear to have lepidopteran faunas commensurate with their modest statures (Lawton 1976). Lichens and algae support numerous species in both the tropics and to a lesser extent in deserts and arid lands (Wagner et al. 2008). Lithosiine arctiines represent a particularly diverse radiation of lichen feeders. Fungi as well are consumed (Rawlins 1984) but, with the exception of the tineids, few groups have diversified. As would be predicted, persistent fungi, such as bracket fungi, have richer caterpillar faunas, but this is another resource dominated by coleopteran larvae and adults.

Coprophagy, detritivory, and saprophagy are not uncommon among caterpillars. Herminiine erebids feed on living and dead plant tissues, with the majority believed to feed on fallen leaves and other detritus, at least in north temperate regions. They can be abundant macro-decomposers in both temperate and tropical forests (Hohn and Wagner 2000). New World calycopidine hairstreak butterflies represent a second radiation of detritivores, with more than 160 species believed to feed principally on fallen leaves, flowers, and fruits, although larvae also feed on living plant tissues (Duarte and Robbins 2010). Tineidae, diverse in species and nearly global in distribution, are notable for their metabolic repertoire, which allows their larvae to digest fur (including wool), feathers, horns, shells, owl pellets, fungi, guano, and other substrates rarely exploited by animals (Davis 1987).

While many lepidopteran lineages have predaceous behaviors, very few are obligate or strict carnivores. Lepidopteran parasitoids are rarer still, and are confined to the Cyclotornidae and Epipyropidae and two closely related zygaenoids whose larvae are external parasitoids of Homoptera (Epstein et al. 1998). Most of the prey of obligatory predators are essentially sessile, such as scale insects and ant and bee brood (Pierce 1995). The largest guild of obligatory predators may be lycaenid ant nest inquilines. More than a dozen lineages of blues and hairstreaks have moved beyond being mutualists to predators that consume larvae and pupae in the brood chamber (Pierce et al. 2002; see Pierce and Dankowicz, Chapter "The Natural History of Caterpillar-Ant Associations", this volume). A few novel cases warrant mention. The snail-eating cosmopterigid *Hyposmocoma molluscivora* first spins a silken net over its intended victim and then enters the shell to consume its prey alive (Rubinoff and Haines 2005). The bagworm, *Perisceptis carnivora* (Psychidae), feeds on ants, spiders, and other small arthropods and then silks the husks of its victims to its case (Davis et al. 2008). Hawaiian *Eupithecia* provide another fantastic case: the caterpillars are sit-and-wait predators that can grab flies and other volant prey that land near their perch (Montgomery 1983, Steve Montgomery pers. comm., Fig. 12a, b). Whitman et al. (1994) and Pierce (1995) treat many additional predatory lineages. Facultative carnivory is far more common and occurs in many lineages. For example, some caterpillars will consume smaller and otherwise vulnerable prey such as another caterpillar in the process of a molt or a teneral pupa. Among Noctuidae, some Heliothinae (including *Helicoverpa* and *Schinia*) and some *Lithophane* are known to be quite carnivorous (Whitman et al. 1994; Hardwick 1996; Wagner et al. 2011). Necrophagy of the cadavers of other con- and heterospecific kin is common among arctiine erebids. This behavior is thought to be driven by a need to ingest pyrrolizidine alkaloids which serve in defense and as components of the male courtship pheromones (Dethier 1937; Bogner and Eisner 1992; Eisner 2003).

The most pervasive form of carnivory among lepidopterans may be cannibalism, in which larvae seemingly have no regard to whether the victim is a sibling or unrelated individual. In most cases, the behavior is associated with a shared food resource that could be overexploited to the detriment of both individuals (see review of drivers and correlates by Whitman et al. 1994). Pieridae that feed on small crucifers, for example, have a high rate of cannibalism (Courtney and Chew 1987; Zago-Braga and Zucoloto 2004). Likewise, caterpillars that only feed on the newest leaves of a vine (e.g., heliconiine butterflies), as well as flowers, seeds, and internally in fruits (many grapholitine Tortricidae, Heliothinae, and Lycaenidae), show a high

Fig. 12 Predaceous Hawaiian *Eupithecia* consuming a termite. *Eupithecia* may be the largest genus of Macrolepidoptera with more 1300 described of species worldwide. They are especially rich in Andean South America. A radiation of eighteen Hawaiian species are thought to be principally predaceous (Montgomery 1983, unpubl. data); facultative predation (including cannibalism) occurs in *Eupithecia* species elsewhere (Wagner 2005)

incidence of cannibalistic behavior (Koptur and Lawton 1988; Hardwick 1996; Richardson et al. 2010; de Nardin and de Araújo 2011). Stem borers are frequently cannibalistic, in part because the chance of finding and safely establishing in a new stem is often low and because when tunnels anastomose one individual is likely to be rendered vulnerable by the constraints of their tunnels: i.e., the larva whose flank or rear is exposed to the mandibles of the second caterpillar is likely to fall victim as happens with some hepialids (Wagner 1985). Thus, in many of the known examples, the phenomenon of cannibalism appears to be linked to the risk of *not* eating one's competitor (Richardson et al. 2010; DLW unpubl. observ.). Interestingly, lycaenid caterpillars, which appear morphologically among the least likely candidates to be predators, include a great many species that eat their siblings (Pierce 1995; Whitman et al. 1994)—typically when their kin is molting, pupal, or in an otherwise vulnerable (immobile) state (DLW unpubl. observ.).

To the above, we can add feeding at sugar solutions. *Hypercompe scribonia* caterpillars occasionally feed at baits of fermenting mixtures of beer, sugar, and rotten fruit (used by collectors and photographers to attract noctuids and other moths). Baer (2018) reports consumption of sugary exudates from extrafloral nectaries in lycaenids, riodinids, and a tropical gelechiid.

Population Dynamics in Brief

Lepidoptera, like other insects, weather heavy bottom-up and top-down pressures and consequently tend to have great fecundities, which routinely range between 30 and a few hundred eggs, with some lineages producing many times this number. A single female of the giant Australian hepialid *Trictena atripalpis* may produce more than 30,000 eggs, which are broadcast during flight (Tindale 1932). Given that as few as a single surviving gravid female is needed to replace a previous generation,

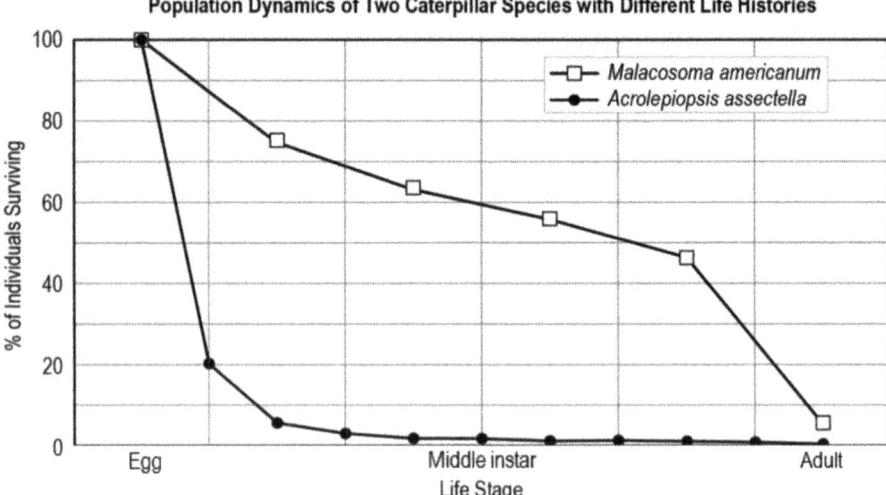

Fig. 13 Caterpillar survivorship curves: type II in *Malacosoma americanum* (Lasiocampidae) and type III in *Acrolepiopsis assectella* (Acrolepididae)

expected mortalities can exceed 99% in fecund lepidopterans—a stunningly high percent sure to surprise many vertebrate biologists. Even small fluctuations around this percentage have the potential to result in population outbreaks or local extirpations; it is our guess that this happens considerably more often than generally appreciated and also speaks to why metapopulation dynamics may be so critical to many insect conservation efforts (Hanski and Thomas 1994; Schultz et al. 2019).

Principal mortality factors include abiotic factors (primarily temperature extremes, but especially freezing temperatures and droughts, i.e., conditions that are too wet or too dry), starvation and plant defenses, pathogens, parasites, parasitoids, and hordes of invertebrate and vertebrate predators. Insects and other fecund invertebrates generally have type III survivorship curves, with greatest mortality coming in the egg and first couple of instars, but type II curves also occur (Price 1997; Schowalter 2017; Fig. 13). In either instance, the caterpillar stage typically experiences the greatest fraction of mortality, as larvae are more numerically abundant than pupae or adults, are relatively sessile, are often the life stage of longest duration, and have the greatest biomass (caloric payoff for natural enemies).

Lepidoptera populations teeter between periods of abundance and scarcity across years and even the generations of a single year. Dramatic swings in populations occur in outbreak species and migrants. Numerous forest pests can reach such high numbers that both their primary and secondary hosts are defoliated. Populations then crash as a result of starvation, responses of natural enemies, or a consequence of unfavorable abiotic conditions, such as extreme winter temperatures and cold, wet springs. Forest outbreak caterpillars tend to be univoltine, polyphagous, spring-feeders that lay their eggs in clusters (Nothnagle and Schultz 1987). Outbreak taxa with summer-feeding caterpillars share similar traits but in addition tend to be gregarious and protected with either physical or chemical defenses (Hunter 1995). A high percentage of the winter- and spring-active geometrids have flightless females, which allows for greater energetic investment in egg production (Wagner and Liebherr 1992) and eliminates the mortality risks associated with flight (Snäll et al. 2007). There is much spatial and temporal variation in such outbreaks, possibly signaling the importance of regional and local microclimate playing important roles in population changes. A single species can have different population dynamics across its range. For example, the forest tent caterpillar (*Malacosoma disstria*) is a boom-and-bust outbreak species northward, but a chronic spring defoliator in the black gum swamps of the southeastern USA (Fitzgerald 1995).

Many lepidopteran crop pests (e.g., *Mythimna unipuncta*, *Peridroma saucia*, and *Spodoptera exigua*, all Noctuidae) are migratory moths that move *en masse*, propelled on the leading edge of storm fronts. Upon arrival, these push through a brood or two and then move on, presumably to avoid natural enemy build-up and diminished food plant availability or quality. This strategy is also shared with many non-pest species: the monarch (*Danaus plexippus*), painted lady (*Vanessa cardui*), and snout butterfly (*Libytheana carinenta*) (all Nymphalidae) have similar population dynamics, moving out of Mexico in the spring, with successive generations pushing farther northward. Smith (1983) hypothesized that inducible plant defenses were the ultimate driver of the famous migration behaviors of sunset moths (*Urania*, Epiplemidae)—that long-distance dispersal during the wet season allowed the caterpillars to find unprotected *Omphalea* (Euphorbiaceae) foliage, which soon became increasingly unpalatable as the plants responded to increasing numbers of feeding caterpillars.

Caterpillar Enemies: Predators and Parasitoids

Principal invertebrate predators of caterpillars include ants, spiders, assassin bugs, predaceous stink bugs, lacewing larvae, ladybugs, beetles, earwigs, sand wasps, vespid, and related wasps. Social wasps and ants can be chronic threats to aboveground feeding caterpillars, with the latter representing a special threat given their ability to recruit to sites of high caterpillar densities. Caterpillars that feed or take refuge on the ground are exposed to additional enemies, including centipedes and ground-dwelling species of ants, spiders, beetles, and still others. The importance of mites that attack eggs and early instars is undoubtedly considerable, but not well quantified.

Caterpillars are a staple of songbirds and comprise the main food for both nestlings and adults of many species (Lack 1950; Laney et al. 2015; Holmes 1980; Singer et al., Chapter "Predators and Caterpillar Diet Breadth: Appraising the Enemy-Free Space Hypothesis", this volume; Di Cecco and Hurlbert Chapter "Caterpillar Patterns in Space and Time: Insights From and Contrasts Between Two Citizen Science Datasets", this volume). As such, birds have been a major selective force in shaping caterpillar phenotypes and behavior, including what, when, and how they eat (Heinrich 1979, 1993, and text below). Rodents, especially mice, are caterpillar and pupa hunters that replace birds in importance as predators of ground- and near-ground-dwelling lepidopterans (Wagner et al. 2011).

The early stages of Lepidoptera are attacked by thousands of different hymenopteran (wasp) and dipteran (fly) parasitoids (see Stireman and Shaw, Chapter "Natural History, Ecology, and Human Impacts on Caterpillar Parasitoids", this volume). Eggs are parasitized by eupelmid, mymarid, platygasterid, and trichogrammatid wasps. Minute microhymenopteran wasps—eulophids, encyrtids, among others—attack first or second instars (as well as later instars). The importance of such early mortality factors is underappreciated and only well characterized for a number of crop and forest pests (Hawkins et al. 1997).

The role of diet in mediating the interactions with natural enemies is currently an area of great ecological importance and almost boundless in scope. A caterpillar's diet can determine how likely it is to be discovered, eaten, or parasitized (Lill et al. 2002), its vulnerability to infection, whether an attack will proceed or be terminated, the nature of the immune response, and more—see Singer et al., Chapter "Predators and Caterpillar Diet Breadth: Appraising the Enemy-Free Space Hypothesis"; Koptur et al., Chapter "Caterpillar Responses to Ant Protectors of Plants"; Lill and Weiss, Chapter "Host Plants as Mediators of Caterpillar-Natural Enemy Interactions"; Smilanch and Muchoney, Chapter "Host Plant Effects on the Caterpillar Immune Response"; Salcido et al., Chapter "Plant-Caterpillar-Parasitoid Natural History Studies Over Decades and Across Large Geographic Gradients Provide Insight into Specialization, Interaction Diversity, and Global Change"; all in this volume.

Morphological, Physiological, and Behavioral Adaptations for Circumventing Bottom-Up and Top-Down Pressures

The daily caterpillar agenda is a simple one: eat and don't get eaten. We first treat adaptations for feeding on plants and then do the same for traits that promote survival in a world replete with enemies. Our treatment is meant to be illustrative, and introduce ecologists to pervasive phenomena, key adaptations, demonstrably adaptive traits, and consider selective pressures that occur across lepidopterans. The anatomical and behavioral adaptions across the order for dealing with predators alone are legion, worthy of their own review or book. Considered collectively, we suggest that they constitute a key set of stratagems that have allowed caterpillars to become the most ecologically successful order of externally feeding herbivores in many terrestrial ecosystems.

Bottom-Up pressures The external structural modifications of caterpillars for feeding on plants (or other substrates) (see also Kaur et al., Chapter "Surface Warfare: Plant Structural Defenses Challenge Caterpillar Feeding", this volume) are rather modest relative to the universe of morphological and behavioral traits that have evolved to deal with the top-down pressures facing caterpillars. Likewise, their modest structural modifications stand in contrast to their impressive physiological abilities to rapidly digest and detoxify a huge range of secondary plant compounds (as well as insecticides and other toxins) (see Groen and Whiteman, Chapter "Ecology and Evolution of Secondary Compound Detoxification Systems in Caterpillars", this volume).

The most dramatic morphological adaptations are exhibited by hypermetamorphic leafmining lineages, most pronounced in Gracillariidae, discussed above. Caterpillars that bore into wood and stems often have prognathous mouthparts, shortened legs, prothoracic shields, and anal plates for the attachment of muscles that facilitate tunneling. Their mandibles are highly sclerotized, melanized, and thickened, with enlarged muscle-packed heads to power them. Proportionately large heads also occur in seed, grass, and many conifer feeders. Bernays et al. (1991) were among the first to note that graminoid-feeders, which must deal with elevated levels of silica, have enlarged mandibles with reduced incisors that are used to clip pieces of host tissue. Convergently, analogous clipping mandibles have evolved in taxa that feed on tough, thickened leaves (Bernays 1998). Conifer-feeding caterpillars also have large heads, especially in neonates and early instars, when thick, toughened needles represent a biomechanical challenge (DLW pers. observ.)

A common theme shared by many larger external feeders, which presumably relates to increasing mobility, is the loss or reduction of anterior prolegs, which allows a caterpillar to loop, rather than crawl. Proleg reductions are especially common in barkresters and species that specialize and move between very young foliage, e.g., *Catocala*, *Zale*, kindred erebids, as well as oncocnemidine noctuids. This strategy is most obvious in the Geometridae, which typically have lost the first three pairs of abdominal prolegs. Enhanced ambulatory abilities have been further extended in many lineages by a lengthening of the abdominal segments—in some geometrids, the length of any one of the anterior abdominal segments can exceed the collective length of the last three abdominal segments (A8–10).

One special aspect of the caterpillar integument that has yet to be studied is their ability to walk over and even consume (Weinhold and Baldwin 2011) the tacky glues used by plants to either exclude or ensnare insects. The latter behavior yields an additional source of nutrients. Tarweeds (Asteraceae), many four o'clocks (Nyctaginaceae), and sundews (Droseraceae)—plant lineages that have sticky secretions that entrap insects—have dietary specialists, with the first two of these supporting small radiations of moths in the American Southwest (DLW unpubl. data).

Behavioral adaptations for circumventing plant defenses are legion. The most widespread of these is simply avoiding plant defenses in space or time (Lawton 1978). Early instars and microlepidopterans can avoid well-defended tissues of their host by feeding on unprotected organs, tissues, or, in the case of miners, cell types. A great many lineages in both tropical and temperate zones, and essentially all xeric communities, synchronize larval development to periods when the larvae will have access to preferred tissues such as buds, young leaves, flowers, and fruits (Coley and

Barone 1996; Endara et al., Chapter "Impacts of Plant Defenses on Host Choice by Lepidoptera in Neotropical Rainforests", this volume).

Dussourd (1993, 2017) reviewed behavioral stratagems that caterpillars use to deactivate plant defenses, emphasizing the disruption of plant vascular and secretory systems and the application of exocrine secretions, and related behaviors, that serve to either elude constitutive defenses or prevent inducible plant responses. Dussourd (2017) reviews an array of fascinating cases of pinching, trenching, vein-cutting, and other behaviors that sabotage the secretory systems of lactiferous plants. Likewise, girdling, clipping, and other behaviors presumably evolved to prevent delivery of inducible defenses. Numerous exocrine secretions, e.g., from the salivary, mandibular, and cervical glands, are employed to inhibit plant defense responses (Felton 2008; Vegliante and Hasenfuss 2012). *Theroa* caterpillars (Notodontidae) have co-opted their acid-producing cervical gland (used widely by heterocampine notodontids in defense) to also incapacitate the laticifers of their euphorb hosts (Dussourd et al. 2019). The caterpillars chew into host laticifers while simultaneously releasing acids from the gland into the plant and in so doing prevent latex delivery to leaf tissues distal to the wound.

A storied case where behavior allows for the exploitation of otherwise protected plants occurs in caterpillars that feed on plants protected by furanocoumarins, which are widespread in Apiaceae and Rutaceae and to a lesser extent in Asteraceae, Fabaceae, Moraceae, and others. Furanocoumarins are reactive cyclic ring compounds that in the presence of ultraviolet light, crosslink DNA and debilitate an individual's ability to function. Generalist caterpillars that lack the ability to deactivate furanocoumarins that are exposed to UV light post-ingestion will likely meet a bad end. Several lepidopteran lineages have evolved workarounds: e.g., they fashion and feed within light-blocking shelters, while others bore into stems and in so doing escape phototoxic repercussions (Berenbaum 1983; Fukui 2001; Lill and Marquis 2007).

Adaptations for Dealing with Top-Down Pressures Much of a caterpillar's phenotype, at least for those that feed externally, has been shaped by the collective selective pressures brought by a species' constellation of predators and parasitoids—its coloration, shape, posture, and texture are quintessentially important to how it is perceived (or, commonly, not perceived) by its enemies (Heinrich 1993; see especially reviews by Salazar and Whitman (2001) and Greeney et al. (2012)). Most are cryptic to avoid detection by visual, tactile, and even olfactory predators (Rothschild 1973; Lederhouse 1990; Heinrich 1993; Stamp and Wilkens 1993). Those that are unpalatable are commonly rendered in bold, aposematic yellows, oranges, reds, black, and white (Rothschild 1973, 1985, 1993; Marquis and Passoa 1989; Bowers 1993). Even this most basic of dichotomies has exceptions: caterpillars can be cryptic at a distance and aposematic at close range—a stratagem shared by many taxa with disruptive coloration (Stamp and Wilkens 1993). Palatable caterpillars can mimic unpalatable ones (Berenbaum 1995; Wagner et al. 2011; Fig. 14). Natural selection may favor warning coloration not only when toxins or a nearby mimicry model is available but also if the plant environment forces caterpil-

Fig. 14 Mimicry in *Acronicta* (Noctuidae) caterpillars: mimics to left and models to right. North American members of the genus *Acronicta* appear to mimic many other protected (and divergently related) caterpillars, including arctiine Erebidae, Lasiocampidae, Limacodidae, Lymantriine Erebidae, and Megalopygidae. Three putative examples are shared: (**a**) *Acronicta americana* and a common model (**b**) *Halysidota tessellaris* (Erebidae: Arctiinae); (**c**) *Acronicta impleta* and a common model (**d**) *Orgyia manto* (Erebidae: Arctiinae); (**e**) *Acronicta radcliffei* and a common model (**f**) *Datana drexelli* (Notodontidae) middle instar. (Image (**f**) courtesy of Michael C. Thomas)

lar apparency: for example, plants with sparse or narrow leaves may make crypsis impossible (Prudic et al. 2007).

The most widespread adaptation for palatable taxa, obvious to any caterpillar hunter, is background matching—something they do exceedingly well, drawing on ontogenetic, plastic, morphological, and behavioral means to blend into their backgrounds. Their preferred resting site seems to be the principal driver for their coloration and shape. At a superficial level, caterpillars that are green rest on leaves or other green tissues (Fig. 15b, c, g); those that are brown or gray rest on bark

Fig. 15 Background matching caterpillars: (**a**) *Bryolymnia viridata* (Noctuidae) a lichen mimic; (**b**) *Sicya morriscaria* (Geometridae) on *Phoradendron juniperinum*; (**c**) *Lithophane lepida* (Noctuidae) on *Pinus rigida*; (**d**) *Plagodis alcoolaria* (Geometridae) on *Betula populifolia*; (**e**) *Catocala ilia* (Noctuidae) on *Quercus rubra*; (**f**) *Nemoria bifilata planuscula* (Geometridae) on *Quercus* sp.; (**g**) *Ianassa lignicolor* (Notodontidae) on *Quercus rubra*

(Figs. 15a, d, f, and 21a–d); flower feeders have stratagems for matching floral colors (Figs. 8 and 16b, d) (Carter and Hargraves 1986; Wagner 2005). The degree to which caterpillars match their background ranges from proximate resemblance in many dietarily generalized taxa to the highly perfected phenotypes of many host plant specialists—involving details of shape, integumental texture, color, reflectance, and choice of resting sites (Figs. 15, 16, and 21a, b)—that attest to the powers

Fig. 16 Caterpillar defensive stratagems. (a), (b) Faux tachinid fly eggs?: *Carboniclava alpicoides* (Notodontidae) (left) and *Prolimacodes badia* (Limacodidae) (right). Some externally feeding caterpillars have a white to creamy spot hypothesized to discourage oviposition—an idea in need of testing, especially given that the *Prolimacodes* caterpillar shown here bears a tachinid egg (toward center) and a similar-sized creamy spot we suggest might serve as a faux egg, in same plane to right. (c), (d) Co-option of flower pigments by flower feeders: *Dichordophora phoenix* (Geometridae) on *Krameria ramosissima* and *Strymon melinus* (Lycaenidae) on *Guaiacum angustifolium*. (Image (b) courtesy of Michael C. Thomas; (c) and (d) courtesy of Berry Nall)

of Darwinian natural selection and the undeniable importance of visual predators, and especially insectivorous birds, in shaping caterpillar phenotypes and behaviors (Schultz 1983; Heinrich 1993). We would go so far as to say that it is the selective pressures of birds and others that have so tuned caterpillars' phenotypes that most externally feeding caterpillars are identifiable to species. Stated differently, a photographic guide to internal or otherwise concealed feeders of a given region, based on larval phenotype alone, would be of more modest utility than a guide to the region's externally feeding, i.e., visually apparent, caterpillar fauna (contrast the images of external and internal feeders in Porter 1997).

Many caterpillars masquerade as inedible or unpalatable objects. Bird-dropping mimicry is common among early instar Papilionidae, a ploy that has independently evolved in Erebidae, Geometridae, Noctuidae, and others (Carter and Hargraves 1986; Suzuki and Sakurai 2015; Wagner 2005). Dead, curling leaves are another common target of caterpillar mimicry. Among the most intriguing, and least studied, are those that masquerade as previously parasitized victims. Caterpillars with white spots that bear a resemblance to tachinid eggs occur in *Prolimacodes* (Limacodidae), Notodontidae, and others (Fig. 16a, b). Whether these in fact discourage oviposition by tachinid females has yet to be experimentally tested. In an analogous fashion, the cocoons of some moths are spun in such a way as to appear to have parasitoid emergence holes or otherwise appear old and inviable (e.g., Epstein 1995).

Several different groups of flower-feeding caterpillars have the ability to match the coloration of the floral tissues that they are consuming. How they do this—presumably by moving pigments from the midgut into the hemocoel and outer tissues of the thorax and abdomen—is neither fully understood nor appreciated. The phenomenon occurs commonly among species of *Eupithecia*, *Nemoria*, and *Dichordophora* (all Geometridae), *Sympistis* (Noctuidae), many Lycaenidae, and others (DLW pers. observ., Fig. 16c, d). Dietary carotenoids facilitate background matching in *Trichoplusia ni* (Noctuidae) larvae—even in this example, the underlying mechanisms behind the different larval phenotypes are not yet understood (Welch et al. 2017).

An abundance of secondary setae has evolved independently across a diverse array of externally feeding caterpillars (and, tellingly, not in internal feeders). Most immediately an abundance of hairlike setae can serve as a physical barrier that protects a caterpillar from many parasitoids and natural enemies of equal size. Most birds avoid hairy caterpillars, but cuckoos are a well-known exception (Barber et al. 2008 and references therein). When the setae are barbed (Erebidae: Arctiinae), especially if deciduous or brittle as in tiger moths, they represent an added danger, especially if they embed in eye tissues, mucosal membranes, and skin (Hossler 2010).

Likewise, integumental spines, chalazae, and scoli physically separate a caterpillar from its normal cast of predators. These integumental outgrowths often bear reverse barbs that catch and impede the entry of an enemy's mandible, leg, or ovipositor. Even the relatively open arrangements of spines in nymphalids and saturniids can distance a caterpillar from many spiders, generalist tachinids, beetles, ants, and others (Bowers 1993).

Sack or case bearers (e.g., Mimallonidae) and internal borers that back out of their case or tunnel to defecate often have an armored anal region that thwarts attack when they are in the business of releasing their feculae. The abdominal terminus of many mimallonids and *Thurberiphaga* (Noctuidae) is extraordinary modified—multiple terminal abdominal segments fuse to yield an armored, flattened plate that blocks entry into the caterpillar's feeding site (Fig. 17). An anal fork or comb that assists in the launching of frass away from the caterpillar's feeding or resting sites occurs in many species (see below, Fig. 4d).

Physiological Responses We focus here on chemical defenses, nearly all of which can be externally deployed substances that discourage the efforts of would-be predators. Internal (immune) responses to microbes and parasitoids are discussed above and in Smilanch and Muchoney (Chapter "Host Plant Effects on the Caterpillar Immune Response", this volume). Many caterpillars regurgitate the contents of the crop as part of their defense repertoire—with some doing so even before they are touched. The expelled fluid tends to be cohesive with appreciable surface tension so that the whole of it also serves as a warning—often it is compositionally distinct

Fig. 17 Armored posterior plates: *Cicinnus melsheimeri* (Mimallonidae) (left) and *Thurberiphaga diffusa* (Noctuidae) (right). In *Cicinnus*, a sack-bearer, segments A8–A10 are fused into a plate that is used to block the posterior end of the caterpillar's case or leaf shelter; in *Thurberiphaga*, a borer in the fruits of wild cotton, segments A9–A10 form a hardened plate used to plug the entrance to its chamber, where it feeds on developing seeds. (Image (b) courtesy of Robert Behrstock)

from newly masticated, undigested leaf tissue (Peterson et al. 1987; Smedley et al. 1993; Wagner pers. observ., Fig. 17). The Australian oecophorid, *Myrascia*, has a foregut diverticulum that stores oils from its eucalyptus hosts that are expelled when alarmed (Common and Bellas 1977).

When an enemy touches the caterpillar, the regurgitant is typically wiped across the point of contact. Many caterpillars recover (imbibe) the fluid after the threat of danger has passed. The crop contents of *Malacosoma americanum* (Lasiocampidae) repel ants; noxious compounds in the regurgitant of other caterpillars can repulse parasitoids (Peterson et al. 1987; Desurmont et al. 2017; see especially Bowers 1993). Caterpillars that use regurgitant in defense may have a larger portion of their digestive system dedicated to their crop (Grant 2006). In some lineages, threatened caterpillars simultaneously exude the contents of the crop and rectum (Brower 1984, Fig. 18). From the latter, the equivalent of a single frass pellet is expelled from the anus; pre-consolidated feculae can be sticky and problematic for some would-be attackers.

The cervical gland or adenosoma (see Fig. 18) of many Noctuoidea and butterflies functions in the manufacture, storage, and release of defensive compounds: acids, alcohols, aldehydes, esters, and terpenes (Osborn and Jaffe 1998; Hallberg and Poppy 2003; Vegliante and Hasenfuss 2012). The gland is especially well developed in the Notodontidae (e.g., Heterocampinae and Nystaleinae) in which much of the anterior portion of the thorax and abdomen is given to the manufacture and storage of formic acid and other minor constituents that are forcibly ejected at would-be attackers (Eisner et al. 1972; Kearby 1975; Attygalle et al. 1993; Eisner 2003). The best-studied caterpillar exocrine structure may be the dorsal osmeterium of Papilionidae, a fleshy horn-like protrusion everted from just behind the head capsule; it also serves a defensive role, delivering a potpourri of propionic and butyric acids (Frankfater et al. 2009). Vegliante and Hasenfuss (2012) discuss additional

Fig. 18 *Datana eileena-perspicua* complex (Notodontidae) on *Arctostaphylos pungens*. Alarmed caterpillars regurgitate and exude fluid from the anus; if left undisturbed the caterpillar may reabsorb both

exocrine glands in Thyrididae, morpho butterflies, and zygaenoids thought to function in defense.

But on the whole, defensive secretions are inexplicably uncommon in caterpillars. The vast majority of caterpillars are unprotected and as a consequence are among the most sought-after prey by birds, other vertebrates, ants, wasps, and others. There are no exocrine secretions widespread across the order. Secretory (primary) setae occur in a few disparate lineages: e.g., Pieridae (Smedley et al. 2002, unpubl data), *Cobubatha* and *Emarginea* (both Noctuidae) (Keegan et al. 2021, Fig. 19), and the hollow-tuft genera of heterocampine Notodontidae (Miller et al. 2021). *Eupackardia calleta* (Saturniidae) secretes a clear fluid, imbued with amines and phenolics, from its chalazae when threatened (Deml and Dettner 1993, frontispiece). Cousins in the Oxyteninae (*Oxytenis*, *Homoeopteryx*, and *Asthenidia*) secrete a sticky substance from the chalazae that discourages ant predation (Aiello and Balcazar 1997, Annette Aiello in litt.). When accosted, *Prolimacodes* (Limacodidae) larvae exude clear droplets from pores below the subdorsal ridge. Reflex bleeding is perplexingly rare; we are aware of only a few examples in the Arctiinae, Geometridae (Fig. 20), and Noctuidae. Caterpillars of the wasp moth *Gymnelia salvini* (Erebidae: Arctiinae) bleed yellow drops when disturbed (Annettte Aiello unpubl. data).

While chalazae, scoli, and an abundance of secondary setae represent physical barriers to would-be enemies, their threat is elevated when impregnated with irritants, toxins, and allergens. Urticating setae—either deciduous or given to breakage such that sections become embedded in unprotected tissues—are found in some lasiocampids (e.g., *Gloveria*); lymantriine erebids (e.g., gypsy and brown-tailed moths, *Lymantria dispar* and *Euproctis chrysorrhoea*, respectively); acronictine

Fig. 19 Secretory setae: (**a**) *Emarginea percara* (Noctuidae) and (**b**) *Cobubatha dividua* (Noctuidae). Secretory setae tend to be more common among early instars and small lepidopterans, e.g., in Notodontidae, although many Pieridae and Saturniidae retain secretory setae through the last instar. Why this is so—the chemical nature of the exudate, and what natural enemies are turned away—are all questions that warrant more study

Fig. 20 Reflex bleeding (each caterpillar was lightly touched with forceps): (**a**) *Somatolophia ectrapelaria* and (**b**) *Meris alticola* (both Geometridae). Both animals are aposematically colored, as might be predicted for reflex bleeders. In each, the hemolymph is brightly colored (and not clear as in caterpillars with secretory setae in Fig. 19)

noctuids (e.g., *Acronicta americana*); many arctiines (e.g., great tiger moth and hickory tussock caterpillar, *Arctia caja* and *Lophocampa caryae*, respectively); and perhaps most famously processionary caterpillars (e.g., *Thaumetopoea pityocampa*) (Notodontidae) (Kawamoto and Kumata 1984; Bowers 1993; Wagner 2005; Hossler 2010). Urticating setae can cause mild to (rarely) severe dermatological reactions when embedded in the skin, eyes, and mucosal membranes. In the eastern USA, a great number of medical cases result from exposure to the hickory tussock moth *Lophocampa caryae* (Erebidae: Arctiinae) (Kuspis et al. 2001; Wagner 2009).

Among the best-protected (and feared) caterpillars are those with stinging setae. Hollow, toxin-filled setae occur in the hemileucine Saturniidae, Limacodidae, and Megalopygidae (Kawamoto and Kumata 1984; Everson et al. 1990; Hossler 2010). The stings of larger caterpillars can be quite uncomfortable and, in rare cases, fatal to humans. Severe stings from *Lonomia*, a gregarious Neotropical hemileucine, can cause deadly internal hemorrhaging if the victim is envenomated by multiple caterpillars (Hossler 2010).

Behavioral Ploys Behaviors to elude top-down pressures are nearly endless in nature; we share some of the most widespread primary (prior to discovery) and secondary (after discovery) defenses common among lepidopterans. We encourage all to consult the review by Greeney et al. (2012), which is anchored to Neotropical caterpillars. One of the most widespread behaviors to avoid predation is to become nocturnal, which, while serving to evade daytime hunters, exposes the lineage to a sweep of nocturnal hunters: mice, snakes, bats, geckos, ants, spiders, and others (Carter and Hargraves 1986; Bernays 1998; Kalka and Kalko 2006; Wagner et al. 2011). The very fact that so many lineages are nocturnal feeders can be taken as a testament of the importance of birds and other visual predators as drivers of caterpillar behavior and evolution. Heinrich (1993) has done much to draw attention to how bird predation has shaped caterpillar phenotypes, how they rest, and even how caterpillars feed. Since birds are able to assess leaf damage as a cue to caterpillar presence, many caterpillars are quick to move away from their feeding site: others feed neatly along a leaf margin or midrib to lower their apparency (Fig. 15g). Caterpillars from at least 12 families clip leaves damaged while feeding and drop these to the ground (Dussourd et al. 2016), ostensibly to eliminate telltale signs of their whereabouts (Heinrich 1993). Additionally, the action might also serve to sabotage a plant's inducible defenses (Dussourd in litt.).

Resting postures, especially among those that are attempting to blend into the background, can add greatly to a caterpillar's likelihood of survival. Twig-mimicking geometrids are decidedly more convincing while resting than when feeding (Fig. 15d). A subset of bark-resting caterpillars even mimic lichens (Fig. 15a, e). Many pine feeders rest with their heads buried at the bottom of a leaf fascicle; these, predictably, have reddish coloration anteriorly that help the caterpillars match the red-orange needle bundle sheathes of their hosts (Fig. 15c). Some notodontids use their own bodies to complete the ragged edge of leaf they simultaneously feed upon (Fig. 15g). The opposite also works: to so contort the body at rest that the caterpillar is rendered unrecognizable as anything edible (Fig. 21e–g). This is what caterpillars do: hide, masquerade, deceive, and blend. The galaxy of examples, and the details of their phenotypes, provide an incontrovertible testament to the powers of natural selection (and the importance of visual predators) in their evolution.

Because volatiles in frass are used by both caterpillar predators and parasitoids to locate prey or hosts (Weiss 2003, 2006; Moraes et al. 2012), how caterpillars defecate and what they do with the frass pellets have enormous fitness consequences. Some distance themselves from their feculae or frass, while others collect their pellets and use these to repel, confuse, block, or otherwise thwart would-be enemies (Weiss 2006). If frass is repellent, it is often gathered into the immediate vicinity of the larvae—either within the shelter (commonly) or affixed to a caterpillar's resting or feeding site (Weiss 2006). Prepupal larvae frequently weave their frass into their shelters or cocoon walls and in so doing better conceal themselves and construct a more formidable refuge. Those that live well off the ground can

Fig. 21 Flash coloration in bark-resting caterpillars and contortionists: (**a**), (**c**) *Catocala aholibah* (Noctuidae) and (**b**), (**d**) *Apotolype blanchardi* (Lasiocampidae). Top row: cryptic nature of these caterpillars at rest; middle row images show their respective venters (same individuals); bark-resters knocked to the ground quickly right themselves. Non-resemblance in resting postures: (**e**) "*Caripeta*" *hilumaria* (Geometridae), perhaps a litter mimic that often hides in leaf curls; (**f**) *Grotella tricolor* (Noctuidae: Grotellinae) on *Boerhavia erecta*; (**g**) *Chalcopasta howardi* (Noctuidae: Stiriinae) on *Hymenothrix wrightii*. In (**f**) and (**g**), the contorted posture more closely resembles a bud, flower, or gall, than a caterpillar

simply release their frass, which falls to the ground and away from their feeding or resting site. Those that live near the ground or within shelters run a much higher risk of detection. Some of these have evolved behaviors that serve to compensate for this challenge: some by moving away from their site of defecation (armyworms and ground-foraging arctiines) or forcibly launching their frass pellets from their feeding and resting sites (Caveney et al. 1998; Weiss 2003, 2006; Moraes et al. 2012). Caterpillars with an anal comb or fork (Fig. 4d) tend to be "frass flickers": included are many shelter-formers (Gelechiidae, Hesperiidae, Tortricidae) and lichen moths (Erebidae: Lithosiini) (Caveney et al. 1998; Weiss 2006). Others flick their posterior abdominal segments at the same moment that the frass is expelled to toss the pellet away. Zygaenoids have a mechanism for forcibly firing their feculae out the anus. An unusual twist occurs in some larentiine geometrids (e.g., *Heterophleps*): late instars grab each pellet with their mandibles and throw their bodies in a way to hurl the pellet as many as 20 body lengths away from the feeding site (Wagner et al. 2002). Weiss (2006) describes these and other varied uses of feculae by caterpillars in her review of insect "fecology."

Upon discovery by a predator, an additional arsenal of secondary defensive strategies is invoked. Some caterpillars leap or fall to the ground; these may remain

Fig. 22 Caterpillar weaponry? (**a**) *Harrisimemna trisignata* retains its head capsules at each molt; when disturbed, these are wielded at points of contact. (**b**) The function of the paddlelike setae in *Acronicta funeralis* has yet to be determined: perhaps they serve to swat away ants, wasps, and other small enemies (Wagner et al. 2011). (Image (**a**) courtesy of Pat Burkett)

motionless or thrash violently, the latter perhaps serving to drive them beneath surface debris (Wagner et al. 2011). Many bark-resters have a boldly colored venter, visible on their way to the ground, that is wholly concealed immediately upon righting themselves (Wagner et al. 2008, Fig. 21a–d). Others drop on a belay line of silk to a position of safety or wait suspended for a spell, before ascending to their shelter or previous feeding site (DeVries 1987; Sugiura and Yamazaki 2006). A great many retain their purchase and adopt a threat display, regurgitate, and/or discharge defense secretions. Startle responses can involve the display of eyespots and the adoption of snake-like postures and, in some Sphingidae, include a whistle or hiss, made by forcing air out the spiracles (Wagner 2005; Janzen et al. 2010; Greeney et al. 2012; Dookie et al. 2017; see Yack, Chapter "Acoustic Defence Strategies in Caterpillars", this volume). Some will flare their mandibles. Many snap at their attackers. Although none inflicts a significant bite that would affect a vertebrate, the brashness and rapidity startle even large vertebrates (humans included).

These same actions can knock away invertebrates of roughly equal size and smaller, and can be especially advantageous for those caterpillars that feed on trees and shrubs, when they are able to propel their enemies to the ground. Caterpillars often throw their head (with or without biting) or flick their rear in an attempt to curb attacks of flies, spiders, and especially ants—with great gain if that ant is a solitary scout. *Acronicta* (Noctuidae) will go so far as to grab ants and hurl them to the ground (Zacharczenko 2017). The head capsule mace of *Harrisimemna* (Noctuidae) (Fig. 22a) appears to be an armament for warding off invertebrate predators and parasitoids. We guess that the paddlelike setae of some caterpillars also serve this function (Fig. 22b).

Silk

Lepidoptera are renowned for their capacity to produce silk, sometimes in copious amounts. Global silk markets, anchored principally to the Oriental silk moth, were valued at more than $19 billion in 2020 and are expected to rise to near $29 billion over the next 5 years (Market Data Forecast Services 2021). The silk gland of caterpillars is a large, paired structure that manufactures and stores silk precursors in a semi-liquid state. While the twisting and convoluted gland may only reach to the sixth abdominal segment (Eaton 1988, Figs. 5 and 6), if the gland were stretched linearly, its full extent might exceed the caterpillar's body length in some taxa. Insect silks are semicrystalline fibers comprised principally of two proteins, sericin and fibroin, that are manufactured and stored as aqueous solutions within the silk glands (Sutherland et al. 2010). In Lepidoptera, these are combined as they are extruded from the labial spinneret (Figs. 2b and 3b, d), which immediately upon secretion, form a strong, light, water-insoluble polymer used for myriad purposes by caterpillars (Craig 1997). We argue here that the ability of Lepidoptera to utilize silk to solve myriad ecological challenges has greatly contributed to the evolutionary success of the order.

In essence, silk is an amino acid polymer and as such requires a large investment of nitrogen (>16% by weight) (Ngô and Bechtold 2018), a limiting element in plant tissues relative to what is required for the metabolism and growth in animals. It is somewhat of an enigma that an element of scarcity, essential for function and growth, would be discharged in such prodigious amounts by caterpillars and that we would find ourselves arguing that the manifold uses of silk by caterpillars represent one of the order's hallmark adaptations.

The importance of cocoons, nests, and shelters constructed with silk (Fig. 23) is worthy of a separate review: silken refuges exclude natural enemies and allow caterpillars to construct and regulate their own microenvironments to prevent desiccation (e.g., in canopy foliage and in myriad xeric communities), exclude water during floods, trap heat and serve as hot houses, and much more (Bernays and Graham 1988; Lill and Marquis 2007; Baer and Marquis 2020; Marquis et al., Chapter "The Impact of Construct Building by Caterpillars on Arthropod Colonists in a World of Climate Change", this volume; see also below). The use of silk to fashion a cocoon in which pupation, and often diapause, will take place is among the most universal uses of silk across the order and, indeed, holometabolous insects (Craig 1997).

In many ways, the silken cases, shelters, and nests of caterpillars are analogous to cocoons but serve the needs of feeding larvae. The majority of microlepidoptera use silk to form a shelter or to line a chamber within the larval feeding substrate; but even among macrolepidopterans, there are sheltering-forming lineages (e.g., some hesperiids, pierids, riodinids, and swallowtails; tent caterpillars; arctiine erebids; many noctuines and notodontids; and geometrids) (Wagner 2005; Wagner et al. 2011). Aquatic crambids use silk to fashion retreats along the surface of submerged plants, rocks, and other objects.

Fig. 23 Some of the manifold uses of silk by caterpillars. (**a**) Nest of *Malacosoma americanum* (Lasiocampidae). (**b**) Case-making microlepidopteran, *Coleophora xyridella* (Coleophoridae). (**c**) Guy line used by many twig-mimicking ennomine geometrids to support the anterior end of the body while at rest; note this line would be nearly invisible to the casual eye, but here has been accentuated by the flash used to take this image. (**d**) Synchlorine geometrids use silk to attach pieces of their host plants (individually chewed free by caterpillar) to dorsal abdominal warts; one can make extraordinary examples by offering captive, flowering-feeding *Synchlora* caterpillars small pieces of colored tissues or flowers with different petal colors. (**e**) Frass chain-refuge of *Adelpha iphiclus* (Nymphalidae); frass chains are fashioned by several genera of biblidine, charaxine, and limenitidine nymphalids; some will additionally weave in necrotic leaf fragments to the leaf vein to fortify their resting site. (**f**) Sensory net of *Brenthia* caterpillar (Choreutidae); silk is deposited wherever the caterpillar walks; immediately adjacent to its head in this image, the caterpillar has chewed an escape hole through which it will dive if any movement is detected on its side of the leaf. (Image (**b**) courtesy of Nelson DeBarros; (**e**) courtesy of Keith Willmott)

Some caterpillars spin communal nests (Fig. 23a). While tent caterpillars are widely known to form nests, communal tent makers also occur among early instar nymphalids (e.g., checkerspots and crescents), pygaerine notodontids, tropical pierids, arctiine erebids (e.g., fall webworm), a few tortricids (e.g., *Archips cerasivorana*), crambids (e.g., *Saucrobotys futilalis*), and others (see Costa and Pierce 1997). Not only do these offer protection from natural enemies but also serve in thermoregulation by acting as greenhouses for digestion and other physiological functions that heat up well above ambient temperatures on cool days (Fitzgerald 1995).

Case-making, where the caterpillar constructs a portable silken case (Fig. 23b), has evolved in no fewer than a dozen moth groups; three particularly speciose lineages include Psychidae, Coleophoridae, and *Hyposmocoma* (Cosmopterigidae). Prepupal Heliozelidae make a mobile case that they affix to their final pupation site. Cases may be made of pure silk or include materials that are woven into the walls: feculae, plant tissues, sand and minute pebbles, twigs, or, in the predaceous psychid *Perisceptis carnivora*, the cadavers of its hapless victims (Davis et al. 2008).

Silk is used for both local and long-distance dispersal. Some caterpillars drop on belay lines to move between leaves, a phenomenon that is especially common during outbreaks when foliage is deteriorating. Among microlepidopterans, it is common to see prepupal larvae dropping from trees on silk lines on their way into leaf litter or soil where pupation will occur. Taxa with flightless females (e.g., some Psychidae, Geometridae, and Lymantriinae) disperse by "ballooning," whereby early instars drop from silken lines and wait for winds to carry them about (Zarlucki et al. 2002; Moore and Hanks 2004).

Silken pads are often spun at a caterpillar's resting site that enable them to secure purchase on smooth leaves and other substrates. These pads or buttons of silk, spun prior to a molt, allow larvae to engage their anal prolegs, anchoring the body (integument) to the silk pad. This helps the next instar free itself of its previous skin over the course of a molt. Likewise, silk buttons, spun by prepupal caterpillars, are used as attachment sites to facilitate pupation and, later, eclosion.

Silk is used in still many other ways. There are caterpillars that lay down silk whenever they walk, which, among other things, allows them to quickly retrace their path, for example, in times of danger. Upon disturbance, scores of microlepidopterans and smaller-bodied macrolepidopteran caterpillars drop (or throw themselves) from the resting or feeding site on a belay line, which can be ascended once danger has passed. Many ennomine geometrids do this preemptively at night, dropping from their perch on a short line of silk, and then returning to their perch at daybreak, thereby thwarting the marauding of nocturnal caterpillar hunters such as spiders, carpenter ants, opilionids, and arboreal carabids and mice (McFarland 1988; Wagner et al. 2002; Wagner 2005). Twig-mimicking geometrids use a fine strand of silk, running from spinneret to twig, as a guy-line to secure their purchase (Fig. 23c). The externally feeding instars of bucculatricids spin molting cocoons that afford protection when transitioning between instars (Braun 1963). Synchlorine Geometridae use silk to attach bits of the flowers that they are eating to their dorsum to camouflage themselves (Wagner et al. 2002; Canfield et al. 2009, Fig. 23d). Tent caterpillars (Lasiocampidae) lay down a silk trail, impregnated with pheromone, that siblings follow to and from the nest (Fitzgerald 1976, 1995; Ruf et al. 2001). In many lineages, frass is collected and silked together to form a larval refugia or frass-encrusted cocoon—see review by Weiss (2006). A curious example of such is that exhibited by early instar Limenitidinae that form a resting "plank" or chain of feculae, most often by reinforcing and/or extending the midrib (or larger secondary vein) of the leaf upon which they are feeding (Freitas and Oliveira 1992; Willmott 2003; Torres et al. 2019, Fig. 23e). Non-feeding caterpillars move onto the frass chain and settle between feeding bouts or to complete a molt. Evidently few predators venture out along these frass chains, turning back before encountering the caterpillar (Freitas and Oliveira 1996). Some microlepidopterans use silk as movement detection networks to learn of approaching natural enemies, in the same way that spiders use their webs to detect movement (Rota and Wagner 2008, Fig. 23f).

Collectively, across instars and lineages, silk has allowed larvae of Lepidoptera to weather challenging abiotic constraints exclude and escape subsets of their natural enemies, disperse, and more. In so doing they exploit and thrive across a broad spectrum of terrestrial (and aquatic) environments to a greater degree than many other insect taxa.

Concluding Remarks

Caterpillars are among the most ecologically important, abundant, and diverse metazoans in terrestrial biomes and account for much of the above-ground insect biomass in forest and shrubland ecosystems, where they are thought to transfer much of the energy from plants to other animals (Janzen 1988; Wagner 2013). Not surprisingly, caterpillars are increasingly the focus of ecological studies, in part because of their diversity and abundance, but also because they are relatively sessile, often present for several weeks during any given generation, and many are identifiable. While extraordinarily diverse in their degree of ecological specialization, external morphology and physiology, internally caterpillars appear structurally generalized: their gut is short, without diverticula or obvious innovation, and most apparently lack dependency on a microbiome. The ability of caterpillars to detoxify, and in some cases sequester, a seemingly endless array of secondary plant compounds and their derivatives has surely contributed greatly to their eco-evolutionary successes. With the exception of the sequestration or modification of plant secondary compounds and myrmecophily-related organs, their defensive chemistry and exocrine systems are not especially impressive. At least in temperate biotas, most species appear rather defenseless and make good fodder for birds and their nestlings, as well as other insectivores. Their most widely employed defensive stratagem is to simply avoid detection. The order's heralded portfolio of primary defense strategies—crypsis, ability to masquerade as unpalatable objects, phenotypic plasticity, shelter construction and countless other behavioral ploys—provides a testament to the remarkable powers of natural selection to shape phenotypes as well as the extraordinary influence of visual predators on the evolution of these and other insects. We submit that the collective abilities of caterpillars to background match, which can involve ontogenetic, plastic, morphological, and behavioral solutions, represents a key adaptation that has allowed the immature stages of Lepidoptera to flourish as exposed, sizeable, external feeders to a greater extent than other insects. We also argue that the ability of caterpillars to produce silk, sometimes in prodigious amounts, has greatly contributed to their evolutionary success, by playing key roles in predator and parasitoid avoidance, ecological engineering, molting processes, dispersal, and more.

Acknowledgments Annette Aiello, Christina Baer, Alex Baranowski, Deane Bowers, Sue Carnahan, David Dussourd, Eric LoPresti, Tanner Matson, Steve Passoa, Moria Robinson, Robert Robbins, Angela Smilanich, Doug Tallamy, David Webb, as well as our editors made helpful suggestions that greatly improved earlier drafts of this chapter. Virginia Wagner contributed the line art for Fig. 2 (reproduced from Wagner 2005). Sam Jaffe of The Caterpillar Lab provided the edited image of the *Pieris* anatomy original painted by Paul Pfurtscheller. Images which greatly added to this effort (in decreasing number) were graciously shared by Berry Nall, Annette Aiello, Mike Thomas, Marc Epstein, Robert Behrstock, Pat Burkett, Jadranka Rota, Gerry Salmon, Doug Tallamy, and Keith Willmott. Alex Baranowski, Jim Roberts, and Abby Ann Sisk provided larvae for study. This work was supported by USFS Co-op Agreement #19DG11420000057 5653380, Earthwatch Institute, and grants from the Richard P. Garmany Fund (Hartford Foundation for Public Giving).

Table 1 Summary of new life history data for Nearctic Macrolepidoptera that appear in this work

Species (Family)	Host	Comments
Geometridae		
"*Caripeta*" *hilumaria* (Geometridae)	Widely polyphagous	My host records mostly from woody plants, e.g., *Ceanothus, Holodiscus, Ribes, Robinia neomexicana*), but also forbs (e.g., *Urtica*)
Dichordophora phoenix (Geometridae)	*Krameria*	Three wild collections and one ex ova collection raised to maturity, all from *Krameria*. Noel McFarland raised an ex ova cohort on a *Rhus* (pers. comm.). *Rhus* unconfirmed from the wild
Somatolophia ectrapelaria (Geometridae)	*Ericameria nauseosa*	First report of host, larval image, and mention of reflex bleeding by larva
Synchlora faseolaria (Geometridae)	*Ericameria ericoides*	Previously reported from *Artemisia californica*. First larval photograph. See also Ferguson (1985)
Noctuidae		
Chalcopasta howardi (Noctuidae)	*Hymenothrix wrightii, Palafoxia texana*	First host association reports and larval image: *Hymenothrix wrightii* in Arizona; *Palafoxia texana* in Rio Grande Valley (Berry Nall); accepting *Florestina tripteris* in captivity (Berry Nall)
Emarginea percara (Noctuidae: Amphipyrinae)	*Phoradendron*	First larval images and mention of secretory setae
Grotella tricolor (Noctuidae)	*Boerhavia erecta*	First discovered and photographed by Robert Behrstock and Karen Lemay in Hereford, AZ
Thurberiphaga diffusa (Noctuidae: Stiriinae)	*Gossypium thurberi*	Larval description and partial life history published by Crumb (1956). Larvae tunnel in ripening cotton bolls (fruits). The first larval images, shared here, were taken by Robert Behrstock

References

Acevedo FE, Peiffer M, Tan CW, Stanley BA, Stanley A, Wang J, Jones AG, Hoover K, Rosa C, Luthe D, Felton G (2017) Fall armyworm-associated gut bacteria modulate plant defense responses. Mol Plant Microbe Interact 30(2):127–137. https://doi.org/10.1094/MPMI-11-16-0240-R

Aiello A (1984) *Adelpha* (Lepidoptera: Nymphalidae): deception on the wing. Psyche 91:1–45

Aiello A, Balcázar MA (1997) The immature stages of *Oxytenis modestia* (Cramer), with comments on the mature larvae of *Asthenidia* and *Homoeopteryx* (Lepidoptera: Saturniidae: Oxyteninae). J Lepid Soc 51:105–118

Akino T, Nakamura KI, Wakamura S (2004) Diet-induced chemical phytomimesis by twig-like caterpillars of *Biston robustum* Butler (Lepidoptera: Geometridae). Chemoecology 14:65–174

Albert AM (1982) Deviations from Dyar's rule in Lithobiidae. Zool Anz 208:192–207

Appel H (1994) The chewing herbivore gut lumen: physicochemical conditions and their impact on plant nutrients, allelochemicals, and insect pathogens. In: Insect-plant interactions, vol V. CRC Press, Boca Raton, pp 209–223

Attygalle AB, Smedley SR, Meinwald J, Eisner T (1993) Defensive secretion of two notodontid caterpillars (*Schizura unicornis, S. badia*). J Chem Ecol 19:2089–2104. https://doi.org/10.1007/BF00979649

Baer CS (2018) Shelter building and extrafloral nectar exploitation by a member of the *Aristotelia corallina* species complex (Gelechiidae) on Costa Rican acacias. J Lepid Soc 72:44–52

Baer CS, Marquis RJ (2020) Between predators and parasitoids: complex interactions among shelter traits, predation and parasitism in a shelter-building caterpillar community. Funct Ecol 34:2186–2198. https://doi.org/10.1111/1365-2435.13641

Barber NA, Marquis RJ, Tori W (2008) Invasive prey impacts distribution of native specialist predators. Ecology 89:2678–2683

Berenbaum MR (1983) Coumarins and caterpillars: a case for coevolution. Evolution 37:163–179

Berenbaum MR (1995) Aposematism and mimicry in caterpillars. J Lepid Soc 49:386–396

Bernays EA (1998) Evolution of feeding behavior in insect herbivores; success seen as different ways to eat without being eaten. Bioscience 48:35–44

Bernays E, Graham BE (1988) On the evolution of host specificity in phytophagous arthropods. Ecology 69:886–892

Bernays E, Janzen DH (1988) Saturniid and sphingid caterpillars: two ways to eat leaves. Ecology 69:1153–1160

Bernays E, Jarzembowski E, Malcolm S (1991) Evolution of insect morphology in relation to plants [and discussion]. Philos Trans R Soc B Biol Sci 333:257–264. https://doi.org/10.1098/rstb.1991.0075

Bernays EA, Chapman RF, Hartmann T (2002) A taste receptor neurone dedicated to the perception of pyrrolizidine alkaloids in the medial galeal sensillum of two polyphagous arctiid caterpillars. Physiol Entomol 27:312–321. https://doi.org/10.1046/j.1365-3032.2002.00304.x

Boege K, Agrawal AA, Thaler JS (2019) Ontogenetic strategies in insect herbivores and their impact on tri-trophic interactions. Curr Opin Insect Sci 32:61–67

Bogner F, Eisner T (1992) Chemical basis of pupal cannibalism in a caterpillar (*Utetheisa ornatrix*). Experientia 48:97–102

Bowers MD (1993) Aposematic caterpillars: life-styles of the warningly colored and unpalatable. In: Stamp NE, Casey TM (eds) Caterpillars: ecological and evolutionary constraints on foraging. Chapman and Hall, New York, pp 331–371

Braun AF (1963) The genus *Bucculatrix* in America North of Mexico (Microlepidoptera). American Entomological Society, Academy of Natural Sciences, Philadelphia

Brower LP (1984) Chemical defence in butterflies. In: Vane-Wright RI, Ackery PR (eds) The biology of butterflies. Academic, New York, pp 109–134

Buchon N, Silverman N, Cherry S (2014) Immunity in *Drosophila melanogaster*—from microbial recognition to whole-organism physiology. Nat Rev Immunol 14:796–810. https://doi.org/10.1038/nri3763

Canfield MR, Changh S, Pierce NE (2009) The double cloak of invisibility: phenotypic plasticity and larval decoration in a geometrid moth, *Synchlora frondaria*, across three diet treatments. Ecol Entomol 34:412–414

Carter D, Hargraves B (1986) Caterpillars of butterflies and moths in Britain and Europe. Watson and Viney Ltd, Alyesbury, p 296

Cassone BJ, Grove HC, Elebute O, Villanueva SMP, LeMoine CMR (2020) Role of the intestinal microbiome in low-density polyethylene degradation by caterpillar larvae of the greater wax moth, *Galleria mellonella*. Proc R Soc B 2872020011220200112

Caveney S, McClean H, Surry D (1998) Faecal firing in a skipper caterpillar is pressure-driven. J Exp Biol 201:121–133

Celorio-Mancera Mde L, Sundmalm SM, Vogel H, Rutishauser D, Ytterberg AJ, Zubarev RA, Janz N (2012) Chemosensory proteins, major salivary factors in caterpillar mandibular glands. Insect Biochem Mol Biol:796–805. https://doi.org/10.1016/j.ibmb.2012.07.008

Chen CY, Mao YB (2020) Research advances in plant–insect molecular interaction. F1000Research 9:198. https://doi.org/10.12688/f1000research.21502.1

Cole BJ (1980) Growth ratios in holometabolous and hemimetabolous insects. Ann Entomol Soc Am 73:489–491

Coley PD, Barone JA (1996) Herbivory and plant defenses in tropical forests. Annu Rev Ecol Syst 27:305–335

Common IFB, Bellas TE (1977) Regurgitation of host-plant oil from a foregut diverticulum in the larvae of *Myrascia megalocentra* and *M. bracteatella* (Lepidoptera: Oecophoridae). J Aust Entomol Soc 16:144–147

Costa JT, Pierce NE (1997) Social evolution in the Lepidoptera: ecological context and communication in larval societies. In: Choe J, Crespi BJ (eds) The evolution of social behavior in insects and arachnids. Cambridge University Press, Cambridge, MA, pp 407–442

Courtney SP, Chew FS (1987) Coexistence and host use by a large community of pierid butterflies: habitat is the template. Oecologia 71:210–220

Craig CL (1997) Evolution of arthropod silks. Annu Rev Entomol 42:231–267

Crumb SE (1956) The larvae of the Phalaenidae. USDA Technical Bulletin 1135. USDA, Washington, DC

Crump D, Silverstein RM, Williams HJ, Fitzgerald TD (1987) Identification of trail pheromone of larva of eastern tent caterpillar *Malacosoma americanum* (Lepidoptera: Lasiocampidae). J Chem Ecol 13:397–402. https://doi.org/10.1007/BF01880088

Davis DR (1987) Tineidae. In: Stehr FW (ed) An introduction to immature insects of North America. Kendall-Hunt Publishing Co, Dubuque, pp 362–365

Davis DR, Stonis JR (2007) A revision of the new world plant-mining moths of the family Opostegidae (Lepidoptera: Nepticuloidea). Smithsonian Institution Scholarly Press, Washington, DC

Davis DR, Quintero DA, Cambra RAT, Aiello A (2008) Biology of a new Panamanian bagworm moth (Lepidoptera: Psychidae) with predatory larvae, and eggs individually wrapped in setal cases. Ann Entomol Soc Am 101:689–702

de Nardin J, de Araújo AM (2011) Kin recognition in immatures of *Heliconius erato phyllis* (Lepidoptera; Nymphalidae). J Ethol 29:499–503

Deml R, Dettner K (1993) Biogenic amines and phenolics characterize the defensive secretion of saturniid caterpillars (Lepidoptera: Saturniidae): a comparative study. J Comp Physiol B 163:123–132

Desurmont GA, Köhler A, Maag D, Laplanche D, Xu H, Baumann J, Demairé C, Devenoges D, Glavan M, Mann L, Turlings TCJ (2017) The spitting image of plant defenses: effects of plant secondary chemistry on the efficacy of caterpillar regurgitant as an anti-predator defense. Ecol Evol 7:6304–6313

Dethier VG (1937) Cannibalism among lepidopterous larvae. Psyche 44:110–115

DeVries PJ (1987) The butterflies of Costa Rica and their natural history. Princeton University Press, Princeton

Dookie A, Young C, Lamothe G, Laura S, Yack J (2017) Why do caterpillars whistle at birds? Insect defence sounds startle avian predators. Behav Process 138. https://doi.org/10.1016/j.beproc.2017.02.002

Dow JAT (1986) Insect midgut function. Adv Insect Physiol 19:187–328

Duarte M, Robbins RK (2010) Description and phylogenetic analysis of the *Calycopidina* (Lepidoptera, Lycaenidae, Theclinae, Eumaeini): a subtribe of detritivores. Rev Bras Entomol 54:45–65

Dussourd DE (1993) Foraging with finesse: caterpillar adaptations for circumventing plant defenses. In: Stamp NE, Casey TM (eds) Caterpillars: ecological and evolutionary constraints on foraging. Chapman and Hall, New York, pp 92–131

Dussourd DE (2017) Behavioral sabotage of plant defenses by insect folivores. Annu Rev Entomol 62:15–34. https://doi.org/10.1146/annurev-ento-031616-035030

Dussourd DE, Peiffer M, Felton GW (2016) Chew and spit: tree-feeding notodontid caterpillars anoint girdles with saliva. Arthropod-Plant Interact 10:143–150

Dussourd DE, Van Valkenburg M, Rajan K, Wagner DL (2019) A notodontid novelty: *Theroa zethus* caterpillars use behavior and anti-predator weaponry to disarm host plants. PLoS One 14:e0218994. https://doi.org/10.1371/journal.pone.0218994

Dyar HG (1890) The number of molts of lepidopterous larvae. Psyche (Stuttg) 5:420–422

Eacock A, Rowland HM, van't Hof AE, Yung CJ, Edmonds N, Saccheri IJ (2019) Adaptive colour change and background choice behaviour in peppered moth caterpillars is mediated by extra-ocular photoreception. Comm Biol 2:286. https://doi.org/10.1038/s42003-019-0502-7

Eaton JL (1988) Lepidopteran anatomy. Wiley, New York

Ehrlich PR, Raven PH (1964) Butterflies and plants: a study in coevolution. Evolution 18:586–608

Eisner T (2003) For love of insects. Belknap Press of Harvard University Press, Cambridge, MA, pp 391–403

Eisner TA, Kluge AF, Carrel JC, Meinwald J (1972) Defense mechanisms of arthropods. XXXIV. Formic acid and acyclic ketones in the spray of a caterpillar. Ann Entomol Soc Am 65:765–766

Epstein ME (1995) False-parasitized cocoons and biology of Aididae (Lepidoptera: Zygaenoidea). Proc Entomol Soc 97:750–756

Epstein ME, Geertsema H, Naumann CM, Tarmann GE (1998) The Zygaenoidea. In: Kristensen NP (ed) Lepidoptera, moths and butterflies, 1: evolution, systematics, and biogeography, Handbook of zoology. Walter de Gruyter, Berlin, pp 159–180

Esperk T, Tammaru T (2006) Determination of female-biased sexual size dimorphism in moths with a variable instar number: the role of additional instars. Eur J Entomol 103:576–586

Everson GW, Chapin JB, Normann SA (1990) Caterpillar envenomations: a prospective study of 112 cases. Vet Hum Toxicol 32:114–119

Felton GW (2008) Caterpillar secretions and induced plant responses. In: Schaller A (ed) Induced plant resistance to herbivory. Springer, Dordrecht. https://doi.org/10.1007/978-1-4020-8182-8_18

Fenton LM, Kennedy Z, Spencer KA (2011) Parasite-induced warning coloration: a novel form of host manipulation. Anim Behav 81:417–422

Ferreira C, Capella AN, Sitnik R, Terra WR (1994) Digestive enzymes in midgut cells, endo- and ectoperitrophic contents, and peritrophic membranes of *Spodoptera frugiperda* (Lepidoptera) larvae. Arch Insect Biochem Physiol 26:299–313

Ferguson DC (1985) Geometroidea: Geometridae (in part). In Dominick RB (ed) The moths of America North of Mexico, fasc. 18.1. Wedge Entomological Research Foundation, Washington, District of Columbia

Fiedler K (1991) Systematic, evolutionary and ecological implications of myrmecophily within the Lycaenidae (Insecta: Lepidoptera: Papilionoidea). Bonn Zool Monogr 31:1–210

Fitzgerald TD (1976) Trail marking by larvae of the eastern tent caterpillar. Science 194:961–963

Fitzgerald TD (1995) The tent caterpillars. Cornell University Press, Ithaca

Forister ML, Basset Y, Coley PD, Diniz IR, Drozd P, Fox M, Glassmire A, Hazen HR, Hrcek J, Jahner JP, Kozubowski TJ, Kursar TA, Lill J, Marquis RJ, Miller SE, Morais HC, Murakami M, Novotny V, Panorska AK, Pardikes N, Ricklefs RE, Singer MS, Smilanich AM, Stireman JO, Wagner DL, Walla T, Weiblen GD, Dyer LA (2015) The global distribution of diet breadth in insect herbivores. Proc Natl Acad Sci 112:442–447

Frankfater C, Tellez MR, Slattery M (2009) The scent of alarm: ontogenetic and genetic variation in the osmeterial gland chemistry of *Papilio glaucus* (Papilionidae) caterpillars. Chemoecology 19:81–96. https://doi.org/10.1007/s00049-009-0013-y

Freitas AV, Oliveira PS (1992) Biology and behavior of the neotropical butterfly *Eunica bechina* (Nymphalidae) with special reference to larval defence against ant predation. J Res Lepid 31:1–11

Freitas AV, Oliveira PS (1996) Ants as selective agents on herbivore biology: effects on the behaviour of a non-myrmecophilous butterfly. J Anim Ecol 65:205–210

Frost C, Hunter MD (2007) Insect herbivores and their frass affect *Quercus rubra* leaf quality and initial stages of subsequent litter decomposition. Oikos 117:13–22

Fukui A (2001) Indirect interactions mediated by leaf shelters in animal–plant communities. Popul Ecol 43:31–40

Gaston KJ (1991) The magnitude of global insect species richness. Conserv Biol 5:283–296. https://doi.org/10.1111/j.1523-1739.1991.tb00140.x

Glądalski M, Bańbura M, Kaliński A, Markowski M, Skwarska J, Wawrzyniak J, Zieliński P, Cyżewska I, Bańbura J (2015) Inter-annual and inter-habitat variation in breeding performance of Blue Tits (*Cyanistes caeruleus*) in central Poland. Ornis Fennica 92:34–42

Glendinning JI, Nelson NM, Bernays EA (2000) How do inositol and glucose modulate feeding in *Manduca sexta* caterpillars? J Exp Biol 203:1299–1315

Glendinning JI, Brown H, Capoor M, Davis A, Gbedemah A, Long E (2001) A peripheral mechanism for behavioral adaptation to specific "bitter" taste stimuli in an insect. J Neurosci 21:3688–3696. https://doi.org/10.1523/JNEUROSCI.21-10-03688.2001

Glendinning JI, Davis A, Ramaswamy S (2002) Contribution of different taste cells and signaling pathways to the discrimination of "bitter" taste stimuli by an insect. J Neurosci 22:7281–7287. https://doi.org/10.1523/JNEUROSCI.22-16-07281.2002

Grant J (2006) Diversification of gut morphology in caterpillars is associated with defensive behavior. J Exp Biol 209:3018–3024. https://doi.org/10.1242/jeb.02335

Greeney HF, Dyer LA, Smilanich AM (2012) Feeding by lepidopteran larvae is dangerous: a review of caterpillars' chemical, physiological, morphological, and behavioral defenses against natural enemies. Invertebrate Surviv 9:7–34

Greene E (1989) A diet-induced developmental polymorphism in a caterpillar. Science 243(4891):643–646. https://doi.org/10.1126/science.243.4891.643

Grunert LW, Clarke JW, Ahuja C, Eswaran H, Nijhout HF (2015) A quantitative analysis of growth and size regulation in *Manduca sexta*: the physiological basis of variation in size and age at metamorphosis. PLoS One 10:e0127988. https://doi.org/10.1371/journal.pone.0127988

Hallberg E, Poppy G (2003) Exocrine glands: chemical communication and chemical defense. In: Kristensen NP (ed) Handbuch der Zoologie: Lepidoptera, moths and butterflies, vol 4, pp 361–375

Hammer TJ, Bowers MD (2015) Gut microbes may facilitate insect herbivory of chemically defended plants. Oecologia 179:1–14. https://doi.org/10.1007/s00442-015-3327-1

Hammer T, Janzen D, Hallwachs W, Jaffe S, Fierer N (2017) Caterpillars lack a resident gut microbiome. Proc Natl Acad Sci U S A 114:9641–9646. https://doi.org/10.2307/26487619

Hammer TJ, Sanders JG, Fierer N (2019) Not all animals need a microbiome. FEMS Microbiol Lett 366:10. https://doi.org/10.1093/femsle/fnz117

Hanski I, Thomas CD (1994) Metapopulation dynamics and conservation: a spatially explicit model applied to butterflies. Biol Conserv 68:167–180

Hardwick DF (1996) A monograph to the North American Heliothentinae (Lepidoptera: Noctuidae). Privately published, Almonte

Hawkins BA, Cornell HV, Hochberg ME (1997) Predators, parasitoids, and pathogens as mortality agents in phytophagous insect populations. Ecology 78:2145–2152

Haylett JM (2000) Distribution of frass produced by larval lepidoptera in a hardwood canopy. MS thesis, Youngstown State University. https://digital.maag.ysu.edu/xmlui/bitstream/handle/1989/6164/b18616343.pdf?sequence=1&isAllowed=y

Heinrich B (1979) Foraging strategies of caterpillars: leaf damage and possible predator avoidance strategies. Oecologia 42:325–337

Heinrich B (1993) How avian predators constrain caterpillar foraging. In: Stamp NE, Casey TM (eds) Caterpillars: ecological and evolutionary constraints on foraging. Chapman and Hall, New York, pp 224–247

Hodges RW, Dominick T, Davis DR, Ferguson DC, Franclemont JG, Munroe, EG, Powell JA (1983) Check list of the Lepidoptera of America North of Mexico, Moths of America north of Mexico. The Wedge Entomological Research Foundation, United States of America

Hohn FM, Wagner DL (2000) Larval substrates of herminiine noctuids (Lepidoptera), macrodecomposers of leaf litter. Environ Entomol 29:207–212

Holmes RT (1980) Resource exploitation patterns and the structure of a forest bird community. Proceedings of the XVII International Ornithology Congress, pp 1056–1062

Hoshizaki D (2012) Fat body. In: Chapman R, Simpson S, Douglas A (eds) The insects: structure and function. Cambridge University Press, Cambridge, MA, pp 132–146. https://doi.org/10.1017/CBO9781139035460.009

Hossler EW (2010) Caterpillars and moths: part II. Dermatologic manifestations of encounters with Lepidoptera. J Am Acad Dermatol 62:13–28

Hunter AF (1995) Ecology, life history, and phylogeny of outbreak and non-outbreak species. In: Cappuccino N, Price PW (eds) Population dynamics: new approaches and synthesis. Academic Press, San Diego, pp 41–64

Hunter M, McNeil J (1997) Host-plant quality influences diapause and voltinism in a polyphagous insect herbivore. Ecology 78:977–986. https://doi.org/10.2307/2265851

Jaenike J (1990) Host specialization in phytophagous insects. Annu Rev Ecol Syst 21:243–273

Janzen DH (1988) Ecological characterization of a Costa Rican dry forest caterpillar fauna. Biotropica 20:120–135

Janzen DH, Hallwach W, Burns JM (2010) A tropical horde of counterfeit predator eyes. PNAS 107:11659–11665

Kalka M, Kalko EKV (2006) Gleaning Bats as Underestimated predators of herbivorous insects: diet of *Micronycteris microtis* (Phyllostomidae) in Panama. J Trop Ecol 22:1–10

Kant MR, Jonckheere W, Knegt B, Lemos F, Liu J, Schimmel BC, Villarroel CA, Ataide LM, Dermauw W, Glas JJ, Egas M, Janssen A, Van Leeuwen T, Schuurink RC, Sabelis MW, Alba JM (2015) Mechanisms and ecological consequences of plant defence induction and suppression in herbivore communities. Ann Bot 115:1015–1051. https://doi.org/10.1093/aob/mcv054

Kawamoto F, Kumata N (1984) Biology and venoms of Lepidoptera. In: Tu AT (ed) Handbook of natural toxins, Vol. 2. Insect poisons, allergens, and other invertebrate venoms. Dekker, New York, pp 291–330

Kearby WH (1975) Variable oakleaf caterpillar [*Heterocampa manteo*] larvae secrete formic acid that causes skin lesions (Lepidoptera: Notodontidae). J Kansas Entomol Soc 48:280–282

Keegan KL, Rota J, Zahiri R, Zilli A, Wahlberg N, Schmidt BC, Lafontaine JD, Goldstein PZ, Wagner DL (2021) Towards a global Noctuidae (Lepidoptera) taxonomy. Insect Syst Biodivers 5:1. https://doi.org/10.1093/isd/ixab005

Kolosov D, O'Donnell MJ (2019) Chapter five—the Malpighian tubules and cryptonephric complex in lepidopteran larvae. Adv Insect Physiol Academic Press 56:165–202. https://doi.org/10.1016/bs.aiip.2019.01.006

Koptur S, Lawton JH (1988) Interactions among vetches bearing extrafloral nectaries, their biotic protective agents, and herbivores. Ecology 69:278–228

Kristensen NP, Scoble MJ, Karsholt O (2007) Lepidoptera phylogeny and systematics: the state of inventorying moth and butterfly diversity. Zootaxa 1668(1):699–747. https://doi.org/10.11646/zootaxa.1668.1.30

Krombein KV, Hurd PD, Smith DR, Burks BD (1979) Catalog of Hymenoptera in America north of Mexico. Smithsonian Institution Press, Washington, DC

Kuspis D, Rawlins J, Krenzelok B (2001) Human exposure to stinging caterpillar: *Lophocampa caryae* exposures. Am J Emerg Med 19:396–398

Lack D (1950) The breeding seasons of European birds. IBIS 92:288–316. https://doi.org/10.1111/j.1474-919X.1950.tb01753.x

Lamas Müller G (2017) The butterflies of Cosñipata: an altitudinal transect study of a megadiverse fauna in southeast Peru. Entomologentagung In: Freising. Abstract, pp 77

Lamas G, McInnis ML, Busby RC (2021) The lycaenid butterfly fauna (Lepidoptera) of Cosñipata, Peru: annotated checklist, elevational patterns, and rarity. Insecta Mundi 0861:1–34

Laney NK, Ayres MP, Stange EE, Sillett TS, Rodenhouse NL, Holmes RT (2015) Breeding timed to maximize reproductive success for a migratory songbird: the importance of phenological asynchrony. Oikos 125:656–666

Lawton JH (1976) The structure of the arthropod community on bracken. Bot J Linn Soc 73:187–216

Lawton JH (1978) Host-plant influences on insect diversity: the effects of space and time. Sympos R Entomol Soc Lond 9:105–125

Lawton JH (1983) Plant architecture and the diversity of phytophagous insects. Annu Rev Entomol 28:23–29

Lawton JH, Schroder D (1977) Effects of plant type, size of geographical range and taxonomic isolation on numbers of insect species associated with British plants. Nature 265:137–140

Lederhouse RC (1990) Avoiding the hunt: primary defenses of lepidopteran caterpillars. In: Insect defense: adaptive mechanisms and strategies of prey and predators. State University of New York Press, Albany, pp 175–189

Lee KP, Wilson K (2006) Melanism in a larval lepidopteran: repeatability and heritability of a dynamic trait. Ecol Entomol 31:196–205. ISSN 1365-2311

Levy SM, Falleiros ÂMF, Moscardi F, Gregório EA, Toledo LA (2004) Morphological study of the hindgut in larvae of *Anticarsia gemmatalis* Hübner (Lepidoptera: Noctuidae). Neotrop Entomol 33:427–431

Levy SM, Falleiros A, Moscardi F, Gregório E, Toledo L (2009) Ultramorphology of digestive tract of *Anticarsia gemmatalis* (Hübner, 1818) (Lepidoptera: Noctuidae) at final larval development. Semina: Ciências Agrárias 29:313–322

Lill JT, Marquis RJ (2007) Microhabitat manipulation: ecosystem engineering by shelter building insects. In: Cuddington KMD, Byers JE, Hastings A, Wilson WG (eds) Ecosystem engineers: concepts, theory, and applications in ecology. Elsevier, San Diego, pp 107–138

Lill JT, Marquis RJ, Ricklefs RE (2002) Host plants influence parasitism of forest caterpillars. Nature 417:170–173

Lin HT, Trimmer B (2010) Caterpillars use the substrate as their external skeleton: a behavior confirmation. Commun Integr Biol 3:471–474. https://doi.org/10.4161/cib.3.5.12560

Lin HT, Slate DJ, Paetsch CR, Dorfmann AL, Trimmer BA (2011) Scaling of caterpillar body properties and its biomechanical implications for the use of a hydrostatic skeleton. J Exp Biol 214:1194–1204

Malicky H (1969) Versuch einer Analyse der ¨okologischen Beziehungen zwischen Lycaeniden (Lepidoptera) und Formiciden (Hymenoptera). Tijdschr Entomol 112:213–298

Malicky H (1970) New aspects on the association between lycaenid larvae (Lycaenidae) and ants (Formicidae, Hymenoptera). J Lepid Soc 24:190–202

Market Data Forecast Services (2021). https://www.marketdataforecast.com/. Accessed May 2021

Marquis RJ, Passoa S (1989) Seasonal diversity and abundance of the herbivore fauna of striped maple *Acer pensylvanicum* L. (Aceraceae) in western Virginia. Am Midl Nat 122:313–320

McFarland N (1988) Portraits of South Australian geometrid moths. Published by Author

McMillan LE, Adamo SA (2020) Friend or foe? Effects of host immune activation on the gut microbiome in the caterpillar *Manduca sexta*. J Exp Biol 223:jeb226662. https://doi.org/10.1242/jeb.226662

Men Q, Wu G (2016) Ultrastructure of the sensilla on larval antennae and mouthparts of the simao pine moth, *Dendrolimus kikuchii* Matsumura (Lepidoptera: Lasiocampidae). Proc Entomol Soc Wash 118:373–381

Miller JS, Wagner DL, Opler PA (2021) Noctuoidea, Notodontidae (part): Heterocampinae, Nystaleinae, Dioptinae, and Dicranurinae. In: Lafontaine JD et al (eds) The moths of North America, fasc. 22.1A (in press)

Mitter C, Farrell B, Wiegmann B (1988) The phylogenetic study of adaptive zones: has phytophagy promoted insect diversification? Am Nat 132:107–128

Mitter C, Davis DR, Cummings MR (2017) Phylogeny and evolution of Lepidoptera. Annu Rev Entomol 62:265–283

Montgomery SL (1983) Carnivorous caterpillars: the behavior, biogeography and conservation of *Eupithecia* (Lepidoptera: Geometridae) in the Hawaiian Islands. GeoJournal 7:549–556. https://doi.org/10.1007/BF00218529

Moore RG, Hanks LM (2004) Aerial dispersal and host plant selection by neonate *Thyridopteryx ephemeraeformis* (Lepidoptera: Psychidae). Ecol Entomol 29:327–335. https://doi.org/10.1111/j.0307-6946.2004.00611.x

Moraes AR, Greeney HF, Oliveira PS, Barbosa EP, Freitas AV (2012) Morphology and behavior of the early stages of the skipper, *Urbanus esmeraldus*, on *Urera baccifera*, an ant-visited host plant. J Insect Sci 12:52. https://doi.org/10.1673/031.012.5201

Morewood WD, Ring RA (1998) Revision of the life history of the High Arctic moth *Gynaephora groenlandica* (Wocke) (Lepidoptera: Lymantriidae). Can J Zool 76:1371–1381

Musser RO, Hum-Musser SM, Eichenseer H, Peiffer M, Ervin G, Murphy JB, Felton GW (2002) Herbivory: caterpillar saliva beats plant defences. Nature 416:599–600

Nall B (2020) American snout (*Libytheana carinenta*) life history. Southern Lepid News 42:212–214

Narango DL, Tallamy DW, Shropshire KJ (2020) Few keystone plant genera support the majority of Lepidoptera species. Nat Commun 11:5751. https://doi.org/10.1038/s41467-020-19565-4

Needham JG (1948) A bucculatricid gall and its hypermetamorphosis. J N Y Entomol Soc LVI:43–50

Needham JG, Frost SW, Tothill BH (1928) Leaf mining Insects. Williams and Watkins Co, Baltimore

Ngô T, Bechtold T (2018) Analysis of the fibroin solution state in calcium chloride/water/ethanol for improved understanding of the regeneration process. Fibres Textile East Eur 26:43–50. https://doi.org/10.5604/01.3001.0012.5174

Nishida K, Robbins RK (2020) One side makes you taller: a mushroom–eating butterfly caterpillar (Lycaenidae) in Costa Rica. Neotrop Biol Conserv 15:463–470

Noor MA, Parnell RS, Grant BS (2008) A reversible color polyphenism in American peppered moth (*Biston betularia cognataria*) caterpillars. PLoS One 3:e3142. https://doi.org/10.1371/journal.pone.0003142

Nothnagle PJ, Schultz JC (1987) What is a forest pest? In: Barbosa P, Schultz JC (eds) Insect outbreaks. Academic, New York, pp 59–80

Osborn F, Jaffe K (1998) Chemical ecology of the defense of two nymphalid butterfly larvae against ants. J Chem Ecol 24:1173–1186

Peters W (1992) Functions of peritrophic membranes. In: Peritrophic membranes, Zoophysiology, vol 30. Springer, Heidelberg. https://doi.org/10.1007/978-3-642-84414-0_6

Peterson SC, Johnson ND, Leguyader JL (1987) Defensive regurgitation of allelochemicals derived from host cyanogenesis by eastern tent caterpillars. Ecology 68:1268–1272

Phalnikar K, Kunte K, Agashe D (2019) Disrupting butterfly caterpillar microbiomes does not impact their survival and development. Proc R Soc B 286:20192438. https://doi.org/10.1098/rspb.2019.2438

Pierce NE (1995) Predatory and parasitic Lepidoptera: carnivores living on plants. J Lepid Soc 49:412–453

Pierce NP, Braby MF, Heath A, Lohman DJ, Mathew J, Rand DB, Travassos MA (2002) The ecology and evolution of ant association in the Lycaenidae (Lepidoptera). Annu Rev Entomol 47:733–771

Pohl GR, Patterson R, Pelham J (2016) Annotated taxonomic checklist of the Lepidoptera of North America, North of Mexico. Working paper published online by the authors at Research Gate.net

Porter J (1997) The colour identification guide to caterpillars of the British Isles. Viking Press, London

Posledovich D, Toftegaard T, Wiklund C, Ehrlén J, Gotthard K (2015) Latitudinal variation in diapause duration and post-winter development in two pierid butterflies in relation to phenological specialization. Oecologia 177:181–190. https://doi.org/10.1007/s00442-014-3125-1

Poulton EB (1892) Further experiments upon the colour-relation between certain lepidopterous larvae, pupae, cocoons, and imagines and their surroundings. Trans Entom Soc Lond 40:293–487

Powell JA (1980) Evolution of larval food preferences in Microlepidoptera. Annu Rev Entomol 25:133–159

Powell JA (1989) Synchronized, mass-emergences of a yucca moth, *Prodoxus y-inversus* (Lepidoptera: Prodoxidae), after 16 and 17 years in diapause. Oecologia 81:490–493. https://doi.org/10.1007/BF00378957

Powell JA, Mitter C, Farrell B (1998) Evolution of larval food preferences in Lepidoptera. In: Kristensen NP (ed) Lepidoptera: moths and butterflies. Vol 1. Evolution, systematics, and biogeography. Handbook of zoology. De Gruyter, New York/Berlin, pp 403–422

Price PW (1997) Insect ecology, 3rd edn. Wiley, New York

Prudic KL, Oliver JC, Sperling FAH (2007) The signal environment is more important than diet or chemical specialization in the evolution of warning coloration. PNAS 104:19381–19386

Ramsay J (1976) The rectal complex in the larvae of Lepidoptera. Philos Trans R Soc Lond Ser B Biol Sci 274:203–226

Rawlins JE (1984) Mycophagy in Lepidoptera. In: Wheeler Q, Blackwell M (eds) Fungus-insect relationships. Perspectives in ecology and evolution. Columbia University Press, New York, pp 382–483

Ray S, Gaffor I, Acevedo FE, Helms A, Chuang WP, Tooker J, Felton GW, Luthe DS (2015) Maize plants recognize herbivore-associated cues from caterpillar frass. J Chem Ecol 41:781–792

Ray S, Basu S, Rivera-Vega LJ, Acevedo FE, Louis J, Felton GW, Luthe DS (2016) Lessons from the far end: caterpillar FRASS-induced defenses in maize, rice, cabbage, and tomato. J Chem Ecol 42:1130–1141

Reavey D (1993) Why body size matters in caterpillars. In: Stamp NE, Casey TM (eds) Caterpillars: ecological and evolutionary constraints on foraging. Chapman and Hall, New York, pp 248–279

Richardson ML, Mitchell RF, Reagel PF, Hanks LM (2010) Causes and consequences of cannibalism in noncarnivorous insects. Annu Rev Entomol 55:39–53

Risley LS, Crossley DA Jr (1988) Herbivore-caused greenfall in the southern Appalachians. Ecology 69:1118–1127

Rota J, Wagner DL (2008) Wormholes, sensory nets and hypertrophied tactile setae: the extraordinary defence strategies of *Brenthia* caterpillars. Anim Behav 76:1709–1713

Rothschild M (1973) Secondary substances and warning colouration in insects. In: Van Emdem HF (ed) Insect-plant relationships. Symp Roy Entomol Soc Lond 6:59–83

Rothschild M (1985) British aposematic Lepidoptera. In: Heath J, Emmet AM (eds) The moths and butterflies of Great Britain and Ireland, Part 2. Harley Books, Colchester, pp 9–62

Rothschild M (1993) Phytochemical selection of aposematic insects. Phytochemistry 33:1037. https://doi.org/10.1016/0031-9422(93)85018-M

Rubinoff D, Haines WP (2005) Web-spinning caterpillar stalks snails. Science 309:575

Ruf C, Costa J, Fiedler K (2001) Trail-based communication in social caterpillars of *Eriogaster lanestris* (Lepidoptera: Lasiocampidae). J Insect Behav 14:231–245

Salazar BA, Whitman DW (2001) Defensive tactics of caterpillars against predators and parasitoids. In: Ananthakrishnan TN (ed) Insect and plant defense dynamics. Science Publishers, Enfield, pp 161–207

Schowalter TD (2017) Insect ecology: an ecosystems approach, 4th edn. Academic Press, New York
Schultz JC (1983) Habitat selection and foraging tactics of caterpillars in heterogeneous trees. In: Denno RF, McClure MS (eds) Variable plants and herbivores in natural and managed systems. Academic, New York, pp 61–90
Schultz CB, Haddad NM, Henry EH, Crone EE (2019) Movement and demography of at-risk butterflies: building blocks for conservation. Annu Rev Entomol 64:167–184
Scoble MJ (1992) The Lepidoptera: form, function and diversity. Oxford University Press, Oxford
Scudder SH (1889) Butterflies of the United States and Canada with special reference to New England, vol III. Cambridge, MA, published by author
Singer MS, Bernays EA, Carriere Y (2002) The interplay between nutrient balancing and toxin dilution in foraging by a generalist insect herbivore. Anim Behav 64:629–643
Smedley SRE, Ehrhardt E, Eisner T (1993) Defensive regurgitation by a noctuid moth larva (*Litoprosopus futilis*). Psyche 100:209–221
Smedley SR, Schroeder FC, Weibel DB, Meinwald J, Lafleur KA, Renwick JA, Rutowski R, Eisner T (2002) Mayolenes: labile defensive lipids from the glandular hairs of a caterpillar (*Pieris rapae*). Proc Natl Acad Sci U S A 99:6822–6827
Smith NG (1983) Host plant toxicity and migration in the dayflying moth *Urania*. Flor Entomol 66:76–85
Smith KW, Smith L (2019) Does the abundance and timing of defoliating caterpillars influence the nest survival and productivity of the Great Spotted Woodpecker *Dendrocopos major*? Bird Study 66:187–197. https://doi.org/10.1080/00063657.2019.1637396
Snäll N, Tammaru T, Wahlberg N, Viidalepp J, Ruohomaki K, Savontaus ML, Huoponen K (2007) Phylogenetic relationships of the tribe Operophterini (Lepidoptera, Geometridae): a case study of the evolution of female flightlessness. Biol J Linn Soc 92:241–252
Song YQ, Sun HZ, Wu JX (2014) Morphology of the sensilla of larval antennae and mouthparts of the oriental fruit moth *Grapholita molesta*. Bull Insectol 67:193–198
Stamp NE, Wilkens RT (1993) On the cryptic side of life: being unapparent to enemies and the consequences for foraging and growth of caterpillars. In: Stamp NE, Casey TM (eds) Caterpillars: ecological and evolutionary constraints on foraging. Chapman and Hall, New York, pp 283–330
Strong DR, Levin DA (1979) Species richness of plant parasites and growth form of their hosts. Am Nat 114:1–22
Strong DR, Southwood JH, Southwood Sir R (1984) Insects on plants: community patterns and mechanisms. Harvard University Press, Cambridge, MA
Sugiura S, Yamazaki K (2006) The role of silk threads as lifelines for caterpillars: pattern and significance of lifeline-climbing behaviour. Ecol Entomol 31:52–57
Sutherland TD, Young JH, Weisman S, Hayashi CY, Merritt DJ (2010) Insect silk: one name, many materials. Annu Rev Entomol 55:171–188
Suzuki TN, Sakurai R (2015) Bent posture improves the protective value of bird dropping masquerading by caterpillars. Anim Behav 105:79–84
Tallamy DW, Shropshire KJ (2009) Ranking lepidopteran use of native versus introduced plants. Conserv Biol 4:941–947. https://doi.org/10.1111/j.1523-1739.2009.01202.x. PMID: 19627321
Tautz J, Markl H (1978) Caterpillars detect flying wasps by hairs sensitive to airborne vibration. Behav Ecol Sociobiol 4:101–110. https://doi.org/10.1007/BF00302564
Taylor CJ, Yak JE (2019) Hearing in caterpillars of the monarch butterfly (*Danaus plexippus*). J Exp Biol 222:jeb211862. https://doi.org/10.1242/jeb.211862
Tindale NBT (1932) Revision of Australian ghost moths (Lepidoptera Homoneura, family Hepialidae). Part I. Rec S Aust Mus 4:497–536
Todd TL (2018) A review of three species-level taxa of the *Anthocharis sara* complex (Lepidoptera: Pieridae: Pierinae: Anthocharidini). Insecta Mundi 1133:1–39
Torres KP, Artieda N, Salazar P, Willmott KR, Hill RI (2019) Life history descriptions of *Adelpha attica attica*, *Adelpha epione agilla*, and *Adelpha jordani* from an eastern Ecuador lowland forest. Trop Lepid Res 29:19–28
Traxler F (1977) General anatomical features of the gypsy moth larva *Lymantria dispar* (Linnaeus) (Lepidoptera: Lymantriidae). J NY Entomol Soc 85:71–97

van Nieukerken EJ (2011) Order Lepidoptera Linnaeus, 1758. In: Zhang ZQ (ed) Animal biodiversity: an outline of higher-level classification and survey of taxonomic richness. Zootaxa 3148:212–221
Vegliante F, Hasenfuss I (2012) Morphology and diversity of exocrine glands in lepidopteran larvae. Annu Rev Entomol 57:187–204
Wagner DL (1985) The biosystematics of the Holarctic Hepialidae, with special emphasis on the *Hepialus californicus* species group. PhD dissertation, University of California, Berkeley, California
Wagner DL (2005) Caterpillars of eastern North America: a guide to identification and natural history. Princeton University Press, Princeton
Wagner DL (2009) The immature stages: structure, function, behavior, and ecology. In: Conner WE (ed) Tiger moths and woolly bears: behavior, ecology, and evolution of the Arctiidae. Oxford University Press, Oxford, pp 31–53
Wagner DL (2013) The biodiversity of moths. In: Levin S et al (eds) Encyclopedia of diodiversity. Academic Press, San Diego, pp 384–403
Wagner DL, Liebherr JK (1992) Flightlessness in insects. Trends Ecol Evol 7:216–220
Wagner DL, Loose JL, Fitzgerald TD, DeBenedictis JA, Davis DR (2000) A hidden past: the hypermetamorphic development of *Marmara arbutiella* (Lepidoptera: Gracillariidae). Ann Entomol Soc Am 93:59–64
Wagner DL, Ferguson DC, McCabe T, Reardon RC (2002) Geometroid caterpillars of Northeastern and Appalachian forests. USFS Technology Transfer Bulletin, FHTET-2001-10. USDA Forest Service, Morgantown, p 239
Wagner DL, Hossler EW, Hossler FE (2006) Not a tiger but a dagger: the larva of *Comachara cadburyi* and reassignment of the genus to the Acronictinae (Lepidoptera: Noctuidae). Ann Entomol Soc Am 99:638–647
Wagner DL, Rota JR, McCabe TL (2008) Larva of *Abablemma* (Lepidoptera: Noctuidae: Hypenodinae) with notes on lichenivory and algivory in Macrolepidoptera. Ann Entomol Soc Am 101:40–52
Wagner DL, Schweitzer DF, Sullivan JB, Reardon RC (2011) Owlet caterpillars of Eastern North America. Princeton University Press, Princeton
Wahlberg M, Wheat CW, Peña C (2013) Timing and patterns in the taxonomic diversification of Lepidoptera (butterflies and moths). PLoS One. https://doi.org/10.1371/journal.pone.0080875
Wang X, Kang L (2014) Molecular mechanisms of phase change in locusts. Annu Rev Entomol 59:225–244
Wang X, Lu H, Shao Y, Zong S (2018) Morphological and ultrastructural characterization of the alimentary canal in larvae of *Streltzoviella insularis* (Staudinger) (Lepidoptera: Cossidae). Entomol Res 48:288–299
Wardhaugh CW (2014) The spatial and temporal distributions of arthropods in forest canopies: uniting disparate patterns with hypotheses for specialisation. Biol Rev 89:1021–1041. https://doi.org/10.1111/brv.12094
Weinhold A, Baldwin IT (2011) Trichome-derived O-acyl sugars are a first meal for caterpillars that tags them for predation. PNAS 108:7855–7859
Weiss MR (2003) Good housekeeping: why do shelter-dwelling caterpillars fling their frass? Ecol Lett 6:361–370
Weiss MR (2006) Defecation behavior and ecology of insects. Annu Rev Entomol 51:635–661
Welch BJ, Obadi OM, Lampert EC (2017) Effects of carotenoid sequestration on a caterpillars cryptic coloration and susceptibility to predation. Entomol Exp Appl 163:177–183. https://doi.org/10.1111/eea.12558
Wesołowski T, Rowiński P (2014) Do blue tits *Cyanistes caeruleus* synchronize reproduction with caterpillar peaks in a primeval forest? Bird Study 61:231–245
Whitman DW, Blum MS, Slansky F Jr (1994) Carnivory in phytophagous insects. In: Ananthakrishnan TN (ed) Functional dynamics of phytophagous insects. Oxford and IBH, New Delhi, pp 161–205

Willmott KR (2003) The genus *Adelpha*: its systematics, biology and biogeography. Scientific Publishers, Gainesville

Zacharczenko BV (2017) Resolving the systematics of Acronictinae (Lepidoptera, Noctuidae), the evolution of larval defenses, and tracking the gain/loss of complex courtship structures in Noctuidae. Doctoral dissertations 1482. https://opencommons.uconn.edu/dissertations/1482

Zago-Braga RC, Zucoloto FS (2004) Cannibalism studies on eggs and newly hatched caterpillars in a wild population of *Ascia monuste* (Godart) (Lepidoptera, Pieridae). Rev Bras Entomol 48:415–420

Zarlucki MP, Clarke AR, Malcolm SB (2002) Ecology and behavior of first instar larval Lepidoptera. Annu Rev Entomol 47:361–393. https://doi.org/10.1146/annurev.ento.47.091201.145220

Zaspel JM, Weller SJ, Epstein ME (2016) Origin of the hungry caterpillar: evolution of fasting in slug moths (Insecta: Lepidoptera: Limacodidae). Mol Phylogenet Evol 94:827–832. https://doi.org/10.1016/j.ympev.2015.09.017

Part II
Impacts of the First Trophic Level on Caterpillar Ecology and Evolution

Surface Warfare: Plant Structural Defenses Challenge Caterpillar Feeding

Ishveen Kaur, Sakshi Watts, Cristina Raya, Juan Raya, and Rupesh Kariyat

First instar *Manduca sexta* caterpillar on leaf surface of silverleaf nightshade *Solanum elaeagnifolium*. Photo by Rupesh Kariyat

Introduction

Plant defenses against herbivores are generally classified into physical and chemical defenses (Howe and Jander 2008). Physical (structural) defenses act as the first line of defense, playing a crucial role in plant-herbivore interactions. Herbivores have to circumvent them to commence feeding before they come in contact with their host plant (Levin 1973; Kariyat et al. 2018). The physical structures on leaves and stems not only make it difficult to grasp, hold, and feed on the plant but also protect the plants against harsh environmental vagaries such as drought stress, evapotranspiration, and solar radiations (Kaur and Kariyat 2020b; Karabourniotis et al. 2020). These defenses against chewing herbivores have been studied at length and have been found to be particularly effective against caterpillars (Krenn 2010). To successfully navigate and feed on their host plants, caterpillars possess chewing and biting mouthparts with strong mandibles that allow them to bite off pieces of leaves, stems, flowers, roots, and/or fruits. Clearly, this ever-increasing pressure is the primary reason for the host plants to evolve specialized and tightly regulated defense mechanisms to reduce the herbivore's impact on their growth, development, and fitness (Kariyat and Portman 2016).

Physical defenses are broadly divided into six main categories including plant waxes, pubescence (trichomes), spinescence (thorns, spines, and prickles), sclerophylly (hardened leaves), raphides (needle-shaped crystals of calcium oxalate or calcium carbonate found in leaves), and latex (Hanley et al. 2007). However, most studies of physical defenses have been focused on their effects against the major group of chewing herbivores, caterpillars, who are the focus of this chapter. Due to space constraints, we also limit the scope of this chapter to two of these major defenses, viz., plant waxes and pubescence.

Plant Waxes

The plant cuticle, which forms the outermost layer of the plant cell wall, is composed of lipids and hydrocarbons. This thin, hydrophobic layer surrounding all aerial plant organs acts as a crucial interface for plant-insect interactions (Jetter et al. 2008). This layer is comprised of two components, cutin (one of the two waxy polymers which provides structural framework) and wax (acts as a hydrophobic layer; also called epicuticular waxes), and both these compounds integrate together to give the leaf surface a three-dimensional structure. Epicuticular waxes are an important component of plant cuticle, protecting plants from various stresses (Chaudhary et al. 2018). These waxes are most commonly composed of straight chains of either saturated or unsaturated aliphatic hydrocarbons consisting of n-alkanes (C_{21}–C_{35}), alkyl esters, fatty acids, primary and secondary alcohols, diols, ketones, and aldehydes (Konno et al. 2006; Jetter et al. 2008). These cuticular waxes play several ecological and physiological functions in mediating plant-insect interactions and can vary both quantitatively and qualitatively among different plant families and among members within a family (Kurtz 1958). For example, in the Solanaceae, these differences are clearly visible through SEM imagery (Fig. 1) with waxes intact and experimentally removed.

Waxes strongly discourage the movement of caterpillars by rendering the surface of leaves slippery (Federle et al. 1997; Whitney and Federle 2013; Figs. 1 and 2). The irregular and prolonged crawling and searching behavior of diamondback moth

Surface Warfare: Plant Structural Defenses Challenge Caterpillar Feeding 67

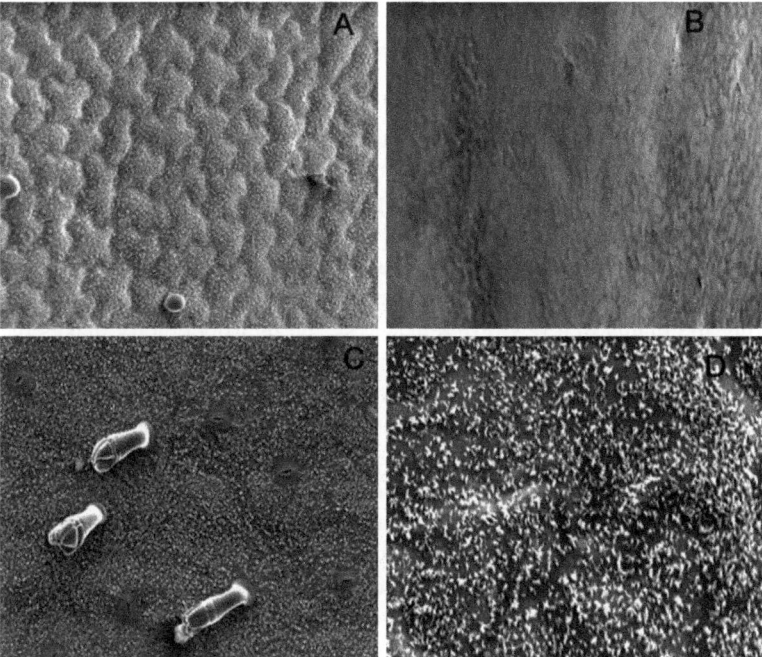

Fig. 1 Scanning electron microscopy images of the leaf surface of (**a**) forest bitterberry (*Solanum anguivi:* Solanaceae) with epicuticular waxes, (**b**) same leaf with waxes removed; (**c**) *Solanum glaucescens* (Solanaceae) at 500X possessing trichomes and waxes, and (**d**) same leaf at more details of wax particles. (Picture credits: Ishveen Kaur and Sakshi Watts)

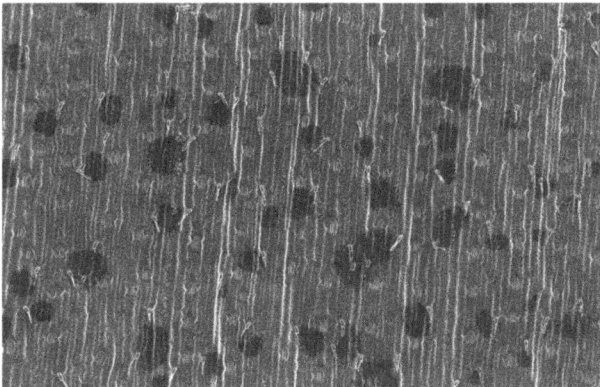

Fig. 2 Scanning electron microscopy image of the leaf surface of sorghum (*Sorghum bicolor:* Poaceae) covered with dense wax crystals at 150X. (Picture credit: Ishveen Kaur)

caterpillars (*Plutella xylostella*: Plutellidae) is observed when caterpillars are placed on cuticular extracts of resistant and susceptible genotypes of canola (*Brassica napus*: Brassicaceae); larvae spend more time crawling and searching food (foraging) than feeding on resistant foliage (Justus et al. 2000; Table 1). A similar increase in walking time and lower rates of feeding were also observed in these caterpillars when put on glossy leaf varieties of collard greens (*Brassica oleracea* var. *viridis*: Brassicaceae), making it difficult for caterpillars to penetrate and grasp the leaf (Stoner 1997; Table 1).

The epicuticular layer also plays a significant role in affecting the intensity and firmness of insect's grip and attachment (Stoner 1990, 1997; Gorb and Gorb 2017), protecting the plants in a few different ways. The hydrophobicity makes the outer surface smooth, which reduces the surface area for insect appendages to embed inside the irregularities or crevices required for gripping and mechanical adherence (Peressadko and Gorb 2004; Gorb et al. 2005; Fig. 2). As a result, larvae have a difficult time penetrating and breaking the leaf epidermis and thus tend to fall from the surface of leaves, delaying access to leaves, making neonate larvae more prone to desiccation. The inability to gain a strong foothold on the leaf surface is also impacted by the wax content as well as by three-dimensional crystalline wax structure formed by integration of waxes on the cutin bed (Stork 1980; Eigenbrode et al. 1996; Barthlott et al. 1998; Eigenbrode and Pillai 1998; Eigenbrode and Jetter 2002; Duetting et al. 2003). The length of wax crystals is another factor that can significantly affect plant-insect interactions (Lewandowska et al. 2020). The thick waxy mat sticks together the mouth parts of larvae, causing them to spend more time on cleaning and preening their mouth parts before they commence feeding (Shelomi et al. 2010), thereby increasing their exposure time to predators (Table 1). Clearly, leaf surface waxes play a significant role as a mechanical barrier in restricting caterpillar feeding across multiple plant families.

Chemical Characteristics of Waxes

Apart from providing a slippery surface to the movement of insects, these waxes also contain a combination of various hydrophobic materials, which can act not only by providing an additional layer of protection to plants as a chemical defense but can also serve as cues for host location (Spencer 1996). They are generally composed of aliphatic chains of alkanes, primary and secondary alcohols, fatty acids, ketones, alkyl esters, and acids (Yang et al. 1993; Jetter et al. 2008; Lewandowska et al. 2020; Table 1). It is particularly significant in caterpillar feeding because plant waxes embedded in this layer are used by caterpillars for their host recognition. This dense waxy layer, often blended with alkanes and alcohols along with the chlorophyll, can provide visual and/or chemical cues to herbivores (Müller 2008). In addition, post host recognition, these compounds can have other functional roles: for instance, free amino acids and soluble carbohydrates in plant waxes of wild leek (*Allium ampeloprasum*; Alliaceae) and maize (*Zea mays*; Poaceae) act as an

Table 1 Examples of positive and negative host plant-caterpillar interactions with epicuticular plant waxes

Plant/family	Characteristics/ chemical composition	Caterpillar species (family)	Effects of wax	References
Brassica oleracea (Brassicaceae)	N-alkane-1-ols, C24 and C25 alcohols	Diamondback moth (Plutella xylostella: Plutellidae)	Increased walking time and more foraging	Eigenbrode and Pillai (1998)
	Triterpenols α and β-amyrin	Diamondback moth (Plutella xylostella: Plutellidae)	Inhibit growth of caterpillars	Sarfaraz et al. (2005)
	Glossy wax	Cabbage butterfly larvae (Pieris rapae: Pieridae)	Increased walking time and failure to commence feeding	Stoner (1997)
		Diamondback moth (Plutella xylostella: Plutellidae)	Resistance by early neonates due to behavioral basis, low pupal weight	Ulmer et al. (2002)
	3-indolyl-methylglucosino-late	Cabbage moth caterpillar (Pieris brassicae: Pieridae)	Host recognition and stimulation of oviposition	van Loon et al. (1992)
Maize (Zea mays: Poaceae)	N-alkanes	Fall army worm (Spodoptera frugiperda; Noctuidae)	Difficulty in establishing on leaf; marked by rapid movement	Eigenbrode and Espelie (1995)
	Lipid extracts	Fall army worm (Spodoptera frugiperda: Noctuidae)	Reduced growth	Yang et al. (1991)
	Myasin (flavone glycoside)	Corn earworm (Helicoverpa zea: Noctuidae)	Antifeedant	Yang et al. (1992)
	Free amino acids, soluble carbohydrates	European corn borer (Ostrinia nubilalis: Pyralidae)	Oviposition preference	Derridj et al. (1996)
	Pentane extract (five n- alkanes)	European corn borer (Ostrinia nubilalis: Pyralidae)	Oviposition stimulant	Udayagiri and Mason (1997)
Potato (Solanum tuberosum: Solanaceae)	Surface lipid extracts	Potato tuber moth (Phthorimaea operculella: Gelechiidae)	Obstruction in movement, spent more time in biting and reduced larval development	Varela and Bernays (1988)

(continued)

Table 1 (continued)

Plant/family	Characteristics/ chemical composition	Caterpillar species (family)	Effects of wax	References
Canola (*Brassica napus*: Brassicaceae);	Heavy waxy layer	Diamondback moth (*Plutella xylostella*: Plutellidae)	Less oviposition activity than glossy sister strains	Justus et al. (2000)
Rapeseed (*Brassica napus*: Brassicaceae)	Heavy waxy coating	Cotton bollworm (*Helicoverpa armigera*: Noctuidae)	Longer first meals due to pre-processing time to remove wax, gumming of mouthparts, more time spent feeding due to thick layer	Shelomi et al. (2010)
White spruce (*Picea glauca*: Pinaceae)	Monoterpenes	Banded leaf roller (*Choristoneura* spp.: Tortricidae)	Feeding deterrent	Daoust et al. (2010)
Sugarcane (*Saccharum alopecuroides*: Poaceae)	Alcohols and carbonyls	African sugarcane borer (*Eldana saccharina* Walker: Pyrallidae)	Resistance against insect herbivore	Rutherford and Staden (1996)
Soyabean (*Glycine max*: Fabaceae)	Volatiles, hexane, and methanol extract	Saltmarsh caterpillar (*Estigmene acrea*: Erebidae)	More time spent on walking	Castrejon et al. (2006)
Carrot (*Daucus carota*: Apiaceae)	Flavonoids glycosides, chlorogenic acid	American swallowtail (*Papilio polyxenes*: Papilionidae)	Oviposition stimulants	Brooks et al. (1996)
Leeks (*Allium ampeloprasum*: Alliaceae)	Free amino acids, soluble carbohydrates (sugars)	European corn borer (*Ostrinia nubilalis*: Pyralidae)	Oviposition stimulant	Derridj et al. (1996)
Mung bean (*Vigna radiata*: Fabaceae)	Alkanes and free acids	Bihar hairy caterpillar (*Spilosoma obliqua* Walker: Arctiidae)	Short-range attractant and oviposition stimulant	Mobarak et al. (2020)
Balsam fir (*Abies balsamea*: Pinaceae)	Leaf waxes	Eastern spruce budworm (*Choristoneura fumiferana*: Tortricidae)	Host detection, oviposition stimulant	Rivet and Albert (1990)
Mulberry (*Morus alba*: Moraceae)	Fatty alcohols hexacosanol and octacosanol	Silk moth (*Bombyx mori*: Bombycidae)	Feeding stimulant	Mori (1982)

(continued)

Table 1 (continued)

Plant/family	Characteristics/ chemical composition	Caterpillar species (family)	Effects of wax	References
Yellow rocketcress (*Barbarea vulgaris*: Brassicaceae)	Glucosinolates	Diamondback moth (*Plutella xylostella*: Plutellidae)	Host recognition and stimulation of oviposition	Hopkins et al. (2009)

oviposition stimulant for European corn borer (*Ostrinia nubilalis*: Pyralidae) (Derridj et al. 1996; Table 1); and the alkane blend present in the wax of leaves of *B. napus* attracts *P. xylostella* moths for oviposition (Barbero 2016; Table 1).

Epicuticular waxes also have semiochemicals that function as sex pheromones and kairomones attracting parasitoids and predators of herbivores post-herbivory, thereby mediating multi-trophic interactions (Rutledge 1996; Dutton et al. 2000). Parasitoids are attracted to the sloughed off lipophilic compounds from their herbivore host on the plant surface, greatly enhancing their foraging success. Similarly, presence of high docosonal (aliphatic alcohol) on leaf surface prevents feeding of tobacco budworm (*Heliothis virescens*: Noctuidae) on tobacco varieties, and sugar and sugar alcohols present on the surface of apples function as kairomones and trigger oviposition in female moths of (*Cydia pomonella*: Tortricidae) (Lombarkia and Derridj 2002).

These waxes also protect plants in many other different ways. Some of the chemicals in wax may also serve as feeding deterrents. Epicuticular waxes also contain low levels of terpenoids, sterols, flavonoids, phenolics, glucosinolates, furanocoumarins, and alkaloids, which have been found to inhibit feeding and oviposition activity of insects and, in many cases, even proven toxic to caterpillars (Eigenbrode and Espelie 1995; Schoonhoven et al. 2005; Städler and Reifenrath 2009; Haliński et al. 2012; Kariyat et al. 2019a).

While considered to be more effective against generalist herbivores which lack co-evolved sequestration or detoxification mechanisms, some of these secondary metabolites (e.g., alkaloids and flavonoids) are highly inducible, such that the concentration of these compounds increases in the wax layer post-herbivory (Müller 2008). For instance, egg deposition by cabbage moth (*Pieris brassicae:* Pieridae) on thale cress (*Arabidopsis thaliana*: Brassicaceae) and brussels sprouts (*Brassica oleracea*: Brassicaceae) alters the wax composition and specific ratio of tetra triacontanoic acid (C34) and tetracosanoic acid (C24), with consequences for enhanced attraction of the egg parasitoid *Trichogramma brassicae* (Hymenoptera: Trichogrammatidae) in the vicinity of host eggs (Blenn et al. 2012). These changes also lead to increased walking and biting time of neonates, as observed in *P. xylostella*, which struggle on the glossy surfaces of cabbage (*B. oleracea*) due to presence of n-alkanes, secondary alcohols, n-alkanoic acids (fatty acids), n-alkane-1-ols, and triterpenoids (Cole and Riggal 1992). And finally, surface waxes also act as a pathway for diffusion of volatiles such as terpenes and glucosinolate derivatives that are diffused through the epicuticular layer when stomata are closed, thus acting as an interface for plant-insect interactions (Müller and Riederer 2005).

Caterpillar Adaptations

In order to evade the hydrophobic and glossy wax layer that restricts movement and reduces grip, caterpillars have evolved a series of physiological and behavioral adaptations. These modifications help them to circumvent waxes and associated defenses and to successfully establish on the plant to commence feeding. For example, *P. brassicae* caterpillars produce water-soluble phenolic compounds that moisten the leaf surface; this later helps the adults to firmly attach eggs to leaves of members of the Brassicaceae (Voigt and Gorb 2009; Fatouros et al. 2012). *Pieris brassicae* has also evolved to ingest and detoxify the glucosinolates present in the heavy wax layer on plants in the Brassicaceae family, thereby enabling caterpillars to grow and develop. In a much more specialized and dramatic adaptation, the giant geometrid (*Biston robustum*: Geometridae) has the ability to alter the chemistry of its outer integument based on the chemicals present in the cuticular wax layer, to resemble the morphology of the plants with respect to surface chemistry, also known as phytomimesis (Akino 2005). This chemical camouflage (mimicking the chemistry of the host plant) allows larvae to hide from their predators. Clearly, while waxes play a significant role as an anti-herbivore defense by both physical and chemical means, caterpillars, especially with a co-evolution history with their host plants, have evolved both behavioral and morphological adaptations to overcome these defenses, an area that still needs to be examined.

Trichomes

Walking on a leaf surface is not an easy task for caterpillars. If waxes do not deter them, their movement is severely obstructed by the dense mat of minute hairs, or trichomes, surrounding them, acting as another physical barrier (Levin 1973; Kariyat et al. 2017, 2018). Although miniature in stature, they play well-established roles in protecting the plants against biotic and abiotic stresses, such as extreme environmental conditions, temperature stresses, detoxification of heavy metals, and high soil salinity (Karabourniotis et al. 2020; Kaur and Kariyat 2020b), and, more importantly, as an anti-herbivore defense (Wagner et al. 2004). Trichomes are broadly divided into two types: non-glandular and glandular trichomes. Non-glandular trichomes are secretion-less sharp and pointed appendages that serve as a physical barrier to the movement of caterpillars (Kariyat et al. 2013a, 2017, 2018, 2019b; Figs. 3, 6, 8, 9a, d, and 10). Glandular trichomes, in addition to acting as physical hindrance, can also have sticky exudates, toxins, and bioactive compounds in their glandular head that deter herbivores via chemical defenses; additionally, the compounds may activate downstream complex defense signaling cascades such as the jasmonic acid pathway (Peiffer et al. 2009; Kaur and Kariyat 2020a, b) (Figs. 3, 4, 5, 8, and 9b, c).

Fig. 3 Artistic representation of the potential defensive roles of trichomes against caterpillars. Trichomes (non-glandular and glandular) can potentially entrap caterpillars, with capacity of glandular trichomes (with glandular bulbs on top) to act as toxins and signaling molecules to activate defense gene expression post-rupturing and stellate trichomes can physically injure the caterpillars. (Illustration by Annette Diaz)

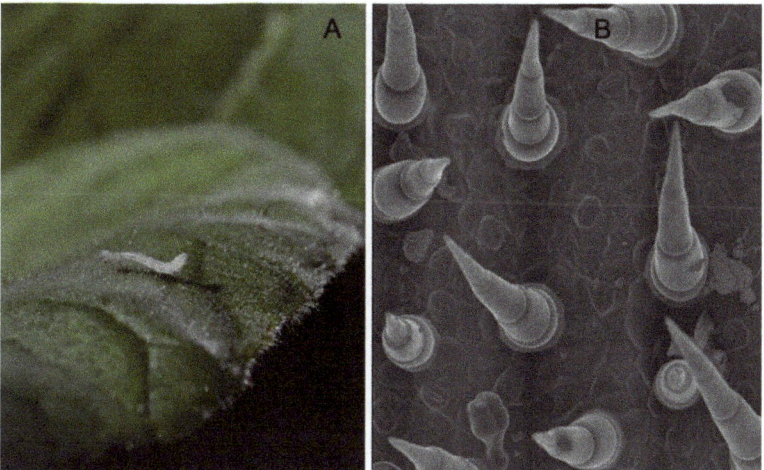

Fig. 4 (**a**) Cabbage looper larva (*Trichoplusia ni:* Noctuidae) struggling to walk on the leaf surface of bottle gourd (*Lagenaria siceraria*: Cucurbitaceae) and (**b**) scanning electron microscopy image of the same leaf possessing both glandular and non-glandular trichomes at 400X. (Picture credit: Rupesh Kariyat and Ishveen Kaur)

Fig. 5 (**a**) Fall army worm (*Spodoptera frugiperda*: Noctuidae) trying to feed on tomato (*Solanum lycopersicon*: Solanaceae) leaf surface covered with a dense mat of glandular and non-glandular trichomes, (**b**) first instar tobacco hornworm (*Manduca sexta:* Sphingidae) caterpillar walking unimpeded on the smooth trichome free surface of pepper (*Capsicum annum:* Solanaceae), (**c**) neonate *M. sexta* making an effort to feed on tobacco (*Nicotiana tabacum:* Solanaceae) leaves covered with glandular trichomes, and (**d**) neonate *M. sexta* struggling to walk on potato (*Solanum tuberosum*: Solanaceae) leaf with glandular trichomes. (Photo credit: Rupesh Kariyat)

Non-glandular trichomes are epidermal outgrowths that provide physical protection from different biotic and abiotic stresses and are devoid of any exudates (Fig. 9a, d). These trichomes start playing their part by disrupting the caterpillar integument with their sharp-pointed ends as soon as young neonates come in contact with a plant (Riddick and Wu 2011) (Figs. 3 and 10) or by preventing their access to the leaf epidermis, resulting in more prolonged foraging than actual feeding (Wilkens et al. 1996; Kariyat et al. 2017, 2018, 2019b; Andama et al. 2020, Kaur and Kariyat 2020b) (Figs. 4, 5, and 10; Table 2). An obvious consequence of this delay in feeding is the prolonged apparency for predators, thus mounting another line of defense even before the whole suite of chemical defenses are activated or induced.

Table 2 Examples demonstrating the role of glandular trichomes in plant defense against caterpillars

Plant species	Trichome type	Caterpillar species	Interaction	References
Garden tomato (*Lycopersicon esculentum*: Solanaceae)	Foliar tetracellular glandular trichomes	Corn earworm (*Helicoverpa zea*: Noctuidae)	Phenolic compounds present in the foliar tetracellular glandular trichomes including flavonol glycoside rutin and some other phenolics as well have antibiotic effects on herbivores, measured by reduction of larval growth	Duffey and Isman (1981)
	Type VI glandular trichomes	Tomato pinworm (*Keiferia lycopersicella*: Gelechiidae)	Glandular exudates provided physical barrier to caterpillars	Lin et al. (1987)
	Foliar glandular trichomes	Tobacco hornworm (*Manduca sexta*: Sphingidae)	Rupturing of foliar glandular trichomes by herbivore contact resulted into the induction of proteinase inhibitor 2 which is involved in defense mechanisms in plants	Peiffer et al. (2009)
	Glandular trichomes	Tobacco hornworm (*Manduca sexta*: Sphingidae)	Trichomes hindered caterpillars from searching for food as caterpillars spent most of the time mowing trichomes	Wilkens et al. (1996)
	Glandular trichomes	Corn earworm (*Helicoverpa zea*: Noctuidae)	Growth of *H. zea* was impaired by higher glandular trichome density.	Tian et al. (2012)
	Glandular trichomes	Tobacco hornworm (*Manduca sexta*: Sphingidae)	A novel recessive mutation called *odorless-2 (od-2)* which altered morphology, density, and chemical composition of glandular trichomes, increased the susceptibility of plants to attack by herbivore, indicating its importance in providing resistance to plants against herbivores	Kang et al. (2010)
	Non-branched glandular trichomes	Tobacco hornworm (*Manduca sexta*: Sphingidae)	Negative effects on larval growth and development of caterpillars were observed due to feeding on trichomes	Kariyat et al. (2019b)

(continued)

Table 2 (continued)

Plant species	Trichome type	Caterpillar species	Interaction	References
L. hirsutum f. *glabratum* Mill	Type VI glandular trichomes	Tomato pinworm (*Keiferia lycopersicella*: Gelechiidae), beet armyworm (*Spodoptera exigua*: Noctuidae)	2-tridecanone and 2-undecanone present in glandular heads of type VI trichomes are major compounds having insecticidal properties	Lin et al. (1987)
Lycopersicon hirsutum Humb. & Bonpl. (LA 361) (Solanaceae)	Type VI glandular trichomes	Tomato pinworm (*Keiferia lycopersicella*: Gelechiidae), beet armyworm (*Spodoptera exigua*: Noctuidae)	Acute toxicity from two unknown sesquiterpenes present in glandular heads of type VI trichomes was found against herbivores	Lin et al. (1987)
Lluttu papa (*Solanum berthaultii*: Solanaceae)	Glandular leaf trichomes	Potato tuber moth (*Phthorimaea operculella*: Gelechiidae)	Deterred oviposition, reduced feeding, longer larval development period, and decreased pupal weight were observed for herbivores feeding on glandular leaf trichomes	Malakar and Tingey (2003)
	Glandular leaf trichomes	Potato tuber moth (*Phthorimaea operculella*: Gelechiidae)	Decreased foliar oviposition and feeding by herbivore, lower pupal weight, and increased mortality were observed	Malakar and Tingey (1999)
	Type A and B glandular trichomes	Potato tuber moth (*Phthorimaea operculella*: Gelechiidae)	Reduced oviposition of herbivore was observed	Horgan et al. (2007)
European crowfoot (*Aquilegia vulgaris*: Ranunculaceae), Pyrenean columbine (*Aquilegia pyrenaica*: Ranunculaceae)	Glandular floral trichomes	General herbivores	Plants were found to have more trichomes in regions with more herbivores. Also, better access by the small insects to the flowers and fruits was observed on the removal of floral trichomes	Jaime et al. (2013)

(continued)

Table 2 (continued)

Plant species	Trichome type	Caterpillar species	Interaction	References
Garden petunia (*Petunia hybrida*: Solanaceae)	Glandular trichomes	African cotton leafworm (*Spodoptera littoralis*: Noctuidae)	*P. hybrida* pleiotropic drug resistance (PhPDR2) type ABC transporters found in trichomes provided resistance to herbivores, since its downregulation decreased levels of petuniasterone and petuniolide (potent toxins against insect herbivores) and increased susceptibility of plants toward herbivore	Sasse et al. (2016)
Solanum tarijense (Solanaceae)	Type A glandular trichomes	Potato tuber moth (*Phthorimaea operculella*: Gelechiidae)	A negative relation between trichome density and oviposition by herbivore has been observed	Horgan et al. (2007, 2009)
Sacred thorn-apple (*Datura wrightii*: Solanaceae)	Glandular and non-glandular trichomes	General herbivores	Overall, the damage to sticky (having more than 95% of glandular trichomes) plants was more than velvety (having less than 5% glandular trichomes) plants, and the sticky plants produced 45% fewer seeds than the velvety plants in the course of providing resistance against herbivores	Elle et al. (1999)
Pelargonium×hortorum	Glandular trichomes	Soybean looper (*Chrysodeixis includens*: Noctuidae)	Lesser hatching of herbivore eggs treated with exudates of glandular trichomes along with higher herbivore mortality was documented. Also, cutting leaf veins before feeding beyond the cuts by last instar caterpillars was observed	Hurley and Dussourd (2014)

Fig. 6 Fifth instar caterpillar of tobacco hornworm (*Manduca sexta*: Sphingidae) (**a**) on trichome-rich surface of outbred horsenettle leaves (*Solanum carolinense*: Solanaceae) plant, (**b**) on low trichome inbred *S. carolinense*, and (**c**) on trichome free pepper (*Capsicum annuum*: Solanaceae) leaves. (Picture credit: Rupesh Kariyat)

Fig. 7 Artistic rendering of *Manduca sexta* caterpillar with peritrophic membrane or gut lining (red color) punctured by undigested trichomes potentially leading to mixing of hemolymph (dark green) and food bolus (light green inside peritrophic membrane; Kariyat et al. 2017). (Illustration by Annette Diaz)

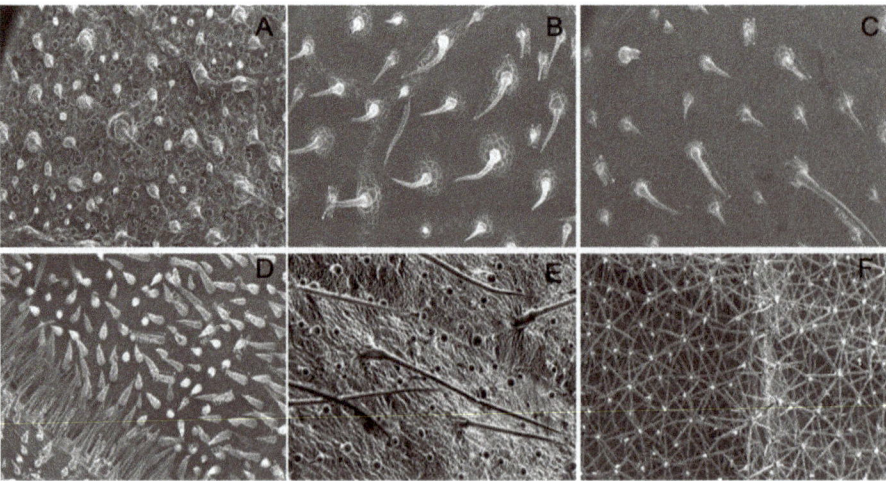

Fig. 8 Scanning electron microscopy images of different types of trichomes present on the leaf surface of (**a**) sunflower (*Helianthus annuus*: Asteraceae) at 70X, (**b**) squash (*Cucurbita pepo*: Cucurbitaceae) at 70X, (**c**) cucumber (*Cucumis sativus*: Cucurbitaceae) at 70X, (**d**) bottle gourd (*Lagenaria siceraria*: Cucurbitaceae), (**e**) potato tree (*Solanum grandiflorum*: Solanaceae) at 100X, and (**f**) Ethiopian eggplant (*Solanum aethiopicum*: Solanaceae) at 60X magnification. (Picture credit: Jesus Chavana, Ishveen Kaur, and Sakshi Watts)

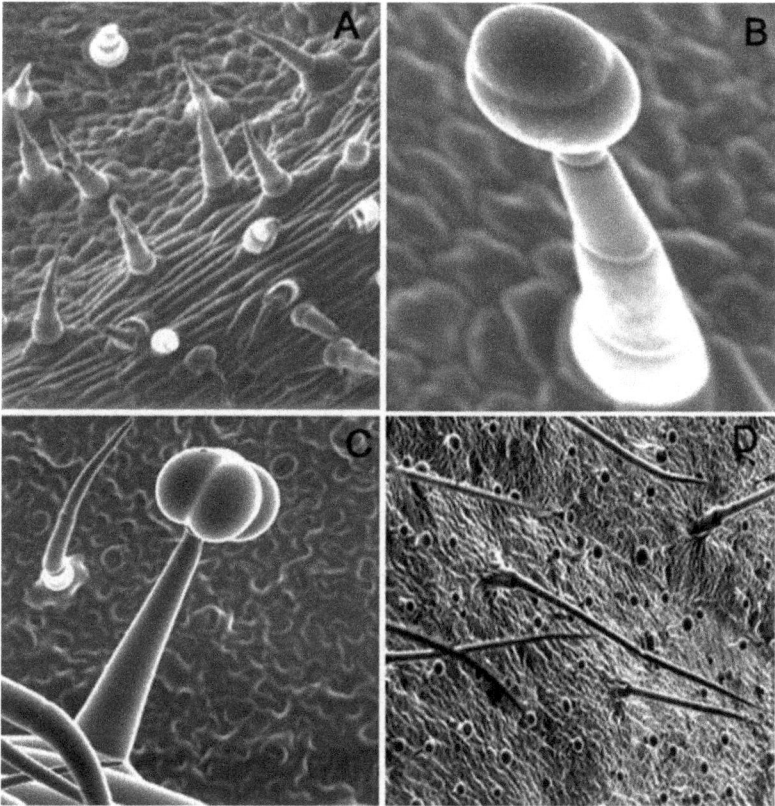

Fig. 9 Scanning electron microscopic images of (**a**) non-glandular trichomes on leaf surface of bottle gourd (*Lagenaria siceraria*: Cucurbitaceae; 100X), (**b** and **c**) glandular trichome on the leaf surface of *L. siceraria* and tomato (*Solanum lycopersicum*: Solanaceae; 400X and 300X, respectively) with bulbous head housing different defensive compounds to ward off herbivory, and (**d**) combination of glandular and non-glandular trichomes on leaf surface of the potato tree (*Solanum grandifolium*: Solanaceae; 100X). (Picture credits: Ishveen Kaur and Sakshi Watts)

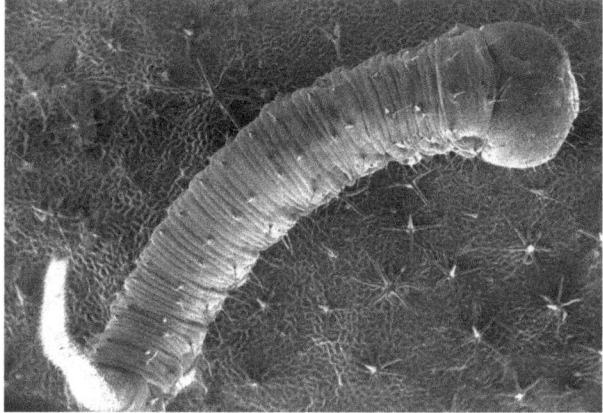

Fig. 10 Scanning electron microscopic image displaying comparative size of stellate trichomes of eggplant (*Solanum melongena*: Solanaceae) and first instar tobacco hornworm (*Manduca sexta*: Sphingidae) caterpillar at 50X. (Picture credit: Sakshi Watts)

Being tender, small, and lacking fully functional defense mechanisms (Zalucki et al. 2002), neonate larvae are even more vulnerable to desiccation and death by starvation or reduced feeding efficiency (Kariyat et al. 2018; Despland 2018) (Figs. 3, 4, and 5; Table 2). For example, neonate tobacco hornworm (*Manduca sexta*; Sphingidae) caterpillars find it difficult to initiate feeding on Solanaceae species possessing trichomes, thus leading to their starvation, desiccation, and ultimately death (Kariyat et al. 2013a, 2017, 2018) (Fig. 5; Table 3). These trichomes act as a strong weapon against herbivores by their mechanical entrapment capabilities, hence impeding the locomotion of these neonates and making them less active and weak by wounding them (Dalin et al. 2008; Peiffer et al. 2009) (Fig. 3). Moreover, they can act as extreme feeding deterrents and physically damaging structures by entangling and impaling soft-bodied insects, in the case of hook-shaped trichomes, as found in French bean (*Phaseolus vulgaris*: Fabaceae) and passionflower (*Passiflora* spp.: Passifloraceae) that can be fatal for herbivores such as sciarid fly (*Bradysia pauper*; Diptera: Sciaridae) and brush-footed butterfly larvae (Heliconiinae: Nymphalidae) (Gilbert 1971; Gepp 1977). Furthermore, the importance of these trichomes as physical structures impeding growth and development of larvae has been clearly demonstrated in *Lycopersicon* spp. (Solanaceae) where neonate *Helicoverpa armigera* suffered high mortality due to their entrapment in non-glandular trichomes (Simmons et al. 2004) (Table 2).

The role of non-glandular trichomes in plant defense has not been fully explored since it is usually assumed that they merely restrict herbivore movement (Figs. 3, 4, 5, 6, and 10). However, recent studies have shown that they can have both pre- and post-ingestive effects, and these effects are also instar specific (Kariyat et al. 2017, 2018) (Fig. 7). In addition to severely restricting feeding of early instars, they can damage later instars, following their ingestion. The undigested stellate trichomes can pierce holes in the peritrophic membrane (gut lining) of caterpillars (Kariyat et al. 2017; Figs. 3, 7, and 8; Table 2), with the dire possibility of food contents in gut mixing with hemolymph (Fig. 7). The peritrophic membrane is known to protect the caterpillar gut from physical and chemical damage, which facilitates digestion. The rupturing of this lining can force hemolymph to ooze inside the gut lining, which would eventually lead to sepsis, thus up-regulating the immune system of caterpillars, leading to diversion of resources toward defense from their growth and development (Pechan et al. 2002; Kariyat et al. 2017) (Fig. 7). Moreover, recent discoveries have shown that the presence of silica (involved in lignification) packaged in non-glandular trichomes of rice (*Oryza sativa*: Poaceae) prevents chewing herbivores from damaging the plant (Andama et al. 2020; Kaur and Kariyat 2020a, b). In one recent examination of histo-chemical and morphological features of non-glandular trichomes, the presence of living cells in the non-glandular trichomes of plants in Lamiaceae and Verbenaceae was confirmed (Tozin et al. 2016). These cells are capable of bioactive compound synthesis, suggesting that non-glandular trichomes are not simply a physical defense but can also assist or enhance plant chemical defenses. Taken together, non-glandular trichomes severely discourage herbivory by affecting mass, feeding and foraging behavior, oviposition of adults,

Table 3 Examples of role of non-glandular trichomes in plant defense against caterpillars

Plant species	Trichome type	Caterpillar/herbivore species	Interaction	References
Mustard (*Brassica nigra*: Brassicaceae)	Non-glandular trichomes	Cabbage whitefly (*Pieris rapae*: Pieridae)	New leaves post damage had 76% more trichomes per unit area than control plants on damage by the herbivore	Traw and Dawson (2002)
	Non-glandular trichomes	Cabbage looper (*Trichoplusia ni*: Noctuidae)	New leaves post damage had 113% more trichomes per unit area than control plants upon damage by the herbivore indicating their role against herbivory	Traw and Dawson (2002)
Jimsonweed (*Datura stramonium*: Solanaceae)	Leaf trichomes	General herbivores	Decrease in the damage caused by herbivory was recorded with increase in the number of leaf trichomes	Valverde et al. (2001)
Horsenettle (*Solanum carolinense*: Solanaceae)	Non-glandular trichomes	Tobacco hornworm (*Manduca sexta*: Sphingidae)	Non-glandular trichomes were found more effective in deterring feeding by first and second instar caterpillars than late instars	Kariyat et al. (2018)
			Feeding on trichomes, besides acting as feeding deterrent, also caused damage to peritrophic membrane (gut lining) of caterpillars	Kariyat et al. (2017)
Silverleaf nightshade (*Solanum elaegnifolium*: Solanaceae)	Non-glandular stellate trichomes	Tobacco hornworm (*Manduca sexta*: Sphingidae)	Feeding on trichomes resulted into reduced mass gain and increased time to pupation by caterpillars	Kariyat et al. (2019b)
Garden tomato (*Lycopersicon esculentum*)	Non-glandular trichomes	Tobacco hornworm (*Manduca sexta*: Sphingidae)	Trichomes hindered caterpillars from searching for food as caterpillars spent most of the time mowing trichomes	Wilkens et al. (1996)

(continued)

Table 3 (Continued)

Plant species	Trichome type	Caterpillar/herbivore species	Interaction	References
Canada nettle (*Laportea canadensis*: Urticaceae)	Stinging trichomes	The red admiral (*Vanessa atalanta*: Nymphalidae)	No significant deterrence in herbivore feeding was observed due to presence of stinging trichomes	Tuberville et al. (1996)
Passionflower (*Passiflora lobata*: Passifloraceae)	Hooked trichomes	Brush-footed butterfly (*Heliconius pachinus*: Nymphalidae)	Trichomes were able to deter non-specialist herbivore (*H. pachinus*)	Cardoso (2008)
Lyre-leaves rock-cress (*Arabidopsis lyrata*: Brassicaceae)	Non-glandular trichomes	Diamondback moth (*Plutella xylostella*: Plutellidae)	For herbivore damage to the plants, significant phenotypic and genetic correlation was found, since the plants with higher trichome density had lower herbivore damage. Although oviposition by herbivore also decreased with increased trichome density, correlation was not significant	Sletvold et al. (2010)
	Non-glandular trichomes	Diamondback moth (*Plutella xylostella*: Plutellidae) and other herbivores	Glabrous (without trichomes) plants had more damage by herbivores than trichome-producing plants	Løe et al. (2007)
Milkweed (*Asclepias syriaca*: Apocynaceae)	Leaf trichomes	*Monarch butterfly* (*Danaus plexippus*: Nymphalidae)	Herbivore abundance was negatively correlated to leaf trichome density	Agrawal (2005)
Passionflower (*Passiflora adenopoda*: Passifloraceae)	Hook-like uncinate trichomes	Heliconine butterflies (*Heliconius* spp.: Nymphalidae)	Hooked trichomes of this species very effectively deterred herbivores	Gilbert (1971)

development, mortality, and fitness (Traw and Dawson 2002; Kariyat et al. 2017, 2018, 2019b) (Table 3).

In the case of glandular trichomes, the trichome head is filled with sticky exudates or secretions, which contain a diversity of defensive compounds (terpenoids, phenolics, alkaloids, and acyl sugars, to name a few) in the secretory cells (Table 2; Fig. 9b, c). In addition, they also have different morphotypes present within the trichomes with specialized functional diversity (Giuliani et al. 2020; Uzelac et al. 2020) as evident from SEM images from different plant families (Figs. 4, 8, and 9). The glandular cells of these trichomes either entangle the herbivores with sticky exudates (Zalucki et al. 2002) or act as toxins (Hare 2005), which can interfere with their metabolic pathways, as soon as herbivores damage/disturb these glandular trichomes. They can synthesize proteinase inhibitors that bind with the digestive enzymes of herbivores affecting the digestive ability of the caterpillar (Peiffer et al. 2009) or release volatile terpenes (Wang et al. 2008; Biswas et al. 2009) that attract predators and parasitoids of the feeding caterpillars. For instance, the chemicals released by glandular trichomes of *Lycopersicon* spp. cause severe irritation and trapping of *Phthorimaea operculella* neonates due to the presence of methyl ketones (2-tridecanone, 2-undecanone) and sesquiterpenes in the globular cells of their trichomes (Lin et al. 1987; Gurr and Grath 2002; Table 2). Similarly, the chemical exudates from glandular trichomes of *L. hirsutum* and *L. pennellii* entrapped larvae of *Helicoverpa armigera*, thereby causing their mortality (Simmons et al. 2004). Glandular trichomes can also activate defense priming through jasmonic acid (Kariyat et al. 2013b), a key phytohormone in plant defense signaling that initiates a wide range of defense responses in plants. In addition to these modes of defenses, exudates from trichomes also render the plant tissues indigestible and unpalatable by housing compounds such as polyphenol oxidase, making it difficult for herbivores to digest the plant tissue. This protective effect is evident by the significant reduction in caterpillar mass gain and pupal mass along with longer development period, evident from reduced growth of *Spodoptera exigua*. The enzyme polyphenol oxidase catalyzes the production of quinones that can further alkylate nucleophilic amino acids, which makes the plant unpalatable (Bhonwong et al. 2009; Fig. 5). Rupturing foliar glandular trichomes of tomato (*L. esculentum*) by *Manduca sexta* induces a proteinase inhibitor, and presence of 2-tridecanone and 2-undecanone in the glandular heads of type VI trichomes in *L. hirsutum* f. *glabratum* has been found to have insecticidal properties against tomato pinworm (*Keiferia lycopersicella*: Gelechiidae) and *S. exigua*, thus highlighting the role of these defenses in plant protection (Lin et al. 1987). Clearly, these glandular trichome exudates act as direct chemical defenses and can significantly reduce caterpillar performance (Malakar and Tingey 1999; Tian et al. 2012; Pradhan and Maradi 2020; Table 2).

In addition, glandular trichomes can act as indirect defenses mediating multitrophic interactions. For example, o-acyl sugars in *Nicotiana tabacum* (Solanaceae) glandular trichomes act as a sugary first meal relished by neonate *M. sexta* caterpillars (Fig. 6); however, post-ingestion of these sugars imparts a distinct scent to larval body and frass that acts as location cues for their natural enemies, such as the rough harvester ant (*Pogonomyrmex rugosus*) (Hymenoptera: Formicidae)

(Weinhold and Baldwin 2011). This ant preys on *M. sexta* eggs and early instar caterpillars. Desert lizards (*Phrynosoma platyrhinos*) (Squamata: Phrynosomatidae) target later instars using the same cues (Stork et al. 2011). Thus, trichome ingestion, which is their first meal, makes them more vulnerable and apparent to their enemies, providing a fascinating example of multitrophic interactions mediated by trichomes.

Caterpillar Adaptations

Specialized strategies to circumvent trichome defenses have evolved in the Lepidoptera, allowing caterpillars to feed and flourish on trichome-rich plants. Caterpillars have evolved morphological, physiological, and behavioral adaptations to minimize trichome defenses, and these caterpillars continue to impose selection pressure on their host plants. For example, before feeding, *M. sexta* caterpillars spend time shaving both glandular and non-glandular trichomes of *Lycopersicon esculentum* (Wilkens et al. 1996). This provides them an opening to the epidermis and then to underlying tissues below the epidermis. Caterpillars have also adapted yet another phenomenon to escape defenses mediated by trichomes, by weaving a network of silk fibers as a pathway to facilitate easier movement on a trichome covered leaf, allowing them to safely reach and remove the protective trichome layer on foliage before reaching their food source (Young and Moffett, 1979; Fordyce and Agrawal, 2001; Yadav and Yack, 2018). Some lepidopterans have been found to produce silk secretions on trichomes to facilitate their easy movement (Rathcke and Poole 1975; Fordyce and Agrawal 2001). Larvae of the specialist herbivore, the zebra longwing butterfly (*Heliconius charithonia*: Nymphalidae) on their host plant passionflower (*Passiflora lobata*: Passifloraceae), interweave silken threads on the hooked trichomes to facilitate their movement. In contrast, generalist herbivores, such as the brush-footed butterfly caterpillar (*Heliconius pachinus*: Nymphalidae) on *P. lobata* and the tiger clearwing butterfly caterpillar (*Mechanitis isthmia*: Nymphalidae) on *Solanum* spp. (Solanaceae), struggle to access the leaf tissue (Gilbert 1971; Cardoso 2008). The studies of lepidopteran larvae and their silk-based defense mechanism collectively suggest that early instar larvae employ behavioral adaptations to reduce their contact with trichomes. This includes moving their head in upward, downward, and side-to-side motions distinct from other caterpillars that produce silk (Hulley 1988). Another unique and interesting phenomenon observed is the structural modification in tarsi and longer tibia of these lepidopterans to facilitate movement without coming in contact with dense trichome surface, but allowing them to insert their legs directly onto the epidermis (Medeiros and Boligon 2007). Similarly, in *H. charithonia* (a specialist herbivore on *Passiflora* spp.), the lateral proleg is sclerotized, helping to prevent the trapping of larvae on trichome hooks. Moreover, the presence of interwoven silken mats to facilitate movement over the trichomes and trichome tips indicates that the caterpillar often bites off the trichome tips to facilitate walking and feeding on plant tissues. In other systems, young neonates often aggregate to collectively feed on trichome-rich or glabrous leaf surfaces. This social facilitation observed in tiger wing butterfly (*Mechanitis menapis*: Nymphalidae) caterpillars while feeding on glandular trichome of *Solanum acerifolium* (Solanaceae) also assists them to continue feeding on leaf tissue with the aid of their siblings (Despland 2019).

Clearly, caterpillars have evolved a variety of behavioral and structural traits that allow them to circumvent this powerful defense. Studies performed on *Manduca sexta* and *Pardasena diversipennis* (Nolidae) on various solanaceous species have found that the caterpillars tend to "shave" trichomes with their mandibles from the leaf surface before they begin feeding on the leaves (Hulley 1988). This behavior is commonly known as mowing. In mowing, the caterpillars first begin feeding at the edge on the upper surface and then use their mandibles to mow the trichomes on the leaf surface, before they eat, thus allowing them to easily gain access to plant material. Also, they start feeding at the edge of upper leaf after mowing the trichomes at the leaf margin, by approaching the leaf at angle of 45 degrees, without touching the trichomes on the abaxial surface of leaf, thus forcing these trichomes to fall under their own weight, thereby removing them (Hulley 1988). While these counter defenses are primarily behavioral or morphological, some caterpillars have evolved detoxification and physiological adaptations so they can ingest trichomes without being affected by their toxins, more critical in the case of glandular trichomes. For example, the larvae of the ello sphinx (*Erinnyis ello*: Sphingidae) feed on spurge nettle (*Cnidoscolus urens*: Euphorbiaceae) by grazing the urticating hairs of the petiole and also by stopping the latex flow into the leaf (Dillon et al. 1983). Additionally, caterpillars often resort to grooming themselves post-feeding, possibly to remove defensive chemicals secreted by the plant. Such defense mechanisms by caterpillars provide them with a greater possibility of survival especially for early instars as their mandibles are small and trichomes could be lethal, causing the loss of prolegs and impaling them (Levin 1973).

Conclusions

Taken together, waxes and trichomes have been well established as an efficient multi-faceted defense strategy in plants against herbivores, especially against lepidopteran larvae. Using Solanaceae as a model plant family and *Manduca sexta* as the model herbivore, we have demonstrated that trichomes are highly inducible by *Manduca sexta* feeding (Kariyat et al. 2013a) and are affected by genetic variation and breeding status of the host plant (Kariyat et al. 2013a). Trichome synthesis and expression are mediated through the jasmonic acid pathway (Kariyat et al. 2012), and caterpillars feeding on trichome-rich species will gain less mass (Kariyat et al. 2018, 2019b) and take more time to begin feeding (Kariyat et al. 2018; Watts and Kariyat 2021), in a density-dependent manner. In addition, these effects are also species specific and vary based on trichome type (Kariyat et al. 2018, 2019b). Trichome-mediated effects are also instar specific where early instars are primarily affected by delayed feeding (Kariyat et al. 2017, 2019b), while late instars that ingest trichomes are affected by their ruptured peritrophic membranes (Kariyat et al. 2017; Kaur and Kariyat 2020a) with possible implications for immune response, development, and survival.

We have been successful in dissecting the genetic mechanisms and developmental regulation underlying trichomes (Chalvin et al. 2020). A wide range of manipulative ecological and physiological studies have successfully determined the role of trichomes as the first line of defense against the voracious caterpillars that continuously attack them. However, new discoveries in this field include the finding that there is metal accumulation in trichomes based on topographic analysis and energy dispersive X-ray (EDX) analysis using scanning electron microscopy (SEM), but these approaches are still in their infancy and require more attention. For example, fortification of trichomes (Mustafa et al. 2018; Hopewell et al. 2021) with metals has been found only in a few families, and the exact functions are unknown. Recently, Andama et al. (2020) demonstrated that trichomes are involved in defense against chewing herbivores in rice (*Oryza sativa*: Poaceae). More research is also warranted in fully understanding potential tradeoffs on trichome induction postherbivory, examining fitness and whether these effects extend to the next generation (Nihranz et al. 2019). The advent of modern analytical chemistry tools also provides us a unique opportunity to examine, identify, and quantify the tremendous variety of compounds present in and on trichomes and their potential role in mediating multitrophic interactions.

References

Agrawal A (2005) Natural selection on common milkweed (*Asclepias syriaca*) by a community of specialized insect herbivores. Evol Ecol Res 7:651–667

Akino T (2005) Chemical and behavioral study on the phytomimetic giant geometer *Biston robustum* Butler (Lepidoptera: Geometridae). Appl Entomol Zool 40:497–505. https://doi.org/10.1303/aez.2005.497

Andama JB, Mujiono K, Hojo Y et al (2020) Nonglandular silicified trichomes are essential for rice defense against chewing herbivores. Plant Cell Environ 43:2019–2032. https://doi.org/10.1111/pce.13775

Barbero F (2016) Cuticular lipids as a cross-talk among ants, plants and butterflies. Int J Mol Sci 17:1966. https://doi.org/10.3390/ijms17121966

Barthlott W, Christoph N, David C, Ditsch F, Meusel I, Theisen I, Wilhelmi H (1998) Classification and terminology of plant epicuticular waxes. J Linn Soc Bot 126:237–260. https://doi.org/10.1111/j.1095-8339.1998.tb02529.x

Bhonwong A, Stout MJ, Attajarusit J, Tantasawat P (2009) Defensive role of tomato polyphenol oxidases against cotton bollworm (*Helicoverpa armigera*) and beet armyworm (*Spodoptera exigua*). J Chem Ecol 35:28–38. https://doi.org/10.1007/s10886-008-9571-7

Biswas KK, Foster AJ, Aung T, Mahmoud SS (2009) Essential oil production: relationship with abundance of glandular trichomes in aerial surface of plants. Acta Physiol Plant 31:13–19. https://doi.org/10.1007/s11738-008-0214-y

Blenn B, Bandoly M, Küffner A, Otte T, Geiselhardt S, Fatouros NE, Hilker M (2012) Insect egg deposition induces indirect defense and epicuticular wax changes in *Arabidopsis thaliana*. J Chem Ecol 38:882–892. https://doi.org/10.1007/s10886-012-0132-8

Brooks JS, Williams EH, Feeny P (1996) Quantification of contact oviposition stimulants for black swallowtail butterfly, *Papilio polyxenes*, on the leaf surfaces of wild carrot, *Daucus carota*. J Chem Ecol 22:2341–2357. https://doi.org/10.1007/bf02029551

Cardoso MZ (2008) Herbivore handling of a plant's trichome: the case of *Heliconius charithonia* (L.) (Lepidoptera: Nymphalidae) and *Passiflora lobata* (Killip) Hutch. (Passifloraceae). Neotrop Entomol 37:247–252. https://doi.org/10.1590/s1519-566x2008000300002

Castrejon F, Virgen A, Rojas JC (2006) Influence of chemical cues from host plants on the behavior of neonate *Estigmene acrea* larvae (Lepidoptera: Arctiidae). Environ Entomol 35:700–707. https://doi.org/10.1603/0046-225x-35.3.700

Chalvin C, Drevenske S, Dron M, Bendahmane A, Boualem A (2020) Genetic control of glandular trichome development. Trends Plant Sci 25(5):477–487. https://doi.org/10.1016/j.tplants.2019.12.025

Chaudhary A, Bala K, Thakur S, Kamboj R, Dumra N (2018) Plant defenses against herbivorous insects. Int J Chem Stud 6(5):681–688

Cole RA, Riggall W (1992) Pleiotropic effects of genes in glossy *Brassica oleracea* resistant to *Brevicoryne brassicae*. In: Proceedings of the 8th international symposium on insect-plant relationships, pp 313–315. https://doi.org/10.1007/978-94-011-1654-1_101

Dalin P, Ågren J, Björkman C, Huttunen P, Kärkkäinen K (2008) Leaf trichome formation and plant resistance to herbivory. In: Schaller A (ed) Induced plant resistance to herbivory. Springer, Dordrecht, pp 89–105. https://doi.org/10.1007/978-1-4020-8182-8_4

Daoust SP, Mader BJ, Bauce E, Despland E, Dussutour A, Albert PJ (2010) Influence of epicuticular-wax composition on the feeding pattern of a phytophagous insect: implications for host resistance. Can Entomol 142:261–270. https://doi.org/10.4039/n09-064

Derridj S, Wu BR, Stammitti L, Garrec JP, Derrien A (1996) Chemicals on the leaf surface, information about the plant available to insects. In: Städler E, Rowell-Rahier M, Bauer R (eds) Proceedings of the 9th international symposium on insect-plant relationships, Series entomologica 53, pp 197–201. https://doi.org/10.1007/978-94-009-1720-0_45

Despland E (2018) Effects of phenological synchronization on caterpillar early-instar survival under a changing climate. Can J For Res 48:247–254. https://doi.org/10.1139/cjfr-2016-0537

Despland E (2019) Caterpillars cooperate to overcome plant glandular trichome defenses. Front Ecol Evol. https://doi.org/10.3389/fevo.2019.00232

Dillon PM, Lowrie S, McKey D (1983) Disarming the "evil woman": petiole constriction by a sphingid larva circumvents mechanical defenses of its host plant, *Cnidoscolus urens* (Euphorbiaceae). Biotropica 15:112. https://doi.org/10.2307/2387953

Duetting PS, Ding H, Neufeld J, Eigenbrode SD (2003) Plant waxy bloom on peas affects infection of pea aphids by *Pandora neoaphidis*. J Invertebr Pathol 84:149–158. https://doi.org/10.1016/j.jip.2003.10.001

Duffey SS, Isman MB (1981) Inhibition of insect larval growth by phenolics in glandular trichomes of tomato leaves. Experientia 37:574–576. https://doi.org/10.1007/bf01990057

Dutton A, Mattiacci L, Dorn S (2000) Plant-derived semiochemicals as contact host location stimuli for a parasitoid of leafminers. J Chem Ecol 26(10):2259–2273. https://doi.org/10.1023/A:1005566508926

Eigenbrode SD, Espelie KE (1995) Effects of plant epicuticular lipids on insect herbivores. Annu Rev Entomol 40:171–194. https://doi.org/10.1146/annurev.en.40.010195.001131

Eigenbrode SD, Jetter R (2002) Attachment to plant surface waxes by an insect predator. ICB Integr Comp Biol 42:1091–1099. https://doi.org/10.1093/icb/42.6.1091

Eigenbrode SD, Pillai SK (1998) Neonate *Plutella xylostella* responses to surface wax components of a resistant cabbage (*Brassica oleracea*). J Chem Ecol 24(10):1611–1627. https://doi.org/10.1023/A:1020812411015

Eigenbrode SD, Castagnola T, Roux M-B, Steljes L (1996) Mobility of three generalist predators is greater on cabbage with glossy leaf wax than on cabbage with a wax bloom. Entomol Exp Appl 81:335–343. https://doi.org/10.1046/j.1570-7458.1996.00104.x

Elle E, van Dam NM, Hare JD (1999) Cost of glandular trichomes, a "resistance" character in *Datura wrightii* Regel (Solanaceae). Evolution 53:22. https://doi.org/10.2307/2640917

Fatouros NE, Lucas-Barbosa D, Weldegergis BT, Pashalidou FG, van Loon JJ, Dicke M, Harvey JA, Rieta G, Huigens ME (2012) Plant volatiles induced by herbivore egg deposition affect insects of different trophic levels. PLoS One. https://doi.org/10.1371/journal.pone.0043607

Federle W, Maschwitz U, Fiala B, Riederer M, Hölldobler B (1997) Slippery ant-plants and skilful climbers: selection and protection of specific ant partners by epicuticular wax blooms in *Macaranga* (Euphorbiaceae). Oecologia 112:217–224. https://doi.org/10.1007/s004420050303

Fordyce JA, Agrawal AA (2001) The role of plant trichomes and caterpillar group size on growth and defence of the pipevine swallowtail *Battus philenor*. J Anim Ecol 70:997–1005. https://doi.org/10.1046/j.0021-8790.2001.00568.x

Gepp J (1977) Hindrance of arthropods by trichomes of bean-plants (*Phaseolus vulgaris* L.). Anz Schädlingskd Pfl Umwelt 50:8–12. https://doi.org/10.1007/BF01993461

Gilbert LE (1971) Butterfly-plant coevolution: has *Passiflora adenopoda* won the selectional race with Heliconiine butterflies? Science 172:585–586. https://doi.org/10.1126/science.172.3983.585

Giuliani C, Bottoni M, Ascrizzi R, Milani F, Papini A, Flamini G, Fico G (2020) *Lavandula dentata* from Italy: analysis of trichomes and volatiles. Chem Biodivers 17(11). https://doi.org/10.1002/cbdv.202000532

Gorb E, Haas K, Henrich A, Enders S, Barbakadze N, Gorb S (2005) Composite structure of the crystalline epicuticular wax layer of the slippery zone in the pitchers of the carnivorous plant *Nepenthes alata* and its effect on insect attachment. J Exp Biol 208:4651–4662. https://doi.org/10.1242/jeb.01939

Gorb EV, Gorb SN (2017) Anti-adhesive effects of plant wax coverage on insect attachment. J Exp Bot 68:5323–5337. https://doi.org/10.1093/jxb/erx271

Gurr GM, McGrath D (2002) Foliar pubescence and resistance to potato moth, *Phthorimaea operculella*, in *Lycopersicon hirsutum*. Entomol Exp Appl 103:35–41. https://doi.org/10.1046/j.1570-7458.2002.00960.x

Haliński ŁP, Paszkiewicz M, Gołębiowski M, Stepnowski P (2012) The chemical composition of cuticular waxes from leaves of the gboma eggplant (*Solanum macrocarpon* L.). J Food Compos Anal 25:74–78. https://doi.org/10.1016/j.jfca.2011.06.004

Hanley ME, Lamont BB, Fairbanks MM, Rafferty CM (2007) Plant structural traits and their role in anti-herbivore defence. PPEES 8:157–178. https://doi.org/10.1016/j.ppees.2007.01.001

Hare JD (2005) Biological activity of acyl glucose esters from *Datura wrightii* glandular trichomes against three native insect herbivores. J Chem Ecol 31:1475–1491. https://doi.org/10.1007/s10886-005-5792-1

Hopewell T, Selvi F, Ensikat H-J, Weigend M (2021) Trichome biomineralization and soil chemistry in Brassicaceae from Mediterranean ultramafic and calcareous soils. Plan Theory 10:377. https://doi.org/10.3390/plants10020377

Hopkins RJ, van Dam NM, van Loon JJA (2009) Role of glucosinolates in insect-plant relationships and multitrophic interactions. Annu Rev Entomol 54:57–83. https://doi.org/10.1146/annurev.ento.54.110807.090623

Horgan FG, Quiring DT, Lagnaoui A, Pelletier Y (2007) Variable responses of tuber moth to the leaf trichomes of wild potatoes. Entomol Exp Appl 125:1–12. https://doi.org/10.1111/j.1570-7458.2007.00590.x

Horgan FG, Quiring DT, Lagnaoui A, Pelletier Y (2009) Effects of altitude of origin on trichome-mediated anti-herbivore resistance in wild Andean potatoes. Flora 204:49–62. https://doi.org/10.1016/j.flora.2008.01.008

Howe GA, Jander G (2008) Plant immunity to insect herbivores. Annu Rev Plant Biol 59:41–66. https://doi.org/10.1146/annurev.arplant.59.032607.092825

Hulley PE (1988) Caterpillar attacks plant mechanical defence by mowing trichomes before feeding. Ecol Entomol 13:239–241. https://doi.org/10.1111/j.1365-2311.1988.tb00351.x

Hurley KW, Dussourd DE (2014) Toxic geranium trichomes trigger vein cutting by soybean loopers, *Chrysodeixis includens* (Lepidoptera: Noctuidae). Arthropod Plant Interact 9:33–43. https://doi.org/10.1007/s11829-014-9348-6

Jaime R, Rey PJ, Alcántara JM, Bastida JM (2013) Glandular trichomes as an inflorescence defence mechanism against insect herbivores in Iberian columbines. Oecologia 172:1051–1060. https://doi.org/10.1007/s00442-012-2553-z

Jetter R, Kunst L, Samuels AL (2008) Composition of plant cuticular waxes. In: Biology of the plant cuticle, pp 145–181. https://doi.org/10.1002/9780470988718.ch4

Justus KA, Dosdall LM, Mitchell BK (2000) Oviposition by *Plutella xylostella* (Lepidoptera: Plutellidae) and effects of phylloplane waxiness. J Econ Entomol 93:1152–1159. https://doi.org/10.1603/0022-0493-93.4.1152

Kang J-H, Liu G, Shi F, Jones AD, Beaudry RM, Howe GA (2010) The tomato odorless-2 mutant is defective in trichome-based production of diverse specialized metabolites and broad-spectrum resistance to insect herbivores. Plant Physiol 154:262–272. https://doi.org/10.1104/pp.110.160192

Karabourniotis G, Liakopoulos G, Nikolopoulos D, Bresta P (2020) Protective and defensive roles of non-glandular trichomes against multiple stresses: structure–function coordination. J For Res 31:1–12. https://doi.org/10.1007/s11676-019-01034-4

Kariyat RR, Portman SL (2016) Plant-herbivore interactions: thinking beyond larval growth and mortality. Am J Bot 103:789–791. https://doi.org/10.3732/ajb.1600066

Kariyat RR, Mena-Alí J, Forry B, Mescher MC, De Moraes CM, Stephenson AG (2012) Inbreeding, herbivory, and the transcriptome of *Solanum carolinense*. Entomol Exp Appl 144:134–144. https://doi.org/10.1111/j.1570-7458.2012.01269.x

Kariyat RR, Balogh CM, Moraski RP, De Moraes CM, Mescher MC, Stephenson AG (2013a) Constitutive and herbivore-induced structural defenses are compromised by inbreeding in *Solanum carolinense* (Solanaceae). Am J Bot 100:1014–1021. https://doi.org/10.3732/ajb.1200612

Kariyat RR, Mauck KE, Balogh CM, Stephenson AG, Mescher MC, De Moraes CM (2013b) Inbreeding in horsenettle (*Solanum carolinense*) alters night-time volatile emissions that guide oviposition by *Manduca sexta* moths. Proc R Soc B 280:20130020. https://doi.org/10.1098/rspb.2013.0020

Kariyat RR, Smith JD, Stephenson AG, De Moraes CM, Mescher MC (2017) Non-glandular trichomes of *Solanum carolinense* deter feeding by *Manduca sexta* caterpillars and cause damage to the gut peritrophic matrix. Proc R Soc B 284:20162323. https://doi.org/10.1098/rspb.2016.2323

Kariyat RR, Hardison SB, Ryan AB, Stephenson AG, De Moraes CM, Mescher MC (2018) Leaf trichomes affect caterpillar feeding in an instar-specific manner. Commun Integr Biol 11:1–6. https://doi.org/10.1080/19420889.2018.1486653

Kariyat RR, Gaffoor I, Sattar S et al (2019a) Sorghum 3-Deoxyanthocyanidin flavonoids confer resistance against corn leaf aphid. J Chem Ecol 45:502–514. https://doi.org/10.1007/s10886-019-01062-8

Kariyat RR, Raya CE, Chavana J, Cantu J, Guzman G, Sasidharan L (2019b) Feeding on glandular and non-glandular leaf trichomes negatively affect growth and development in tobacco hornworm (*Manduca sexta*) caterpillars. Arthropod-Plant Interact 13:321–333. https://doi.org/10.1007/s11829-019-09678-z

Kaur I, Kariyat RR (2020a) Eating barbed wire: direct and indirect defensive roles of non-glandular trichomes. Plant Cell Environ 43:2015–2018. https://doi.org/10.1111/pce.13828

Kaur J, Kariyat RR (2020b) Role of trichomes in plant stress biology. In: Evolutionary ecology of plant-herbivore interaction, pp 15–35. https://doi.org/10.1007/978-3-030-46012-9_2

Konno K, Nakamura M, Tateishi K, Wasano N, Tamura Y, Chikara H, Hattori M, Koyama A, Ono H, Kohno K Tateishi M, Wasano K (2006) Various ingredients in plant latex: their crucial roles in plant defense against herbivorous insects. Plant Cell Physiol, vol 47. Great Clarendon St, Oxford OX2 6DP, England, pp S48–S48

Krenn HW (2010) Feeding mechanisms of adult lepidoptera: structure, function, and evolution of the mouthparts. Annu Rev Entomol 55:307–327. https://doi.org/10.1146/annurev-ento-112408-085338

Kurtz EB (1958) A survey of some plant waxes of southern Arizona. JAOCS 35:465–467. https://doi.org/10.1007/bf02539916

Levin DA (1973) The role of trichomes in plant defense. Q Rev Biol 48:3–15. https://doi.org/10.1086/407484

Lewandowska M, Keyl A, Feussner I (2020) Wax biosynthesis in response to danger: its regulation upon abiotic and biotic stress. New Phytol 227:698–713. https://doi.org/10.1111/nph.16571

Lin SY, Trumble JT, Kumamoto J (1987) Activity of volatile compounds in glandular trichomes of *Lycopersicon* species against two insect herbivores. J Chem Ecol 13:837–850. https://doi.org/10.1007/bf01020164

Løe G, Toräng P, Gaudeul M, Ågren J (2007) Trichome production and spatiotemporal variation in herbivory in the perennial herb *Arabidopsis lyrata*. Oikos 116:134–142. https://doi.org/10.1111/j.2006.0030-1299.15022.x

Lombarkia N, Derridj S (2002) Incidence of apple fruit and leaf surface metabolites on *Cydia pomonella* oviposition. Entomol Exp Appl 104:79–87. https://doi.org/10.1046/j.1570-7458.2002.00993.x

Malakar R, Tingey WM (1999) Resistance of *Solanum berthaultii* foliage to potato tuberworm (Lepidoptera: Gelechiidae). J Econ Entomol 92:497–502. https://doi.org/10.1093/jee/92.2.497

Malakar R, Tingey WM (2003) Glandular trichomes of *Solanum berthaultii* and its hybrids with potato deter oviposition and impair growth of potato tuber moth. Entomol Exp Appl 94:249–257. https://doi.org/10.1046/j.1570-7458.2000.00627.x

Medeiros L, Boligon DS (2007) Adaptations of two specialist herbivores to movement on the hairy leaf surface of their host, *Solanum guaraniticum* Hassl (Solanaceae). Rev Bras Entomol 51:210–216. https://doi.org/10.1590/s0085-56262007000200011

Mobarak SH, Koner A, Mitra S, Mitra P, Barik A (2020) The importance of leaf surface wax as short-range attractant and oviposition stimulant in a generalist Lepidoptera. J Appl Ecol 44:616–631. https://doi.org/10.1111/jen.12769

Mori M (1982) n-Hexacosanol and n-octacosanol: feeding stimulants for larvae of the silkworm, *Bombyx mori*. J Insect Physiol 28:969–973. https://doi.org/10.1016/0022-1910(82)90114-7

Müller C (2008) Resistance at the plant cuticle. In: Schaller A (ed) Induced plant resistance to herbivory. Springer, Dordrecht, pp 107–129. https://doi.org/10.1007/978-1-4020-8182-8_5

Müller C, Riederer M (2005) Plant surface properties in chemical ecology. J Chem Ecol 31:2621–2651. https://doi.org/10.1007/s10886-005-7617-7

Mustafa A, Ensikat H-J, Weigend M (2018) Mineralized trichomes in Boraginales: complex microscale heterogeneity and simple phylogenetic patterns. Ann Bot 121:741–751. https://doi.org/10.1093/aob/mcx191

Nihranz CT, Kolstrom RL, Kariyat RR et al (2019) Herbivory and inbreeding affect growth, reproduction, and resistance in the rhizomatous offshoots of *Solanum carolinense* (Solanaceae). Evol Ecol 33:499–520. https://doi.org/10.1007/s10682-019-09997-w

Pechan T, Cohen A, Williams WP, Luthe DS (2002) Insect feeding mobilizes a unique plant defense protease that disrupts the peritrophic matrix of caterpillars. PNAS 99:13319–13323. https://doi.org/10.1073/pnas.202224899

Peiffer M, Tooker JF, Luthe DS, Felton GW (2009) Plants on early alert: glandular trichomes as sensors for insect herbivores. New Phytol 184:644–656. https://doi.org/10.1111/j.1469-8137.2009.03002.x

Peressadko A, Gorb SN (2004) When less is more: experimental evidence for tenacity enhancement by division of contact area. J Adhes 80:247–261. https://doi.org/10.1080/00218460490430199

Pradhan K, Maradi RM (2020) Plant glandular trichomes: the natural pesticide factories. Biotica Res Today 2(8):713–716

Rathcke BJ, Poole RW (1975) Coevolutionary race continues: butterfly larval adaptation to plant trichomes. Science 187:175–176. https://doi.org/10.1126/science.187.4172.175

Riddick EW, Wu Z (2011) Lima bean–lady beetle interactions: hooked trichomes affect survival of *Stethorus punctillum* larvae. BioControl 56:55–63. https://doi.org/10.1007/s10526-010-9309-7

Rivet M-P, Albert PJ (1990) Oviposition behavior in spruce budworm *Choristoneura fumiferana* (Clem.) (Lepidoptera: Tortricidae). J Insect Behav 3:395–400. https://doi.org/10.1007/bf01052116

Rutherford RS, van Staden J (1996) Towards a rapid near-infrared technique for prediction of resistance to sugarcane borer *Eldana saccharina* walker (Lepidoptera: Pyralidae) using stalk surface wax. J Chem Ecol 22:681–694. https://doi.org/10.1007/bf02033578

Rutledge CE (1996) A survey of identified kairomones and synomones used by insect parasitoids to locate and accept their hosts. Chemoecology 7:121–131. https://doi.org/10.1007/bf01245964

Santos Tozin LR, de Melo Silva SC, Rodrigues TM (2016) Non-glandular trichomes in Lamiaceae and Verbenaceae species: morphological and histochemical features indicate more than physical protection. N Z J 54:446–457. https://doi.org/10.1080/0028825x.2016.1205107

Sarfraz M, Keddie AB, Dosdall LM (2005) Biological control of the diamondback moth, *Plutella xylostella*: a review. Biocontrol Sci Tech 15:763–789. https://doi.org/10.1080/09583150500136956

Sasse J, Schlegel M, Borghi L et al (2016) *Petunia hybrid* PDR2 is involved in herbivore defense by controlling steroidal contents in trichomes. Plant Cell Environ 39:2725–2739. https://doi.org/10.1111/pce.12828

Schoonhoven LM, Van Loon B, Van Loon JJ, Dicke M (2005) Insect-plant biology. Oxford University Press on demand University of Arizona, London

Shelomi M, Perkins LE, Cribb BW, Zalucki MP (2010) Effects of leaf surfaces on first-instar *Helicoverpa armigera* (Hübner) (Lepidoptera: Noctuidae) behaviour. Aust J Entomol 49:289–295. https://doi.org/10.1111/j.1440-6055.2010.00766.x

Simmons AT, Gurr GM, McGrath D et al (2004) Entrapment of *Helicoverpa armigera* (Hubner) (Lepidoptera: Noctuidae) on glandular trichomes of *Lycopersicon* species. Aust J Entomol 43:196–200. https://doi.org/10.1111/j.1440-6055.2004.00414.x

Sletvold N, Huttunen P, Handley R, Kärkkäinen K, Ågren J (2010) Cost of trichome production and resistance to a specialist insect herbivore in *Arabidopsis lyrata*. Evol Ecol 24:1307–1319. https://doi.org/10.1007/s10682-010-9381-6

Spencer JL (1996) Waxes enhance *Plutella xylostella* oviposition in response to sinigrin and cabbage homogenates. Entomol Exp Appl 81:165–173. https://doi.org/10.1111/j.1570-7458.1996.tb02028.x

Städler E, Reifenrath K (2009) Glucosinolates on the leaf surface perceived by insect herbivores: review of ambiguous results and new investigations. Phytochem Rev 8:207–225. https://doi.org/10.1007/s11101-008-9108-2

Stoner KA (1990) Glossy leaf wax and plant resistance to insects in *Brassica oleracea* under natural infestation. Environ Entomol 19:730–739. https://doi.org/10.1093/ee/19.3.730

Stoner KA (1997) Behavior of neonate imported cabbageworm larvae (Lepidoptera: Pieridae) under laboratory conditions on collard leaves with glossy or normal waxi. J Entomol Sci 32:290–295. https://doi.org/10.18474/0749-8004-32.3.290

Stork N (1980) Role of waxblooms in preventing attachment to brassicas by the mustard beetle, *Phaedon cochleariae*. Entomol Exp Appl 28:100–107. https://doi.org/10.1111/j.1570-7458.1980.tb02992.x

Stork WF, Weinhold A, Baldwin IT (2011) Trichomes as dangerous lollipops: do lizards also use caterpillar body and frass odor to optimize their foraging? Plant Signal Behav 6:1893–1896. https://doi.org/10.4161/psb.6.12.18028

Tian D, Tooker J, Peiffer M, Chung SH, Felton GW (2012) Role of trichomes in defense against herbivores: comparison of herbivore response to woolly and hairless trichome mutants in tomato (*Solanum lycopersicum*). Planta 236:1053–1066. https://doi.org/10.1007/s00425-012-1651-9

Traw BM, Dawson TE (2002) Differential induction of trichomes by three herbivores of black mustard. Oecologia 131:526–532. https://doi.org/10.1007/s00442-002-0924-6

Tuberville TD, Dudley PG, Pollard AJ (1996) Responses of invertebrate herbivores to stinging trichomes of *Urtica dioica* and *Laportea canadensis*. Oikos 75:83. https://doi.org/10.2307/3546324

Udayagiri S, Mason CE (1997) Epicuticular wax chemicals in *Zea mays* influence oviposition in *Ostrinia nubilalis*. J Chem Ecol 23:1675–1687. https://doi.org/10.1023/b:joec.0000006443.72203.f7

Ulmer B, Gillott C, Woods D, Erlandson M (2002) Diamondback moth, *Plutella xylostella* (L.), feeding and oviposition preferences on glossy and waxy *Brassica rapa* (L.) lines. Crop Prot 21:327–331. https://doi.org/10.1016/s0261-2194(02)00014-5

Uzelac B, Stojičić D, Budimir S (2020) Glandular trichomes on the leaves of *Nicotiana tabacum*: morphology, developmental ultrastructure, and secondary metabolites. In: Ramawat K, Ekiert H, Goyal S (eds) Plant cell and tissue differentiation and secondary metabolites, Reference series in phytochemistry. Springer, Cham, pp 25–61. https://doi.org/10.1007/978-3-030-30185-9_1

Valverde PL, Fornoni J, Núñez-Farfan J (2001) Defensive role of leaf trichomes in resistance to herbivorous insects in *Datura stramonium*. J Evol Biol 14:424–432. https://doi.org/10.1046/j.1420-9101.2001.00295.x

van Loon JJ, Blaakmeer A, Griepink FC, van Bleek TA, Schoonhoven LM, de Groot A (1992) Leaf surface compound from *Brassica oleracea* (Cruciferae) induces oviposition by *Pieris brassicae* (Lepidoptera: Pieridae). Chemoecology 3:39–44. https://doi.org/10.1007/bf01261455

Varela LG, Bernays EA (1988) Behavior of newly hatched potato tuber moth larvae, *Phthorimaea operculella* Zell. (Lepidoptera: Gelechiidae), in relation to their host plants. J Insect Behav 1:261–275. https://doi.org/10.1007/bf01054525

Voigt D, Gorb S (2009) Egg attachment of the asparagus beetle *Crioceris asparagi* to the crystalline waxy surface of *Asparagus officinalis*. Proc R Soc B 277:895–903. https://doi.org/10.1098/rspb.2009.1706

Wagner GJ, Wang E, Shepherd R (2004) New approaches for studying and exploiting an old protuberance, the plant trichome. Ann Bot 93:3–11. https://doi.org/10.1093/aob/mch011

Wang G, Tian L, Aziz N, Broun P, Dai X, He J, King A, Zhao PX, Dixon RA (2008) Terpene biosynthesis in glandular trichomes of hop. Plant Physiol 148:1254–1266. https://doi.org/10.1104/pp.108.125187

Watts S, Kariyat R (2021) Picking sides: feeding on the abaxial leaf surface is costly for caterpillars. Planta. https://doi.org/10.1007/s00425-021-03592-6

Weinhold A, Baldwin IT (2011) Trichome-derived O-acyl sugars are a first meal for caterpillars that tags them for predation. PNAS 108:7855–7859. https://doi.org/10.1073/pnas.1101306108

Whitney HM, Federle W (2013) Biomechanics of plant–insect interactions. Curr Plant Biol 16:105–111. https://doi.org/10.1016/j.pbi.2012.11.008

Wilkens RT, Shea GO, Halbreich S, Stamp NE (1996) Resource availability and the trichome defenses of tomato plants. Oecologia 106:181–191. https://doi.org/10.1007/bf00328597

Yadav C, Yack JE (2018) Immature stages of the masked birch caterpillar, *Drepana arcuata* (Lepidoptera: Drepanidae) with comments on feeding and shelter building. J Insect Sci 18(1):18. https://doi.org/10.1093/jisesa/iey006

Yang G, Isenhour DJ, Espelie KE (1991) Activity of maize leaf cuticular lipids in resistance to leaf-feeding by the fall armyworm. Fla Entomol 74:229. https://doi.org/10.2307/3495301

Yang G, Wiseman BR, Espelie KE (1992) Cuticular lipids from silks of seven corn genotypes and their effect on development of corn earworm larvae [*Helicoverpa zea* (Boddie)]. J Agric Food Chem 40:1058–1061. https://doi.org/10.1021/jf00018a030

Yang G, Espelie KE, Wiseman BR, Isenhour DJ (1993) Effect of corn foliar cuticular lipids on the movement of fall armyworm (Lepidoptera: Noctuidae) neonate larvae. Fla Entomol 76:302. https://doi.org/10.2307/3495730

Young AM, Moffett MW (1979) Studies on the population biology of the tropical butterfly *Mechanitis isthmia* in Costa Rica. Am Midl Nat 101:309. https://doi.org/10.2307/2424596

Zalucki MP, Clarke AR, Malcolm SB (2002) Ecology and behavior of first instar larval Lepidoptera. Annu Rev Entomol 47:361–393. https://doi.org/10.1146/annurev.ento.47.091201.145220

Impacts of Plant Defenses on Host Choice by Lepidoptera in Neotropical Rainforests

María-José Endara, Dale Forrister, James Nicholls, Graham N. Stone, Thomas Kursar, and Phyllis Coley

Letys mycerina (Erebidae) feeding on young leaves of *Inga thibaudiana* in Los Amigos, Peru. Credit: María-José Endara

M.-J. Endara
Grupo de Investigación en Biodiversidad, Medio Ambiente y Salud-BIOMAS, Universidad de las Américas, Quito, Ecuador

School of Biological Sciences, University of Utah, Salt Lake City, UT, USA
e-mail: majo.endara@utah.edu

D. Forrister · T. Kursar · P. Coley (✉)
School of Biological Sciences, University of Utah, Salt Lake City, UT, USA
e-mail: coley@biology.utah.edu

J. Nicholls
Institute of Evolutionary Biology, University of Edinburgh, Edinburgh, UK

CSIRO, Australian National Insect Collection, Black Mountain Labs, Acton, ACT, Australia

G. N. Stone
Institute of Evolutionary Biology, University of Edinburgh, Edinburgh, UK

Introduction

In forests, insect herbivores and their host plants are major components of the community. The study of their interactions is essential for understanding the mechanisms promoting and maintaining species diversity and niche differentiation in both trophic levels (Becerra 2015). Theory has long predicted that the evolution of plant antiherbivore defenses and insect counter-adaptations is the driver of trait diversification and coevolution (Ehrlich and Raven 1964). This arms race has also been implicated in mechanisms of coexistence (Becerra 1997; Lewinsohn and Roslin 2008; Kursar et al. 2009; Marquis et al. 2016; Maron et al. 2019). For example, recent work suggests that herbivores play a key role in maintaining the high local diversity of rainforests by preventing most plant species from becoming abundant (Comita et al. 2014). Species with different defenses do not share herbivores and therefore can coexist, promoting high local diversity (Janzen 1970; Becerra 2007, Fine et al. 2013; Comita et al. 2014; Coley and Kursar 2014, Salazar et al. 2016a, b; Vleminckx et al. 2018; Forrister et al. 2019). For herbivores, host plant specialization has been regarded as one of the main mechanisms promoting insect diversity, as specialized partitioning of plant resources may allow more herbivore species to coexist (Novotny et al. 2006; Lewinsohn and Roslin 2008). Thus, there is an intricate relationship between plant traits and host plant choice, yet we still do not fully understand the impacts of plant defenses on host choice nor the consequences at ecological and evolutionary time scales for herbivorous insects. In this chapter, we will review the relative effect of different plant defenses on host plant choice by Lepidoptera caterpillars. We will focus on insect herbivores associated with the Neotropical tree genus *Inga* (Fabaceae) in the rainforests of Central and South America, a region where the diversity of plants and invertebrates is among the highest in the terrestrial world and where the arms race coevolution may be particularly pronounced.

With more than 300 species, *Inga* is an ecologically important and abundant genus in Neotropical rainforests. *Inga* is among those genera with the greatest congeneric species richness at a given site, with more than 40 species together contributing 6% of the stems occurring in 25 ha in Ecuador (Valencia et al. 2004). As a system for study of plant-herbivore interactions, the genus *Inga* is exceptional in that within a single genus, we can find a broad range of defensive traits and a diverse assemblage of herbivores. Throughout the Amazon and Panama, lepidopteran larvae are the dominant group of herbivores attacking *Inga*, both in terms of numbers and damage caused (Kursar et al. 2006; Coley et al. 2018). Thus, for this chapter, we will focus mainly on lepidopteran herbivores due to their importance.

Herbivores in Neotropical Forests Prefer Young Expanding Leaves

In tropical rainforests, one of the most prevalent patterns in herbivory is the difference in damage between expanding and mature leaves. The typical leaf lifespan in the understory is between 2 and 4 years (Coley and Barone 1996). Nevertheless,

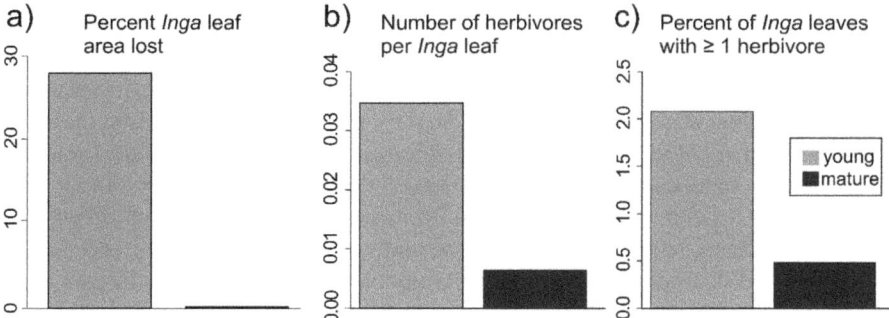

Fig. 1 Comparison of herbivore attack on young expanding leaves vs mature leaves in *Inga*. (**a**) Average percent leaf area lost for ten species of *Inga* on Barro Colorado Island, Panama; (**b**) number of caterpillar herbivores per leaf found feeding on young and mature leaves for 15 species of *Inga* in Yasuní National Park, Ecuador. (**c**) Percentage of leaves where at least one herbivore was found feeding for fifteen species of *Inga* in Yasuní National Park, Ecuador. Data are for saplings that were visited monthly from February 2000 to November 2004 on Barro Colorado Island, Panama, and from September 2018 to February 2020 in Yasuní, Ecuador

across saplings of many genera, more than 75% of the damage that accrues during the lifetime of a leaf occurs during the ephemeral period of leaf expansion (1 to 3 weeks) (Coley and Barone 1996; Kursar and Coley 2003). For example, for ten species of *Inga* on Barro Colorado Island, Panama, 26% of leaf area was lost during leaf expansion, versus 0.23% of lost leaf area for mature leaves (Fig. 1a). This pattern is also reflected in herbivore host preferences. During a period of 18 months, in the Yasuní National Park in the Ecuadorian Amazon, the number of caterpillars associated with 15 species of *Inga* was 5.4 times greater on young than on mature leaves (Fig. 1b). Similarly, occurrence of lepidopteran larvae was 4.3 times greater on young expanding leaves than on mature leaves (Fig. 1c). Thus, herbivores prefer young over mature leaves. The greater tenderness and nutritive value of young leaves (Kursar and Coley 2003) may allow caterpillar herbivores to grow faster and thereby minimize the amount of time they are vulnerable to predation and parasitism (Benrey and Denno 1997; van Nouhuys and Lei 2004).

The preference of Lepidoptera for expanding leaves and the high rates of damage they inflict suggest that young leaves are under strong selective pressure to invest in anti-herbivore defenses. Mature leaves are tough and high in fiber (Coley and Kursar 2003), a very effective physical defense. Because young leaves cannot lignify cell walls until they stop expanding, they must rely on defenses other than toughness (see below). It remains unclear, however, how leaf age-associated differences are translated into herbivore host choice. We hypothesize that given that young and mature leaves are so different, each leaf age must be associated with a different assemblage of herbivores.

Host Plant Selection by Herbivores and Plant Defensive Traits

For the rest of the chapter, we will focus only on the defenses of young expanding leaves, given that they are the most preferred by herbivores and therefore the most relevant for host choice. Key young leaf defenses in *Inga* can be grouped into six evolutionarily independent axes: (a) secondary metabolites, (b) density and length of trichomes, (c) diameter of extra-floral nectaries and the number and identity of ants visiting them, (d) chlorophyll content and rate of leaf expansion, and the (e) synchrony and (f) timing of young leaf production (Kursar et al. 2009; Endara et al. 2017, 2018a). All defense axes contribute to host choice by herbivores (Table 1, Endara et al. 2017). In addition to being independent of each other, each defense category shows substantial variation across *Inga* species, with closely related species being more different than expected by chance (Kursar et al. 2009). Consequently, plant traits, more than plant phylogeny, determine host choice by herbivores (Endara et al. 2017).

Table 1 Summary of the relationships between herbivore assemblages and *Inga* host defensive traits. The percentages in parentheses indicate the increase (↑) or decrease (↓) in the probability of occurrence for every unit of change in the host defensive trait. Blank cells indicate no significant effect

	Lepidopteran herbivore family		
Inga host defensive trait	Riodinidae	Gelechioidea	Noctuidae (including Erebidae)
Chemistry (includes phenolics, saponins, and amines)	Avoid hosts with tyrosine (↓ 32%) and tyramine gallates (↓ 51%)	Avoid hosts with tyrosine gallates (↓ 24%) Prefer host with saponins (↑ 179%)	Prefer hosts with tyrosine (↑ 86%) and tyramine gallates (↑ 168%)
Trichomes (includes length and density)	Positive	Negative	
Developmental (includes chlorophyll content and leaf expansion rate)	Prefer hosts with a high rate of expansion (↑ 70%)	Prefer hosts with a slow rate of expansion (↓ 49%)	
Biotic (includes extrafloral nectary size and ants visiting the nectaries)	Positive		
Timing in leaf production			Prefer hosts that flush leaves at certain times of the year
Synchrony in leaf production			Prefer synchronous hosts (↑ 170%)

Chemical Defenses

Although we refer to chemistry as a single defensive class, in reality it includes thousands of compounds. Each compound can vary independently, leading to an almost infinite number of potential niche axes for herbivore species to occupy (Coley and Kursar 2014). We have catalogued over 9,000 compounds from *Inga*, including non-protein amino acids, flavonoids, flavan-3-ols, and saponins (Lokvam et al. 2004; Lokvam and Kursar 2005; Brenes-Arguedas et al. 2006; Lokvam et al. 2007). *Inga* also overexpresses L-tyrosine, an essential amino acid, which is toxic to herbivores at the elevated concentrations found in young *Inga* leaves (Coley et al. 2019; Lokvam et al. 2006). Together, these soluble chemical defenses are 1.9 times greater in expanding leaves than in mature leaves, constituting 46% of the total dry weight (DW) of a young leaf (Wiggins et al. 2016). Furthermore, chemical defensive profiles between expanding and mature leaves are qualitatively different, with mature leaves showing higher intraspecific variation than expanding leaves (Wiggins et al. 2016). In fact, there is a consistent chemical phenotype among young leaves within the same species, despite variation in the environment or genotype (Bixenmann et al. 2011; Sinimbu et al. 2012; Endara et al. 2018b). Higher levels of intraspecific variation among mature leaves should select for broad diet breadth of herbivores (i.e., generalists), while defense canalization and greater investment in young leaves should select for higher specialization of herbivores. Although we have not compared host range for herbivores on mature versus expanding leaves, our results suggest that across the Amazon and Panama, most caterpillar species feed on young leaves of only 1–3 species of *Inga* at a site, even though 30–50 *Inga* species may be available (Fig. 2). Thus, lepidopteran herbivores associated with young *Inga* leaves are mainly specialists.

Secondary metabolites in *Inga* are strongly correlated with host choice and performance of herbivores in laboratory and field experiments (Endara et al. 2015, 2017; Coley et al. 2019; Forrister et al. 2019). Bioassays using artificial diets were conducted in Utah with *Heliothis virescens* (Noctuidae), a generalist herbivore of tropical origin, and in Panama with *Phoebis philea* (Pieridae), which feeds on *Cassia* (Fabaceae) but not on *Inga*. Bioassays showed that all extracts and fractions

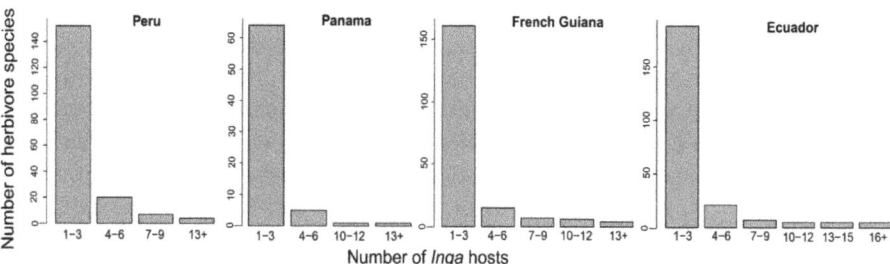

Fig. 2 Number of *Inga* species per species of Lepidoptera from Los Amigos, Peru; Barro Colorado Island, Panama; Nouragues, French Guiana; and Tiputini, Ecuador

of secondary metabolites were toxic (Coley et al. 2005; Lokvam and Kursar 2005; Lokvam et al. 2006; Coley et al. 2018). In addition to secondary metabolites, one clade of 17 species overexpresses tyrosine in young expanding leaves (Coley et al. 2019). At these concentrations (5% DW to 29% DW), tyrosine is highly toxic to *Heliothis virescens*. In laboratory bioassays, we found that concentrations of 3.8% DW of tyrosine reduced larval growth by 50% compared to controls ($p < 0.01$), whereas concentrations of 10% DW (the mean concentration found in the young leaves) reduced growth to 2% of controls ($p < 0.001$) and produced high mortality (Lokvam et al. 2007). In Peru, larvae of a specialist herbivorous sawfly (Argidae) were presented with fresh young leaves from two chemotypes of *I. capitata* in a Petri dish choice experiment. This experimental approach separated the effects of field traits such as habitat, phenology, and ant protection from secondary metabolites and nutritional content. Although the chemotypes were quite similar, the sawflies significantly preferred the chemotype on which they were most commonly found in the field (Endara et al. 2015). This suggests that even small differences in chemistry matter for host choice.

Chemical Defenses Affect Host Choice of Lepidoptera

At the community level, our studies have shown a key role for *Inga* defensive chemistry in structuring herbivore assemblages both within (Endara et al. 2015, 2017; Forrister et al. 2019) and across sites (Table 1, Endara et al. 2018a; Coley et al. 2019). For 38 species of *Inga* coexisting at a single site in the Peruvian Amazon, phylogenetically controlled analyses have shown that chemistry alone explained 30% of the total variation in the assemblage of lepidoptera larvae associated with young leaves. Host plant species that were more similar in chemistry were also more similar in their assemblage of herbivores (Endara et al. 2017). For sympatric host plants with similar chemistry, 20% of herbivore species were shared, whereas dissimilar host plants had no herbivores in common. Secondary metabolites have also been found to play a key role in structuring local assemblages of herbivores associated with other species-rich genera such as *Bursera*, *Ficus*, *Protium*, and *Piper* (Becerra 2007; Volf et al. 2017; Salazar et al. 2018; Richards et al. 2015; Massad et al. 2017).

Across sites, it has proven more difficult to determine the role of chemistry, or any host plant trait, on the assembly of herbivore communities. Our work with *Inga* is one of the few to measure defense traits and test their effect on herbivore assemblages across wide geographic ranges. For *Inga*, across four communities that span Panama and the Amazon Basin in Ecuador, Peru, and French Guiana, plant defensive chemistry is the main predictor structuring sawfly larvae associations (Hymenoptera, Argidae; Endara et al. 2018a). We expect that a similar pattern will emerge for Lepidoptera.

Although we have not examined large spatial effects of the entire suite of *Inga* secondary metabolites on lepidopteran herbivores, we found that the overexpression of tyrosine and its derivatives constrains host plant selection by herbivores (Table 1). For larvae of the family Lycaenidae, expression of tyrosine and tyramine gallates in the young leaves decreased the probability of host association by 32% and 51%, respectively. Larvae of the superfamily Gelechioidea avoid hosts that express tyrosine and its derivatives (Coley et al. 2019). In contrast, for Noctuidae moths in Peru, these compounds are among the main positive predictors for host plant association (Endara et al. 2017), illustrating the point that chemical traits that function initially as defenses can become oviposition cues for specialist herbivores (Lankau 2007; Reudler Talsma et al. 2008). Thus, the association of different herbivore families with different classes of secondary metabolites suggests that each family may have a different set of adaptations for handling saponins, tyrosine, and its galloylated derivatives, with some groups being specialists at detoxifying these chemicals.

Chemical Diversity Affects Diversity and Abundance of Lepidoptera

Each *Inga* species invests in a large diversity of compounds, with a median of 316 unique secondary metabolites observed in a single young *Inga* leaf. This value changes from one species to another, with some species investing in as many as 1485 unique secondary metabolites. Variation in compound diversity across species can potentially affect variation in the species richness and abundance of insect herbivores. In fact, theory has predicted that the richness of insect herbivore assemblages is linked to the chemical richness of their host plants (Richards et al. 2015; Volf et al. 2019). In this regard, useful insights can come from studies that include assembly of host plants that are coexisting in sympatry and are exposed to the same pool of herbivores.

Preliminary analyses for assemblies of herbivores associated with sympatric *Inga* species show relationships between the richness of chemical compound classes and the abundance and richness of herbivore species. Here, chemical class richness is defined as the total number of classes present in the following categories: phenolic compounds (33 classes), saponins (1 class), and metabolites containing amines (3 classes) (Endara et al. 2018b). We also quantified gravimetrically the percent DW investment in metabolites. For *Inga* trees coexisting at the Tiputini Biological Station in the Ecuadorian Amazon, species with higher richness of chemical classes were attacked by fewer lepidopteran herbivores ($R^2 = -0.42$, $p = 0.02$, Fig. 3a). Furthermore, the impact of chemical class richness differentially affected insect herbivores with contrasting diet breadths. A higher richness of chemical classes was negatively associated with the richness of specialist insect herbivores (lepidopteran herbivores with 1–3 hosts; $R^2 = -0.40$, $p = 0.007$, Fig. 3b). There was no significant correlation for more generalist herbivores (>3 hosts). Similar patterns were found for coexisting *Inga* species at Los Amigos Biological Station at the Peruvian Amazon, where hosts with higher chemical richness were also associated with less

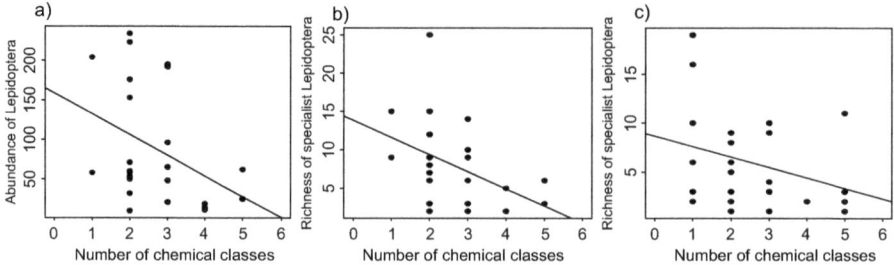

Fig. 3 Relationship between chemical richness and lepidopteran herbivore abundance and richness in the genus *Inga*. (**a**) Number of chemical classes vs the number of individuals of lepidopteran herbivores in Tiputini, Ecuador ($R^2 = -0.49$, $p = 0.02$), (**b**) number of chemical classes vs number of specialist herbivore species in Tiputini, Ecuador ($R^2 - 0.40$, $p = 0.007$). (**c**) Number of chemical classes vs number of specialist herbivore species in Los Amigos, Peru ($R^2 = -0.30$, $p = 0.006$). Specialist herbivores are defined as those lepidopteran larvae associated with ≤3 *Inga* host species

species-rich assemblages of specialized herbivores ($R^2 = -0.30$, $p = 0.006$, Fig. 3c), although no relationship with herbivore abundance was found. At both sites, there was no significant correlation between the degree of investment in metabolites and the abundance or species richness of herbivores.

These results agree with the hypothesis that greater chemical diversity of host plants should have a stronger effect on specialist than generalist insect herbivores (Root 1975), given that constraints on diet specialization have led specialized herbivores to be more limited in the types of defensive chemicals that they can overcome or circumvent (Jaenike 1990). For coexisting *Piper* species at La Selva in Costa Rica, higher diversity of high-volatility compounds also showed a greater negative effect on specialist herbivores than on generalist herbivores (Salazar et al. 2016a, b; Massad et al. 2017, but see Richards et al. 2015). Consequently, chemically diverse hosts would benefit from reduced herbivore pressure, since diverse mixtures of metabolites would allow them to be defended against a wide variety of insect enemies. Together, these findings support the defensive nature of secondary metabolites and the idea that community-wide chemical diversity influences plant-insect interactions, including aspects of species diversity and composition of herbivore assemblies (Coley et al. 2005; Kursar et al. 2009; Coley et al. 2018; Becerra 2015; Salazar et al. 2018).

Trichomes

Trichomes, hairs on the upper and lower surfaces of the leaf, play a pivotal role in plant defense against herbivores (Tian et al. 2012). In particular, foliar trichome density negatively influences herbivore populations by physically hindering insect movement and behavior and/or through toxic chemicals they produce or release

(Agrawal 1999; Horgan et al. 2009; Kessler and Baldwin 2002; Lill et al. 2006; Peiffer et al. 2009; Tian et al. 2012). Some *Inga* species have trichomes on the surface of their young leaves, with those leaves that have a higher density of trichomes also having longer trichomes ($R^2 = 0.79$, $p < 0.001$, Endara et al. 2017). The covariance between these two traits may result from a mechanism of maximizing defense. Interspecific variation in the length and density of trichomes explained a significant variation of the herbivore assemblage on different *Inga* hosts at Los Amigos in Peru ($R^2 adj = 0.06$, $p = 0.02$). In particular, for species of the moth superfamily Gelechioidea, trichome length significantly reduced the odds of occurrence for their larvae (proportional odds estimate for trichome length= 0.69, 95% CI (1.05 to 0.45)). Furthermore, correlations between trichomes and herbivore associations need not be negative; for example, *Inga* host associations of butterflies in the family Riodinidae correlate positively with trichome density (Table 1, $R^2 adj = 0.12$, $p = 0.06$, Endara et al. 2017). This positive association is a possible example of specialist herbivores using a defense that they have circumvented as a cue in host choice (e.g., Lankau 2007; Reudler Talsma et al. 2008).

Extrafloral Nectaries and Ant Attraction

Inga is characterized by the presence of nectaries on leaves (i.e., extrafloral nectaries) that produce nectar and attract protective ants (Koptur 1984) only during the short period of leaf expansion (Fig. 4a). Across Central and South America, both traits are highly correlated in *Inga*, with larger nectaries receiving higher rates of ant visitation ($R^2 = 0.12$, $p \leq 0.001$, Coley et al. 2018). Variation in foliar nectar production across *Inga* species also results in variation in ant visitation (Bixenmann et al. 2013), and this variation seems to affect herbivore host selection. In Peru, phylogenetically controlled analyses show that herbivore assemblage similarity across *Inga* hosts correlates positively with increasing similarity in the number of ants visiting the extrafloral nectaries (Mantel $r = -0.21$, $p = 0.02$). Furthermore, species of *Inga* that are defended by similar species of ants are also associated with similar assemblages of lepidopteran herbivores (Mantel $r = 0.25$, $p = 0.09$).

Although ants commonly prey on caterpillars, the probability of occurrence of riodinids on host plants substantially increases with the number of ants visiting a nectary (Table 1). For every unit of increase in the mean number of ants, the odds of occurrence for Riodinidae species on *Inga* increased by 22 times (proportional odds estimate for ants= 22.14, 95% CI (221.3 to 2.73)). This apparently counterintuitive correlation is explained by the fact that the larvae of many riodinid species are tended by ants (myrmecophily, Fiedler 1991, Pierce et al. 2002) in exchange for larval secretions that contain carbohydrates and amino acids (Pierce and Mead 1981; Pierce 1985). A strong positive effect of ants on riodinid host plant choice is thus expected (Fig. 4b).

Fig. 4 (**a**) Extrafloral nectaries on young leaves of *Inga auristellae* being visited by ants in Tiputini, Ecuador. Credit: Thomas Kursar. (**b**) An ant (*Ectatomma tuberculatum*) tending a *Synargis* sp. (Riodinidae) on *Inga thibaudiana* in Los Amigos, Peru. Credit: Maria Jose Endara. (**c**) A Gelechiidae larva webbing young leaves of *Inga bourgonii* in Tiputini, Ecuador. Credit: Thomas Kursar. d) Sawfly larvae (Argidae sp.) feeding on young leaves of *Inga capitata* in Los Amigos, Peru. Credit: Maria Jose Endara

Chlorophyll Content and Rate of Leaf Expansion

Inga species show large variation in the expansion rate and chlorophyll content of young leaves consistent with adaptations to reduce the impact of herbivory (Kursar & Coley 1992a, b, c). Within the genus *Inga*, these traits are highly negatively correlated, probably as a result of evolutionary trade-offs (Kursar & Coley 1992a, b, c; Coley et al. 2018). The species that escape herbivory by expanding their leaves rapidly fuel this growth by delaying the development of chloroplasts (delayed greening, Fig. 4a) until the leaf is fully expanded and defended by toughness (Kursar & Coley 1992a, b, c; Kursar and Coley 2003). Since these correlated traits are related to the development of the expanding leaf and fall into an independent axis of defense variation from other traits (Endara et al. 2017), we recognized this defense strategy as a developmental defense (Kursar et al. 2009).

Developmental defenses influence the time frame for which leaves are both tender and nutritious, and as such, they constitute an important driver for host plant choice by herbivores. For example, modeling of the probability of occurrence for larvae of the moth superfamily Gelechioidea across widely separated communities in Central and South America suggests that these herbivores prefer host plants with a relatively low rate of leaf expansion (Table 1). For every unit of increase in the rate of leaf expansion, the odds of occurrence for Gelechioidea species on *Inga* decreased by half (proportional odds estimate for leaf expansion rate = 0.61, 95% CI (0.88 to 0.42)). Gelechioidea have a wide range of feeding habits, with *Inga*-associated larvae mostly showing leaf-mining, webbing, and scraping habits (Fig. 4c). Variation in leaf development could be particularly important for these intimate feeding strategies because they need more time for successful development and are confined to a single leaf during their entire larval stage. The three-dimensional structure of leaf mines and webs is both dependent on the structure of leaf tissues, and rapidly expanding tissues may not be compatible with suitable structures, microhabitats, or nutrient supply (Pincebourde and Casas 2006; Ayabe et al. 2018; Aoyama and Ohshima 2019).

In contrast, myrmecophilous Riodinidae larvae (Fig. 4b) are more likely to occur on *Inga* trees with rapidly expanding leaves (Table 1), with the odds of occurrence almost doubling for every unit of increase in the rate of leaf expansion (proportional odds estimate for leaf expansion rate= 1.7, 95% CI (2.6 to 1.08)). It has been suggested that ant-tended riodinids have been selected to feed upon nitrogen-rich plants in order to meet the energetic requirements for their own development as well as the secretion of amino acids for the attendant ants (Pierce 1985). Leaf expansion rate is positively correlated with the nitrogen content of expanding leaves. Thus, riodinids prefer *Inga* species with rapidly expanding leaves because they have more nitrogen.

Synchrony and Timing of Young Leaf Production: Phenological Defenses

Synchronization and timing of leaf production define windows of food availability for insect herbivores. Adaptive modification of these traits has been shown to be important in antiherbivore defense by Feeny (1976), Aide (1993), Coley and Kursar (1996), and more recently Lamarre et al. (2014). High synchrony in new leaf production, particularly if it is community wide, can exceed the ability of the herbivores that are present at that time to eat all the leaves. It could also impose tight phenological tracking on specialist herbivores. Although the phenology of tropical herbivores has been little studied, previous work has suggested that peaks of abundance for insect herbivores, particularly for those with narrow host ranges, may reflect the availability of their principal food resource, young leaves (Checa et al. 2009; Kishimoto-Yamada et al. 2010; Kishimoto-Yamada and Itioka 2015). Since young leaves are an ephemeral resource, it seems clear that food availability is limited. Field surveys often show a tight correlation between peaks of leaf production and peaks of herbivory (Murali and Sukumar 1993; Lamarre et al. 2014), suggesting that insect herbivores track production of their key food resource, especially of plants in which production of young leaves is highly synchronized. In that case, individual plants that flush early are favored by suffering less herbivory, probably because fewer specialist herbivores are present at the beginning of a peak of flushing (Murali and Sukumar 1993).

Our studies in the Peruvian Amazon have shown that the phenology of *Inga* leaf production is an important predictor for host plant selection by lepidopteran herbivores (Endara et al. 2017). In particular, for larvae of the moth family Erebidae, *Inga* with flushing peaks in June–July and October–November were preferred over species that flushed during other times of the year (Table 1, R^2adj = 0.13, p = 0.04). In addition, Erebidae larvae were mainly associated with *Inga* hosts that are highly synchronous (Table 1), with their probability of occurrence more than doubling for every unit of increase in the degree of synchrony in leaf production (proportional odds estimate for synchrony in leaf production= 2.7, 95% CI (5.58 to 0.97)). Thus, timing of flushing and synchrony in leaf production are important variables for host plant selection by herbivores.

Constraints on Host Specialization

The "arms race" paradigm predicts that interactions between plants and insect herbivores may drive diversification and trait evolution in both groups, leading to phylogenetic signal in plant defenses and in host plant use by insect herbivores (Ehrlich and Raven 1964). Our findings in *Inga* are consistent with the hypothesis that *Inga*-herbivore interactions drive defense trait evolution, but in contrast to the classic phylogenetic signal paradigm, we found closely related *Inga* species to present substantial differences in defenses (Kursar et al. 2009). This suggests that herbivore selective pressure has promoted rapid and divergent evolution of anti-herbivore traits.

In addition, although the Ehrlich and Raven paradigm predicts host plant use by insect herbivores to be phylogenetically conserved (Ehrlich and Raven 1964; Brooks and McLennan 2002), we found Lepidoptera on *Inga* to shift between distantly related hosts, disrupting any signal of codiversification (Endara et al. 2017). Rather than a phylogenetically structured arms race model of reciprocal coevolution, our results support recruitment of herbivore assemblages through a process of ecological sorting based on *Inga* host defenses (Endara et al. 2017; Endara et al. 2018a). This is exemplified by the three most abundant families of Lepidoptera associated with *Inga* trees in Peru. Closely related herbivores in each of the Gelechioidea, Erebidae, and Riodinidae are associated with *Inga* species that share similar defenses rather than close phylogenetic relationships. These results imply that herbivores are tracking resources for which their behaviors, morphologies, and physiologies are to some extent pre-adapted (Janzen 1985; Agosta and Klemens 2008). Adult Lepidoptera must synchronize oviposition with the timing of leaf production, and larvae must match growth rates with the leaf expansion phase; handle nutritional, chemical, physical, and biotic defenses; as well as minimize predation. Switches to novel hosts with divergent defenses would require simultaneous changes in many of these herbivore adaptations (Brooks and McLennan 2002). It appears that closely related lepidopteran herbivores are similar in this complex set of adaptations and consequently constrained to feed on hosts with similar defenses.

A more detailed phylogenetic analysis with another group of insects associated with *Inga* suggests that the tracking of defenses in evolutionary time might be a general pattern for specialized herbivores (Endara et al. 2018a). *Inga*-associated sawflies (Hymenoptera: Argidae) are highly specialized insect herbivores, whose larvae feed on expanding leaves of only one to two host plant species (Fig. 4d). Sawflies often sequester or modify toxic host compounds for use in anti-predator defense (Eisner et al. 1974; Petre et al. 2007; Boevé et al. 2013) and are often highly dependent on the host plant chemistry.

Across Panama and the Amazon basin, more than 90% of diversification events of *Inga*-associated sawflies involved shifts between *Inga* species with similar defensive chemistry, regardless of phylogenetic relatedness. Although most of these events occurred in allopatry, host switching in sympatry also involved chemically similar *Inga*. For example, within French Guiana sawfly MOTU (molecular operational taxonomic unit) 9 attacked *Inga jenmanii*, and a sister taxon MOTU 10 attacked *I. obidensis*. Both host plant species have a defensive chemistry based on galloylated tyrosine phenolics, but are not closely related phylogenetically (Endara et al. 2018a). There were few sawfly examples of host shifts to novel hosts that expressed different classes of chemical defenses, suggesting that ecological speciation is much rarer than defense tracking. Thus, in a manner analogous to Lepidoptera, sawfly diversification (often allopatric) seems largely constrained to colonization of chemically similar hosts for which they have appropriate adaptations.

The macroevolutionary patterns we observe with insect herbivores appear to reflect genetic and developmental constraints, but not in the classic sense. Traditionally, the use of host plants by phytophagous insects has been considered conserved at both ecological and evolutionary time scales, with closely related herbivores associated with closely related plants across multiple levels of phylogenetic divergence. However, insects appear to shift hosts much more frequently than expected (Agosta 2006; Janz 2011), which was initially considered to be evidence of high lability in the evolution of host association. However, analytical advances, such as phylogenetic structural models, that allow the combination of phylogenetic and host plant trait information (e.g., Hadfield 2010) have shown that, in reality, herbivore traits involved in host plant choice are evolving slowly, and host shifts depend more on existing host-choice traits. This suggests that plant defenses determine the extent of host choice in both ecological and evolutionary time scales. Therefore, improving our understanding of the ecology and evolution of plant-herbivore interactions will require close attention to host plant defenses.

Plant Traits Influence Herbivore Interactions with the Third Trophic Level

Herbivores not only are challenged by plant defenses but are also confronted with attack by the third trophic level. In lowland tropical rainforests, rates of predation and parasitism for larval Lepidoptera can be very high. Surveys in the Amazon using clay caterpillars found rates of attack to be 65%/caterpillar/day primarily due to arthropod predators, a rate almost ten-fold higher than in temperate forests (Roslin et al. 2017; Coley et al. 2018). Rates of attack to clay caterpillars are similar to those of cryptic, undefended real caterpillars (Richards and Coley 2007). Parasitism averages ~20% in both tropical and temperate systems (Dyer & Coley 2002) with some tropical studies finding that up to 43% of larvae are parasitized (Connahs et al. 2011).

Several plant nutritional and defensive traits appear to influence the vulnerability of larvae to the third trophic level. Although release of volatile organic compounds following damage has been shown to attract predators and parasitoids (Aartsma et al. 2017), we have no evidence of this in *Inga*. Instead, extra-floral nectaries serve to attract predators, especially ants (Bixenmann et al. 2011 & 2013). We suggest that the relative rarity of mature leaf feeders in tropical forests may be because the high tannin and low nutritional content of mature leaves extends developmental times, thereby prolonging the period when they could be attacked by natural enemies (Coley et al. 2006). As a consequence, mature leaf feeders have more defenses such as hairs, gregarious behavior, and warning colors and are rejected by ants in feeding trials. If mature leaf feeders are defended with chemicals, we predict that these will be synthesized by the caterpillar, as tannins, the most common compound class in mature leaves, are not feasible to sequester. In contrast, caterpillars that feed on fast-expanding young leaves have a short period before the leaf toughens and becomes unpalatable, so they must grow fast. The high nitrogen and water contents and low chemical defenses permit fast larval growth. They tend to be cryptic and highly palatable to ants in feeding trials. In contrast, caterpillars feeding on slow expanders have intermediate growth rates and apparently invest more in defense, as they are less preferred by ants. Because slow-expanding leaves have high concentrations of chemicals, including low molecular weight molecules, caterpillars may have the opportunity to sequester the compounds of the host plant. Thus, the type of plant defense may shape the growth and defense strategies of herbivores and, in turn, their susceptibility to the third tropic level.

Global Climate Change Will Affect Insect Herbivores Through Changes Experienced by Their Host Plants

A growing body of studies has documented the impact that climate change and extreme weather events are having on plants and animals. This impact is predicted to be particularly disruptive for multispecies interactions (Tylianakis et al. 2008), and to increase with the level of specialization in interaction networks (Salcido et al. 2020), since these interactions are susceptible to the phenology, physiology, and behavior of multiple species. In this review, we showed how the tight ecological and evolutionary relationships that exist between plant defenses and host plant choice have shaped the high levels of host specialization we observe in tropical forests. Thus, it would not be surprising if plant-herbivore interactions in the tropics were to be at a particularly high risk of perturbation by climate change (Tylianakis et al. 2008).

Many global change studies have focused on temperate ecosystems (Feeley et al. 2017), but only a handful have dealt with herbivory (Tylianakis et al. 2008). Increases in temperature and the length of the dry season can directly influence herbivores' survival and development, but changes experienced by their host plants will also have impacts (Coley 1998, Cornelissen 2011). Below we discuss how two major plant traits, defensive chemistry and phenology of young leaf production, might change. We have no data on how other defenses, such as nectar production at extra-floral nectaries, leaf expansion rates, and trichome density, would respond, but we suspect they may show less change.

In general, the predicted increase in depositions of CO_2 is expected to modify plant quality and defensive chemistry. A surge in carbon availability for plant tissue induces an increase in the C/N ratio, producing a "nitrogen dilution effect" (Coley 1998; Bidart-Bouzat and Imeh-Nathaniel 2008; Robinson et al. 2012; Welti et al. 2020) that decreases the nutritional quality of leaves for herbivores (Lincoln et al. 1986; Robinson et al. 2012). The excess in carbon for plant growth can also be shunted into secondary metabolites, producing an increase in carbon-based defensive compounds such as terpenes and phenolics (reviewed in Coley 1998; Tylianakis et al. 2008, Ryan et al. 2010). Nevertheless, a decline in C-based compounds has been detected as well (Vanette and Hunter 2011; Decker et al. 2019).

Variations in plant tissue associated with higher atmospheric CO_2 have been related to alterations in herbivore performance and host choice preferences. Although herbivores can compensate for reduction of nutritional quality in plant leaves by higher consumption rates (Stiling and Cornelissen 2007), they still show substantial decreases in relative growth rates and pupal weight (Stiling and Cornelissen 2007), as well as extended developmental times (Goverde and Erhardt 2002; Smith and Jones 2002). These patterns are particularly exacerbated in chewing herbivores, such as caterpillars (Bidart-Bouzat and Imeh-Nathaniel 2008; Cornelissen 2011). Compensatory feeding has also been observed for specialized herbivores that sequester toxic chemicals from their hosts, presumably to maintain the appropriate concentration of sequestered compounds (Decker et al. 2019). In addition, herbivores have responded to variations in plant tissue quality by switching hosts or the plant parts on which they prefer to feed (Williams et al. 1997; Agrell et al. 2005). Such alterations in herbivore development and behavior have the potential to make herbivores more vulnerable to predation and parasitism (Stiling et al. 1999; Decker et al. 2019), with a subsequent reduction in herbivore diversity and abundance (Cornelissen 2011). Studies that span multiple decades have already documented widespread declines in Lepidoptera abundance for temperate regions

(Thomas et al. 2004; Schultz et al. 2017) and in generalist caterpillars and parasitoids in intact tropical forests (Salcido et al. 2020).

Changes in climatic factors, such as temperature and rainfall patterns, and the associated shift in phenological synchrony of life history events between interacting species, are among the most pronounced consequences of rapid environmental change (Bidart-Bouzat and Imeh-Nathaniel 2008; Yang and Rudolf 2010; Cornelissen 2011). For insect herbivores, synchronization with the phenology of their host plants is crucial. Key stages in the life cycles of herbivorous insects, such as egg deposition, diapause, migration, and possibly mating, must be synchronized with the availability of their principal food resource, expanding leaves. In temperate regions, global warming has advanced budburst timing, producing a phenological mismatch between plants and their insect herbivores (Bidart-Bouzat and Imeh-Nathaniel 2008; Yang and Rudolf 2010; Cornelissen 2011). The result is the alteration of tightly coevolved species interactions, whose broad implications are still not well understood (Yang and Rudolf 2010; Burgess et al. 2018).

In tropical forests, the climatic drivers of leaf production are still unclear (Cleland et al. 2007; Girardin et al. 2016; Hubert-Wagner et al. 2017). In the Amazon, leaf production in aseasonal forests (where no substantial moisture stress is experienced during the whole year, as in the northwest Amazon) has been shown to be precipitation-driven (Asner et al. 2000; Girardin et al. 2016, but see Hubert-Wagner et al. 2017). In forests with a marked dry season (e.g., southeastern Amazonian forests), leaf production is more sensitive to solar radiation (Bi et al. 2015; Hubert-Wagner et al. 2017). Because shifts in precipitation and radiation, such as an increase in drought events in the Amazon (Li et al. 2006), are predicted to occur in concert with rising global temperatures (Cleland et al. 2007), phenological synchrony between plants and their insect herbivores is likely to be particularly sensitive to climate change. As discussed earlier, insect herbivores closely track young leaf production of their host plants in Amazonian forests, and extensive mismatches between the phenology of hosts and their herbivores could have far-reaching effects.

Because climate-induced changes can have pervasive effects on a range of plant traits, as well as directly on the many species of herbivores, it is difficult to make precise predictions of community responses. Furthermore, climate change could influence the abundance of host plant species at a given site. We suggest that generalist herbivores may be the most resilient to changes in plant chemistry, phenology, and abundance as they are able to feed on a variety of hosts. However, most insect herbivores on *Inga* (Fig. 2) and in the tropics in general (Forister et al. 2015) are quite specialized, so shifts in host plant traits could have more severe impacts for tropical herbivores.

Conclusions

In this chapter, we used the Neotropical tree genus *Inga* and its associated insect herbivores to illustrate the tight relationship that exists between plant defenses and host plant choice by herbivores and their role in shaping the high levels of host specialization observed in tropical forests. Related herbivores show a strong signal of feeding on *Inga* species with similar defenses rather than similar ancestry. Furthermore, *Inga* species with more classes of secondary metabolites are fed on by a fewer number of specialist herbivore species. Many studies on host range for insect herbivores at ecological and evolutionary levels focus on the role of host plant phylogeny and do not include information on host defensive traits. We argue that hypotheses exploring the role of host range in herbivore ecology and evolution should incorporate host defenses, or "host resources" sensu Brooks and McLennan (2002), including food availability. They should also incorporate phylogenetic relationships between species in each trophic level (Endara et al. 2018a).

Although the diversity of insect herbivores and their host plants in the Neotropics is among the highest in the terrestrial world, there is an evident lack of long-term and multi-site studies. This has slowed understanding of the factors structuring herbivore assemblages at a single site, as well as our understanding of the processes shaping host association and species divergence at regional scales. Similarly, much global change research has been geographically and taxonomically biased toward temperate ecosystems, despite the acknowledgement that plant-herbivore interactions in the tropics might be at a higher risk due to global change perturbations. Thus, we need comprehensive long-term studies in the tropics that include plants, insects, species traits, and the effect of multiple drivers of global environmental change if we want to make reliable predictions of the effect of climate change on species interactions and its escalated effect for the entire community.

Acknowledgments The authors are indebted to the many excellent field and lab assistants who were essential for data collection; to Toby Pennington and Kyle Dexter for the *Inga* phylogeny; to governments of Peru, Ecuador, Panama, and French Guiana for granting research and export permits; and to colleagues around the world for inspiration and feedback. TAK and PDC are grateful to the National Science Foundation for funding (DEB 0234936, DEB 0640630, DEB 0108150, and Dimensions of Biodiversity DEB-1135733) and M.J.E to the Secretaría Nacional de Educación Superior, Ciencia, Tecnología e Innovación del Ecuador (SENESCYT) for funding.

References

Aartsma Y, Bianchi FJJA, van der Werf W, Poelman EH, Dicke M (2017) Herbivore-induced plant volatiles and tritrophic interactions across spatial scales. New Phytologist 216:1054–1063

Agosta SJ (2006) On ecological fitting, plant-insect associations, herbivore host shifts, and host plant selection. Oikos 114:556–565

Agosta SJ, Klemens JA (2008) Ecological fitting by phenotypically flexible genotypes: implications for species associations, community assembly and evolution. Ecol Lett 11:1123–1134

Agrawal AA (1999) Induced responses to herbivory in wild radish: effects on several herbivores and plant fitness. Ecology 80:1713–1723

Agrell A, Kopper B, McDonald EP, Lindroth RL (2005) CO_2 and O_3 effects on host plant preferences of the forest tent caterpillar (*Malacosoma disstria*). Glob Change Biol 11:588–599

Aide TM (1993) Patterns of leaf development and herbivory in a tropical understory community. Ecology 74:455–466

Aoyama H, Ohshima H (2019) Changing leaf geometry provides a refuge from a parasitoid for a leaf miner. Zool Sci 36(1):31–37

Asner GP, Alan RT, Braswell BH (2000) Satellite observation of El Niño effects on Amazon Forest phenology and productivity. Geophys Res Lett 27(7):981–984

Ayabe Y, Minoura T, Hijii N (2018) Oviposition site selection by a lepidopteran leafminer in response to heterogeneity of leaf surface conditions: structural traits and microclimates. Ecol Entomol 42(3):294–305

Becerra JX (1997) Insects on plants: Macroevolutionary chemical trends in host use. Science 276(5310):253–256

Becerra JX (2007) The impact of herbivore-plant coevolution on plant community structure. P Natl Acad Sci USA 104(18):7483–7488

Becerra JX (2015) On the factors that promote the diversity of herbivorous insects and plants in tropical forests. P Natl Acad Sci USA 112(19):6098–6103

Benrey B, Denno RF (1997) The slow-growth-high-mortality hypothesis: a test using the cabbage butterfly. Ecology 78:987–999

Bi J, Knyazikhin Y, Choi S, Park T, Barichivich J, Ciais P et al (2015) Sunlight mediated seasonality in canopy structure and photosynthetic activity of Amazonian rainforests. Environ Res Lett 10(6):064014

Bidart-Bouzat MG, Imeh-Nathaniel A (2008) Global change effects on plant chemical defenses against insect herbivores. J Integr Plant Biol 50:1339–1354

Bixenmann RJ, Coley PD, Kursar TA (2011) Is extrafloral nectar production induced by herbivores or ants in a tropical facultative ant-plant mutualism? Oecologia 165(2):417–425

Bixenmann RJ, Coley PD, Kursar TA (2013) Developmental changes in direct and indirect defenses in the young leaves of the neotropical tree genus *Inga* (Fabaceae). Biotropica 45(2):175–184

Boevé JL, Blank SM, Meijer G, Nyman T (2013) Invertebrate and avian predators as drivers of chemical defensive strategies in tenthredinid sawflies. BMC Evol Biol 13:198

Brenes-Arguedas T, Horton MW, Coley PD, Lokvam J, Waddell RA, Meizoso-O'Meara BE, Kursar TA (2006) Contrasting mechanisms of secondary metabolite accumulation during leaf development in two tropical tree species with different leaf expansion strategies. Oecologia 149:91–100

Brooks DR, McLennan DA (2002) The nature of diversity: an evolutionary voyage of discovery. The University of Chicago Press, Chicago

Burgess MD, Smith KW, Leech D, Pearce-Higgins JW, Branston CJ, Briggs K, Clark JR, Evan KL, du Feu CR, Nager RG, Sheldon BC, Smith J, Whytock RC, Willis SG, Phillimore AB (2018) Tritrophic phenological match–mismatch in space and time. Nat Ecol Evol 2(6):970–975

Checa MF, Barragán A, Rodríguez J, Christman M (2009) Temporal abundance of butterfly communities (Lepidoptera: Nymphalidae) in the Ecuadorian Amazonia and their relationship with climate. Ann Soc Entomol Fr (n.s) 45(4):470–486

Cleland EE, Chuine I, Menzel A, Mooney HA, Schwartz MD (2007) Shifting plant phenology in response to global change. Trends Ecol Evol 22(7):357–365

Coley PD (1998) The effects of climate change on plant-herbivore interactions in moist tropical rainforests. Clim Change 39: 455-472.

Coley PD, Barone JA (1996) Herbivory and plant defenses in tropical forests. Annu Rev Ecol Syst 27:305–335

Coley PD, Kursar TA (2014) On tropical forests and their pests. Science 343(6166):35–36

Coley PD, Kursar TA (1996) Anti-herbivore defenses of young tropical leaves: Physiological constraints and ecological tradeoffs. In Mulkey SS, Chazdon RL and Smith AP (Eds) Tropical Forest Plant Ecophysiology, Chapman and Hall, NY, pp 305–336.

Coley PD, Kursar TA (2003) Reading the tree leaves. Nat Hist 112:12–12.

Coley PD, Lokvam J, Rudolph K, Bromberg K, Sackett TE, Wright L, Brenes-Arguedas T, Dvorett D, Ring S, Clark A, Baptiste C, Pennington RT, Kursar TA (2005) Divergent defensive strategies of young leaves in two Neotropical species of *Inga*. Ecology 86:2633–2643

Coley PD, Bateman ML, Kursar TA (2006) The effects of plant quality on caterpillar growth and defense against natural enemies. Oikos 115:219–228

Coley PD, Endara MJ, Kursar TA (2018) Consequences of interspecific variation in defenses and herbivore host choice for the ecology and evolution of Inga, a speciose rainforest tree. Oecologia 187(2):361–376

Coley PD, Endara MJ, Ghabash G, Kidner CA, Nicholls JA, Pennington RT, Mills AG, Soule AJ, Lemes MR, Stone GN, Kursar TA (2019) Macroevolutionary patterns in overexpression of tyrosine: an anti-herbivore defence in a speciose tropical tree genus, *Inga* (Fabaceae). J Ecol 107:1620–1632

Comita LS, Queenborough SA, Murphy SJ, Eck JL, Xu KY, Krishnadas M, Beckman N, Zhu Y (2014) Testing predictions of the Janzen-Connell hypothesis: a meta-analysis of experimental evidence for distance- and density-dependent seed and seedling survival. J Ecol 102(4):845–856

Connahs H, Aiello A, Van Bael S, Rodríguez-Castañeda G (2011) Caterpillar abundance and parasitism in a seasonal dry forest versus wet tropical forest of Panama. J Trop Ecol 27:51–58

Cornelissen T (2011) Climate change and its effects on terrestrial insects and herbivory patterns. Neotrop Entomol 40(2):155–163

Decker LE, Soule AJ, de Roode JC, Hunter MD (2019) Phytochemical changes in milkweed induced by elevated CO_2 alter wing morphology but not toxin sequestration in monarch butterflies. Funct Ecol 33(3):411–421

Dyer LA, Coley PD (2002) Tritrophic interactions in tropical and temperate communities. In: Tscharntke T, Hawkins B (Eds) Multitrophic level interactions. Cambridge University Press, pp 67–68.

Ehrlich PR, Raven PH (1964) Butterflies and plants – a study in coevolution. Evolution 18(4):586–608

Eisner T, Johnessee JS, Carrel J, Meinwald J (1974) Defensive use by an insect of a plant resin. Science 184:996–999

Endara MJ, Weinhold A, Cox JE, Wiggins NL, Coley PD, Kursar TA (2015) Divergent evolution in antiherbivore defences within species complexes at a single Amazonian site. J Ecol 103(5):1107–1118

Endara MJ, Coley PD, Ghabash G, Nicholls JA, Dexter KG, Donoso DA, Stone GN, Pennington RT, Kursar TA (2017) Coevolutionary arms race versus host defense chase in a tropical herbivore-plant system. P Natl Acad Sci USA 114(36):E7499–E7505

Endara MJ, Nicholls JA, Coley PD, Forrister DL, Younkin GC, Dexter KG, Kidner CA, Pennington RT, Stone GN, Kursar TA (2018a) Tracking of host defenses and phylogeny during the radiation of Neotropical *Inga*-feeding sawflies (Hymenoptera; Argidae). Front Plant Sci 9:16

Endara MJ, Coley PD, Wiggins NL, Forrister DL, Younkin GC, Nicholls JA, Pennington RT, Dexter KG, Kidner CA, Stone GN, Kursar TA (2018b) Chemocoding as an identification tool where morphological- and DNA-based methods fall short: *Inga* as a case study. New Phytol 218:847–858

Feeley KJ, Stroud JT, Perez TM (2017) Most 'global' reviews of species' responses to climate change are not truly global. Div Distrib 23:231–234

Feeny P (1976) Plant apparency and chemical defense. Recent Adv Phytochem 10:1–40

Fiedler K (1991) Systematic, evolutionary, and ecological implications of myrmecophily within the Lycaenidae (Insecta: Lepidoptera: Papilionoidea). Bonn Zool Monog 31:1–210

Fine PVA, Metz MRM, Lokvam J, Mesones I, Ayarza Zuniga JM, Lamarre GPA, Vásquez Pilco M, Baraloto C (2013) Insect herbivores, chemical innovation, and the innovation of habitat specialization in Amazonian trees. Ecology 94:1764–1775.

Forister ML et al (2015) Global distribution of diet breadth in insect herbivores. Proc Nat Acad Sci 112:442–447. https://doi.org/10.1073/pnas.1423042112

Forrister DL, Endara MJ, Younkin GC, Coley PD, Kursar TA (2019) Herbivores as drivers of negative density dependence in tropical forest saplings. Science 363(6432):1213–1216

Girardin CAJ, Malhi Y, Doughty CE, Metcalfe DB, Meir P, del Aguila-Pasquel J et al (2016) Seasonal trends of Amazonian rainforest phenology, net primary productivity, and carbon allocation. Global Biogeochem Cy 30(5):700–715

Goverde M, Erhardt A (2002) Effects of elevated CO_2 on development and larval food-plant preference in the butterfly *Coenonympha pamphilus* (Lepidoptera, Satyridae). Global Change Biol 91(1):74–83

Hadfield J (2010) MCMC methods for multi-response generalized linear mixed models: The MCMCglmm R package. J Stat Softw 33(2)

Horgan FG, Quiring DT, Lagnaoui A, Pelletier Y (2009) Effects of altitude of origin on trichome-mediated anti-herbivore resistance in wild Andean potatoes. Flora 204:49–62

Jaenike J (1990) Host specialization in phytophagous insects. Ann Rev Ecol Syst 21:243–273

Janz N (2011) Ehrlich and Raven revisited: Mechanisms underlying codiversification of plants and enemies. Ann Rev Ecol Evol Syst 42:71–89

Janzen DH (1970) Herbivores and the number of tree species in the tropical forests. Am Nat 104: 501–528.

Janzen DH (1985) On ecological fitting. Oikos 45:308–310

Kessler A, Baldwin IT (2002) Plant responses to insect herbivory: the emerging molecular analysis. Ann Rev Plant Biol 53:299–328

Kishimoto-Yamada K, Itioka T (2015) How much have we learned about seasonality in tropical insect abundance since Wolda (1988). Entomol Sci 18:407–419

Kishimoto-Yamada K, Itioka T, Sakai S (2010) Seasonality in light-attracted chrysomelid populations in a Bornean rainforest. Insect Conserv Div 3:266–277

Koptur S (1984) Experimental evidence for defense of *Inga* (Mimosoideae) saplings by ants. Ecology 65:1787–1793

Kursar TA, Coley PD (1992a) Delayed development of the photosynthetic apparatus in tropical rainforest species. Funct Ecol 6:411–422

Kursar TA, Coley PD (1992b) The consequences of delayed greening during leaf development for light absorption and light use efficiency. Plant Cell Environ 15:901–909

Kursar TA, Coley PD (1992c) Delayed greening in tropical leaves: An antiherbivore defense? Biotropica 24:256–262

Kursar TA, Coley PD (2003) Convergence in defense syndromes of young leaves in tropical rainforests. Biochem Syst Ecol 31:929–949

Kursar TA, Wolfe BT, Epps MJ, Coley PD (2006) Food quality, competition, and parasitism influence feeding preference in a Neotropical lepidopteran. Ecology 87:3058–3069

Kursar TA, Dexter KG, Lokvam J, Pennington RT, Richardson JE, Weber MG, Murakami ET, Drake C, McGregor R, Coley PD (2009) The evolution of antiherbivore defenses and their contribution to species coexistence in the tropical tree genus *Inga*. P Natl Acad Sci USA 106(43):18073–18078

Lamarre GPA, Mendoza I, Fine PVA, Baraloto C (2014) Leaf synchrony and insect herbivory among tropical tree habitat specialists. Plant Ecol 215:209–220

Lankau RA (2007) Specialist and generalist herbivores exert opposing selection on a chemical defense. New Phyt 175(1):176–184

Lewinsohn TM, Roslin T (2008) Four ways towards tropical herbivore megadiversity. Ecol Lett 11:398–416

Li W, Fu R, Dickinson RE (2006) Rainfall and its seasonality over the Amazon in the 21st century as assessed by the coupled models for the IPCC AR4. J Geophys Res 111:D02111

Lill JT, Marquis RJ, Forkner RE, Le Corff J, Holmberg N, Barber NA (2006) Leaf pubescence affects distribution and abundance of generalist slug caterpillars (Lepidoptera: Limacodidae). Environ Entomol 35:797–806

Lincoln DE, Couvet D, Sionit N (1986) Response of an insect herbivore to host plants grown in carbon dioxide enriched atmospheres. Oecologia 69:556–560

Lokvam J, Kursar TA (2005) Divergence in structure and activity of phenolic defenses in two co-occurring *Inga* species. J Chem Ecol 31:2563–2580

Lokvam J, Coley PD, Kursar TA (2004) Cinnamoyl glucosides of catechin and dimeric procyanidins from young leaves of *Inga umbellifera* (Fabaceae). Phytochemistry 65:351–358

Lokvam J, Brenes-Arguedas T, Lee JS, Coley PD, Kursar TA (2006) Allelochemic function for a primary metabolite: The case of L-tyrosine hyper-production in *Inga umbellifera* (Fabaceae). Am J Bot 93:1109–1115

Lokvam J, Clause TP, Grapov D, Coley PD, Kursar TA (2007) Galloyl depsides of tyrosine from young leaves of *Inga laurina*. J Nat Prod 70(1):134–136

Maron JL, Agrawal AA, Schemske DW (2019) Plant–herbivore coevolution and plant speciation. Ecology:e02704

Marquis RJ, Salazar D, Baer C, Reinhardt J, Priest G, Barnett K (2016) Ode to Ehrlich and Raven or how herbivorous insects might drive plant speciation. Ecology 97(11):2939–2951

Massad T, Martins de Morae M, Philbin C, Oliveira C, Cebrian G, Yamaguchi L, Jeffrey C, Dyer L, Richards L, Kato M (2017) Similarity in volatile communities leads to increase herbivory and greater tropical forest diversity. Ecology 98:1750–1756

Murali KS, Sukumar R (1993) Leaf flushing phenology and herbivory in a tropical dry deciduous forest, southern India. Oecologia 94:114–119

Novotny V, Drozd P, Miller SE, Kulfan M, Janda M, Basset Y, Weiblen GD (2006) Why are there so many species of herbivorous insects in tropical rainforests? Science 313(5790):1115–1118

Peiffer M, Tooker JF, Luthe DS, Felton GW (2009) Plants on early alert: glandular trichomes as sensors for insect herbivores. New Phytologist 184:644–656

Petre CA, Detrain C, Boevé JL (2007) Anti-predator defence mechanisms in sawfly larvae of Arge (Hymenoptera, Argidae). J Insect Physiol 53:668–675

Pierce NE (1985) Lycaenid butterflies and ants: selection for nitrogen-fixing and other protein-rich food plants. Am Nat 125:888–895

Pierce NE, Mead PS (1981) Parasitoids as selective agents in the symbiosis between lycaenid butterfly and ants. Science 211:1185–1187

Pierce NE, Braby MF, Heath A, Lohman DJ, Mathew J, Rand DB, Travassos MA (2002) The ecology and evolution of ant association in the Lycaenidae (Lepidoptera). Ann Rev Entomol 47:733–771

Pincebourde S, Casas J (2006) Multitrophic biophysical budgets: thermal ecology of an intimate herbivore insect-plant interaction. Ecol Monogr 76(2):175–194

Reudler Talsma JH, Biere A, Harvey JA, van Nouhuys S (2008) Oviposition cues for a specialist butterfly – plant chemistry and size. J Chem Ecol 34(9):1202–1212

Richards LA, Coley PD (2007) Seasonal and habitat differences affect the impact of food and predation on herbivores: a comparison between gaps and understory of a tropical forest. Oikos 116:31–40

Richards LA, Lee DA, Forister ML, Smilanich AM, Dodson CD, Leonard MD, Jeffrey CS (2015) Phytochemical diversity drives plant-insect community diversity. P Natl Acad Sci USA 112(35):10973–10978

Robinson EA, Ryan GD, Newman JA (2012) A meta-analytical review of the effects of elevated CO_2 on plant–arthropod interactions highlights the importance of interacting environmental and biological variables. New Phytol 194:321–336

Root RB (1975) Organization of a plant-arthropod association in simple and diverse habitats: the fauna of collards (*Brassica oleracea*). Ecol Monogr 43:95–120

Roslin T, Hardwick B, Novotny V, Petry WK, Andrew NR, Asmus A, Barrio IC, Basset Y, Boesing AL et al (2017) Higher predation risk for insect prey at low latitudes and elevations. Science 356(6339):742–744

Ryan GD, Rasmussen S, Newman JA (2010) Global atmospheric change and trophic interactions: Are there any general responses? In: Baluška F, Ninkovic V (eds) Plant communications from an ecological perspective. Springer, Berlin/Heidelberg, pp 179–214

Salazar D, Jaramillo A, Marquis RJ (2016a) The impact of plant chemical diversity on plant-herbivore interactions at the community level. Oecologia 181(4):1199–1208

Salazar D, Jaramillo MA, Marquis RJ (2016b) Chemical similarity and local community assembly in the species rich tropical genus *Piper*. Ecology 97(11):3176–3183

Salazar D, Lokvam J, Mesones I, Ayarza Zuñiga JM, Vásquez Pilco M, de Valpine P, Fine PVA (2018) Origin and maintenance of chemical diversity in a species-rich tropical tree lineage. Nature Ecol Evol 2:983–990

Salcido DM, Forister ML, Garcia Lopez H, Dyer LA (2020) Loss of dominant caterpillar genera in a protected tropical forest. Sc Rep UK 10:422

Schultz CB, Brown LM, Pelton E, Crone EE (2017) Citizen science monitoring demonstrates dramatic declines of monarch butterflies in western North America. Biol Cons 214:343–346

Sinimbu G, Coley PD, Lemes MR, Lokvam J, Kursar TA (2012) Do the antiherbivore traits of developing leaves in the Neotropical tree *Inga paraensis* (Fabaceae) vary with light availability? Oecologia 170(3):669–676

Smith PHD, Jones TH (2002) Effects of elevated CO_2 on the chrysanthemum leaf-miner *Chromatomyia syngenesiae*: a greenhouse study. Global Change Biol 4(3):287–291

Stiling P, Cornelissen T (2007) How does elevated carbon dioxide (CO_2) affect plant-herbivore interactions? A field experiment and a meta-analysis of CO_2-mediated changes on plant chemistry and herbivore performance. Global Change Biol 13:1823–1842

Stiling P, Rossi AM, Hungate B, Dijkstra P, Hinkle CR, Knott WM, Drake B (1999) Decreased leaf-miner abundance in elevated CO2: reduced leaf quality and increased parasitoid attack. Ecol App 9:240–244

Thomas JA, Telfer MG, Roy DB, Preston CD, Greenwood JJD, Asher J, Fox R, Clarke RT, Lawton JH (2004) Comparative losses of British butterflies, birds, and plants and the global extinction crisis. Science 303:1879–1881

Tian D, Tooker J, Pfeiffer M, Ho Chung S, Felton GW (2012) Role of trichomes in defense against herbivores: comparison of herbivore response to woolly and hairless trichome mutants in tomato (*Solanum lycopersicum*). Planta 236:1053–1066

Tylianakis JM, Didham RK, Bascompte J, Wardle DA (2008) Global change and species interactions in terrestrial ecosystems. Ecol Lett 11(12):1351–1363

Valencia R, Condit R, Foster RB, Romoleroux K, Munoz GV, Svenning JC, Magärd E, Bass M, Losos EC, Balsev H (2004) Yasuní forest dynamics plot, Ecuador. In: Losos EC, Leigh EG (eds) Tropical forest diversity and dynamism: findings from a large-scale plot network. University of Chicago Press, Chicago, pp 609–620

van Nouhuys S, Lei GC (2004) Parasitoid and host metapopulation dynamics: the influences of temperature mediated phenological asynchrony. J Anim Ecol 73:526–535

Vleminckx J, Salazar D, Fortunel C, Mesones I, Dávila N, Lokvam J, Beckley K, Baraloto C, Fine PVA (2018) Divergent secondary metabolites and habitat filtering both contribute to tree species coexistence in the Peruvian Amazon. Front Plant Sci 9:836

Volf M, Segar ST, Miller SE, Isua B, Sisol M, Aubona G, Simek P, Moos M, Laitila J, Kim J, Zima J et al (2017) Community structure of insect herbivores is driven by conservatism, escalation and divergence of defensive traits in *Ficus*. Ecol Lett 21:83–92

Volf M, Salminen JP, Segar ST (2019) Evolution of defences in large tropical plant genera: perspectives for exploring insect diversity in a tri-trophic context. Curr Opin Insect Sci 32:91–97

Wagner FH, Hérault B, Rossi V, Hilker T, Maeda EE, Sanchez A et al (2017) Climate drivers of the Amazon forest greening. PLoS ONE 12(7):e0180932

Welti EA, Roeder KA, de Beurs KM, Joern A, Kaspari M (2020) Nutrient dilution and climate cycles underlie declines in a dominant insect herbivore. Proc Nat Acad Sci 117:7271–7275

Wiggins NL, Forrister DL, Endara MJ, Coley PD, Kursar TA (2016) Quantitative and qualitative shifts in defensive metabolites define chemical defense investment during leaf development in *Inga*, a genus of tropical trees. Ecol Evol 6(2):478–492

Williams RS, Lincoln DE, Thomas RB (1997) Effects of elevated CO2-grown loblolly pine needles on the growth, consumption, development, and pupal weight of red-headed pine sawfly larvae reared within open-topped chambers. Global Change Biol 3:501–511

Yang LH, Rudolf VHW (2010) Phenology, ontology and the effects of climate change on the timing of species interactions. Ecol Lett 13:1–10

Ecology and Evolution of Secondary Compound Detoxification Systems in Caterpillars

Simon C. Groen and Noah K. Whiteman

Monarch caterpillar (*Danaus plexippus*) on showy milkweed (*Asclepias speciosa*) in Oakland, CA. Photo by Noah K. Whiteman.

Introduction

Ecological specialization generates and maintains biological diversity through evolutionary divergence between populations and subsequent coexistence between species (Allio et al. 2021; Braby and Trueman 2006; Gloss et al. 2016; Wiens et al. 2015). Dietary specialization typifies the life histories of most Lepidoptera (Forister et al. 2015), nearly all species of which are herbivorous (Wagner and Hoyt, Chapter "On Being a Caterpillar: Structure, Function, Ecology, and Behavior"). This form of ecological specialization is driven by both bottom-up (host plant quality and

S. C. Groen (✉)
Department of Nematology, University of California, Riverside, CA, USA
e-mail: simon.groen@ucr.edu

N. K. Whiteman
Department of Integrative Biology, University of California, Berkeley, CA, USA

defenses) and top-down (enemies) selective forces (Lawton and McNeill 1979; Bernays and Graham 1988). In either case, specialization revolves around so-called plant secondary compounds – those chemicals not typically required for primary plant growth, maintenance, and reproduction – although some clearly are used by plants as signaling molecules within defense pathways (Clay et al. 2009). Plants produce an enormous diversity of secondary chemicals, and the *raison d'être* of many of these is that they function as toxic anti-feedants (Fraenkel 1959). A paradox is that these same toxins can become co-opted by specialized arthropods, including Lepidoptera, as host-finding cues, feeding/oviposition stimulants (or antistimulants, in the case of compounds to which the insect is not adapted), and defensive mechanisms for the arthropods themselves. The biology of lepidopteran larvae (caterpillars) has played a central role in the development of the field of coevolution. Foundational papers on the topic, including ones by Ehrlich and Raven (1964) and Berenbaum (1983), focus on patterns of host use in caterpillars as they relate to secondary chemistry.

Dietary specialization in Lepidoptera requires the ability to mitigate the toxic effects of these secondary compounds, which we broadly define as detoxification. In this chapter, we focus on detoxification strategies deployed by specialized caterpillars for exemplar toxins at two ends of the mode of action spectrum: cardiac glycosides (CGs) and glucosinolates (GSLs). Studies of these two classes of toxins have been foundational for our understanding of plant-caterpillar interactions (Fig. 1).

One mode-of-action strategy for plant toxins is to target highly conserved essential proteins or even specific amino acid residues found in animals but not in plants. The targeting of proteins used in nervous and circulatory systems is particularly widespread. Among such toxins, the best studied are the CGs, which bind to the first extracellular loop of the sodium/potassium ATPase (Na+/K+-ATPase; Fig. 1b). CGs contain three structures: a steroid core, a 5-(cardenolides) or 6-(bufadienolides) membered lactone ring, and sugar residue(s). These toxins evolved in ca. 60 genera from 12 plant families as well as in toads (Bufonidae) and fireflies (Lampyridae; Agrawal et al. 2012). Because plant genomes do not encode a copy of the Na+/K+-ATPase, they do not suffer from its toxic effects.

The process of detoxification in all animals, not just insects, can be divided into three phases of xenobiotic metabolism: phase I is the functionalization step of detoxification characterized by oxidation, hydrolysis, and reduction reactions; phase II is the conjugation step in which lipophilic compounds are converted into more hydrophilic ones to facilitate excretion or sequestration; and in phase III excretion takes place (Amezian et al. 2021; Nakata et al. 2006). As we will discuss later, strategies to detoxify CGs that involve proteins active in these phases have evolved in several insects. However, an important alternative strategy in Danainae butterflies and other herbivores specialized on CG-producing plants involves target site insensitivity (TSI). TSI describes a biophysical phenomenon in which the toxic ligand fails to bind (or binds poorly) to the target site owing to "alteration in structure or accessibility" (Berenbaum 1986 citing Brooks 1976). Several insects have evolved to sequester CGs from their host plants in response to pressure from the

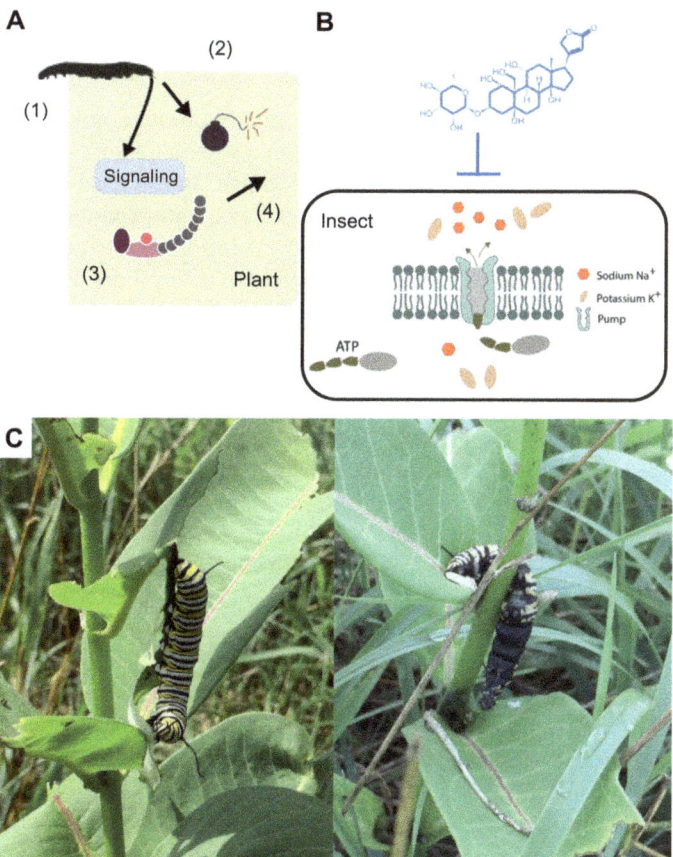

Fig. 1 (**a**) Upon attack by caterpillars (1), plants activate defense responses. In the case of Brassicaceae species, a reservoir of aliphatic glucosinolates (GSLs) is turned into toxic isothiocyanates (ITCs), activating the "mustard oil bomb" (2). In plants from all families, an intricate signaling network regulates production of heightened levels of defensive chemicals on top of a pre-made reservoir of stored chemicals. Such plant immune responses are activated after plants recognize the onset of attack through cell-surface and intracellular receptors (3). Brassicaceae species in the genus *Erysimum* produce toxic cardiac glycosides (CGs) in addition to producing GSLs (Züst et al. 2020). CGs are further produced by milkweeds and other Apocynaceae, plus species in 11 other plant families (Agrawal et al. 2012; 4). (**b**) CGs derive their toxicity from blocking activity of the caterpillars' sodium/potassium ATPases (Na+/K+-ATPases). (**c**) Caterpillars of the monarch butterfly engage in leaf vein-cutting or laticifer clipping behavior. On the left, a caterpillar cut the main mid-vein of a milkweed leaf and can now feed on a leaf with impaired defensive capabilities. On the right, a caterpillar died from exposure to CG-rich latex before it could disable this highly effective defensive barrier. (Cartoons by Simon C. Groen and Sophie Zaaijer, photos by Simon C. Groen)

third trophic level, in some cases by co-opting through gene duplication phase III drug transporters that originally evolved to remove CGs, which we will elaborate upon below.

At the other end of the spectrum, many plants produce non-toxic precursor glucoside molecules that are hydrolyzed, upon tissue damage, to toxic antiherbivore compounds by one or more β-glucosidases stored elsewhere (Fig. 1a). However, this reaction can yield toxins that are also auto-toxic to plants (Morant et al. 2008). Cyanogenic glycosides and their evolutionary derivatives, the GSLs, are well-studied examples of precursors relevant to caterpillars, as are iridoid and benzoxazinoid glucosides. GSLs are found only in plant species of the Brassicales and in the distantly related tropical tree genera *Drypetes* and *Putranjiva* (Malpighiales: Putranjivaceae; Rodman et al. 1998). As such, GSLs are used as host-finding/oviposition cues and feeding stimulants for many specialized insects. Interactions between GSLs and *Pieris* spp. gave rise to the field of chemical ecology, owing to Verschaffelt's 1910 study in which GSLs painted on non-host leaves stimulated feeding by *Pieris* spp. caterpillars, the first *bona fide* experiment showing that plant secondary compounds could be co-opted in this way. Some of the more toxic hydrolysis products of aliphatic GSLs, derived primarily from the amino acid methionine, are the isothiocyanates (ITCs), which give wasabi and other mustards their peppery and pungent taste. ITCs are general toxins that widely target nucleophilic residues such as exposed cysteine and lysine residues in proteins as well as DNA. In this case, full TSI could not evolve because the toxin is so promiscuous. Instead, a common strategy in Brassicaceae-feeding insects is to "disarm" the mustard oil bomb by preventing the formation of the ITCs through desulfation of the GSLs (e.g., in *Plutella* spp.) or diversion of hydrolysis products to nitriles (e.g., in *Pieris* spp.); this has occurred in both cases through a process of gene duplication and neofunctionalization (see references below). In generalists, or more recently derived specialists, the main route of GSL detoxification is a metabolically expensive strategy: the use of phase II detoxification enzymes (specifically glutathione *S*-transferases). Bacterial symbionts are able to hydrolyze ITCs, potentially facilitating colonization of GSL-bearing plants. Indolic GSLs, derived from the amino acid tryptophan, do not form stable ITCs, but rather are hydrolyzed into compounds that are oxidized by phase I enzymes. Thus, four of the principle means of detoxification (TSI, modification via phase I, conjugation via phase II, and excretion via phase III enzymes) can be subsumed by CGs and GSLs and will now be the subject of more detail.

We will use these two toxin classes to illustrate the different mechanisms by which caterpillars interact with toxins in general but will extend our discussion to other life stages, toxins, and plant-insect interactions to indicate potentially general mechanisms or to supplement known gaps in knowledge of how caterpillars interact with GSLs and CGs. We will start by providing an overview of functionally described proximate mechanisms of detoxification in Lepidoptera and then use this as a platform for diving into what is known about ultimate evolutionary patterns of Lepidoptera in response to their plant hosts.

Proximate Mechanisms of Detoxification

Resistance to host plant toxins evolves through different behavioral, physical, and physiological mechanisms including avoidance of toxin ingestion, reduced penetration through surface membranes such as the cuticle and gut lining, TSI, and active detoxification through metabolic enzymes (Li et al. 2007). These mechanisms often can be found in combination, providing a multi-tiered protection against toxins (Beran et al. 2018).

Behavioral

Studies across Lepidoptera and beyond have established functional roles for members of at least five chemoreceptor gene families in mediating behavioral avoidance of, or attraction to, plant odors and tastants that act as chemical signals. Chemical sensing starts through binding of an external ligand (e.g., a plant volatile) to receptor proteins that are located in the dendritic membrane of chemosensory neurons, such as those found in antennae (peripheral events). This interaction is then translated into an electrical cue to the central nervous system. Most of the chemoreceptors expressed in insect sensory organs are members of three main families, the gustatory, ionotropic, and odorant receptors (GRs, IRs, and ORs, respectively; Depetris-Chauvin et al. 2015). Added to these are receptors from the transient receptor potential (Trp) and degenerin/epithelial sodium channel (DEG/ENaC) or pickpocket (ppk) families, as well as the insect orphan G-protein-coupled DmXR protein (Depetris-Chauvin et al. 2015). Although members of these latter families are tightly involved in chemoreception in the main genetic model insect, the "fruit" fly *Drosophila melanogaster* (Benton et al. 2006, 2009; Matsuura et al. 2009; Mitri et al. 2009; Scott et al. 2001; Zelle et al. 2013), DEG/ENaC and DmXR orthologs have not yet been functionally described in Lepidoptera. Because of this, we will not discuss these further.

Olfactory Receptors

Insects detect a wide set of plant volatiles through expressing ORs in olfactory sensory neurons. OR function relies on an obligate partner, Orco, which is an OR itself (Benton et al. 2006). Indeed, knocking out Orco with CRISPR gene editing leads to largely disrupted foraging and oviposition behaviors of juvenile and adult moths toward host plants, as was observed for the silkmoth *Bombyx mori* (Bombycidae), the tobacco hawkmoth *Manduca sexta* (Sphingidae), and the Egyptian cotton leafworm *Spodoptera littoralis* (Noctuidae; Fandino et al. 2019; Koutroumpa et al. 2016; Liu et al. 2017). In one moth species, the importance of ORs in host plant detection was narrowed down to the level of an individual OR: CRISPR knockout individuals for Or42 in the cotton bollworm *Helicoverpa armigera* (Noctuidae) were impaired for host detection because they could not sense phenylacetaldehyde (Guo et al. 2021). ORs also form one of the mechanisms through which at least adult insects may perceive ITCs. In the diamondback moth *Plutella xylostella* (Plutellidae), ITCs stimulate oviposition by gravid females, and this response relies on the combined activity of Or35 and Or49 (Liu et al. 2020).

Ionotropic Receptors

A second class of receptors involved in sensing a wide set of plant volatiles is that of the IRs, which do not depend on Orco function (Benton et al. 2009). There is currently no evidence for a role of IRs in mediating caterpillar responses to ITCs or other volatile chemicals emitted from plants. However, a functional genetic study in *M. sexta* observed that adult females are deterred from ovipositing on two host plants, *Nicotiana attenuata* and *Datura wrightii*, when plants are already occupied by a feeding caterpillar from the same species or another such as *S. littoralis* (Zhang et al. 2019a). This avoidance behavior is displayed upon detection of the caterpillar frass-emitted carboxylic acids 3-methylpentanoic acid and hexanoic acid and mediated through Ir8a, which was verified through abolishing Ir8a function using CRISPR (Zhang et al. 2019a).

Gustatory Receptors

With one recent exception involving *Pieris rapae* and GSLs (Yang et al. 2021a, b), the GRs that insect taste sensilla express have only been functionally described in Lepidoptera when sensing chemicals not considered defensive chemicals. In *Plutella xylostella*, which specializes on GSL-containing plants, caterpillars made foraging decisions partially based on sensing the canonical plant hormones brassinolide and 24-epibrassinolide via Gr34 (Yang et al. 2020). This was functionally verified through RNA interference/RNA silencing (RNAi) of *Gr34* expression (Yang et al. 2020). That GRs can have dramatic effects on plant acceptance or rejection by caterpillars was demonstrated for larvae of the mulberry (*Morus alba*) specialist *B. mori*, where knocking out *Gr66* with CRISPR led to the acceptance of a wide variety of plant species unrelated to mulberry when foraging. This stood in stark contrast to foraging patterns of wild-type *B. mori* caterpillars, which retained a strong feeding preference for mulberry (Zhang et al. 2019c).

Transient Receptor Potential Channels

One of the main mechanisms by which insects and other animals may sense ITCs and other, often bitter, electrophilic plant compounds with deterrent effects is through transient receptor potential (Trp) channels (Kang et al. 2010). Functional genetic studies in *D. melanogaster* revealed the Trp channels TrpA1 and Painless to be involved in sensing ITCs, as knockout mutant flies showed a reduction in aversive responses to ITCs (Al-Anzi et al. 2006; Kang et al. 2010). Although more studies are needed in Lepidoptera, for now at least, we know that TrpA1 and Painless are expressed in sensory organs of the Brassicaceae specialist *P. rapae* (Mao et al. 2020) and that one of the "model" ITCs, allyl ITC (AITC), activates the TrpA1 channel in the generalist *Helicoverpa armigera* (Wei et al. 2015). Furthermore, there is functional evidence that TrpA1 is involved in tasting bitter compounds in caterpillars of *Manduca sexta* (Afroz et al. 2013).

Non-receptor Chemosensory Gene Families

Before reaching a herbivore's chemoreceptor, plant compounds travel through the lymph that fills the sensilla housing the dendrites of chemosensory neurons. This sensillar lymph contains a variety of water-soluble proteins, including members of two closely related families, the odorant-binding proteins (OBPs) and chemosensory proteins (CSPs; Vieira and Rozas 2011). Although these proteins are highly abundant, much about them is still unknown. Most likely, OBPs and CSPs mediate the solubilization and transport of generally hydrophobic odorants through the sensillar lymph and thereby regulate the sensitivity of the olfactory system (Leal, 2013; Vieira and Rozas 2011).

OBPs and CSPs typically contain six and four positionally conserved cysteine residues, respectively, which could have particular ecological relevance in Brassicaceae specialists such as *Plutella xylostella*. The exposed cysteines could make OBPs vulnerable to attack by reactive electrophiles such as the ITCs that mustard plants produce. A study of another Brassicaceae specialist, the fly *Scaptomyza flava* (Drosophilidae), observed a striking loss of OBPs (Gloss et al. 2019b). Losses were particularly apparent within the Plus-C OBP subfamily whose member genes encode six additional cysteine residues compared to other OBPs (Zhou et al. 2004), which might render them even more vulnerable to ITCs. Loss of OBPs may in this scenario contribute to a lower sensitivity of Brassicaceae specialists to the deterrent effects of ITCs.

On the other hand, OBPs and CSPs may have a detoxification function in the strict sense if they can remove harmful ligands such as ITCs from the peripheral nervous system. Moreover, expression of OBP and CSPs is not restricted to the olfactory tissues; they may also participate in detoxification of plant defensive chemicals in other tissues such as the gut (Bautista et al. 2015), although this still awaits experimental support (Pelosi et al. 2018). Such potential multiple functions in xenobiotic responses make it difficult to formulate predictions for how OBPs may evolve in response to the presence of host plant-derived ITCs. When characterizing the genomes of Lepidoptera that are Brassicaceae specialists, such as *Plutella xylostella*, and those of Lepidoptera that are not, such as the monarch butterfly (*Danaus plexippus*), there is no obvious difference in the number of OBPs in their genomes: 38 and 32, respectively (Cai et al. 2020; You et al. 2013; Zhan et al. 2011).

A similar pattern was visible for the CSP gene family, with 31 CSPs for *P. xylostella* and 34 CSPs for the monarch (You et al. 2013; Zhan et al. 2011).

While there is at least some mechanistic knowledge of how caterpillars sense potential host plants from the Brassicaceae that give rise to ITCs, virtually nothing is known about how herbivores sense plants that store less reactive toxins such as CGs (Agrawal et al. 2021). It will be fascinating to find out more about the molecular mechanisms that give rise to complex adaptive behaviors such as the leaf vein-cutting behavior displayed by larvae of the monarch and several other herbivores of milkweeds, including the milkweed tussock moth *Euchaetes egle* (Arctiidae) (Dussourd and Eisner 1987). Via a process of elimination, a series of experiments suggested that polar (water-soluble) CGs or non-CG chemicals might stimulate this behavior in monarch caterpillars (Helmus and Dussourd 2005). Deactivating the latex-containing canals in veins of milkweed leaves (which contain concentrated CGs) reduces exposure to toxic CGs, making this a life or death matter (Fig. 1c).

Prevention of Defense Response Induction

While behaviors such as selection of host plants and tissues as well as laticifer clipping are effective ways to avoid or, in the case of certain specialist herbivores, perhaps seek exposure to toxic plant defensive chemicals, there are further mechanisms that have evolved to prevent activation of plant defenses upon engagement of lepidopterans with host plants. Through expressing enzymes with immuno-suppressive effects on the host plant, caterpillars could actively stop plants from inducing toxin production upon feeding. One widespread mechanism is for caterpillars to produce glucose oxidase in their saliva (Eichenseer et al. 2010). Glucose oxidase is the most highly abundant salivary enzyme in *H. zea* and other caterpillars, converting D-glucose and molecular oxygen to D-gluconic acid and hydrogen peroxide (Musser et al. 2002). The hydrogen peroxide in turn elicits a burst of salicylic acid (SA) production by the host plant, which suppresses the synthesis of higher levels of defensive chemicals through interference with plant defensive signaling by jasmonic acid (JA) and ethylene (Fig. 1a; Diezel et al. 2009). JA/SA antagonism and its modulation by ethylene likely evolved in the last common ancestor of angiosperms (Groen and Whiteman 2014; Thaler et al. 2012a, b). The conserved nature of JA/SA antagonism may partially explain the pattern that caterpillars of highly polyphagous species were more likely to possess relatively high levels of glucose oxidase activity than caterpillars from more specialized species (Eichenseer et al. 2010).

Another mechanism of preventing plant production of defensive chemicals is to evade molecular detection of attack by plant receptor proteins that survey plant cells (Fig. 1a; Ngou et al. 2021; Yuan et al. 2021a, b). A particularly well-studied example can be found in the interaction between cowpea (*Vigna unguiculata*) and caterpillars. This plant activates production of defensive chemicals upon recognition of so-called inceptin-related peptides, present in caterpillar oral secretions, which are peptides derived from chloroplastic ATP synthase γ-subunit proteins (Schmelz et al. 2012; Steinbrenner et al. 2020). While these active inceptins are generated when caterpillars of generalist herbivores such as the fall armyworm *Spodoptera frugiperda* (Noctuidae) are attacking cowpea, they are not generated when larvae of the legume-specializing velvet bean caterpillar (*Anticarsia gemmatalis*) feed on the plant. A functional screen of inceptin amino acid building blocks identified that unlike the main inceptin found in all other Lepidoptera examined (Vu-In; +ICDINGVCVDA−), the oral secretions of *A. gemmatalis* caterpillars predominantly contained an inactive, C-terminal truncated peptide (Vu-In−A; +ICDINGVCVD−), which also functioned as an effective antagonist of Vu-In-induced responses (Schmelz et al. 2012).

Diversion Strategies for Precursor Toxins

If defensive chemicals are already stored constitutively, as is the case for the mustard oil bomb and other toxins that are released upon β-glucosidase-mediated hydrolysis of stored precursor glucoside molecules, an alternative strategy to prevent toxin formation is to modify the precursors or divert the hydrolytic process. Prevention of ITC formation could have strong effects on caterpillar survival, growth, and development time, as was shown definitively for the small cabbage white *Pieris rapae* in feeding experiments with microencapsulated formulations of allyl ITC, its precursor allyl GSL, and myrosinase (Agrawal and Kurashige 2003).

One effective way through which several specialists on Brassicaceae disarm the mustard oil bomb and prevent ITC formation is to remove the sulfate group in GSLs using sulfatase enzymes (GSSs; Ratzka et al. 2002). This removal renders myrosinases ineffective, as they cannot use desulfo-GSLs as substrates and are competitively inhibited by sulfate (Ratzka et al. 2002). This mechanism has evolved in *Plutella xylostella*, whose genome encodes three GSSs with distinct expression patterns and substrate specificity patterns in response to dietary GSLs (Heidel-Fischer et al. 2019). Two of these gene copies evolved under positive selection while acquiring their new GSL desulfation capabilities (Heidel-Fischer et al. 2019). As a further testament to the importance of GSSs for *P. xylostella* fitness when feeding on GSL-containing host plants, larvae experienced reduced survival and slower development when GSSs were knocked out using CRISPR (Chen et al. 2020).

A second diversion mechanism of the mustard oil bomb has evolved in the pierid butterflies. Upon caterpillar feeding and concomitant GSL degradation, nitrile-specifier proteins (NSPs) in the gut redirect the GSL hydrolysis reaction away from formation of ITCs and toward formation of nitriles, which are subsequently excreted with the feces (Wittstock et al. 2004, Wheat et al. 2007). The genes involved in GSL and ITC production in Brassicales plants and members of the NSP gene family in pierids show evidence of evolving in an escalating evolutionary arms race pattern (Berenbaum and Feeny 1981). Key innovations are linked to gene and genome duplications and shifts in diversification rates, followed by gradual changes in trait complexity that appear to have been facilitated by allelic turnover (Edger et al. 2015).

Physical Barriers (Peritrophic Membrane)

The peritrophic membrane or matrix (PM) (see Wagner and Hoyt, Chapter "On Being a Caterpillar: Structure, Function, Ecology, and Behavior") is a semipermeable chitinous matrix that lines the midgut of caterpillars and most other insects. The PM not only serves to protect the midgut epithelium from microorganisms and mechanical damage, but also from large plant defensive chemicals such as CGs, including the highly polar CG digitoxin (Barbehenn 1999, 2001). A study in *Helicoverpa zea* observed that the PM reduced hydrogen peroxide in the midgut, acting as a physical antioxidant (Summers and Felton 1996).

The PM in insects is formed through binding between chitin fibrils and PM proteins with multiple chitin binding domains (CBDs). Multi-CBD chitin binding proteins form the two major types of structural proteins in the PM alongside the insect intestinal mucin proteins. While CRISPR knockout mutants for mucin proteins in the cabbage looper *Trichoplusia ni* did not perform worse when fed a diet of GSL-containing cabbage leaves than wild-type caterpillars (Wang and Wang 2020), mucin proteins are involved in protecting caterpillars of *Plutella xylostella* against the harmful effects of terpenoids such as (3E)-4,8-dimethyl-1,3,7-nonatriene (DMNT; Chen et al. 2021). DMNT repressed expression of *PxMucin* in the larval midgut, and knock-down of this gene led to PM rupture and caterpillar death. These harmful effects of DMNT were both direct and indirect, since DMNT-induced damage to the PM led to further costly imbalances in the midgut microbiome of caterpillars (Chen et al. 2021).

Another constituent protein of the PM is the chitin-binding protein Peritrophin A. Insect herbivores show enhanced expression of this gene when jasmonic acid-mediated defensive signaling and production of reactive oxygen species are active (Groen et al. 2016; Mittapalli et al. 2007; Whiteman et al. 2011). The chitin fibrils and glycoproteins present in the PM are further targeted by a group of carbohydrate-binding proteins known as lectins. Indeed, a study dissecting the PM from the *Spodoptera littoralis* midgut showed distinct abnormalities in the PM with disrupted microvilli structures owing to lectin binding (Vandenborre et al. 2011).

A second set of important physical barriers are transepithelial diffusion barriers such as septate junctions in the midgut and the hemolymph (or blood)-brain barrier (BBB), which is also known as the perineurium (Petschenka et al. 2013). Septate junctions limit solute passage through intercellular spaces in epithelia. One of the proteins that has been implicated in maintaining junctional activity is the Na+/K+-ATPase β subunit encoded by the gene *Nrv2* in *D. melanogaster* (Paul et al. 2003, 2007). This epithelial barrier function is independent of its role in Na+/K+-ATPase pump activity. The presence of the junctions, combined with a lack of active uptake mechanisms for hydrophilic substances, which cannot permeate lipid bilayer membranes passively, can provide at least some protection against polar CGs such as ouabain (Dobler et al. 2015; Petschenka et al. 2013; Rubin et al. 1983). However, to prevent lipophilic defensive chemicals (e.g., the apolar CGs digoxin and digitoxin) from penetrating the midgut and the BBB, active detoxification mechanisms that counteract passive diffusion of the compounds through the lipid bilayers are necessary, which we will discuss below.

Target Site Insensitivity

Physiological investigations of the monarch butterfly provided early evidence of the existence of a Na+/K+-ATPase (the target of CGs) with dramatically lowered sensitivity (increased resistance) to CGs (Holzinger et al. 1992; Holzinger and Wink 1996). Molecular investigations demonstrated that this insensitivity may be explained in the monarch butterfly, at least in part, by an amino acid substitution of asparagine for histidine at position 122 (N122H) of the Na+/K+-ATPase's alpha subunit (Holzinger et al. 1992; Holzinger and Wink 1996). This form of molecular substitution that alters the toxin's binding potential to the enzyme is called "target site insensitivity" (TSI). By screening all Na+/K+-ATPase transmembrane domains involved in CG binding, a pair of studies detected the presence of the same substitution in five distantly related insect species representing a total of at least four independent origins across a phylogenetic distance of 300 million years (Dobler et al. 2012; Zhen et al. 2012). Remarkably, these screens also identified other amino acid substitutions associated with TSI of the Na+/K+-ATPase to CGs.

However, it was unknown if these substitutions could be sufficient for conferring resistance at the whole-organism level in a way that is beneficial, i.e., adaptive, for the animal. A follow-up study embarked on reconstructing possible mutational paths linked to CG insensitivity by comparing protein sequences of the CG binding site between the monarch butterfly and other animals with CG-rich diets to those of animals not regularly encountering dietary CGs (Weinreich et al. 2006, Karageorgi et al. 2019). Many evolutionary paths involved mutations in binding site residues 111, 119, and 122 (Karageorgi et al. 2019). A subset of these paths, including the monarch's, were then introduced into the genome of *D. melanogaster* through single-base edits using CRISPR (Gratz et al. 2014; Lin et al. 2014; Port et al. 2014; Groen and Whiteman 2016; Karageorgi et al. 2019). Since *D. melanogaster* is not specialized on a CG-rich diet, Karageorgi and co-workers (2019) tested whether the mutations conferred CG insensitivity at the neurophysiological and whole-organism levels.

A series of fly lines was engineered that represents steps in the evolution of CG insensitivity as observed in the lineages of the monarch butterfly and other CG-resistant species (Karageorgi et al. 2019). Mutating residues Q111 and N122 caused nervous system dysfunction, and co-introduction of A119S limited these deleterious side effects (Karageorgi et al. 2019). At the neurophysiological and whole-organism levels, flies with insensitivity mutations at sites 111 and 122 were highly resistant to CGs, just as the monarch is. Again, co-introducing A119S was important by enhancing the resistance-conferring effects of these insensitivity mutations (Karageorgi et al. 2019). Overall, residue S119 unlocked adaptive paths to resistance through interactive effects (epistasis) with sites 111 and 122 (Weinreich et al. 2006, Karageorgi et al. 2019), a result confirmed independently (Taverner et al. 2019).

TSI is a particularly effective strategy in response to toxins with narrow target ranges such as CGs, where a single or few TSI mutations have the potential for producing large fitness consequences. However, toxins such as ITCs, other reactive electrophiles, and reactive oxygen species have a wide target range, and it is unknown if insensitivity of at least some of the target sites has the potential to evolve in response to such toxins.

We explored whether this could be the case by taking a comparative genomics approach for a Brassicaceae-specialized herbivorous fly, *Scaptomyza flava*. For such a comparative analysis, we could not work with lepidopteran herbivores because herbivory evolved too long ago and the availability of genomic data is still relatively limited (Groen and Whiteman 2016). In the analysis we used data from *D. melanogaster* and further leveraged available protein biochemistry data from human biomedical science studies where interactions between GSL breakdown products and target proteins were studied functionally. We find that *S. flava* orthologs of genes that encode proteins targeted by GSL breakdown products in humans evolve faster than orthologs of human genes that do not encode such proteins (Fig. 2). It will be interesting to see if similar polygenic patterns of presumptive TSI have evolved in lepidopteran specialists on Brassicaceae such as *Pieris* spp. and *Plutella* spp.

Detoxification

Alongside the behavioral changes, structural barriers, immunosuppressive mechanisms, and TSI to prevent or negate the toxic effects of plant defensive chemicals, caterpillars may actively detoxify and metabolize these compounds through a conserved set of enzyme families. These enzymes are active not only at the interface of plant cells and caterpillar mouthparts as part of the insect's saliva (Rivera-Vega et al. 2017a,b) but also in tissues such as the gut, the BBB, and the Malpighian tubules (Li et al. 2007).

The three phases of detoxification in animals, as defined earlier, are each characterized by the activity of certain ubiquitous enzyme families, and we will review these below. Caterpillars of different species harbor distinct subsets of these enzyme families, and in most cases specific plant defensive chemicals can only be metabolized by a small number of detoxification enzymes (Heidel-Fischer and Vogel 2015).

Over the last 10–20 years, genomics and transcriptomics studies have provided evermore comprehensive insights into xenobiotic metabolism of caterpillars. One comparative genomics study found that among lepidopteran species feeding on

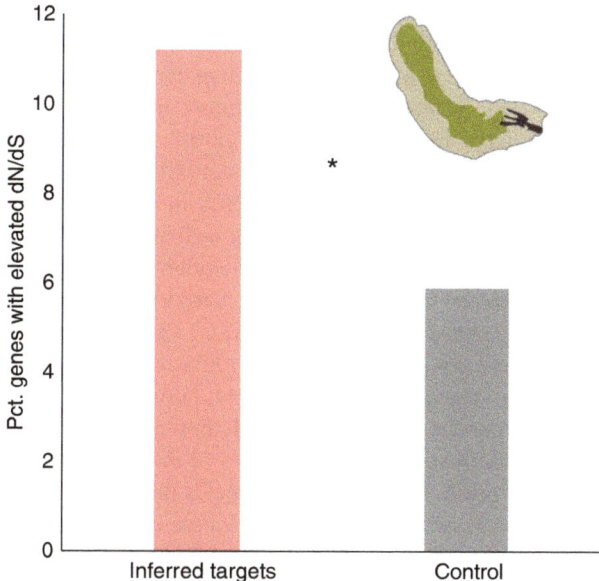

Fig. 2 Genes encoding proteins putatively targeted by GSL breakdown products display accelerated evolution in the Brassicaceae-specialized herbivorous fly *Scaptomyza flava*. Inferred putative targets of GSL breakdown products in *S. flava* and its one-to-one orthologs in *D. melanogaster* relatives were determined via orthology with human proteins that have functionally verified interactions with these products using the PantherDB database (Mi et al. 2013). Then, for each set of single-copy orthologous *Scaptomyza* and *Drosophila* genes, amino acid sequences from five species were aligned in MUSCLE: *S. flava*, *D. grimshawi*, *D. mojavensis*, *D. virilis*, and *D. melanogaster* (Gloss et al. 2019b). Using PAML v4.5's codeml module (Yang 2007), branch site tests for accelerated ratios of the number of non-synonymous substitutions per non-synonymous site (dN) to the number of synonymous substitutions per synonymous site (dS), dN/dS, were run for all terminal branches (Yang 1998), which has been described in more detail previously (Gloss et al. 2014). We define "accelerated" as being part of the top 5% tail of dN/dS values. Asterisk indicates a significant difference ($P < 0.05$) in the number of putative targets of GSL breakdown products with accelerated ratios of dN/dS (inferred targets) versus the number of putative non-targets with accelerated ratios of dN/dS (control) using a chi-square test. (Cartoon of *S. flava* larva by Sophie Zaaijer)

photosynthesizing plant tissue, highly polyphagous species had higher numbers of genes encoding cytochrome P450 monooxygenase (CYP450; phase I), carboxyl/choline esterase (CCE; phase I), and glutathione *S*-transferase (GST; phase II) genes (Gloss et al. 2019a; Rane et al. 2019). These genes are collectively among the most important in detoxification sensu stricto because they transform toxins into less toxic molecules.

Comparative gene expression studies in which transcriptomes have been sequenced in caterpillars reared on genetically manipulated crucifer plants, such as the model plant *Arabidopsis thaliana*, have shown how generalists and specialists appear to use different strategies to try to cope with the mustard oil bomb. In the tobacco budworm *Heliothis virescens*, a generalist, 3,747 transcripts were differentially expressed when feeding on plants with intact GSL production compared to engineered plants with disrupted production, whereas only 254 transcripts were differentially regulated in a specialist, the large cabbage white *Pieris brassicae* (Schweizer et al. 2017). Moreover, twice as many transcripts were upregulated rather than downregulated in *H. virescens*, while these proportions were similar (i.e., 50:50) in *P. brassicae*. Several canonical detoxification genes were strongly induced in *H. virescens* by the presence of GSLs in host plants (up to 30-fold), including 17 CYP450s and 9 CCEs (phase I), as well as 7 ABC transporters (phase III; Schweizer et al. 2017). In *P. brassicae*, on the other hand, a member of the NSP gene family, known to divert GSL breakdown toward less toxic nitriles (see above), was regulated by GSLs, plus a homologue of *GSTD1* (Schweizer et al. 2017), which efficiently catalyzes the conjugation of reduced glutathione (GSH) with ITCs in the dipteran herbivore *Scaptomyza nigrita* (see below; Gloss et al. 2014).

Although similar experiments with genetically engineered host plants are not yet possible for milkweed herbivores, transcriptomes of monarch caterpillars reared on *Asclepias curassavica* and *A. incarnata*, two species that differ substantially in CG concentrations, have been measured. Monarch larvae differentially expressed several hundred genes when feeding on these different hosts, including numerous phase I, II, and III detoxification genes, suggesting that these genes play a role in monarch toxin resistance and sequestration (Tan et al. 2019a, b).

Transcription of xenobiotic metabolism genes is regulated by a signaling network with at least five different pathways through it, each initiated by different classes of receptors: (1) the membrane-localized G protein-coupled receptors; (2) cyclic adenosine 3´,5´-monophosphate (cAMP)-response element binding protein (CREB), which is a bZIP family transcription factor and requires phosphorylation by environment-responsive mitogen-activated protein kinase (MAPK) cascades to initiate signaling; (3) Cap'n'collar isoform C/Kelch-like ECH associated protein 1 (CncC/Keap1), which is another bZIP family transcription factor and ortholog of Nuclear factor erythroid-derived 2-related factor 2 (Nrf2) found in mammals; (4) the basic helix-loop-helix (bHLH)-Per-ARNT-Sim (PAS) domain-class transcription factor aryl hydrocarbon receptor (AhR), which heterodimerizes with the AhR nuclear translocator (ARNT) before binding to xenobiotic response elements (XRE) in target gene promoters to activate their transcription; and 5) the nuclear receptor (NR) superfamily transcription factor Hormone receptor-like in 96 (HR96), which is related to genes encoding the Steroid and Xenobiotic Receptor (SXR) and Constitutive Androstane Receptor (CAR) in vertebrates (Amezian et al. 2021; Li et al. 2021a, b). Both CAR and SXR may translocate to the nucleus upon activation and subsequently dimerize with Retinoid-X-Receptor (RXR) to enhance target gene transcription (Amezian et al. 2021).

Which receptors initiate signaling depends partly on the solubility of the plant defensive chemicals the insect encounters. ITCs and another GSL breakdown product, indol-3-carbinol (I3C), are relatively lipophilic, and after passing through the cell membrane, they can elicit a burst of reactive oxygen species (ROS) directly or indirectly. This in turn can activate transcription through CncC/Keap1 (Nrf2) interaction with the antioxidant response element (ARE) in promoters of downstream detoxification genes such as CYP450s and GSTs (Chen et al. 2018; Giraudo et al. 2015; Li et al. 2021a, b).

CGs, on the other hand, occur in a range of polarities and, therefore, solubilities. In addition to being perceived through their inhibition of the Na+/K+-ATPase, there are hints they could be perceived by intracellular receptors, which may depend on the solubility of individual CGs. Although polar, water-soluble plant defensive compounds, including several alkaloids such as nicotine, cannot passively diffuse through membranes and may thus be perceived by membrane-localized receptors such as GPCR (Amezian et al. 2021; Li et al. 2021a, b; Yang et al. 2020), polar CGs have not been connected with this mechanism. Certain polar compounds, including the CG ouabain, can be actively transported into cells via transmembrane transporters such as organic anion transporter peptides (Groen et al. 2017; Wink 2018). Polar CGs, along with lipophilic membrane-permeable CGs such as digitoxin, might then be perceived by intracellular receptors. However, while in mammals the relatively polar CG digoxin interacted with the nuclear receptor RORγT, this was not the case for its distant ortholog in insects, the steroid-sensing receptor DH3/Hr3 (Ahmed et al. 2020; Huh et al. 2011). Genetic screening in the model insect *D. melanogaster* may be the most efficient way forward for identifying if there is an intracellular receptor for CGs in insects in addition to the Na+/K+-ATPase at the cell membrane (Groen and Whiteman 2016).

We will now go into more depth regarding the multiple families of canonical insect xenobiotic metabolism genes.

Phase I: Oxidation, Hydrolysis, Reduction

Here, the goal is to provide an overview of the role that phase I enzymes, principally CYP450s, play in mediating detoxification of plant secondary compounds encountered by lepidopteran larvae. We then narrow our discussion to focus on their role in CG and GSL detoxification.

CYP450s are membrane-localized enzymes with important roles in metabolizing a variety of chemicals, ranging from steroid hormones to fatty acids to vitamins. The monooxygenases achieve this by adding oxygen atoms to target chemicals, using heme as a co-factor. A critical part of the heme group is an iron atom, which is activated by a conserved cysteine residue (Feyereisen 2012). The oxygenated substrates typically become more water-soluble and more amenable to being targeted by enzymes in subsequent phases of the detoxification process (which is why they are called phase I).

CYP450s are critical for successful detoxification of a range of plant defensive chemicals, and particularly well-studied members of the CYP450 family in this regard are those of the CYP6 clade. Members of the CYP6AE clade show a bloom (expansion in gene number through duplications) in Lepidoptera (Dermauw et al. 2020). Silencing or knocking out CYP6AE genes in the cotton bollworm *H. armigera* impairs caterpillar tolerance toward the cotton toxin gossypol (Mao et al. 2007) and the furanocoumarin xanthotoxin that is found in plants from the Rutaceae and

Apiaceae (Wang et al. 2018), respectively. In particular, CYP6AE19 was shown to metabolize xanthotoxin, but not as efficiently as the P450 CYP6B1 from the black swallowtail *Papilio polyxenes*, a specialist on furanocoumarin-containing plants (Wang et al. 2018). *P. polyxenes* caterpillars can tolerate dietary furanocoumarin concentrations of up to 1% using CYP6B1, its paralogue CYP6B3, and other CYP6Bs as detoxifying enzymes (Berenbaum and Zangerl 1993; Cohen et al. 1992; Hung et al. 1995; Wen et al. 2003). CYP6B1 and -3 probably evolved toward subfunctionalization under independent purifying selection after the duplication event that gave rise to both, and now display different efficiencies with which they metabolize different types of furanocoumarin (Wen et al. 2006). A similar pattern of subfunctionalization under selection apparently occurred in the parsnip webworm *Depressaria radiella* (formerly *D. pastinacella*), which has an even narrower host range (restricted to Apiaceae) than *P. polyxenes*, with at least two CYP450s (CYP6AE89 and CYP6AB3) efficiently metabolizing a variety of different furanocoumarins (Calla et al. 2020; Li et al. 2004a, b; Mao et al. 2006, 2007a, 2008). Going in the other direction, away from specialization and toward more generalized host plant ranges, substrate specificities of CYP450s in *Papilio* spp. were broader in the oligophagous species *P. multicaudatus* than in the specialist *P. polyxenes* and broader still in the polyphagous species *P. glaucus* and *P. canadensis*; this was linked to the relative abundance of furanocoumarin-producing plants in the diet (Li et al. 2003; Mao et al. 2007b).

In the context of handling toxic GSL breakdown products, it appears that CYP6B enzymes can process I3C as a substrate, which is one of the major derivatives of indole GSLs. Caterpillars of the generalist moth *H. virescens* showed enhanced transcription of CYP6B8 and several other CYP6AE and CYP6AB genes after encountering GSLs, including I3C (Schweizer et al. 2017). Comparison of the homolog of CYP6B8 in another generalist, *H. zea* (which has a wide host range and occasionally encounters GSLs), with CYP6B1 from the Rutaceae and Apiaceae specialist *P. polyxenes* (which practically never encounters GSL-producing plants), showed that while CYP6B1 did not metabolize the indole GSL breakdown product I3C, CYP6B8 did (Li et al. 2004a, b). CYP6B8 further metabolized a number of other chemically diverse plant defensive compounds including quercetin, flavone, chlorogenic acid, rutin, and xanthotoxin (Li et al. 2004a, b). The latter compound is one of the defensive chemicals abundant in hosts of *P. polyxenes*, and indeed, CYP6B1 of the specialist had a 30-fold higher metabolic clearance rate toward xanthotoxin than CYP6B8 (Li et al. 2004a, b), pointing to a trade-off between breadth and efficiency in terms of substrate handling for these CYP450s.

There is some evidence that CGs may also be substrates for CYP450s (Marty and Krieger 1984). However, the identity of individual CYP450s that may metabolize CGs in caterpillars from the monarch and other milkweed herbivores is currently unknown. Two studies that compared transcriptomes of monarch caterpillars reared on host plants with different CG profiles revealed suites of CYP450s that were differentially expressed (Gonzalez-De-la-Rosa et al. 2020; Tan et al. 2019a), potentially narrowing down the set of candidate CYP450s that may be involved in processing CGs. It has recently been established that an enzymatic reduction step is critical for detoxification of the toxic CG voruscharin, produced by one of the monarch's main host plants *Asclepias curassavica* (Agrawal et al. 2021). After a first non-enzymatic step in which voruscharin is converted to uscharidin, a step facilitated by the alkaline pH of the gut milieu (Berenbaum 1980), this compound is then enzymatically reduced to the more polar and less toxic CGs calactin and calotropin (Agrawal et al. 2021; Marty and Krieger 1984; Seiber et al. 1980). Oxidoreductases such as CYP450s are candidates for carrying out this step, as indeed, CYP450s such as the Halloween genes have well-studied roles in facilitating molecular alterations of plant-derived steroids that are chemically related to CGs to synthesize molting hormones (Gilbert 2004; Seiber et al. 1980).

The carboxyl/cholinesterases (CCEs) form another functionally diverse superfamily of enzymes. These hydrolyze carboxylic esters to their component alcohols and acids. Although CCEs have been studied less intensively than P450s, evidence has been found for a role of CCEs in targeting host plant defensive chemicals. In caterpillars of *Depressaria radiella*, CCEs are involved in processing plant-derived aliphatic esters in the midgut (Zangerl et al. 2012). Furthermore, in adults of the generalist moth *Spodoptera littoralis*, two CCE genes, *SlCXE7* and *SlCXE10*, were found to degrade the plant volatile (Z)-3-hexenyl acetate in the antennae, but it is unclear which of these genes could have a role in processing volatile cues in the larval stage as well (Durand et al. 2010, 2011). It is further unknown if CCEs could be involved in processing GSLs, GSL breakdown products, or CGs. However, transcriptomic studies have identified a number of CCEs that are responsive to the presence of dietary GSLs in *Heliothis virescens* (Schweizer et al. 2017) and to host plants with different CG contents in the monarch (Gonzalez-De-la-Rosa et al. 2020; Tan et al. 2019a).

Phase II: Conjugation

In the second phase, the products of the first phase or, often, the toxins themselves are conjugated to other molecules. The enzymes that catalyze these reactions are various transferases such as GSTs, many of which are regulated by the Keap1-Nrf2-ARE signaling pathway. Perhaps their best-studied detoxification mechanism is the conjugation reaction with GSH. Conjugation neutralizes reactive nucleophile sites of plant defensive chemicals. It can further increase their solubility in water, thereby facilitating their excretion from cells in phase III.

GST-mediated detoxification can happen through the metabolism of secondary products generated from other detoxification enzymes (phase II). It can also occur directly during phase I as an alternative to P450- or CCE-mediated detoxification. Despite their central role in processing a range of plant defensive chemicals, GSTs appear not to have undergone a gene family-wide expansion in the Lepidoptera (You et al. 2015).

GSTs play an important role in the detoxification of ITCs in caterpillars of generalist species that have not evolved specialized mechanisms to prevent ITC formation, such as GSL desulfation through GSSs in *Plutella* spp. and diversion of GSL breakdown toward nitriles under the influence of NSPs in *Pieris* spp. Although mechanistic evidence is still being gathered, it appears that GST-mediated ITC detoxification occurs via a series of enzymatic steps known from mammalian studies as the mercapturic acid pathway (Traka and Mithen 2009). This pathway starts with activity of GSTs, generating ITC conjugates with GSH, cysteinylglycine (CysGly), and Cys, which end up as conjugates with an N-acetylcysteine group through the action of N-acetyltransferases (Traka and Mithen 2009). The last step deserves particular attention. While ITCs leave the mammalian body in urine and bile as N-acetylcysteine conjugates, such conjugates have not been observed in caterpillar frass, despite detection of all conjugates from intermediate steps in the pathway (Jeschke et al. 2016, 2017, 2021; Schramm et al. 2012). It is currently unclear if lepidopteran genomes do not encode the required enzymes, whether such enzymes are perhaps not expressed at the caterpillar stage, or if the enzymatic reaction may be impeded by the relatively high pH of the caterpillar midgut milieu (Berenbaum 1980; Schramm et al. 2012).

Thus far, ITC detoxification via GSTs and the mercapturic acid pathway has been studied in a variety of generalists (e.g., *Helicoverpa armigera*, *Mamestra brassicae*, *Spodoptera* spp., *Trichoplusia ni*) and Brassicaceae specialists, but also in a specialist on legumes: *Anticarsia gemmatalis*. A comparative study of GST activity in response to the presence of dietary ITCs showed that in the highly polyphagous species *Spodoptera frugiperda*, GSTs metabolize a wide range of ITCs (Wadleigh and Yu 1988). This range becomes progressively narrower in GSTs of *T. ni*, which is less polyphagous and metabolizes only allyl and benzyl ITC, and *A. gemmatalis*, which does not typically encounter ITCs and metabolizes only benzyl ITC. These comparisons suggest that GST substrate specificity may evolve according to the proportion of GSL-containing plant material in the diet (Wadleigh and Yu 1988).

This study and subsequent studies further identified that GST levels are induced, not only when ITCs are present in the diet, but also when indole GSL-derived I3C and indole-3-acetonitrile are present in the diet (Li et al. 2007; Wadleigh and Yu 1988). In the generalist *Spodoptera litura*, expression of the epsilon-class GST (*Slgste1*) in the midgut was responsive to the formation of ROS induced by I3C (Chen et al. 2018). Induction of expression was regulated by binding of SlNrf2 to an antioxidant response *cis*-regulatory element in the *Slgste1* promoter. This was functionally verified through RNAi on *SlNrf2*: caterpillars with silenced *SlNrf2* showed reduced expression of *Slgste1*, lower levels of peroxidase reactions by GSTs, and reduced cell viability in response to treatment with I3C (Chen et al. 2018).

Although a specialist such as *Pieris rapae* does not rely mainly on GST- and GSH-dependent detoxification to handle dietary GSLs, it may have additional adaptations to prevent oxidative damage that could still be induced by non-ITC breakdown products of GSLs. *P. rapae* individuals show genetic variation in/near *Glyoxalase 1* (*Glo1*), encoding a lactoyl-GSH lyase that is linked to caterpillar performance on *Arabidopsis thaliana* plants (Nallu et al. 2018). As part of the glyoxalase pathway, Glo1 neutralizes toxic by-products of metabolism, using GSH in the process.

In addition to clade-specific defensive chemicals such GSLs, GSTs have also been found to provide protection against more widely occurring toxins. The compound 12-oxophytodienoic acid (12-OPDA), which is part of the jasmonate family and also acts as a signaling molecule (Groen et al. 2013), has a reactive α,β-unsaturated carbonyl structure. It easily adds cellular nucleophiles, making OPDA potentially toxic for herbivores. The glutathione *S*-transferase GST16 inactivates 12-OPDA in the insect gut by isomerization to inactive *iso*-OPDA in *Helicoverpa armigera* (Shabab et al. 2014), and GST family members perform the same function in a suite of other generalist moth larvae (Dabrowska et al. 2009).

A more recently identified family of genes acting in phase II detoxification is that of the UDP-glycosyltransferases (UGTs; Ahn et al. 2012). UGTs may catalyze conjugation of sugars with lipophilic plant defensive chemicals, which increases water solubility of the toxins and makes it easier for them to be processed further in subsequent phases of detoxification. UGTs show lineage-specific expansions within the Lepidoptera and appear to play an important role in the xenobiotic response (Ahn et al. 2012).

While not yet studied in the context of GSLs and ITCs, a role for UGTs has been identified for caterpillar detoxification of three other classes of toxins that share certain properties with ITCs. The first class is represented by capsaicin from peppers (*Capsicum* spp.), which in mammals and *D. melanogaster* is perceived by Trp receptors as are ITCs (Li et al. 2020a, b). Although it is unknown if Trp receptors are involved in capsaicin perception in Lepidoptera as well, capsaicin does have a deterrent effect on feeding and oviposition in *Helicoverpa* spp. moths (Ahn et al. 2011a). Interestingly, these species all appear to employ UGT-mediated glucosylation as a means of capsaicin detoxification, including not only the generalists *H. armigera* and *H. zea* but also the specialist *H. assulta*, despite the latter showing higher capsaicin tolerance levels (Ahn et al. 2011a,b).

The second class is exemplified by the sesquiterpene dimer gossypol from cotton, which, not unlike ITCs, is able to cross membranes passively as an apolar chemical, deriving its toxicity from damaging amino acids in proteins. Gossypol toxicity occurs through interaction between its highly reactive aldehyde groups and amino acids, while six phenolic hydroxyl groups lend it additional toxicity. Enzymatic essays with insect cells expressing UGT41B3 and UGT40D1 from the generalist *Helicoverpa armigera* showed that these UGTs can glycosylate gossypol to diglycosylated gossypol isomers, a process which may be involved in detoxification in vivo (Krempl et al. 2016).

The third class is formed by benzoxazinoid glycosides, which are produced by a subset of monocots, including maize. Benzoxazinoids are indole-derived defensive chemicals whose aglucone breakdown products delay caterpillar growth and survival. *Spodoptera frugiperda* detoxifies these aglucones through UGT-mediated reglucosylation. In the process, the chemical is inverted compared to its original benzoxazinoid glycoside state as found in the host plant. This inverted glucosylation ensures that the benzoxazinoids cannot be turned into the toxic aglucone form by either plant or insect ß-glucosidases again, making the detoxification strategy effective for enhancing caterpillar fitness (Maag et al. 2014; Wouters et al. 2014).

In all of these examples, more work will be necessary to narrow down the mechanistic involvement of UGTs to the levels of individual genes and the enzymes for which they code. Lastly, UGTs are enriched in the transcriptome of monarch caterpillars compared to the transcriptomes of the pupal and adult life stages (Ranz et al. 2020). Although it was speculated that these UGTs may have a role in the detoxification of milkweed host toxins such as CGs, this has not yet been studied functionally (Ranz et al. 2020).

Phase III: Excretion

Enzymatic reactions in phases I and II make plant defensive chemicals available for the last phase of the detoxification process, if they were not already available as water-soluble compounds. In this last phase, phase III, the compounds become substrates of several diverse sets of transporters from multiple gene families and subfamilies. Activity of these transporters is particularly important in three tissue types where they shunt away plant defensive chemicals and/or their processed derivatives: the gut, the BBB, and the Malpighian tubules. We will now focus on two classes of transporters that are expressed in all three of these tissues.

The first class is formed by the multidrug transporters (Mdrs), which are also known as P-glycoproteins and B-type ABC transporters (Dermauw and Van Leeuwen 2014). Tissue-specific gene expression measurements and staining with Mdr-specific antibodies detected the presence of Mdrs in the midguts of generalist herbivores as well as CG-adapted insects (Dobler et al. 2015; Petschenka et al. 2013). Mdr expression is further enriched in the Malpighian tubules (Chahine and O'Donnell 2009; Dow and Davies 2006), where efflux capacity increases dramatically upon toxin exposure (Chahine and O'Donnell 2009). The regulation of Mdr expression appears to be coordinated with that of genes involved in earlier phases of xenobiotic detoxification (Chahine and O'Donnell 2011). Lastly, Mdrs are expressed in the BBB across all of the animal kingdom (Hindle and Bainton 2014). Physiological assays, complemented with reverse genetic studies, have established that Mdrs act as active diffusion barriers to apolar CGs such as digoxin in Lepidoptera, other insects, and vertebrates (Gozalpour et al. 2013; Petschenka et al. 2013; Groen et al. 2017).

Interestingly, knockout mutants of *Mdr50* in *D. melanogaster* are compromised in their digoxin resistance (Groen et al. 2017). The putative monarch orthologs show interesting properties: (1) the monarch orthologs appear to have undergone a bloom compared to orthologs in caterpillars that do not regularly encounter dietary CGs (Fig. 3); and (2) expression of these genes is upregulated on a diet containing CG-rich milkweeds (Gonzalez-de-la-Rosa et al. 2020). If the role of Mdr50 is conserved in the monarch butterfly, this might provide a mechanism for the monarch to minimize exposure to apolar CGs by reducing their entry from the midgut to the hemolymph. Excluding apolar CGs such as the thiazolidine ring-containing voruscharin from the hemolymph could have important fitness consequences. This CG is the most abundant CG in one of the monarch's main milkweed hosts, *Asclepias*

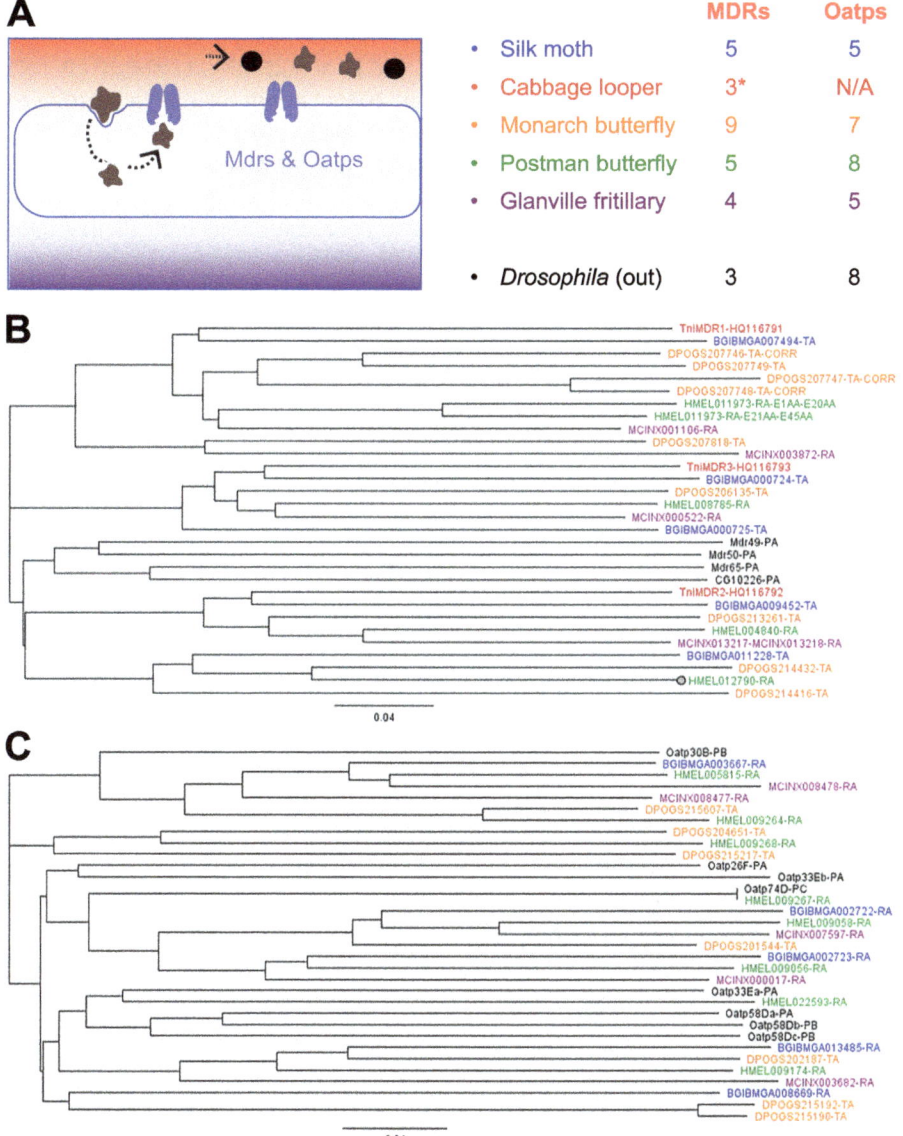

Fig. 3 (**a**) In addition to TSI-conferring substitutions in the Na+/K+-ATPase, monarch caterpillars may resist CG toxicity by excluding CGs (black and brown compounds) from the sensitive nervous tissue by ABC transporters and organic anion transporting polypeptides (purple transmembrane proteins) that are mainly active in the midgut, blood-brain barrier (depicted), and Malpighian tubules. This mechanism is particularly important for protecting the nervous tissue (purple area) against apolar CGs, which can cross membranes passively, by transporting these back into the hemolymph (red area), whereas polar CGs (black) can be kept out to some extent through tight junctions between cells. (**b**) Orthologs of *D. melanogaster* Mdr50 (a B-type ABC transporter) may have experienced a gene bloom in the monarch butterfly (*Danaus plexippus*) relative to the silk moth (*Bombyx mori*), the cabbage looper *Trichoplusia ni*, the postman butterfly (*Heliconius melpomene*), and the Glanville fritillary (*Melitaea cinxia*). The asterisk at the cabbage looper indicates that its genome may encode more than three Mdrs. (**c**) A duplication was also detected for the monarch ortholog of Oatp33Eb (see text for details). (Cartoon by Sophie Zaaijer)

curassavica, accounting for 40% of leaf CGs, and its abundance was negatively correlated with caterpillar growth (Agrawal et al. 2021). It will also be interesting to study Mdrs more closely in caterpillars of species such as *Empyreuma pugione* and *Daphnis nerii*. These species specialize on CG-bearing plants, but do not have known TSI substitutions in their Na+/K+-ATPases. Indeed, in vitro analyses of enzyme activity in the presence of increasing CG concentrations indicate that their Na+/K+-ATPases are highly sensitive to CGs (Petschenka and Dobler 2009; Petschenka et al. 2012, 2013; Petschenka and Agrawal 2015). This sensitivity suggests they may have evolved alternative mechanisms of handling dietary CGs, which may include efflux through Mdrs (Petschenka et al. 2013).

A second class of transporters is formed by the organic anion transporting polypeptides (Oatps). Many of these transporters show strong expression in the BBB and midgut (Hagenbuch and Stieger 2013; Hindle and Bainton 2014), while some are highly expressed in the Malpighian tubules (Torrie et al. 2004). Like Mdrs, the expression of Oatps is coordinated with that of other enzymes involved in xenobiotic detoxification. Besides their role in this process, Oatps are also involved in the metabolism and efflux of endogenous solutes (Dow and Davies 2006). In vitro and in vivo reverse genetic screens on *D. melanogaster* established that a subset of Oatps prevent polar CGs such as ouabain from interfering with Na+/K+-ATPase function (Groen et al. 2017; Torrie et al. 2004). The Oatps provide a baseline level of protection against CGs in insects not specializing on CG-containing diets. These transporters may have provided a substrate for natural selection to work upon in insects that transitioned to feeding on CG-producing host plants (Groen et al. 2017).

Although the Oatp family and the superfamily of solute carrier transporters they belong to, the SLC22 organic cation/anion/zwitterion transporters, underwent an expansion in the Lepidoptera (Denecke et al. 2020), the absolute number of Oatps does not appear to have changed in the monarch (Fig. 3). However, there has been a duplication of the monarch ortholog of *D. melanogaster* Oatp33Eb, and a fly knockout mutant of Oatp33Eb (an Oatp that is typically expressed in the gut system) showed the lowest lethal dose of ouabain of several Oatp mutants compared to wild-type flies (Groen et al. 2017). It will be interesting to find out if these monarch Oatps are indeed involved in dealing with dietary CGs.

Which transporters allow herbivores on Brassicaceae to expel ITCs and other GSL breakdown products has not been determined. However, evidence from biomedical studies suggests that instead of B-type ABC transporters (P-glycoproteins or Mdrs), it is likely the C- and G-type ABC transporters that may be important. Like B-type transporters, the C-type transporters are full ABC transporters with at least 12 transmembrane domains and a nucleotide-binding domain that has ATPase activity (Dermauw and Van Leeuwen 2014). In human cells, Multidrug resistance protein1 (MRP1 or ABCC1) mediates efflux of AITC, BITC, PEITC, and sulforaphane as conjugates with GSH and cysteinylglycine (Callaway et al. 2004; Hu and Morris 2004; Zhang and Callaway 2002), whereas its subfamily relative MRP2 (ABCC2) transports the GSH-conjugated form of PEITC (Ji and Morris 2005a).

Unlike B- and C-type transporters, Breast cancer resistance protein (BCRP or ABCG2) is a half transporter, and besides the nucleotide binding domain with

ATPase activity, it contains only six transmembrane domains. BCRP transports the unchanged form of PEITC, without conjugation to molecules such as GSH (Ji and Morris 2005b). Future functional studies may find out if B-, C-, and/or G-type transporters may be involved in GSL detoxification in caterpillars as well.

Microbial Interactions

With important caveats (e.g., that many caterpillar individuals may lack a resident gut microbiome), microbes associated with caterpillars and their immediate host plants may have important modulating effects on the different mechanisms caterpillars use for dealing with plant defensive chemicals.

Chewing herbivores could benefit from microbes through at least two mechanisms. One is through the sometimes immunosuppressive effects of microbes on the host plant when deposited via oral secretions (regurgitant derived from the foregut) or the saliva (Grant 2006). Experiments with the Colorado potato beetle (*Leptinotarsa decemlineata*) demonstrated that larvae benefitted from the suppressive effects of oral secretions containing *Pseudomonas* and *Enterobacter* spp. bacteria on antiherbivore defenses in one of the host plants, tomato (*Solanum lycopersicum*; Chung et al. 2013). Immunosuppression by bacteria in oral secretions has more recently also been found to occur for *Spodoptera frugiperda* caterpillars, particularly when the herbivores deposited *Pantoea* spp. bacteria on tomato host plants (Acevedo et al. 2017). It is not yet known if bacteria in caterpillar saliva, as opposed to regurgitant oral secretions (Grant 2006), could also influence the outcome of plant-herbivore interactions. However, it is interesting that salivary glands of *Trichoplusia ni* are enriched for a distinct bacterial flora compared to other organs that open directly into the digestive system, including the mandibular glands, the Malpighian tubules, and the midgut itself, and that *Pseudomonas* bacteria were one of the enriched genera (Lawrence et al. 2020).

A second mechanism of microbial effects on caterpillar fitness, and one that has been studied somewhat more extensively, is through modification of plant defensive chemicals by enzymes derived from microbes (Mason et al. 2019a). At an extreme, entire microbes become internalized in herbivore cells in an endosymbiotic relationship. More commonly, however, single microbial genes end up in the herbivore genome through horizontal gene transfer (Hansen and Moran 2014). In this scenario, a microbe-herbivore association becomes fixed and microbe-produced detoxification enzymes are now indirectly derived from microbes (Mason et al. 2019a). This has happened relatively frequently in clades of herbivores such as piercing/sucking insects and chelicerates (Hansen and Moran 2014; Wybouw et al. 2018; Greenhalgh et al. 2020). In the whitefly *Bemisia tabaci*, the herbivore genome even had a host plant-derived phenolic glucoside malonyltransferase gene incorporated that allows detoxification of phenolic glycosides (Xia et al. 2021). Genomic analysis of three lepidopteran herbivores (*Bombyx mori*, *Heliconius melpomene*, and *Danaus plexippus*) revealed that horizontal transfer events had occurred ca. 12 times per species and that at least some of the genes with putative origins from bacteria or fungi were transferred prior to the formation of many herbivore species (Sun et al. 2013). Several of the genes encode enzymes that are potentially involved in metabolizing amino acids, starch, and sugar, and some might be involved in detoxification of host plant defensive chemicals (Li et al. 2011). In one well-studied example, all lepidopteran genomes examined contain orthologs of bacterial β-cyanoalanine synthase/cysteine synthase (*CAS/CYS*) genes, which is probably the result of an ancient horizontal gene transfer event from methylobacteria in the ancestor of all Lepidoptera (Wybouw et al. 2014, 2016). Caterpillars of a variety of species show inducible CAS activity upon encountering plant-produced cyanide in their diet. Functional studies in the Brassicaceae specialist *Pieris rapae* showed that CAS enzymes convert this toxic defensive chemical via a cross-reaction with cysteine into the less toxic products β-cyanoalanine and hydrogen sulfide (Witthohn and Naumann 1987; Meyers and Ahmad 1991; Stauber et al. 2012; Van Ohlen et al. 2016).

Yet, many of the relevant associations between microbes and caterpillars fall toward the more plastic/labile end of the spectrum (Mason et al. 2019a). Unlike herbivores with piercing/sucking mouthparts (Hansen and Moran 2014), caterpillars appear to lack a resident gut microbiome (Hammer et al. 2017). They probably derive a large proportion of their gut microbiome from their diet (Hammer et al. 2017) and may even obtain much of it from the soil (Hannula et al. 2019). In addition to this lack of specificity in caterpillar gut microbiomes, there also remains much to be discovered about whether and how caterpillars may receive benefits from microbes in dealing with host plant defenses (Hammer and Bowers 2015). Observations on fitness outcomes of interactions between caterpillars and internal, non-disease causing microbes show a continuum from positive, to neutral, to negative. Caterpillars of *Anticarsia gemmatalis* showed better survivorship and growth when their gut microbiome was left intact (Visôtto et al. 2009), while suppressing gut bacteria had no detectable effect on fitness in *Manduca sexta* (Hammer et al. 2017). A negative effect of gut microbes was observed in *Spodoptera frugiperda* caterpillars feeding on maize plants. When a defensive protease (Mir1-CP) produced by maize damaged the peritrophic matrix, gut bacteria from the genera *Enterobacter*, *Enterococcus*, and *Klebsiella* then penetrated this protective barrier, invaded the hemocoel, and exacerbated the negative fitness consequences of the maize protease on the caterpillars (Mason et al. 2019b). It will be fascinating to see if such interactive effects of plant defenses and microbial infections occur more generally.

Several studies have assessed mechanisms of how gut microbes may affect detoxification of ITCs and GCs. Although more work on caterpillars is needed, experiments across various species of chewing insects (and humans) have identified bacteria that metabolize these defensive chemicals. Among the gut microbiota of the cabbage stem flea beetle *Psylliodes chrysocephala*, the bacterial genera *Pantoea*, *Pseudomonas*, and *Acinetobacter* were associated with degradation of ITCs (Shukla and Beran 2020). However, only *Pantoea* spp. had measurable effects on ITC detoxification in follow-up experiments (Shukla and Beran 2020), despite the fact that strains of *Pseudomonas* bacteria produce enzymes that detoxify ITCs (Fan et al.

2011) and can suppress plant defenses locally and systemically (Groen et al. 2013, 2016). Separate studies on the human gut microbiome identified that the bacterium *Eggerthella lenta* carries a "CG reductase" operon that metabolizes CGs (Koppel et al. 2018). Taken together, these studies show that gut microbes have the potential to play a role in ITC and CG detoxification, but much more research will be needed to determine if the microbiome may perform similar functions in the guts of caterpillars that feed on toxic host plants.

Ultimate Causes of the Evolution and Maintenance of Detoxification Mechanisms

A salient discussion of the genomic and phenotypic targets of selection associated with how herbivorous insects interact with plant defensive chemicals requires identification of the agents of selection. Selection on insect herbivores is applied by both bottom-up agents (e.g., the host plants that are fed on) and top-down agents (e.g., predators and parasites; Price et al. 1980). Comparison between different species of herbivores and between herbivores and their non-herbivorous relatives can reveal genotypic and phenotypic signatures of selective pressure by each of these agents.

Bottom-Up Agents of Selection

Host plant species are typically polymorphic for the production of defensive chemicals, and the same is true for many counter-adaptations in insects. Such coinciding patterns of trait distributions are hypothesized to be the consequence of coevolutionary dynamics (Flor 1956; Ehrlich and Raven 1964; Karasov et al. 2014; Stahl et al. 1999).

These dynamics can be subdivided into distinct classes according to several criteria, a main one being if dynamics show directionality or whether instead they are fluctuating (Hall et al. 2020; Woolhouse et al. 2002). When directionality is present, the dynamics often resemble "arms races," which may, for example, result in escalation of plant defensive chemical production over generations and counter-adaptations by herbivores (Ehrlich and Raven 1964; Dawkins and Krebs 1979; Kareiva 1999; Van Valen 1973). As part of arms race dynamics, successive selective sweeps are likely to occur, purging alleles that are non-adaptive in the participating species. However, depending on fitness costs associated with evolving traits and the genetic architecture of these traits, polymorphisms can be maintained over short to longer periods of time. When polymorphisms are stably maintained, the dynamics appear as "trench warfare" (Stahl et al. 1999). On the other hand, costs may also drive selection and evolutionary dynamics to fluctuate, favoring different traits or trait values during different episodes of selection (Hall et al. 2020). This could result in fluctuations in the frequencies of alleles involved in regulating the traits (Speed et al. 2015).

The presence and nature of fitness costs associated with traits under selection can thus play an important role in determining which type of coevolutionary dynamics populations of herbivores and their host plants will follow over time. On the plant side, the production of toxins can be constrained by several different types of costs: 1) opportunity costs may arise if toxin production in early life stages diminishes subsequent plant growth vigor and competitive ability (Coley et al. 1985; Züst et al. 2011); 2) metabolic costs are incurred when toxins are produced (Bekaert et al. 2012; Gershenzon 1994); 3) allocation costs may cause growth and/or reproduction to be reduced when limited resources are spent on toxin production (Simms 1992); 4) toxin production can carry genetic costs depending on the presence and level of genetic correlation with other traits, for example, via genetically hardwired signaling networks (Groen et al. 2020; Züst and Agrawal 2017); and 5) production of toxins effective against one herbivore genotype may have negative fitness consequences on interactions with other genotypes or other species and thus carry ecological costs. For example, producing toxins effective against a generalist herbivore may harm mutualistic interactions with pollinators or increase plant susceptibility to specialist herbivores (Strauss et al. 1999). Although fitness costs have been notoriously difficult to measure (Bergelson and Purrington 1996; Koricheva 2002), it appears that at least in some environmental contexts, GSL and CG production may incur costs to plants (Stowe and Marquis 2011; Züst et al. 2015).

On the herbivore side, the types of costs associated with detoxification can be divided into similar classes. While in plants costs and benefits of toxin production will be influenced by the probability of encountering certain herbivores, costs and benefits of detoxification in herbivores will be influenced by the chance that dietary toxins will be encountered (Després et al. 2007). Perhaps they have not received as much attention from scientists in terms of theoretical framework development and experimental work as the costs on the plant side (Després et al. 2007; Karban and Agrawal 2002).

Behavioral avoidance of toxin ingestion by searching for hosts or tissues with lower toxin levels comes with opportunity costs in the form of spending time searching or actively manipulating the host plant to subvert activation of defenses. These costs will increase as well-defended plants increase in population frequency (Després et al. 2007; Karban and Agrawal 2002). Another set of costs that increase as hosts produce more toxins are the metabolic and allocation costs as herbivores spend energy on detoxification (Després et al. 2007). Costs of handling plant toxins have thus far been established for several toxin-herbivore combinations in the Lepidoptera, including GSLs in *Pieris rapae* and *Helicoverpa armigera* (Agrawal and Kurashige 2003; Wang et al. 2021; Jeschke et al. 2021), nicotine in *Spodoptera eridania* (Cresswell et al. 1992), furanocoumarins in *Depressaria pastinacella* (Berenbaum and Zangerl 1994), and CGs in the monarch (Seiber et al. 1980; Zalucki et al. 2001; Agrawal 2005; Rasmann et al. 2009; Tao et al. 2016; Agrawal et al. 2021).

As a general pattern, herbivores combine several of the mechanisms described in the previous section to deal with plant defensive chemicals: e.g., behavioral avoidance of toxin ingestion is regularly associated with enzymatic detoxification. The monarch combines laticifer clipping behavior with enzymatic detoxification of and TSI to CGs (Agrawal et al. 2021; Dussourd and Eisner 1987; Marty and Krieger 1984; Seiber et al. 1980), while a generalist herbivore on Brassicaceae such as *Helicoverpa armigera* combines searching for low-level GSL areas of leaves with GSL detoxification via the mercapturic acid pathway (Jeschke et al. 2021; Shroff et al. 2008). However, it is unknown if such trait co-occurrences arise from environment-imposed, phenotypic, or genetic constraints (Després et al. 2007). Theoretical modeling has shown that such combined strategies may confer fitness advantages when traits are associated with ever-rising costs and the probability of ingesting certain toxins is low (Vacher et al. 2005). Genetic costs may be particularly pronounced when TSI-conferring mutations evolve, especially when the target proteins of toxins are active in the nervous system. TSI-conferring mutations can incur costs when they lower the efficiency of a protein in the herbivore (Després et al. 2007). We have observed this in experiments with *D. melanogaster*, when substitutions conferring TSI of the Na+/K+-ATPase to CGs that have evolved in the monarch and other specialists on milkweeds were introduced in flies (Karageorgi et al. 2019; Taverner et al. 2019). While the substitutions heightened insect resistance to CGs, they also appear to have caused pleiotropic nervous system defects. These potential defects were ameliorated through epistasis when accompanied by a facilitating or compensatory substitution near the TSI-conferring substitutions in the first extracellular loop of the Na+/K+-ATPase (Karageorgi et al. 2019; Taverner et al. 2019). Contrary to other toxin resistance traits, the costs of TSI are fixed, i.e., they do not change with the probability of dietary toxin ingestion (Després et al. 2007). However, these costs can be modulated through epistatic interactions with genetic variation elsewhere in the herbivore genome and by environmental fluctuations.

A second general pattern is that costs and benefits of toxin resistance traits in herbivores can be phenotypically plastic. Generalists, and to a lesser extent specialists, are presented with highly variable levels and diverse combinations of toxins both across and within host plant species (Després et al. 2007). The within-species variability is partially under genetic control by the plant and partially by factors such as the plant's phenological stage and fluctuations in biotic and abiotic factors it encounters. To the extent that this is controlled by genetics (Fig. 1a), such variability may be an evolved plant strategy that follows the moving target theory or, perhaps more likely, the optimal defense theory, since it is thought to increase costs for the herbivore to acclimate its gut milieu and other traits as cocktails of dietary toxins change in composition, causing the herbivore population to always be chasing moving fitness optima (Wetzel and Thaler 2016; Li et al. 2020a, b). A study with artificial diets with variable levels of the furanocoumarin xanthotoxin presented to caterpillars of the generalist *Trichoplusia ni* provides support for this notion that toxin level variability suppresses herbivore performance (Pearse et al. 2018).

In response to variable toxin levels in host plants, generalists have evolved toxin-induced avoidance behaviors and both constitutive and induced production of detoxification enzymes (Després et al. 2007). TSI, on the other hand, is typically restricted to specialist herbivores that use it alongside more generalized toxin resistance mechanisms. The use of more than one resistance mechanism may confer robustness to the efforts of specialists to deal with host plant toxins. This strategy might also prevent specialization from becoming an evolutionary "dead end" if host plant populations dwindle, in which case shifts to novel host plants might be necessary (Termonia et al. 2001).

Examples from specialists on CG-producing plants illustrate how herbivore adaptations to the presence of certain toxins in their host plants may facilitate shifts to other plant species producing those toxins. Our reconstructions of host plant usage of herbivorous insects revealed that in three independent instances among the Coleoptera, Diptera, and Lepidoptera, close relatives of specialists on CG-producing plants in the Apocynaceae were feeding on *Solanum* spp. (Solanaceae) plants (Fig. 4; Begon 1975; Brown 1987; Schoville et al. 2018). Intriguingly, the species feeding on *Solanum* spp. hosts all possess one or more substitutions in the Na+/K+-ATPase that confer TSI to CGs (Karageorgi et al. 2019). While only a subset of Solanaceae plant species are known to produce CGs like the Apocynaceae, Solanaceae produce saponins such as glycoalkaloids and steroidal glycosides (Pomilio et al. 2008), and there is some evidence that these may inhibit Na+/

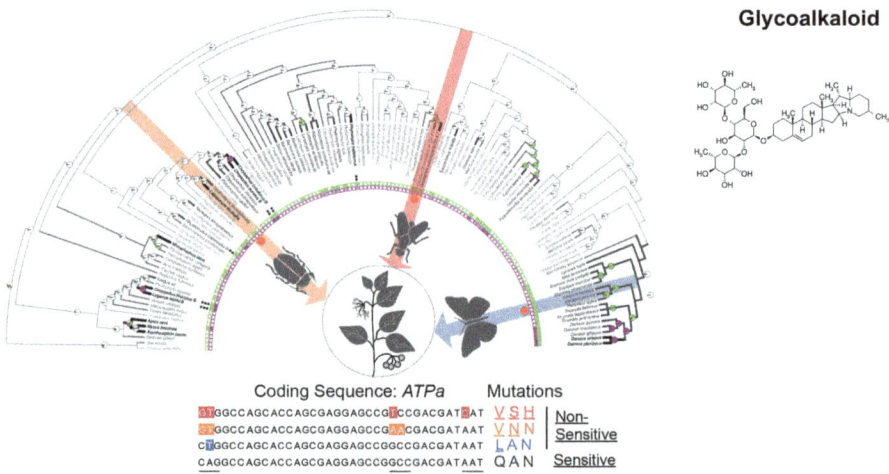

Fig. 4 Mutations in three codons of the Na+/K+ATPase alpha subunit gene *ATPa* (highlighted in the sequences above the sequence of *D. melanogaster* as a reference species without target site insensitivity mutations in the bottom) have evolved at least three times (red dots) in insects from different orders that feed on plant species of the nightshade family (Solanaceae) (center). These insects are weakly or completely non-sensitive to the steroidal toxins that the plants produce. The known species are the nymphalid butterfly *Mechanitis polymnia* (blue, with mutations causing codon changes to amino acids L, A, and N at positions 111, 119, and 122 of the Na+/K+ATPase alpha subunit, respectively, appearing as if the mutations were introduced into the *D. melanogaster* sequence), the "fruit" fly *D. subobscura* (red, with mutations causing codon changes to amino acids V, S, and H), and the Colorado potato beetle *Leptinotarsa decemlineata* (orange, with mutations causing codon changes to amino acids V, N, and N). (Cartoon by Sophie Zaaijer)

K+-ATPase as well (Blankemeyer et al. 1995). This sets up a potential mechanism of cross-resistance that could facilitate host switches between Solanaceae and CG-producing Apocynaceae plants, which may be facilitated further by the activity of conserved, generalized toxin resistance mechanisms such as the expression of multidrug transporters in the midgut and BBB of all of these species (Fig. 3; Dobler et al. 2015; Groen et al. 2017). Indeed, the milkweed butterfly clade (Danainae) is the sister group of Ithomiinae, which are specialists on the Solanaceae. The latter clade includes *Mechanitis polymnia*, which has a somewhat CG-insensitive Na+/K+-ATPase (Petschenka et al. 2013; Karageorgi et al. 2019). It is likely that host switching between Solanaceae and Apocynaceae has occurred (Brown 1987).

Although fluctuating dynamics and "trench warfare" dynamics are yet to be studied in the context of plant-herbivore interactions (Gloss et al. 2013), dynamics that resemble arms races have been examined in several plant-lepidopteran herbivore study systems. Among them are the well-studied interactions between Brassicaceae plants and their herbivore communities, which include *Pieris* spp. (Edger et al. 2015; Griese et al. 2021); between milkweeds and their herbivore communities, including monarch and queen butterflies (Agrawal and Fishbein 2008; Agrawal et al. 2021); and between wild parsnip and the herbivores *Depressaria pastinacella* and *Papilio polyxenes* (Berenbaum and Feeny 1981; Berenbaum and Zangerl 1998).

Potential mechanisms for how arms race dynamics may lead to co-diversification between herbivore and host plant species have been studied in most detail for the pierid butterflies and their Brassicales host plants (Edger et al. 2015). Here we review the herbivore side of these interactions. *Pieris* spp. contain the NSPs, which are part of the NSP-like gene family that also includes the NSP paralog, the major allergen (MA) protein. These proteins are unique to pierids and are related to the single domain major allergen (SDMA) proteins, which are generally expressed in the gut systems of caterpillars (Fischer et al. 2008). Like NSPs, MAs can also disarm the mustard oil bomb (Edger et al. 2015). Based on experimental work and comparative analyses, it appears that the *Pieris* spp. maintain a breadth of potential host plant species while specializing on a smaller subset of hosts through gene duplications and subsequent sub- or neo-functionalization of NSPs and MAs. While NSPs show more stable expression, they have experienced positive selection related to specialization on different host plants with unique GSL profiles (Heidel-Fischer et al. 2010; Okamura et al. 2019a,b). MAs showed GSL-inducible expression but were more evolutionarily stable and are perhaps involved in detoxification of those GSLs that are produced more commonly among the host plants of the Pieridae (Okamura et al. 2019a,b). Like the NSPs, the horizontally transferred CAS/CYS enzymes also underwent further duplication in *Pieris* spp. and other species feeding on cyanogenic plants compared to lepidopteran species not feeding on such plants (Li et al. 2021a, b). This may have further facilitated the ability of *Pieris* spp. to handle the formation of equimolar levels of cyanide upon the breakdown of GSLs to nitriles (Steiner et al. 2018). In particular, the number of BSAS genes encoding the CAS/CYS enzymes showed a stepwise increase as species specialized onto Brassicaceae host plants with BSAS2, which shows high affinity for cyanide, generally present in all Lepidoptera; while BSAS3 is restricted to the Pieridae, and

BSAS1 is restricted even further to the Pierinae (Herfurth et al. 2017). The CAS/CYS enzymes may be complemented in their role of cyanide detoxification by two rhodaneses, which may add robustness to the detoxification process. The rhodanese-encoding genes, *TST1* and *TST2*, differ in their expression, subcellular localization, and kinetic properties and are the result of a rhodanese family expansion in the Pieridae (Herfurth et al. 2017; Steiner et al. 2018).

However, such arms race dynamics between Brassicales specialists and their host plants are not a given: *Plutella xylostella*'s genome encodes three GSSs that stem from duplications of insect arylsulfatases (Heidel-Fischer et al. 2019). Each GSS has distinct expression patterns in response to dietary GSLs and mediates desulfation of different types of GSLs with varying efficiency. Rather than showing signatures of arms race coevolution early after duplication from an arylsulfatase gene and evolving in a stepwise manner, copies of GSS genes neofunctionalized in parallel under positive selection caused by the herbivore's host shift to GSL-producing plants while gaining their novel detoxification functions (Heidel-Fischer et al. 2019).

Interestingly, aside from *D. radiella*, all Lepidoptera in these examples are multivoltine (Hazel 1977; Berenbaum and Zangerl 1991; Brower 1998; Fei et al. 2014; Agrawal 2017; Moranz and Rahman et al. 2019). The herbivores thus have the potential to evolve faster than their host plants, which have no more than one generation per year. This discrepancy sets up an apparent paradox: how are host plants able to prevent losing out in these arms races? A first potential reason might be that defense or, alternatively, loss of susceptibility is relatively more straightforward for the host plant than using a plant as a host is for the herbivore (Thompson 1986). For example, the most abundant sterol in herbivorous insects is cholesterol, but insects rely on plant-produced sterols to synthesize it. Changes in sterol profiles may not have apparent fitness consequences in host plants (Corbin et al. 2001) but could provide effective loss of susceptibility to herbivores, with relative cholesterol levels and larval survival deteriorating the most in a host plant specialist (Jing et al. 2012, 2013). A second potential reason is that escalation of arms races comes with the production of novel defenses by host plants, and being able to combine defensive traits may give plants an evolutionary advantage (Gilman et al. 2012; Speed et al. 2015). A third potential reason is that coevolution can be diffuse. For example, because of its migratory lifestyle, the monarch butterfly encounters multiple species of milkweed hosts. This may pose a limitation to the monarch for evolving more efficient mechanisms of handling the CGs and other toxins produced by any one milkweed species (Agrawal et al. 2021). A fourth potential reason is that herbivores are attacked by natural enemies in the form of pathogens, parasites, parasitoids, and predators, and top-down control by these organisms may dampen the negative effects that herbivore populations may have on host plant populations. It is possible that natural selection becomes less efficient if effective population sizes are reduced. Finally, interactions between the first and third trophic levels lead to trade-offs that prevent herbivores from adapting strictly to plant defenses. We will now take a more detailed look at the effects of these top-down agents of selection on herbivore-plant interactions.

Top-Down Agents of Selection

Organisms that are natural enemies of caterpillars and other lepidopteran life stages not only form independent agents of selection by consuming their prey partially or wholly (Bernays 1997) but also influence caterpillar fitness in conjunction with bottom-up, host plant-derived agents of selection (Bernays and Graham 1988; Lill et al. 2002; Thaler et al. 2012a, b; Kaplan et al. 2014; Singer et al. 2014). For these effects to occur, caterpillars do not need to experience attack directly; even the perceived threat of attack may cause caterpillars, including *Pieris rapae* and the monarch, to become less efficient at dealing with plant defensive chemicals (Lund et al. 2020; Lee et al. 2021). In addition, plant toxin level variability may not only affect herbivore performance from the bottom-up but may influence top-down selection as well. *Trichoplusia ni* caterpillars ingesting higher dietary levels of the furanocoumarin xanthotoxin were attacked at lower rates by the parasitoid wasp *Copidosoma floridanum* (Paul et al. 2020). Interactive top-down and bottom-up effects can even be modulated further by viruses, microbes, and parasites of the natural enemies, showing the ecological complexities (Harvey et al. 2003; Tan et al. 2018).

Among Brassicales specialist herbivores, the effects of plant-produced GSLs and their breakdown products on multi-trophic interactions appear to be species dependent. For example, the performance of an endoparasitoid *Diadegma semiclausum* was negatively correlated with GSL concentrations, as the wasp developed better when caterpillars of its host *Plutella xylostella* were actively detoxifying GSLs via desulfation (Sun et al. 2020). In contrast, the performance of the endoparasitoid *Hyposoter ebeninus* was positively correlated with higher GSL concentrations of the Brassicaceae plants that their hosts, caterpillars of *Pieris rapae* and *Spodoptera exigua*, were feeding on (Kos et al. 2012). The authors speculated that this may have been caused by negative effects of plant GSLs on caterpillar immunity against the parasitoid (see also Smilanich and Muchoney, Chapter "Host Plant Effects on the Caterpillar Immune Response").

Interactive effects between host plant defensive chemicals and the insect immune system were also invoked to explain population-specific patterns of selection on immunity genes in the monarch butterfly (Tan et al. 2019a, b). While the North American population of monarchs predominantly uses the common milkweed *Asclepias syriaca* as larval host plant, caterpillars of monarch populations outside North America typically feed on other milkweed species, including *A. curassavica*. This species and other alternative milkweed hosts outside North America contain higher CG concentrations. Such elevated CG levels are known to affect the success rate of infection by the protozoan parasite *Ophryocystis elektroscirrha* (Sternberg et al. 2012; Gowler et al. 2015; Tao et al. 2016) and may also influence performance of other pathogens, predators, and parasites of the monarch (Brower et al. 1967, 1968). The use of dietary CGs in defense against attack could in principle lead to relaxation of selection on the monarch's immune system genes, especially when their expression is accompanied by costs (de Roode et al. 2013; Gerardo et al. 2010; Parker et al. 2011).

One mechanism by which the monarch and many other specialist herbivores on a variety of host plants minimize fitness losses from attack by natural enemies is through sequestration of plant defensive chemicals (see Bowers, Chapter "Sequestered Caterpillar Chemical Defenses: From "Disgusting Morsels" to Model Systems"). However, sequestration of these chemicals comes with a set of challenges. For chewing herbivores such as caterpillars, this is particularly true in the case of sequestering plant-produced, non-toxic precursor glucoside molecules such as GSLs that are hydrolyzed by plant-derived β-glucosidases upon herbivore feeding. Herbivores would need to leave GSLs intact if they are to evolve GSL storage and the ability to set up their own mustard oil bomb. Perhaps not surprisingly, the first well-studied instance of GSL sequestration was for an aphid species that specializes on Brassicaceae, *Brevicoryne brassicae* (Kazana et al. 2007). As a piercing-sucking herbivore, it can leave at least the aliphatic GSL intact, allowing it to store GSLs in its body. It further produces its own myrosinase enzyme in separate compartments, which is brought in contact with the GSLs upon wounding, thereby effectively setting itself up as a "booby trap" to predators and parasites. However, chewing herbivores, including caterpillars, may not have easy access to this option, given the amount of tissue disruption they bring about. Sequestration of intact and/or modified GSL by chewing herbivores has thus far only been reported outside Lepidoptera: in larvae of the sawfly *Athalia rosae* (Hymenoptera; Müller et al. 2001) and in the flea beetle *Phyllotreta armoraciae* (Coleoptera; Sporer et al. 2021). In the sawfly, GSL breakdown in the gut appears to be prevented by rapid GSL uptake across the epithelium, which may be facilitated by low activity of plant myrosinases in the anterior gut (Abdalsamee et al. 2014). The flea beetle appears to employ similar mechanisms and may have an additional mechanism to reduce activity of plant myrosinases in the gut to trace levels (Sporer et al. 2021). Intriguingly, *P. armoraciae* can supercharge GSL sequestration via 13 putative sugar porters in the major facilitator superfamily (MFS) that import GSLs (Yang et al. 2021a, b). These proteins, dubbed glucosinolate-specific transporters (GTRs), show expression predominantly in the Malpighian tubules, and silencing them via RNAi showed that GTR activity in the tubules enabled the beetles to sequester high GSL levels in their hemolymph (Yang et al. 2021a, b). Characterization of sugar transporters has started in the moths *Bombyx mori* and *Helicoverpa armigera* (Govindaraj et al. 2016; Yuan et al. 2021a, b), and it will be interesting to see their characterization in Brassicaceae-specializing lepidopterans such as *Pieris* spp. and *Plutella* spp. It could also be fruitful to study ABC transporters in caterpillars of Brassicaceae specialists since at least one of these broad-spectrum transporters, the C-type ABC transporter MRP, has already been shown to mediate toxin sequestration in another beetle, *Chrysomela populi* (Strauss et al. 2013).

While Brassicales specialists such as *Pieris brassicae* and *P. rapae* do not appear to sequester intact GSLs (Müller et al. 2003), *P. brassicae* caterpillars do show attack-induced production of an intensely green regurgitant (that likely contains high levels of nitriles), which has been shown to act as a deterrent to *Myrmica rubra* ants. These observations further suggest that nitriles may have a defensive role for *P. brassicae* and could come with adaptive benefits (Müller et al. 2003). Sequestration of nitriles might even bring more benefits to herbivores in some interactions with natural enemies than the ability to release ITCs. When GSL desulfation in *Plutella xylostella* was disrupted by silencing its *GSS* genes via RNAi, the caterpillars systemically accumulated ITCs (Sun et al. 2019). Not only did the ITCs impair caterpillar development, but the larvae were still efficiently captured and eaten by the lacewing *Chrysoperla carnea*, a predator able to degrade ingested ITCs via the mercapturic acid pathway (Sun et al. 2019).

Specialists on CG-producing plants may have easier paths to evolve sequestration since these dietary toxins come to the herbivores in stable form. A series of different studies over the last 50 years using a variety of approaches have elucidated an important part of the genetic, molecular, and physiological mechanisms underlying CG sequestration in the monarch. Several studies with monarch butterflies reared on milkweeds (including *Asclepias curassavica* and *A. fruticosa* as host plants) demonstrated that the monarch may selectively avoid sequestration of more toxic apolar CGs such as voruscharin, a compound to which its Na+/K+-ATPase is sensitive, despite the monarch's TSI mutations. The monarch preferentially sequesters the less toxic polar CGs calotropin and calactin, compounds to which the TSI mutations provide >50-fold relative increase in resistance (Reichstein et al. 1968; Roeske et al. 1976; Seiber et al. 1980, 1983; Cheung et al. 1988; Groeneveld et al. 1990; Malcolm 1990; Nelson 1993; Malcolm 1995; Petschenka et al. 2018; Jones et al. 2019; Agrawal et al. 2021). The monarch achieves this biased sequestration in part through converting voruscharin into calotropin and calactin via non-enzymatic and enzymatic steps (Agrawal et al. 2021; Marty and Krieger 1984; Seiber et al. 1980) and through transporting CGs via as-of-yet unknown carriers (Frick and Wink 1995). New experimental work should identify these CG carriers in the monarch; past studies have identified a set of candidate carriers. Kowalski and co-workers recently identified that the B-type ABC transporters ABCB1-3 may allow the dogbane beetle *Chrysochus auratus*, a specialist on the CG-producing plant *Apocynum cannabinum*, to sequester calotropin and other CGs (Kowalski et al. 2020). Interestingly, the most efficient transporters of calotropin were ABCB2 and -3, which are most closely related to *D. melanogaster* Mdr50 (Groen et al. 2017). It is precisely in orthologs of Mdr50 that we observed a gene bloom in the monarch genome (Fig. 3). From data produced by several population genetic/genomic studies, it can be observed that the monarch population does not seem to show genetic variation for the TSI mutations (Aardema et al. 2012; Zhan et al. 2014, Pierce et al. 2016), but does show genetic variation for sequestration (Freedman et al. 2020). It will be interesting to see if this genetic variation may be found in and around genes that code for CG detoxification enzymes, CG carriers, and/or other proteins that may be involved in sequestration.

Evolution of the substitutions in the monarch's Na+/K+-ATPase that confer TSI to many, but not all, CGs appears to have followed arms race dynamics (Aardema et al. 2012; Petschenka et al. 2013; Petschenka and Agrawal 2015; Pierce et al. 2016). The latest escalation (at least as far as major effect substitutions in the first extracellular loop are concerned) was the addition of substitution N122H. This step was most likely linked to CG sequestration, rather than merely coping with the toxins as part of the diet (Petschenka and Agrawal 2015). N122H was not necessary for protecting caterpillars against CG toxicity when toxins were ingested with the diet, but the substitution allowed tolerance of CGs when hemolymph with sequestered CGs from the monarch was injected into the body cavity (Petschenka and Agrawal 2015). Intriguingly, not all CG-sequestering lepidopteran species have evolved accompanying TSI substitutions. For example, larvae of several species of arctiid moths sequester CGs, but their Na+/K+-ATPases do not harbor TSI substitutions (Petschenka et al. 2012; Petschenka and Agrawal 2015). This suggests that costs of N122H and other TSI substitutions may be high and would need to be offset by compensatory mechanisms and/or ecological benefits. Our own and our collaborators' work with *D. melanogaster* has shown that the monarch's TSI substitutions indeed come with substantial costs in the form of imbalances in nervous system functioning (Karageorgi et al. 2019). Exactly how the monarch nullifies these deleterious side effects is unknown, but one mechanism is the evolution of a facilitating substitution in the form of A119S that offsets the negative pleiotropic consequences of N122H (Karageorgi et al. 2019). For sequestration to evolve, other (potential) costs need to be overcome. Agrawal and colleagues recently measured significant CG sequestration costs for monarch caterpillars that were evident in reduced growth rates (Agrawal et al. 2021). Slower growth may have been caused by the burden of energetic costs that selective detoxification and transport mechanisms may incur (Després et al. 2007). Ultimately, the sum total of all costs needs to be lower than the ecological benefits of sequestration in the form of lower predation rates, which will depend on local environmental constraints (Després et al. 2007). Reduced predation due to sequestration is certainly possible for the monarch in at least some locations and conditions as several studies with natural enemies have shown (Brower et al. 1967, 1968), and this fits within a more general pattern that toxin-sequestering specialists are measurably better defended against predation than generalists (Zvereva and Kozlov 2016).

A meta-analysis of 159 publications on the costs and benefits of toxin accumulation in herbivores further revealed that chemical defenses were generally beneficial when herbivores are threatened by generalist predators, but not when threatened by specialist predators or generalist and specialist parasitoids (Zvereva and Kozlov 2016) (see also Singer et al., Chapter "Predators and Caterpillar Diet Breadth: Appraising the Enemy-Free Space Hypothesis"). Furthermore, chemical defenses were more effective against vertebrate predators, particularly birds, compared to invertebrate predators (Zvereva and Kozlov 2016). Studies with different types of natural enemies of the monarch show patterns that are broadly consistent with this (Brower et al. 1967, 1968, 1985; Fink and Brower 1981; Fink et al. 1983; Brower and Calvert 1985; Brower 1988; Glendinning et al. 1988; Glendinning and Brower 1990; Glendinning 1992; Koch et al. 2003; Rafter et al. 2013; Hermann et al. 2019; Stenoien et al. 2019).

One important mechanism through which sequestering species may enhance the benefits of sequestration is evolving aposematism (see also Bowers, Chapter "Sequestered Caterpillar Chemical Defenses: From "Disgusting Morsels" to Model

Systems"). The monarch and other sequestering specialist herbivores have evolved warning coloration as a corollary to their accumulation of protective toxins that serves to advertise the herbivores' toxicity and can prevent attacks from happening, especially when vertebrate predators are a threat (Zvereva and Kozlov 2016). A first population genomic study has identified part of the genetic basis of the monarch's orange-and-black warning coloration (Zhan et al. 2014). Future studies may more fully characterize the genetic architecture of the monarch's CG detoxification- and sequestration-related traits and determine the extent of genetic correlation with its aposematic colorations. In this way, the evolution of the monarch's mechanisms to deal with bottom-up and top-down selection pressures can be understood more completely.

The herbivores on which we have focused, the Brassicaceae specialist pierid butterflies and the milkweed butterflies, and their mechanisms of handling host plant-produced toxins are fitting illustrations of broader patterns concerning the role of defensive chemical detoxification and sequestration for caterpillars to navigate interactions with selective agents at lower and higher trophic levels. A meta-analysis of 112 studies found that effect sizes of top-down selection pressures were generally larger than those of bottom-up selection pressures (Vidal and Murphy 2018). However, for specialist chewing herbivores, this pattern was turned upside down, which suggests that mechanisms such as the sequestration of host plant defensive chemicals in defense against natural enemies could have alleviated top-down selection pressures. An illustration of this pattern was found in a study of the insect community around *Brassica nigra* and *B. oleracea* plants: in this community, where specialist herbivores were more abundant than generalists, bottom-up selection had a larger influence on herbivore abundance than top-down selection (Kos et al. 2011).

Finally, it is interesting to contemplate the role that climate change may play in influencing the ecology and evolution of detoxification phenotypes *sensu lato*. For example, experimental increases in temperature raised cardenolide levels in foliage of *A. curassavica*, a species now widespread in the southern USA, that may be causing a reduction in the proportion of migrating monarchs (Faldyn et al. 2018). There is some concern that, owing to the fitness reduction monarchs experienced when feeding on plants grown in experimentally warmed conditions, these butterflies could become caught in an ecological trap. Adult female monarchs in the southern USA prefer to oviposit on *A. curassavica*, and as the climate warms, so too should cardenolide levels rise in these plants. Although higher cardenolide levels tend to enhance protection from natural enemies, there are also costs to sequestration, and overall, this could reduce average fitness of monarchs in these populations. Unconsidered by Faldyn et al. (2018) is the potential role for an evolutionary response in such scenarios. Adaptation in the populations of monarchs facing higher cardenolide concentrations owing to warming conditions could produce any variety of adaptations, including reduced preference for *A. curassavica*, mitigation of the higher cardenolide levels physiologically, and/or increased resistance or tolerance of cardenolides that are particularly toxic. On the other hand, higher temperatures directly reduce fitness as well (York and Oberhauser 2002). This one example highlights the difficulty in predicting the impacts of climate change at the plant-insect nexus. More research in this area is certainly needed, especially in the area of adaptation per se.

References

Aardema ML, Zhen Y, Andolfatto P (2012) The evolution of cardenolide-resistant forms of Na+, K+-ATPase in Danainae butterflies. Mol Ecol 21:340–349
Abdalsamee MK, Giampà M, Niehaus K et al (2014) Rapid incorporation of glucosinolates as a strategy used by a herbivore to prevent activation by myrosinases. Insect Biochem Mol Biol 52:115–123
Acevedo FE, Peiffer M, Tan CW et al (2017) Fall armyworm-associated gut bacteria modulate plant defense responses. Mol Plant-Microbe Interact 30:127–137
Afroz A, Howlett N, Shukla A et al (2013) Gustatory receptor neurons in *Manduca sexta* contain a TrpA1-dependent signaling pathway that integrates taste and temperature. Chem Senses 38:605–617
Agrawal AA (2005) Natural selection on common milkweed (*Asclepias syriaca*) by a community of specialized insect herbivores. Evol Ecol Res 7:651–667
Agrawal AA (2017) Monarchs and milkweed. Princeton University Press, Princeton
Agrawal AA, Fishbein M (2008) Phylogenetic escalation and decline of plant defense strategies. Proc Natl Acad Sci USA 105:10057–10060
Agrawal AA, Kurashige NS (2003) A role for isothiocyanates in plant resistance against the specialist herbivore *Pieris rapae*. J Chem Ecol 29:1403–1415
Agrawal AA, Petschenka G, Bingham RA et al (2012) Toxic cardenolides: chemical ecology and coevolution of specialized plant–herbivore interactions. New Phytol 194:28–45
Agrawal AA, Böröczky K, Haribal M et al (2021) Cardenolides, toxicity, and the costs of sequestration in the coevolutionary interaction between monarchs and milkweeds. Proc Natl Acad Sci USA 118:e2024463118
Ahmed SMH, Maldera JA, Krunic D et al (2020) Fitness trade-offs incurred by ovary-to-gut steroid signalling in *Drosophila*. Nature 584:415–419
Ahn SJ, Badenes-Pérez FR, Heckel DG (2011a) A host-plant specialist, *Helicoverpa assulta*, is more tolerant to capsaicin from *Capsicum annuum* than other noctuid species. J Insect Physiol 57:1212–1219
Ahn SJ, Badenes-Pérez FR, Reichelt M et al (2011b) Metabolic detoxification of capsaicin by UDP-glycosyltransferase in three *Helicoverpa* species. Arch Insect Biochem Physiol 78:104–118
Ahn SJ, Vogel H, Heckel DG (2012) Comparative analysis of the UDP-glycosyltransferase multigene family in insects. Insect Biochem Mol Biol 42:133–147
Andrew D, Gloss Anna C, Nelson DB, Goldman-Huertas NK, Whiteman (2013) Maintenance of genetic diversity through plant–herbivore interactions. Current Opinion in Plant Biology 16(4):443–450 https://doi.org/10.1016/j.pbi.2013.06.002
Ana, Depetris-Chauvin D, Galagovsky Y, Grosjean (2015) Chemicals and chemoreceptors: ecologically relevant signals driving behavior in *Drosophila*. Frontiers in Ecology and Evolution 310.3389/fevo.2015.00041
Anja S, Strauss Sven, Peters Wilhelm, Boland Antje, Burse (2013) ABC transporter functions as a pacemaker for sequestration of plant glucosides in leaf beetles. eLife 210.7554/eLife.01096
Al-Anzi B, Tracey WD Jr, Benzer S (2006) Response of *Drosophila* to wasabi is mediated by painless, the fly homolog of mammalian TRPA1/ANKTM1. Curr Biol 16:1034–1040
Allio R, Nabholz B, Wanke S et al (2021) Genome-wide macroevolutionary signatures of key innovations in butterflies colonizing new host plants. Nat Commun 12:354

Allison K., Hansen Nancy A., Moran (2014) The impact of microbial symbionts on host plant utilization by herbivorous insects. Molecular Ecology 23(6):1473–1496 https://doi.org/10.1111/mec.12421

Amezian D, Nauen R, Le Goff G (2021) Transcriptional regulation of xenobiotic detoxification genes in insects-an overview. Pestic Biochem Physiol 174:104822

Barbehenn RV (1999) Non-absorption of ingested lipophilic and amphiphilic allelochemicals by generalist grasshoppers: The role of extractive ultrafiltration by the peritrophic envelope. Arch Insect Biochem Physiol 42:130–137

Barbehenn RV (2001) Roles of peritrophic membranes in protecting herbivorous insects from ingested plant allelochemicals. Arch Insect Biochem Physiol 47:86–99

Bautista MAM, Bhandary B, Wijeratne AJ et al (2015) Evidence for trade-offs in detoxification and chemosensation gene signatures in *Plutella xylostella*. Pest Manag Sci 71:423–432

Begon M (1975) The relationships of *Drosophila obscura* Fallen and *D. subobscura* Collin to naturally-occurring fruits. Oecologia 20:255–277

Bekaert M, Edger PP, Hudson CM et al (2012) Metabolic and evolutionary costs of herbivory defense: systems biology of glucosinolate synthesis. New Phytol 196:596–605

Benton R, Sachse S, Michnick SW et al (2006) Atypical membrane topology and heteromeric function of *Drosophila* odorant receptors in vivo. PLoS Biol 4:e20

Benton R, Vannice KS, Gomez-Diaz C et al (2009) Variant ionotropic glutamate receptors as chemosensory receptors in *Drosophila*. Cell 136:149–162

Beran F, Sporer T, Paetz C et al (2018) One pathway is not enough: The cabbage stem flea beetle *Psylliodes chrysocephala* uses multiple strategies to overcome the glucosinolate-myrosinase defense in its host plants. Front Plant Sci 9:1754

Berenbaum MR (1980) Adaptive significance of midgut pH in larval Lepidoptera. Am Nat 115:138–146

Berenbaum MR (1983) Coumarins and caterpillars: a case for coevolution. Evolution 37:163–179

Berenbaum MR (1986) Target site insensitivity in insect-plant interactions. In: Molecular aspects of insect-plant associations. Springer, Boston, pp 257–272

Berenbaum M, Feeny P (1981) Toxicity of angular furanocoumarins to swallowtail butterflies: escalation in a coevolutionary arms race? Science 212:927–929

Berenbaum MR, Zangerl AR (1991) Acquisition of a native hostplant by an introduced oligophagous herbivore. Oikos:153–159

Berenbaum MR, Zangerl AR (1993) Furanocoumarin metabolism in *Papilio polyxenes*: biochemistry, genetic variability, and ecological significance. Oecologia 95:370–375

Berenbaum MR, Zangerl AR (1994) Costs of inducible defense: protein limitation, growth, and detoxification in parsnip webworms. Ecology 75:2311–2317

Berenbaum MR, Zangerl AR (1998) Chemical phenotype matching between a plant and its insect herbivore. Proc Natl Acad Sci USA 95:13743–13748

Bergelson J, Purrington CB (1996) Surveying patterns in the cost of resistance in plants. Am Nat 148:536–558

Bernays EA (1997) Feeding by lepidopteran larvae is dangerous. Ecol Entomol 22:121–123

Bernays EA, Graham M (1988) On the evolution of host specificity in phytophagous arthropods. Ecology 69:886–892

Blankemeyer JT, Atherton R, Friedman M (1995) Effect of potato glycoalkaloids alpha-chaconine and alpha-solanine on sodium active transport in frog skin. J Agric Food Chem 43:636–639

Braby MF, Trueman JWH (2006) Evolution of larval host plant associations and adaptive radiation in pierid butterflies. J Evol Biol 19:1677–1690

Brooks GT (1976) Penetration and distribution of insecticides. In: Wilkinson CF (ed) Insecticide biochemistry and physiology. Plenum Publishing Corporation, New York, pp 3–60

Brower LP (1988) Avian predation on the monarch butterfly and its implications for mimicry theory. Am Nat 131:S4–S6

Brower LP, Calvert WH (1985) Foraging dynamics of bird predators on overwintering monarch butterflies in Mexico. Evolution 39:852–868

Brower LP, Van Zandt BJ, Corvino JM (1967) Plant poisons in a terrestrial food chain. Proc Natl Acad Sci USA 57:893–898
Brower LP, Ryerson WN, Coppinger LL et al (1968) Ecological chemistry and the palatability spectrum. Science 161:1349–1350
Brower LP, Horner BE, Marty MA et al (1985) Mice (*Peromyscus maniculatus, P. spicilegus*, and *Microtus mexicanus*) as predators of overwintering monarch butterflies (*Danaus plexippus*) in Mexico. Biotropica 17:89–99
Brown KS Jr (1987) Chemistry at the Solanaceae/Ithomiinae interface. Ann Missouri Bot Gard 74:359–397
Cai LJ, Zheng LS, Huang YP et al (2020) Identification and characterization of odorant binding proteins in the diamondback moth, *Plutella xylostella*. Insect Sci. https://doi.org/10.1111/1744-7917.12817
Calla B, Wu WY, Dean CAE et al (2020) Substrate-specificity of cytochrome P450-mediated detoxification as an evolutionary strategy for specialization on furanocoumarin-containing hostplants: CYP6AE89 in parsnip webworms. Insect Mol Biol 29:112–123
Callaway EC, Zhang Y, Chew W et al (2004) Cellular accumulation of dietary anticarcinogenic isothiocyanates is followed by transporter-mediated export as dithiocarbamates. Cancer Lett 204:23–31
Chahine S, O'Donnell MJ (2009) Physiological and molecular characterization of methotrexate transport by Malpighian tubules of adult *Drosophila melanogaster*. J Insect Physiol 55:927–935
Chahine S, O'Donnell MJ (2011) Interactions between detoxification mechanisms and excretion in Malpighian tubules of Drosophila melanogaster. J Exp Biol 214:462–468
Chen S, Lu M, Zhang N et al (2018) Nuclear factor erythroid-derived 2–related factor 2 activates glutathione S-transferase expression in the midgut of *Spodoptera litura* (Lepidoptera: Noctuidae) in response to phytochemicals and insecticides. Insect Mol Biol 27:522–532
Chen W, Dong Y, Saqib HSA et al (2020) Functions of duplicated glucosinolate sulfatases in the development and host adaptation of *Plutella xylostella*. Insect Biochem Mol Biol 119:103316
Chen C, Chen H, Huang S et al (2021) Volatile DMNT directly protects plants against *Plutella xylostella* by disrupting peritrophic matrix barrier in midgut. elife 10:e63938
Cheung HA, Nelson CJ, Watson TR (1988) New glucoside conjugates and other cardenolide glycosides from the monarch butterfly reared on *Asclepias fruticosa* L. J Chem Soc Perkin Trans 1:1851–1857
Chung SH, Rosa C, Scully ED et al (2013) Herbivore exploits orally secreted bacteria to suppress plant defenses. Proc Natl Acad Sci USA 110:15728–15733
Clay NK, Adio AM, Denoux C et al (2009) Glucosinolate metabolites required for an *Arabidopsis* innate immune response. Science 323:95–101
Cohen MB, Schuler MA, Berenbaum MR (1992) A host-inducible cytochrome P-450 from a host-specific caterpillar: molecular cloning and evolution. Proc Natl Acad Sci USA 89:10920–10924
Coley PD, Bryant JP, Chapin FS (1985) Resource availability and plant antiherbivore defense. Science 230:895–899
Corbin DR, Grebenok RJ, Ohnmeiss TE et al (2001) Expression and chloroplast targeting of cholesterol oxidase in transgenic tobacco plants. Plant Physiol 126:1116–1128
Cresswell JE, Merritt SZ, Martin MM (1992) The effect of dietary nicotine on the allocation of assimilated food to energy metabolism and growth in fourth-instar larvae of the southern armyworm, *Spodoptera eridania* (Lepidoptera: Noctuidae). Oecologia 89:449–453
Dąbrowska P, Freitak D, Vogel H et al (2009) The phytohormone precursor OPDA is isomerized in the insect gut by a single, specific glutathione transferase. Proc Natl Acad Sci USA 106:16304–16309
Dawkins R, Krebs JR (1979) Arms races between and within species. Proc R Soc B 205:489–511
de Roode JC, Lefèvre T, Hunter MD (2013) Self-medication in animals. Science 340:150–151
Denecke SM, Driva O, Luong HNB et al (2020) The identification and evolutionary trends of the solute carrier superfamily in arthropods. Genome Biol Evol 12:1429–1439

Dermauw W, Van Leeuwen T (2014) The ABC gene family in arthropods: comparative genomics and role in insecticide transport and resistance. Insect Biochem Mol Biol 45:89–110

Dermauw W, Van Leeuwen T, Feyereisen R (2020) Diversity and evolution of the P450 family in arthropods. Insect Biochem Mol Biol 127:103490

Després L, David JP, Gallet C (2007) The evolutionary ecology of insect resistance to plant chemicals. Trends Ecol Evol 22:298–307

Diezel C, von Dahl CC, Gaquerel E et al (2009) Different lepidopteran elicitors account for crosstalk in herbivory-induced phytohormone signaling. Plant Physiol 150:1576–1586

Dobler S, Dalla S, Wagschal V et al (2012) Community-wide convergent evolution in insect adaptation to toxic cardenolides by substitutions in the Na, K-ATPase. Proc Natl Acad Sci USA 109:13040–13045

Dobler S, Petschenka G, Wagschal V et al (2015) Convergent adaptive evolution–how insects master the challenge of cardiac glycoside-containing host plants. Entomol Exp Appl 157:30–39

Dow JA, Davies SA (2006) The Malpighian tubule: rapid insights from post-genomic biology. J Insect Physiol 52:365–378

Durand N, Carot-Sans G, Chertemps T et al (2010) Characterization of an antennal carboxylesterase from the pest moth *Spodoptera littoralis* degrading a host plant odorant. PLoS ONE 5:e15026

Durand N, Carot-Sans G, Bozzolan F et al (2011) Degradation of pheromone and plant volatile components by a same odorant-degrading enzyme in the cotton leafworm, *Spodoptera littoralis*. PLoS ONE 6:e29147

Dussourd DE, Eisner T (1987) Vein-cutting behavior: insect counterploy to the latex defense of plants. Science 237:898–901

Edger PP, Heidel-Fischer HM, Bekaert M et al (2015) The butterfly plant arms-race escalated by gene and genome duplications. Proc Natl Acad Sci USA 112:8362–8366

Ehrlich PR, Raven PH (1964) Butterflies and plants: a study in coevolution. Evolution:586–608

Eichenseer H, Mathews MC, Powell JS et al (2010) Survey of a salivary effector in caterpillars: glucose oxidase variation and correlation with host range. J Chem Ecol 36:885–897

Faldyn MJ, Hunter MD, Elderd BD (2018) Climate change and an invasive, tropical milkweed: an ecological trap for monarch butterflies. Ecology 99:1031–1038

Fan J, Crooks C, Creissen G et al (2011) Pseudomonas sax genes overcome aliphatic isothiocyanate–mediated non-host resistance in *Arabidopsis*. Science 331:1185–1188

Fandino RA, Haverkamp A, Bisch-Knaden S et al (2019) Mutagenesis of odorant coreceptor Orco fully disrupts foraging but not oviposition behaviors in the hawkmoth *Manduca sexta*. Proc Natl Acad Sci USA 116:15677–15685

Fen, Mao Wan-jun, Lu Yi, Yang Xiaomu, Qiao Gong-yin, Ye Jia, Huang (2020) Identification Characterization and Expression Analysis of TRP Channel Genes in the Vegetable Pest Pieris rapae. Insects 11(3):192-10.3390/insects11030192

Fei M, Gols R, Harvey JA (2014) Seasonal phenology of interactions involving short-lived annual plants, a multivoltine herbivore and its endoparasitoid wasp. J Anim Ecol 83:234–244

Feyereisen R (2012) Insect *CYP* genes and P450 enzymes. In: Insect molecular biology and biochemistry. Academic, pp 236–316

Fink LS, Brower LP (1981) Birds can overcome the cardenolide defence of monarch butterflies in Mexico. Nature 291:67–70

Fink LS, Brower LP, Waide RB et al (1983) Overwintering monarch butterflies as food for insectivorous birds in Mexico. Biotropica 15:151–153

Fischer HM, Wheat CW, Heckel DG et al (2008) Evolutionary origins of a novel host plant detoxification gene in butterflies. Mol Biol Evol 25:809–820

Flor HH (1956) The complementary genic systems in flax and flax rust. In: Advances in genetics, vol 8. Academic, pp 29–54

Forister ML, Novotny V, Panorska AK et al (2015) The global distribution of diet breadth in insect herbivores. Proc Natl Acad Sci USA 112:442–447

Fraenkel GS (1959) The raison d'etre of secondary plant substances. Science 129:1466–1470

Freedman MG, Jason C, Ramírez SR et al (2020) Host plant adaptation during contemporary range expansion in the monarch butterfly. Evolution 74:377–391

Frick C, Wink M (1995) Uptake and sequestration of ouabain and other cardiac glycosides in *Danaus plexippus* (Lepidoptera: Danaidae): evidence for a carrier-mediated process. J Chem Ecol 21:557–575

Gerardo NM, Altincicek B, Anselme C et al (2010) Immunity and other defenses in pea aphids, Acyrthosiphon pisum. Genome Biol 11:1–17

Gershenzon J (1994) Metabolic costs of terpenoid accumulation in higher plants. J Chem Ecol 20:1281–1328

Gilbert LI (2004) *Halloween* genes encode P450 enzymes that mediate steroid hormone biosynthesis in *Drosophila melanogaster*. Mol Cell Endocrinol 215:1–10

Gilman RT, Nuismer SL, Jhwueng DC (2012) Coevolution in multidimensional trait space favours escape from parasites and pathogens. Nature 483:328–330

Giraudo M, Hilliou F, Fricaux T et al (2015) Cytochrome P450s from the fall armyworm (*Spodoptera frugiperda*): responses to plant allelochemicals and pesticides. Insect Mol Biol 24:115–128

Glendinning JI (1992) Effectiveness of cardenolides as feeding deterrents to *Peromyscus* mice. J Chem Ecol 18:1559–1575

Glendinning JI, Brower LP (1990) Feeding and breeding responses of five mice species to overwintering aggregations of the monarch butterfly. J Anim Ecol:1091–1112

Glendinning JI, Mejia AA, Brower LP (1988) Behavioral and ecological interactions of foraging mice (*Peromyscus melanotis*) with overwintering monarch butterflies (*Danaus plexippus*) in Mexico. Oecologia 75:222–227

Gloss AD, Vassao DG, Hailey AL et al (2014) Evolution in an ancient detoxification pathway is coupled with a transition to herbivory in the Drosophilidae. Mol Biol Evol 31:2441–2456

Gloss AD, Groen SC, Whiteman NK (2016) A genomic perspective on the generation and maintenance of genetic diversity in herbivorous insects. Annu Rev Ecol Evol Syst 47:165–187

Gloss AD, Abbot P, Whiteman NK (2019a) How interactions with plant chemicals shape insect genomes. Curr Opin Insect Sci 36:149–156

Gloss AD, Dittrich ACN, Lapoint RT et al (2019b) Evolution of herbivory remodels a *Drosophila* genome. bioRxiv. https://doi.org/10.1101/767160

Gonzalez-De-la-Rosa PM, Loustalot-Laclette MR, Abreu-Goodger C et al (2020) Differential gene expression reflects larval development and survival of monarch butterflies on different milkweed hosts. bioRxiv. https://doi.org/10.1101/2020.09.05.284489

Govindaraj L, Gupta T, Esvaran VG et al (2016) Genome-wide identification, characterization of sugar transporter genes in the silkworm *Bombyx mori* and role in *Bombyx mori* nucleopolyhedrovirus (BmNPV) infection. Gene 579:162–171

Gowler CD, Leon KE, Hunter MD et al (2015) Secondary defense chemicals in milkweed reduce parasite infection in monarch butterflies, *Danaus plexippus*. J Chem Ecol 41:520–523

Gozalpour E, Wittgen HG, van den Heuvel JJ et al (2013) Interaction of digitalis-like compounds with p-glycoprotein. Toxicol Sci 131:502–511

Grant JB (2006) Diversification of gut morphology in caterpillars is associated with defensive behavior. J Exp Biol 209:3018–3024

Gratz SJ, Ukken FP, Rubinstein CD et al (2014) Highly specific and efficient CRISPR/Cas9-catalyzed homology-directed repair in *Drosophila*. Genetics 196:961–971

Greenhalgh R, Dermauw W, Glas JJ et al (2020) Genome streamlining in a minute herbivore that manipulates its host plant. elife 9:e56689

Griese E, Caarls L, Bassetti N et al (2021) Insect egg-killing: a new front on the evolutionary arms-race between brassicaceous plants and pierid butterflies. New Phytol 230:341–353

Groen SC, Whiteman NK (2014) The evolution of ethylene signaling in plant chemical ecology. J Chem Ecol 40:700–716

Groen SC, Whiteman NK (2016) Using *Drosophila* to study the evolution of herbivory and diet specialization. Curr Opin Insect Sci 14:66–72

Groen SC, Whiteman NK, Bahrami AK et al (2013) Pathogen-triggered ethylene signaling mediates systemic-induced susceptibility to herbivory in *Arabidopsis*. Plant Cell 25:4755–4766

Groen SC, Humphrey PT, Chevasco D et al (2016) *Pseudomonas syringae* enhances herbivory by suppressing the reactive oxygen burst in *Arabidopsis*. J Insect Physiol 84:90–102

Groen SC, LaPlante ER, Alexandre NM et al (2017) Multidrug transporters and organic anion transporting polypeptides protect insects against the toxic effects of cardenolides. Insect Biochem Mol Biol 81:51–61

Groeneveld HW, Steijl H, Van Den Berg B et al (1990) Rapid, quantitative HPLC analysis of *Asclepias fruticosa* L, *Danaus plexippus* L. cardenolides. J Chem Ecol 16:3373–3382

Guo M, Du L, Chen Q et al (2021) Odorant receptors for detecting flowering plant cues are functionally conserved across moths and butterflies. Mol Biol Evol 38:1413–1427

Hagenbuch B, Stieger B (2013) The SLCO (former SLC21) superfamily of transporters. Mol Asp Med 34:396–412

Hall AR, Ashby B, Bascompte J et al (2020) Measuring coevolutionary dynamics in species-rich communities. Trends Ecol Evol 35:539–550

Hammer TJ, Bowers MD (2015) Gut microbes may facilitate insect herbivory of chemically defended plants. Oecologia 179:1–14

Hammer TJ, Janzen DH, Hallwachs W et al (2017) Caterpillars lack a resident gut microbiome. Proc Natl Acad Sci USA 114:9641–9646

Hannula SE, Zhu F, Heinen R et al (2019) Foliar-feeding insects acquire microbiomes from the soil rather than the host plant. Nat Commun 10:1254

Harvey JA, van Dam NM, Gols R (2003) Interactions over four trophic levels: foodplant quality affects development of a hyperparasitoid as mediated through a herbivore and its primary parasitoid. J Anim Ecol 72:520–531

Hazel WN (1977) The genetic basis of pupal colour dimorphism and its maintenance by natural selection in *Papilio polyxenes* (Papilionidae: Lepidoptera). Heredity 38:227–236

Heidel-Fischer HM, Vogel H (2015) Molecular mechanisms of insect adaptation to plant secondary compounds. Curr Opin Insect Sci 8:8–14

Heidel-Fischer HM, Vogel H, Heckel DG et al (2010) Microevolutionary dynamics of a macroevolutionary key innovation in a lepidopteran herbivore. BMC Evol Biol 10:60

Heidel-Fischer HM, Kirsch R, Reichelt M et al (2019) An insect counteradaptation against host plant defenses evolved through concerted neofunctionalization. Mol Biol Evol 36:930–941

Helmus MR, Dussourd DE (2005) Glues or poisons: which triggers vein cutting by monarch caterpillars? Chemoecology 15:45–49

Herfurth AM, Ohlen MV, Wittstock U (2017) β-Cyanoalanine synthases and their possible role in pierid host plant adaptation. Insects 8:62

Hermann SL, Blackledge C, Haan NL et al (2019) Predators of monarch butterfly eggs and neonate larvae are more diverse than previously recognised. Sci Rep 9:14304

Hindle SJ, Bainton RJ (2014) Barrier mechanisms in the *Drosophila* blood-brain barrier. Front Neurosci 8:414

Holzinger F, Wink M (1996) Mediation of cardiac glycoside insensitivity in the monarch butterfly (*Danaus plexippus*): role of an amino acid substitution in the ouabain binding site of Na+, K+-ATPase. J Chem Ecol 22:1921–1937

Holzinger F, Frick C, Wink M (1992) Molecular basis for the insensitivity of the Monarch (*Danaus plexippus*) to cardiac glycosides. FEBS Lett 314:477–480

Hu K, Morris ME (2004) Effects of benzyl-, phenethyl-, and α-naphthyl isothiocyanates on P-glycoprotein- and MRP1-mediated transport. J Pharm Sci 93:1901–1911

Huh JR, Leung MW, Huang P et al (2011) Digoxin and its derivatives suppress TH 17 cell differentiation by antagonizing RORγt activity. Nature 472:486–490

Hung CF, Harrison TL, Berenbaum MR et al (1995) CYP6B3: a second furanocoumarin-inducible cytochrome P450 expressed in *Papilio polyxenes*. Insect Mol Biol 4:149–160

Jeschke V, Gershenzon J, Vassão DG (2016) A mode of action of glucosinolate-derived isothiocyanates: detoxification depletes glutathione and cysteine levels with ramifications on protein metabolism in *Spodoptera littoralis*. Insect Biochem Mol Biol 71:37–48

Jeschke V, Kearney EE, Schramm K et al (2017) How glucosinolates affect generalist lepidopteran larvae: growth, development and glucosinolate metabolism. Front Plant Sci 8:1995

Jeschke V, Zalucki JM, Raguschke B et al (2021) So much for glucosinolates: a generalist does survive and develop on brassicas, but at what cost? Plants 10:962

Ji Y, Morris ME (2005a) Transport of dietary phenethyl isothiocyanate is mediated by multidrug resistance protein 2 but not P-glycoprotein. Biochem Pharmacol 70:640–647

Ji Y, Morris ME (2005b) Membrane transport of dietary phenethyl isothiocyanate by ABCG2 (breast cancer resistance protein). Mol Pharm 2:414–419

Jing X, Grebenok RJ, Behmer ST (2012) Plant sterols and host plant suitability for generalist and specialist caterpillars. J Insect Physiol 58:235–244

Jing X, Grebenok RJ, Behmer ST (2013) Sterol/steroid metabolism and absorption in a generalist and specialist caterpillar: Effects of dietary sterol/steroid structure, mixture and ratio. Insect Biochem Mol Biol 43:580–587

Jones PL, Petschenka G, Flacht L et al (2019) Cardenolide intake, sequestration, and excretion by the monarch butterfly along gradients of plant toxicity and larval ontogeny. J Chem Ecol 45:264–277

Kareiva P, (1999) Coevolutionary arms races: Is victory possible?. Proceedings of the National Academy of Sciences 96(1):8–10 10.1073/pnas.96.1.8

Kang K, Pulver SR, Panzano VC et al (2010) Analysis of *Drosophila* TRPA1 reveals an ancient origin for human chemical nociception. Nature 464:597–600

Kaplan I, McArt SH, Thaler JS (2014) Plant defenses and predation risk differentially shape patterns of consumption, growth, and digestive efficiency in a guild of leaf-chewing insects. PLoS ONE 9:e93714

Karageorgi M, Groen SC, Sumbul F et al (2019) Genome editing retraces the evolution of toxin resistance in the monarch butterfly. Nature 574:409–412

Karasov TL, Kniskern JM, Gao L et al (2014) The long-term maintenance of a resistance polymorphism through diffuse interactions. Nature 512:436–440

Karban R, Agrawal AA (2002) Herbivore offense. Annu Rev Ecol Syst 33:641–664

Kazana E, Pope TW, Tibbles L et al (2007) The cabbage aphid: a walking mustard oil bomb. Proc R Soc B 274:2271–2277

Koch RL, Hutchison WD, Venette RC et al (2003) Susceptibility of immature monarch butterfly, *Danaus plexippus* (Lepidoptera: Nymphalidae: Danainae), to predation by *Harmonia axyridis* (Coleoptera: Coccinellidae). Biol Control 28:265–270

Koppel N, Bisanz JE, Pandelia ME et al (2018) Discovery and characterization of a prevalent human gut bacterial enzyme sufficient for the inactivation of a family of plant toxins. elife 7:e33953

Koricheva J (2002) Meta-analysis of sources of variation in fitness costs of plant antiherbivore defenses. Ecology 83:176–190

Kos M, Broekgaarden C, Kabouw P et al (2011) Relative importance of plant-mediated bottom-up and top-down forces on herbivore abundance on *Brassica oleracea*. Funct Ecol 25:1113–1124

Kos M, Houshyani B, Wietsma R et al (2012) Effects of glucosinolates on a generalist and specialist leaf-chewing herbivore and an associated parasitoid. Phytochemistry 77:162–170

Koutroumpa FA, Monsempes C, François MC et al (2016) Heritable genome editing with CRISPR/Cas9 induces anosmia in a crop pest moth. Sci Rep 6:29620

Kowalski P, Baum M, Körten M et al (2020) ABCB transporters in a leaf beetle respond to sequestered plant toxins. Proc R Soc B 287:20201311

Krempl C, Sporer T, Reichelt M et al (2016) Potential detoxification of gossypol by UDP-glycosyltransferases in the two heliothine moth species *Helicoverpa armigera* and *Heliothis virescens*. Insect Biochem Mol Biol 71:49–57

Lawrence SD, Novak NG, Shao J et al (2020) Cabbage looper (*Trichoplusia ni* Hübner) labial glands contain unique bacterial flora in contrast with their alimentary canal, mandibular glands, and Malpighian tubules. Microbiol Open 9:e994

Lawton JH, McNeill S (1979) Between the devil and the deep blue sea: on the problem of being a herbivore. In: Merson R, Turner BD, Taylor LR (eds) Population dynamics. Symposium of the British Ecological Society. Blackwell, Oxford, pp 223–244

Leal WS (2013) Odorant reception in insects: roles of receptors, binding proteins, and degrading enzymes. Annu Rev Entomol 58:373–391

Lee ZA, Baranowski AK, Preisser EL (2021) Auditory predator cues affect monarch (*Danaus plexippus*; Lepidoptera: Nymphalidae) development time and pupal weight. Acta Oecol 111:103740

Li W, Schuler MA, Berenbaum MR (2003) Diversification of furanocoumarin-metabolizing cytochrome P450 monooxygenases in two papilionids: specificity and substrate encounter rate. Proc Natl Acad Sci USA 100:14593–14598

Li W, Zangerl AR, Schuler MA et al (2004a) Characterization and evolution of furanocoumarin-inducible cytochrome P450s in the parsnip webworm, *Depressaria pastinacella*. Insect Mol Biol 13:603–613

Li X, Baudry J, Berenbaum MR et al (2004b) Structural and functional divergence of insect CYP6B proteins: from specialist to generalist cytochrome P450. Proc Natl Acad Sci USA 101:2939–2944

Li X, Schuler MA, Berenbaum MR (2007) Molecular mechanisms of metabolic resistance to synthetic and natural xenobiotics. Annu Rev Entomol 52:231–253

Li ZW, Shen YH, Xiang ZH et al (2011) Pathogen-origin horizontally transferred genes contribute to the evolution of lepidopteran insects. BMC Evol Biol 11:356

Li D, Halitschke R, Baldwin IT et al (2020a) Information theory tests critical predictions of plant defense theory for specialized metabolism. Sci Adv 6:eaaz0381

Li Y, Bai P, Wei L et al (2020b) Capsaicin functions as *Drosophila* ovipositional repellent and causes intestinal dysplasia. Sci Rep 10:9963

Li X, Deng Z, Chen X (2021a) Regulation of insect P450s in response to phytochemicals. Curr Opin Insect Sci 43:108–116

Li Y, Zhou Y, Jing W et al (2021b) Horizontally acquired cysteine synthase genes undergo functional divergence in lepidopteran herbivores. Heredity. https://doi.org/10.1038/s41437-021-00430-z

Lill JT, Marquis RJ, Ricklefs RE (2002) Host plants influence parasitism of forest caterpillars. Nature 417:170–173

Lin S, Staahl BT, Alla RK et al (2014) Enhanced homology-directed human genome engineering by controlled timing of CRISPR/Cas9 delivery. elife 3:e04766

Liu Q, Liu W, Zeng B et al (2017) Deletion of the *Bombyx mori* odorant receptor co-receptor (BmOrco) impairs olfactory sensitivity in silkworms. Insect Biochem Mol Biol 86:58–67

Liu XL, Zhang J, Yan Q et al (2020) The molecular basis of host selection in a crucifer-specialized moth. Curr Biol 30:4476–4482

Lund M, Brainard DC, Coudron T et al (2020) Predation threat modifies *Pieris rapae* performance and response to host plant quality. Oecologia 193:389–401

Maag D, Dalvit C, Thevenet D et al (2014) 3-β-D-Glucopyranosyl-6-methoxy-2-benzoxazolinone (MBOA-N-Glc) is an insect detoxification product of maize 1,4-benzoxazin-3-ones. Phytochemistry 102:97–105

Malcolm SB (1990) Chemical defence in chewing and sucking insect herbivores: plant-derived cardenolides in the monarch butterfly and oleander aphid. Chemoecology 1:12–21

Malcolm SB (1995) Milkweeds, monarch butterflies and the ecological significance of cardenolides. Chemoecology 5:101–117

Mao W, Rupasinghe S, Zangerl AR et al (2006) Remarkable substrate-specificity of CYP6AB3 in *Depressaria pastinacella*, a highly specialized caterpillar. Insect Mol Biol 15:169–179

Mao YB, Cai WJ, Wang JW et al (2007) Silencing a cotton bollworm P450 monooxygenase gene by plant-mediated RNAi impairs larval tolerance of gossypol. Nat Biotechnol 25:1307–1313

Mao W, Rupasinghe SG, Zangerl AR et al (2007a) Allelic variation in the *Depressaria pastinacella* CYP6AB3 protein enhances metabolism of plant allelochemicals by altering a proximal surface residue and potential interactions with cytochrome P450 reductase. J Biol Chem 282:10544–10552

Mao W, Schuler MA, Berenbaum MR (2007b) Cytochrome P450s in *Papilio multicaudatus* and the transition from oligophagy to polyphagy in the Papilionidae. Insect Mol Biol 16:481–490

Mao W, Zangerl AR, Berenbaum MR et al (2008) Metabolism of myristicin by *Depressaria pastinacella* CYP6AB3v2 and inhibition by its metabolite. Insect Biochem Mol Biol 38:645–651

Marty MA, Krieger RI (1984) Metabolism of uscharidin, a milkweed cardenolide, by tissue homogenates of monarch butterfly larvae, *Danaus plexippus* L. J Chem Ecol 10:945–956

Mason CJ, Jones AG, Felton GW (2019a) Co-option of microbial associates by insects and their impact on plant–folivore interactions. Plant Cell Environ 42:1078–1086

Mason CJ, Ray S, Shikano I et al (2019b) Plant defenses interact with insect enteric bacteria by initiating a leaky gut syndrome. Proc Natl Acad Sci USA 116:15991–15996

Matsuura H, Sokabe T, Kohno K et al (2009) Evolutionary conservation and changes in insect TRP channels. BMC Evol Biol 9:228

Meyers DM, Ahmad S (1991) Link between L-3-cyanoalanine synthase activity and differential cyanide sensitivity of insects. Biochim Biophys Acta 1075:195–197

Mi H, Muruganujan A, Casagrande JT et al (2013) Large-scale gene function analysis with the PANTHER classification system. Nat Protoc 8:1551–1566

Mitri C, Soustelle L, Framery B et al (2009) Plant insecticide L-canavanine repels *Drosophila* via the insect orphan GPCR DmX. PLoS Biol 7:e1000147

Mittapalli O, Sardesai N, Shukle RH (2007) cDNA cloning and transcriptional expression of a peritrophin-like gene in the Hessian fly, *Mayetiola destructor* [Say]. Arch Insect Biochem Physiol 64:19–29

Morant AV, Jørgensen K, Jørgensen C et al (2008) β-Glucosidases as detonators of plant chemical defense. Phytochemistry 69:1795–1813

Moranz R, Brower LP (1998) Geographic and temporal variation of cardenolide-based chemical defenses of queen butterfly (*Danaus gilippus*) in northern Florida. J Chem Ecol 24:905–932

Müller C, Agerbirk N, Olsen CE et al (2001) Sequestration of host plant glucosinolates in the defensive hemolymph of the sawfly *Athalia rosae*. J Chem Ecol 27:2505–2516

Müller C, Agerbirk N, Olsen CE (2003) Lack of sequestration of host plant glucosinolates in *Pieris rapae* and *P. grarricae*. Chemoecology 13:47–54

Musser RO, Hum-Musser SM, Eichenseer H et al (2002) Caterpillar saliva beats plant defences. Nature 416:599–600

Nakata K, Tanaka Y, Nakano T et al (2006) Nuclear receptor-mediated transcriptional regulation in Phase I, II, and III xenobiotic metabolizing systems. Drug Metab Pharmacokinet 21:437–457

Nallu S, Hill JA, Don K et al (2018) The molecular genetic basis of herbivory between butterflies and their host plants. Nat Ecol Evol 2:1418–1427

Nelson C (1993) A model for cardenolide and cardenolide glycoside storage by the monarch butterfly. In: Zalucki MP, Malcolm SB (eds) Biology and conservation of the monarch butterfly. Los Angeles County Museum of Natural History, Los Angeles, pp 83–90

Ngou BPM, Ahn HK, Ding P et al (2021) Mutual potentiation of plant immunity by cell-surface and intracellular receptors. Nature 592:110–115

Okamura Y, Sato A, Tsuzuki N et al (2019a) Molecular signatures of selection associated with host plant differences in *Pieris* butterflies. Mol Ecol 28:4958–4970

Okamura Y, Sato A, Tsuzuki N et al (2019b) Differential regulation of host plant adaptive genes in *Pieris* butterflies exposed to a range of glucosinolate profiles in their host plants. Sci Rep 9:7256

Parker BJ, Barribeau SM, Laughton AM et al (2011) Non-immunological defense in an evolutionary framework. Trends Ecol Evol 26:242–248

Paul SM, Ternet M, Salvaterra PM et al (2003) The Na+/K+ ATPase is required for septate junction function and epithelial tube-size control in the *Drosophila* tracheal system. Development 130:4963–4974

Paul SM, Palladino MJ, Beitel GJ (2007) A pump-independent function of the Na, K-ATPase is required for epithelial junction function and tracheal tube-size control. Development 134:147–155

Paul RL, Pearse IS, Ode PJ (2020) Fine-scale plant defense variability increases top-down control of an herbivore. Funct Ecol. https://doi.org/10.1111/1365-2435.13808

Pearse IS, Paul R, Ode PJ (2018) Variation in plant defense suppresses herbivore performance. Curr Biol 28:1981–1986

Pelosi P, Iovinella I, Zhu J et al (2018) Beyond chemoreception: diverse tasks of soluble olfactory proteins in insects. Biol Rev 93:184–200

Petschenka G, Agrawal AA (2015) Milkweed butterfly resistance to plant toxins is linked to sequestration, not coping with a toxic diet. Proc R Soc B 282:20151865

Petschenka G, Dobler S (2009) Target-site sensitivity in a specialized herbivore towards major toxic compounds of its host plant: the Na+ K+-ATPase of the oleander hawk moth (*Daphnis nerii*) is highly susceptible to cardenolides. Chemoecology 19:235–239

Petschenka G, Offe JK, Dobler S (2012) Physiological screening for target site insensitivity and localization of Na+/K+-ATPase in cardenolide-adapted Lepidoptera. J Insect Physiol 58:607–612

Petschenka G, Pick C, Wagschal V et al (2013) Functional evidence for physiological mechanisms to circumvent neurotoxicity of cardenolides in an adapted and a non-adapted hawk-moth species. Proc R Soc B 280:20123089

Petschenka G, Fei CS, Araya JJ et al (2018) Relative selectivity of plant cardenolides for Na+/K+-ATPases from the monarch butterfly and non-resistant insects. Front Plant Sci 9:1424

Pierce AA, de Roode JC, Tao L (2016) Comparative genetics of Na+/K+-ATPase in monarch butterfly populations with varying host plant toxicity. Biol J Linn Soc 119:194–200

Pomilio AB, Falzoni EM, Vitale AA (2008) Toxic chemical compounds of the Solanaceae. Nat Prod Commun 3:1934578X0800300420

Port F, Chen HM, Lee T et al (2014) Optimized CRISPR/Cas tools for efficient germline and somatic genome engineering in *Drosophila*. Proc Natl Acad Sci USA 111:E2967–E2976

Price PW, Bouton CE, Gross P et al (1980) Interactions among three trophic levels: influence of plants on interactions between insect herbivores and natural enemies. Annu Rev Ecol Syst 11:41–65

Rafter JL, Agrawal AA, Preisser EL (2013) Chinese mantids gut toxic monarch caterpillars: avoidance of prey defence? Ecol Entomol 38:76–82

Rahman MM, Zalucki MP, Furlong MJ (2019) Diamondback moth egg susceptibility to rainfall: effects of host plant and oviposition behavior. Entomol Exp Appl 167:701–712

Rane RV, Ghodke AB, Hoffmann AA et al (2019) Detoxifying enzyme complements and host use phenotypes in 160 insect species. Curr Opin Insect Sci 31:131–138

Ranz JM, González PM, Clifton BD et al (2020) A *de novo* genome assembly, gene annotation, and expression atlas for the monarch butterfly *Danaus plexippus*. bioRxiv. https://doi.org/10.1101/2020.09.19.304162

Rasmann S, Johnson MD, Agrawal AA (2009) Induced responses to herbivory and jasmonate in three milkweed species. J Chem Ecol 35:1326–1334

Ratzka A, Vogel H, Kliebenstein DJ et al (2002) Disarming the mustard oil bomb. Proc Natl Acad Sci USA 99:11223–11228

Reichstein TV, Von Euw J, Parsons JA et al (1968) Heart poisons in the monarch butterfly. Science 161:861–866

Rivera-Vega LJ, Acevedo FE, Felton GW (2017a) Genomics of Lepidoptera saliva reveals function in herbivory. Curr Opin Insect Sci 19:61–69

Rivera-Vega LJ, Galbraith DA, Grozinger CM et al (2017b) Host plant driven transcriptome plasticity in the salivary glands of the cabbage looper (*Trichoplusia ni*). PLoS ONE 12:e0182636

Rodman JE, Soltis PS, Soltis DE et al (1998) Parallel evolution of glucosinolate biosynthesis inferred from congruent nuclear and plastid gene phylogenies. Am J Bot 85:997–1006

Roeske CN, Seiber JN, Brower LP et al (1976) Milkweed cardenolides and their comparative processing by monarch butterflies (*Danaus plexippus* L.). In: Biochemical interaction between plants and insects. Springer, Boston, pp 93–167

Rubin AL, Stirling CE, Stahl WL (1983) 3H-ouabain binding autoradiography in the abdominal nerve cord of the hawk moth, *Manduca sexta*. J Exp Biol 104:217–230

Schmelz EA, Huffaker A, Carroll MJ et al (2012) An amino acid substitution inhibits specialist herbivore production of an antagonist effector and recovers insect-induced plant defenses. Plant Physiol 160:1468–1478

Schoville SD, Chen YH, Andersson MN et al (2018) A model species for agricultural pest genomics: the genome of the Colorado potato beetle, *Leptinotarsa decemlineata* (Coleoptera: Chrysomelidae). Sci Rep 8:1931

Schramm K, Vassão DG, Reichelt M et al (2012) Metabolism of glucosinolate-derived isothiocyanates to glutathione conjugates in generalist lepidopteran herbivores. Insect Biochem Mol Biol 42:174–182

Schweizer F, Heidel-Fischer H, Vogel H et al (2017) *Arabidopsis* glucosinolates trigger a contrasting transcriptomic response in a generalist and a specialist herbivore. Insect Biochem Mol Biol 85:21–31

Scott K, Brady R Jr, Cravchik A et al (2001) A chemosensory gene family encoding candidate gustatory and olfactory receptors in *Drosophila*. Cell 104:661–673

Seiber JN, Tuskes PM, Brower LP et al (1980) Pharmacodynamics of some individual milkweed cardenolides fed to larvae of the monarch butterfly (*Danaus plexippus* L.). J Chem Ecol 6:321–339

Seiber JN, Lee SM, Benson JM (1983) Cardiac glycosides (cardenolides) in species of *Asclepias* (Asclepiadaceae). In: Handbook of natural toxins. CRC Press, pp 43–83

Shabab M, Khan SA, Vogel H et al (2014) OPDA isomerase GST 16 is involved in phytohormone detoxification and insect development. FEBS J 281:2769–2783

Shroff R, Vergara F, Muck A et al (2008) Nonuniform distribution of glucosinolates in *Arabidopsis thaliana* leaves has important consequences for plant defense. Proc Natl Acad Sci USA 105:6196–6201

Shukla SP, Beran F (2020) Gut microbiota degrades toxic isothiocyanates in a flea beetle pest. Mol Ecol 29:4692–4705

Simon C., Groen I, Ćalić Z, Joly-Lopez AE., Platts JY, Choi M, Natividad K, Dorph WM., Mauck B, Bracken Carlo Leo U., Cabral A, Kumar RO., Torres R, Satija G, Vergara A, Henry SJ., Franks Michael D., Purugganan (2020) The strength and pattern of natural selection on gene expression in rice. Nature 578(7796) 572-576 10.1038/s41586-020-1997-2

Simms EL (1992) Costs of plant resistance to herbivory. In: Fritz RS, Simms EL (eds) Plant resistance to herbivores and pathogens: ecology, evolution, and genetics. University of Chicago Press, Chicago, pp 392–425

Singer MS, Lichter-Marck IH, Farkas TE et al (2014) Herbivore diet breadth mediates the cascading effects of carnivores in food webs. Proc Natl Acad Sci USA 111:9521–9526

Speed MP, Fenton A, Jones MG et al (2015) Coevolution can explain defensive secondary metabolite diversity in plants. New Phytol 208:1251–1263

Sporer T, Körnig J, Wielsch N et al (2021) Hijacking the mustard-oil bomb: how a glucosinolate-sequestering flea beetle copes with plant myrosinases. Front Plant Sci 12:831

Stahl EA, Dwyer G, Mauricio R et al (1999) Dynamics of disease resistance polymorphism at the *Rpm1* locus of *Arabidopsis*. Nature 400:667–671

Stauber EJ, Kuczka P, Van Ohlen M et al (2012) Turning the 'mustard oil bomb' into a 'cyanide bomb': aromatic glucosinolate metabolism in a specialist insect herbivore. PLoS ONE 7:e35545

Steinbrenner AD, Muñoz-Amatriaín M, Chaparro AF et al (2020) A receptor-like protein mediates plant immune responses to herbivore-associated molecular patterns. Proc Natl Acad Sci USA 117:31510–31518

Steiner AM, Busching C, Vogel H et al (2018) Molecular identification and characterization of rhodaneses from the insect herbivore *Pieris rapae*. Sci Rep 8:10819

Stenoien CM, Meyer RA, Nail KR et al (2019) Does chemistry make a difference? Milkweed butterfly sequestered cardenolides as a defense against parasitoid wasps. Arthropod-Plant Interact 13:835–852

Sternberg ED, Lefèvre T, Li J et al (2012) Food plant derived disease tolerance and resistance in a natural butterfly-plant-parasite interactions. Evolution 66:3367–3376

Stowe KA, Marquis RJ (2011) Costs of defense: correlated responses to divergent selection for foliar glucosinolate content in *Brassica rapa*. Evol Ecol 25:763–775

Strauss SY, Siemens DH, Decher MB et al (1999) Ecological costs of plant resistance to herbivores in the currency of pollination. Evolution 53:1105–1113

Summers CB, Felton GW (1996) Peritrophic envelope as a functional antioxidant. Arch Insect Biochem Physiol 32:131–142

Sun BF, Xiao JH, He SM et al (2013) Multiple ancient horizontal gene transfers and duplications in lepidopteran species. Insect Mol Biol 22:72–87

Sun R, Jiang X, Reichelt M et al (2019) Tritrophic metabolism of plant chemical defenses and its effects on herbivore and predator performance. elife 8:e51029

Sun R, Gols R, Harvey JA et al (2020) Detoxification of plant defensive glucosinolates by an herbivorous caterpillar is beneficial to its endoparasitic wasp. Mol Ecol 29:4014–4031

Tan CW, Peiffer M, Hoover K et al (2018) Symbiotic polydnavirus of a parasite manipulates caterpillar and plant immunity. Proc Natl Acad Sci USA 115:5199–5204

Tan WH, Acevedo T, Harris EV et al (2019a) Transcriptomics of monarch butterflies (*Danaus plexippus*) reveals that toxic host plants alter expression of detoxification genes and down-regulate a small number of immune genes. Mol Ecol 28:4845–4863

Tan WH, Mongue AJ, de Roode JC et al (2019b) Population genomics reveals complex patterns of immune gene evolution in monarch butterflies (*Danaus plexippus*). bioRxiv. https://doi.org/10.1101/620013

Tao L, Hoang KM, Hunter MD et al (2016) Fitness costs of animal medication: antiparasitic plant chemicals reduce fitness of monarch butterfly hosts. J Anim Ecol 85:1246–1254

Taverner AM, Yang L, Barile ZJ et al (2019) Adaptive substitutions underlying cardiac glycoside insensitivity in insects exhibit epistasis *in vivo*. elife 8:e48224

Termonia A, Hsiao TH, Pasteels JM et al (2001) Feeding specialization and host-derived chemical defense in Chrysomeline leaf beetles did not lead to an evolutionary dead end. Proc Natl Acad Sci USA 98:3909–3914

Thaler JS, Humphrey PT, Whiteman NK (2012a) Evolution of jasmonate and salicylate signal crosstalk. Trends Plant Sci 17:260–270

Thaler JS, McArt SH, Kaplan I (2012b) Compensatory mechanisms for ameliorating the fundamental trade-off between predator avoidance and foraging. Proc Natl Acad Sci USA 109:12075–12080

Thompson JN (1986) Constraints on arms races in coevolution. Trends Ecol Evol 1:105–107

Torrie LS, Radford JC, Southall TD et al (2004) Resolution of the insect ouabain paradox. Proc Natl Acad Sci USA 101:13689–13693

Traka M, Mithen R (2009) Glucosinolates, isothiocyanates and human health. Phytochem Rev 8:269–282

Vacher C, Brown SP, Hochberg ME (2005) Avoid, attack or do both? Behavioral and physiological adaptations in natural enemies faced with novel hosts. BMC Evol Biol 5:60

Van Ohlen M, Herfurth AM, Kerbstadt H et al (2016) Cyanide detoxification in an insect herbivore: Molecular identification of β-cyanoalanine synthases from *Pieris rapae*. Insect Biochem Mol Biol 70:99–110

Van Valen L (1973) A new evolutionary law. Evol Theory 1:1–30

Vandenborre G, Smagghe G, Van Damme EJ (2011) Plant lectins as defense proteins against phytophagous insects. Phytochemistry 72:1538–1550

Verschaffelt E (1910) The cause determining the selection of food in some herbivorous insects. Proc R Acad Amst 13:536–542

Vidal MC, Murphy SM (2018) Bottom-up vs. top-down effects on terrestrial insect herbivores: a meta-analysis. Ecol Lett 21:138–150

Vieira FG, Rozas J (2011) Comparative genomics of the odorant-binding and chemosensory protein gene families across the Arthropoda: origin and evolutionary history of the chemosensory system. Genome Biol Evol 3:476–490

Visôtto LE, Oliveira MGA, Guedes RNC et al (2009) Contribution of gut bacteria to digestion and development of the velvetbean caterpillar, *Anticarsia gemmatalis*. J Insect Physiol 55:185–191

Wadleigh RW, Yu SJ (1988) Detoxification of isothiocyanate allelochemicals by glutathione transferase in three lepidopterous species. J Chem Ecol 14:1279–1288

Wang S, Wang P (2020) Functional redundancy of structural proteins of the peritrophic membrane in *Trichoplusia ni*. Insect Biochem Mol Biol 125:103456

Wang H, Shi Y, Wang L et al (2018) *CYP6AE* gene cluster knockout in *Helicoverpa armigera* reveals role in detoxification of phytochemicals and insecticides. Nat Commun 9:4820

Wang P, Vassão DG, Raguschke B et al (2021) Balancing nutrients in a toxic environment: the challenge of eating. Insect Sci. https://doi.org/10.1111/1744-7917.12923

Wei JJ, Fu T, Yang T et al (2015) A TRPA1 channel that senses thermal stimulus and irritating chemicals in *Helicoverpa armigera*. Insect Mol Biol 24:412–421

Weinreich DM, Delaney NF, DePristo MA et al (2006) Darwinian evolution can follow only very few mutational paths to fitter proteins. Science 312:111–114

Wen Z, Pan L, Berenbaum MR et al (2003) Metabolism of linear and angular furanocoumarins by *Papilio polyxenes* CYP6B1 co-expressed with NADPH cytochrome P450 reductase. Insect Biochem Mol Biol 33:937–947

Wen Z, Rupasinghe S, Niu G et al (2006) CYP6B1 and CYP6B3 of the black swallowtail (*Papilio polyxenes*): adaptive evolution through subfunctionalization. Mol Biol Evol 23:2434–2443

Wetzel WC, Thaler JS (2016) Does plant trait diversity reduce the ability of herbivores to defend against predators? The plant variability–gut acclimation hypothesis. Curr Opin Insect Sci 14:25–31

Wheat CW, Vogel H, Wittstock U et al (2007) The genetic basis of a plant–insect coevolutionary key innovation. Proc Natl Acad Sci USA 104:20427–20431

Whiteman NK, Groen SC, Chevasco D et al (2011) Mining the plant–herbivore interface with a leafmining *Drosophila* of *Arabidopsis*. Mol Ecol 20:995–1014

Wiens JJ, Lapoint RT, Whiteman NK (2015) Herbivory increases diversification across insect clades. Nat Commun 6:8370

Wink M (2018) Plant secondary metabolites modulate insect behavior-steps toward addiction? Front Physiol 9:364

Witthohn K, Naumann CM (1987) Cyanogenesis – a general phenomenon in the Lepidoptera? J Chem Ecol 13:1789–1809

Wittstock U, Agerbirk N, Stauber EJ et al (2004) Successful herbivore attack due to metabolic diversion of a plant chemical defense. Proc Natl Acad Sci USA 101:4859–4864

Woolhouse ME, Webster JP, Domingo E et al (2002) Biological and biomedical implications of the co-evolution of pathogens and their hosts. Nat Genet 32:569–577

Wouters FC, Reichelt M, Glauser G et al (2014) Reglucosylation of the benzoxazinoid DIMBOA with inversion of stereochemical configuration is a detoxification strategy in lepidopteran herbivores. Angew Chem Int Ed 53:11320–11324

Wybouw N, Dermauw W, Tirry L et al (2014) A gene horizontally transferred from bacteria protects arthropods from host plant cyanide poisoning. elife 3:e02365

Wybouw N, Pauchet Y, Heckel DG et al (2016) Horizontal gene transfer contributes to the evolution of arthropod herbivory. Genome Biol Evol 8:1785–1801

Wybouw N, Van Leeuwen T, Dermauw W (2018) A massive incorporation of microbial genes into the genome of *Tetranychus urticae*, a polyphagous arthropod herbivore. Insect Mol Biol 27:333–351

Xia J, Guo Z, Yang Z et al (2021) Whitefly hijacks a plant detoxification gene that neutralizes plant toxins. Cell 184:1693–1705

Yang Z (1998) Likelihood ratio tests for detecting positive selection and application to primate lysozyme evolution. Mol Biol Evol 15:568–573

Yang Z (2007) PAML 4: phylogenetic analysis by maximum likelihood. Mol Biol Evol 24:1586–1591

Yang K, Gong XL, Li GC et al (2020) A gustatory receptor tuned to the steroid plant hormone brassinolide in *Plutella xylostella* (Lepidoptera: Plutellidae). elife 9:e64114

Yang J, Guo H, Jiang N-J et al (2021a) Identification of a gustatory receptor tuned to sinigrin in the cabbage white butterfly *Pieris rapae*. PLoS Genet

Yang ZL, Nour-Eldin HH, Hänniger S et al (2021b) Sugar transporters enable a leaf beetle to accumulate plant defense compounds. Nat Commun 12:2658

York HA, Oberhauser KS (2002) Effects of duration and timing of heat stress on monarch butterfly (*Danaus plexippus*) (Lepidoptera: Nymphalidae) development. J Kansas Entomol Soc 75:290–298

You M, Yue Z, He W et al (2013) A heterozygous moth genome provides insights into herbivory and detoxification. Nat Genet 45:220–225

You Y, Xie M, Ren N et al (2015) Characterization and expression profiling of glutathione S-transferases in the diamondback moth, *Plutella xylostella* (L.). BMC Genomics 16:152

Yuan M, Jiang Z, Bi G et al (2021a) Pattern-recognition receptors are required for NLR-mediated plant immunity. Nature 592:105–109

Yuan YY, Xin YC, Han JL et al (2021b) Functional characterization of a novel, highly expressed ion-driven sugar antiporter in the thoracic muscles of *Helicoverpa armigera*. Insect Sci. https://doi.org/10.1111/1744-7917.12908

Zalucki MP, Brower LP, Alonso MA (2001) Detrimental effects of latex and cardiac glycosides on survival and growth of first-instar monarch butterfly larvae *Danaus plexippus* feeding on the sandhill milkweed *Asclepias humistrata*. Ecol Entomol 26:212–224

Zangerl AR, Liao LH, Jogesh T et al (2012) Aliphatic esters as targets of esterase activity in the parsnip webworm (*Depressaria pastinacella*). J Chem Ecol 38:188–194

Zelle KM, Lu B, Pyfrom SC et al (2013) The genetic architecture of degenerin/epithelial sodium channels in *Drosophila*. G3 3:441–450

Zhan S, Merlin C, Boore JL et al (2011) The monarch butterfly genome yields insights into long-distance migration. Cell 147:1171–1185

Zhan S, Zhang W, Niitepold K et al (2014) The genetics of monarch butterfly migration and warning colouration. Nature 514:317–321

Zhang Y, Callaway EC (2002) High cellular accumulation of sulphoraphane, a dietary anticarcinogen, is followed by rapid transporter-mediated export as a glutathione conjugate. Biochem J 364:301–307

Zhang J, Bisch-Knaden S, Fandino RA et al (2019a) The olfactory coreceptor IR8a governs larval feces-mediated competition avoidance in a hawkmoth. Proc Natl Acad Sci USA 116:21828–21833

Zhang ZJ, Zhang SS, Niu BL et al (2019b) A determining factor for insect feeding preference in the silkworm, *Bombyx mori*. PLoS Biol 17:e3000162

Zhen Y, Aardema ML, Medina EM et al (2012) Parallel molecular evolution in an herbivore community. Science 337:1634–1637

Zhou JJ, Huang W, Zhang GA et al (2004) "Plus-C" odorant-binding protein genes in two *Drosophila* species and the malaria mosquito *Anopheles gambiae*. Gene 327:117–129

Züst T, Joseph B, Shimizu KK et al (2011) Using knockout mutants to reveal the growth costs of defensive traits. Proc R Soc B 278:2598–2603

Züst T, Rasmann S, Agrawal AA (2015) Growth–defense tradeoffs for two major anti-herbivore traits of the common milkweed *Asclepias syriaca*. Oikos 124:1404–1415

Züst T, Strickler SR, Powell AF et al (2020) Independent evolution of ancestral and novel defenses in a genus of toxic plants (*Erysimum*, Brassicaceae). elife 9:e51712

Zvereva EL, Kozlov MV (2016) The costs and effectiveness of chemical defenses in herbivorous insects: a meta-analysis. Ecol Monogr 86:107–124

Sequestered Caterpillar Chemical Defenses: From "Disgusting Morsels" to Model Systems

M. Deane Bowers

To Theodore Sargent and Lincoln Brower, two talented biologists who shared with all of us their vast knowledge of both larval and adult Lepidoptera. They are missed.

Euphydryias phaeton larva (Nymphalidae) on white turtlehead, *Chelone glabra* (Plantaginaceae). (Photo: M. Deane Bowers)

Introduction

As is clear from the chapters in this book, caterpillars have served as the inspiration and basis for many historically critical steps forward in our understanding of ecology, behavior, and evolution. For example, relevant to this chapter on caterpillar sequestration and unpalatability, in his work on the importance of coloration in sexual selection, Darwin was stymied by the bright colors of many caterpillars, which, of course, cannot reproduce. This led to a series of experiments and observations in the late nineteenth century, showing that this bright coloration advertised the fact that these caterpillars were unpalatable to predators (see below), thus reconciling Darwin's dilemma. In 1930, Fisher published *The Genetical Theory of Natural Selection*, in which he used gregarious caterpillars as a way to resolve the conundrum of how warningly colored, unpalatable insects could evolve. He noted that in gregarious, toxic caterpillars, if an individual loses its life, its siblings will be protected; although, "… the selective potency of the avoidance of brothers will of course be only half as great as if the individual itself were protected; against this is to be set the fact that it applies to the whole of a possibly numerous brood" (Fisher 1930, p. 178). Here is stated an early incarnation of the theory of kin selection, as well as the resolution of a paradox: how warning coloration could evolve.

Other important advancements in our understanding of plant and insect ecology and evolution have involved lepidopteran caterpillars. For example, the concept of coevolution was developed based on the relationships between butterflies and the chemistry of their larval host plants (Ehrlich and Raven 1964), and our ideas about the role of plant secondary compounds as important plant defenses and their subsequent use by insects as defenses were also based on research with caterpillars (Slater 1877; Haase 1896; Vershaffelt 1910; Fraenkel 1959, 1969). Although other groups of phytophagous insects, such as grasshoppers, bees, and beetles, have played important roles in the development and testing of theory and practice in ecology and evolution, lepidopteran caterpillars are clearly the stars.

While all caterpillars are caught in the middle between their host plants and their natural enemies, those that are unpalatable due to sequestering chemical defense compounds from their host plants are particularly notable: not only do they acquire nutrition from their host plants, but they also acquire chemical compounds that serve as defenses against higher trophic levels: predators, parasitoids, and pathogens. As a result, not only is the nutritional quality of a potential host plant important for larval growth and development, but the content of chemical compounds that can be sequestered is important as well. Thus selective pressures on sequestering caterpillars come both from the host plant (traits such as host plant choice, toxin tolerance, and toxin detoxification) and from natural enemies (levels of sequestered compounds and efficacy against different types of enemies) (Price et al. 1980; Ode 2006; Fordyce and Nice 2008). In the past few decades, researchers have come a long way in understanding how the bottom-up and top-down pressures impact sequestering herbivores; but there is still a great deal to learn.

Plants produce an incredible diversity of secondary metabolites (Hegnauer 1962-1996; Rosenthal and Berenbaum 1991). While these compounds may serve several primary roles, such as antioxidants or UV filters (and lines between primary and secondary compounds are blurred, Erb and Kliebenstein 2020), their primary role is as a protection against plant enemies, including herbivores and pathogens (Fraenkel 1959, 1969). While these secondary metabolites may serve as effective deterrents or toxins to many herbivores, certain species can tolerate these compounds, and may actually use them as feeding or oviposition stimulants (Schoonhoven et al. 2006; Nishida 2014). After ingestion by herbivores, these plant secondary metabolites may undergo a number of fates: they may be eliminated intact; they may be metabolized with any of a number of effective enzyme systems (Heckel 2014) and the metabolites absorbed (e.g., sugars that are the result of metabolism may be absorbed and used (Pasteels et al. 1983)) or excreted; or they may be sequestered, either as the intact compound or as a metabolite (Duffey 1980; Rimpler 1991; Bowers 1993; Nishida 2002; Opitz and Müller 2009; Dobler et al. 2011). Sequestration can be defined as the selective uptake, transport, and storage of chemical compounds from host plants or prey (see Heckel 2014). While this review will deal with sequestration of plant secondary compounds (also called plant allelochemicals or plant secondary metabolites) by caterpillars from their host plants, sequestration of defensive compounds is not unique to invertebrates. Poison dart frogs provide a fascinating system in which sequestration of defensive compounds by this diverse clade of frogs is from their invertebrate prey (e.g., Saporito et al. 2012). In addition, caterpillars can defend themselves via other chemical methods (Bowers 1993), such as the production of hairs or spines that produce urticating or toxic compounds (Diaz 2005; Battisti et al. 2011) and the osmeteria of swallowtail caterpillars that produce a variety of different unpalatable compounds (Honda 1981, 1983; Frankfater et al. 2009). Although potentially potent defenses, those will not be covered in this chapter.

The goals of this chapter are to first provide some historic perspective on studies of caterpillar palatability and unpalatability and then to review some of the more recent work on caterpillar sequestration and defense against natural enemies. In this latter context, I will use caterpillars sequestering one group of plant secondary metabolites, the iridoid glycosides, as a model system, discuss variation in sequestration among different caterpillar species and how host plant species can affect sequestration, and then consider how global change, such as introduced plants, elevated nitrogen and carbon dioxide, and use of pesticides can affect caterpillar sequestration and their multitrophic interactions.

Historic Studies of Caterpillar Palatability/Unpalatability

Initial descriptions of unpalatability in insects focused on *adult* Lepidoptera. In 1862, Bates proposed that butterflies in what he designated as the Danaidae and Heliconiidae (the latter including an assemblage of taxa later shown to include three different groups: ithomiines, heliconiines, and acraeines) were unpalatable, writing "There is nothing apparent in their structure or habits which could render them safe from persecution by the numerous insectivorous animals which are constantly on the watch in the same parts of the forest which they inhabit. It is probable they are unpalatable to insect enemies" (Bates 1862, p. 510). Five years later, Wallace (1867b) tackled the question of brightly colored *caterpillars*, when, in a meeting of the Royal Entomological Society of London, he asked members to help him "clear up a difficult point": this being Darwin's conclusion that bright coloration in animals was due to sexual selection. Wallace noted that larval Lepidoptera were an exception to this rule and, "could not owe their gaudy attire to sexual selection."

Wallace wrote that "Just as certain moths were agreeable and others distasteful to birds, so also he did not doubt that certain larvae were agreeable and others distasteful to birds, but distastefulness alone would be insufficient to protect a larva unless there were some outward sign to indicate to its would-be destroyer that his contemplated prey would prove a disgusting morsel, and so deter him from attack" (Wallace 1867b p. lxxx–lxxxi).

To obtain more information about caterpillar palatability and coloration, Wallace inserted a notice in the March 23, 1867, issue of *The Field, The Country Gentleman's Newspaper*, in which he asked the help of readers to make observations that would be "of great interest to Mr. Darwin and myself." In this notice, entitled "Caterpillars and Birds" (Fig. 1), he asked readers to collect observations on bird acceptance or rejection of different caterpillar species. He asked for observations on which caterpillars birds ate or rejected; for readers to offer birds as many different caterpillar species as they could collect and observe bird responses; or to put caterpillars in the garden, "in a soup plate or other vessel, which must be placed in a larger vessel of water, so that the creatures cannot escape, and then after a few hours note which have been taken and which left." At the end of this note, he added, "this question has an important bearing on the whole theory of the origin of the colours of animals, and especially of insects" (Wallace 1867a, Fig. 1). This may have been one of the earliest, if not the earliest, call to arms of citizen scientists.

This request resulted in several sets of such observations, some of which were reported in meetings of the Royal Entomology Society. In 1869, Weir and Butler published back-to-back articles addressing Wallace's request and reporting their results with a variety of different caterpillar species. Weir (1869) tested a variety of adults, some pupae, and larvae of several species and found three of them to be rejected: *Diloba caeruleocephala* L. (Noctuidae) (feeds on deciduous trees and shrubs, especially *Sorbus, Prunus spinosa, Crataegus*), *Zygaena (Anthrocera) filipindulae* (Zygaenidae) (feeds on *Lotus corniculatus* and *L. pentaphyllum* (Fabaceae)), and *Cucullia verbasci* (Noctuidae) (feeds on *Verbascum* (Scrophulariaceae), which contains iridoid glycosides). Butler (1869) found *Abraxas grossulariata* (Geometridae) (feeds on *Ribes rubrum, R. nigrum, Prunus spinosa, Crataegus, Corylus, Euonymus europaeus, Salix*) and *Halia vauaria* (now *Macaria wauaria*) (Geometridae; feeds on *Ribes*) to be unpalatable. For certain of these species, we now know not only that they are unpalatable, but the likely source of their unpalatability; for example, *Zygaena filipendulae* sequesters cyanogenic glycosides (Zagrobelny et al. 2004, 2018; Zabrobelny and Møller 2011), and *C. verbasci* is likely to sequester iridoid glycosides (Bowers, pers. obs.).

Slater (1877), in his paper "On the food of gaily-coloured caterpillars," acknowledged these earlier observations by "modern entomologists" (Slater 1877, p. 205) and went on to be the first to suggest that these "…strikingly-coloured insects, not otherwise specially protected, will be found to feed upon poisonous plants, or upon such as, though not poisonous, possess unpleasant, or at least very powerful odours or flavours" (op.cit., p. 205). These observations led to numerous observations and experiments during the late 1800s and early 1900s about the palatability or unpalatability of many different caterpillar species (Poulton 1887, 1890 summarize many of the experiments done to date), using a variety of animals as predators, including

THE FIELD, THE COUNTRY GENTLEMAN'S NEWSPAPER. [MARCH 23, 1867.

CATERPILLARS AND BIRDS.

SIR,—May I be permitted to ask the co-operation of your readers in making some observations during the coming spring and summer, which are of great interest to Mr Darwin and myself. I will first state what observations are wanted, and then explain briefly why they are wanted. A number of our smaller birds devour quantities of caterpillars, but there is reason to suspect that they do not eat all alike. Now we want direct evidence as to which species they eat and which they reject. This may be obtained in two ways. Those who keep insectivorous birds, such as thrushes, robins, or any of the warblers (or any other that will eat caterpillars), may offer them all the kinds they can obtain, and carefully note (1) which they eat, (2) which they refuse to touch, and (3) which they seize but reject. If the name of the caterpillar cannot be ascertained, a short description of its more prominent characters will do very well, such as whether it is hairy or smooth, and what are its chief colours, especially distinguishing such as are green or brown from such as are of bright and conspicuous colours, as yellow, red, or black. The food plant of the caterpillar should also be stated when known. Those who do not keep birds, but have a garden much frequented by birds, may put all the caterpillars they can find in a soup plate or other vessel, which must be placed in a larger vessel of water, so that the creatures cannot escape, and then after a few hours note which have been taken and which left. If the vessel could be placed where it might be watched from a window, so that the kind of birds which took them could also be noted, the experiment would be still more complete. A third set of observations might be made on young fowls, turkeys, guineafowls, pheasants, &c., in exactly the same manner.

Now the purport of these observations is to ascertain the law which has determined the colouration of caterpillars. The analogy of many other insects leads us to believe that all those which are green or brown, or of such speckled or mottled tints as to resemble closely the leaf or bark of the plant on which they feed, or the substance on which they usually repose, are thus to some degree protected from the attacks of birds and other enemies. We should expect, therefore, that all which are thus protected would be greedily eaten by birds whenever they can find them. But there are other caterpillars which seem coloured on purpose to be conspicuous, and it is very important to know whether they have another kind of protection, altogether independent of disguise, such as a disagreeable odour and taste. If they are thus protected, so that the majority of birds will never eat them, we can understand that to get the full benefit of this protection they should be easily recognised, should have some outward character by which birds would soon learn to know them and thus let them alone; because if birds could not tell the eatable from the uneatable till they had seized and tasted them, the protection would be of no avail, a growing caterpillar being so delicate that a wound is certain death. If, therefore, the eatable caterpillars derive a partial protection from their obscure and imitative colouring, then we can understand that it would be an advantage to the uneatable kinds to be well distinguished from them by bright and conspicuous colours.

I may add that this question has an important bearing on the whole theory of the origin of the colours of animals, and especially of insects. I hope many of your readers may be thereby induced to make such observations as I have indicated, and if they will kindly send me their notes at the end of the summer, or earlier, I will undertake to compare and tabulate the whole, and to make known the results, whether they confirm or refute the theory here indicated. ALFRED R. WALLACE.
9, St. Mark's-crescent, Regent's Park, N.W.

Fig. 1 Alfred R. Wallace's notice in *The Field, the Country Gentleman's Newspaper*, in which he asked for readers to do experiments to investigate the acceptability or not of various caterpillar species. Accessed from http://wallace-online.org

lizards (Poulton 1887; Pritchett 1903; Eltringham 1909), birds (Butler 1869; Weir 1869, 1870; Poulton 1887), frogs (Poulton 1887), and spiders (Poulton 1887).

It was not until the 1960s and early 1970s that experimental work on butterfly and, to a lesser extent, caterpillar palatability and chemical defense began again in earnest. The early experiments by Jane Van Zandt Brower (JVZ Brower 1958a, b, c) and Lincoln Brower (Brower et al. 1968; Brower 1969) and the chemical and behavioral experiments by Miriam Rothschild and Thomas Reichstein (e.g., Rothschild et al. 1970) were the beginnings of what is now a burgeoning field of study, the field of chemical ecology. These pioneers, notably two of them women (J. VZ Brower and M. Rothschild) in a field dominated by men, accompanied by others investigating plant secondary compounds and their importance for herbivorous insects (see below), opened the doors to what is now an exploding area of investigation, from both basic and applied perspectives.

Caterpillar Sequestration and Unpalatability

Sequestration of plant secondary metabolites by caterpillars (and other insects) has been the focus of a number of reviews (e.g., Duffey 1980; Bowers 1993; Nishida 2002; Opitz and Müller 2009; Dobler et al. 2011; Heckel 2014). In an extensive and detailed review, Opitz and Müller (2009) document over 250 species of insects that sequester plant secondary compounds: these belonging to six orders: Coleoptera, Diptera, Hemiptera, Hymenoptera, Lepidoptera, and Orthoptera. Although this number does not make sequestration a "common phenomenon among herbivorous insects" (Petschenka and Agrawal 2016), it is also true that new examples of sequestration are frequently being discovered. For example, a few example reports since Opitz and Müller (2009) include the Death's Head Hawkmoth, *Acherontia atropos* (Sphingidae), which sequesters the alkaloid atropine (an anticholinergic agent) from *Atropa belladonna* (Solanaceae) (Kubinova et al. 2014); colchicine (an alkaloid and a mitotic inhibitor) sequestration by larvae of *Polytela gloriosae* (Noctuidae), the Lily Moth, from the *Gloriosa* lily, *Gloriosa superba* (Colchicaceae) (Sajitha et al. 2019); and sequestration of the iridoid glycoside, antirrhinoside, by larvae of *Calophasia lunula* (Noctuidae), the toadflax defoliator, from dalmatian toadflax (*Linaria dalmatica*, Plantaginaceae) (Jamieson and Bowers 2010). Thus, it is likely that the numbers of sequestering species will continue to increase as more species are investigated.

A number of different classes of plant secondary metabolites are sequestered by caterpillars (Nishida 2002; Opitz and Müller 2009; Heckel 2014). Especially well-studied are the alkaloids (e.g., Hartmann and Ober 2000 Wink 2019), cardenolides (Rothschild et al. 1970; Dobler et al. 2011; Petschenka et al. 2013), cyanogenic glycosides (Engler-Chaouat and Gilbert 2007; Zagrobelny and Moller 2011), glucosinolates (Müller 2009; Winde and Wittstick 2011; Halkier and Gershenzon 2006) and iridoid glycosides (Bowers 1991, 1993; Dobler et al. 2011). Other classes of compounds that are sequestered include grayanoids (Nishida 2002), terpenoids

(Opitz and Müller 2009), cycasin (Rothschild et al. 1986; Bowers and Larin 1989; Nash et al. 1992), and phenolics (Hesbacher et al. 1995; Scott et al. 2014; Scott Chialvo et al. 2018).

In some, but not most, cases, data from chemical analyses and feeding experiments with potential predators are both available, thus allowing linkage of palatability/unpalatability with levels of chemical sequestration; but this information is not always available nor feasible. This is certainly an area inviting further investigation. For example, early experiments on caterpillar palatability to predators showed that certain species were acceptable while others were avoided, but the basis of predator rejections was not determined (see above) and more recent experiments (e.g., Dyer 1995) did not identify the chemical basis of caterpillar rejection. Similarly, while sequestration of various plant secondary metabolites has been documented in a variety of species (Opitz and Müller 2009), the consequences for palatability of those insects are not always known, and almost never to the full range of possible natural enemies.

Iridoid Glycosides as a Model System

The iridoid glycosides are terpenoid-derived compounds found in over 50 families of plants (Hegnauer 1973; Bowers 1991; Jensen 1991; Rimpler 1991). Because of their medical importance (Dinda 2019), the last few decades have seen increased emphasis on the discovery and isolation of iridoid glycosides; thus in 1980, there were about 500 iridoid glycosides known (El-Naggar and Beal 1980) and a recent book (Dinda 2019) reported over 3000 iridoids! Many traditional medicinal plants apparently owe their properties to iridoid glycosides, and increasing interest in the pharmacological value of these compounds has led to rapid discovery of new compounds (Dinda 2019 and references therein). Indeed, their wide range of bioactivities supports a number of therapeutic possibilities for these compounds (Tundis et al. 2008; Dinda 2019), and the use of valerian (*Valeriana officinalis*, Valerianaceae) as a treatment for insomnia has been known for decades (Shinjyo et al. 2020).

Probably the first suggestion of the importance of iridoid glycosides for caterpillars was a paper by Nayar and Fraenkel (1963), in which they suggested that "catalposides" (a mixture of iridoid glycosides found in *Catalpa* spp., Bignoniaceae) served as feeding stimulants for larvae of the catalpa sphinx, *Ceratomia catalpae* (Sphingidae). Slightly later, Hegnauer (1973, Volume 6, page 352) in his classic series on plant secondary chemistry, *Chemotaxonomie der Pflanzen*, suggested that iridoid glycosides were important in the host plant relationships of checkerspot butterflies (*Euphydryas* spp.). In 1979, Bowers suggested that these compounds were also important in the unpalatability of caterpillars and adults of butterflies in the genus *Euphydryas* (Nymphalidae). The anti-feedant properties of iridoid glycosides were first illustrated with the iridoid glycoside, ipolamiide, which was shown to be a feeding deterrent to a generalist caterpillar, *Spodoptera littoralis* (Noctuidae), as well as two generalist grasshopper species (Bernays and DeLuca 1981). Sequestration of iridoid glycosides was first described by Bowers and Puttick (1986) in larvae of three species that specialize on plants containing iridoid glycosides, *Euphydryas phaeton* (Nymphalidae, the Baltimore checkerspot), *Junonia coenia* (Nymphalidae, the buckeye), and *C. catalpae*, and by Stermitz et al. (1986) for adults of another checkerspot species, *E. anicia* (Nymphalidae, the anicia checkerspot).

Experiments demonstrating unpalatability of larvae of taxa that sequester iridoid glycosides are less common than chemical analysis; however, there have been several. In early experiments, Bowers (1980, 1981) showed that adult checkerspots in the genus *Euphydryas* were unpalatable to blue jays, *Cyanocitta cristata*, and that larvae of *E. phaeton* were also unpalatable to these birds (Bowers 1980). Later experiments showed that adult checkerspots were also unpalatable to gray jays, *Perisoreus canadensis* (Bowers and Farley 1990). Subsequent experiments with larvae of the buckeye (*J. coenia*) showed that ants (Dyer and Bowers 1996), spiders (Theodoratus and Bowers 1999; Strohmeyer et al. 1998), paper wasps (Stamp 2001), and praying mantis (Fig. 2, Bowers and Massa unpublished) found larvae of the buckeye to be unpalatable; however, unpalatability depended on the host plant species on which caterpillars had fed. In general, both vertebrate and invertebrate

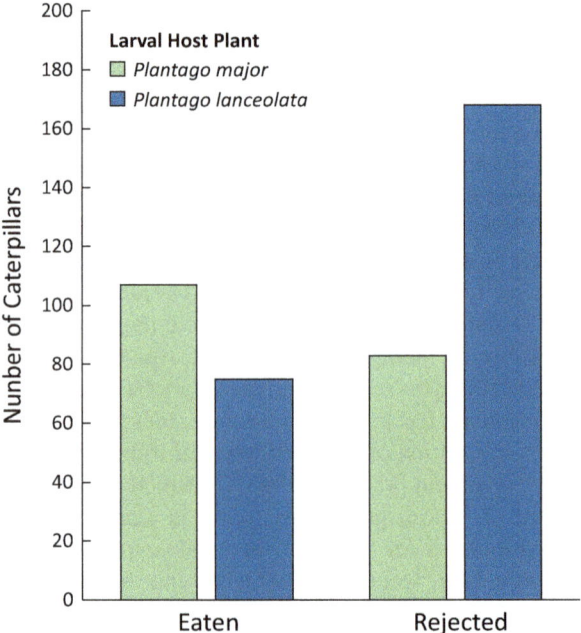

Fig. 2 Praying mantis (third instar) behavior toward third instar larvae of the buckeye (*Junonia coenia*) when they were reared on *Plantago major* (contains only aucubin and in relatively low amounts of 0.5–2% dry mass) or *P. lanceolata* (contains aucubin and catalpol in amounts from 5% to 12% dry mass) throughout their development. Caterpillar chemical content was not measured. Mantis ate significantly more larvae reared on *P. major* than those reared on *P. lanceolata* ($\chi^2 = 5.63$, P < 0.025) and rejected significantly more larvae reared on *P. lanceolata* than those reared on *P. major* ($\chi^2 = 28.78$; P < 0.001). Unpublished data from Massa and Bowers

predators behave similarly toward larvae sequestering iridoid glycosides: they find them distasteful and reject those sequestering higher levels of these compounds (Bowers 1980, 1981; Bowers and Farley 1990; op.cit.). Variation in palatability and chemical defense among individuals within a species can be due not only to the host plant on which a caterpillar feeds (Bowers 1980; Theodoratus and Bowers 1999), but also to the ontogenetic stage of the particular plant species used as food (Quintero et al. 2014; Quintero and Bowers 2018).

The ability to sequester iridoid glycosides has been documented in four different orders (reviewed in Rimpler 1991): Hemiptera (Nishida and Fukami 1989), Hymenoptera (Bowers et al. 1993), Coleoptera (Willinger and Dobler 2001; Baden et al. 2011), and Lepidoptera (Rimpler 1991; Bowers 1993; Fig. 3, Table 1) and the grasshopper, *Romalea guttata*, can sequester the non-glycosidic iridoid, nepetalactone (Blum et al. 1987). Within the Lepidoptera, sequestration has been documented in several different families (Rimpler 1991; Table 1), including both butterflies and moths. For iridoid glycosides, not all can be sequestered; some are metabolized or broken down, and these metabolites are excreted (Rimpler 1991). For some iridoid glycosides a more complex compound is metabolized into one that can be sequestered. Such is the case with the catalpa sphinx, *Ceratomia catalpae* (Sphingidae), in which the iridoid glycoside, catalposide, is broken down into catalpol and a metabolite, and the catalpol sequestered (Bowers 2003). Thus the host plant compounds may not be directly mirrored in the sequestering caterpillar.

While most species shown to sequester iridoid glycosides are specialists on plants containing these compounds, a few of these species are quite general in their feeding habits, for example, *Spilosoma congrua* (Erebidae, Arctiinae) (Robinson et al. 2002) and the painted lady, *Vanessa cardui* (Nymphalidae) (Robinson et al. 2002). For both of these species, iridoid glycoside sequestration occurs when they feed on the introduced weed, narrow-leaved plantain (*Plantago lanceolata*, Plantaginaceae) (Bowers and Stamp 1997; Lampert et al. 2014), although levels of sequestered iridoids are lower than those of most specialists (Fig. 3).

Caterpillar Sequestration: Comparisons Across Species

Although sequestration is found in many different lepidopteran groups, the ability of different species to sequester is quite variable, even in closely related species. The monarch (*Danaus plexippus*, Nymphalidae) and the queen (*Danaus gilippus*) butterflies served as early models for the study of unpalatability and sequestration. JVZ Brower (1958a) showed that both monarch and queen adults were unpalatable, although monarchs were more unpalatable than queens; later, L Brower showed that larvae of the monarch were unpalatable (Brower et al. 1967). Chemical analysis and more direct comparisons across species showed that queen butterflies sequestered lower amounts of cardenolides than monarchs when reared on the same host plant species and that the palatability of these two species differed (Cohen 1985; Malcolm 1991).

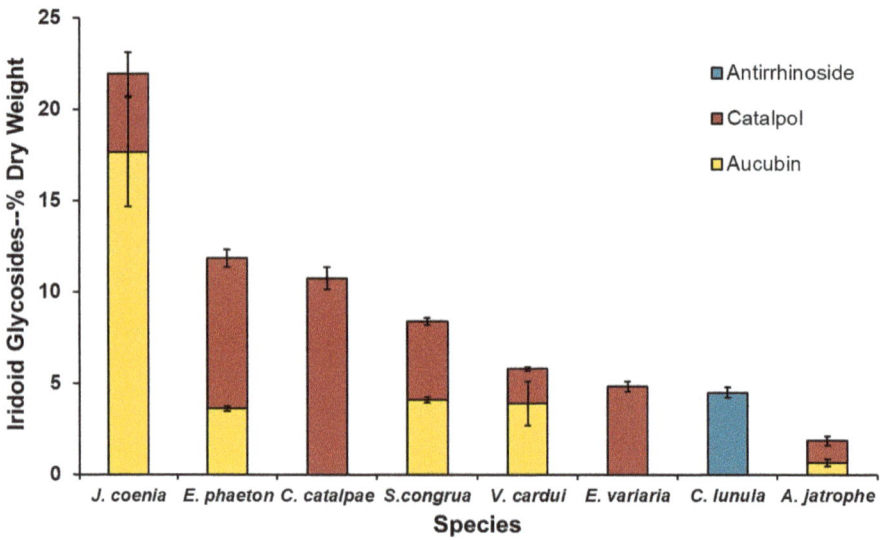
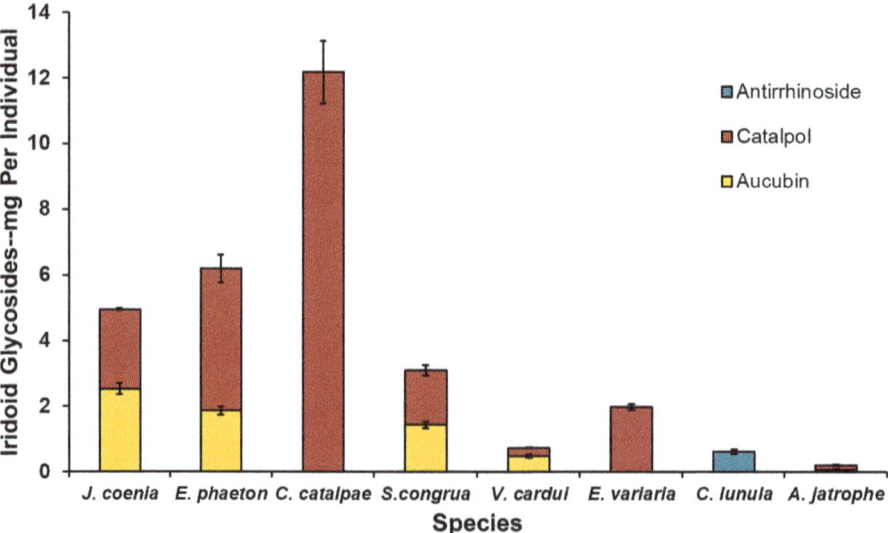

Fig. 3 Comparison of sequestration of iridoid glycosides in eight lepidopteran species. All measurements were made on newly molted larvae in the last larval instar before pupation; this was the fifth instar in all but *E. phaeton*, in which it was the sixth instar. Table 2 shows the larval host plants on which larvae were fed. Iridoid glycosides were quantified using gas chromatography. Sample sizes range from 10 to 16. A. Iridoid glycosides as percent dry weight of larvae. B. Iridoid glycosides as milligrams per individual

Table 1 Lepidopterans shown in Fig. 4 and their taxonomic affiliations and the host plants on which larvae were reared to obtain sequestration data

Species	Family	Tribe	Host plant (family)	References
Junonia coenia	Nymphalidae	Junoniini	*Plantago lanceolata* (Plantaginaceae)	Knerl and Bowers (2013)
Euphydryas phaeton	Nymphalidae	Melitaeini	*Plantago lanceolata* (Plantaginaceae)	Bowers (unpublished data)
Vanessa cardui	Nymphalidae	Nymphalini	*Plantago lanceolata* (Plantaginaceae)	Lampert et al. (2014)
Anartia jatrophae	Nymphalidae	Victoriini	*Plantago lanceolata* (Plantaginaceae)	Knerl and Bowers (2013)
Ceratomia catalpae	Sphingidae	Sphingini	*Catalpa bignonioides* (Bignoniaceae)	Bowers (2003)
Spilosoma congrua	Erebidae	Arctiini	*Plantago lanceolata* (Plantaginaceae)	Bowers and Stamp (1997)
Eucaterva variaria	Geometridae	Ourapterygini	*Chilopsis linearis* (Bignoniaceae)	Bowers (unpublished data)
Calophasia lunula	Noctuidae	Oncocnemidini	*Linaria dalmatica* (Plantaginaceae)	Jamieson and Bowers (2010)

Sequestration of cardenolides has been examined more broadly in danaines; although most of this research is focused on adult cardenolide content, larval stages have been investigated as well. A comparison of larvae of three danaine species, which differ in the resistance of their sodium/potassium ATPase (NA^+/K^+-ATPase) to cardenolides (*Euploea core*, not resistant; *Danaus gilippus*, intermediate resistance; *D. plexippus*, highly resistant; Petschenka and Agrawal 2015), showed that these species also differ in their ability to sequester cardenolides (Petschenka and Agrawal 2015). The resistance of this enzyme is due to specific amino acid substitutions that alter its cardenolide binding properties (termed target site specificity) (Dobler et al. 2012; Petschenka and Agrawal 2015). Thus monarchs, which have a highly resistant NA^+/K^+-ATPase, have an amino acid substitution that prevents cardenolides from effectively binding to the enzyme (Vaughan and Jungreis 1977; Holzinger et al. 1992). Larvae were reared on eight different milkweed species and, for all host plants, *E. core* sequestered no detectable cardenolides, *D. gilippus* was intermediate, and *D. plexippus* sequestered the highest amounts (Petschenka and Agrawal 2015). Thus, larvae of these three species vary considerably in their ability to sequester cardenolides.

Similarly, heliconiine species sequestering cyanogenic glycosides may also vary in their ability to sequester these compounds, with some species being very efficient and others being poor sequesterers (Engler-Chaout and Gilbert 2007; Sculfort et al. 2020). Host plant levels of cyanogenic glycosides were also important in

Fig. 4 Caterpillars for which data are shown in Fig. 3. (**a**) *Euphydryas phaeton* (Nymphalidae), the Baltimore checkerspot. Photo by M.D. Bowers. (**b**) *Junonia coenia* (Nymphalidae), the buckeye. Photo by M.D. Bowers. (**c**) *Eucaterva variaria* (Geometridae). Photo by M.S. Singer. (**d**) *Vanessa cardui* (Nymphalidae). Photo by K. Hernandez. (**e**) *Spilosoma congrua* (Erebidae), the agreeable tiger moth. Photo by M.D. Bowers. (**f**) *Ceratomia catalpae* (Sphingidae). Photo by M.D. Bowers. (**g**) *Calophasia lunula* (Noctuidae), the toadflax defoliator. Photo by M. Jamieson. H. *Anartia jatrophae* (Nymphalidae), the white peacock. Photo by N. Muchoney

determining levels found in the insects. Furthermore, heliconiine larvae may produce cyanogenic glycosides de novo, and there may be a trade-off between sequestration and de novo production of these compounds (Engler-Chaout and Gilbert 2007).

The ability to sequester iridoid glycosides has arisen in a number of different lepidopteran taxa, occurring in five different families: Nymphalidae, Erebidae, Noctuidae, Geometridae, and Sphingidae (Rimpler 1991; Table 1). In order to directly compare the efficiency with which larvae of eight different species, representing these five families, sequester iridoid glycosides, data were compiled for iridoid glycoside content of newly molted last instar larvae. Newly molted larvae were used to minimize any plant material being found in the gut. Iridoid glycosides are polar compounds and are sequestered in the insects' hemolymph (Bowers 2003). Five of these species were reared on the same host plant species, *Plantago lanceolata* (Plantaginaceae) (Table 1), a common weed introduced into North America about 200 years ago (Cavers et al. 1980). This plant species contains primarily two iridoid glycosides, aucubin and catalpol (Bowers et al. 1992a, b). The others were reared on the host plant on which they most commonly feed: *C. catalpae* was reared on *Catalpa bignonioides* (Bignoniaceae), which contains primarily catalpol and catalposide (von Poser et al. 2000; Bowers 2003) (only catalpol is found in larvae); *Eucaterva variaria* was reared on *Chilopsis linearis*, which contains catalpol and esters of catalpol (von Poser et al. 2000) (only catalpol is found in larvae, Bowers unpublished data); *Calophasia lunula* was reared on *Linaria dalmatica*, which contains antirrhinoside, linarioside, and several other iridoid glycosides (Handjieva et al. 1993) (only antirrhinoside is found in larvae, Jamieson and Bowers 2010). In some cases, such as *C. catalpae* and *E. variaria* above, as well as others (e.g., Gardner and Stermitz 1988; Kelly and Bowers 2018), caterpillars are converting esters of catalpol, such as catalposide, 6-isovanillylcatalpol (Gardner and Stermitz 1988), and scutellarioside (Kelly and Bowers 2018) into catalpol (see also Rimpler 1991).

This compilation shows that the ability to sequester iridoid glycosides can vary immensely and over an order of magnitude, from very high levels (15–25% dry mass, 6–12 mg per caterpillar), such as found in *J. coenia* and *C. catalpae*, to very low levels, such as found in (Nymphalidae) (2% dry mass, 0.2 mg per caterpillar) (Fig. 3). Data are presented as both percent dry mass and total milligrams per caterpillar because both of these measures are important in the interactions of the caterpillars with enemies. The concentration of compounds will be important for enemies that attack, but may not ingest, the entire caterpillar—higher concentrations in a drop or two of hemolymph will provide greater deterrence. The total milligrams per caterpillar will be important for enemies that ingest an entire larva—a higher dose could lead to greater toxicity or a stronger physiological response. In contrast to what is known for other groups of compounds, such as cardenolides, in which levels of sequestration among different species are linked to the sensitivity of the targeted enzyme to these compounds (Petschenka and Agrawal 2015), little is known about what regulates sequestrative ability in iridoid glycoside-sequestering species.

Caterpillar Sequestration and Unpalatability: Importance of Host Plant Variation

On what plant species, in which plant population, and on which individual plant, ontogenetic stage, and plant part a caterpillar feeds can determine levels of caterpillar sequestration. Likewise, insect features, such as species, ontogenetic stage, genotype, and interactions of hostplant chemistry and nutrient content are also determinants of levels of compounds that are sequestered (Bowers 1993). Amounts of compounds in the host plants, however, may not directly predict levels that are sequestered; thus, for cardenolides, monarch caterpillars are more efficient at sequestering when levels of cardenolides are low (Jones et al. 2019). Variation in chemical content of host plant species and the consequences of that variation for sequestering herbivores have been most extensively examined in milkweeds (Apocynaceae) and monarchs (e.g., Roeske et al. 1976; Jones et al. 2019; Zust et al. 2019). Milkweed cardenolide content can vary by orders of magnitude (e.g., Zust et al. 2019), and this variation results in monarchs that vary in their cardenolide content (Malcolm 1991) as well as their palatability to both vertebrate (Brower et al. 1967) and invertebrate (Rayor 2004) predators, and their susceptibility to parasites (Sternberg et al. 2012). Similarly, in members of the genus *Heliconius*, in which caterpillars feed on plants in the Passifloraceae and sequester cyanogenic glycosides, host plant species can influence the amounts of cyanogenic glycosides sequestered (Hay-Roe and Nation 2007; Engler-Chaouat and Gilbert 2007). Chemical defense in *Heliconius*, however, is complicated by the fact that cyanogenic glycosides may also be synthesized de novo (Nahrstedt and Davis 1983; Wray et al. 1983; Engler-Chaouat and Gilbert 2007).

For insects sequestering iridoid glycosides, host plant species may also determine levels of sequestered compounds, as well as which specific compounds are sequestered (Gardner and Stermitz 1988; Rimpler 1991; Dobler et al. 2011). For example, Gardner and Stermitz (1988) and L'Empereur and Stermitz (1990) showed that two different populations of the checkerspot, *Euphydryas anicia* (Nymphalidae) in Colorado, sequestered different iridoid glycosides, even though the same two host plant species (*Castilleja integra*, Orobanchaceae, and *Besseya alpina* Plantaginaceae) were available in both populations. Specifically, they found that most butterflies in one population (Red Hill) contained large amounts of one iridoid glycoside, macfadienoside, that was found only in one host plant species, *C. integra*, while butterflies at the other site (Michigan Hill) did not contain this compound, indicating that they did not use *C. integra* at this site. The effects of this difference in sequestration for interaction with natural enemies were not examined, however.

The influence of host plant species on iridoid glycoside sequestration may vary depending on the caterpillar species examined. For example, in a comparison of three caterpillar species that sequester iridoid glycosides, larvae were reared on *Plantago lanceolata* or *P. major* (Lampert and Bowers 2010); *P. lanceolata* contains primarily two iridoid glycosides, aucubin and catalpol (Bowers and Stamp 1993), whereas *P. major* contains only aucubin (Barton and Bowers 2006, and references therein). Amounts of iridoid glycosides in these two plant species also vary considerably, with *P. major* containing 0.5–2.0% dry weight aucubin (Barton and Bowers 2006) and *P. lanceolata* containing amounts of aucubin and catalpol combined as high as 10–12% dry weight (Bowers and Stamp 1993). The three caterpillar species compared were the buckeye, *J. coenia* (a specialist on iridoid glycoside containing plants), and two generalist arctiines, the agreeable tiger moth, *Spilosoma congrua* (Erebidae, Arctiini) and the saltmarsh caterpillar, *Estigmene acraea* (Erebidae, Arctiini). For two of the three species, *J. coenia* and *S. congrua*, levels of iridoid glycosides were lowest when larvae fed on *P. major* and highest when larvae fed on *P. lanceolata*. However, for *E. acraea*, levels of iridoid glycosides were quite low, and host plant species did not affect iridoid glycoside levels (Lampert and Bowers, 2010).

Within a plant species, populations may vary in their chemical content (e.g., Darrow and Bowers 1997; Jamieson and Bowers 2010; Moore et al. 2013; Pellessier et al. 2014; Glassmire et al. 2016; Hahn and Maron 2016), which may then influence sequestration and chemical defense of insects feeding on them. For example, population variation in cyanogenic glycoside content of the passionflower species, *Passiflora biflora*, was reflected in the cyanogenic glycoside content of *Heliconius erato* feeding on those plants (Mattila et al. 2020). In contrast, variation in sequestered aristolochic acids in larvae of the pipevine swallowtail (*Battus philenor*, Papilionidae) depended more on the larval family line than on variation in host plant (*Aristolochia erecta*, Aristolochiaceae) aristolochic acids (DiMarco et al. 2012). Thus, the consequences of plant chemical variation for sequestration may be quite different in different caterpillar species.

Plants may also change dramatically in their chemical content as they develop (Bowers et al. 1992a, b; Darrow and Bowers 1997; Boege and Marquis 2005; Barton and Koricheva 2010; Boege et al. 2019), and these ontogenetic changes may interact with ontogenetic trajectories in herbivores to determine caterpillar chemical defenses (Quintero et al. 2014; Boege et al. 2019; Jones et al. 2019). For example, later instar larvae of the monarch accumulated greater total amounts of cardenolides, although there were different patterns for caterpillar body tissues compared to hemolymph (Jones et al. 2019). There was actually a decrease in body tissue cardenolides over development; however, hemolymph cardenolides showed more complex patterns, with a significant increase in hemolymph cardenolides between the fourth and fifth instars (Jones et al. 2019). In larvae of the buckeye, sequestration of iridoid glycosides increases with larval stage; however, an experiment with *P. lanceolata* showed that plant ontogenetic stage interacted with larval instar such that variation in levels of iridoid glycosides sequestered by different larval instars changed as a function of host plant ontogenetic stage (Quintero and Bowers 2018). Thus, variation in sequesterable host plant secondary metabolites may have important consequences for caterpillar sequestration, but the dynamics of this relationship may not always be clear-cut.

Caterpillar Sequestration and Chemical Defense: Consequences of Anthropogenic Change to the Environment

Introduced Plants

The introduction of exotic plants into novel ecosystems can have profound effects on the native inhabitants. Such introductions are increasing and they are relatively common, with widespread impacts on native lepidopteran herbivores (Graves and Shapiro 2003). These introductions can generate novel interactions with native species, with resulting effects on interactions with other organisms, as well as community structure. Incorporation of these novel host plants into the diet of native caterpillars can affect their population dynamics and interactions with higher trophic levels, such as predators, parasitoids, and pathogens. For insect species that sequester chemical compounds from their host plants, novel host plants may alter both quality (which compounds are sequestered) and quantity (amounts of compounds that are sequestered) of sequestered compounds, as well as affect whether or not compounds are sequestered at all (e.g., Knerl and Bowers 2013).

Asclepias curassavica (Apocynaceae), commonly known as tropical milkweed or blood flower, has been introduced into North America and is commonly grown in gardens throughout the southern United States (Malcolm 2018). It is a favored host plant of monarch butterflies and is used extensively in experiments with monarchs (e.g., Brower et al. 1967; Faldyn et al. 2018; Decker et al. 2019; Jones et al. 2019) because it is easy to grow and larvae perform well on it (Malcolm 2018). It is relatively high in cardenolides compared to many native North American milkweeds and thus caterpillars, and the resulting butterflies, reared on it are well-protected against predators and pathogens (De Roode et al. 2008; Tan et al. 2019). On the negative side, however, *A. curassavica* can provide a food resource that is available all year round, thus breaking reproductive diapause (Majewska and Altizer 2019) and reducing the propensity of monarchs to migrate (Malcolm 2018). Feeding on this species can also increase disease prevalence in populations that are not migrating (Satterfield et al. 2015). Thus, feeding on this introduced milkweed can have both positive (increased chemical defense against predators, reduced susceptibility to a parasite) and negative (breaking reproductive diapause and reduced migratory propensity) effects on monarchs.

Plantago lanceolata and *P. major* are two common weeds introduced into North America in the eighteenth and seventeenth centuries (respectively) (Cavers et al. 1980; Hawthorne 1974). A number of native caterpillar taxa have incorporated these species into their diets (Robinson et al. 2002). Several of these caterpillar species sequester iridoid glycosides from these plants, and which of these two plants caterpillars feed upon can influence their sequestration (e.g., Lampert and Bowers 2010) and their susceptibility to natural enemies (Theodoratus and Bowers 1999; Fig. 2), likely due to the differences in the amounts and specific iridoid glycosides that are sequestered from these two plant species. Wolf spiders (*Lycosa carolinensis*) found caterpillars of the buckeye, *J. coenia*, to be more unpalatable when reared on *P.*

lanceolata than when reared on *P. major* (Theodoratus and Bowers 1999). Similarly, praying mantis (*Mantis religiosa*) showed similar behaviors, accepting many fewer buckeye caterpillars that had been reared on *P. lanceolata* compared to those reared on *P. major* and rejecting more of the *P. lanceolata* reared individuals (Fig. 2).

The Baltimore checkerspot, *Euphydryas phaeton* (Nymphalidae), as well as other species of *Euphydryas* in North America, have also incorporated *P. lanceolata* into their diets (Stamp 1979; Bowers 1980; Thomas et al. 1987; Bowers et al. 1992a, b; Haan et al. 2018, 2021). For *E. phaeton*, use of this novel host plant, compared to use of the native host plant, turtlehead (*Chelone glabra*, Plantaginaceae), has a number of negative effects. Feeding on *P. lanceolata* results in more palatable caterpillars and butterflies (Bowers 1980) and reduced immune response and greater prevalence of an entomopathogenic virus (Muchoney et al. in review). However, population growth rates of *E. phaeton* can be higher when they are feeding on *P. lanceolata* (Brown et al. 2017). In a different checkerspot species, use of *P. lanceolata* by populations of *Euphydryas editha* in California may result in extinction of populations that come to depend on this novel host plant species (Singer and Parmesan 2018). In contrast, use of *P. lanceolata* by the endangered *E. editha taylori* in Washington state has positive effects, with early instar larvae in the field performing better on this introduced species than on its native Orobanchaceae host plants, *Castilleja levisecta* and *C. hispida* (Haan et al. 2018). Early instar larvae feeding on *P. lanceolata* also had higher levels of iridoid glycosides than larvae feeding on either of the native host plant species (Haan et al. 2021). Thus, even within this relatively small genus, the effects of this introduced plant may vary considerably.

A recent incorporation of *P. lanceolata* into the diet of the native white peacock butterfly, *Anartia jatrophae* (Nymphalidae), resulted in larvae being able to sequester iridoid glycosides from this novel host plant, with fourth instar larvae containing a mean of about 4% dry mass iridoid glycosides (Knerl and Bowers 2013). While this is substantially lower than what was found in larvae of the buckeye, *J. coenia* (Knerl and Bowers 2013; Fig. 3), it could be sufficient to protect these larvae from enemies (Knerl and Bowers 2013). Similarly, larvae of the painted lady, *Vanessa cardui*, can sequester low amounts of iridoid glycosides when they feed on *P. lanceolata* (Fig. 3), but the efficacy of these sequestered compounds in protecting the larvae from attack has not been investigated.

Other Types of Anthropogenic Change

As discussed above, introduced plants may have important effects on caterpillar sequestration and the interaction of these caterpillars with their natural enemies; these effects can be positive or negative. Other components of human-induced changes to the environment can also be important for caterpillar sequestration; increased levels of carbon dioxide, increasing nitrogen deposition, application of herbicides and pesticides, and changes in precipitation and thermal regime may influence caterpillar sequestration indirectly, via changes in plant chemical content or phenology, or directly, by changing caterpillar feeding rates or metabolism (Robinson et al. 2012; Jamieson and Bowers 2012; Jamieson et al. 2017; Schultz et al. 2016; Hamann et al. 2021). Relatively few studies have examined the direct or indirect effects of such anthropogenic changes on insect chemical defenses of any kind, much less caterpillar sequestration, although authors may allude to such effects (e.g., Veteli et al. 2002). Those studies that have been conducted do not reveal a consistent picture; this is clearly an area worthy of increased attention.

Increased soil nitrogen, through atmospheric deposition or agricultural run-off, can alter plant secondary metabolite concentrations (reviewed in Throop and Lerdau 2004; Jamieson et al. 2017), and a number of different studies have looked at the consequences of increased soil nitrogen for plant chemistry and caterpillar performance (reviewed in Hunter 2016; Li et al. 2016; Jamieson et al. 2017). Many fewer, however, have examined the effects on sequestration. For iridoid glycosides, fertilization was shown to decrease iridoid glycosides in *Plantago lanceolata*, and caterpillars fed on fertilized *P. lanceolata* sequestered about four times less iridoid glycosides than those fed on unfertilized plants (Prudic et al. 2005). In addition, the relative proportion of the two sequestered compounds also changed in caterpillars fed on fertilized plants: the proportion of total iridoid glycosides sequestered that was catalpol was about 75% of total iridoid glycosides when caterpillars were fed on fertilized plants and only 20% of total when fed on unfertilized plants (Prudic et al. 2005). In another iridoid-sequestering caterpillar, the toadflax defoliator, *Calophasia lunula* (Erebidae), nitrogen addition reduced iridoid glycosides in the host plant, the invasive weed, dalmatian toadflax (*Linaria dalmatica*, Plantaginaceae), and caterpillars feeding on these fertilized plants also had lower iridoid glycosides (Jamieson and Bowers 2012). For monarchs, the effects of nitrogen fertilization appear to be more complex: Tao et al. (2014) found no effect of nitrogen fertilization on cardenolide concentrations in monarch butterflies, but the sequestration efficiency (the proportion of ingested defense that is retained by the herbivore (Bowers and Collinge 1992)) was significantly lower when caterpillars were fed fertilized plants.

Elevated carbon dioxide levels can also have important consequences for trophic interactions, through both direct and indirect effects on insect and plant physiology, as well as insect behavior (Jamieson et al. 2017). In only a few cases, however, have insect defenses been examined in this context, and more rarely has sequestration been examined. In monarchs feeding on milkweeds, effects of elevated CO_2 on milkweed cardenolides differed among milkweed species, with only one species, *Asclepias curassavica*, showing a significant effect: surprisingly, cardenolides were reduced by about 50% in plants grown under elevated (760 ppm) compared to ambient (400 ppm) CO_2 levels (Fig. 1 in Decker et al. 2019). Despite this difference in plant chemistry, there were no differences in cardenolide concentrations of wings of monarchs fed on *A. curassavica* plants grown under ambient versus elevated CO_2 (Decker et al. 2019). However, the efficiency with which larvae sequestered cardenolides (cardenolides sequestered per unit cardenolide available in host plants) was higher when fed on milkweeds grown under elevated CO_2 (Decker et al. 2019).

Thus, when cardenolides are lower (as when plants are grown under elevated CO_2), monarch larvae are more efficient at sequestering them, even though the insects are still lower in total cardenolides.

The application of pesticides and herbicides can also alter plant secondary chemistry (Lydon and Duke 1989); however, the effects of such applications on sequestering caterpillars have only rarely been examined. One such study examined the effects of the application of grass-specific herbicides (graminicides) on larval sequestration of the checkerspot, *Euphydryas colon* (Schultz et al. 2016). Application of graminicides can be effective in managing exotic grasses (Marushia and Allen 2011) and be beneficial for maintaining and restoring butterfly habitat (Blake et al. 2011). However, there may be non-target, negative effects of these graminicides, and different butterfly species show different responses (LaBar and Schultz 2012). In the Schultz et al. (2016) experiment, larvae were fed on *Plantago lanceolata* and then separated from the host plant and exposed directly to three different graminicides or were untreated (controls). When caterpillars entered diapause (fourth instar), a set of larvae were analyzed for iridoid glycoside sequestration. Results showed that, although exposure to graminicides did not affect overall amounts of iridoid glycosides sequestered, graminicide treatment did alter the relative proportions of the two compounds sequestered (aucubin and catalpol): specifically, caterpillars treated with graminicides had 1.5 to two times the amount of aucubin as catalpol, while the control caterpillars had about 1.5 times more catalpol than aucubin. Earlier experiments have shown that catalpol appears to be the more toxic of these two iridoid glycosides (Bowers and Puttick 1988; Puttick and Bowers 1988); thus caterpillars exposed to graminicides could be less toxic to enemies because they have lower catalpol contents.

Conclusions and Future Directions

Studies of caterpillars and their interactions with their host plants and enemies have served as important cornerstones for the fields of ecology and evolution. Continuing to explore these and other aspects of the chemical ecology of multitrophic interactions is crucial to our understanding of the complex drivers of these interactions, their impacts across trophic levels, and their role in community and ecosystem interactions. For caterpillars that are chemically defended by sequestering compounds from the plants on which they feed, there are still many questions to be answered. The examples proposed below are only a few of the many fascinating questions that remain to be addressed using sequestering caterpillars; there are certainly many more.

1. **Why are some species able to sequester a particular class of compounds, but even closely related species cannot?**

For most classes of compounds, we know relatively little about what mechanisms allow sequestration in some species but not others; and in sequestering species, what prevents autotoxicity (with cardenolides being an exception (e.g., Dobler et al. 2012)). For example, the Catalpa sphinx, *Ceratomia catalpae*, sequesters iridoid glycosides, but its congener, *C. undulosa*, cannot (Lampert and Bowers 2014); similarly, *Spilosoma congrua* sequesters iridoid glycosides, but its congeners *S. virginica* and *S. latipennis* do not (Bowers and Stamp 1997). Identifying the physiological, biochemical, and molecular drivers of sequestration is ultimately key to determining how different species deal with plant secondary metabolites.

2. **How does sequestration of chemical compounds correspond to efficacy of defense against enemies?**

As previously noted, there are relatively few studies where both palatability to predators, or effectiveness against parasitoids or pathogens, and identification and quantification of caterpillar chemical defenses are directly linked, in either the field or the lab. Caterpillars are attacked by multiple natural enemies, and the primary enemies attacking different life stages can vary dramatically (Hawkins et al. 1997); furthermore, the effectiveness of chemical defenses may vary among these different enemies. Other issues to consider are that enemies may respond to caterpillar chemical defenses in different ways. They may respond in a dose-dependent manner, such that the higher the levels of defense, the more deterrent an individual or group of individuals will be. Alternatively, enemies may show a threshold effect, in which above a certain level, it will not matter how much an individual sequesters. Because different types of enemies may differ in their mode of response and thus their strength as selective agents, understanding these features of different types of enemies is an important component of assessing selection on caterpillar chemical defenses.

Experiments investigating the effectiveness of chemical defenses against different types of enemies are also needed. For example, bird (Fink et al. 1983) and mouse (Glendinning 1990, 1993) species differ in their response to cardenolides. In addition, parasitoids may be able to deal with high levels of defense compounds in their caterpillar prey that predators cannot tolerate (Lampert et al. 2010); thus sequestering caterpillars may serve as safe havens for parasitoids (Dyer and Gentry 1999; Smilanich et al. 2009). In an experiment using three different types of predators (ants, predatory wasps, bugs), Dyer (1997) found that each predator was influenced by a different set of caterpillar defenses. Understanding how different enemies respond to sequestered compounds, how variation in the amounts and kinds of these compounds is important for these enemies, and how these enemies might serve as selective agents in natural populations is certainly a productive research endeavor.

3. **What are the anticipated consequences of global change for caterpillar sequestration and chemical defense?**

Levels of host plant secondary metabolites are influenced by both natural and anthropogenic forces; how such changes affect sequestering caterpillars is complex. The direct impacts of such changes on the physiology of the caterpillars themselves are relatively little studied. For example, there may be trade-offs between allocation of resources to sequestration and other functions, such as the immune response (Smilanich et al. 2009), which could be altered by changing environments. And there may be other impacts on caterpillar, and ultimately adult, features. In monarchs, wing morphology is altered when caterpillars feed on plants raised under elevated carbon dioxide levels, and these changes are impacted by both host plant species and whether larvae are infected with a protozoan parasite, *Ophryocystis elektroscirrha* (Decker et al. 2018). For monarchs, a migrating species, these changes in wing morphology can impact migratory ability (Decker et al. 2018).

As discussed earlier in this chapter, human alterations to the environment, such as increased introductions of exotic plants, elevated nitrogen and carbon-dioxide levels, changing thermal environments, and use of insecticides and herbicides, may all have consequences for caterpillar sequestration and potentially for caterpillar interactions with their host plants, as well as their enemies. Because of the importance of such multitrophic interactions for structuring communities and, ultimately, ecosystems, attention to how such environmental changes impact caterpillar sequestration, either via changes to host plant chemistry or via changes in caterpillar physiology, is increasingly important. We are just beginning to understand some of the consequences of such changes for caterpillar sequestration, but only a few systems have been studied in this regard; investigation of many more is certainly of great importance.

Acknowledgments Many thanks to the wonderful students, post-docs and colleagues who, over many years, have contributed to my thoughts about the behavior, ecology, and evolution of unpalatable caterpillars. I appreciate the comments from Adrian Carper, the Bowers-Resasco lab group, and three anonymous reviewers. I thank Greg Hill for German translation, Bob Marquis for the citizen science observation and Katherine Hernandez, Mary Jamieson, Nadya Muchoney, and Mike Singer for use of their photographs. I especially thank Suzanne Koptur and Bob Marquis for the invitation to contribute to this volume. In this chapter, much of the research from my laboratory was funded by the National Science Foundation and the United States Department of Agriculture. While writing this chapter, research was funded by NSF DEB-1929522.

References

Baden CU, Geier T, Franke S, Dobler S (2011) Sequestered iridoid glycosides – highly effective deterrents against ant predators? Biochem Syst Ecol 39:897–901
Barton K, Bowers MD (2006) Neighbor species differentially alter resistance phenotypes in *Plantago*. Oecologia 150:442–452
Barton KE, Koricheva J (2010) The ontogeny of plant defense and herbivory: characterizing general patterns using meta-analysis. Am Nat 175:481–493
Bates HW (1862) Contributions to an insect fauna of the Amazon Valley. Lepidoptera: Heliconidae. Trans Linn Soc Lond 23:495–566
Battisti A, Holm G, Fagrell B, Larsson S (2011) Urticating hairs in arthropods: their nature and medical significance. Annu Rev Entomol 56:203–220
Bernays E, DeLuca C (1981) Insect anti-feedant properties of an iridoid glycoside: ipolamiide. Experientia 37:1289–1290
Blake RJ, Woodcock BA, Westbury DB, Sutton P, Potts SG (2011) New tools to boost butterfly habitat quality in existing grass buffer strips. J Insect Conserv 15:221–232
Blum MS, Whitman DW, Severson RF, Arrendale RF (1987) Herbivores and toxic plants: evolution of a menu of options for processing allelochemicals. Int J Trop Insect Sci 8:459–463

Boege K, Marquis RJ (2005) Facing herbivory as you grow up: the ontogeny of resistance in plants. Trends Ecol Evol 20:441–448

Boege K, Agrawal AA, Thaler JS (2019) Ontogenetic strategies in insect herbivores and their impact on tri-trophic interactions. Curr Op Insect Sci 32:61–67

Bowers MD (1979) Unpalatability as a defense strategy of checkerspot butterflies with special reference to *Euphydryas phaeton* (Nymphalidae). Dissertation, University of Massachusetts

Bowers MD (1980) Unpalatability as a defense strategy of *Euphydryas phaeton* (Lepidoptera, Nymphalidae). Evolution 34:586–600

Bowers MD (1981) Unpalatability as a defense strategy of western checkerspot butterflies (*Euphydryas*, Nymphalidae). Evolution 35:367–375

Bowers MD (1991) Iridoid glycosides. In: Rosenthal GA, Berenbaum MR (eds) Herbivores: their interactions with secondary plant metabolites, vol 2. Academic Press, San Diego, pp 297–325

Bowers MD (1993) Aposematic caterpillars: lifestyles of the unpalatable and warningly colored. In: Stamp NE, Casey T (eds) Caterpillars: ecological and evolutionary constraints on foraging. Chapman and Hall, New York, pp 331–371

Bowers MD (2003) Defensive chemistry and ecology of the Catalpa Sphinx, *Ceratomia catalpae* (Sphingidae). J Chem Ecol 29:2359–2367

Bowers, MD, Collinge, SK (1992) Sequestration and metabolism of iridoid glycosides by larvae of the Buckeye, *Junonia coenia* (Nymphalidae). J Chem Ecol 18:817–831

Bowers MD, Farley S (1990) The behavior of grey jays, *Perisoreus canadensis*, towards palatable and unpalatable Lepidoptera. Anim Behav 39:699–705

Bowers MD, Larin Z (1989) Acquired chemical defense in the lycaenid butterfly, *Eumaeus atala*. J Chem Ecol 15:1133–1146

Bowers MD, Puttick GM (1986) Fate of ingested iridoid glycosides in lepidopteran herbivores. J Chem Ecol 12:169–178

Bowers MD, Puttick GM (1988) Response of generalist and specialist insects to qualitative allelochemical variation. J Chem Ecol 14:319–334

Bowers MD, Stamp NE (1993) Effect of plant age, genotype, and herbivory on *Plantago* performance and chemistry. Ecology 74:1778–1791

Bowers MD, Stamp NE (1997) Fate of hostplant iridoid glycosides in larvae of the Nymphalidae and Arctiidae. J Chem Ecol 23:2955–2965

Bowers MD, Stamp NE, Collinge SK (1992a) Early stage of host range expansion by a specialist herbivore, *Euphydryas phaeton* (Nymphalidae). Ecology 73:526–536

Bowers MD, Collinge SK, Gamble SE, Schmitt J (1992b) Effects of genotype, habitat, and seasonal-variation on iridoid glycoside content of *Plantago lanceolata* (Plantaginaceae) and the implications for insect herbivores. Oecologia 91:201–207

Bowers MD, Boockvar K, Collinge SK (1993) Iridoid glycosides of *Chelone glabra* (Scrophulariaceae) and their sequestration by larvae of a sawfly, *Tenthredo grandis* (Tenthredinidae). J Chem Ecol 19:815–823

Brower JVZ (1958a) Experimental studies of mimicry in some North American butterflies. Part I. The Monarch, *Danaus plexippus* and viceroy, *Limenitis archippus archippus*. Evolution 12:32–47

Brower JVZ (1958b) Experimental studies of mimicry in some North American butterflies. Part II. *Battus philenor* and *Pailiio troilus, P. polyxenes* and *P. glaucus*. Evolution 12:123–136

Brower JVZ (1958c) Experimental studies of mimicry in some North American butterflies. Part III. *Danaus gilippus berenice* and *Limenitis archippus floridensis*. Evolution 12:273–285

Brower LP (1969) Ecological chemistry. Sci Am 220:22–29

Brower LP, Brower JVZ, Corvino JM (1967) Plant poisons in a terrestrial food chain. Proc Natl Acad Sci 57:893–898

Brower LP, Ryerson WN, Coppinger LL, Glazier SC (1968) Ecological chemistry and the palatability spectrum. Science 161:1349–1350

Brown LM, Breed GA, Severns PM, Crone EE (2017) Losing a battle but winning the war: moving past preference-performance to understand native herbivore-novel host plant interactions. Oecologia 183:441–453

Butler AG (1869) Remarks upon certain caterpillars, etc., which are unpalatable to their enemies. Trans R Entomol Soc London 1869:27–29

Cavers PB, Bassett IJ, Crompton CW (1980) The biology of Canadian weeds: 47. *Plantago lanceolata* L. Can J Plant Sci 60:1269–1282

Cohen JA (1985) Differences and similarities in cardenolide contents of queen and monarch butterflies in Florida and the ecological and evolutionary implications. J Chem Ecol 11:85–103

Darrow K, Bowers MD (1997) Phenological and population variation in iridoid glycosides of *Plantago lanceolata* (Plantaginaceae). Biochem Syst Ecol 25:1–11

de Roode JC, Pedersen AB, Hunter MD, Altizer A (2008) Host plant species affects virulence in monarch butterfly parasites. J Anim Ecol 77:120–126

Decker LE, deRoode JC, Hunter MD (2018) Elevated atmospheric concentrations of carbon dioxide reduce monarch tolerance and increase parasite virulence by altering the medicinal properties of milkweeds. Ecol Lett 21:1353–1363

Decker LE, Soule AJ, deRoode JC, Hunter MD (2019) Phytochemical changes in milkweed induced by elevated CO_2 alter wing morphology but not toxin sequestration in monarch butterflies. Funct Ecol 33:411–421

Diaz JH (2005) The evolving global epidemiology, syndromic classification, management, and prevention of caterpillar envenoming. Am J Trop Med Hyg 72:347–357

Dimarco RD, Nice CC, Forcyce JA (2012) Family matters: effect of host plant variation in chemical and mechanical defenses on a sequestering specialists herbivore. Oecologia 170:687–693

Dinda B (2019) Pharmacology and applications of naturally occurring iridoids. Springer Nature, Cham

Dobler S, Petschenka G, Pankoke H (2011) Coping with toxic plant compounds – the insect's perspective on iridoid glycosides and cardenolides. Phytochemistry 72:1593–1604

Dobler S, Dalla S, Wagschal V, Agrawal AA (2012) Community-wide convergent evolution in insect adaptation to toxic cardenolides by substitutions in the NA,K-ATPase. Proc Natl Acad Sci 109:13040–13045

Duffey SS (1980) Sequestration of plant natural products by insects. Annu Rev Entomol 25:447–477

Dyer LA (1995) Tasty generalists and nasty specialists? Antipredator mechanisms in tropical lepidopteran larvae. Ecology 76:1483–1496

Dyer LA (1997) Effectiveness of caterpillar defenses against three species of invertebrate predators. J Res Lepid 34:48–68

Dyer LA, Bowers MD (1996) The importance of sequestered iridoid glycosides as a defense against an ant predator. J Chem Ecol 22:1527–1539

Dyer LA, Gentry G (1999) Predicting natural-enemy responses to herbivores in natural and managed systems. Ecol Appl 9:402–408

Ehrlich PR, Raven P (1964) Butterflies and plants: a study on coevolution. Evolution 18:586–608

El-Naggar LJ, Beal JL (1980) Iridoids: a review. J Nat Prod 43:649–707

Eltringham H (1909) A account of some experiments on the edibility of certain lepidopterous larvae. Trans R Entomol Soc Lond 1909:471–478

Engler-Chaouat HS, Gilbert LE (2007) De novo synthesis vs. sequestration: negatively correlated metabolic traits and the evolution of host plant specialization in cyanogenic butterflies. J Chem Ecol 33:25–42

Erb M, Kliebenstein DJ (2020) Plant secondary metabolites as defenses, regulators, and primary metabolites: the blurred functional trichotomy. Plant Physiol 184:39–52

Faldyn MJ, Hunter MD, Elderd BD (2018) Climate change and invasive, tropical milkweed: an ecological trap for monarch butterflies. Ecology 99:1031–1038

Fink L, Brower L, Waide R, Spitzer P (1983) Overwintering monarch butterflies as food for insectivorous birds in Mexico. Biotropica 15:151–153

Fisher RA (1930) The genetical theory of natural selection. Clarendon Press, Oxford

Fordyce JA, Nice CC (2008) Antagonistic, stage-specific selection on defensive chemical sequestration in a toxic butterfly. Evolution 62:1610–1617

Fraenkel G (1959) The raison d'etre of secondary plant substances. Science 139:1466–1470

Fraenkel G (1969) Evaluation of our thoughts on secondary plant substances. Entomol Exp Appl 12:473–486

Frankfater C, Tellez MR, Slattery M (2009) The scent of alarm: ontogenetic and genetic variation in the osmeterial gland chemistry of *Papilio glaucus* (Papilionidae) caterpillars. Chemoecology 19:81–96

Gardner DR, Stermitz FR (1988) Host plant utilization and iridoid glycoside sequestration by *Euphydryas anicia* (Lepidoptera: Nymphalidae). J Chem Ecol 14:2147–2168

Glassmire AE, Jeffrey CS, Forister ML, Parchman TL, Nice CC, Jahner JP, Wilson JS, Walla TR, Richards LA, Smilanich AM, Leonard MD, Morrison CR, Simbana W, Salagaje LA, Dodson GD, Miller JS, Tepe EJ, Villamarin-Cortez S, Dyer LA (2016) Intraspecific phytochemical variation shapes community and population structure for specialist caterpillars. New Phytol 212:208–219

Glendinning JI (1990) Responses of three mouse species to deterrent chemicals in the monarch butterfly. II. Taste tests using intact monarchs. Chemoecology 1:124–130

Glendinning JI (1993) Comparative feeding responses of the mice *Peromyscus melanotis, P. aztecus, Reithrodontomys sumichrasti*, and *Microtus mexicanus* to overwintering monarch butterflies in Mexico. In: Malcolm SB, Zalucki MP (eds) Biology and conservation of the monarch butterfly. Natural History Museum of Los Angeles County, Los Angeles, pp 323–333

Graves SD, Shapiro A (2003) Exotics as host plants of the California butterfly fauna. Biol Conserv 110:413–433

Haan NL, Bakker JD, Dunwiddie P, Linders MJ (2018) Instar-specific effects of host plants on survival of endangered butterfly larvae. Ecol Entomol 43:742–753

Haan NL, Bowers MD, Bakker JD (2021) Preference, performance, and chemical defense in an endangered butterfly using novel and ancestral host plants. Sci Rep 11:992

Haase E (1896) Researches on mimicry on the basis of a natural classification of the Papilionidae. Pt II. Transl. by CM Childs. Nagel, Stuttgard

Hahn PG, Maron JL (2016) A framework for predicting intraspecific variation in plant defense. Trends Ecol Evol 31:646–656

Halkier BA, Gershenzon J (2006) Biology and biochemistry of glucosinolates. Annu Rev Plant Biol 57:303–333

Hamann E, Blevins C, Franks SJ, Jameel MI, Anderson JT (2021) Climate change alters plant-herbivore interactions. New Phytol 229:1894–1910

Handjieva NV, Elieva EI, Spassov SL, Popov SS, Duddeck H (1993) Iridoid glycosides from *Linaria* species. Tetrahedron 49:9261–9266

Hartmann T, Ober D (2000) Biosynthesis and metabolism of pyrrolizidine alkaloids in plants and specialized insect herbivores. Top Curr Chem 209:207–243

Hawkins BA, Cornell HV, Hochberg ME (1997) Predators, parasitoids, and pathogens as mortality agents in phytophagous insect populations. Ecology 78:2145–2152

Hawthorne WR (1974) The biology of Canadian weeds. 4. *Plantago major* and *P. rugelii*. Can J Plant Sci 54:383–396

Hay-Roe MM, Nation J (2007) Spectrum of cyanide toxicity and allocation in *Heliconius erato* and *Passiflora* host plants. J Chem Ecol 33:319–329

Heckel DG (2014) Insect detoxification and sequestration strategies. Annu Plant Rev 47:77–114

Hegnauer R (1962-1996) Chemotaxonomies der Pflanzen. Vols I–X. Berkhauser Verlag, Basel/Stuttgart

Hesbacher S, Giez I, Embacher G, Fiedler K, Max W, Trawoger A, Turk R, Lange OL, Proksch P (1995) Sequestration of lichen compounds by lichen-feeding members of the Arctiidae (Lepidoptera). J Chem Ecol 21:2079–2089

Holzinger F, Frick C, Wink M (1992) Molecular basis for the insensitivity of the monarch (*Danaus plexippus*) to cardiac glycosides. FEBS Lett 314:477–480

Honda K (1981) Larval osmeterial secretions of the swallowtails (*Papilio*). J Chem Ecol 7:1089–1113

Honda K (1983) Defensive potential of components of the larval osmeterial secretion of papilionid butterflies against ants. Physiol Entomol 8:173–179

Hunter MD (2016) The phytochemical landscape: linking trophic interactions and nutrient dynamics. Princeton University Press, Princeton

Jamieson M, Bowers MD (2010) Iridoid glycoside variation in the invasive plant dalmatian toadflax, *Linaria dalmatica* (Plantaginaceae) and sequestration by the biological control agent *Calophasia lunula* (Noctuidae). J Chem Ecol 36:70–79

Jamieson M, Bowers MD (2012) Plant-mediated effects of soil nitrogen enrichment on a chemically defended specialist herbivore, *Calophasia lunula*. Ecol Entomol 37:300–308

Jamieson MA, Burkle LA, Manson JS, Runyon JB, Trowbridge AM, Zientek J (2017) Global change effects on plant-insect interactions: the role of phytochemistry. Curr Opin Insect Sci 23:70–80

Jensen SR (1991) Plant iridoids, their biosynthesis and distribution in angiosperms. In: Harborne JB, Tomas-Barberan FA (eds) Ecological chemistry and biochemistry of plant terpenoids. Clarendon Press, Oxford, pp 133–158

Jones PL, Petschenka G, Flecht L, Agrawal AA (2019) Cardenolide intake, sequestration, and excretion by the monarch butterfly along gradients of plant toxicity and larval ontogeny. J Chem Ecol 45:264–277

Kelly CA, Bowers MD (2018) Host plant iridoid glycosides mediate herbivore interactions with natural enemies. Oecologia 188:491–500

Knerl A, Bowers MD (2013) Incorporation of an introduced weed into the diet of a native butterfly: consequences for preference, performance and chemical defense. J Chem Ecol 39:1313–1321

Kubinova R, Svajdleka E, Kulovana T (2014) Accumulation of secondary metabolites in larvae of Death's-head hawk moth (*Acherontia atropos*) is dependent on food composition. Chem List 108:1145–1148

L'Empereur KM, Stermitz FR (1990) Iridoid glycoside content of *Euphydryas anicia* (Lepidoptera, Nymphalidae) and its major hostplant, *Besseya plantaginea* (Scrophulariaceae) at a high-plains Colorado site. J Chem Ecol 16:187–197

LaBar CC, Schultz CB (2012) Investigating the role of herbicides in controlling invasive grasses in prairie habitats: effects on non-target butterflies. Nat Areas J 32:177–189

Lampert EC, Bowers MD (2010) Host plant influences on iridoid glycoside sequestration of generalist and specialist caterpillars. J Chem Ecol 36:1101–1104

Lampert EC, Bowers MD (2014) Incompatibility between plant-derived defensive chemistry and immune response of two sphingid herbivores. J Chem Ecol 41:85–92

Lampert EC, Dyer LA, Bowers MD (2010) Caterpillar chemical defense and parasitoid success: *Cotesia congregata* parasitism of *Ceratomia catalpae*. J Chem Ecol 36:992–998

Lampert EC, Dyer LA, Bowers MD (2014) Dietary specialization and the effects of plant species on potential multitrophic interactions of three species of nymphaline caterpillars. Entomol Exp Appl 153:207–216

Li F, Dudley TL, Chen B, Chang X, Liang L, Peng S (2016) Responses of tree and insect herbivores to elevated nitrogen inputs: a meta-analysis. Acta Oecol 77:160–167

Lydon J, Duke SO (1989) Pesticide effects on secondary metabolism of higher plants. Pest Manag Sci 25:361–373

Majewska AA, Altizer S (2019) Exposure to non-native tropical milkweed promotes reproductive development in migratory monarch butterflies. Insects 10:253. https://doi.org/10.3390/insects10080253

Malcolm S (1991) Cardenolide-mediated interactions between plants and herbivores. In: Rosenthal GA, Berenbaum MR (eds) Herbivores: their interactions with secondary plant metabolites, vol 1. Academic Press, San Diego, pp 251–295

Malcolm S (2018) Anthropogenic impacts on mortality and population viability of the monarch butterfly. Annu Rev Entomol 63:277–302

Marushia RG, Allen EB (2011) Control of exotic annual grasses to restore native forbs in abandoned agricultural land. Restor Ecol 19:45–54

Mattila ALK, Jiggins CD, Poedal OH, Montejo-Kovacevich G, de Castro E, WO MM, Baquet C, Saastamoinen M (2020) High evolutionary potential in the chemical defenses of an aposematic *Heliconius butterfly*. BioRxiv. preprint posted January 15, https://doi.org/10.1101/2020.01.14.905950

Moore BD, Andrew RL, Kulheim C, Foley WJ (2013) Explaining intraspecific diversity in plant secondary metabolites in an ecological context. New Phytol 201:733–750

Muller C (2009) Interactions between glucosinolate- and myrosinase-containing plants and the sawfly *Athalia rosae*. Phytochem Rev 8:121–134

Nahrstedt A, Davis RH (1983) Occurrence, variation and biosynthesis of the cyanogenic glucosides linamarin and lotaustralin in species of the Heliconiini (Insecta: Lepidoptera). Comp Biochem Physiol Part B Biochem 75:65–73

Nash RJ, Bell EA, Ackery PR (1992) The protective role of cycasin in *Cycad*-feeding Lepidoptera. Phytochemistry 31:1955–1957

Nayar JK, Fraenkel G (1963) The chemical basis of host selection in the catalpa sphinx, *Ceratomia catalpae* (Lepidoptera, Sphingidae). Ann Entomol Soc Am 56:119–122

Nishida R (2002) Sequestration of defensive substances from plants by Lepidoptera. Annu Rev Entomol 47:57–92

Nishida R (2014) Chemical ecology of insect-plant interactions: ecological significance of plant secondary metabolites. Biosci Biotechnol Biochem 78:1–13

Nishida R, Fukami H (1989) Host plant iridoid-based chemical defense of an aphid, *Acyrthosiphon nipponicus*, against ladybird beetles. J Chem Ecol 15:1837–1845

Ode PJ (2006) Plant chemistry and natural enemy fitness: effects on herbivores and natural enemy interactions. Annu Rev Entomol 51:163–185

Opitz SEW, Müller C (2009) Plant chemistry and insect sequestration. Chemoecology 19:117–154

Pasteels JM, Rowell-Rahier M, Braekman JC, Dupoint A (1983) Salicin from host plant as precursor of salicylaldehyde in defensive secretion of chrysomeline larvae. Physiol Entomol 8:307–314

Pellissier L, Rober A, Bilat J, Rasmann S (2014) High elevation *Plantago lanceolata* plants are less resistant to herbivory than their low elevation conspecifics: is it just temperature? Ecography 37:950–959

Petschenka G, Agrawal AA (2015) Milkweed butterfly resistance to plant toxins is linked to sequestration, not coping with a toxic diet. Proc R Soc B 282:20151865. https://doi.org/10.1098/rspb.2015.1865

Petschenka G, Agrawal AA (2016) How herbivores coopt plant defenses: natural selection, specialization, and sequestration. Curr Opin Insect Sci 14:17–24

Petschenka G, Fandrich S, Sander N, Wagschal V, Boppre M, Dobler S (2013) Stepwise evolution of resistance to toxic cardenolides vial genetic substitutions in the NA$^+$/K$^+$-ATPase of milkweed butterflies (Lepidoptera: Danaini). Evolution 67:2753–2761

Poulton EB (1887) The experimental proof of the protective value of colour and markings in insects with reference to their vertebrate enemies. Proc Zool Soc London 1887:191–274

Poulton EB (1890) The colours of animals. D Appleton and Co, New York

Price PC, Bouton CE, Gross PC, Bouton E, Gross P, McPheron BA, Thompson JN, Weis AE (1980) Interactions among three trophic levels: influence of plants on interactions between insect herbivores and natural enemies. Annu Rev Ecol Syst 11:41–65

Pritchett AH (1903) Some experiments in feeding lizards with protectively colored insects. Biol Bull 5:271–287

Prudic KL, Oliver JC, Bowers MD (2005) Soil nutrient effects on oviposition preference, larval performance and chemical defense of a specialist insect herbivore. Oecologia 143:578–587

Puttick GM, Bowers MD (1988) Effect of qualitative and quantitative variation in allelochemicals on a generalist insect: iridoid glycosides and the southern armyworm. J Chem Ecol 14:335–351

Quintero C, Bowers MD (2018) Plant and herbivore ontogeny interact to shape the preference, performance and chemical defense of a specialist herbivore. Oecologia 187:401–441

Quintero C, Lampert EC, Bowers MD (2014) Time is of the essence: direct and indirect effects of plant ontogenetic trajectories on higher trophic levels. Ecology 95:2589–2602

Rayor LS (2004) Effects of monarch larval host plant chemistry and body size on *Polistes* wasp predation. In: Oberhauser KS, Solensky MJ (eds) Monarch butterfly biology and conservation. Cornell University Press, Ithaca, pp 39–46

Rimpler H (1991) Sequestration of iridoids by insects. In: Harborne JB, Tomas-Barberan FA (eds) Ecological chemistry and biochemistry of plant terpenoids. Clarendon Press, Oxford, pp 314–330

Robinson GS, Ackery PR, Kitching I, Beccaloni G, Hernandez L (2002) Hostplants of moth and butterfly caterpillars North of Mexico. Memoirs of the American Entomological Institute, Gainesville

Robinson EA, Ryan GC, Newman JA (2012) A meta-analytical review of the effects of elevated CO_2 on plant-arthropod interactions highlights the importance of interacting environmental and biological variables. New Phytol 194:321–336

Roeske CN, Seiber JN, Brower LP, Moffitt CM (1976) Milkweed cardenolides and their comparative processing by monarch butterflies (*Danaus plexippus* L.). Rec Adv Phytochem 10:93–167

Rosenthal GA, Berenbaum MR (eds) (1991) Herbivores: their interactions with secondary plant metabolites. Academic, San Diego

Rothschild M, Reichstein T, von Euw J, Aplin R, Harman RRM (1970) Toxic Lepidoptera. Toxicon 8:293–299

Rothschild M, Nash RJ, Bell EA (1986) Cycasin in the endangered butterfly *Eumaeus atala florida*. Phytochemistry 25:1853–1854

Sajitha TP, Siva R, Manjunatha BL, Rajai P, Navdeep G, Kavita D, Ravikanth G, Shaanker RU (2019) Sequestration of the plant secondary metabolite, colchicine, by the noctuid moth *Polytela gloriosae* (Fab.). Chemoecology 29:135–142

Saporito RA, Connelly MA, Spande TF, Garraffo HM (2012) A review of chemical ecology in poison frogs. Chemoecology 22:159–168

Satterfield DA, Maerz JC, Altizer S (2015) Loss of migratory behavior increases infection risk for a butterfly host. Proc R Soc B 282:20141734

Schoonhoven LM, van Loon JJA, Dicke M (2006) Insect-plant biology, 2nd edn. Oxford University Press, Oxford

Schultz CB, Zemaitis JL, Thomas CC, Bowers MD, Crone EE (2016) Non-target effects of grass-specific herbicides differ among species, chemicals and host plants in *Euphydryas* butterflies. J Insect Conserv 20:867–877

Scott Chialvo CH, Chialvo P, Holland JD, Anderson TJ, Breinholt JW, Kawahara KY, Zhou X, Liu S, Zaspel JM (2018) A phylogenomic analysis of lichen-feeding tiger moths uncovers evolutionary origins of host chemical sequestration. Mol Phylogenet Evol 121:23–34

Scott CH, Zaspel JM, Chialvo P, Weller SJ (2014) A preliminary molecular phylogenetic assessment of the lichen moths (Lepidoptera: Erebidae: Arctiinae: Lithosiini) with comments on palatability and chemical sequestration. Syst Entomol 39:286–303

Sculfort O, de Castro EP, Kozak KM, Bak S, Elias M, Nay B, Llaurens V (2020) Variation of chemical compounds in wild Heliconiini reveals ecological factors involved in the evolution of chemical defenses in mimetic butterflies. Ecol Evol 10:2677–2694

Shinjyo N, Waddell G, Green J (2020) Valerian root in treating sleep problems and associated disorders—a systematic review and meta-analysis. J Evid Based Integr Med 25:1–31

Singer MC, Parmesan C (2018) Lethal trap created by adaptive evolutionary response to an exotic resource. Nature 557:238–241

Slater JW (1877) On the food of gaily coloured caterpillars. Trans R Entomol Soc Lond 1877:205–209

Smilanich AM, Dyer LA, Chambers JQ, Bowers MD (2009) Immunological cost of chemical defence and the evolution of herbivore diet breadth. Ecol Lett 12:612–621

Stamp NE (1979) New oviposition plant for *Euphydryas phaeton* (Nymphalidae). J Lepid Soc 33:203–204

Stamp NE (2001) Effects of prey quantity and quality on predatory wasps. Ecol Entomol 26:292–301

Stermitz FR, Gardner DR, Odendaal FJ, Ehrlich PR (1986) *Euphydryas anicia* (Lepidoptera: Nymphalidae) utilization of iridoid glycosides from *Castilleja* and *Besseya* (Scrophulariaceae) host plants. J Chem Ecol 12:1455–1468

Sternberg ED, Levevre T, Li J, Fernandez de Castillejo CL, Li H, Hund D, de Roode JC (2012) Food plant-derived disease tolerance and resistance in a natural butterfly-plant-parasite interaction. Evolution 66:3367–3376

Strohmeyer HH, Stamp NE, Jarzomski CM, Bowers MD (1998) Prey species and prey diet affect growth of invertebrate predators. Ecol Entomol 23:68–79

Tan W-H, Acevedo T, Harris EV, Alcaide TY, Walters JR, Hunter MD, Gerardo NM, de Roode JC (2019) Transcriptomics of monarch butterflies (*Danaus plexippus*) reveals that toxic host plants alter expression of detoxification genes and down-regulate a small number of immune genes. Mol Ecol 28:4845–4863

Tao L, Berns AR, Hunter MD (2014) Why does a good thing become too much? Interactions between foliar nutrients and toxins determine performance of an insect herbivore. Funct Ecol 28:190–196

Theodoratus DH, Bowers MD (1999) Effect of sequestered iridoid glycosides on prey choice of the prairie wolf spider, *Lycosa carolinensis*. J Chem Ecol 25:283–295

Thomas CD, Ng D, Singer MC, Mallet JLB, Parmesan C, Billington HL (1987) Incorporation of a European weed into the diet of a north American herbivore. Evolution 41:892–901

Throop HL, Lerdau MT (2004) Effects of nitrogen deposition on insect herbivory: implications for community and ecosystem processes. Ecosystems 7:109–133

Tundis R, Loizzo M, Menichini F, Statti G, Menichini F (2008) Biological and pharmacological activities of iridoids: recent developments. Mini Rev Med Chem 8:399–420

Vaughan GL, Jungreis AM (1977) Insensitivity of lepidopteran tissues to ouabain: physiological mechanisms for protection from cardiac glycosides. J Insect Physiol 23:585–589

Verschaffelt E (1910) The cause determining the selection of food in some herbivorous insects. Proc Acad Sci Amsterdam 1:536–542

Veteli TO, Kuokkanen K, Julkenen-Titto R, Roininen H, Tahavanainen J (2002) Effects of elevated CO_2 and temperature on plant growth and herbivore defensive chemistry. Glob Chang Biol 8:1240–1252

Von Poser GL, Schripsema J, Henriques AT, Jensen SR (2000) Th distribution of iridoids in Bignoniaceae. Biochem Syst Ecol 28:351–366

Wallace AR (1867a) Caterpillars and birds. Field, The Country Gentleman's Newspaper 29 (743): 206. Accessed from http://wallace-online.org

Wallace AR (1867b) Untitled. Proc R Entomol Soc London 1867:lxxx–lxxxi

Weir JJ (1869) On insects and insectivorous birds; and especially on the relation between the color and edibility of Lepidoptera and their larvae. Trans R Entomol Soc Lond 1869:21–26

Weir JJ (1870) Further observations on the relations between the color and edibility of Lepidoptera and their larvae. Trans R Entomol Soc Lond 1870:337–339

Willinger G, Dobler S (2001) Selective sequestration of iridoid glycosides from their host plants in *Longitarsus* flea beetles. Biochem Syst Ecol 29:335–346

Winde I, Wittstock U (2011) Insect herbivore counteradaptations to the plant glucosinolate-myrosinase system. Phytochemistry 72:1566–1575

Wink M (2019) Quinolizidine and pyrrolizidine alkaloid chemical ecology – a mini-review on their similarities and differences. J Chem Ecol 45:109–115

Wray V, Davis RH, Nahrstedt A (1983) Biosynthesis of cyanogenic glycosides in butterflies and moths: incorporation of valine and isoleucine into linamarin and lotaustralin by *Zygaena* and *Heliconius* species (Lepidoptera). Zeit Für Naturfor C 38:583–588

Zagrobelny M, Møller BL (2011) Cyanogenic glucosides in the biological warfare between plants and insects: the burnet moth-birdsfoot trefoil model system. Phytochemistry 72:1585–1592

Zagrobelny M, Pinheiro De Castro EC, Moller BL, Bak S (2018) Cyanogenesis in arthropods: From chemical warfare to nuptial gifts. *Insects* 9: 51; https://doi.org/10.3390/insects9020051

Zust T, Petschenka G, Hastings A, Agrawal AA (2019) Toxicity of milkweed leaves and latex: chromatographic quantification versus biological activity of cardenolides in 16 *Asclepias* species. J Chem Ecol 45:50–60

Part III
Impacts of the Third Trophic Level on Caterpillar Ecology and Evolution

Acoustic Defence Strategies in Caterpillars

Jayne E. Yack

Late instar *Drepana arcuata*. (Photo by Tom Eisner, presented to Jayne Yack)

Introduction

Caterpillars have many enemies, including invertebrate predators and parasitoids (e.g. wasps, flies, mantids, stink bugs, dragonflies, ants, and spiders) and vertebrate predators (e.g. bats, birds, lizards, rodents, toads) (Heinrich 1993; Montllor and Bernays 1993; Wagner 2005; Kalka and Kalko 2006; Greeney et al. 2012; Sugiura 2020). While vulnerable in their soft exoskeletons and with limited options for escape, they are not exactly helpless. In fact, caterpillars are well recognized for their many antipredator strategies, including crypsis, mimesis, deimatic displays, urticating and poisonous spines and bristles, irritating sprays, warning coloration,

J. E. Yack (✉)
Department of Biology, Nesbitt Biology Building, Carleton University, Ottawa, ON, Canada
e-mail: jayneyack@cunet.carleton.ca

thrashing, dropping, and shelter building (Lederhouse 1990; Gentry and Dyer 2002; Greeney et al. 2012; Sugiura 2020). Research on antipredator tactics has focused on those operating in the visual and chemical realms, and comparatively less is understood about acoustic defences. Do caterpillars use sounds and vibrations to detect or repel enemies, or to avoid detection? Arguably caterpillars should be exploiting airborne sounds and solid-borne vibrations to avoid attack. Their enemies generate a diversity of acoustic signals and cues that can provide information to assess risk. Also, considering that many enemies of caterpillars have hearing capabilities, producing acoustic signals should be effective in communicating with these enemies. Additionally, caterpillars have limited visual capabilities, but live in rather complex vibroacoustic environments, being substrate-bound organisms (see Yack and Yadav 2021). These points considered, one is hard-pressed to explain the few examples of caterpillar acoustic defences in reviews covering the topics of acoustic defences in Lepidoptera (e.g. Minet and Surlykke 2003), acoustic defences in insects (e.g. Conner 2014), insect defences (e.g. Evans and Schmidt 1990; Ruxton et al. 2004; Sugiura 2020), or caterpillar defences (Lederhouse 1990; Stamp and Casey 1993; Gentry and Dyer 2002; Greeney et al. 2012). Still, there have been reports, many dating back to the 1800s, of caterpillars producing sounds or responding to them, purportedly, in the context of defence. In the past few decades, alongside a growing awareness of the importance of near-field sounds and solid-borne vibrations in insect communication, there has been an increasing number of experimental studies confirming that caterpillars live in rather complex acoustic environments. This chapter reviews the literature on caterpillar acoustic defences to gain an appreciation for the taxonomic diversity and functions of hearing and sound production in the context of defence, and to propose future lines of investigation.

First it is important to define the terms used in this chapter to discuss acoustic stimuli, how they are detected in insects, and how they might be relevant to an insect prey. Broadly speaking, acoustic events are vibrations transmitted through any elastic medium (Windmill and Jackson 2016). Vibrations travelling through air and water are commonly referred to as 'sounds', whereas those transmitted through solids such as plant material, silk, waxes, or soil are commonly referred to as 'vibrations', 'substrate-borne vibrations', or 'solid-borne vibrations'. Airborne vibrations are further categorized as 'far-field' and 'near-field' sounds, which describe the pressure and displacement components of sound respectively. Far-field sounds are pressure waves transmitted over long distances and detected by pressure detectors such as tympanal ears found in many adult insects and most vertebrates. Near-field sounds, resulting from the displacement component of a vibrating source, typically are transmitted over shorter distances (within a few meters) and are restricted to lower frequencies (less than 2 kHz). Near-field sound receptors have been described in adult and juvenile insects and include lightweight receivers such as hairs (i.e., trichoid sensilla) and antennae. Vibrations propagated through solids are used by insects in a variety of contexts. The sensory organs best known for vibration reception are subgenual organs in adults of some insect orders (see Yack 2016). In this Chapter, I use the terms *sound* to mean airborne vibrations in general, and *near- and far-field sounds* to distinguish between the displacement and pressure components respectively. I use the terms *vibrations* or *solid-borne vibrations* to describe waves transmitted through solids. For more in-depth discussions of the nomenclature associated with acoustic vibrations and sensory receptors in insects, see Hill (2008, 2014), Hill and Wessel (2016), Lakes-Harlan and Strauss (2014), Windmill and Jackson (2016), and Yack (2004, 2016).

Acoustic stimuli relevant to a discussion on caterpillar defences include those that arise from predators and parasitoids, as well as non-predators (conspecifics and heterospecifics), and these stimuli can be categorized as signals or cues. Here I use the term *cue* to refer to sounds and vibrations that have not evolved in the context of communication (i.e., they have not evolved to alter the behaviour of an intended recipient). Acoustic cues in this context include sounds and vibrations generated as a consequence of movement (e.g. flying, walking, digging). On the other hand, *signal* is used to describe a sound or vibration that evolved in the context of communication (i.e., conveying a message to an intended receiver). The intended receiver could be oneself (e.g. echolocation) or another recipient (e.g. alarm or mating call). For further discussion of the nomenclature relating to signals, cues, and communication, see Maynard-Smith and Harper (2003) and Yack et al. (2020).

Acoustic Antipredator Strategies in Insects

What acoustic strategies do insect prey use to avoid attack? To address this question, I have broadly categorized acoustic anti-predator strategies into acoustic crypsis, sound and vibration detection, and sound and vibration production. These categories are outlined below and in Fig. 1 with representative examples drawn from insects in general, and then further discussed in the context of caterpillar defences in sections "Acoustic crypsis in caterpillars", "Sound and vibration reception in caterpillars", and "Generating sounds and vibrations in caterpillars" of this chapter.

Crypsis can be defined as any trait, whether visual, chemical, tactile, electric, or acoustic, that minimizes the probability of being detected when potentially detectable by a predator (Conner 2014). Acoustic crypsis includes the following strategies: (i) reducing sounds that predators could use to locate prey. For example, some insects shut down advertisement or mating calls in the presence of a predator (e.g. Faure and Hoy 2000; Greenfield and Baker 2003; Hamel and Cocroft 2019) or cease movement to avoid being detected by vibration cues (e.g. Djemai et al. 2001; Takanashi et al. 2016); (ii) altering the physical characteristics of sound (e.g. amplitude, frequency) to be less conspicuous to an enemy (e.g. Nakano et al. 2008); (iii) rendering oneself inconspicuous to echolocating predators by reducing the amplitude of the echo through morphological features (e.g. Zeng et al. 2011); and (iv) blending into the background acoustically to avoid being detected or recognized as prey (e.g. Rydell 1998). Acoustic crypsis is believed to be an understudied defence strategy in insects (Conner 2014). The topic of acoustic crypsis in caterpillars is discussed in section "Acoustic crypsis in caterpillars".

> **Acoustic Anti-predator Strategies in Insects**
>
> **Acoustic Crypsis**
> A trait that minimizes the probability of being detected when potentially detectable by a predator or parasitoid.
>
> > **Examples in Insects**
> > - Reducing acoustic cues available to predators by ceasing movement, feeding, singing
> > - Adjusting sound characteristic (e.g. amplitude, frequency) to be out of predator hearing range
> > - Match or blend into background to be inconspicuous to echolocating predators
>
> **Sound and Vibration Detection**
> Receiving acoustic signals or cues of predators, or alarm calls or recruitment signals of non-predators.
>
> > **Examples in Insects**
> > - Detecting acoustic cues produced by an approaching predator (e.g. crawling vibrations, flight sounds, rustling leaves) (Cue by predator)
> > - Hearing echolocation calls or vibrational sounding signals (Signal by predator)
> > - Eavesdropping on communication calls, songs of predators (Signal by predator)
> > - Detecting alarm or recruitment signals by non-predators (Signal by non-predator)
>
> **Sound and Vibration Production**
> Producing sounds or vibrations that are directed at the predator to stop an attacker, or directed at non-predators to recruit help or coordinate defenses.
>
> > **Examples in Insects**
> > - Acoustic aposematism (Directed at predator)
> > - Deimatic displays (Directed at predator)
> > - Mimicry (Directed at predator)
> > - Jamming or interference (Directed at predator)
> > - Alarm signals (Directed at non-predator)
> > - Recruitment signals (Directed at non-predator)

Fig. 1 An overview of different acoustic defence strategies employed by insects, including acoustic crypsis, detecting sounds and vibrations, and generating sounds and vibrations. Examples (or lack thereof) of these strategies employed by caterpillars are discussed in the text

Detecting sounds and vibrations can be important for insect prey. Relevant sounds and vibrations generated by predators include incidental cues resulting from movement (e.g. wings flapping, leaves rustling, crawling). For example, some butterflies detect the flight sounds of insectivorous birds (Mikhail et al. 2018), and moths detect the rustling leaf sounds of foraging birds (Jacobs et al. 2008). Prey also attend to communication signals (e.g. advertisement songs, echolocation calls) of predators to assess risk. Many flying adult insects, including moths and butterflies, have evolved tympanal ears to detect the echolocation calls of bats (Hoy 1992; Miller and Surlykke 2001; Yack et al. 2007; Conner and Corcoran 2012; Yager 2012; Pollack 2016). There are no confirmed examples, to the best of my knowledge, of insects eavesdropping on the social calls of their predators, although this hypothesis has been proposed to explain hearing in some butterflies (Ribaric and Gogala 1996; Mikhail et al. 2018) and is a common strategy for assessing risk in vertebrate prey (see Yack et al. 2020). Relevant sounds and vibrations produced by non-predators include alarm and recruitment signals. There are several examples of adult social insects detecting and responding to the alarm calls of non-predators (Hunt and Richard 2013). The topic of caterpillar 'hearing' in the context of avoiding attack is discussed in section "Sound and vibration reception in caterpillars".

Insects also can generate sounds and vibrations when under attack or threat of attack. Such signals have been called distress, alarm, warning, and defence signals (Alexander 1967; Masters 1980; Conner 2014; Bura et al. 2016). Defence sounds directed at a predator may function as aposematic warning signals, deimatic displays, interference signals, or mimics of sounds advertising danger (Conner 2014; Low et al. 2021). Acoustic defence signals can also be directed at non-predators, such as conspecifics or heterospecifics, and these function primarily to warn kin, or recruit help from others (Cocroft and Hamel 2010; Hunt and Richard 2013). Despite the widespread occurrence of defence sounds and vibrations among insects, their survival benefits are not well understood (Conner 2014; Low et al. 2021). The topic of caterpillar sound and vibration production in the context of defence is discussed in section "Generating sounds and vibrations in caterpillars".

A review on the topic of acoustically mediated defences in caterpillars is due, for a couple of reasons. First, the subject has not previously been the focus of a review, although some aspects of the topic have been addressed in reviews on vibratory communication in insects (Yack 2016), vibratory communication in caterpillars (Yack and Yadav 2021), vibratory-mediated predator prey interactions in insects (Virant-Doberlet et al. 2019), and insect defence sounds (Low et al. 2021). Second, there have been an increasing number of experimental examples of acoustically mediated communication in caterpillars over the past two decades. It is now apparent that larval insects inhabit complex vibro-acoustic environments and attend to sounds and vibrations in a diversity of contexts, including territoriality and spacing (e.g. Yack et al. 2001; Fletcher et al. 2006; Scott et al. 2010; Yack et al. 2014), obtaining food (e.g. Ishay et al. 1974; McIver and Beech 1986), recruitment and coordinating group activities (e.g. Fletcher 2007, 2008; Yadav et al. 2017), mimicry to exploit resources (e.g. Travassos and Pierce 2000; Sala et al. 2014), and avoiding enemies (e.g. Castellanos and Barbosa 2006; Low 2008; Roberts 2017; Taylor and Yack 2019). In the majority of reports on larval acoustics, the sounds and vibrations are not easily detected by humans without the aid of recording equipment such as laser vibrometers and specialized microphones. However, with increasing awareness of the importance of vibro-acoustic communication in insects and the broader availability of specialized recording instruments, more examples are being reported for acoustic communication in juvenile insects.

Acoustic Crypsis in Caterpillars

Cryptic silence is to the ear what cryptic appearance is to the eye. The silence of which I speak is not a passive condition- a mere absence of sound. It is an active quality....
Cott (1940)

Predators and parasitoids of caterpillars use different sensory modalities, including their acoustic senses, to identify and locate prey. For example, stink bugs and parasitoid wasps eavesdrop on chewing and crawling movements of caterpillars (Pfannenstiel et al. 1995; Meyhofer et al. 1997), and bats use echolocation and passive listening to locate prey (Kalka and Kalko 2006; Wilson and Barclay 2006; Geipel et al. 2013; Page and Bernal 2020). Conceivably, caterpillars have evolved strategies to render themselves acoustically cryptic to their enemies. They have been shown to avoid both invertebrate and vertebrate predators by reducing movement and freezing (e.g. Heinrich 1993; Montllor and Bernays 1993) (Table 1). Although it is often assumed that this is a strategy to avoid visually hunting predators, reduction of movement could also render caterpillars acoustically cryptic. For example, the masked birch caterpillar (*Drepana arcuata*) ceased activities (chewing, movement) when approached by a predatory stink bug that uses vibrations to locate prey (Guedes et al. 2012). The apple leaf miner *Phyllonorycter malella* stops feeding and remains immobile in the presence of a parasitoid wasp *Sympiesis sericeicornis* that uses vibrations to locate its host (Meyhöfer et al. 1997). Other strategies that caterpillars could employ to render themselves acoustically cryptic would be to restrict feeding and movement activities to times of the day when predators are not hunting, to acoustically match their backgrounds to avoid detection by echolocating predators, or to mask vibrations caused by their activities by selecting noisy backgrounds. Hiding acoustically from predators and parasitoids that use sound and vibratory cues to identify and locate prey is a likely strategy used by caterpillars, and deserves further research attention.

Sound and Vibration Reception in Caterpillars

The fact that a considerable number of species is now known to respond [to sound]... suggests that the response to sound is characteristic of many, perhaps all, caterpillars.
(Minnich 1936).

Several species of caterpillars have been reported to respond behaviourally to air- or solid-borne vibrations. The sensory mechanisms used for sound and vibration reception in caterpillars, however, remain mostly unknown. Only in two species have acoustic receptors been experimentally confirmed to the best of my knowledge, and these are both trichoid sensilla used to detect near-field sounds (Markl and Tautz 1975; Taylor and Yack 2019). Receptors of solid-borne vibrations have not yet been identified in caterpillars despite the many confirmed examples of vibration reception based on behavioural experiments (Yack and Yadav 2021). Structures

Table 1 Sound an vibration detection in a defensive context based on behavioural responses

Taxon			Acoustic stimulus/stimuli		S/V[a]	Behavioural response	Proposed function	Reference(s)
Superfamily	Family	Species	Source					
Bombycoidea	Saturniidae	*Actias Luna*	Tuning forks (~200–1000 Hz)		S	Head retraction	N/A	Minnich (1936)
Bombycoidea	Saturniidae	*Automeris io*	Tuning forks (~200–1000 Hz)		S	Cessation of movement, body contraction	N/A	Minnich (1936)
Bombycoidea	Saturniidae	*Hylesia nigricans*	Playback sound of wing beat of wasp or insectivorous bird		S	Head flicking, ultrasound emission	Stimulate defensive aggregation	Breviglieri and Romero (2019)
Bombycoidea	Saturniidae	*Hylesia sp.*	Human voice, music, wasp flight		S	Jerking of anterior body	Avoiding parasites	Hogue (1972)
Bombycoidea	Saturniidae	*Samia cecropia* (*Hyalolophora cecropia*)	Tuning forks (~200–1000 Hz)		S	Head retraction	N/A	Minnich (1936)
Bombycoidea	Saturniidae	*Telea polyphemus* (*Antheraea polyphemus*)	Tuning forks (~200–1000 Hz)		S	Head retraction	N/A	Minnich (1936)
Drepanoidea	Drepanidae	*Drepana arcuata*	Crawling vibrations of predatory stink bug		V	Cessation of activity, signalling	Predator avoidance, deterrent	Guedes et al. (2012)
Geometroidea	Geometridae	*Semiothisa aemulataria*	Leaf vibration induced by wasps, stink bugs, 200 Hz sounds		V	Dropping by silk thread	Predator avoidance	Castellanos and Barbosa (2006)
Geometroidea	Geometridae	Species unidentified	Tuning forks (~200–1000 Hz)		S	Freezing, body contraction	N/A	Minnich (1936)
Gracillarioidea	Gracillariidae	*Phyllonorycter blancardella*	Vibrations simulating ovipositor insertions		V	Ceasing movement, evasive behaviour	Avoid parasitism	Djemai et al. (2001)

(continued)

Table 1 (continued)

Taxon			Acoustic stimulus/stimuli			Behavioural response	Proposed function	Reference(s)
Superfamily	Family	Species	Source		S/V[a]			
Gracillarioidea	Gracillariidae	*Phyllonorycter malella*	Insertion of ovipositor		V	Cease feeding or movement, increase movement	Avoid parasitism	Meyhöfer et al. (1997)
Lasiocampoidea	Lasiocampidae	*Bombyx quercus* (*Lasiocampa quercus*)	Shouting, whistling, hand clapping		S	Jerking heads and bodies	N/A	Johnson (1893)
Lasiocampoidea	Lasiocampidae	*Malacosoma pluviale* (*M. californicum*)	Simulated flight sound of parasitic fly		S	Head flicking	Avoid parasitism	Myers and Smith (1978)
Noctuoidea	Erebidae	*Callimorpha dominula*	Human voices		S	Lateral body thrashing	N/A	Tutt (1893)
Noctuoidea	Erebidae	*Empyreuma pugione*	100–350 Hz tones		S	Twitched violently	Avoiding predator or parasitoid	Conner and Wilson (2009)
Noctuoidea	Erebidae	*Euchaetias egle*	Tuning forks (~200–1000 Hz)		S	Cease movement, jerking anterior or posterior body laterally	N/A	Minnich (1936)
Noctuoidea	Erebidae	*Nemeophila plantaginis*	Human voices		S	Lateral thrashing	N/A	Tutt (1893)
Noctuoidea	Erebidae	Species unidentified	Tuning forks (~200–1000 Hz)		S	Cease movement, freezing, body contraction	N/A	Minnich (1936)
Noctuoidea	Noctuidae	*Barathra brassicae* (*Mamestra Brassicae*)	Flight sound of parasitoid (100–160 Hz), 40–1000 Hz sound		S	Cease movement, squirming, dropping	Avoid detection, escape	Markl and Tautz (1975), Tautz and Markl (1978)
Noctuoidea	Noctuidae	*Mamestra picta*	Tuning forks (~200–1000 Hz)		S	Anterior lateral thrashing	N/A	Minnich (1936)

Noctuoidea	Noctuidae	*Spodoptera exigua*	Honeybee flight	S	Stops feeding	Avoid detection	Tautz and Rostas (2008)
Noctuoidea	Notodontidae	*Cerura borealis*	175–180 Hz and 330–360 Hz tones	S	Rearing head, moving tentacles, extending flagella	Avoiding parasitoids	White et al. (1983)
Noctuoidea	Notodontidae	*Cerura scitiscripta (Tecmessa scitiscripta)*	Human voice	S	Thrashing anterior and posterior ends	Defence reaction	Klots (1969)
Noctuoidea	Notodontidae	*Datana ministra, D. perspicua*	512 and 728 Hz tones	S	Anterior and posterior flicking, lateral thrashing	N/A	Abbott (1927)
Papilionoidea	Nymphalidae	*Basilarchia arthemis (Limenitis arthemis)*	Tuning forks (~200–1000 Hz)	S	Cease movement, freezing, body contraction	N/A	Minnich (1936)
Papilionoidea	Nymphalidae	*Danaus plexippus*	Buzzing sounds of flying bumble bees and wasps, aircraft, human voices	S	Ducking, agitation of tubercles, twitching, volatile release	Antipredator display	Rothschild and Bergström (1997)
Papilionoidea	Nymphalidae	*Danaus plexippus*	50–900 Hz tones	S	Contraction, freezing, vertical anterior flicking	Predator or parasitoid avoidance	Taylor and Yack (2019)
Papilionoidea	Nymphalidae	*Danaus plexippus*	Tuning forks (~200–1000 Hz)	S	Anterior jerking	N/A	Minnich (1936)
Papilionoidea	Nymphalidae	*Vanessa antiopa (Nymphalis antiopa)*	Tuning forks (~200–1000 Hz), piano, human voice, organ, violin	S	Body contraction, head retraction, cessation of locomotion	N/A	Minnich (1925)

(continued)

Table 1 (continued)

Taxon			Acoustic stimulus/stimuli				Proposed function	Reference(s)
Superfamily	Family	Species	Source		S/V[a]	Behavioural response		
Papilionoidea	Nymphalidae	*Vanessa atalanta*	Tuning forks (~200–1000 Hz)		S	Cessation of movement, freezing, body contraction	N/A	Minnich (1936)
Papilionoidea	Nymphalidae	*Vanessa cardui*	Tuning forks (~200–1000 Hz)		S	Cessation of movement, freezing, body contraction	N/A	Minnich (1936)
Papilionoidea	Nymphalidae	*Vanessa urticae* (*Aglais urticae*)	Human voice		S	Lateral thrashing	N/A	Tutt (1893)
Papilionoidea	Papilionidae	*Papilio polyxenes*	Tuning forks (~200–1000 Hz)		S	Cessation of movement, freezing, body contraction	N/A	Minnich (1936)
Papilionoidea	Papilionidae	*Papilio turnus*	Tuning forks (~200–1000 Hz)		S	Cessation of movement, freezing, body contraction	N/A	Minnich (1936)
Papilionoidea	Pieridae	*Colias philodice*	Tuning forks (~200–1000 Hz)		S	Cessation of movement, freezing, body contraction	N/A	Minnich (1936)

[a]S/V = Sound/Vibration. The acoustic stimulus is reported as being either sound (airborne) or vibration (solid-borne) based on either what is proposed by the authors of the published paper, or what is inferred by the source of the stimulus. This does not mean that the caterpillar is necessarily responding to that form of stimulus however. For example, if a caterpillar responds to the sounds of a tuning fork, it could conceivably be receiving these vibrations via air (near-field receptors) or solids (vibration receptors), particularly if the receptor is undescribed

proposed to detect solid-borne vibrations in caterpillars include antennae (Dethier 1941), various scolopidia located in the thorax or abdomen (Hasenfuss 1992), and trichoid sensilla on the prolegs (Rosi-Denadai 2018). A possible mechanism for detecting far-field airborne sounds is through resonating internal structures (see Shaw 1994 for discussion of these in adult cockroaches), although currently there is no evidence for caterpillars detecting far-field sounds. In the context of avoiding predators, there are several reports of caterpillars responding to acoustic cues generated by flying, crawling, and ovipositing insect enemies. Also, there is limited evidence that caterpillars detect signals of non-predators to coordinate defences. These are discussed below.

Detecting Sounds and Vibrations Generated by Predators

Detection of Near-Field Sounds

Reports on caterpillars responding to sound date back more than 200 years (reviewed in Minnich 1936). Sound sources evoking responses have included tuning forks, highway noise, jet aircraft, human voices, flying insects, human voices, various musical instruments, hand clapping, and tones played from speakers (e.g. Tutt 1893; Johnson 1893; Minnich 1925, 1936; Abbott 1927; Hogue 1972; Markl and Tautz 1975; Myers and Smith 1978; Rothschild and Bergström 1997; Davis et al. 2018) (Table 1). Behavioural responses to these sounds include flicking different body parts, freezing, body contraction, squirming, increased heart rate, cessation of movement, and dropping from silk threads. Despite the numerous reports over the past two centuries, there has been little formal research on the adaptive significance and sensory mechanisms associated with caterpillar 'hearing'. The best studied species to date include larvae of the cabbage moth *Mamestra brassicae* (Noctuidae) and the monarch butterfly *Danaus plexippus* (Nymphalidae).

Cabbage moth (*Mamestra brassicae*) larvae respond to sounds in a number of ways, including ceasing locomotion, contracting, squirming, and dropping from the substrate (Markl and Tautz 1975). Caterpillars respond to pure tones between 40 and 1000 Hz, with best sensitivity at 100–600 Hz (Markl and Tautz 1975). Eight thoracic trichoid sensilla function as near-field sound receptors (Markl and Tautz 1975; Tautz 1977, 1978). Tautz and Markl (1978) demonstrated that defensive behaviours were evoked by flight sounds of the parasitoid wasp *Dolichovespula media* (Vespidae). When the wasp flies close to the larva, the sensilla are deflected, evoking defensive responses in the caterpillar. Responses varied with loudness of the sounds, with low amplitude sounds causing larvae to freeze, and higher amplitude sounds causing them to squirm and drop from the plant (Tautz and Markl 1978). Experimental tests involving sensory ablation showed that significantly more larvae were attacked if their sensilla had been removed compared to controls with intact sensilla (Tautz and Markl 1978). The resonance frequency of sensilla is ~150 Hz, matching the wingbeat frequency of *D. media* (Tautz and Markl 1978).

This comprehensive series of experiments confirmed that trichoid sensilla function in hearing, and that an adaptive function of hearing is to detect parasitoid wasps.

Monarch butterfly (*Danaus plexippus*) caterpillars were first formally tested for their hearing capabilities by Minnich (1936), who reported that larvae responded to tuning forks by freezing, contracting, and jerking their anterior ends. Rothschild and Bergström (1997) subsequently reported that monarch caterpillars responded to sounds of passing aircraft, 'buzzing' sounds, and human voices, by making 'sudden ducking or twitching movements'. More recently, Davis et al. (2018) demonstrated that the sounds of traffic noise caused monarch larvae to increase their heart rates. Taylor and Yack (2019) conducted a series of experiments to characterize the behavioural responses to sounds and their tuning characteristics, and to identify the primary hearing organs. Late instar (4th and 5th) larvae responded to pure tone sounds by freezing, contracting, and vertically flicking their thorax (Fig. 2a). These responses were evoked by sound frequencies ranging between 50 and 900 Hz, with best sensitivity at 100–200 Hz. Sound amplitude affected the type of response, with low amplitudes causing freezing and contraction, and higher amplitudes evoking vigorous dorsoventral flicks (Taylor 2009; Taylor and Yack 2019). This result suggests that caterpillars respond differently as the enemy approaches, first by freezing to purportedly render themselves acoustically or visually cryptic and then by flicking their bodies and tubercles to knock off the predator or prevent egg laying by a parasitoid. Caterpillars were shown to habituate to sounds upon repeated exposures. This result has implications for conservation of monarch butterflies, as a larva's ability to detect and respond to enemies could be compromised in the presence of anthropogenic noise. The primary sensory receptors were confirmed, using ablation experiments, to comprise a pair of prothoracic trichoid sensilla. It is not known whether these receptors are homologous to those in *M. brassicae*, but given that these species are distantly related and exhibit different responses to sound, it is probable that sound reception resulted from convergent evolution. It is proposed that monarchs evolved hearing in response to flight sounds of aerial predators such as wasps and tachinid flies. However, experiments with live predators or parasitoids have not yet been conducted.

Behavioural responses to sounds have been reported for caterpillar species belonging to several lepidopteran taxa (Table 1). In addition to the above-mentioned examples, there are reports of tent caterpillars responding to sounds by flicking their heads and dropping from their tents (Myers and Smith 1978; Taylor 2009), notodontid larvae thrashing tentacles (White et al. 1983), and gregarious saturniid caterpillars raising their heads and generating alarm calls (Breviglieri and Romero 2019). Given the diversity of behavioural responses observed in distantly related species, and the likelihood that near-field receptors are presumably relatively 'easy' to evolve, it would not be surprising if sound reception evolved multiple times in caterpillars. Trichoid sensilla that detect air currents and near-field sounds tend to be long (>500 µm) and filiform in shape (see Keil 1997), and these probably evolve as specializations of the many trichoid sensilla (i.e. innervated hairs) that cover the bodies of larval Lepidoptera. Future studies on 'hearing' should involve testing of more species for their behavioural responses to sound, either played through

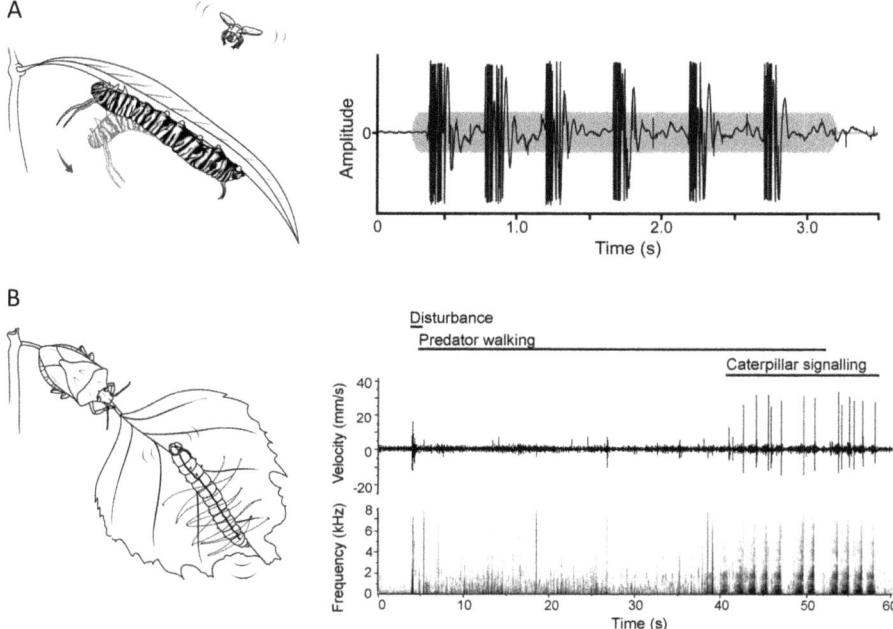

Fig. 2 Defensive sound and vibration detection in caterpillars. (**A**) Monarch (*Danaus plexippus*) caterpillars detect low frequency sounds (~100–500 Hz) by freezing, contracting, and flicking their anterior bodies. Sound frequencies match those generated by a flying insect predator or parasitoid. The waveform illustrates a late instar caterpillar responding to a sound by dorsally flicking. The dark part of the trace shows a laser vibrometer recording from a milkweed leaf, showing six consecutive flicks of the caterpillar. The grey part of the trace indicates the timing of a 300 ms, 200 Hz sound played to the caterpillar through a speaker (adapted from Taylor and Yack 2019). (**B**) The masked birch caterpillar (*Drepana arcuata*) responds to plant-borne vibration cues resulting from the walking movements of a predatory stink bug (*Podisus* sp.). Caterpillars typically first freeze upon detecting the predator, and then signal when the predator is close. The trace shown is a laser vibrometer recording from a birch leaf upon which the caterpillar is residing in its silk shelter. Crawling vibrations of the stink bug are shown to occur prior to the caterpillar signalling. The disturbance marks the application of the predator to the leaf twig (adapted from Guedes et al. 2012)

speakers in the lab, or in the presence of live predators. Experiments should also focus on identifying the receptors, and ascertaining the capabilities of these receptors for encoding sound frequency, amplitude, and direction, characteristics that could convey information about the location and type of enemy.

Detection of Solid-Borne Vibrations

Caterpillars are substrate-bound organisms residing primarily on or within plants (e.g. leaves, twigs) or silk (e.g. silk mats on leaves, or tents). Consequently, solid-borne vibrations generated by predators or parasitoids should be important for risk assessment. Predator-generated vibrations could include cues resulting from the predator approaching the prey, signals used for hunting (i.e. echolocation), or signals used for communicating with others such as conspecifics. These types of cues and signals could conceivably be transmitted to the prey directly through the solid substrate, or indirectly, by air-borne sounds vibrating the substrate (see Caldwell 2014; Yack 2016). While there are many sources of vibrations of potential interest to caterpillar prey, there are currently few experimentally confirmed examples of caterpillars using these vibrations to assess risk. Late instar masked birch caterpillars (*Drepana arcuata*) (Drepanidae) respond to low-frequency crawling vibrations generated by an approaching predatory stink bug (*Podisus* sp.) (Guedes et al. 2012) (Fig. 2b). The caterpillars respond differently depending on the distance of the approaching enemy, by first ceasing activity, and then signalling when the predator is close or has launched an attack. *Semiothisa aemulataria* (Geometridae) larvae respond to plant-borne vibrations resulting from approaching predators (stink bugs, wasps) (Castellanos and Barbosa 2006). Late instar larvae could distinguish between vibrations caused by different sources, including predators, non-predators, and abiotic sources. They also showed evidence of being able to discriminate between vibrations of wasps and stinkbugs by escaping on different lengths of silk thread. Apple leaf miners *Phyllonorycter malella* (Gracillariidae) stop feeding upon detecting vibrations produced by a parasitic wasp *Sympiesis sericeicornis* (Eulophidae) inserting its ovipositor into a mine; the caterpillar resumes feeding only after the parasitoid leaves (Meyhöfer et al. 1997).

There are likely to be many instances of caterpillars using vibrations generated by predators and parasitoids. To document these cases, predator-prey interactions should be staged on natural substrates while recording with vibration sensors to assess what vibrations are available to caterpillars. Playing back these vibrations to prey can be helpful in assessing a prey's response to these vibrations, although vibratory playbacks can be methodologically complex (Cocroft et al. 2014). Identifying the receptors used for vibratory sensing is also needed to better understand vibratory-mediated risk assessment in caterpillars.

Detecting Sounds and Vibrations Made by Non-predators

Caterpillars communicate with non-predators, including conspecifics and heterospecifics, to coordinate defences against their common predators (Costa 2006). For example, some species form assemblages with conspecifics to enhance warning displays, or to build protective shelters. Other species form relationships with ants for protection. The roles of sound and vibration detection in caterpillars remain to be tested in most cases, but there are a few inferential examples. Breviglieri and Romero (2019) report that social *Hylesia nigricans* (Saturniidae) caterpillars respond to ultrasonic alarm signals generated by conspecifics that are being attacked by wasp and bird predators, and these sounds are proposed to function in coordinating group defences. Other examples of caterpillars detecting non-predator sounds and vibrations in the context of defence are discussed in section "Sending signals to non-predators", which focuses on signal generation in caterpillars.

Summary of Defensive Sound and Vibration Detection

Although caterpillars lack tympanal ears that are commonly used in adult Lepidoptera for avoiding predators, they have nonetheless evolved mechanisms to detect near-field sounds and solid-borne vibrations to avoid attack. Eavesdropping on the acoustic cues produced by enemies has been reported numerous times, and is thought to be widespread among caterpillars. It is very likely that caterpillars also use sounds and vibrations to coordinate defences in social groups, and possibly, to eavesdrop on the communication signals of predators such as insectivorous birds. Future lines of investigation should involve staging interactions with natural predators while recording air- and solid-borne vibrations to gain a better appreciation for the cues and signals available to larvae during these interactions. Once hearing and vibration reception has been confirmed based on behavioural experiments, the sensory mechanisms involved should be identified. Also, it is worth considering the effects of anthropogenic noises, both airborne and vibratory, on the abilities of caterpillars to respond to predators.

Generating Sounds and Vibrations in Caterpillars

> *The larva of the North American Saturnian moth Telea polyphemus can, in the third and fourth stages, by rubbing the powerfully constructed mandibles against each other produce a tolerably loud, tapping sound, which is audible at the distance of several meters. That here is question of a means of intimidation is not to be doubted, for if the larva is left in peace it keeps perfectly quiet, but when the larva-cage is touched, or the larvae are taken out, they make this peculiar tapping sound, resembling the ticking of a watch.* (Federley 1905).

Defence sounds in Lepidoptera are taxonomically widespread and highly variable with respect to the types of sounds and mechanisms of sound production. In adults, these sounds, which primarily function to warn, frighten, or confuse echolocating bats, are well documented (reviewed in Minet and Surlykke 2003; Conner and Corcoran 2012; Greenfield 2014; Kawahara and Barber 2015). Comparatively less is understood about defensive sounds in juveniles, even though disturbance sounds have been documented for both pupae and larvae of many species (e.g. Hinton 1948; Devries 1991a; Bura et al. 2016; Dolle et al. 2018). Caterpillars conceivably would benefit from communicating acoustically with their vertebrate predators (i.e. birds, rodents, bats, frogs, and lizards) which have well-developed hearing. Other caterpillars, such as those living in social groups, or those attended by ants, could benefit from sending recruitment or alarm signals to gain protection or coordinate a defence. Caterpillar sound production in the context of defence is discussed below under two categories: caterpillars that send signals to predators (section "Sending signals to predators") and those that send signals to non-predators (section "Sending signals to non-predators").

Sending Signals to Predators

Many species of silk and hawkmoth (Bombycoidea) caterpillars produce sounds upon being disturbed (Table 2). Earlier reports variously described these sounds in different species as 'singing' (Reed 1868), 'tcep or tceep' (Sanborn 1868), 'squeaking' (Packard 1904), 'crepitating noise' (Packard 1904), 'ticking of a watch' (Federley 1905), and 'crackling-rasping noises' (Heinrich 1979), although the functions of these sounds remained untested. More recently, these sounds have been shown to be widespread throughout the superfamily, variable in their signal characteristics in mechanisms, and to function in defence (see Brown et al. 2007; Bura et al. 2009; Bura et al. 2011; Bura et al. 2012; Bura et al. 2016; Dookie et al. 2017; Rosi-Denadai et al. 2018; Sugiura et al. 2020). In a study of 61 species of late instar larvae, Bura et al. (2016) showed that 31% of species and 45% of genera produced sounds following simulated attacks. Four distinct sound types and respective mechanisms were reported: clicking (mandibular stridulation), chirping (mandibular stridulation), whistling (forced air out of spiracles), and vocalizing (forced air out of buccal cavity). It is proposed that these sounds are directed primarily at vertebrate predators, and function as warning signals (acoustic aposematism), as startle displays, or to mimic alarm calls of a predator. Acoustic aposematism is predicted to occur in species that use a chemical defence, with sounds preceding or accompanying chemical release. Bura et al. (2016) demonstrated that in species with 'high' chemical scores (i.e. chemical production through regurgitation or release from scoli occurring promptly following attack), sound production preceded or accompanied chemical release. These sounds tend to be short-duration clicks or chirps and are proposed to warn the predator of an impending defence. In a study using live predators, Brown et al. (2007) showed that clicking *Antheraea polyphemus* caterpillars (Fig. 3a) survived attacks by chickens following sound production, and mice were repelled by the chemical regurgitant. Other sounds are proposed to startle predators. These sounds tend to be loud and long in duration, such as whistles and vocalizations, and are not typically associated with a chemical defence. Bura et al. (2016) demonstrated that species with low chemical scores (i.e. rarely produced a chemical following multiple attacks) tended to produce sounds with these characteristics. Trials with live predators showed that whistles of the walnut sphinx, *Amorpha juglandis* (Fig. 3b), caused yellow warblers (*Dendroica petechia*) and red winged black birds (*Agelaius phoeniceus*) to escape by diving or flying away (Bura et al. 2011; Dookie et al. 2017). Another hypothesis explaining the function of caterpillar defence sounds is mimicry. The whistles of *A. juglandis* caterpillars resemble the 'seet' warning calls of insectivorous birds, and it was proposed that these sounds mimic the alarm calls of avian predators (Dookie et al. 2017).

Table 2 Caterpillars producing sounds and vibrations in a defensive context

Taxon (of sender)			Acoustic signal			Target animal (proposed) Predator, non-predator, parasitoid	Function (proposed)	Reference(s)
Superfamily	Family	Species	Descriptors of sound, mechanism	NF/FF/V^a (proposed)				
Adeloidea	Heliozelidae	*Antispila nysaefoliella*	Abdominal movements	V		Parasitoids	Defence against parasitoids	Low (2008)
Bombycoidea	Saturniidae	*Actias Luna*	Stridulation, mandibles, clicking	FF		Predators	Aposematism	Bura et al. (2016)
Bombycoidea	Saturniidae	*Antheraea pernyi*	Stridulation, mandibles, clicking	FF		Predators	Aposematism	Bura et al. (2016)
Bombycoidea	Saturniidae	*Antheraea polyphemus*	Stridulation, mandibles, clicking	FF		Predators	Aposematism	Brown et al. (2007), Bura et al. (2016)
Bombycoidea	Saturniidae	*Callosamia promethea*	Stridulation, mandibles, clicking	FF		Predators	Aposematism	Bura et al. (2016)
Bombycoidea	Saturniidae	*Calosaturnia mendocino*	Stridulation, mandibles, chirping	FF		Predators	Aposematism	Bura et al. (2016)
Bombycoidea	Saturniidae	*Citheronia lobesis*	Stridulation, mandibles, chirping	FF/V		Predators	Aposematism	Bura et al. (2016)
Bombycoidea	Saturniidae	*Hylesia nigricans*	Ultrasound emission	FF		Non-predators (conspecifics)	Recruitment	Breviglieri and Romero (2019)
Bombycoidea	Saturniidae	*Rhodinia fugax*	Forced air, spiracles, whistling	FF		Predators	Startle display	Sugiura et al. (2020)
Bombycoidea	Saturniidae	*Saturnia pyri*	Stridulation, mandibles, chirping	FF		Predators	Aposematism	Bura et al. (2016, 2009)
Bombycoidea	Saturniidae	*Schausiella santarosensis*	Stridulation, mandibles, chirping	FF		Predators	Aposematism	Bura et al. (2016)

(continued)

Table 2 (continued)

Taxon (of sender)			Acoustic signal		NF/FF/V[a] (proposed)	Target animal (proposed) Predator, non-predator, parasitoid	Function (proposed)	Reference(s)
Superfamily	Family	Species	Descriptors of sound, mechanism					
Bombycoidea	Sphingidae	*Acherontia atropos*	Stridulation, mandibles, clicking		FF	Predators	Aposematism	Bura et al. (2016)
Bombycoidea	Sphingidae	*Amorpha juglandis*	Forced air, spiracles, whistling		FF	Predators	Startle display	Bura et al. (2011, 2016)
Bombycoidea	Sphingidae	*Amphion floridensis*	Forced air, buccal cavity, vocalizing		FF	Predators	Startle display	Bura et al. (2016), Rosi-Denadai et al. (2018)
Bombycoidea	Sphingidae	*Eumorpha satellitia*	Stridulation, mandibles, clicking		FF	Predators	Aposematism	Bura et al. (2016)
Bombycoidea	Sphingidae	*Langia zenzeroides*	Forced air, spiracles, whistling		FF	Predators	Defence against invertebrate attackers	Sugiura and Takanashi (2018)
Bombycoidea	Sphingidae	*Manduca pellenia*	Stridulation, mandibles, clicking		FF	Predators	Aposematism	Bura et al. (2016)
Bombycoidea	Sphingidae	*Manduca sexta*	Stridulation, mandibles, clicking		FF	Predators	Aposematism	Bura et al. (2012, 2016)
Bombycoidea	Sphingidae	*Nyceryx magna*	Forced air, buccal cavity, vocalizing		FF	Predators	Startle display	Bura et al. (2016)
Bombycoidea	Sphingidae	*Pachygonidia drucei*	Forced air, buccal cavity, vocalizing		FF	Predators	Startle display	Bura et al. (2016)
Bombycoidea	Sphingidae	*Phyllosphingia dissimilis*	Forced air, spiracles, whistling		FF	Predators	Startle display	Bura et al. (2016)

				FF			
Bombycoidea	Sphingidae	*Sphecodina abbottii*	Forced air, buccal cavity, vocalizing		Predators	Startle display	Packard (1904), Heinrich (1979), Bura et al. (2016)
Drepanoidea	Drepanidae	*Drepana arcuata*	Percussion, stridulation mandibles, anal oars	V	Predators (stink bug)	Deterrent to predators	Guedes et al. (2012)
Gelechioidea	Oecophoridae (Lypusidae)	*Diurnea fagella*	Thoracic leg hook scraping on leaf surface	V	Arthropods and parasitoids	Refuge defence	Hunter (1987)
Noctuoidea	Notodontidae	*Cerura* sp. *Dicranura* sp.	Stridulation, mandibles	V	NA	Intimidation	Federley (1905)
Papilionoidea	Lycaenidae	*Cupido argiades Jalmenus evagoras Lycaena phaleas Maculinea alcon Scolitintides orion Thereus pedusa* etc.	N/A	V	Non-predator (ants)	Increase survival	DeVries (1990), (1991a), Travassos and Pierce (2000), Riva et al. (2017)
Papilionoidea	Lycaenidae	*Arhopala madytus*	Stridulation, abdominal segment 5 and 6	V	Non-predator (ant)	Protection from ants	Hill (1993)
Papilionoidea	Lycaenidae	*Hypolycaena othona*	Rhythmic muscle contractions, croaking	V	N/A	Maybe alarm calls to ants	Fiedler (1992a)

(continued)

Table 2 (continued)

Taxon (of sender)			Acoustic signal		Target animal (proposed) Predator, non-predator, parasitoid	Function (proposed)	Reference(s)
Superfamily	Family	Species	Descriptors of sound, mechanism	NF/FF/V[a] (proposed)			
Papilionoidea	Lycaenidae	*Lycaena dispar, L. helle, L. phlaeas*	N/A	N/A	Natural enemies	Signal and repel natural enemies	Riva et al. (2017)
Papilionoidea	Lycaenidae	*Surendra florimel*	N/A	V/FF	Non-predator (ant), other predator	Protection from ants, distress calls	Fiedler (1992b)
Papilionoidea	Riodinidae	*Eurybia* sp.	N/A	V	Non-predator (ants)	Increase survival	DeVries (1990, 1991a)
Papilionoidea	Riodinidae	*Juditha molpe Nymphidium* sp. *Synargis mycone Thisbe irenea Theope* sp. Etc.	Stridulation, vibratory papillae, and epicranial granulations	V	Non-predator (ants)	Increase survival	DeVries (1990, 1991a, 1991b)

[a]NF/FF/V = Near-field sound, far-field sound, vibration (solid-borne). These describe the proposed (or suggested) mode of transmission to the receiver

Fig. 3 Caterpillars that generate sounds and vibrations to avoid attack. (**A**) Aposematic warning sounds. When attacked, the silkmoth caterpillar *Antheraea polyphemus* (Saturniidae) produces clicking sounds by stridulating its mandibles. These sounds are followed by regurgitation and function as warning sounds (see Brown et al. 2007). The waveform and spectrogram show a train of clicks following a simulated attack to the caterpillar with blunt forceps. (**B**) Startle sounds. The walnut sphinx *Amorpha juglandis* produces whistles by forcing air out of its eighth abdominal spiracles. These sounds have been shown to startle avian predators (see Bura et al. 2011; Dookie et al. 2017). The oscillogram and spectrogram show a train of five whistles following a simulated attack with blunt forceps. (**C**) Vibratory recruitment signals. Parasitic larvae of the butterfly *Scolitantides orion* generate acoustic signals to recruit ants for protection. The oscillogram and spectrogram show a train of acoustic signals generated by the larva. The mechanism of signal production is unknown. Sounds and photographs for A and B are from the Yack lab, and for C are provided courtesy of Francesca Barbero and Marco Gherlenda

Some caterpillars are proposed to generate solid-borne vibrations directed at an attacker, although the functional significance of these vibratory signals is not clear. Tupelo leaf miners, *Antispila nysaefoliella*, generate vibrations when disturbed by a parasitoid (Low 2008). The vibrations were described as 'ticks' and 'rattles' produced by specialized structures on the abdomen. Signalling is proposed to interfere with foraging in parasitic wasps, although this hypothesis remains untested. The masked birch caterpillar, *Drepana arcuata*, generates drumming and scraping vibrations when pursued by a stink bug predator (*Podisus* sp.), and these signals appeared to stop the attack (Guedes et al. 2012). It is possible that the vibrations signal to the predator that the prey is unprofitable to pursue. *Diurnea fagella* larvae produce vibratory signals by scraping a hook on their thoracic leg against the leaf surface (Hunter 1987). It was postulated that signalling is directed at intruding spiders, but the anti-predator functions of these signals were not tested.

Sending Signals to Non-predators

Caterpillars also send acoustic signals to non-predators in the context of defence. These signals are primarily vibratory, and function, or are proposed to function, as alarm or recruitment signals to coordinate a defence or recruit others for protection. Currently, the majority of examples involve myrmecophilic (ant-loving) butterfly larvae. Larvae of Lycaenidae and Riodinidae butterflies can generate vibratory signals to recruit and maintain relationships with ants in mutualistic, commensalistic, and parasitic relationships (reviewed in Devries 1991a; Riva et al. 2017; Schönrogge et al. 2017; Casacci et al. 2019) (Table 2). Lycaenidae larvae can produce a variety of vibrations described as pulses, drums, grunts, and hisses to communicate with ants (Travassos and Pierce 2000; Schönrogge et al. 2017). While the mechanisms of signal production are not well understood, one such mechanism involves an abdominal stridulatory apparatus (Hill 1993). Other Lycaenidae species have been described to produce vibrations by a 'shivering' behaviour, which is probably a form of tremulation (e.g. Devries 1991a). Riodinidae larvae also produce vibrations to call to ant hosts. Many species generate signals using a stridulatory mechanism comprising two structures: vibratory papillae and epicranial granulations (Devries 1990, 1991a). Vibratory papillae are grooved rod-like appendages located on the prothorax that strike against textured (granulated) surfaces on the head to produce vibrations as the head oscillates. One of the benefits that caterpillars gain from their relationships with ants is protection from predators and parasitoids. In species that use vibratory signals to gain acceptance into the ant colony, the anti-predator benefits derived from calling are indirect, as being tended by ants provides protection from predators and parasitoids (Pierce et al. 1987; Devries 1991b). However, in cases where myrmecophilous species live within the ant territory but outside of the ant nest, larvae generate vibratory and chemical signals to recruit ants for protection (Schönrogge et al. 2017; Casacci et al. 2019). For example, *Scolitantides orion* (Fig. 3c) calls to ants for protection when disturbed (Riva et al. 2017; Barbero pers. comm.), and *Hypolycaena othona*, although only weakly associated with ants, signals upon disturbance (Fiedler 1992a), presumably to gain protection by ants.

Other examples of caterpillars proposed to communicate acoustically with non-predators to gain protection include early instar *D. arcuata* that signal to recruit conspecifics to build protective shelters (Yadav et al. 2017; Yadav and Yack 2018), and early instar *H. nigricans* that produce airborne sounds to coordinate a group defence (Breviglieri and Romero 2019). However, the antipredator benefits of these signals have not been experimentally validated to date.

Summary of Caterpillar Defensive Sound Production

Despite having soft bodies that limit their capabilities for producing sounds and vibrations, caterpillars have evolved an impressive variety of acoustic defence signals. These can be directed at a predator and function as warning, startle, or mimicry signals, or they can be directed at non-predators to recruit help or coordinate defences. Many of these signals are inconspicuous to humans without the assistance of specialized recording instruments, and it is expected that there are many undocumented examples.

Conclusions and Future Research

When considering caterpillar defence strategies, hearing and sound production do not immediately come to mind. Yet, evidenced by the examples discussed here, it is clear that acoustic antipredator strategies are taxonomically widespread and functionally diverse in caterpillars. Still, we have just scratched the surface in our understanding of this topic. The following lines of investigation are recommended for future studies: 1. Behavioural responses to low frequency sounds (less than 2 kHz) should be documented across different taxa and developmental stages of larval instars. Low frequency sounds simulate those of flying insect predators and parasitoids that impose significant selection pressures on caterpillars to evolve near-field sound and vibration receptors, and this form of hearing is likely to have evolved multiple times. 2. Recordings of air- and solid-borne vibrations from natural substrates (i.e. host plants, silk shelters) should be performed while videotaping predator/parasitoid and caterpillar prey interactions. 3. Sound and vibration receptor mechanisms should be identified using neurophysiological, morphological, and behavioural experiments. Given the diversity of acoustic signals and cues that are detectable by caterpillars, as well as the diversity of behavioural responses exhibited, it would not be surprising to see a diversity of sound and vibration receptors resulting from convergent evolution. 4. Hypotheses explaining the functions and evolution of defence sounds in Bombycoidea caterpillars require further testing using experiments with live predators and comparative phylogenetic analyses. Research on acoustically mediated defences is key for gaining a comprehensive understanding of the survival strategies of caterpillars, but also has some practical applications. For example, as anthropogenic noise may impair a caterpillar's ability to detect an enemy, it is important to understand what a caterpillar 'hears' for conservation purposes. On the flip side, sounds and vibrations can be implemented in pest management, as caterpillars have been shown to respond to acoustic signals and cues by ceasing movement and feeding, and dropping from host plants.

Acknowledgements I am grateful to Mairelys Naranjo for the help in reviewing literature and preparing the tables, to Kaylen Brzezinski for the line drawings in Fig. 2, to Francesca Barbero for discussions on myrmecophilic caterpillars, and to F. Barbero and Marco Gherlenda for contributing a photo and recording of *Scolitantides orion* (Fig. 3c). I also thank the reviewers for their constructive comments. This work was funded by the Natural Science and Engineering Council of Canada Discovery Grants (2014-05947 and 2020-07056) to JEY.

References

Abbott CE (1927) The reaction of *Datana* larvae to sounds. Psyche 34:129–133. https://doi.org/10.1155/1927/90746

Alexander RD (1967) Acoustical communication in arthropods. Annu Rev Entomol 12:495–526. https://doi.org/10.1146/annurev.en.12.010167.002431

Breviglieri CPB, Romero GQ (2019) Acoustic stimuli from predators trigger behavioural responses in aggregate caterpillars. Austral Ecol 44:880–890. https://doi.org/10.1111/aec.12757

Brown SG, Boettner GH, Yack JE (2007) Clicking caterpillars: acoustic aposematism in *Antheraea polyphemus* and other Bombycoidea. J Exp Biol 210:993–1005. https://doi.org/10.1242/jeb.001990

Bura VL, Fleming AJ, Yack JE (2009) What's the buzz? Ultrasonic and sonic warning signals in caterpillars of the great peacock moth (*Saturnia pyri*). Naturwissenschaften 96:713–718. https://doi.org/10.1007/s00114-009-0527-8

Bura VL, Rower VG, Martin PR, Yack JE (2011) Whistling in caterpillars (*Amorpha juglandis*, Bombycoidea): sound-producing mechanism and function. J Exp Biol 214:30–37. https://doi.org/10.1242/jeb.046805

Bura VL, Hnain AK, Hick JN, Yack JE (2012) Defensive sound production in the tobacco hornworm, *Manduca sexta* (Bombycoidea: Sphingidae). J Insect Behav 25:114–126. https://doi.org/10.1007/s10905-011-9282-8

Bura VL, Kawahara AY, Yack JE (2016) A comparative analysis of sonic defences in Bombycoidea caterpillars. Sci Rep 6:31469. https://doi.org/10.1038/srep31469

Caldwell MS (2014) Interactions between airborne sound and substrate vibration in animal communication. In: Cocroft R, Gogala M, Hill P, Wessel A (eds) Studying vibrational communication, Animal signals and communication, vol 3. Springer, Berlin/Heidelberg, pp 65–92. https://doi.org/10.1007/978-3-662-43607-3_6

Casacci LP, Bonelli S, Balletto E, Barbero F (2019) Multimodal signaling in myrmecophilous butterflies. Front Ecol Evol 7:454. https://doi.org/10.3389/fevo.2019.00454

Castellanos I, Barbosa P (2006) Evaluation of predation risk by a caterpillar using substrate-borne vibrations. Anim Behav 72:461–469. https://doi.org/10.1016/j.anbehav.2006.02.005

Cocroft RB, Hamel JA (2010) Vibrational communication in the "other insect societies": a diversity of ecology, signals and signal functions. In: O'Connell-Rodwell CE (ed) The use of vibrations in communication: properties, mechanisms and function across taxa. Transworld Research Network, Kerala, pp 47–68

Cocroft RB, Hamel J, Su Q, Gibson J (2014) Vibrational playback experiments: challenges and solutions. In: Cocroft R, Gogala M, Hill P, Wessel A (eds) Studying vibrational communication, Animal signals and communication, vol 3. Springer, Berlin/Heidelberg, pp 249–274. https://doi.org/10.1007/978-3-662-43607-3_13

Conner WE (2014) Adaptive sounds and silences: acoustic anti-predator strategies in insects. In: Hedwig B (ed) Insect hearing and acoustic communication. Animal signals and communication, vol 1. Springer, Berlin/Heidelberg, pp 65–79. https://doi.org/10.1007/978-3-642-40462-7_5

Conner WE, Corcoran AJ (2012) Sound strategies: the 65-million-year-old battle between bats and insects. Annu Rev Entomol 57:21–39. https://doi.org/10.1146/annurev-ento-121510-133537

Conner WE, Wilson R (2009) Caterpillar talk. In: Conner WE (ed) Tiger moths and woolly bears: behavior, ecology, and evolution of the Arctiidae. Oxford University Press, New York, pp 173–176

Costa JT (2006) The other insect societies. Harvard University Press, Cambridge MA

Cott HB (1940) Adaptive coloration in animals. Methuen, London

Davis AK, Schroeder H, Yeager I, Pearce J (2018) Effects of simulated highway noise on heart rates of larval monarch butterflies, *Danaus plexippus*: implications for roadside habitat suitability. Biol Lett 14:20180018. https://doi.org/10.1098/rsbl.2018.0018

Dethier VG (1941) The function of the antennal receptors in lepidopterous larvae. Biol Bull 80:403–414. https://doi.org/10.2307/1537725

DeVries PJ (1990) Enhancement of symbioses between butterfly caterpillars and ants by vibrational communication. Science 248:1104–1106. https://doi.org/10.1126/science.248.4959.1104

DeVries PJ (1991a) Call production by Myrmecophilous Riodinid and Lycaenid butterfly caterpillars (Lepidoptera): morphological, acoustical, functional, and evolutionary patterns. Amer Mus Novitates 3025:1–23

DeVries PJ (1991b) Mutualism between *Thisbe irenea* butterflies and ants, and the role of ant ecology in the evolution of larval-ant associations. Biol J Linn Soc 43:179–195. https://doi.org/10.1111/j.1095-8312.1991.tb00592.x

Djemai I, Casas J, Magal C (2001) Matching host reactions to parasitoid wasp vibrations. Proc R Soc B 268:2403–2408. https://doi.org/10.1098/rsbp.2001.1811

Dolle P, Klein P, Fischer OW, Schnitzler H, Gilbert LE, Boppré M (2018) Twittering pupae of papilionid and nymphalid butterflies (Lepidoptera): novel structures and sounds. Annu Entomolo Soc Am 111:341–354. https://doi.org/10.1093/aesa/say029

Dookie AL, Young CA, Lamothe G, Schoenle LA, Yack JE (2017) Why do caterpillars whistle at birds? Insect defence sounds startle avian predators. Behav Process 138:58–66. https://doi.org/10.1016/j.beproc.2017.02.002

Evans DL, Schmidt JO (eds) (1990) Insect defenses. State University of New York Press, Albany

Faure P, Hoy R (2000) The sounds of silence: cessation of singing and song pausing are ultrasound-induced acoustic startle behaviors in the katydid *Neoconocephalus ensiger* (Orthoptera; Tettigoniidae). J Comp Physiol A 186:129–142. https://doi.org/10.1007/s003590050013

Federley H (1905) Sound produced by Lepidopterous larvae. J N Y Entomol Soc 13:109–110

Fiedler K (1992a) Notes on the biology of *Hypolycaena othona* (Lepidoptera: Lycaenidae) in West Malaysia. Nachr Entomol Vereins Apollo 13:65–92

Fiedler K (1992b) The life-history of *Surendra florimel* Doherty 1889 (Lepidoptera:Lycaenidae) in West Malaysia. Nachr Entomol Vereins Apollo 113:107–135

Fletcher LE (2007) Vibrational signals in a gregarious sawfly larva (*Perga affinis*): group coordination or competitive signalling? Behav Ecol Sociobiol 61:1809–1821. https://doi.org/10.1007/s00265-007-0414-2

Fletcher LE (2008) Cooperative signaling as a potential mechanism for cohesion in a gregarious sawfly larva, *Perga affinis*. Behav Ecol Sociobiol 62:1127–1138. https://doi.org/10.1007/s00265-007-0541-9

Fletcher LE, Yack JE, Fitzgerald TD, Hoy RR (2006) Vibrational communication in the cherry leaf roller caterpillar *Caloptilia serotinella*. J Insect Behav 19:1–18. https://doi.org/10.1007/s10905-005-9007-y

Geipel I, Jung K, Kalko EKV (2013) Perception of silent and motionless prey on vegetation by echolocation in the gleaning bat *Micronycteris microtis*. Proc R Soc B 280:20122830. https://doi.org/10.1098/rspb.2012.2830

Gentry GL, Dyer LA (2002) On the conditional nature of neotropical caterpillar defenses against their natural enemies. Ecology 83:3108–3119. https://doi.org/10.1890/0012-9658(2002)083[3108:OTCNON]2.0.CO;2

Greeney HF, Dyer LA, Smilanich AM (2012) Feeding by lepidopteran larvae is dangerous : a review of caterpillars' chemical, physiological, morphological, and behavioral defenses against natural enemies. Invertebr Surviv J 9:7–34

Greenfield MD (2014) Acoustic communication in the nocturnal Lepidoptera. In: Hedwig B (ed) Insect hearing and acoustic communication. Animal signals and communication, vol 1. Springer, Berlin/Heidelberg, pp 81–100. https://doi.org/10.1007/978-3-642-40462-7_6

Greenfield MD, Baker M (2003) Bat avoidance in non-aerial insects: the silence response of signaling males in an acoustic moth. Ethology 109:427–442. https://doi.org/10.1046/j.1439-0310.2003.00886.x

Guedes RNC, Matheson SM, Frei B, Smith ML, Yack JE (2012) Vibration detection and discrimination in the masked birch caterpillar (*Drepana arcuata*). J Comp Physiol A 198:325–335. https://doi.org/10.1007/s00359-012-0711-8

Hamel JA, Cocroft RB (2019) Maternal vibrational signals reduce the risk of attracting eavesdropping predators. Front Ecol Evol 7:204. https://doi.org/10.3389/fevo.2019.00204
Hasenfuss I (1992) Morphology, evolution, and taxonomic importance of supposed web-vibration receptors in the larvae of butterflies (Lepidoptera: Pyraloidea and Gelechioidea). Entomol Generalis 18:43–54. https://doi.org/10.1127/entom.gen/18/1993/43
Heinrich B (1979) Foraging strategies of caterpillars. Oecologia 42:325–337. https://doi.org/10.1007/BF00346597
Heinrich B (1993) How avian predators constrain caterpillar foraging. In: Stamp NE, Casey TM (eds) Caterpillars: ecological and evolutionary constraints on foraging. Chapman and Hall, New York, pp 224–247
Hill CJ (1993) The myrmecophilous organs of *Arhopala madytus* Fruhstorfer (Lepidoptera: Lycaenidae). J Aust Entomol Soc 32:283–288. https://doi.org/10.1111/j.1440-6055.1993.tb00587.x
Hill PSM (2008) Vibrational communication in animals. Harvard University Press, Cambridge
Hill PSM (2014) Stretching the paradigm or building a new? Development of a cohesive language for vibrational communication. In: Cocroft R, Gogala M, Hill P, Wessel A (eds) Studying vibrational communication, Animal signals and communication, vol 3. Springer, Berlin/Heidelberg, pp 13–30. https://doi.org/10.1007/978-3-662-43607-3_2
Hill PSM, Wessel A (2016) Biotremology. Curr Biol 26:R181–R191. https://doi.org/10.1016/j.cub.2016.01.054
Hinton HE (1948) Sound production in Lepidopterous pupae. The Entomologist 81:254–269
Hogue CL (1972) Protective function of sound perception and gregariousness in *Hylesia* larvae (Saturniidae: Hemileucinae). J Lepid Soc 26:33–34
Hoy RR (1992) The evolution of hearing in insects as an adaptation to predation from bats. In: Webster DB, Popper AN, Fay RR (eds) The evolutionary biology of hearing. Springer, New York, pp 115–129. https://doi.org/10.1007/978-1-4612-2784-7_8
Hunt JH, Richard FJ (2013) Intracolony vibroacoustic communication in social insects. Insect Soc 60:403–417. https://doi.org/10.1007/s00040-013-0311-9
Hunter MD (1987) Sound production in larvae of *Diurnea fagella* (Lepidoptera: Oecophoridae). Ecol Entomol 12:355–357. https://doi.org/10.1111/j.1365-2311.1987.tb01015.x
Ishay J, Motro A, Gitter S, Brown MB (1974) Rhythms in acoustical communication by the oriental hornet, *Vespa orientalis*. Anim Behav 22:741–744. https://doi.org/10.1016/S0003-3472(74)80026-6
Jacobs DS, Ratcliffe JM, Fullard JH (2008) Beware of bats, beware of birds: the auditory responses of eared moths to bat and bird predation. Behav Ecol 19:1333–1342. https://doi.org/10.1093/beheco/arn071
Johnson AJJ (1893) Sensibility of larvae to sound. Entomol Rec 4:240–241
Kalka M, Kalko EKV (2006) Gleaning bats as underestimated predators of herbivorous insects: diet of *Micronycteris microtis* (Phyllostomidae) in Panama. J Trop Ecol 22:1–10. https://doi.org/10.1017/S0266467405002920
Kawahara AY, Barber JR (2015) Tempo and mode of antibat ultrasound production and sonar jamming in the diverse hawkmoth radiation. Proc Natl Acad Sci 112:6407–6412. https://doi.org/10.1073/pnas.1416679112
Keil TA (1997) Functional morphology of insect mechanoreceptors. Microsc Res Tech 39:506–531
Klots AB (1969) Audition by *Cerura* larvae (Lepidoptera: Notodontidae). J N Y Entomol Soc 77:10–11
Lakes-Harlan R, Strauß J (2014) Functional morphology and evolutionary diversity of vibration receptors in insects. In: Cocroft R, Gogala M, Hill P, Wessel A (eds) Studying vibrational communication, Animal signals and communication, vol 3. Springer, Berlin/Heidelberg, pp 277–302. https://doi.org/10.1007/978-3-662-43607-3_14
Lederhouse RC (1990) Avoiding the hunt: primary defenses of lepidopteran caterpillars. In: Evans D, Schmidt J (eds) Insect defenses: adaptive mechanisms and strategies of prey and predators. State University of New York Press, Albany, pp 175–189

Low C (2008) Seismic behaviours of a leafminer, *Antispila nysaefoliella* (Lepidoptera: Heliozelidae). Fla Entomol 91:604–609. https://doi.org/10.1653/0015-4040-91.4.604

Low ML, Naranjo M, Yack JE (2021) Survival sounds in insects: diversity, function and evolution. Front Ecol Evol. https://doi.org/10.3389/fevo.2021.641740

Markl H, Tautz J (1975) The sensitivity of hair receptors in caterpillars of *Barathra brassicae* L. (Lepidoptera, Noctuidae) to particle movement in the sound field. J Comp Physiol 99:79–87. https://doi.org/10.1007/BF01464713

Masters WM (1980) Insect disturbance stridulation: characterization of airborne and vibrational components of the sound. J Comp Physiol A 135:259–268. https://doi.org/10.1007/BF00657254

Maynard-Smith J, Harper D (2003) Animal signals. Oxford University Press, Oxford

Mclver S, Beech M (1986) Prey finding behavior and mechanosensilla of larval *Toxorhynchites brevipalpis* Theobald (Diptera:Culicidae). Int J Insect Morphol Embryol 15:213–225. https://doi.org/10.1016/0020-7322(86)90059-0

Meyhöfer R, Casas J, Dorn S (1997) Vibration-mediated interactions in a host-parasitoid system. Proc R Soc B 264:261–266. https://doi.org/10.1098/rspb.1997.0037

Mikhail A, Lewis JE, Yack JE (2018) What does a butterfly hear? Physiological characterization of auditory afferents in *Morpho peleides* (Nymphalidae). J Comp Physiol A 204:791–799. https://doi.org/10.1007/s00359-018-1280-2

Miller LA, Surlykke A (2001) How some insects detect and avoid being eaten by bats: tactics and countertactics of prey and predator. Bioscience 51:570–581

Minet J, Surlykke A (2003) Auditory and sound producing organs. In: Kristensen NP (ed) Handbook of zoology: Vol IV Arthropoda: Insecta. Part 36. Lepidoptera, moths and butterflies, vol 2. W G de Gruyter, New York, pp 289–323

Minnich DE (1925) The reactions of the larvae of *Vanessa antiopa* Linn. To sounds. J Exp Zool 42:443–468. https://doi.org/10.1002/jez.1400420404

Minnich DE (1936) The responses of caterpillars to sounds. J Exp Zool 72:439–453. https://doi.org/10.1002/jez.1400720305

Montllor CB, Bernays EA (1993) Invertebrate predators and caterpillar foraging. In: Stamp NE, Casey TM (eds) Caterpillars: ecological and evolutionary constraints on foraging. Chapman and Hall, New York, pp 170–202

Myers JH, Smith JNM (1978) Head flicking by tent caterpillars: a defensive response to parasite sounds. Can J Zool 56:1628–1631. https://doi.org/10.1139/z78-225

Nakano R, Skals N, Takanashi T, Surlykke A, Koike T, Yoshida K, Maruyama H, Tatsuki S, Ishikawa Y (2008) Moths produce extremely quiet ultrasonic courtship songs by rubbing specialize scales. PNAS 105:11812–11817. https://doi.org/10.1073/pnas.0804056105

Packard AS (1904) Sound produced by a japanese saturnian caterpillar. J N Y Entomol Soc 12:92–93

Page RA, Bernal XE (2020) The challenge of detecting prey: private social information use in predatory bats. Funct Ecol 34:344–363. https://doi.org/10.1111/1365-2435.13439

Pfannenstiel RS, Hunt RE, Yeargan KV (1995) Orientation of a hemipteran predator to vibrations produced by feeding caterpillars. J Insect Behav 8:1–9. https://doi.org/10.1007/BF01990965

Pierce NE, Kitching RL, Buckley RC, Taylor MFJ, Benbow KF (1987) The costs and benefits of cooperation between the Australian lycaenid butterfly, *Jalmenus evagoras*, and its attendant ants. Behav Ecol Sociobiol 21:237–248. https://doi.org/10.1007/BF00292505

Pollack GS (2016) Hearing for defense. In: Pollack G, Mason A, Popper A, Fay R (eds) Insect hearing, Springer handbook of auditory research, vol 55. Springer, Cham, pp 81–98. https://doi.org/10.1007/978-3-319-28890-1_4

Reed E (1868) A musical larva. Can Entomol 1:40–41. https://doi.org/10.4039/Ent140-5

Ribarič D, Gogala M (1996) Acoustic behaviour of some butterfly species of the genus *Erebia* (Lepidoptera: Satyridae). Acta Entomol Slov 4:5–12

Riva F, Barbero F, Bonelli S, Balletto E, Casacci LP (2017) The acoustic repertoire of lycaenid butterfly larvae. Bioacoustics 26:77–90. https://doi.org/10.1080/09524622.2016.1197151

Roberts D (2017) Mosquito larvae can detect water vibration patterns from a nearby predator. Bull Entomol Res 107:499–505. https://doi.org/10.1017/S0007485316001140
Rosi-Denadai CA (2018) How do caterpillars detect vibration? Proleg sensory hairs as vibration receptors in *Drepana arcuata* (Drepanidae) and *Trichoplusia ni* (Noctuidae). MSc thesis, Carleton University, Ottawa, Canada. https://doi.org/10.22215/etd/2018-12859
Rosi-Denadai CA, Scallion ML, Merrett CG, Yack JE (2018) Vocalization in caterpillars: a novel sound-producing mechanism for insects. J Exp Biol 221:jeb169466. https://doi.org/10.1242/jeb.169466
Rothschild M, Bergström G (1997) The monarch butterfly caterpillar (*Danaus plexippus*) waves at passing hymenoptera and jet aircraft - are repellent volatiles released simultaneously? Phytochemistry 45:1139–1144. https://doi.org/10.1016/S0031-9422(97)00138-6
Ruxton GD, Sherratt TN, Speed MP (2004) Avoiding attack: the evolutionary ecology of crypsis, warning signals and mimicry. Oxford University Press, Oxford
Rydell J (1998) Bat defense in lekking ghost swifts (*Hepialus humuli*), a moth without ultrasonic hearing. Proc R Soc Lond B 265:1373–1376. https://doi.org/10.1098/rspb.1998.0444
Sala M, Casacci LP, Balletto E, Bonelli S, Barbero F (2014) Variation in butterfly larval acoustics as a strategy to infiltrate and exploit host ant colony resources. PLoS One 9:e94341. https://doi.org/10.1371/journal.pone.0094341
Sanborn FG (1868) Musical larvae. Can Entomol 1:48
Schönrogge K, Barbero F, Casacci LP, Settele J, Thomas JA (2017) Acoustic communication within ant societies and its mimicry by mutualistic and socially parasitic myrmecophiles. Anim Behav 134:249–256. https://doi.org/10.1016/j.anbehav.2016.10.031
Scott JL, Kawahara AY, Skevington JH, Yen S-H, Sami A, Smith ML, Yack JE (2010) The evolutionary origins of ritualized acoustic signals in caterpillars. Nat Commun 1:1–9. https://doi.org/10.1038/ncomms1002
Shaw SR (1994) Detection of airborne sound by a cockroach 'vibration detector': a possible missing link in insect auditory evolution. J Exp Biol 193:13–47
Stamp NE, Casey TM (eds) (1993) Caterpillars: ecological and evolutionary constraints on foraging. Chapman and Hall, New York
Sugiura S (2020) Predators as drivers of insect defenses. Entomol Sci 23:316–337. https://doi.org/10.1111/ens.12423
Sugiura S, Takanashi T (2018) Hornworm counterattacks: defensive strikes and sound production in response to invertebrate attackers. Biol J Linn Soc 123:496–505. https://doi.org/10.1093/biolinnean/blx156
Sugiura S, Takanashi T, Kojima W, Kajiura Z (2020) Squeaking caterpillars: independent evolution of sonic defense in wild silkmoths. Ecology 101(10):e03112. https://doi.org/10.1002/ecy.3112
Takanashi T, Fukaya M, Nakamuta K, Skals N, Nishino H (2016) Substrate vibrations mediate behavioral responses via femoral chordotonal organs in a cerambycid beetle. Zool Lett 2:18
Tautz J (1977) Reception of medium vibration by thoracal hairs of caterpillars of *Barathra brassicae* L. (Lepidoptera, Noctuidae) I mechanical properties of the receptor hairs. J Comp Physiol 118:13–31. https://doi.org/10.1007/BF00612334
Tautz J (1978) Reception of medium vibration by thoracal hairs of caterpillars of *Barathra brassicae* L. (Lepidoptera, Noctuidae) II response characteristics of the sensory cell. J Comp Physiol 125:67–77. https://doi.org/10.1007/BF00656832
Tautz J, Markl H (1978) Caterpillars detect flying wasps by hairs sensitive to airborne vibration. Behav Ecol Sociobiol 4:101–110. https://doi.org/10.1007/BF00302564
Tautz J, Rostas M (2008) Honeybee buzz attenuates plant damage by caterpillars. Curr Biol 18:R1125–R1126. https://doi.org/10.1016/j.cub.2008.10.038
Taylor CJ (2009) Hearing in larvae of the monarch butterfly, *Danaus plexippus*, and selected other Lepidoptera. MSc thesis, Carleton University, Ottawa, Canada
Taylor CJ, Yack JE (2019) Hearing in caterpillars of the monarch butterfly (*Danaus plexippus*). J Exp Biol 222:jeb211862. https://doi.org/10.1242/jeb.211862
Travassos MA, Pierce NE (2000) Acoustics, context and function of vibrational signaling in a lycaenid butterfly ant mutualism. Anim Behav 60:13–26. https://doi.org/10.1006/anbe.1999.1364

Tutt JW (1893) Note on sensibility of larvae to sound. Entomol Rec 4:241

Virant-Doberlet M, Kuhelj A, Polajnar J, Sturm R (2019) Predator-prey interactions and eavsdropping in vibrational communication networks. Front Ecol Evol 7:203

Wagner DL (2005) Caterpillars of Eastern North America. Princeton University Press, Princeton

White TR, Weaver JS, Agee HR (1983) Response of *Cerura borealis* (Lepidoptera: Notodontidae) larvae to low-frequency sound. Ann Entomol Soc Am 76:1–5. https://doi.org/10.1093/aesa/76.1.1

Wilson JM, Barclay RMR (2006) Consumption of caterpillars by bats during an outbreak of western spruce budworm. Am Midl Nat 155:244–249. https://doi.org/10.1674/0003-0031(2006)155[0244:COCBBD]2.0.CO;2

Windmill JFC, Jackson JC (2016) Mechanical specializations of insect ears. In: Pollack G, Mason A, Popper A, Fay R (eds) Insect hearing, Springer handbook of auditory research, vol 55. Springer, Cham, pp 125–157. https://doi.org/10.1007/978-3-319-28890-1_6

Yack JE (2004) The structure and function of auditory chordotonal organs in insects. Microsc Res Tech 63:315–337. https://doi.org/10.1002/jemt.20051

Yack JE (2016) Vibrational signaling. In: Pollack GS, Mason AC, Popper AN, Fay RR (eds) Insect hearing, Springer handbook of auditory research, vol 55. Springer, Cham, pp 99–123. https://doi.org/10.1007/978-3-319-28890-1_5

Yack JE, Yadav C (2021) Chapter 20: Vibratory sensing and communication in caterpillars. In: Hill PSM, Mazzoni V, Stritih Peljhan N, Virant-Doberlet M, Wessel A, Janik EM, McGregor P (eds) Biotremology: physiology, ecology and evolution, Animal signals and communication, vol. X. Springer, Berlin/Heidelberg, p xx–xx. (in production)

Yack JE, Smith ML, Weatherhead PJ (2001) Caterpillar talk: acoustically mediated territoriality in larval Lepidoptera. Proc Natl Acad Sci U S A 98:11371–11375. https://doi.org/10.1073/pnas.191378898

Yack JE, Kalko EKV, Surlykke A (2007) Neuroethology of ultrasonic hearing in nocturnal butterflies (Hedyloidea). J Comp Physiol A 193:577–590. https://doi.org/10.1007/s00359-007-0213-2

Yack JE, Gill S, Drummond-Main C, Sherratt TN (2014) Residency duration and shelter quality influences signalling displays in a territorial caterpillar. Ethology 120:354–364. https://doi.org/10.1111/eth.12210

Yack JE, Raven BH, Leveillee MB, Naranjo M (2020) What does an insect hear? Reassessing the role of hearing in predator avoidance with insights from vertebrate prey. Integr Comp Biol 60:1036–1057. https://doi.org/10.1093/icb/icaa097

Yadav C, Yack JE (2018) Immature stages of the masked birch caterpillar, *Drepana arcuata* (Lepidoptera: Drepanidae) with comments on feeding and shelter building. J Insect Sci 18:1–9. https://doi.org/10.1093/jisesa/iey006

Yadav C, Guedes RNC, Matheson SM, Timbers TA, Yack JE (2017) Invitation by vibration: recruitment to feeding shelters in social caterpillars. Behav Ecol Sociobiol 71:51. https://doi.org/10.1007/s00265-017-2280-x

Yager DD (2012) Predator detection and evasion by flying insects. Curr Opin Neurobiol 22:201–207. https://doi.org/10.1016/j.conb.2011.12.011

Zeng J, Xiang N, Jiang L, Jones G, Zheng Y, Liu B, Zhang S (2011) Moth wing scales slightly increase absorbance of bat echolocation calls. PLoS One 6:e27190. https://doi.org/10.1371/journal.pone.0027190

Natural History and Ecology of Caterpillar Parasitoids

John O. Stireman III and Scott R. Shaw

Head-on caterpillar of *Eumorpha pandorus* (Sphingidae; ventral side up), covered with the silk cocoons of a microgastrine braconid (probably *Cotesia* species), with an egg of a tachinid fly on its face, photographed in northern Arkansas. Photo: Kenji Nishida, Wakayama, Japan.

J. O. Stireman III (✉)
Department of Biological Sciences, Wright State University, Dayton, OH, USA

S. R. Shaw
UW Insect Museum, Department of Ecosystem Science and Management, University of Wyoming, Laramie, WY, USA

© The Author(s), under exclusive license to Springer Nature Switzerland AG 2022
R. J. Marquis, S. Koptur (eds.), *Caterpillars in the Middle*, Fascinating Life Sciences, https://doi.org/10.1007/978-3-030-86688-4_8

Introduction

The larval stages of Lepidoptera face many hazards in their race to reach maturity, including both abiotic insults of heat, cold, wind, and storms, and biotic affronts such as plant defenses, predators, and pathogens. However, parasitoid insects are among the most important and well-studied enemies of caterpillars. Parasitoids are a special category of secondary consumers that complete their development on or in a single "host" individual (in this case, a caterpillar), killing it in the process (Godfray 1994). Parasitoid insects are distinguished from parasites such as lice and fleas, which may exploit resources from multiple "host" individuals and typically do not kill their hosts (although the line between parasitoid and parasite is sometimes difficult to discern). The vast majority of Lepidoptera species are host to one or more parasitoid species, and some caterpillar species may be attacked by dozens of parasitoid species (e.g., Sabrosky and Reardon 1976; Delucchi 1982). Although for some lepidopteran species predators may collectively be a greater source of mortality throughout larval development (Remmel et al. 2011), parasitoids are often responsible for a significant fraction of deaths across immature life stages (Hawkins et al. 1997), with parasitism frequencies ranging as high as 80% or more in certain cases (e.g., Clausen 1978; Hawkins 1994). Like many parasites, parasitoids interact with their hosts intimately, often developing inside a living host over an extended period. Due to this intimacy, they are often specialized on a taxonomically narrow range of host species, frequently a single species (especially in the tropics; e.g., Smith et al. 2007, 2008; Arias-Penna et al. 2019). This high specificity and the close correspondence between parasitoids and hosts has attracted considerable attention from theoretical ecologists and applied entomologists, since parasitoids have served as important models for population ecology theory (e.g., Hassell 2000) and are a key focus of biological control efforts (e.g., Waage and Hassell 1982). Here, we focus on the parasitoid enemies of caterpillars, examining their diverse natural histories, their complex interactions with caterpillars, and their responses to anthropogenic environmental change.

Parasitoid Life History Diversity

The parasitoid habit represents a distinct natural phenomenon with particular shared challenges that shape parasitoid morphology, physiology, and behavior. Parasitoid adults must locate host habitats, assess quality of suitable hosts, and lay eggs on, into, or near the selected host individual. Both adults and immatures must cope with morphological, physiological, and behavioral defenses of hosts, and the mature larva must seek out or create a protective site for pupation (or take advantage of the caterpillar's pupation retreat). In these selective pressures, insect parasitoids are not unlike the caterpillars they parasitize, except that their hosts tend to be much smaller and more mobile than the plant hosts of caterpillars and their physiological interactions are quite different.

Parasitoids of caterpillars are able to locate and attack their hosts in a diverse array of microhabitats. While many parasitoids attack exposed externally feeding caterpillars directly, various cryptic-feeding caterpillars such as shelter-builders, leaf rollers, stem borers, leaf miners, and gall-inducers are all subject to attack. In fact, these concealed feeders tend to experience higher parasitism rates and host larger parasitoid complexes than many exposed taxa (Hawkins 1994). In addition, wood boring and subterranean caterpillars may be host to parasitoids that have evolved specialized strategies to locate and attack such hidden hosts (Chandra and Gupta 1977; Hinz and Short 1983; Zong et al. 2012). Parasitoids of most caterpillars develop internally as endoparasitoids; however, ectoparasitism, where the immature parasitoid develops outside the host body, is fairly common and widespread among the parasitic wasps (Quicke 1997, 2015; Gibson et al. 1997). In addition to being endo- or ectoparasitic, parasitoids are often categorized into two main groups based on their developmental strategies – idiobionts and koinobionts (Askew and Shaw 1986). Members of the former group halt the development of the host at the time of parasitism (typically via injecting a paralyzing venom), whereas the latter allow their hosts to continue to develop and grow for some time, often delaying their own larval development until the host has achieved sufficient size before the parasitoids consume them. However, boundaries between these strategies are blurred in some instances (Quicke 1997), and they may not be as clearly applicable outside the parasitic wasps, as flies and other taxa lack paralyzing venoms yet often differ in developmental schedules and synchronization from koinobiont wasps (Dindo 2011). Parasitoids also vary in the stage of host attacked. Our focus here is on caterpillars, but parasitoids are known from every developmental stage of lepidopterans including eggs, all larval instars, pupae, and even adults, albeit rarely (McCabe 1998; Pereira et al. 2015). Parasitoids frequently emerge from a life stage different than the stage attacked (e.g., egg-larval and larval-pupal parasitoids). Finally, parasitism of other parasitoids, or hyperparasitism, is quite common in many caterpillar systems, so in this regard caterpillars may not always be "in the middle" but may just be a necessary layer in a more complex, multilayered trophic "sandwich."

Parasitoid Taxonomic Diversity

It has been estimated that approximately 15% of all insects possess parasitoid lifestyles (Godfray 1994). This is likely a substantial underestimate, as recent studies suggest that the diversity of some parasitoid groups is considerably greater than current described species suggest (e.g., Quicke 2015; Stireman et al. 2017; Forbes et al. 2018; Burington et al. 2020). Each of the major holometabolous insect orders is known to contain species with parasitoid lifestyles (Eggleton and Belshaw 1992), and important parasitoid taxa are also found in other phyla (namely, nematodes and fungi; Eggleton and Gaston 1990). However, most parasitoid enemies of caterpillars belong to two major insect orders: Hymenoptera (wasps) and Diptera (flies). In Hymenoptera, the parasitoid habit appears to have arisen once (the stinging nest-provisioning wasps, as well as the social wasps, ant, and bees all evolved from parasitic ancestors; Peters et al. 2017) and characterizes a broad paraphyletic grade of lineages ranging from the parasitoid sawfly superfamily Orussoidea, to the hyperdiverse Ichneumonoidea, and to the unusual hyperparasitic Trigonalidae and many others. In terms of diversity, parasitism (or "parasitoidism") is the dominant life history among hymenopterans. In contrast, the parasitoid habit appears to have arisen numerous times within the true flies (Diptera), with at least 21 families exhibiting parasitoid lifestyles, each representing at least one and often many independent derivations of the life history (Eggleton and Belshaw 1992; Feener and Brown 1997). Despite their many origins, few dipteran parasitoids have successfully colonized caterpillars as hosts, and the vast majority belong to a single family, Tachinidae.

In this chapter, we provide an overview of the natural history and ecology of some prominent parasitoid enemies of caterpillars and consider briefly how their diversity and interactions may be influenced by the environmental changes our world is currently experiencing. Our specific goals are to (1) review the natural histories and ecologies found in major hymenopteran and dipteran parasitoid groups (namely, Ichneumonoidea, Chalcidoidea, and Tachinidae); (2) examine their community patterns and roles in influencing the ecology and evolution of caterpillars; and (3) evaluate how major anthropogenic environmental changes, including habitat loss and fragmentation, invasive species, and global warming, affect parasitoids and their interactions with hosts. Given their vast diversity and ecological importance, and the voluminous scientific literature concerning them, we can provide but a glimpse into the world of parasitoids and touch on only a few of the many interesting and important aspects of their natural history and ecology. More exhaustive explorations of parasitoid biology and ecology can be found in Waage and Greathead (1986), Godfray (1994), Hawkins and Sheehan (1994), Feener and Brown (1997), Quicke (1997, 2015), and Stireman et al. (2006), among other works.

Hymenopteran Parasitoids of Caterpillars

Before delving into the rich diversity of parasitoid wasps (Figs. 1, 2, and 3) associated with caterpillars, it is worthwhile to point out that all Hymenoptera share a distinctive genetic characteristic: haplodiploid sex determination (Gauld and Bolton 1988; Quicke 2015). Simply stated, this means that the sex of an individual wasp is determined by whether or not an egg is fertilized. If the wasp egg is fertilized, it develops (eventually) into a diploid female wasp. However, if the egg is unfertilized, it develops into a haploid male wasp. Most readers are probably familiar with the theory of kin selection and the possible relevance of haplodiploid sex determination to the origins of eusocial behavior in bees, ants, and wasps (Wilson 2005; Foster et al. 2006). However, few have considered the adaptive value of haplodiploid sex determination for the non-social Hymenoptera. Haplodiploidy evolved long before social behavior and it is prevalent throughout all the diverse lineages of parasitoid Hymenoptera. So how might it benefit parasitoids? Consider a tiny female wasp in

Fig. 1 Examples of cocoons of braconid parasitoids of caterpillars. (**a**) *Cotesia* sp. cocoons on a sphingid carcass (FL: SA Marshall). (**b**) Microgastrine braconid cocoons on *Prochoerodes lineola* (Geometridae) (OH: JOS). (**c**) A microgastrine cocoon on an unidentified geometrid (MO: SA Marshall). (**d**) Microgastrine braconid cocoons on an unidentified erebid (Ecuador: SA Marshall). (**e**) Gregarious meteorine braconid cocoons suspended from a twig (Brazil: SA Marshall)

a vast tropical forest, searching for a mate and for caterpillars to parasitize, and the tiny males out there searching for females. In a large complex world, it may be much more difficult for a male and female wasp to find each other and mate, than for a female to locate (more numerous) caterpillars. With other insects, such as parasitic flies, an unmated female might lay her eggs as they develop, but they are infertile and die. However, a parasitoid wasp can develop and lay eggs at times when male wasps cannot be found and those eggs can develop into male wasps. Thereby, an unmated female wasp has the capacity to flood the local environment with male wasps at precisely the time when males are scarce. Therefore, haplodiploidy, one of the most basic characteristics of the order Hymenoptera, gives these insects a key advantage over other groups of insect parasitoids: they can exist and survive locally at lower population densities.

Fig. 2 Adults and pupae of Ichneumonoidea. (**a**) An adult of the chelonine braconid *Chelonus* (Brazil: SA Marshall). (**b**) An agathidine braconid (*Agathis malvacearum* Latreille) ovipositing into a *Metzneria lappella* (L.) caterpillar (Gelechiidae) concealed inside a cocklebur inflorescence (*Xanthium strumarium* L.) (ONT Canada: SA Marshall). (**c**) A campoplegine ichneumonid cocoon attached to the carcass of a sphingid caterpillar (OH: JOS). (**d**) A fluffy mass of microgastrine braconid cocoons concealing the noctuid host carcass from which they emerged (OH: JOS). (**e**) A notodontid caterpillar (*Heterocampa guttivitta* (Walker)) "guarding" the microgastrine braconid parasitoids that emerged from it (OH: JOS)

In parasitoid species that utilize sparsely distributed and difficult to locate hosts, inbreeding is common due to "local mate competition" (Godfray 1994). Many eggs might be oviposited into one host caterpillar, increasing the chances that emerging wasps will mate with their own siblings. Haplodiploid sex determination and control of egg fertilization allows female wasps to determine the sex of their offspring as eggs are laid, allowing them to skew sex ratios such that just barely sufficient numbers of males are produced. Female wasps can also assess the quality of

Fig. 3 Ichneumonid and eulophid parasitoids. (a) A campoplegine ichneumonid spinning its cocoon and "wearing" the remains of its host (*Adelpha tracta* (Butler): Nymphalidae) (Costa Rica: K Nishida). (b) The finished campoplegine cocoon from 3A (K Nishida). (c) Another campoplegine ichneumonid cocoon and host carcass (*Hypena madefactalis* Guenée: Erebidae) (OH: JOS). (d) Eulophid larvae (Chalcidoidea: Eulophidae) clustered on a notodontid caterpillar (*Misogada unicolor* (Packard)) (OH: JOS). (e) The noctuid caterpillar (*Sympistis badistriga* (Grote)) with a group of ectoparasitic eulophid larvae (probably *Euplectrus*) riding atop it

individual hosts, and preferentially place female eggs into higher quality hosts that will provide more grand-offspring (Godfray 1994; Quicke 1997).

The order Hymenoptera first arose in the early Mesozoic (Triassic Period). The parasitoid lineages of Hymenoptera evolved and diversified in the Jurassic Period, well before the surge in diversity of flowering plants (angiosperms) and associated lepidopteran plant-feeders in the late Cretaceous Period, and onwards over the last 70 million years or so (Grimaldi and Engel 2005; Shaw 2014). The ancestral lineage of parasitic Hymenoptera was likely associated with wood-boring beetle or sawfly larvae, probably for tens of millions of years before they later diversified into the various independent radiations of parasitoid lineages that attack caterpillars. The modern success of the entire order Hymenoptera is due, in large part, to the hyperdiversity of parasitoid wasp lineages (especially the massive superfamilies Ichneumonoidea, Chalcidoidea, and Platygastroidea), and their great species richness is due, in large measure, to their successful exploitation of Lepidoptera as hosts. These lineages are so vast and variable that what follows can only skim the surface with selected examples. While we try to give the reader a broad overview of caterpillar parasitoid biologies, the examples that follow are admittedly weighted somewhat toward discussion of the Braconidae and the Tachinidae, due to the particular areas of specialization of the two coauthors.

Diversity: Superfamily Ichneumonoidea

The massive superfamily Ichneumonoidea is one of the most species-rich lineages of insects, containing at least 60,000 described species and estimated to comprise several times this figure (Quicke 2015). Ichneumonoidea are divided into three families: the two large families Braconidae and Ichneumonidae, and one small primitive Australian family, the Trachypetidae (Quicke et al. 2020). Braconids and ichneumonids are extraordinarily diverse in their host associations, utilizing a wide range of insect orders and even spiders, but we focus on lineages that attack caterpillars (or other stages of Lepidoptera). Collectively, the ichneumonoid lineages associated with caterpillars display a wide range of parasitic strategies. Some are ectoparasitic (as in many Eulophidae (Chalcidoidea) parasitoids of caterpillars, Fig. 3d, e), while many others are endoparasitic and are adapted to cope with the caterpillar's internal milieu of physiological defenses. Some are idiobionts, arresting the development of the host with a paralytic venom, while many others are koinobionts that allow the host to continue developing for some time before consuming them (Shaw 2004; Quicke 2015). Some are solitary, each developing in isolation in a single caterpillar, while many others develop gregariously, with multiple larvae (sometimes many hundreds) sharing a single caterpillar. Among the gregarious feeders there are two strategies: most just quickly insert many eggs into one host caterpillar, but some have evolved polyembryony (where one egg clones its embryo into multiple individuals, sometimes producing hundreds from a single egg) (Parker 1931). Many ichneumonoids are larval-larval parasitoids (both attacking and emerging from the caterpillar stage). Many others, however, are egg-larval parasitoids, larval-pupal parasitoids, or pupal-pupal parasitoids. While most exit the host caterpillar after they have completed their feeding, many others have evolved the capacity to pupate inside the host's remains (Shaw 2006). Most ichneumonoids are primary parasitoids (feeding on the caterpillar host), but many Ichneumonidae are hyperparasitoids (merely using the caterpillar as a vessel to locate other parasitoids inside it, then using them as hosts) (Araujo et al. 2018). While many attack exposed leaf-feeding caterpillars, many others specialize in attacking concealed-feeders, such as leaf miners, leaf rollers, stem borers, or even cryptic and hidden cocooned pupal stages (e.g., Fig. 2b).

While the superfamily Ichneumonoidea displays a wide range of parasitoid strategies and behaviors, the two major families, Braconidae and Ichneumonidae, do not display these variations equally (Gauld and Bolton 1988; Shaw and Huddleston1991; Hanson and Gauld 1995; Quicke 1997, 2015). Braconidae comprise relatively more larval-larval caterpillar parasitoids and more egg-larval parasitoids, as well as most of the leaf-miner parasitoids. Furthermore, only rarely are Braconidae hyperparasitic (Kula et al. 2012) and only rarely do they attack pupae (although many emerge from cocoons of pre-pupal caterpillars). By contrast, the Ichneumonidae include many larval-pupal or pupal-pupal lepidopteran parasitoids, and some very diverse hyperparasitoid lineages that use caterpillars as a primary host but actually feed on other parasitoids inside the caterpillar, such as other ichneumonids, braconids, or tachinids. Given the wide range of ways for living on or inside caterpillars displayed by these two hyperdiverse families, it is worth discussing each of them in more detail.

Family Braconidae

At least 13 major lineages of Braconidae have diversified as larval-larval specialists attacking caterpillars (Agathidinae, Cardiochilinae, Euphorinae tribe Meteorini, Exothecinae, Gnamptodontinae, Homolobinae, Hormiinae, Ichneutinae tribe Muesebeckiini, Macrocentrinae, Microgastrinae, Miracinae, Orgilinae, and Rogadinae). Two other subfamilies (Braconinae and Doryctinae) have broad host ranges encompassing several orders of insects, most often beetle larvae, but contain many species that will utilize concealed-feeding caterpillars (Quicke 2015). Perhaps the most successful lineage of larval-larval caterpillar parasitoids is the braconid subfamily Microgastrinae, with some 2999 named species worldwide and estimated to comprise 30,000 to 50,000 species when undescribed species are included (Rodriguez et al., 2013; Fernandez-Triana et al., 2020). These koinobiont endoparasitoids may develop as solitary larvae in small caterpillars or gregariously in larger caterpillars. Mature microgastrine larvae typically exit the host caterpillar and spin their silk cocoons on the surface of the host carcass (as in Figs. 1a–d and 2d, e). The caterpillar hosts of Microgastrinae species are stunningly diverse and include almost all Lepidoptera (except Micropterigoidea, Eriocranioidea, Hepialoidea and Nepticuloidea).

The basal leaf-mining microlepidopterans are not, however, immune from attack by braconid parasitoids. At least 10 hymenopteran families include members that attack leaf miners (eight of them in the Chalcidoidea, Gates et al. 2002), and at least eight lineages of Braconidae have evolved minute body sizes and the capacity to utilize leaf-mining and other cryptic-feeding caterpillars (Whitfield and Wagner 1991). These include Nepticulidae, Eriocraniidae, Coleophoridae, Bucculatricide, Gracillariidae, Lyonetiidae, and many others. Braconids attacking such tiny caterpillars include some Cheloninae (tribe Adeliini) (Shimbori et al. 2019), some

Exothecinae, Gnamptodontinae, Hormiinae, Ichneutinae (tribe Muesebeckiini), Miracinae, some Orgilinae, and some Rogadinae (tribe Stiropiini) (Shaw 1983; van Achterberg 1983; Whitfield and Wagner 1991). The adeliines were formerly treated as subfamily Adeliinae but have recently been reclassified as a tribe of Cheloninae (Kittel et al. 2016; Shimbori et al. 2019). Some leaf-miner parasitoids are quite specialized, such as adeliines and gnamptodontines that are koinobiont endoparasitoids of exclusively Nepticulidae larvae (Gauld and Bolton 1988; Whitfield 1988). By contrast, others possess broad host ranges, such as species of the exothecine genus *Colastes* that are ectoparasitic idiobionts. These are highly polyphagous, eating whatever larvae they find in mines on a particular plant including Nepticulidae, Tischeriidae, Momphidae, Eriocraniidae, Heliozelidae, and Gracillariidae, and even leaf-mining larvae of Coleoptera, Diptera, and tenthredinoid Hymenoptera (Shaw 1983).

Another highly successful group of caterpillar parasitoids, notable for its varied (and sometimes spectacular) cocoon-spinning behaviors (Shaw and Nishida 2005; Stigenberg and Ronquist 2011), is the subfamily Euphorinae, tribe Meteorini, genus *Meteorus* (formerly subfamily Meteorinae). The name refers to distinctive pendant cocoons of solitary *Meteorus* species, which are suspended from a long silk thread, reminiscent of the tail of a meteor streaking across the night sky. Most *Meteorus* larvae feed as koinobiont endoparasitoids inside caterpillars, spinning their distinctive cocoons as they emerge from the caterpillar host (Shaw and Huddleston 1991; Zitani and Shaw 2002). Meteorines are parasitoids of at least 15 lepidopteran families (Jones and Shaw 2012; Aguirre et al. 2014; Aguirre and Shaw 2014a, b; Yu 2014). However, Aguirre et al. (2015) reported that more than 70% of host records for Neotropical meteorines were from just five families: Erebidae (Arctiinae), Noctuidae, Pyralidae, Nymphalidae, and Megalopygidae. While it has been presumed that the suspended cocoons of solitary *Meteorus* species are an adaptation to avoiding crawling predators such as ants, it has also been observed that these cocoons are commonly targeted by hyperparasites such as mesochorine ichneumonids (Simmonds 1947; Zitani and Shaw 2002; Shaw unpublished data). Many *Meteorus* species have evolved gregariousness, with as many as 250 wasps developing from a single sphingid larva (Zitani et al. 1998). Such gregarious species often cooperate to form communal cocoon masses (Fig. 1e), sometimes suspended from a jointly spun multi-threaded cord, sometimes covered by jointly spun masses of loose protective threads like baling wire, and, most remarkably, sometimes cooperating to spin communal radially symmetrical cocoon masses covered by a spherical sheath (which may protect 90% of the cocoon surfaces from direct attack by hyperparasitoids) (Zitani and Shaw 2002).

The Rogadinae are another hyperdiverse lineage of endoparasitic koinobiont caterpillar specialists that evolved another quite novel approach to pupation. While the larvae of most koinobiont endoparasitic wasp species rapidly exit the remains of the host to safely spin their cocoons elsewhere, Rogadinae species efficiently consume all edible body contents then pupate inside the host caterpillar's exoskeletal remains, which shrink and harden (Shaw and Huddleston 1991). Rogadinae contain many species and genera, but by far the most commonly encountered are species in the genus *Aleiodes* (Shaw et al. 1997; Areekul-Butcher and Quicke 2011; Shimbori and Shaw 2014; van Achterberg et al. 2020; Shaw et al. 2020), collectively found worldwide and with more than 200 named species in the New World (Yu et al. 2012; Garro et al. 2017; Shaw et al. 2020). *Aleiodes* are sometimes called "mummy wasps" (Shaw 2006) due to their characteristic habit of pupating inside the remains of the host caterpillar, which shrinks and dries into a caterpillar "mummy" (Shaw 1983, 1994; Shaw and Huddleston 1991; Shaw et al. 1997; Shaw 2006; Zaldívar–Riverón et al. 2008; Shimbori and Shaw 2014). While morphological identification of Rogadinae species can be challenging, the mummified remains of the host caterpillar (conveniently preserved for study by the parasitoid's unique behavior) often provide clear evidence of the rogadine species identity (Shaw 2006; Shimbori and Shaw 2014). Rogadines commonly use caterpillars that are in the middle of their developmental cycle: usually second and third instars, and the caterpillar host is killed before it reaches the ultimate instar (Shaw and Huddleson 1991). While rogadines utilize hosts as diverse as Limacodidae, Zygaenidae, Lycaenidae, and Riodinidae, the vast majority are solitary parasitoids developing in small Geometridae, Erebidae, and Noctuidae caterpillars; a few develop gregariously in large Lasiocampidae or Noctuidae caterpillars (Fortier 2000).

While gregarious development may not be the norm for Rogadinae, another braconid subfamily, the Macrocentrinae, are known for it. More particularly, macrocentrines are notable for their interesting method of accomplishing gregariousness: by polyembryony. While other braconids (such as microgastrines, meteorines, and a few rogadines) achieve gregariousness by inserting multiple eggs into a caterpillar, the macrocentrines inject a single egg that divides clonally to produce multiple, sometimes numerous, same-sex embryos that ultimately develop into a cohort of genetically identical wasps (Parker 1931). It appears that some species achieve mixed broods of males and females by ovipositing more than one egg, each of which develops polyembryonically into many individuals. In some cases, a *Macrocentrus* develops as a solitary individual as the result of a single embryo dominating over many others that fail (Finney et al. 1947). Macrocentrines are delicate-bodied, long-legged nocturnally active wasps that specialize in attacking cryptic-feeding caterpillars of families such as Oecophoridae, Gelechiidae, Tortricidae, Sesiidae, Pyralidae, and some Noctuidae (Gauld and Bolton 1988; Shaw and Huddleston 1991).

Species of the subfamily Homolobinae are solitary koinobiont endoparasitoids of Geometridae, Erebidae, and Noctuidae. They share with many (if not most) species of the abovementioned Meteorini, Rogadinae, and Macrocentrinae another key innovation for utilizing caterpillars as hosts: nocturnal behavior. While many caterpillars, especially tropical ones, feed at night to elude day-active predators, females of these braconids specialize in hunting for caterpillars at night. This allows these wasps to make better use of chemical cues, such as kairomones released by caterpillar feeding, silk, and frass, as well as the sounds of caterpillar feeding to locate these insects (Quicke 2015). These nocturnal wasps tend to have quite large eyes and ocelli, are usually pale-colored, and are commonly attracted to lights (van Achterberg 1979; van Achterberg and Shaw 2009).

The hyperdiverse koinobiont endoparasitoid braconid subfamily Cheloninae are sometimes called "turtle-shelled" wasps because of their distinctive carapace-like abdominal dorsum. Chelonine wasps are specialists on small, concealed caterpillars, including species of Tortricidae, Pyralidae, Crambidae, and Plutellidae (Shaw and Huddleston 1991; Shaw 2017; Dadelahi et al. 2018). Chelonine females are commonly seen walking rapidly and erratically over leaf surfaces (Fig. 2a). Chelonines are egg-larval parasitoids, that is, wasps which parasitize lepidopteran eggs but delay their development and emerge later from caterpillars (Kainoh and Tamaki 1982). Even more remarkably, chelonines actually insert their egg into the embryonic caterpillar inside its egg (Vance 1932), such that many young caterpillars are "born" (hatch from their egg) already parasitized. The chelonine egg hatches into a larva with massive mandibles that kills any competing parasitoid larvae inside the host, and then the chelonine molts again into a hibernating form that diapauses until the caterpillar has reached its ultimate stage. The host caterpillar is allowed to form its pupation retreat and spin a silk cocoon before the chelonine larva breaks diapause and begins feeding and rapidly growing, ultimately killing the prepupa before pupation (Vance 1932; Kainoh and Tamaki 1982). Chelonines are thus exquisitely adapted to take advantage of the ease of locating exposed host eggs, as well as allowing the host caterpillar to attain maximum biomass and create a protective retreat within which to pupate.

Egg parasitoids. It is worth observing that while many braconids oviposit into lepidopterous eggs, no known ichneumonoids develop as true egg parasitoids (developing entirely inside one egg). However, Chalcidoidea (Trichogrammatidae) and Platygastroidea (Scelionidae) contain a great diversity of minute egg parasitoids ("nanohymenoptera"). Technically these are not "caterpillar parasitoids" since they live entirely inside the lepidopteran egg. Nevertheless, they have a profound impact on the diversity and abundance of caterpillars. Among the Scelionidae, which collectively attack eggs of species in many insect orders, perhaps the most frequently encountered are species of the large genus *Telenomus*, which prefer eggs of Lepidoptera. Egg parasitoids are discussed further in the Chalcidoidea section (below).

Family Ichneumonidae

Like the braconids, several ichneumonid lineages include successful larval-larval parasitoids of caterpillars (Broad et al. 2018). These caterpillar-eaters include koinobiont endoparasitoids among the Anomaloninae, Banchinae, Campopleginae (formerly Porizontinae; Figs. 2c and 3c), Cremastinae, some Ichneumoninae, Metopiinae, Ophioninae, and a few Ctenopelmatinae (Townes 1969; Townes and Townes 1978; Gauld and Bolton 1988; Quicke 1997). These subfamilies tend to be broadly polyphagous in their host ranges, and while there are several braconid subfamilies that are exclusively parasitoids of Lepidoptera, in Ichneumonidae there are only a few that are dedicated Lepidoptera specialists (Cremastinae, Ichneumoninae, and Metopiinae) (Gauld et al. 2002). Some of these, such as the ichneumonine genus *Alomya*, have managed to do things that no braconid has accomplished:

attacking root-feeding larvae of Hepialidae (Hinz and Short 1983). Others, such as the banchine *Lissonotus*, have long ovipositors and are able to drill deep into woody branches and roots to reach cryptic-feeding cossid larvae (Chandra and Gupta 1977; Zong et al. 2012). Many of these wasps, especially in Ichneumoninae and Metopiinae, are quite large insects, commonly brightly marked with aposematic patterns and may mimic stinging wasps (Heinrich 1977).

Some campoplegine ichneumonids are reminiscent of the rogadine braconids in that they "mummify" the caterpillar remains (Townes 1969; Gauld and Bolton 1988). However, campoplegines accomplish this in a very different way. In Rogadinae, the wasp larva spins little silk, and the caterpillar mummy is supported by hardening of the host cuticle. In contrast, the mummy produced by a campoplegine ichneumonid is supported by a thick, oval cocoon spun of coarse silk. In some campoplegines, such as species of *Hyposoter*, the wasp cocoon is not fully enclosed inside the caterpillar remains but is visible laterally (the caterpillar cuticle is merely stuck over the top of the wasp's cocoon like a pelt; Fig. 3a, b).

Unlike braconids, which rarely feed upon or emerge from lepidopteran pupae, the Ichneumonidae contain several lineages of larval-pupal parasitoids (Anomaloninae and Ichneumoninae) and ones that specialize in parasitizing the pupal stage of Lepidoptera directly (Ichneumoninae, some Cryptinae, and some Pimplinae) (Gauld 1991). Location of pupae has been facilitated by chemodetection of silk, and in some ichneumonid lineages (e.g., Pimplini) this has allowed host range expansion from Lepidoptera to other organisms enclosed in silk, such as spider eggs (Gauld and Fitton 1984; Gauld and Bolton 1988; Quicke 2015). The subfamily Ichneumoninae, regarded as the second-largest ichneumonid subfamily, is unique among the Ichneumonidae in being (so far as is known) entirely restricted to using Lepidoptera as hosts. Throughout the Ichneumoninae, whether they attack and enter the caterpillar or its pupa, the parasitoid emerges from the pupal stage.

Some tryphonine ichneumonids have evolved a novel method for securing their eggs to the host caterpillar. Unlike other ectoparasitoids that typically glue their egg to the host's cuticle, tryphonines have uniquely stalked eggs. The stalk and anchoring hook move down a channel of the ovipositor while the egg is pulled along externally. The female wasp injects the anchoring base of the egg stalk, like a harpoon, into the host's thoracic cuticle, attaching it securely in an area where the caterpillar cannot remove it with its mandibles. The ichneumonid subfamily Tryphoninae includes three tribes that largely attack caterpillars (Oedemopsini, Phytodietini, and Sphinctini). Most are solitary ectoparasitic idiobionts, but a few develop gregariously.

Hyperparasitism: More Layers in the Trophic Sandwich

The Ichneumonidae also provide diverse examples of a phenomenon seldom seen in the Braconidae: hyperparasitism. Hyperparasitoids attack and feed upon other parasitoid species (primary parasitoids) inside a carrier insect (Gauld and Bolton 1988; Hanson and Gauld 1995; Quicke 1997). Within the Ichneumonidae several lineages are specialized as hyperparasitoids or contain hyperparasitic species, including the Mesochorinae, Eucerotinae, some Pimplinae, and some Cryptinae. While technically these are not caterpillar parasitoids (they are just using the caterpillar as a living space), hyperparasitoids are worth considering briefly to illustrate the complexity of caterpillar-parasitoid interactions. There are even (in the Chalcidoidea) hyperparasitoids of hyperparasitic species (tertiary parasites), so in some cases there may well be primary, secondary, and tertiary parasitic species, all living inside leaf-feeding caterpillars.

The small and curious ichneumonid subfamily Eucerotinae provides a distinctive example of hyperparasitism. *Euceros* females lay tiny stalked eggs on the surfaces of the leaves of plants (reminiscent of some Tachinidae). These eggs hatch into an active planidium-type larva, which attaches itself to a passing caterpillar. It may feed on this "carrier-host" but the eucerotine larva does not develop completely unless the caterpillar is attacked by another parasitoid, such as a campoplegine ichneumonid, in which case the eucerotine larva enters the other ichneumonid larva and completes its feeding as a hyperparasite.

The more than 900 species of mesochorine ichneumonids (Araujo et al. 2018) are obligatory endoparasitic hyperparasitoids (Dasch 1971, 1974). They are particularly diverse in the neotropics (Gauld et al. 2002) and often emerge from pupal stages of the other primary parasitoids mentioned elsewhere in this chapter, e.g., silk cocoons of microgastrine or meteorine wasps, from mummified caterpillars occupied by rogadine braconids or campoplegine ichneumonids, or tachinid puparia (Shaw, personal observation).

Perhaps the most fascinating strategy of hyperparasitism inside caterpillars involves the wasp family Trigonalidae (or Trigonalyidae; Weinstein and Austin, 1991; Carmean 1995; Krauth and Williams 2006). Trigonalid females lay minute eggs on the ventral surfaces of leaves, often in rows along the leaf edges where they might be consumed by caterpillars (Gauld and Bolton 1988; Murphy et al. 2009). A single female trigonalid wasp may lay thousands (as many as 10,000) of these microtype eggs (Clausen 1940; Carmean and Kimsey 1998). Once inside the digestive tract, the trigonalid larva chews through the gut wall and enters the caterpillar's hemocoel. Only rarely do trigonalids develop as a primary parasitoid (Raff 1934). Most often the young trigonalid larva will not develop further unless the caterpillar "carrier" is further parasitized by an ichneumonoid wasp, or a tachinid fly, in which case it develops as an endoparasitic hyperparasitoid. Even more remarkable, if the infected "carrier" caterpillar is captured and macerated by a foraging vespid social wasp and returned to the nest to be fed to developing vespid larvae, the trigonalid can be acquired trophically a second time and develop in the vespid larva (Clausen 1940; Gauld and Bolton 1988; Carmean 1991). Trigonalid wasps are seldom seen unless they are reared from an infected caterpillar (Smith et al. 2012), a moth pupa, an ichneumonoid cocoon, or a tachinid fly puparium (Clausen 1940). Perhaps even most remarkably, many tropical trigonalids are Batesian mimics of vespid social wasps, mimicking species from which they may have developed (Hanson and Gauld 1995; Carmean and Kimsey 1998; Smith et al. 2012).

Superfamily Chalcidoidea

The hymenopteran superfamily Chalcidoidea rivals the Ichneumonoidea in species-richness, abundance, and biological diversity (Pérez-Benavides et al. 2020; Ghahari et al. 2021). With over 25,000 named species (Pérez-Benavides et al. 2020), Chalcidoidea species comprise approximately 33% of the known parasitoid Hymenoptera worldwide. The number of undescribed species is so great that it is conservatively estimated that at least 60,000 to 100,000 Chalcidoidea species may exist (Grissell and Schauff 1997). The exceptional morphological and behavioral diversity among Chalcidoidea has resulted in this superfamily being classified into far more families than the Ichneumonoidea: there are 19 chalcidoid families, which are divided into over 90 subfamilies and about 300 genera (Heraty et al. 2013; Noyes 2021; Pérez-Benavides et al. 2020).

While many chalcidoid species are associated with caterpillars, species of the Chalcidoidea collectively have much broader host range than the Ichneumonoidea, including at least 13 orders of insects (including some such as dragonflies, fleas, twisted-wing parasites, and thrips, which are never utilized by Ichneumonoidea). Hosts of Chalcidoidea even include a variety of non-insect hosts such as spider egg sacks, ticks and gall-forming mites, pseudoscorpions, and even gall-forming nematodes (Grissell and Schauff 1997). A much higher proportion of Chalcidoidea species are gall-formers or have lost the parasitic habit and become secondarily phytophagous upon gall or seed tissues. Among those Chalcidoidea utilizing lepidopteran hosts, relatively few are direct parasitoids of caterpillars only, and many are either strictly egg parasitoids or direct pupal parasitoids (Quicke 1997). In contrast, while some Braconidae are egg-larval parasitoids, there are no strict egg parasitoids in either the Ichneumonidae or Braconidae; Braconidae rarely attack pupae, and Ichneumonidae are more likely to be larval-pupal parasitoids than strict pupal parasites (Quicke 2015). While no single chalcidoid family specializes exclusively on Lepidoptera, the principal families that include caterpillar-attacking lineages are briefly touched on below.

The Chalcididae are among the best-known chalcidoids due to their comparatively large body sizes, striking color patterns, and toothed or spined hind legs with outlandishly swollen femurs. Chalcididae use a wide range of hosts (including Diptera, Hymenoptera, Coleoptera, and Neuroptera), but most are solitary primary koinobiont endoparasitoids of caterpillars, or hyperparasitic upon Tachinidae or Braconidae in caterpillars (Gordh et al. 1999). Chalcidids associated with Lepidoptera may emerge from either the fully grown caterpillar or the pupa (Narendran and Amareswara Rao 1987). Some Chalcididae are of interest as parasitoids of forest pests, such as *Brachymeria intermedia* (Nees), which attacks caterpillar of the gypsy moth, *Lymantria dispar* (L.), although not in sufficient numbers for effective biological control (Gauld and Bolton 1988; Drost 1991). Some members of the Torymidae and Encyrtidae have similar biology as hyperparasitoids of assorted primary parasitoids inside caterpillars, and these species are among the wasps most easily recognized as hyperparasitoids when they emerge from caterpillars mummified by rogadine Braconidae. Since Rogadinae are the only primary parasitoids initiating such distinctive mummification of caterpillars, any chalcidoids emerging from mummified caterpillar must be hyperparasitoids of a rogadine braconid inhabiting that host (Shaw 2006, and personal observations).

The Eulophidae comprise an exceptionally diverse family using a wide range of hosts, including Lepidoptera (Hansson 2002). Most eulophids are primary ectoparasitoids of concealed-feeding larva, with many species attacking leaf-mining hosts. However, many species of *Eulophus* and *Euplectrus* are gregarious ectoparasitoids of exposed-feeding caterpillars, making them among the most apparent and best-known chalcidoid parasitoids of caterpillars (Fig. 3d, e). When they complete their ectoparasitic feeding, eulophid larvae pupate either on the host remains or attached to the substrates nearby. While most Chalcidoidea differ from the Ichneumonoidea in pupating as exposed exarate pupae rather than spinning cocoons, some *Euplectrus* species are exceptional in spinning loose silk cocoons, not from labial gland silk as in most Hymenoptera, but from silk produced by the Malpighian tubules (Gauld and Bolton 1988; Quicke 1997). Some Eulophidae have been successfully employed as biological control agents for forest pest caterpillars, such as *Chrysocharis laricinellae* (Ratzeburg), which has been introduced to North America to suppress *Coleophora laricella* (Hübner) (Coleophoridae), the larch casebearer (Ryan et al. 1987; Ryan 1990, 1997). However, recent studies of forest defoliation patterns from 1962 to 2018 indicate that the effectiveness of this parasitoid at regulating its host is decreasing in response to climate warming trends (Ward et al. 2020).

The Encyrtidae are another large family with a broad range of hosts, but in contrast to Eulophidae, encyrtids are mostly primary endoparasitoids, although some are hyperparasitic (Gordh et al. 1999). Encyrtids may be either larval parasitoids or egg parasitoids. Some Encyrtidae species are highly gregarious and produce large numbers of offspring by polyembryony (notably the genus *Copidosoma*; Guerrieri and Noyes 2005). Caterpillars parasitized by such polyembryonic species often are so stuffed full of small encyrtid larvae that the caterpillar's cuticle is stretched thin and the parasitic larvae inside are easily visible through the host's translucent cuticle (Gauld and Bolton 1988). In some gregarious species, certain individual encyrtid larvae are larger than others and may act as soldiers, seeking out and destroying competing parasitic species present in the same host (Gordh et al. 1999). While many encyrtids are valued as beneficial biological control agents, in some cases where their host is an endangered or threatened species, they may pose a threat to insect conservation. The endangered *Papilio homerus* butterfly, for example, may suffer up to 77% egg mortality due to attack by Encyrtidae (Garraway et al. 2008).

Unlike encyrtids, which only sometimes attack insect eggs, the families Mymaridae and Trichogrammatidae are obligatory, mostly solitary but sometimes gregarious, idiobiont egg-parasitoids (Nagarkatti and Nagaraja 1977; Schauff 1984; Noyes and Valentine 1989; Quicke 1997). Both families have broad host ranges including many insect orders (Huber 1986; Doutt and Vigianni 1968), but Trichogrammatidae include many species that target eggs of Lepidoptera. Both Mymaridae and Trichogrammatidae share key adaptations for exploiting eggs as hosts, including microscopic body sizes, adult body shapes that conform to the form of host eggs, and larval specializations for rapidly and effectively usurping the insect egg microhabitat. Of these, the evolution of the "mymariform" first instar is perhaps the most striking (Jackson 1961; Gauld and Bolton 1988; Yoshimoto 1990; Quicke 1997). The mymariform larva is highly modified, conical-headed, and long-tailed form, capable of swimming about inside the host egg, destroying the caterpillar embryo, fighting, and killing other parasitoid larvae to avoid superparasitism and multiple parasitism (Jackson 1961; Boivin and Baaren 2008; Nénon et al. 2011). Trichogrammatid larvae exhibit hypermetamorphosis, with mymariform larvae molting and transforming to hymenopteriform larvae that complete their development by feeding on the remains of the caterpillar embryo and egg yolk. The capacity to develop from egg to adult wasp by feeding only on the contents within a single lepidopteran egg, by necessity, requires the evolution of ultra-small body sizes. Indeed, the Trichogrammatidae and Mymaridae include the smallest adult insects and the smallest flying animals, with some having bodies less than 300 μm in length (Doutt and Vigianni 1968; Schauff 1984; Noyes and Valentine 1989; Quicke 1997; Ghahari et al. 2021). Along with minute body sizes, trichogrammatids have evolved uniquely reduced internal anatomy. One trichogrammatid species has the smallest known insect nervous system, with adults having fewer than 7600 neurons. They are the only animals of any kind known to have neurons lacking cell nuclei (Polilov, 2012). Trichogrammatid use of various Lepidoptera eggs as hosts has drawn considerable interest for biological control applications (Nagarkatti and Nagaraja 1977). Many trichogrammatids are phoretic; that is, they find and attach themselves to the body of an adult female moth or butterfly and ride around with her until she lays eggs, which are then attacked (Fatouros and Huigens 2012). One Neotropical trichogrammatid species is phoretic on the bodies of *Caligo* owl butterflies, sometimes with as many as 250 individual wasps attached to one butterfly (Malo 1961).

Some Chalcidoidea species differ from ichneumonoids in how they achieve contact with the host caterpillar. While among Ichneumonoidea the eggs are almost always oviposited either onto or into the host insect, species of Perilampidae lay their eggs on plants or other substrates in the general environment of the host insect (Heraty and Darling 1984). These hatch into a mobile "planidium" type first instar larva, which actively seeks out a caterpillar (or other primary host insect), chews into the host, and attacks the larvae of Tachinidae, Ichneumonidae, or Braconidae found inside, as a hyperparasitoid (Hanson and Gauld 1995; Quicke 1997). The perilampid larva is an obligatory hyperparasitoid and may persist as an undeveloped planidium throughout development into the adult moth if the caterpillar remains unparasitized by other wasps or flies (Purrington, 1979).

While most Chalcidoidea use a broad range of hosts at the family level, many individual species are highly specialized on a narrow set of hosts, often a single species (Noyes 2021). More rarely, species may exhibit extreme levels of opportunistic dietary generalism. For example, the pteromalid wasp *Dibrachys cavus* (Walker) has long been regarded as the most polyphagous parasitoid wasp known (Grissell and Schauff 1997). While this species has been reared from hundreds of caterpillar species across a range of families, its documented host range also includes hundreds of additional insect taxa belonging to other orders including Diptera, Hymenoptera,

Coleoptera, Neuroptera, and even spiders (Noyes 2021). Although careful research has revealed this species to comprise a complex of four or more morphologically cryptic sibling species (Peters and Baur 2011), the host ranges of such pteromalids remain stunningly broad compared to other hymenopterous parasitoid lineages. Similarly, broad host ranges are exhibited by some Eupelmidae species such as *Eupelmus vesicularis* (Retzius) (Grissell and Schauff 1997; Gibson and Fusu 2016; Noyes 2021).

While there is much more to say about the diversity and biology of Chalcidoidea species, as well as hymenopteran parasitoids of Lepidoptera belonging to other lineages (such as scelionine Platygastroidea), our review must necessarily be constrained. Therefore, readers seeking more information about the Chalcidoidea (and other parasitoid Hymenoptera) are referred to other sources such as Clausen (1940), Gauld and Bolton (1988), Hanson and Gauld (1995), Grissell and Schauff (1997), Gordh et al. (1999), Quicke (1997, 2015), and Ghahari et al. (2021).

Dipteran Parasitoids of Caterpillars

The non-hymenopteran, or "other" parasitoids (Stireman 2016), comprise a diverse and disparate assemblage of taxa including various families of beetles and flies and a few groups of Lepidoptera and Neuropteroidea (Eggleton and Belshaw 1992). As mentioned previously, however, relatively few of these groups have been able to take advantage of the diverse and plentiful resource that caterpillars offer. Among Diptera, only tachinid flies, sarcophagid flies, bee flies (Bombyliidae), and a few scattered taxa in other fly families (e.g., Phoridae, Chloropidae) have been recorded as parasitoids of Lepidoptera (Eggleton and Belshaw 1992; Yeates and Greathead 1997). Records of lepidopteran parasitism by non-Hymenoptera are even sparser outside of the Diptera, with only a handful of records from Lepidoptera (Pyralidae, Epipyropidae), Neuroptera (Mantispidae), and Coleoptera (Cleridae) (Eggleton and Belshaw 1992). Tachinidae ("bristle flies") are clearly the dominant group of non-hymenopteran parasitoids, with almost 8600 described species (O'Hara et al. 2020), about three fifths of which use caterpillars as hosts (Cerretti 2010). Lepidoptera dominate as hosts for the large and diverse subfamilies Exoristinae and Tachininae (Fig. 4e), as well as some lineages of Dexiinae such as the Voriini grade of taxa (Stireman et al. 2019). Due to their dominance among caterpillar parasitoids, we focus on Tachinidae (Fig. 4) in our brief examination of the biology of non-hymenopteran parasitoids, with occasional notes regarding other taxa.

Unlike the hymenopteran parasitoids, which have ancient origins and likely radiated onto early lepidopteran lineages as they themselves radiated, tachinid flies are a recent clade, arising perhaps 30–35 mya (Cerretti et al. 2017; Stireman et al. 2019). Remarkably, the breadth of tachinid hosts rivals that of all the parasitic wasps, with species recorded from at least 13 arthropod orders (Stireman et al. 2006). Their ancestral host associations are uncertain, but the basal-most branching, extant lineages are beetle parasitoids, suggesting multiple subsequent colonizations of caterpillar hosts (Cerretti et al. 2014; Stireman et al. 2019). Thus, tachinids

Fig. 4 Illustrations of tachinid fly life cycles. (**a**) Tachinid eggs (probably *Winthemia* sp. on a *Nephelodes minians* Guenée caterpillar (bronzed cutworm: Noctuidae) (OH: JOS). (**b**) A tachinid puparium (*Actia interrupta* Curran) above the remains of its tortricid caterpillar host (likely *Choristoneura*) (OH: JOS). (**c**) A dying monarch caterpillar (*Danaus plexippus* (L.)) being devoured by tachinid larvae (*Lespesia archippivora* (Riley)) (OH: JOS). (**d**) A close-up of the posterior end of a tachinid larva (from 4C) with posterior spiracles protruding from the monarch larvae it is consuming (OH: JOS). (**e**) A spiny, caterpillar-attacking, "hedgehog" tachinid fly (*Adejeania* sp.) from Peru with enlarged, cigar-like palpi (SA Marshall)

diversified after most of the major lineages of Lepidoptera had already radiated and had been colonized by many groups of Ichneumonoidea and other parasitoid wasp lineages. How they were able to successfully colonize the already occupied niches of caterpillar parasitoids, disperse all over the world and into nearly every terrestrial habitat, and diversify into so many varied forms in such a short time is unclear. What is clear is that the family represents one of the most successful and rapidly diversifying clades of caterpillar parasitoids.

Stage Attacked

Tachinid flies as a rule attack larval stages of their lepidopteran hosts. None is known to be an egg parasitoid, nor to attack pupae, though they may often emerge from pupae. Other dipteran groups are known as predators on eggs or egg masses of various insect groups including Lepidoptera, but most of the fly lineages in which the parasitoid habit evolved are too large to develop in all but the largest of lepidopteran eggs (excepting Phoridae). Lepidopteran pupae are attacked by several dipteran parasitoid groups, notably certain Sarcophagidae and Bombyliidae, which can be significant enemies of economically important species such as the forest tent caterpillar (*Malacosoma disstria*; Roland and Taylor 1997) and pine processionary moth (*Thaumetopoea pityocampa*; Battisti et al. 2000), respectively. A number of sarcophagids, as well as phorid flies, recorded from caterpillars and lepidopteran pupae are facultative parasitoids, attacking previously injured or incapacitated hosts (Feener and Brown 1997). As mentioned previously, there are a few reports of dipteran parasitoids emerging from adult moths and butterflies (Smith 1981; Greeney and Stireman 2002 – note that this was likely *Helicobia* sp., rather than *Arachnidomyia aldrichi*); however it is unclear whether the adult stage was attacked or the parasitoid persisted from a previous stage. Most non-hymenopteran parasitoids, including all tachinids, develop internally as endoparasitoids, but some other dipteran groups tend to develop externally and, as with the Hymenoptera, in these groups the lines between parasitoid and predator can be blurred (Yeates and Greathead 1997).

Most caterpillar attacking tachinids (and other dipteran parasitoids) use "macromoths" and butterflies as hosts. In particular, they are often reared from members of the families Erebidae, Geometridae, Noctuidae, Nymphalidae, Sphingidae, and Saturniidae (Arnaud 1978; Tschorsnig 2017). This is probably due to the relative large size of most tachinid species. However, many smaller tachinid taxa, such as species belonging to the tribes Siphonini, Graphogastrini, and Blondeliini, frequently attack exposed and concealed-feeding caterpillars belong to the "microlepidopteran" grade of Ditrysia including Pyralidae, Crambidae, Tortricidae, Oecophoridae, Elachistidae, Gelechiidae, and Gracillariidae, among others (Tschorsnig 2017).

Mode of Attack

All tachinids, and several other groups of dipteran parasitoids, are endoparasitoids, developing inside the living host. However, unlike their wasp counterparts, they ancestrally lack piercing ovipositors and thus must gain entrance into the host caterpillar by other means. Many species deposit eggs (or larvae in the case of Sarcophagidae) externally on the host (Fig. 4a), and the newly hatched larvae must burrow their way inside through the host's exoskeleton. This provides an opportunity for caterpillars to defend themselves by biting off parasitoid eggs or larvae (or molting their cuticle and the eggs along with it), which may explain why so often tachinid eggs are observed on the head capsules or thoraces of caterpillars where they are out of reach of the mandibles. Many lineages of tachinids, as well as lineages of other dipteran parasitoids, have secondarily evolved piercing structures derived from abdominal sclerites that they use to inject eggs into the host (Feener and Brown 1997; Blaschke et al. 2018). This is seen in tachinid genera such as *Eucelatoria*, *Blondelia*, and *Phorocera* that use abdominal piercers to inject eggs into caterpillars, at least some of which appear to pierce leaf rolls to parasitize concealed caterpillars (Burington 2017).

In a substantial departure from most hymenopteran parasitoids, adult females of many tachinid (and bombyliid) parasitoids of caterpillars do not contact the host at all. Rather, it is the larva or egg that contacts the host. Some taxa, such as members of the tribe Tachinini (the "hedgehog" flies; Fig. 4e), possess ambushing larvae. After hatching from eggs laid on suitable substrates, the larvae lie in wait for a passing caterpillar that they can latch on to and burrow into (Clausen 1940; Herting 1960; O'Hara 1985). Other tachinids, such as Leskiini (e.g., *Leskia*, *Genea*), certain Polideini (e.g., *Chrysotachina*, *Lypha*), and other taxa (e.g., *Lydella*) possess mobile, planidial larvae that seek out their caterpillar hosts. These taxa tend to attack caterpillars concealed in leaf rolls, webs, stems, or other structures that are inaccessible to adult females. Finally, a diverse clade of almost entirely caterpillar parasitoids, the Goniini, possess highly specialized microtype eggs that are eaten by the host. These tiny, hard-shelled eggs, which are laid on leaves or other substrates frequented by hosts (Herting 1960; Mondor and Roland 1998; Ichicki et al. 2012), contain fully developed larvae that hatch in the gut and burrow into the host hemocoel after they are ingested. Goniines and other tachinids with "indirect" oviposition strategies tend to be associated with concealed caterpillars (Hrcek et al. 2013), but many species are known to attack exposed feeders (Stireman et al. 2017), indicating that this strategy cannot be viewed simply as an adaptation for attacking physically inaccessible hosts. Similar strategies are employed by the rare trigonalid wasp hyperparasitoids and caterpillar-parasitizing nematodes (Poinar 1979; Murphy et al. 2009). The latter group, consisting mostly of the family Mermithidae, can be important parasitoids in lowland tropical forest (Gentry and Dyer 2002).

Immature Development

Because they generally lack piercing ovipositors and associated accessory gland products, dipteran parasitoids cannot paralyze their hosts during oviposition. For this reason, koinobiont and idiobiont life history categorizations do not generally apply to parasitoids outside the Hymenoptera (Dindo 2011). Furthermore, non-hymenopteran parasitoids generally cannot inject substances to incapacitate host immune defenses such as venom proteins, polydnaviruses, or other viruses (Pennachio and Strand 2006; Beckage and Drezen 2012; Colinet et al. 2013). Therefore, they have had to evolve other means by which to avoid host immune defenses. Tachinid larvae of at least some groups are able to avoid host encapsulation, wherein host hemocytes adhere to foreign intruders and form multilayered suffocating sheath around them, through the formation of respiratory funnels. These structures, formed by melanized concretions of host hemocytes, envelop the posterior of the larvae, but allow the posterior spiracles to maintain contact with outside air via the body wall or through host tracheae (e.g., Michalkova et al. 2009; Valigurova et al. 2014; Fig. 4b, c). At least some polyphagous tachinids (e.g., *Compsilura concinnata* (Meigen)) appear to avoid host immune defenses during early development by situating themselves between the peritrophic membrane and gut wall (Ichiki and Shima 2003). Recent evidence suggests that other species are able to "cloak" themselves in host tissues to hide from encapsulating hemocytes (Yamashita et al. 2019). Tachinid larvae may exit the host to pupate in soil, leaf litter, or other concealed locations or they may pupate inside the host remains. They lack the ability to produce silk cocoons, and instead the pupa is protected inside a puparium, the inflated, sclerotized, last larval instar cuticle (Fig. 4e).

Tachinids and other dipteran parasitoids may be solitary or gregarious. In some cases, only a single fly larva can complete development in a host, and if supernumerary (additional) larvae are present, they are killed directly or indirectly (via competition) by the most vigorous individual. In other cases, several to dozens of larvae can coexist within a host and complete development, although sometimes at a cost in terms of body size (Wilson et al. 2020). Unlike many parasitic wasps, there is little evidence that tachinid flies have the ability to discriminate between parasitized and unparasitized hosts even when the costs of superparasitism (i.e., laying eggs on a host that has already been parasitized) are high (Belshaw 1994; Caron et al. 2010, although see López et al. 1995).

Ecology and Host Relations of Caterpillar Parasitoids

The multitude of ecological relationships between caterpillars and their parasitoids defy any meaningful summary. Each species of parasitoid possesses a particular (and sometimes fluid) set of hosts and means of host location, selection, parasitism, and development. Likewise, different caterpillar species possess particular sets of strategies to avoid or defend against parasitoids, leading to distinct ecological interactions. Furthermore, these interactions take place in a diverse ecological web of host plants, predators, pathogens, and competitors and result in varied selective pressures and ecological consequences. We touch on just of few aspects of these complex relationships.

Community Patterns

One rather simple way we can gain insight into parasitoid-host relationships and their impact is by examining parasitism rates across communities of caterpillars from rearing studies. Such studies generally recover overall parasitism frequencies varying between 10% and 30% (Table 1). However, comparisons among such studies can be problematic due to different methods and approaches. For example, if early instar caterpillars are sampled, then potential parasitism of later instars may be missed; on the other hand, if only late instars are sampled, early stage parasitoids (mostly Hymenoptera) will be overlooked. Furthermore, rearing success is often quite low, with half or more of caterpillars failing to produce an adult moth or parasitoid (Hrcek et al. 2013; Stireman et al. 2017). It is unclear what proportion of these may have hosted parasitoids. Recently developed methods of metabarcoding or metagenomics, by which parasitoids can be detected inside hosts via their DNA, offer a potentially insightful way to circumvent some of the problems associated with rearing (Condon et al. 2014; Wirta et al. 2014; Hrcek and Godfray 2015; Kitson et al. 2019; Sow et al. 2019). Interaction webs of whole communities of caterpillars and their parasitoids can be reconstructed in this manner. However, in order to move beyond operational taxonomic units, this approach requires a thorough and accurate reference database of parasitoid sequence data. In addition, it still suffers from the

Table 1 Total estimated parasitism rates and percent parasitism by wasps and tachinid flies of forest caterpillar communities derived from broad rearing studies for which overall frequencies were available. (Percentages may not sum due to unknown parasitoids or parasitoids of other taxa)

Location	Total %	% Hym	% Tach	References
Greenland	20.9[a]	18.4[a]	2.4[a]	Wirta et al. (2014)/ Várkonyi and Roslin (2013)
	15.5	15.2	0.3	
Canada	17.6	11.10	6.5	Lill et al. (2002)
CAN: ONT	20	5.5	14.5	Timms et al. (2012)
USA: MA	18.0[b]	–	–	Schaffner and Griswold (1934)
USA: CN	24.5	9.7	14	Farkas and Singer (2013)
USA: MD	22.6	18.40	~4.2	Barbosa et al. (2004)
USA: OH	21.2	15.5	5.7	Stireman et al. (unpub. data)
USA: VA	16.2	7.40	8.8	Petrice et al. (2004)
USA: MO	15.4	–	–	Le Corff et al. (2000)
USA: AZ	18.4	3.70	14.7	Stireman and Singer (2003a, b)
Costa Rica: Guanacaste	~11	3-4	7.0	Janzen (1995)
Costa Rica: La Selva	31.0	12.1	14.3	Gentry and Dyer (2002)
Ecuador: Yanayacu	28.8	19.90	8.2	Stireman et al. (2017)
Papua New Guinea	11.6	7.1	4.5	Hrcek et al. (2013)

[a]Based on DNA metabarcoding
[b]Based on number of collections that produced parasitoids, which overestimates true parasitism frequency

issue of missing potential future parasitism at the time of collection. Furthermore, metabarcoding methods may detect potential parasitism (i.e., parasitoid immatures that are encapsulated or otherwise fail to develop), as well as realized parasitism (i.e., that which would kill the caterpillar and result in an adult parasitoid), which may lead to inappropriate conclusions regarding host-parasitoid associations and realized parasitism frequency. The marriage of these two methods, rearing and metagenomics, is particularly promising, as both potential and realized associations can be revealed and the "space" between these may be a particularly rich source of information about caterpillar-parasitoid interactions and coevolution (Wirta et al. 2014).

It might be predicted that parasitism rates should increase with decreasing latitude, as has been observed for pressure from predators (Roslin et al. 2017). On the other hand, parasitoids may be less diverse and represent a relatively weaker component of the enemy community in the tropics, due in part to enhanced pressure by enemies such as ants (e.g., Novotny et al. 2006). One of the few comparative studies of caterpillar communities found that parasitism rates across sites in the Americas did not exhibit a clear latitudinal pattern (Stireman et al. 2006; also see Table 1). Rather, parasitism frequency varied with environmental variability, with overall rates decreasing as variability in precipitation increased – especially for Hymenoptera (Stireman et al. 2006). A similar pattern has been observed with altitude in New Guinea: predation pressure decreased with increasing altitude, but parasitism exhibited no altitudinal pattern (Libra et al. 2019; though see Koptur 1985).

A survey of caterpillar parasitism rates from rearing studies suggests that the relative preponderance of hymenopteran and dipteran parasitoids can vary strongly across communities. Some studies have recorded greater parasitism by Tachinidae (e.g., Stireman and Singer 2003a; Farkas and Singer 2013), others have shown a clear dominance of Hymenoptera (Le Corff et al. 2000; Lill et al. 2002; Stireman et al. 2017), and still others have found similar proportions of parasitism by the two groups (Gentry and Dyer 2002; Petrice et al. 2004). This variation in relative parasitism may be due to variation in the taxonomic or ecological types of caterpillars sampled. For example, concealed feeders are expected to harbor relatively more Hymenoptera, which have piercing ovipositors with which to pierce protective plant tissues (Hrcek et al. 2013). Macrolepidoptera, in contrast, may experience relatively high parasitism rates by Tachinidae, which tend to be larger bodied on average than parasitoid wasps and require greater host resources. Differences in parasitoid composition may also vary with particular habitats or ecosystems. For example, Hymenoptera may tend to dominate in high latitude communities (Hawkins 1994; Wirta et al. 2014), whereas Tachinidae appear to comprise a relatively higher fraction of caterpillar parasitoids in the tropics (Hawkins 1994; Gentry and Dyer 2002), perhaps due to greater impacts of resource fragmentation or "nasty hosts" on wasps (Janzen 1981; Gauld et al. 1992; Burington et al. 2020). At a local level, hymenopteran koinobiont parasitoids may be relatively more susceptible to disturbance due to their generally greater host-specificity.

Relatively few comprehensive parasitoid-host food webs have been constructed for caterpillar communities. Analyses of parasitoid-caterpillar networks for particular parasitoid clades have been published (e.g., Stireman and Singer 2003a; Stireman et al. 2017) and parasitoid complexes of particular caterpillars (especially pests) are frequently documented, but few have examined entire parasitoid communities (e.g., Lill et al. 2002; Timms et al. 2012; Hrcek et al. 2013; Wirta et al. 2014). This is due in part to the narrow taxonomic focus of most parasitoid researchers on particular families, tribes, and lower taxa, which is a consequence of the enormous diversity of parasitoids. One cannot know them all. Clearly, we need greater collaboration among parasitoid researchers, and among parasitoid and caterpillar researchers to understand parasitoid-caterpillar interaction networks. Innovations in molecular data acquisition are helping to solve this problem by allowing anyone to identify parasitoid and caterpillar OTUs via DNA barcodes, and thus species connections (e.g., Hrcek et al. 2011, Hrcek and Godfray 2015), hopefully not at the expense of taxonomic expertise.

Predictors of Susceptibility to Parasitoids

Hidden in the overall percentages of parasitism in Table 1 are widely varying frequencies for particular caterpillar species, from taxa that appear all but immune to parasitoids, with extremely low frequencies of parasitism, to those experiencing intense pressure from many parasitoid species with a majority of individuals parasitized. One general finding from rearing studies of caterpillar communities is that taxa that appear to be well-defended from predators tend to experience higher rates of parasitism by parasitoids. For example, gregarious caterpillars and concealed feeders experience greater parasitism than solitary and exposed feeders (Dyer and Gentry 1999; Gentry and Dyer 2002; Stireman and Singer 2003a; Stireman et al. 2017). Furthermore, hairy caterpillars and those with deterrent extracts have been found to exhibit elevated parasitism frequencies (Gentry and Dyer 2002; Stireman and Singer 2003a, b). The richness of parasitoid complexes of caterpillars also appears to be associated with morphological and behavioral defenses (e.g., hairiness, gregariousness; Stireman and Singer 2003a).

These patterns have led several authors to hypothesize that parasitoids preferentially use well-defended hosts in order to gain "enemy-free space" (Jeffries and Lawton 1984) from potential predators of their hosts (Gentry and Dyer 2002; Murphy et al. 2014). Negative associations between predation and parasitoid mortality of caterpillars in both temperate (Murphy et al. 2014) and tropical (Baer and Marquis 2020) forests support this hypothesis. These associations likely represent behavioral adaptations on the part of the parasitoids (i.e., parasitoids attracted to defended hosts have been selected for). However, if parasitized caterpillars are more susceptible to predators than unparasitized ones, the association could be artifactual, at least in part.

This is not to say that caterpillars are defenseless against parasitoids. Caterpillars have evolved myriad physiological, physical, and behavioral defenses against their enemies (e.g., Gross 1993; Montllor and Bernays 1993; Dyer 1995, 1997; Greeney et al. 2012), and much recent work has focused on understanding how enemy pressures, in combination with bottom-up pressures from plants, have shaped caterpillar life histories, morphology, behavior, and community structure (Greeney et al. 2012). For example, defensive behaviors such as dropping and biting, as well as regurgitating crop contents, are associated with lower parasitism rates by some parasitoids (Gentry and Dyer 2002; Desurmont et al. 2017). Furthermore, physical defenses that are effective against predators such as hairs and constructed shelters can also be effective against parasitoids (Kageyama and Sugiura 2016; LoPresti and Morse 2013, respectively). Although caterpillars with predator-deterrent extracts appear to experience elevated parasitism rates in some communities (Gentry and Dyer 2002), a number of studies have shown negative effects of sequestered toxins on parasitoid success (Singer and Stireman 2003, Singer et al. 2004; Ode 2006). Dissection and genetic metabarcoding studies confirm that many parasitoid eggs and larvae perish inside the host such that not all parasitized caterpillars produce adult parasitoids (e.g., Ahmadou et al. 2019). Indeed, host physiological immune defenses are potent anti-parasitoid weapons of caterpillars, explaining considerable variation in parasitism frequencies (Smilanich et al. 2009). However, as mentioned previously, these defenses are often suppressed, circumvented, or avoided by parasitoids. Recent studies have revealed the existence of tradeoffs between the efficacy of caterpillar immune defenses and that of plant-derived chemical defenses against parasitoids (Hansen et al. 2017; Slinn et al. 2018) (see also Smilanich and Muchovey, Chapter "Host Plant Effects on the Caterpillar Immune Response", this volume). Further studies of such interactions promise to be an insightful area of research.

Parasitoid Host Ranges

The number of host species attacked by parasitoid species varies widely. Many species of caterpillar parasitoids have been reared from only a single host species and typically only a particular developmental stage of that host. Other parasitoids have been reared from literally hundreds of host species (e.g., *Compsilura concinnata* (Arnaud 1978) and *Dibrachys microgastri* (Bouche) Noyes 2021). As with caterpillars, parasitoid host ranges form a complex continuum, and categorization as "specialists" and "generalists" is somewhat arbitrary (Forister et al. 2012). In practice, species using one or a few host species belonging to a single family are often considered specialists while those with hosts belonging more than one family are considered generalists (e.g., Fernandez-Triana et al. 2014b). Assessing the breadth of hosts used by a parasitoid species can be difficult because one must survey all potential hosts, and the more times a parasitoid is reared the more potential hosts it may be reared from (Eggleton and Gaston 1992; Stireman et al. 2017). Furthermore, host ranges may vary geographically due to varying distributions of suitable or preferred hosts. On the other hand, some parasitoids are occasionally reared from atypical hosts – apparent "accidents" in which a novel host proves permissive to parasitoid development. Thus, several authors have pointed out the importance of distinguishing between the realized host range (all recorded hosts) and the "usual" host range (i.e., those that are typically attacked; Shaw 1994; Quicke 2015).

Host associations are unknown for the majority of parasitoid species. This lack of knowledge not only hampers our understanding of parasitoid evolution and ecology, but also how parasitoids shape the evolution of their hosts and how parasitoid-caterpillar-plant networks are structured. Documenting host-associations through rearing studies is laborious and time consuming, as well as difficult to fund for non-pest host insects. As mentioned previously, metabarcoding can accelerate our understanding of parasitoid-host food webs, and it is now even possible in some cases to identify the host of an adult parasitoid with metabarcoding approaches (Rougerie et al. 2011; Haran et al. 2018). The increasing application of genetic data (e.g., DNA barcodes), combined with classical morphological and host data, is providing new insight into parasitoid host ranges and host-parasitoid interaction webs. Hundreds of new parasitoid species have recently been described through the efforts of major tropical caterpillar rearing programs (e.g., Dyer et al. 2007; Janzen and Hallwachs 2009) combined with DNA barcoding initiatives (e.g., Fernandez-Triana et al. 2014a, b; Arias-Penna et al. 2019; Fleming et al. 2019). Inclusion of DNA and host data has often revealed the existence of morphologically cryptic species that are associated with a narrow range of hosts. For example, Smith et al. (2007) found that a set of 15 apparently generalist tachinid species consisted of at least 73 distinct mtDNA lineages, many of which were specialized on a narrow set of host caterpillars. Findings are similar for the even more specialized parasitoid wasps (Smith et al. 2008). Fernandez-Triana et al. (2014b) found that 90% of 186 new *Apantales* species were monophagous or oligophagous (using hosts from at most a single family), and they suggested that at least some of the apparently polyphagous species likely represent more specialized species complexes. Likewise, 96% of recently described *Glyptapantales* (N = 136) were found to use at most a single host family (Arias-Penna et al. 2019).

Host ranges of parasitoids are shaped by a complex array of ecological and evolutionary interactions. Traditionally, the determinants of parasitoid host ranges have been considered a function of host phylogeny and host ecology (Askew and Shaw 1986). Effects of host phylogeny are thought to be due to shared physiology, morphology, and behavior. Physiological constraints on host range can be a function of adaptations to particular host immune defenses, tolerance of host toxins (either innate or acquired), and synchronization of development (Belshaw 1994). Morphology can limit host ranges through size (e.g., hosts may be too small to support development) or because structural defenses (e.g., hairs, thick cuticle, crypsis) limit parasitoid access. Behaviors such as biting, thrashing, dropping, fleeing, and regurgitating can also exclude potential hosts from a parasitoid's options. Host feeding ecology can limit parasitoid use in a variety of ways in terms of where, when, and what caterpillars feed on, as well as how they feed on it. For example, many caterpillars feed nocturnally when many parasitoids and predators are inactive, and hide in inaccessible, sheltered locations during the day (e.g., soil, leaf litter, leaf rolls, under bark; Wagner 2005). At least some parasitoids have evolved means to surmount each of these constraints, and thus, host ranges (or patterns of host use) can be viewed as the result of an evolutionary interplay between traits of hosts and parasitoids in a particular ecological context.

The particular hosts used by parasitoid species determine the cues they use in host location, as selection is expected to hone the host location process toward use of the most detectable and most reliable cues (Vet and Dicke 1992). However, the process of, and cues used for, host location can also shape patterns of host use. There is abundant evidence that parasitism risk varies with caterpillar host plants (e.g., Barbosa et al. 2001; Lill et al. 2002; Farkas and Singer 2013) and parasitoids are well known to use volatile organic compounds released by plants, often due to herbivore feeding, to locate suitable hosts (Turlings et al. 1990; Vet and Dicke 1992). In many cases it appears that the habitat of a caterpillar (e.g., host plant, feeding niche, pupation site) is an important determinant of whether it is included in a parasitoid's host range (Shaw 1994). This is especially likely for parasitoid taxa in which host physiology does not appear to strongly limit host ranges, such as many Tachinidae (Stireman et al. 2006, 2009). Indeed, it has been suggested that host specificity in dipteran parasitoids is more a function of events leading up to oviposition (e.g., host searching) than interactions after oviposition (host physiological defenses; Feener and Brown 1997), and this extends to many hymenopteran parasitoids as well. It is seemingly paradoxical that parasitoid host ranges both determine and are determined by the manner in which they locate hosts, but host use and host searching/selection are dynamic phenomena that feed back on one another via natural selection (as in phytophagous insects). For example, if two caterpillar species produce overlapping kairomones, perhaps because they feed on the same host plant, a typical parasitoid of one of them may encounter and attack the other species as well. If the secondary host is conducive to parasitoid development, selection can favor physiological adaptions for increased performance and perhaps relaxation of or divergence in host location and selection cues, expanding host range. Subsequently, parasitoid populations on each host may diverge and, if barriers allow, eventually resolve into distinct species (Shaw 2002).

Some parasitoid lineages may be locked into tight coevolutionary races with a particular host lineage, speciating in concert, but this appears to be rare. Most current models for the evolution of host associations in phytophagous insects assume that host ranges are dynamic (e.g., Janz et al. 2006; Janz and Nylin 2008; Fordyce 2010; though see Hardy and Otto 2014), varying across space and expanding and contracting over evolutionary time. This is likely true for most of their parasitoids as well (Shaw 2002; Quicke 2015). Certain parasitoid lineages appear to consist of large clades of relatively specialized species (e.g., Smith et al. 2006, 2008), where host expansions appear to be associated with speciation. However, in many groups, polyphagous and oligophagous (or monophagous) species are phylogenetically intermingled (e.g., Shaw 1994, 2002; Stireman 2005; Smith et al. 2007), suggesting either transitory stages of polyphagy or persistent polyphagous lineages repeatedly "spinning off" more specialized lineages (Stireman 2005).

Parasitoids in the Anthropocene

A growing number of studies are pointing to widespread declines in insect diversity and biomass (Sanchez-Bayo and Wyckhuys 2019; Montgomery et al. 2020; Wagner et al. 2021). Parasitoids may be particularly susceptible to environmental change because they occupy relatively high trophic levels and many species possess highly specialized niches (LaSalle and Gauld 1991; Holt et al. 1999). These traits make them well suited as early-warning indicator species of ecosystem health; however, they have rarely been utilized in this regard. As a group, parasitoids are so poorly known that we have little idea how many species have already gone extinct or may currently be teetering on the brink due to anthropogenic changes to the environment. Even for most described species, we have little more than a name and a morphological description (or conversely DNA sequence BINs with host and habitat data but no names). It is certain that vast numbers of species have disappeared in recent centuries in heavily impacted environments such as Brazil's Atlantic Forest, Indonesian rainforests, tropical dry forests of the Americas, Madagascar, and temperate forests and grasslands of North America and Europe. Today, parasitoids are facing an increasingly severe array of anthropogenic threats including habitat destruction, degradation, and fragmentation, agricultural intensification, invasive species, and climate change (Tylianakis and Binzer 2014). As important enemies of phytophagous insects such as caterpillars, the consequences of their diminishment and loss may be profound (Salcido et al. 2020).

Habitat Loss and Fragmentation

Loss and degradation of suitable habitat is probably the most serious environmental threat facing parasitoid insects as well as most other terrestrial animals on Earth (LaSalle and Gauld 1991; Pereira et al. 2010; Sanchez-Bayo and Wyckhuys 2019). Forests, prairies, wetlands, and other habitats are being destroyed, fragmented, and degraded at an increasingly rapid rate by agricultural expansion and land use change fueled by increasing per capita consumption of a growing human population (Tilman et al. 2001). Due to their relatively small size and concomitantly large population sizes, it might be expected that parasitoid species should be resistant to extinction. However, current knowledge of host use, although incomplete, suggests most parasitoids attack a narrow range of hosts, often caterpillars that are themselves restricted to certain host plants, and thus, their populations can be quite sensitive to disturbance of lower trophic levels. Even apparently polyphagous species of parasitoids may be specialized with respect to particular habitats (Strand and Obrycki 1996; Stireman and Singer 2003b). LaSalle and Gauld (1991) predicted that the widespread extinction of parasitic Hymenoptera species would have devastating effects on ecological stability and community balance due to their roles as keystone species and the consequent cascading effects as species are lost. More recently, studies of other animal taxa have shown that those with specialized niches are most likely to suffer from habitat loss, fragmentation, and/or degradation, leading to a dominance of generalists and more diffuse ecological interactions (Davies et al. 2004; Matthews et al. 2014).

In addition to the simple loss of habitat area, most natural habitats are becoming increasingly fragmented into isolated patches with adverse consequences for biodiversity and ecosystem function (Haddad et al. 2015). Due to their life history, in which free-living adults must locate small hosts (and mates) in a large and complex world, most parasitoid taxa are relatively vagile and can potentially disperse long distances between patches of habitat (e.g., Couchoux et al. 2016). However, even strong flying parasitoids like tachinid flies exhibit evidence of declines in diversity in small habitat fragments, which is exacerbated by increasing isolation (Inclán et al. 2014). Despite strong dispersal abilities, reliance on specific host populations makes parasitoids particularly susceptible to declines and local extinction in fragmented habitats (van Nouhuys 2005; Gravel et al. 2011; Martinson and Fagan 2014; Stireman and Singer 2018). For example, fragmentation of the Chaco Serrano forest in Argentina has resulted in a decline in parasitoid biodiversity and lower parasitism rates of lepidopteran leaf miners (Fenoglio et al. 2012). Consistent with the prediction that impacts of habitat loss and fragmentation should be most evident among specialized parasitoids, dipteran leaf miners in the Chaco Serrano forests, which tend to be attacked by relatively polyphagous parasitoids, did not exhibit strong reductions in parasitism or parasitoid diversity in smaller fragments (Fenoglio et al. 2012). Similarly, among externally feeding forest caterpillars, specialists exhibited declining parasitism rates with diminishing forest fragment size, presumably due to negative impacts of fragmentation on specialized parasitoids, whereas generalist feeders showed no such pattern (Anderson et al. 2019). Recently, it has been found that declines in biodiversity due to habitat fragmentation are exacerbated by ecosystem decay (Chase et al. 2020), wherein altered ecological interactions cause demographic instability of populations, increasing the risk of local extinction, which may then lead to even greater instability.

Agricultural Intensification

Although several factors may have contributed to the apparent widespread declines of insect abundance and diversity in recent decades (Sanchez-Bayo and Wyckhuys 2019; Wagner et al. 2021), a number of studies point to agricultural intensification as a primary culprit (Hallmann et al. 2017). Intense management practices such as removal of peripheral vegetation (e.g., hedgerows), soil disturbance, and heavy application of fertilizers, herbicides, and pesticides, all have the potential to negatively impact parasitoid populations and their ability to help regulate phytophagous insect populations (e.g., Jonsson et al. 2012). These negative effects have been recognized in the biological control literature for decades (e.g., Landis and Menalled 1998; Landis et al. 2000; Tilman et al. 2001; Tscharntke et al. 2005), and attempts to mitigate them have been applied in many Integrated Pest Management programs. Yet, agricultural practices are changing at a rapid pace with the development and spread of genetically modified crops, increased mechanization and efficiency, and novel, long lasting, systemic pesticides. We are only beginning to understand the impacts of these changes and may only now be seeing the consequences of agricultural intensification and the conversion of vast areas of our landscapes into ecological deserts of potentially toxic monocultures (Hallmann et al. 2017).

Several studies have shown that agricultural intensification is associated with lower diversity and diminished biological control services of caterpillar parasitoids (e.g., Jonsson et al. 2012; Letourneau et al. 2012; Gonzalez et al. 2020). Generally, parasitoids tend to be more susceptible to insecticides than their hosts (Croft and Brown 1975), and even pesticides that are not directly lethal may have sublethal detrimental effects on parasitoid development and behavior (Desneux et al. 2007). Conversely, less intensive (organic) farming has been shown to increase parasitoid diversity at both local and landscape scales relative to conventional agriculture (Inclán et al. 2015). Still, even in organic farms, with their somewhat more stringent regulations on pesticides and herbicides, loss of surrounding natural vegetation has been shown to result in lower parasitoid diversity and lower parasitism frequencies of potential lepidopteran pests (Letourneau et al. 2012, 2015). Simplification of the larger landscape such as that associated with large-scale agriculture has been shown to result in more homogenous communities composed of more generalist taxa (Gámez-Virués et al. 2015).

Intensive agricultural practices may not only create agricultural landscapes that contribute little to biodiversity and ecosystem services, but their effects may extend far beyond the field margins due to runoff and drift of fertilizer, herbicides, and pesticides (Godfray et al. 2015), as well as their potential to act as biodiversity sinks for mobile insect populations (Kautz and Gardiner 2019). Given the expansion and/or intensification of agriculture necessary to feed the world's burgeoning human population (Tilman et al. 2001), preserving diverse parasitoid communities and the ecological services they provide in agroecosystems and surrounding areas will be challenging. It will require the maintenance of non-crop habitats, mitigation of non-target effects of agricultural chemicals, and an emphasis on sustainable, low disturbance practices.

Invasive Species

The impacts of invasive species on parasitoids may be significant but have been relatively little studied. At the primary producer level, invasive plants may outcompete and displace native host plants of caterpillars, which may in turn negatively impact parasitoid populations (Carvalheiro et al. 2010; Bezemer et al. 2014; López-Núñez et al. 2017). If their host caterpillars are able to colonize novel invasive plants, parasitoids may not be able to locate them effectively (i.e., they may gain a degree of enemy-free space) or to tolerate the novel secondary chemicals ingested by their hosts. Invasive herbivores can also negatively impact native parasitoids, particularly those that decimate host plant populations (and thus herbivore populations) in their introduced ranges, such as the gypsy moth, *Lymantria dispar* (though see Timms et al. 2012), emerald ash borer, *Agrilus planipennis* Fairmaire, and the cactus moth, *Cactoblastis cactorum* (Berg). Furthermore, invasive herbivores can interfere with native parasitoid-host interactions by disrupting plant volatile cues used by parasitoids in host location (Martorana et al. 2017; Desurmont et al. 2018).

Parasitoids introduced inadvertently or purposefully for biological control may also act as invasive species and can have substantial impacts on populations of non-target hosts (Parry 2009; Myers and Cory 2017). The tachinid *Compsilura concinnata* provides an example of such impacts. This fly was introduced repeatedly as a biological control agent against gypsy moths in North America (Elkinton and Boettner 2012) despite having an extremely broad host range in its native range (Tschorsnig 2017). Rearing studies have indicated that *C. concinnata* is exploiting many species of native forest caterpillars in its introduced range and may be responsible for widespread declines in wild silk moths in the eastern United States (Boettner et al. 2000; Elkinton and Boettner 2012). A similar situation may be occurring with the braconid *Cotesia glomerata* (L.) which was introduced to control the cabbage white, *Pieris rapae* (L.), but has contributed to marked declines in populations of the native mustard white, *P. oleracea* (Harris) (van Driesche et al. 2004; Keeler et al. 2006). Introduced parasitoids may also negatively affect other native parasitoids via competition or hyperparasitism. In New Zealand, the introduced tachinid *Trigonospila brevifacies* (Hardy) has become the dominant parasitoid of a leaf-rolling guild of native caterpillars and may be displacing native parasitoid species (Munro and Henderson 2002). In Hawaii, the parasitoid fauna of native lepidopterans is almost completely dominated (97%) by intentionally and unintentionally introduced hymenopteran parasitoids, with unknown consequences for the native caterpillar and parasitoid fauna (Henneman and Memmot 2001). In an interesting twist, the aforementioned introduced pierid parasitoid *Co. glomerata* is apparently being replaced by another introduced species *Co. rubecula* (Marshall) (van Driesche 2008).

Climate Change

Evidence is mounting that climate change may have dramatic consequences for parasitoids of caterpillars especially via disruption of their interactions with hosts. As mentioned previously, parasitism frequencies of caterpillar communities exhibit a negative relationship with increasing precipitation variability across geographic sites (Stireman et al. 2005) with ominous implications for the future as climate predictions call for an increasing frequency of extreme weather events, droughts, and floods. Such effects, combined with habitat fragmentation and agricultural intensification, appear to be having substantial negative impacts on species diversity and interaction diversity of parasitoids and their caterpillar hosts in a well-studied tropical site in Costa Rica (Salcido et al. 2020).

One of the most obvious consequences of climate change for parasitoids is the creation of potential phenological mismatches with their hosts (Jeffs and Lewis 2013; Visser and Gienapp 2019). If hosts and parasitoids differ in developmental rates, respond differentially to environmental cues in developmental transitions (e.g., adult eclosion), or vary in their plasticity to respond to climate change (e.g., multivoltinism), this could lead to declines in parasitism and potential outbreaks of caterpillars and other herbivores (Forrest 2016; Chidawanyika et al. 2019; Ward et al. 2020). On the other hand, temporal or spatial refuges of caterpillars could be compromised as parasitoids shift phenologically and/or geographically, potentially leading to declines in or extinction of lepidopteran populations. This may be the case in Britain, where the nymphalid parasitizing tachinid *Sturmia bella* (Meigen) has dramatically expanded its range northward, potentially contributing to declines in the small tortoiseshell butterfly (*Aglais urticae* (L.); Gripenberg et al. 2011; Jeffs and Lewis 2013). Finally, dramatic and lasting impacts of weather anomalies on caterpillar populations such as late freezes and summer droughts (Marquis et al. 2019) may ramify upward to parasitoids and result in local extinction.

Idiobiont parasitoid species often have broad host ranges relative to koinobiont taxa; therefore, they may be relatively more likely to persist in modified habitats. Idiobionts may also gain advantages with a warming climate. While koinobionts are better adapted to exploit the overwintering behavior of their host insects, idiobiont species are limited to overwintering along with hosts in more exposed habitats or as adults (Hance et al. 2007). Consequently, idiobionts are comparatively more sensitive to harsh winter conditions, and consistent with this, they comprise a smaller proportion of lepidopteran parasitoids as one moves toward the arctic (Timms et al. 2013, 2016; Kankaanpää et al. 2020). As the arctic warms due to anthropogenic sources of greenhouse gases, it is predicted that idiobionts will comprise an increasing proportion of the parasitoids of caterpillar communities (Kankaanpää et al. 2020)

The potential effects of climate change on caterpillar-parasitoid interactions are manifold and difficult to predict even in simple systems. For example, Dyer et al. (2013) observed complex, chemistry-mediated effects of elevated temperature and CO_2 in an experimental tritrophic plant-caterpillar-parasitoid system. They found that elevated temperature disrupted developmental synchrony of the parasitoid and host, leading to parasitoid extinction, while elevated CO_2 indirectly lowered parasitism rates by decreasing plant quality and increasing caterpillar development time. In agroecosystems, differing thermal tolerances of caterpillars and parasitoids and effects of temperature on parasitoid success may alter host parasitoid interactions and lead to diminished regulatory impacts of parasitoids (Hance et al. 2007; Stireman and Singer 2018). Effects of climate change extend beyond impacts on particular parasitoid-host pairs or tri-trophic chains but influence the structure and dynamics of entire interaction webs of parasitoids and their hosts (Tylianakis and Binzer 2014). Despite the increasingly rapid rate of climate change, we are only beginning to understand what these effects may be.

Consequences of Parasitoid Loss

In considering the many anthropogenic environmental threats that parasitoids face and their likelihood of being among the first components of plant-caterpillar-enemy tritrophic systems to be lost, we might ask: what are the consequences of this loss? We know so little about most natural tritrophic systems and communities that this question is difficult to answer, but several lines of evidence suggest that the consequences could be dramatic. For example, the widespread phenomenon of "enemy release" in introduced herbivores in both natural and managed systems belies the important role of parasitoids in controlling herbivore populations (i.e., such herbivores appear to be "released," but they are not escaping more generalized invertebrate and vertebrate predators). The frequent success of biological control programs using parasitoids also provides evidence of their potential impact in natural systems (Greathead and Greathead 1992). Thus, we might expect that the loss of parasitoids or declines in parasitism may lead to more frequent and more severe irruptions of caterpillar populations with potentially adverse consequences for agricultural and natural ecosystems.

Local losses of parasitoids will also lead to simplified food webs of constituent species (Tylianakis and Binzer 2014) that may be less resilient and more susceptible to disturbance (Dunne et al. 2002; Tylianakis et al. 2010). Such simplified food webs may be more likely to collapse under increasing environmental pressures of climate change and habitat fragmentation. Characteristics of ecological interaction networks are expected to change as specialist parasitoids are lost and generalists, which are less susceptible to habitat loss and degradation, become more dominant (Clavel et al. 2011). Food web connectivity and linkage density should increase with a preponderance of generalist taxa, but vulnerability (mean number of parasitoid species per caterpillar species), interaction evenness, and compartmentalization will likely decrease. Increases in generalist parasitoids may also lead to greater levels of apparent competition among caterpillars via their shared enemies (Holt 1977). These changes in food web structure may have further consequences for food web stability and resistance to disturbance (e.g., Montoya et al. 2006).

Some of us find the life histories of parasitoids fascinating. We marvel at their incredible diversity, their complex multitrophic interactions with caterpillars and their host plants, and the remarkable adaptations and counter-adaptations of parasitoids and their hosts. Others may have less charitable views, possibly finding them creepy or disgusting. Nevertheless, hopefully all can understand the incredibly important ecological roles parasitoids play in natural and managed systems as key enemies of caterpillars and other phytophagous insects (as well as pollinators and as prey). We also hope that all can come to appreciate the fascinating stories of their ecology, evolution, and diversification. We look forward to further unraveling more of these complex stories and exploring how parasitoids are responding to anthropogenic environmental changes to Earth's ecosystems.

Acknowledgments We would like to thank Suzanne Koptur and Robert Marquis for inviting us to contribute to this edited volume and providing valuable feedback as well as three reviewers (including RJ Marquis and JB Whitfield) whose comments served to improve this chapter. We also thank Steve Marshall and Kenji Nishida for granting us permission to publish their photos. Stireman was supported in part by NSF DEB 1442134 and Shaw in part by DEB 14-42110 during the preparation of this chapter. Any opinions, findings, and conclusions expressed are those of the authors and do not necessarily reflect the views of the National Science Foundation.

References

Aguirre H, Shaw SR (2014a) *Meteorus* Haliday (Hymenoptera: Braconidae) parasitoids of Pyralidae: description and biology of two new species and first record of *Meteorus desmiae* Zitani, 1998 from Ecuador. J Nat Hist 48:2375–2388

Aguirre H, Shaw SR (2014b) Neotropical species of *Meteorus* Haliday (Hymenoptera: Braconidae: Meteorinae) parasitizing Arctiinae (Lepidoptera: Noctuoidea: Erebidae). Zootaxa 3779:353–367

Aguirre H, Shaw SR, Berry JA, de Sassi C (2014) Description and natural history of the first micropterous *Meteorus* species: *M. orocrambivorus* sp. n. (Hymenoptera, Braconidae, Euphorinae), endemic to New Zealand. J Hymenopt Res 38:45–57

Aguirre H, de Almeida LP, Shaw SR, Sarmiento CE (2015) An illustrated key to Neotropical species of the genus *Meteorus* Haliday (Hymenoptera, Braconidae, Euphorinae). Zookeys 489:33–94

Ahmadou S, Brevaoult T, Benoit L, Chapis M-P, Galan M, Couer d'acier A, Delvare G, Sembene M, Haran J (2019) Deciphering host-parasitoid interactions and parasitism rates of crop pests using DNA metabarcoding. Sci Rep 9:3646

Anderson RM, Dallar NM, Pirtel NL, Connors CJ, Mickley J, Bagchi R, Singer MS (2019) Bottom-up and top-down effects of forest fragmentation differ between dietary generalist and specialist caterpillars. Front Ecol Evol 7:452

Araujo RO, Vivallo F, Santo BF (2018) Ichneumonid wasps of the subfamily Mesochorinae: new replacement names, combinations, and an updated key to the World genera. Zootaxa 4521:52–60

Areekul-Butcher B, Quicke DLJ (2011) Revision of *Aleiodes* (*Hemigyroneuron*) parasitic wasps (Hymenoptera: Braconidae: Rogadinae) with reappraisal of subgeneric limits, descriptions of new species and phylogenetic analysis. J Nat Hist 45:1403–1476

Arias-Penna CD, Whitfield JB, Janzen DH, Hallwachs W, Dyer LA, Smith MA, Hebert PDN, Fernández-Triana JL (2019) A species-level taxonomic review and host associations of *Glyptapanteles* (Hymenoptera, Braconidae, Microgastrinae) with an emphasis on 136 new reared species from Costa Rica and Ecuador. ZooKeys 890:1–685

Arnaud PH Jr (1978) A host-parasite catalog of North American Tachinidae (Diptera). USDA Misc Publ 1319. 860 pp

Askew RR, Shaw MR (1986) Parasitoid communities: their size, structure, and development. In: Waage JK, Greathead D (eds) Insect parasitoids. Academic, London, pp 225–264

Baer CS, Marquis RJ (2020) Between predators and parasitoids: Complex interactions among shelter traits, predation and parasitism in a shelter-building caterpillar community. Funct Ecol 00:1–13. https://doi.org/10.1111/1365-2435.13641

Barbosa P, Segarra AE, Gross P, Caldas A, Ahlstrom K, Carlson RW, Ferguson DC, Grissell EE, Hodges RW, Marsh PM, Poole RW, Schauff ME, Shaw SR, Whitfield JB, Woodley NE (2001) Differential parasitism of macrolepidopteran herbivores on two deciduous tree species. Ecology 82:698–704

Barbosa P, Tammaru T, Caldas A (2004) Is parasitism of numerically dominant species in macrolepidopteran assemblages independent of their abundance? Basic Appl Ecol 5:357–366

Battisti A, Bernardi M, Ghiraldo C (2000) Predation by the hoopoe (*Upupa epops*) on pupae of *Thaumetopoea pityocampa* and the likely influence on other natural enemies. Biocontrol 45:311–323

Beckage NE, Drezen J-M (eds) (2012) Parasitoid viruses: symbionts and pathogens. Elsevier, London

Belshaw R (1994) Life history characteristics of Tachinidae (Diptera) and their effect on polyphagy. In: Hawkins BA, Sheehan W (eds) Parasitoid community ecology. Oxford University Press, Oxford, pp 145–162

Bezemer TM, Harvey JA, Cronin JT (2014) Response of native insect communities to invasive plants. Annu Rev Entomol 59:119–141

Blaschke JS, Stireman JO III, O'Hara JE, Cerretti P, Moulton JK (2018) Molecular phylogenetics and piercer evolution in the bug-killing flies (Diptera: Tachinidae: Phasiinae). Syst Entomol 43:218–238

Boettner GH, Elkinton JS, Boettner CJ (2000) Effects of a biological control introduction on three nontarget native species of saturniid moths. Conserv Biol 14:1798–1806

Boivin G, van Baaren J (2008) The role of larval aggression and mobility in the transition between solitary and gregarious development in parasitoid wasps. Ecol Lett 3:469–474

Broad GR, Shaw MR, Fitton MG (2018) Ichneumonid Wasps (Hymenoptera: Ichneumonidae): their Classification and Biology. Handbook Ident Br Insec 7(12) 418 pp

Burington ZL (2017) Evolution and biogeography of the tachinid flies with focus on the tribe Blondeliini (Insecta: Diptera: Tachinidae). PhD dissertation, Wright State University

Burington ZL, Inclán-Luna DJ, Pollet M, Stireman JO III (2020) Latitudinal patterns in tachinid parasitoid diversity (Diptera: Tachinidae): a review of the evidence. Insect Cons Divers 13:419–431

Carmean D (1991) Biology of the Trigonalyidae (Hymenoptera), with notes on the vespine parasitoid *Bareogonalos canadensis*. NZ J Zool 18:209–214

Carmean D (1995) Trigonalyidae. In: Hanson PE, Gauld ID (eds) The Hymenoptera of Costa Rica. Oxford University Press, pp 187–192

Carmean D, Kimsey L (1998) Phylogenetic revision of the parasitoid wasp family Trigonalidae (Hymenoptera). Syst Entomol 23:35–76

Caron V, Myers J, Gillespie D (2010) The failure to discriminate: Superparasitism of *Trichoplusia ni* Hübner by a generalist tachinid parasitoid. Bull Entomol Res 100:255–261

Carvalheiro LG, Buckley YM, Memmott J (2010) Diet breadth influences how the impact of invasive plants is propagated through food webs. Ecology 91:1063–1074

Cerretti P (2010) I tachinidi della fauna italiana (Diptera Tachinidae), con chiave interattiva dei generi ovest-paleartici. Vol. I. Centro Nazionale Biodiversita Forestale – Verona. Cierre Edizioni, Verona 573 pp

Cerretti P, O'Hara JE, Wood DM, Shima H, Inclán DJ, Stireman JO III (2014) Signal through the noise? Phylogeny of the Tachinidae (Diptera) as inferred from morphological evidence. Syst Entomol 39:335–353

Cerretti P, Stireman JO III, Pape T, Marinho MAT, Rognes K, Grimaldi D (2017) First definitive fossil of an oestroid fly (Diptera: Calyptratae: Oestroidea) and the dating of oestroid divergences. PLoS One 12:e0182101

Chandra G, Gupta VK (1977) Ichneumonologia Orientalis Part VII: the tribes Lissonotini and Banchini (Hymenoptera: Ichneumonidae: Banchinae). Orient Insect Monogr 7:1–290

Chase JM, Blowes SA, Knight TM, Gerstner K, May F (2020) Ecosystem decay exacerbates biodiversity loss with habitat loss. Nature 584:238–243

Chidawanyika F, Mudavanhu P, Nyamukondiwa C (2019) Global climate change as a driver of bottom-up and top-down factors in agricultural landscapes and the fate of host-parasitoid interactions. Front Ecol Evol 7:1–13

Clausen CP (1940) Entomophagous insects. McGraw-Hill Book Company, New York

Clausen CP (1978) Introduced parasites and predators of arthropod pests and weeds: a review
Clavel J, Julliard R, Devictor V (2011) Worldwide decline of specialist species: toward a global functional homogenization? Front Ecol Environ 9:222–228
Colinet D, Mathe-Hubert H, Allemand R, Gatti J-L, Poirie M (2013) Variability of venom components in immune suppressive parasitoid wasps: from a phylogenetic to a population approach. J Insect Physiol 59:205–212
Condon MA, Scheffer SJ, Lewis ML, Wharton R, Adams DC, Forbes AA (2014) Lethal interactions between parasites and prey increase niche diversity in a tropical community. Science 343:1240–1244
Couchoux C, Seppä P, van Nouhuys S (2016) Strong dispersal in a parasitoid wasp overwhelms habitat fragmentation and host population dynamics. Mol Ecol 25:3344–3355
Croft BA, Brown AWA (1975) Responses of arthropod natural enemies to insecticides. Annu Rev Entomol 20:285–335
Dadelahi S, Shaw SR, Aguirre H, de Almeida LF (2018) A taxonomic study of Costa Rican *Leptodrepana* with descriptions of twenty-four new species (Hymenoptera: Braconidae: Cheloninae). Zookeys 750:59130
Dasch CE (1971) Ichneumon-flies of North America north of Mexico: 6. Mesochorinae. Volume 16. Mem Am Entomol Inst 16:1–376
Dasch CE (1974) Neotropic Mesochorinae. Volume 22. Mem Am Entomol Inst 22:1–509
Davies KF, Margules CR, Lawrence JF (2004) A synergistic effect puts rare, specialized species at greater risk of extinction. Ecology 85:265–271
Delucchi V (1982) Parasitoids and hyperparasitoids of *Zeiraphera diniana* [Lep, Tortricidae] and their pole in population control in outbreak areas. Entomophaga 27:77–92
Desneux N, Decourtye A, Delpuech J-MN (2007) The sublethal effects of pesticides on beneficial arthropods. Annu Rev Entomol 52:81–106
Desurmont GA, Köhler A, Maag D et al (2017) The spitting image of plant defenses: effects of plant secondary chemistry on the efficacy of caterpillar regurgitant as an anti-predator defense. Ecol Evol 7:6304–6313
Desurmont GA, Guiguet A, Turlings T (2018) Invasive insect herbivores as disrupters of chemically-mediated tritrophic interactions: effects of herbivore density and parasitoid learning. Biol Invasions 20:195–206
Dindo ML (2011) Tachinid parasitoids: are they to be considered as koinobionts? Biocontrol 56:249–255
Doutt RL, Viggiani G (1968) The classification of the Trichogrammatidae (Hymenoptera: Chalcidoidea). Proc Calif Acad Sci 35:477–586
Drost YC (1991) Development of oviposition behavior of *Brachymeria intermedia*, a parasitoid of the gypsy moth, *Lymantria dispar*. PhD dissertation, University of Massachusetts, Amherst
Dunne JA, Williams RJ, Martinez ND (2002) Network structure and biodiversity loss in food webs: robustness increases with connectance. Ecol Lett 5:558–567
Dyer LA (1995) Tasty generalists and nasty specialists? A comparative study of antipredator mechanisms in tropical lepidopteran larvae. Ecology 76:1483–1496
Dyer LA (1997) Effectiveness of caterpillar defenses against three species of invertebrate predators. J Res Lep 35:1–16
Dyer LA, Gentry G (1999) Larval defensive mechanisms as predictors of successful biological control. Ecol Appl 9:402–408
Dyer LA, Singer MS, Lill JT, Stireman JO, Gentry GL, Marquis RJ, Ricklefs RE, Greeney HF, Wagner DL, Morais HC, Diniz IR, Kursar TA, Coley PD (2007) Host specificity of Lepidoptera in tropical and temperate forests. Nature 448:606–700
Dyer LA, Richards LA, Short SA, Dodson CD (2013) Effects of CO_2 and temperature on tritrophic interactions. PLoS ONE 8:e62528
Eggleton P, Belshaw R (1992) Insect parasitoids: an evolutionary overview. Philos Trans R Soc Lond Ser B Biol Sci 337:1–20

Eggleton P, Gaston KJ (1990) "Parasitoid" species and assemblages: convenient definitions or misleading compromises? Oikos 59:417–421

Eggleton P, Gaston KJ (1992) Tachinid host ranges: a reappraisal (Diptera: Tachinidae). Entomol Gaz 43:139–143

Elkinton JS, Boettner GH (2012) Benefits and harm caused by the introduced generalist tachinid, *Compsilura concinnata*, in North America. BioControl 57:277–288

Farkas TE, Singer MS (2013) Can caterpillar density or host-plant quality explain host-plant-related parasitism of a generalist forest caterpillar assemblage? Oecologia 173:971–983

Fatouros NE, Huigens ME (2012) Phoresy in the field: natural occurrence of *Trichogramma* egg parasitoids on butterflies and moths. BioControl 57:493–502

Feener DH Jr, Brown BV (1997) Diptera as parasitoids. Annu Rev Entomol 42:73–97

Fenoglio MS, Srivastava D, Valladares G, Cagnolo L, Salvo A (2012) Forest fragmentation reduces parasitism via species loss at multiple trophic levels. Ecology 93:2407–2420

Fernandez-Triana J, Janzen D, Hallwachs W, Whitfield J, Smith M, Kula R (2014a) Revision of the genus *Pseudapanteles* (Hymenoptera, Braconidae, Microgastrinae), with emphasis on the species in Area de Conservación Guanacaste, northwestern Costa Rica. ZooKeys 446:1–82

Fernandez-Triana J, Whitfield J, Rodriguez J, Smith M, Janzen D, Hajibabaei M, Burns J, Solis A, Brown J, Cardinal S, Goulet H, Hebert P (2014b) Review of *Apanteles sensu stricto* (Hymenoptera, Braconidae, Microgastrinae) from Area de Conservación Guanacaste, northwestern Costa Rica, with keys to all described species from Mesoamerica. ZooKeys 383:1–565

Fernandez-Triana J, Shaw MR, Boudreault C, Beaudin M, Broad GR (2020) Annotated and illustrated checklist of Microgastrinae parasitoid wasps (Hymenoptera: Braconidae). Zookeys 920:1–1089

Finney GL, Flanders SE, Smith HS (1947) Mass culture of *Macrocenturs ancylivorus* and its host, the potato tuber moth. Hilgardia 17:437–483

Fleming A, Wood D, Smith M, Dapkey T, Hallwachs W, Janzen D (2019) Twenty-two new species in the genus *Hyphantrophaga* Townsend (Diptera: Tachinidae) from Area de Conservación Guanacaste, with a key to the species of Mesoamerica. Biodivers Data J 7:e29553

Forbes AA, Bagley RK, Beer M, Hippee AC, Widmayer HA (2018) Quantifying the unquantifiable: why Hymenoptera – not Coleoptera – is the most speciose animal order. BMC Ecol 18:21

Fordyce JA (2010) Host shifts and evolutionary radiations of butterflies. Proc R Soc B 277:3735–3743

Forister ML, Dyer LA, Singer MS, Stireman JO, Lill JT (2012) Revisiting the evolution of ecological specialization, with emphasis on insect-plant interactions. Ecology 93:981–991

Forrest JRK (2016) Complex responses of insect phenology to climate change. Curr Opin Insect Sci 17:49–54

Fortier JC (2000) Description of a new gregarious species of *Aleiodes* Wesmael (Hymenoptera: Braconidae: Rogadinae). J Hymenopt Res 9:288–291

Foster KR, Wenseleers T, Ratnieks FLW (2006) Kin selection is the key to altruism. Trends Ecol Evol 21:57–60

Gámez-Virués S, Perović D, Gossner M et al (2015) Landscape simplification filters species traits and drives biotic homogenization. Nat Commun 6:8568

Garraway EB, Bailey AJA, Freeman BE, Parnell JR, Emmel TC (2008) Insect conservation and islands. Springer, Dordrecht, pp 189–203

Garro LS, Shimbori EM, Penteado-Dias AM, Shaw SR (2017) Four new species of *Aleiodes* (Hymenoptera: Braconidae: Rogadinae) from the Neotropical Region. Can Entomol 149:560–573

Gates MW, Heraty JM, Schauff ME, Wagner DL, Whitfield JB, Wahl DB (2002) Survey of parasitic Hymenoptera on leaf miner in California. J Hymenopt Res 11:213–270

Gauld ID (1991) The Ichneumonidae of Costa Rica, 1. Introduction, keys to subfamilies, and keys to the species of the lower Pimpliform subfamilies Rhyssinae, Poemeniinae, Acaenitinae and Cylloceriinae. Mem Am Entomol Inst 47:1–589

Gauld I, Bolton B (1988) The Hymenoptera. Oxford University Press, British Museum of Natural History
Gauld ID, Fitton MG (1984) An introduction to the Ichneumonidae of Australia. British Museum of Natural History, London
Gauld ID, Gaston KJ, Janzen DH (1992) Plant allelochemicals, tritrophic interactions and the anomalous diversity of tropical parasitoids: the "nasty" host hypothesis. Oikos 65:353–357
Gauld ID, Sithole R, Gomez JU, Godoy C (2002) The Ichneumonidae of Costa Rica, 4. Mem Am Entomol Inst 66:1–768
Gentry G, Dyer L (2002) On the conditional nature of Neotropical caterpillar defenses against their natural enemies. Ecology 83:3108–3119
Ghahari H, Gibson GAP, Viggiani G (2021) Chalcidoidea of Iran (Insecta: Hymenoptera). CAB International, London. 480 pp
Gibson GA, Fusu L (2016) Revision of the Palaearctic species of *Eupelmus* (*Eupelmus*) Dalman (Hymenoptera: Chalcidoidea: Eupelmidae). *Zootaxa* 4081:1–331
Gibson GAP, Huber JT, Woolley JB (eds) (1997) Annotated Keys to the Genera of Nearctic Chalcidoidea (Hymenoptera). NRC Research Press, Ottawa
Godfray HCJ (1994) Parasitoids: behavioral and evolutionary ecology. Princeton University Press, Princeton
Godfray HCJ, Blacquiere T, Field LM, Hails RS, Potts SG, Raine NE, Vanbergen AJ, McLean AR (2015) A restatement of recent advances in the natural science evidence base concerning neonicotinoid insecticides and insect pollinators. Proc R Soc Lond B Biol Sci 282:20151821
González E, Landis DA, Knapp M, Valladares G (2020) Forest cover and proximity decrease herbivory and increase crop yield via enhanced natural enemies in soybean fields. J Appl Ecol 57:2296–2306
Gordh G, Legner LE, Caltagirone LE (1999) Chapter 15, biology of parasitic Hymenoptera. In: Bellows TS, Fisher TW (eds) Handbook of biological control. Academic, pp 355–381
Gravel D, Massol F, Canard E, Mouillot D, Mouquet N (2011) Trophic theory of island biogeography. Ecol Lett 14:1010–1016
Greathead DJ, Greathead AH (1992) Biological control of insect pests by insect parasitoids and predators: the BIOCAT database. Biocontrol News Info 13:61N–68N
Greeney HF, Stireman JO III (2002) Emergence of parasitic flies from adult *Actinote diceus* (Nymphalidae: Acraeinae) in Ecuador. J Lep Soc 55:79–80
Greeney HF, Dyer LA, Smilanich AM (2012) Feeding by lepidopteran larvae is dangerous: a review of caterpillars' chemical, physiological, morphological, and behavioral defenses against natural enemies. Invertebr Surviv J 9:7–34
Grimaldi D, Engel MS (2005) The evolution of the insects. Cambridge University Press, Cambridge
Gripenberg S, Hamer NIA, Brereton TOM, Roy DB, Lewis OT (2011) A novel parasitoid and a declining butterfly: cause or coincidence? Ecol Entomol 36:271–281
Grissell EE, Schauff ME (1997) Chapter 3. Superfamily Chalcidoidea. In: Gibson GA, Huber JT, Wooley JB (eds) Annotated keys to the genera of Nearctic Chalcidoidae. NRC Research Press, Ottawa, pp 45–116
Gross P (1993) Insect behavioral and morphological defenses against parasitoids. Annu Rev Entomol 38:251–273
Guerrieri E, Noyes J (2005) Revision of the European species of *Copidosoma* Ratzeburg (Hymenoptera: Encyrtidae), parasitoids of caterpillars (Lepidoptera). Syst Entomol 30:97–174
Haddad NM, Brudvig LA, Clobert J et al (2015) Habitat fragmentation and its lasting impact on Earth's ecosystems. Sci Adv 1:e1500052
Hallmann CA, Sorg M, Jongejans E et al (2017) More than 75 percent decline over 27 years in total flying insect biomass in protected areas. PLoS One 12:e0185809
Hance T, van Baaren J, Vernon P, Boivin G (2007) Impact of extreme temperatures on parasitoids in a climate change perspective. Annu Rev Entomol 52:107–126
Hansen AC, Glassmire AE, Dyer LA, Smilanich AM (2017) Patterns in parasitism frequency explained by diet and immunity. *Ecography* 40:803–805

Hanson PE, Gauld ID (1995) The Hymenoptera of Costa Rica. Oxford University Press, Oxford

Hansson C (2002) Eulophidae of Costa Rica (Hymenoptera, Chalcidoidea). Mem Am Entomol Inst 67:1–290

Haran J, Delvare G, Vayssieres JF, Benoit L, Cruaud P, Rasplus J-Y, Cruaud A (2018) Increasing the utility of barcode databases through high-throughput sequencing of amplicons from dried museum specimens, an example on parasitic Hymenoptera (Braconidae). Biol Control 122:93–100

Hardy NB, Otto SP (2014) Specialization and generalization in the diversification of phytophagous insects: tests of the musical chairs and oscillation hypotheses. Proc R Soc B 281:20132960

Hassell MP (2000) Host-parasitoid population dynamics. J Anim Ecol 69:543–566

Hawkins BA (1994) Patterns and process in host–parasitoid interactions. Cambridge University Press, Cambridge

Hawkins BA, Sheehan W (eds) (1994) Parasitoid community ecology. Oxford University Press, Oxford

Hawkins BA, Cornell HV, Hochberg ME (1997) Predators, parasitoids, and pathogens as mortality agents in phytophagous insect populations. Ecology 78:2145–2152

Heinrich G (1977) Ichneumoninae of Florida and neighboring states (Hymenoptera: Ichneumonidae, subfamily Ichneumoninae). Vol. 9 in Arthropods of Florida and neighboring land areas, Florida Department of Agriculture & Consumer Services, pp 1–350

Henneman ML, Memmott J (2001) Infiltration of a Hawaiian community by introduced biological control agents. Science 293:1314–1316

Heraty JM, Darling DC (1984) Comparative morphology of the planidial larvae of Eucharitidae and Perilampidae (Hymenoptera: Chalcidoidea). Syst Entomol 9(3):309–328

Heraty JM, Burks RA, Cruaud A, Gibson GA, Liljeblad P, Munro J (2013) A phylogenetic analysis of the megadiverse Chalcidoidea (Hymenoptera). Cladistics 29:466–542

Herting B (1960) Biologie der westpalarktischen Raupenfliegen. Dipt., Tachinidae. Monogr Angew Entomol 16, 188 pp

Hinz R, Short J (1983) Life history and systematic position of the European *Alomya* species. Entomol Scand 14:462–466

Holt RD (1977) Predation, apparent competition, and the structure of prey communities. Theor Popul Biol 12:197–229

Holt RD, Lawton JH, Polis GA, Martinez ND (1999) Trophic rank and the species-area relationship. Ecology 80:1495–1504

Hrcek J, Godfray HCJ (2015) What do molecular methods bring to host–parasitoid food webs? Trends Parasitol 31:30–35

Hrcek J, Miller SE, Quicke DLJ, Smith MA (2011) Molecular detection of trophic links in a complex insect host–parasitoid food web. Mol Ecol Resour 11:786–794

Hrcek J, Miller SE, Whitfield JB, Shima H, Novotny V (2013) Parasitism rate, parasitoid community composition and host specificity on exposed and semi-concealed caterpillars from a tropical rainforest. Oecologia 173:521–532

Huber JT (1986) Systematics, biology, and hosts of the Mymaridae and Mymarommatidae (Insecta: Hymenoptera). Entomography 4:185–243

Ichiki R, Shima H (2003) Immature Life of *Compsilura concinnata* (Meigen) (Diptera: Tachinidae). Ann Entomol Soc Am 96:161–167

Ichiki R, Ho G, Wajnberg E, Kainoh Y, Tabata J, Nakamura S (2012) Different uses of plant semiochemicals in host location strategies of the two tachinid parasitoids. Naturwissenschaften 99:687–694

Inclán DJ, Cerretti P, Marini L (2014) Interactive effects of area and connectivity on the diversity of tachinid parasitoids in highly fragmented landscapes. Landsc Ecol 29:879–889

Inclán DJ, Cerretti P, Gabriel D, Benton TG, Sait SM, Kunin WE, Gillespie MAK, Marini L (2015) Organic farming enhances parasitoid diversity at the local and landscape scales. J Appl Ecol 52:1102–1109

Jackson DJ (1961) Observations on the biology of *Caraphractus cinctus* Walker (Hymenoptera: Mymaridae), a parasitoid of the eggs of Dytsicidae (Coleoptera). 2. Immature stages and seasonal history with a review of mymarid larvae. Parasitology 51:269–294

Janz N, Nylin S (2008) The oscillation hypothesis of host-plant range and speciation. In: Tilmon K (ed) Specialization, speciation, and radiation: the evolutionary biology of herbivorous insects. University of California Press, Berkeley, pp 203–215

Janz N, Nylin S, Wahlberg N (2006) Diversity begets diversity: host expansions and the diversification of plant-feeding insects. BMC Evol Biol 6:4

Janzen DH (1981) The peak in North American ichneumonid species richness lies between 38-degrees and 42-degrees-N. Ecology 62:532–537

Janzen DH (1995) The caterpillars and their parasitoids of a tropical dry forest. Tachinid Times 8:1–3

Janzen DH, Hallwachs W (2009) Dynamic database for an inventory of the macrocaterpillar fauna, and its food plants and parasitoids, of Area de Conservacion Guanacaste (ACG), northwestern Costa Rica. http://janzen.sas.upenn.edu. Accessed Dec 2020

Jeffries MJ, Lawton JH (1984) Enemy-free space and the structure of ecological communities. Biol J Linn Soc 23:269–286

Jeffs CT, Lewis OT (2013) Effects of climate warming on host–parasitoid interactions. Ecol Entomol 38:209–218

Jones GZ, Shaw SR (2012) Ten new species of *Meteorus* (Braconidae: Hymenoptera) from Ecuador reared at the Yanayacu Biological Center for Creative Studies. Zootaxa 3547:1–23

Jonsson M, Buckley HL, Case BS, Wratten SD, Hale RJ, Didham RK (2012) Agricultural intensification drives landscape-context effects on host–parasitoid interactions in agroecosystems. J Appl Ecol 49:706–714

Kageyama A, Sugiura S (2016) Caterpillar hairs as an anti-parasitoid defence. Sci Nat 103:86

Kainoh Y, Tamaki Y (1982) Searching behavior and oviposition of the egg-larval parasitoid, *Ascogaster reticulatus* Watanabe (Hymenoptera: Braconidae). Appl Entomol Zool 17:194–206

Kankaanpää T, Vesterinen E, Hardwick B et al (2020) Parasitoids indicate major climate-induced shifts in arctic communities. Glob Chang Biol 00:1–20. https://doi.org/10.1111/gcb.15297

Kautz AR, Gardiner MM (2019) Agricultural intensification may create an attractive sink for Dolichopodidae, a ubiquitous but understudied predatory fly family. J Insect Conserv 23:453–465

Keeler MS, Chew FS, Goodale BC, Reed JM (2006) Modelling the impacts of two exotic invasive species on a native butterfly: top-down vs. bottom-up effects. J Anim Ecol 75:777–788

Kitson JJM, Hahn C, Sands RJ, Straw NA, Evans DM, Lunt DH (2019) Detecting host-parasitoid interactions in an invasive Lepidopteran using nested tagging DNA-metabarcoding. Mol Ecol 28:471–483

Kittel RN, Austin AD, Klopfstein SA (2016) Molecular and morphological phylogenetics of chelonine parasitoid wasps (Hymenoptera: Braconidae), with a critical assessment of divergence time estimations. Mol Phylogenet Evol 101:224–241

Koptur S (1985) Alternative defenses against herbivores in *Inga* (Fabaceae: Mimosoideae) over an elevational gradient. Ecology 66:1639–1650

Krauth SJ, Williams AH (2006) Notes on *Taeniogonalos gundlachii* (Hymenoptera: Trigonalidae) from Wisconsin. Great Lakes Entomol 39:54–58

Kula RR, Dix-Luna O, Shaw SR (2012) Review of *Ilatha* Fischer (Hymenoptera: Braconidae: Alysiinae) including descriptions of six new species and a key to species. Proc Entomol Soc Wash 114:293–328

Landis DA, Menalled FD (1998) Ecological considerations in the conservation of effective parasitoid communities in agricultural systems. In: Barbosa P (ed) Conservation biological control. Academic, New York, pp 101–121

Landis DA, Wratten SD, Gurr GM (2000) Habitat management to conserve natural enemies of arthropod pests in agriculture. Annu Rev Entomol 45:175–201

LaSalle J, Gauld ID (1991) Parasitic Hymenoptera and the biodiversity crisis. Redia 74:315–334

Le Corff J, Marquis RJ, Whitfield JB (2000) Temporal and spatial variation in a parasitoid community associated with the herbivores that feed on Missouri *Quercus*. Environ Entomol 29:181–194

Letourneau DK, Allen SGB, Stireman JO III (2012) Perennial habitat fragments, parasitoid diversity and parasitism in ephemeral crops. J Appl Ecol 49:1405–1416

Letourneau DK, Kula RR, Sharkey MJ, Stireman JO III (2015) Habitat eradication and cropland intensification may reduce parasitoid diversity and natural pest control services in annual crop fields. Elementa 3:000069

Libra M, Tulai S, Novotny V, Hrcek J (2019) Elevational contrast in predation and parasitism risk to caterpillars in a tropical rainforest. Entomol Exp Appl 167:922–931

Lill JT, Marquis RJ, Ricklefs RE (2002) Host plants influence parasitism of forest caterpillars. Nature 417:170–173

López R, Ferro DN, Van Driesche RG (1995) Two tachinid species discriminate between parasitized and non-parasitized hosts. Ent Exp Appl 74:37–45

López-Núñez FA, Heleno RH, Ribeiro S, Marchante H, Marchante E (2017) Four-trophic level food webs reveal the cascading impacts of an invasive plant targeted for biocontrol. Ecology 98:782–793

LoPresti EF, Morse DH (2013) Costly leaf shelters protect moth pupae from parasitoids. Arthro-Plant Interact 7:445–453

Malo F (1961) Phoresy in *Xenufens* (Hymenoptera: Trichogrammatidae). A parasite of *Caligo eurylochus* (Lepidoptera: Nymphalidae). J Econ Entomol 54:465–466

Marquis RJ, Lill JT, Forkner RE, Le Corff J, Landoski J, Whitfield JB (2019) Declines and resilience of leaf chewing insects on Missouri oaks following spring frost and summer drought. Front Ecol Evol 7:396

Martinson HM, Fagan WF (2014) Trophic disruption: a meta-analysis of how habitat fragmentation affects resource consumption in terrestrial arthropod systems. Ecol Lett 17:1178–1189

Martorana L, Foti MC, Rondoni G, Conti E, Colazza S, Peri E (2017) An invasive insect herbivore disrupts plant volatile-mediated tritrophic signalling. J Pest Sci 90:1079–1085

Matthews TJ, Cottee-Jones HE, Whittaker RJ (2014) Habitat fragmentation and the species–area relationship: a focus on total species richness obscures the impact of habitat loss on habitat specialists. Divers Distrib 20:1136–1146

McCabe TL (1998) Dipterous parasitoids from adults of moths (Lepidoptera). Entomol News 109:325–328

Michalková V, Valigurová A, Dindo ML, Vanhara J (2009) Larval morphology and anatomy of the parasitoid *Exorista larvarum* (Diptera: Tachinidae), with an emphasis on cephalopharyngeal skeleton and digestive tract. J Parasitol 95:544–554

Mondor EB, Roland J (1998) Host searching and oviposition by *Leschenaultia exul*, a tachinid parasitoid of the forest tent caterpillar, *Malacosoma disstria*. J Insect Behav 11:583–592

Montgomery GA, Dunn RR, Fox R, Jongejans E, Leather SR, Saunders ME, Shortall CR, Tingley MW, Wagner DL (2020) Is the insect apocalypse upon us? How to find out. Biol Conserv 108327

Montllor CB, Bernays EA (1993) Invertebrate predators and caterpillar foraging. In: Stamp NE, Casey TM (eds) Caterpillars: ecological and evolutionary constraints on foraging. Chapman & Hall, New York, pp 170–202

Montoya JM, Pimm SL, Solé RV (2006) Ecological networks and their fragility. Nature 442:259–264

Munro VMV, Henderson IM (2002) Nontarget effect of entomophagous biocontrol: shared parasitism between native lepidopteran parasitoids and the biocontrol agent *Trigonospila brevifacies* (Diptera: Tachinidae) in foresthabitats. Environ Entomol 31:388–396

Murphy SM, Lill JT, Smith DR (2009) A scattershot approach to host location: uncovering the unique life history of the trigonalid hyperparasitoid *Orthogonalys pulchella* (Cresson). Am Entomol 55:82–87

Murphy SM, Lill JT, Bowers MD, Singer MS (2014) Enemy free space for parasitoids. Environ Entomol 43:1465–1474

Myers JH, Cory JS (2017) Biological control agents: invasive species or valuable solutions? In: Vilà M, Hulme P (eds) Impact of biological invasions on ecosystem services, Invading nature – Springer series in invasion ecology, vol 12. Springer, Cham, pp 191–202

Nagarkatti S, Nagaraja H (1977) Biosystematics of *Trichogramma* and Trichogrammatoidea species. Annu Rev Entomol 22:157–176

Narendran TC, Amareswara Rao S (1987) Biosystematics of Chalcididae (Chalcidoidea: Hymenoptera). Proc Ind Acad Sci (Anim Sci) 96:543–550

Nénon J-P, Boivin G, Le Lannic J, van Baaren J (2011) Functional morphology of the mymariform and sacciform larvae of the egg parasitoid *Anaphes victus* Huber (Hymenoptera: Mymaridae). Can J Zool 73:996–1000

Novotny V, Drozd P, Miller SE, Kulfan M, Janda M et al (2006) Why are there so many species of herbivorous insects in tropical rainforests? Science 313:1115–1118

Noyes JS (2021) Universal Chalcidoidea database. Available at: http://www.nhm.ac.uk/chalcidoids

Noyes JS, Valentine EW (1989) Mymaridae (Insecta: Hymenoptera) – introduction and review of genera. Fauna NZ 17:1–95

O'Hara JE (1985) Oviposition strategies of the Tachinidae, a family of beneficial parasitic flies. Univ Alberta Agric For Bull 8:31–34

O'Hara JE, Henderson SJ, Wood DM (2020) Preliminary checklist of the Tachinidae of the world. Version 2.0. PDF document, 1039 pages. Available at: http://www.nadsdiptera.org/Tach/WorldTachs/Checklist/Worldchecklist.html. Accessed 4 Sept 2020

Ode PJ (2006) Plant chemistry and natural enemy fitness: effects on herbivore and natural enemy interactions. Annu Rev Entomol 51:163–185

Parker HL (1931) *Macrocentrus gifuensis* Ashmead, a polyembryonic braconid parasite in the European corn borer. Tech Bull USDA 230:1–32

Parry D (2009) Beyond Pandora's box: quantitatively evaluating non-target effects of parasitoids in classical biological control. Biol Invasions 11:47–58

Pennacchio F, Strand MR (2006) Evolution of developmental strategies in parasitic Hymenoptera. Annu Rev Entomol 51:233–258

Pereira HM, Leadley PW, Proença V et al (2010) Scenarios for global biodiversity in the 21st century. Science 330:1496–1501

Pereira FF, Kassab SO, Ferreira Calado VR, Vargas EL, de Oliveira HN, Zanuncio JC (2015) Parasitism and Emergence of *Tetrastichus howardi* (Hymenoptera: Eulophidae) on *Diatraea saccharalis* (Lepidoptera: Crambidae) Larvae, Pupae and Adults. Fla Entomol 98:377–380

Pérez-Benavides AL, Fernando Hernández-Baz F, González JM, Romero-Nápoles J, Hanson PE, Zaldívar-Riverón A (2020) Integrative taxonomy to assess the species richness of chalcidoid parasitoids (Hymenoptera) associated to Bruchinae (Coleoptera: Chrysomelidae) from Mexico. Rev Mex Biodivers 91:e913492

Peters RS, Baur H (2011) A revision of the *Dibrachys cavus* species complex (Hymenoptera: Chalcidoidea: Pteromalidae). Zootaxa 2937:1–30

Peters RS, Krogmann L, Mayer C et al (2017) Evolutionary history of the Hymenoptera. Curr Biol 27:1013–1018

Petrice TR, Strazanac JS, Butler L (2004) A survey of hymenopteran parasitoids of forest Macrolepidoptera in the central Appalachians. J Econ Entomol 97:451–459

Poinar GO (1979) Nematodes for biological control of insects. CRC Press, Boca Raton

Polilov AA (2012) The smallest insects evolve anucleate neurons. Arthrop Struct Dev:4129–4134

Purrington FF (1979) Biology of the hyperparasitic wasp *Perilampus similis* (Hymenoptera: Perilampidae). Great Lakes Entomol 12:63–66

Quicke DLJ (1997) Parasitic wasps. Chapman and Hall/Springer, Dordrecht

Quicke DLJ (2015) The braconid and ichneumonid parasitoid wasps: biology, systematics, evolution and ecology. Wiley-Blackwell, London

Quicke DLJ, Austin AD, Fagan-Jeffries EP, Hebert PDN, Butcher BA (2020) Recognition of the Trachypetidae stat.n. as a new extant family of Ichneumonoidea (Hymenoptera), based on molecular and morphological evidence. Syst Entomol 45:771–782

Raff JW (1934) Observations on sawflies of the genus *Perga*, with notes on some reared primary parasites of the families Trigonalidae, Ichneumonidae, and Tachinidae. Trans Proc Roy Soc Vict 47:54–77

Remmel T, Davison J, Tammaru T (2011) Quantifying predation on folivorous insect larvae: the perspective of life-history evolution. Biol J Linn Soc 104:1–18

Rodriguez JJ, Fernandez Triana J, Whitfield JB, Smith MA, Erwin TL (2013) Extrapolations from field studies and known faunas converge on much higher estimates of world microgastrine parasitoid wasp species richness. Ins Conserv Divers 6:530–536

Roland J, Taylor PD (1997) Insect parasitoid species respond to forest structure at different spatial scales. *Nature* 386:710–713

Roslin T, Hardwick B, Novotny V et al (2017) Higher predation risk for insect prey at low latitudes and elevations. Science 356:742–744

Rougerie R, Smith MA, Fernandez-Triana J, Lopez-Vaamonde C, Ratnasingham S, Hebert PD (2011) Molecular analysis of parasitoid linkages (MAPL): gut contents of adult parasitoid wasps reveal larval host. Mol Ecol 20:179–186

Ryan RB (1990) Evaluation of biological control: introduced parasites of larch casebearer (Lepidoptera: Coleophoridae) in Oregon. Environ Entomol 19:1873–1881

Ryan RB (1997) Before and after evaluation of biological control of the larch casebearer (Lepidoptera: Coleophoridae) in the Blue Mountains of Oregon and Washington, 1972–1995. Environ Entomol 26:703–715

Ryan RB, Tunnock S, Ebel FW (1987) The larch casebearer in North America. J For 85:33–39

Sabrosky CW, Reardon RC (1976) Tachinid parasites of the gypsy moth, *Lymantria dispar*, with keys to adults and puparia. Misc Pubs Entomol Soc Am 10:1–126

Salcido DM, Forister ML, Garcia Lopez H, Dyer LA (2020) Loss of dominant caterpillar genera in a protected tropical forest. Sci Rep 10:422

Sanchez-Bayo F, Wyckhuys KAG (2019) Worldwide decline of the entomofauna: a review of its drivers. Biol Conserv 232:8–27

Scaffner JV, Griswold CL (1934) Macrolepidoptera and their parasites reared from field collections in the northeastern part of the United States. USDA Miscellaneous Publication, Washington

Schauff ME (1984) The Holarctic genera of Mymaridae (Hymenoptera: Chalcidoidea). Mem Entomol Soc Wash 12:1–67

Shaw MR (1983) On the evolution of endoparasitism: the biology of some genera of Rogadinae (Braconidae). In: Gupta VK (ed) Studies on the Hymenoptera: a collection of articles on Hymenoptera commemorating the 70th birthday of Henry K. Townes, vol 20. Contributions of the American Entomological Institute, pp 307–328

Shaw MR (1994) Parasitoid host ranges. In: Hawkins BA, Sheehan W (eds) Parasitoid community ecology. Oxford University Press, Oxford, pp 111–144

Shaw MR (2002) Host ranges of *Aleiodes* species and an evolutionary hypothesis. In: Melika G, Thuroczy C (eds) Parasitic wasps: evolution, systematics, biodiversity and biological control. Agroinform Kiado, Budapest, pp 321–327

Shaw SR (1995) In: Hanson P, Gauld ID (eds) Chapter 12.2, Braconidae. The Hymenoptera of Costa Rica. Oxford University Press, London, pp 431–463

Shaw SR (2004) Essay on the evolution of adult-parasitism in the subfamily Euphorinae (Hymenoptera: Braconidae). Proc Russ Entomol Soc St Petersburg 75:1–15

Shaw SR (2006) *Aleiodes* wasps of eastern forests: a guide to parasitoids and associated mummified caterpillars, Technology Transfer Series. Forest Health Technology Enterprise Team, Morgantown, 121 pp

Shaw SR (2014) Planet of the bugs. University of Chicago Press, Chicago

Shaw SR (2017) Subfamily Cheloninae. In: Wharton RA, Marsh PM, Sharkey MJ (eds) Identification manual of the New World Genera of the Family Braconidae (Hymenoptera), 2nd ed, vol 1. International Society of Hymenopterists Special Publication, Washington, DC, pp 198–207

Shaw MR, Huddleston T (1991) Classification and biology of braconid wasps. Handbooks for the identification of British insects, Volume 7, Part 11. Royal Entomological Society of London, London, 126 pp

Shaw SR, Nishida K (2005) A new species of gregarious *Meteorus* (Hymenoptera: Braconidae) reared from caterpillars of *Venadicodia caneti* (Lepidoptera: Limacodidae) in Costa Rica. Zootaxa 1028:49–60

Shaw SR, Marsh PM, Fortier JC (1997) Revision of North American *Aleiodes* (Part 1): the *pulchripes* Wesmael species-group in the New World (Hymenoptera: Braconidae, Rogadinae). J Hymenopt Res 6:10–35

Shaw SR, Shimbori EM, Penteado-Dias AM (2020) A revision of the *Aleiodes bakeri* (Brues) species subgroup of the *A. seriatus* species group with the descriptions of 18 new species from the Neotropical Region. Zookeys 964:41–107

Shimbori EM, Shaw SR (2014) Twenty-four new species of *Aleiodes* Wesmael from the eastern Andes of Ecuador with associated biological information (Hymenoptera, Braconidae, Rogadinae). Zookeys 405:1–81

Shimbori EM, Bortoni MA, Shaw SR, Soussa-Gessner C d S, Cerântola P d CM, Penteado-Diaz AM (2019) Revision of the new World genera *Adelius* Haliday and *Paradelius* DeSaeger (Hymenoptera: Braconidae: Cheloninae: Adeliini). Zootaxa 4571:151–200

Simmonds FJ (1947) The biology of *Phytodietus pulcherrimus* (Cress.) (Ichneumonidae, Tryphoninae), parasitic on *Loxostege sticticalis* L. in North America. Parasitology 38:150–156

Singer MS, Stireman JO III (2003) Does anti-parasitoid defense influence host-plant selection by a generalist caterpillar? Oikos 100:554–562

Singer MS, Rodrigues D, Stireman JO III, Carriere Y (2004) Comparing bi-trophic and tri-trophic causes of host use in a phytophagous insect. Ecology 85:2747–2753

Slinn HL, Richards LA, Dyer LA, Hurtado P, Smilanich AM (2018) Across multiple species, Phytochemical diversity and herbivore diet breadth have cascading effects on herbivore immunity and parasitism in a tropical model system. Front Plant Sci 9:656

Smilanich AM, Dyer LA, Gentry GL (2009) The insect immune response and other putative defenses as effective predictors of parasitism. Ecology 90:1434–1440

Smith KGV (1981) A tachinid (Diptera) larva in the abdomen of an adult moth (Geometridae). Entomol Gaz 32:174–176

Smith MA, Woodley NE, Janzen DH, Hallwachs W, Hebert PDN (2006) DNA barcodes reveal cryptic host-specificity within the presumed polyphagous members of a genus of parasitoid flies (Diptera: Tachinidae). Proc Natl Acad Sci U S A 103:3657–3662

Smith MA, Wood DM, Janzen DH, Hallwachs W, Hebert PDN (2007) DNA barcodes affirm that 16 species of apparently generalist tropical parasitoid flies (Diptera, Tachinidae) are not all generalists. Proc Natl Acad Sci U S A 104:4967–4972

Smith MA, Rodriguez JJ, Whitfield JB, Deans AR, Janzen DH, Hallwachs W, Hebert PDN (2008) Extreme diversity of tropical parasitoid wasps exposed by iterative integration of natural history, DNA barcoding, morphology, and collections. Proc Natl Acad Sci U S A 105:12359–12364

Smith DR, Janzen DH, Hallwachs W, Smith AM (2012) Hyperparasitoid wasps (Hymenoptera, Trigonalidae) reared from dry forest and rain forest caterpillars of Area de Conservación Guanacaste, Costa Rica. J Hymenopt Res 29:119–144

Sow A, Brévault T, Benoit L, Chapuis M-P, Galan M, Coeur D'acier A, Delvare G, Sembène M, Haran J (2019) Deciphering host-parasitoid interactions and parasitism rates of crop pests using DNA metabarcoding. Sci Rep 9:1–12

Stigenberg J, Ronquist F (2011) Revision of the Western Palearctic Meteorini (Hymenoptera, Braconidae), with a molecular characterization of hidden Fennoscandian species diversity. Zootaxa 3084:1–95

Stireman JO III (2005) The evolution of generalization? Parasitoid flies and the perils of inferring host range evolution from phylogenies. J Evol Biol 18:325–336

Stireman JO III (2016) Community ecology of the "other" parasitoids. Curr Opin Insect Sci 14:87–93

Stireman JO III, Singer MS (2003a) Determinants of parasitoid-host associations: insights from a natural tachinid-lepidopteran community. Ecology 84:296–310

Stireman JO III, Singer MS (2003b) What determines host range in parasitoids? An analysis of a tachinid parasitoid community. Oecologia 135:629–638

Stireman JO III, Singer MS (2018) Tritrophic niches of insect herbivores in an era of rapid environmental change. Curr Opin Insect Sci 29:117–125

Stireman JO III, Dyer LA, Janzen DH, Singer MS, Lill JT, Marquis RJ et al (2005) Climatic unpredictability and parasitism of caterpillars: implications of global warming. Proc Natl Acad Sci U S A 102:17384–17387

Stireman JO III, O'Hara JE, Wood DM (2006) Tachinidae: evolution, behavior, and ecology. Annu Rev Entomol 51:525–555

Stireman JO III, Greeney HF, Dyer LA (2009) Species richness and host associations of Lepidoptera-attacking Tachinidae in the northeast Ecuadorian Andes. J Insect Sci 9:39

Stireman JO III, Dyer LA, Greeney HF (2017) Specialized generalists? Food web structure of a tropical tachinid-caterpillar community. Insect Conserv Divers 10:367–384

Stireman JO III, Cerretti P, O'Hara JE, Moulton JK (2019) Molecular phylogeny and evolution of world Tachinidae. Mol Phylogenet Evol 139:106358

Strand MR, Obrycky JJ (1996) Host specificity of insect parasitoids and predators. Bioscience 46:422–429

Tilman D, Fargione J, Wolff B, D'Antonio C, Dobson A, Howarth R, Schindler D, Schlesinger WH, Simberloff D, Swackhamer D (2001) Forecasting agriculturally driven global environmental change. Science 292:281–284

Timms LL, Walker SC, Smith SM (2012) Establishment and dominance of an introduced herbivore has limited impact on native host-parasitoid food webs. Biol Invasions 14:229–244

Timms LL, Bennett AMR, Buddle CM, Wheeler TA (2013) Assessing five decades of change in a high Arctic parasitoid community. Ecography 36:1227–1235

Timms LL, Schwarzfeld M, Sääksjärvi IE (2016) Extending understanding of latitudinal patterns in parasitoid wasp diversity. Insect Conserv Divers 9:74–86

Townes HK (1969) Genera of Ichneumonidae, Part 3 (Lycorininae, Banchinae, Scolobatinae, Porizontinae). Mem Am Entomol Inst 13:1–307

Townes HK, Townes M (1978) Ichneumon flies of America north of Mexico: 7. Subfamily Banchinae, tribes Lissonotini and Banchini. Mem Am Entomol Inst 26:1–614

Tscharntke T, Klein A-M, Kruess A, Steffan-Dewenter I, Thies C (2005) Landscape perspectives on agricultural intensification and biodiversity – ecosystem service management. Ecol Lett 8:857–874

Tschorsnig H-P (2017) Preliminary host catalogue of Palaearctic Tachinidae (Diptera). http://www.nadsdiptera.org/Tach/WorldTachs/CatPalHosts/Home. Accessed Sept 2020

Turlings TCJ, Tumlinson JH, Lewis WJ (1990) Exploitation of herbivore-induced plant odors by host seeking parasitic wasps. Science 250:1251–1253

Tylianakis JM, Binzer A (2014) Effects of global environmental changes on parasitoid–host food webs and biological control. Biol Control 75:77–86

Tylianakis J, Laliberté E, Nielsen A, Bascompte J (2010) Conservation of species interaction networks. Biol Conserv 143:2270–2279

Valigurová A, Michalková V, Koník P, Dindo M, Gelnar M, Vaňhara J (2014) Penetration and encapsulation of the larval endoparasitoid *Exorista larvarum* (Diptera: Tachinidae) in the factitious host *Galleria mellonella* (Lepidoptera: Pyralidae). Bull Entomol Res 104:203–212

van Achterberg C (1979) A revision of the subfamily Zelinae auct. (Hymenoptera, Braconidae). Tijdschr Entomol 122:241–479

van Achterberg C (1983) Revisionary notes on the subfamily Gnamptodontinae, with description of eleven new species (Hymenoptera, Braconidae). Tijdschr Entomol 126:25–57

van Achterberg C, Shaw SR (2009) A new species of the genus *Homolobus* from Ecuador (Hymenoptera: Braconidae: Homolobinae). Zool Meded 83:805–810 van Achterberg C, Shaw MR, Quicke DLJ (2020) Revision of the western Palaearctic species of *Aleiodes* Wesmael (Hymenoptera, Braconidae, Rogadinae). Part 2: revision of the A. apicalis group. Zookeys 919:1–259

van Driesche RG (2008) Biological control of *Pieris rapae* in New England: Host suppression and displacement of *Cotesia glomerata* by *Cotesia rubecula* (Hymenoptera: Braconidae). Fla Entomol 91:22–25

van Driesche RG, Nunn C, Pasqual A (2004) Life history pattern, host plants, and habitat determinants of population survival of *Pieris napi oleracea* interacting with an introduced braconid parasitoid. Biol Control 29:278–287

van Nouhuys S (2005) Effects of habitat fragmentation at different trophic levels in insect communities. Ann Zool Fenn 42:433–447

Vance AM (1932) The biology and morphology of the braconid *Chelonus annulipes* Wesmael, a parasite of the European corn borer. Tech Bull USDA 233:1–28

Várkonyi G, Roslin T (2013) Freezing cold yet diverse: dissecting a high-arctic parasitoid community associated with Lepidoptera hosts. Can Entomol 145:193–218

Vet LEM, Dicke M (1992) Ecology of infochemical use by natural enemies in a tritrophic context. Annu Rev Entomol 47:141–172

Visser ME, Gienapp P (2019) Evolutionary and demographic consequences of phenological mismatches. Nat Ecol Evol 12:879–885

Waage J, Greathead D (1986) Insect parasitoids. In: 13th symposium of the Royal Entomological Society of London, Academic, London

Waage J, Hassell M (1982) Parasitoids as biological control agents – a fundamental approach. Parasitology 84:241–268

Wagner DL (2005) Caterpillars of Eastern North America. Princeton University Press, Princeton. 512 pp

Wagner DL, Grames EM, Forister ML, Berenbaum MR, Stopak D (2021) Insect decline in the Anthropocene: death by a thousand cuts. Proc Natl Acad Sci 118:e2023989118

Ward SF, Aukema BH, Fei S, Liebhold AM (2020) Warm temperatures increase population growth of a nonnative defoliator and inhibit demographic responses by parasitoids. Ecology 101:e03156

Weinstein P, Austin AD (1991) The host relationships of trigonalid wasps (Hymenoptera: Trigonalyidae), with a review of their biology and catalogue to world species. J Nat Hist 25:399–433

Whitfield JB (1988) Two new species of *Paradelius* (Hymenoptera: Braconidae) of North America with biological notes. Pan-Pac Entomol 64:313–319

Whitfield JB, Wagner DL (1991) Annotated key to the genera of Braconidae (Hymenoptera) attacking leafmining Lepidoptera in the holarctic region. J Nat Hist 25:733–754

Wilson EO (2005) Kin selection as the key to altruism: its rise and fall. Soc Res 72:159–166

Wilson JK, Ruiz L, Davidowitz G (2020) Within-host competition drives energy allocation trade-offs in an insect parasitoid. PeerJ 8:e8810

Wirta HK, Hebert PDN, Kaartinen R, Prosser SW, Várkonyi G, Roslin T (2014) Complementary molecular information changes our perception of food web structure. Proc Natl Acad Sci U S A 111:1885–1890

Yamashita K, Zhang K, Ichiki R, Nakamura S, Furukawa S (2019) Novel host immune evasion strategy of the endoparasitoid *Drino inconspicuoides*. Bull Entomol Res 109:643–648

Yeates DK, Greathead D (1997) The evolutionary pattern of host use in the Bombyliidae (Diptera): a diverse family of parasitoid flies. Biol Linn Soc 60:149–185

Yoshimoto CM (1990) A review of the genera of New World Mymaridae (Hymenoptera: Chalcidoidea). In: Flora and fauna handbook, vol 7. Sandhill Crane Press Inc, Gainsville, pp 1–166

Yu DS (2014) Taxapad. Home of Ichneumonoidea

Yu DS, van Achterberg K, Horstmann K (2016) World Ichneumonoidea 2011. Taxonomy, Biology, Morphology and Distribution. Taxapad.com. Canada

Zaldívar-Riverón A, Shaw MR, Sáez AG, Mori M, Belokoblylskij SA, Shaw SR, Quicke DLJ (2008) Evolution of the parasitic wasp subfamily Rogadinae (Braconidae): phylogeny and evolution of lepidopteran host ranges and mummy characteristics. BMC Evol Biol 8:1–20

Zitani NM, Shaw SR (2002) From meteors to death stars: variations on a silk thread. Am Entomol 48:228–225

Zitani NM, Shaw SR, Janzen DH (1998) Systematics of Costa Rican *Meteorus* (Hymenoptera: Braconidae: Meteorinae) species lacking a dorsope. J Hymenopt Res 7:182–208

Zong S-X, Sheng M-L, Luo Y-Q, Lu C-K (2012) *Lissonota holcocerica* Sheng sp.n (Hymenoptera: Ichneumonidae) parasitizing *Holcocerus hippophaecolus* (Lepidoptera: Cossidae) from China. J Insect Sci 112:1–7

Predators and Caterpillar Diet Breadth: Appraising the Enemy-Free Space Hypothesis

Michael S. Singer, Riley M. Anderson, Andrew B. Hennessy, Emily Leggat, Aditi Prasad, Sydnie Rathe, Benjamin Silverstone, and Tyler J. Wyatt

Polistes comanchus (Vespidae) wasp chewing its unidentified caterpillar victim to separate the body from the gut. (Photo: Michael S. **Singer**)

M. S. Singer (✉) · R. M. Anderson · A. B. Hennessy · E. Leggat · A. Prasad · S. Rathe
B. Silverstone · T. J. Wyatt
Department of Biology, Wesleyan University, Middletown, CT, USA
e-mail: msinger@wesleyan.edu

Introduction

Caterpillars have featured prominently in generating and testing theory about the evolutionary ecology of dietary specialization. For nearly a century, entomologists, ecologists, and evolutionary biologists have asked and tried to answer the question of why most insect herbivores have diets restricted to a small set of plant species, commonly from the same family or genus (Brues 1924). This taxonomic specificity of insect herbivore diets is what we mean by dietary specialization in this chapter. Although many explanations for dietary specialization by insect herbivores have been hypothesized (e.g., Hardy et al. 2020), the conventional answer attributes it to selection from plant defenses, especially plant secondary metabolites that are unique to particular plant taxa (e.g., Dethier 1954; Fraenkel 1959; Ehrlich and Raven 1964; Krieger et al. 1971; Thompson 1988; Jaenike 1990; Cornell and Hawkins 2003; Rothwell and Holeski 2020). Sometimes called the physiological efficiency hypothesis, the conventional explanation posits that dietary specialization enables enhanced growth, survival, and reproduction on specific host plants (Rothwell and Holeski 2020). This fitness benefit is thought to arise from specific counter-adaptations to specific plant defenses that generalist herbivores lack. If counter-adapted "offense" traits (Karban and Agrawal 2002), such as enzymatic detoxification, cause increased herbivore fitness on particular plant phenotypes (e.g., species or chemically similar species) at the expense of fitness on others, the resulting fitness trade-offs would favor the evolution of dietary specialist phenotypes of the herbivores (e.g., Futuyma and Moreno 1988; Rausher 1988; Joshi and Thompson 1995; Fry 1996, 2003; Agrawal et al. 2010; Gompert and Messina 2016).

The focus of this chapter is the challenge to this conventional explanation from research taking a tri-trophic, rather than bi-trophic perspective (Price et al. 1980; Strong 1988; Bernays and Graham 1988). As a lightning rod for the debate, Bernays and Graham (1988) made the most pointed critique. They argued that evidence did not strongly support the physiological efficiency hypothesis. Experiments found neither physiological efficiency advantages of dietary specialists (but see Rothwell and Holeski 2020) nor clear growth performance trade-offs of herbivores among diets of alternative host plants (Bernays and Graham 1988; Hardy et al. 2020). It was also known that some dietary specialist herbivores are capable of feeding on a wider range of host plants in the laboratory than the relatively restricted set they use in nature, suggesting that other ecological or behavioral factors are responsible for their restricted diets (e.g., Janzen 1985). In addition, Bernays and Graham (1988) pointed to evidence that plant secondary metabolites have more varied effects on herbivores than toxicity (e.g., deterrence without toxicity). This point questioned the prediction from plant-herbivore coevolution theory that plant chemical defenses select against dietary generalist herbivores lacking specialized counteradaptations to their toxicity (Cornell and Hawkins 2003; Rothwell and Holeski 2020). It is possible that chemical deterrence, for example, indicates a poor host plant for reasons other than its food quality to the herbivore.

The other reason proposed by Bernays and Graham (1988) and Bernays (1988, 1989) is the risk of mortality from dietary generalist predators, such as ants, vespid wasps, spiders, and birds, which might select for herbivores to specialize on plants offering them the best anti-predator defenses (see Plate 1). They reasoned that such generalist predators are major sources of mortality for insect herbivores, thus providing a possible source of strong selection. This argument applies especially well to caterpillars and other immature insects, as predation of immatures guarantees zero reproductive fitness for those individuals. Consequently, selection should be strong for traits that prevent predation during immature stages. Bernays and Graham (1988) posited that predation, not plant defense, imposes fitness trade-offs on herbivores, thus selecting for herbivore phenotypes that can both eat a specific plant *and* use the host plant in ways that reduce the risk of predation (i.e., acquiring enemy-free space, Jeffries and Lawton 1984). Selection from *generalist* predators is a key part of this enemy-free space hypothesis (Bernays 1989), although this point has sometimes been overlooked. While Bernays (1989) did not specify how general a predator's diet must be to select for narrow diet breadth in herbivores, the rationale was that predators with broad diets would selectively avoid well-defended prey while enemies with specialized diets, such as many parasitoids, would be adapted to overcome defenses of their specific prey or hosts.

Plate 1 How is dietary specialization linked to anti-predator defense? (**a**) *Polistes comanchus* (Vespidae) wasp chewing its unidentified caterpillar victim to separate the body from the gut. (**b**) *Phaeoura* sp. (probably *cristifera*) (Geometridae) exhibiting specific camouflage on its host plant. (**c**) *Cucullia lilicina* (Noctuidae) is a dietary specialist with warning coloration, possibly an advertisement of sequestered allelochemicals; this coloration might also function as camouflage from a distance. (All photos by Michael S. Singer)

As bold as the enemy-free space hypothesis was, it provoked more rhetoric than empirical testing, as our literature search will attest (see below). Importantly for this volume, much of the initial testing involved measuring generalist predator responses to caterpillar species of varying diet breadths (Bernays 1988, 1989; Bernays and Cornelius 1989; Dyer and Floyd 1993; Dyer 1995, 1997). These studies tested the prediction that dietary specialist herbivores should have better anti-predator defenses than their generalist counterparts. Although these studies provided valuable comparative perspectives on caterpillar diet breadth and predation risk in select geographic locations, few similar studies followed. The limited testing of the enemy-free space hypothesis is unfortunate because the role of anti-predator defenses as a driver of dietary specialization in insect herbivores has far-reaching implications for ecology and evolutionary biology (Singer and Stireman 2005; Futuyma and Agrawal 2009; Vidal and Murphy 2018a). In short, it implies that the broader tri-trophic community drives the origin and maintenance of dietary specialization in herbivorous insects, offering a unique perspective on the evolutionary and ecological diversification of this mega-diverse group (Singer and Stireman 2005, Wiens et al. 2015; Vidal and Murphy 2018a, Hardy et al. 2020). From a practical perspective, a more thorough understanding of the ecological niches (including diet) of herbivorous insects could also inform the conservation and management of biodiversity in this era of rapid environmental change (Stireman and Singer 2018).

To be clear, our focus is on enemy-free space as a possible explanation for dietary specialization of caterpillars and other herbivores. However, predators and parasites can also drive host shifts and host expansions by insect herbivores (Berdegue et al. 1996; Stamp 2001; Murphy 2004; Vidal and Murphy 2018b). Therefore, enemy-free space can explain patterns of host-plant use other than dietary specialization per se, i.e., through host shifts and expansions. This point is important because it shows the potentially broad explanatory power of the enemy-free space hypothesis compared to various alternatives, such as the physiological efficiency hypothesis.

Given the potential explanatory power of the enemy-free space hypothesis, it is ripe for a review and appraisal. One of our goals is to place Bernays' hypothesis into our contemporary understanding of ecology and evolution of plant-insect interactions, which generally accepts the importance of tri-trophic interactions (Abdala-Roberts et al. 2019). Some similar research efforts have already been made. For example, Mooney et al. (2012) formulated the tri-trophic interactions hypothesis by integrating the enemy-free space hypothesis with the physiological efficiency and the slow-growth/high-mortality hypotheses. The meta-analysis by Vidal and Murphy (2018a) compares the effect sizes of bottom-up and top-down forces on insect herbivores, confirming the contemporary view that plants and carnivores exert effects of similar magnitude on the performance of the herbivores. One of their analyses compares these effects between dietary specialist and generalist herbivores. Although they found no difference in top-down effect sizes between specialist and generalist herbivores, they did not specifically test the prediction that dietary specialization entails reduced mortality from *generalist* predators. Here we test this prediction with a quantitative analysis, and review the literature on the enemy-free space hypothesis as well as various mechanisms for it. Following this review of the evidence, we appraise and refine the hypothesis to stimulate further study.

Qualitative and Quantitative Review of the Evidence

We reviewed as much literature on the enemy-free space hypothesis as we could find (985 sources), beginning with the seminal paper by Bernays and Graham (1988). We reviewed all 903 English-language citations of this paper listed in the ISI Web of Science and/or Google Scholar as of 1 June 2020. We also scoured the bibliographies of the 281 most relevant papers from that search to arrive at the 985 sources. The most relevant papers were those that addressed the issue of insect herbivore diet breadth in relation to predation or parasitism (excluding pathogens) with more than a passing phrase and citation. To analyze the evidence for the enemy-free space hypothesis, we sought articles that measured predation or parasitism among insect herbivores of varied diet breadth. From the 37 papers that fit this criterion, we restricted our quantitative analysis to those (15) for which we could obtain effect sizes of the enemies on the herbivores.

Qualitative Evidence

Many studies show that predation depends on insect herbivore diet breadth, but others found no such effect. The pattern seems to depend in part on the type of predator (addressed by the quantitative analysis). For example, several studies report evidence that dietary specialist insect herbivores gained enemy-free space advantages over their generalist counterparts. The most compelling of these used experimental methods to compare many specialist and generalist insect herbivore species subjected to predation or predator foraging responses in field or greenhouse settings (e.g., Bernays 1988, 1989; Bernays and Cornelius 1989; Dyer and Floyd 1993; Dyer 1995, 1997; Singer et al. 2014; Bosc et al. 2018). This set of studies encompasses a variety of ecosystems, geographic regions, and predator taxa, suggesting that the anti-predator benefits of dietary specialization are quite general. However, studies with these criteria were not universally supportive of the prediction that generalist predators target generalist herbivores (e.g., Singer et al. 2019). Another set of studies used observational methods to compare mortality from parasitoids in specialist and generalist insect herbivores (mostly caterpillars) collected from their natural communities (e.g., Gentry and Dyer 2002; Stireman and Singer 2003; Anderson et al. 2019). These studies report no evidence for enemy-free space advantages of dietary specialist herbivores. Numerous other studies compared much smaller numbers of herbivore species, typically in laboratory experiments with arthropod predators or parasitoids, and the evidence from these studies is variable. Below, we delve into some of these studies to address the mechanistic basis of the enemy-free space hypothesis (see Mechanisms of enemy-free space advantages to specialists).

Quantitative Evidence

To test the main prediction proposed by Bernays and Graham (1988), we tested if dietary specialist herbivores gain a greater survival advantage over generalist herbivores against generalist enemies compared to specialist enemies. We tested this prediction by comparing effect sizes across studies, but the limited dataset precluded a formal meta-analysis (sensu Koricheva et al. 2013). To calculate effect sizes for each study, we followed the methods of Vidal and Murphy (2018a). In short, log response ratios (ln(mean treatment/mean control)) were used when mean counts or densities were reported. However, the majority of studies (13/15) reported the effects of predators as percent mortality or survival. Following Vidal and Murphy (2018a), we used the percent survival as the effect size in these cases. When studies compared multiple species of varying diet breadth and individually reported these values, we were able to extract an effect size for each comparison (e.g., Oppenheim and Gould 2002; Vencl et al. 2005). More commonly, studies reported survival or mortality of multiple specialist and generalist herbivore species without making direct comparisons. In these cases, we averaged the values for all specialists and all the generalists to generate two values. In some studies, values for all specialists and all generalists were presented as aggregate values, and we used these values in our calculations. In all studies, we used the authors' diet breadth classifications. We note that "specialist" and "generalist" are relative terms (Forister et al. 2012), and we make no attempt here to create an absolute scale of diet breadth. Following Rothwell and Holeski (2020), we made direct comparisons between dietary specialist and generalist herbivores *within* each study by using a single effect size calculated as the difference in effect size (LRR or % survival) between specialist and generalist insect herbivores. We refer to this metric as Δeffect size. Positive values of Δeffect size indicate that dietary specialists had greater survival relative to dietary generalists within a study. Negative values of Δeffect size indicate the opposite, and values at or near zero indicate that specialists and generalists had similar survival rate within a study. Most studies did not report standard errors or sample sizes; including only those that did would have drastically reduced the number of useable studies. Therefore, we were unable to use more sophisticated meta-analysis methods (e.g., Lajeunesse 2015). We consider our quantitative analysis a first step at evaluating the literature, but it is not definitive or exhaustive. To test the hypothesis that dietary specialization provides protection from predation, we modeled Δeffect size as a function of enemy type using a one-way ANOVA weighted by the number of herbivore species used in a study. We categorized natural enemy type as either predator or parasitoid because assigning diet breadth to the enemies was not straightforward. However, as parasitoids are typically more host-specific than opportunistic predators such as ants, vespid wasps, and birds, we categorized parasitoids as specialist predators in contrast to generalist predators. Additionally, to test the predicted but overlooked explanatory power of including enemy diet breadth, we compared the survival values of dietary specialist and generalist herbivores without accounting for enemy type using a Welch two-sample t-test. All analyses were performed in R, version 3.5.1 (R Core Team 2018).

We extracted 18 effect sizes from 15 studies. The majority of the studies used larval Lepidoptera as the insect herbivores (12/15), with only 3 other insect orders represented (Coleoptera 1/15, Diptera 1/15, and Hemiptera 1/15). Of the 18 effect sizes, 12 were greater than 0, meaning dietary specialists had greater survival than generalists in most studies. Of the studies in which dietary specialists had greater survival than generalists, 6 were from predator experiments, and 4 were from parasitoid experiments. The analysis shows that dietary specialist insect herbivores gained a significantly greater survival advantage over generalists when enemies were distinguished as predators or parasitoids in the model ($\beta = 0.387$, std.err. = 0.151, $P = 0.021$). Specifically, dietary specialist herbivores were more likely than generalist herbivores to escape natural enemy attack when those natural enemies were generalist predators as opposed to parasitoids. The results were similar when we restricted the data to Lepidoptera ($\beta = 0.406$, std.err. = 0.175, $P = 0.038$). Including fixed effects of herbivore order and/or experiment type did not improve the models. As reported by Vidal and Murphy (2018a), specialist and generalist herbivores did not differ in their levels of enemy-free space when enemy type was not accounted for ($t = 0.689$, $P = 0.496$). This result demonstrates that including some measure of enemy diet breadth is necessary to detect general support for the enemy-free space hypothesis.

Mechanisms of Enemy-Free Space Advantages to Specialists

To review evidence for various mechanisms, we examined all the articles returned in our search that compared dietary specialist and generalist insect herbivores in terms of allelochemical sequestration, specific camouflage, efficiency of feeding decisions, and suppression of herbivore-induced plant volatiles. With the exception of chemical sequestration, for which a recent meta-analysis has already been conducted (Zvereva and Kozlov 2016), the relatively small number of cases in each of these categories of mechanisms precluded quantitative analysis. We chose the first two categories because they were most extensively discussed by Bernays and Graham (1988) and Bernays (1988, 1989). We investigated papers on the efficiency of feeding decisions because of its emphasis in follow-up work by Bernays and colleagues (reviewed in Bernays 2001). Lastly, we looked at suppression of herbivore-induced plant volatiles due to its more recent attention in the field of tri-trophic interactions. Other possible mechanisms discussed by Bernays and Graham (1988), such as morphological adaptations to living on the plant surface, have received less attention. Our review of these mechanisms below is not exhaustive, but rather highlights those studies we found to be especially informative.

Sequestration of Plant Allelochemicals

Bernays and Graham (1988) suggested that a key advantage of dietary specialization is superior protection from generalist predators via sequestration of plant allelochemicals. There is now extensive evidence supporting this mechanism. Our literature search identified many examples and comparisons of generalist and specialist herbivores that utilize their host-plant's allelochemicals to produce or enhance their antipredator defenses (e.g., Dyer 1995; Camara 1997; Traugott and Stamp 1997; Leuthardt et al. 2013; Lampert et al. 2014; Katsanis et al. 2016; Zvereva and Kozlov 2016; Zvereva et al. 2018). Dyer (1995) was the first to rigorously examine the anti-predator traits underlying the anti-predator advantage of dietary specialists. He tested the palatability of 70 Costa Rican caterpillar species to the predatory ant *Paraponera clavata* to explore the effectiveness of various anti-predator traits. Prey chemistry, quantified by consumption of herbivore extracts by *P. clavata*, was the best predictor of rejection, and diet breadth was the second best. Specialist caterpillars were rejected more often, and, importantly, species within the same family that seemingly only differed based on diet breadth were better protected if they were specialists. These results argue that dietary specialization conferred anti-predator benefits to caterpillars mainly because specialists possess superior chemical defenses.

Dietary specialization may allow herbivores to strengthen their anti-predator defenses through sequestration either by enhancing existing chemical defenses (Jones et al. 1989; Engler-Chaouat and Gilbert 2007; Zvereva et al. 2017; Zvereva et al. 2018) or enabling them to evolve chemical defenses that generalists may be unable to utilize (Lampert et al. 2014; Kumar et al. 2014; Zvereva and Kozlov 2016; Katsanis et al. 2016). For example, sequestration is effective in bolstering existing defenses when utilized in place of or in addition to de novo synthesis of natural herbivore chemical defenses (Engler-Chaouat and Gilbert 2007; Zvereva et al. 2017, 2018). Engler-Chaouat and Gilbert (2007) compared the abilities of *Passiflora*-generalist and -specialist *Heliconius* butterflies to sequester cyanogens from *Passiflora*. While *Heliconius* butterflies naturally produce cyanogens as defensive compounds, specialists have traded some ability to synthesize cyanogens de novo for the ability to sequester them. Experiments showed that monophagous specialist butterflies in the *H. sara-sapho* clade sequestered cyanogens at a rate seven times higher than *Passiflora*-generalist species. Yet, when fed on *Passiflora* species from which specialists cannot sequester cyanogens, they contained significantly lower levels of cyanogens than generalists. This finding also raises an important question about the role of natural selection from predators on dietary specialization in *Heliconius* butterflies. Does the dependence of anti-predator defenses on specific host plants drive the evolution of dietary specialization?

Katsanis et al. (2016) show the anti-predator benefit of sequestration coupled with physiological specialization in a comparison of aphid species that differ in diet breadth but share the same host plant. First, laboratory experiments showed that the specialist species (*Brevicoryne brassicae*) had a 14% greater sequestration rate of glucosinolates than the generalist (*Myzus persicae*). Subsequently, they found that coccinellid predators were threefold more likely to attack the generalist aphids than the specialists. While both generalists and specialists are able to sequester glucosinolates, these chemicals require myrosinase to produce biologically active and toxic compounds. Given that these generalist aphids do not contain endogenous myrosinase like specialists do, they do not receive the full defensive benefits of the sequestered glucosinolates. Therefore, the tri-trophic effect of plant defenses on predators was stronger and more consistent when transferred through the specialist herbivore than the generalist. The strong differences in aphid survival in the presence of predators, combined with no difference in aphid performance in their absence, suggest that the primary benefit of restricting diet breadth is related more to predator deterrence than to plant-herbivore interactions alone.

Often, even low levels of allelochemical sequestration may cause predators to discriminate between generalist and specialist herbivores as predicted by the enemy-free space hypothesis. Choice assays conducted in a lab by Kumar et al. (2014) showed that the wolf spider (*Camptocosa parallela*) preferred generalist *Spodoptera exigua* over specialist *Manduca sexta* caterpillars when both were fed artificial diets containing nicotine. As nicotine is a common alkaloid in many of its natural host plants, *M. sexta* is less sensitive to nicotine than *S. exigua*, which must oxidize some nicotine into the less toxic cotinine. *Manduca sexta* exhales nicotine when consuming plant material, which deters predators, while incorporating only 0.6% of this ingested nicotine into its hemolymph. This small amount of sequestration increases the nicotine concentration of *M. sexta* by ca. 15 micrograms per ml hemolymph compared to *S. exigua*. But in combination with *M. sexta's* tolerance and exhalation of nicotine, this slight difference gives the specialist an advantage in deterring the wolf spider. Another example comes from greenhouse choice assays in which the extracts of larval cuticle and surface extracts from the caterpillar *Uresiphita reversalis* were highly deterrent to both wasps (*Mischocyttarus flavitarsis*) and ants (*Iridomyrmex humilis*). Only about 1% of the plant compounds ingested by *U. reversalis* need to be sequestered to reach a concentration in the cuticle that is deterrent to generalist predators (Montllor et al. 1991).

When low levels of sequestration effectively deter generalist predators, the provided protection can create further selection for increased sequestration and dietary specialization when interacting species coevolve or ecological circumstances change. Such a scenario might explain the escalation of cardenolide sequestration in the caterpillars of danaine milkweed butterflies (Petschenka and Agrawal 2015). Herbivores utilizing small amounts of plant defense chemicals are not toxic, but if they reach the threshold for deterrence, they may make generalist predators more likely to opt for other prey species. For example, specialization by individuals on certain host plants can increase an herbivore species' pre-existing ability to sequester secondary compounds and thus to effectively deter predators. Jones et al. (1989) treated the dietary generalist grasshopper *Romalea guttata* with three diets: (1) restricted diet of only wild onions, (2) a broader plant diet resembling the grasshopper's natural diet, and (3) artificial diet not including the secondary compounds *R. guttata* would naturally encounter. The wild-onion only diet resulted in significantly higher sequestration than both the natural and artificial alternatives. Furthermore, when ant predators of *R. guttata* were given a choice between grasshoppers from the three diet groups, they were most strongly deterred by the secretions of the wild-onion fed grasshoppers. This example of a dietary generalist species suggests the potential for evolutionary transitions from broader to narrower diets mediated by selection from predators for increasingly efficient sequestration of plant defensive compounds (Montllor et al. 1991; Kelly and Bowers 2018).

Zvereva and Kozlov's (2016) meta-analysis of the effectiveness of chemical defenses of insect herbivores provides broad support for the enemy-free space hypothesis. They found that specialist herbivores had more effective anti-predator defenses than generalist herbivores, primarily due to their superior sequestration ability. The meta-analysis also suggests that specialist and generalist enemies exert opposing selection on herbivore diet breadths. Selection from parasitoids may oppose selection from generalist predators for narrow herbivore diet breadth: several studies show that chemicals sequestered by specialist insect herbivores can be used by specialist parasitoids and predators as search cues (Köpf et al. 1997; Zvereva and Rank 2004; Zvereva et al. 2016). These sequestered chemicals may also make the herbivore a superior host for parasitoids because a chemically protected host will also protect the parasitoids developing inside it (Murphy et al. 2014). Alternatively, the sequestered chemicals may constrain the herbivore's immunological defenses against parasitoids (e.g., Smilanich et al. 2009). Either way, specialized sequestering herbivores face a potential trade-off between reduced mortality from generalist natural enemies and increased susceptibility to specialized natural enemies (Lampert et al. 2014; Zvereva and Kozlov 2016; Ali and Agrawal 2017; Kelly and Bowers 2018).

Specific Camouflage

Bernays and Graham (1988) also suggested that dietary specialist herbivores might gain an anti-predator advantage over generalists by possessing superior camouflage. While this prediction has not been extensively tested (Sandoval and Nosil 2005), the limited evidence tends to support it. Before we examine this evidence, we note the updated terminology since 1988 (see also Koptur et al., Chapter "Caterpillar Responses to Ant Protectors of Plants"). Current work in this field defines camouflage as the close resemblance of an organism to its background (Ruxton et al. 2018). Crypsis, which has historically been used synonymously with camouflage, refers to camouflage that limits the detectability of the camouflaged organism. The relatively new concept of masquerade (Skelhorn et al. 2010) refers to cases of camouflage in which the predator detects the prey but does not recognize it as prey (Ruxton et al. 2018).

Few direct tests have evaluated the contention that specific camouflage can mediate the enemy-free space advantage of dietary specialization. Singer et al. (2014) measured the strength of bird predation on 41 different caterpillar species that varied in diet breadth and anti-predator traits, such as morphological camouflage.

Dietary specialization predicted reduced bird predation and superior camouflage (for the set of camouflaged caterpillar species), measured as the latency to detection by human observers. A companion study found higher values of latency to detection (more effective camouflage) to be among the strongest predictors of reduced bird predation for caterpillars (Lichter-Marck et al. 2015). Both of these studies also showed evidence for a behavioral component of camouflage, measured as the fidelity of each caterpillar species to hide on specific parts of the host plant. This measure, termed stereotypy, was also associated with dietary specialization (more host-specific caterpillar species exhibited more stereotypy, Singer et al. 2014) as well as with reduced bird predation (Lichter-Marck et al. 2015). The latter study also showed that increased expression of behavioral defensive responses of caterpillars (e.g., thrashing, biting, locomotion) was associated with increased bird predation. This set of studies indicates that specific camouflage can be an important mediator of dietary specialization and protection from predators. The process by which predators exert selection on camouflaged phenotypes is not addressed in the work described above.

On this point, several studies do link selection by predators to specific camouflage and host-plant use in insect herbivores. Perhaps the most well-studied example is that of *Timema* stick insects (e.g., Sandoval 1994; Sandoval and Nosil 2005). *Timema podura* uses plant species in the genera *Ceanothus* and *Adenostoma*, and green morphs are specifically camouflaged on *Ceanothus* while brown morphs are specifically camouflaged on *Adenostoma*. Predation experiments reveal differential predation by birds targeting background mismatches between herbivore morph and host plant (Sandoval and Nosil 2005). The same phenomenon is recorded for the species *T. cristinae* on *Ceanothus* and *Adenostoma*, except with striped morphs and unstriped morphs, respectively (Sandoval and Nosil 2005). Remarkably, experimental data reveal that all morphs of both species have higher fecundity feeding on *Ceanothus*, but all morphs exhibit host preference toward the plant on which they better obtain enemy-free space via specific camouflage despite the fecundity trade-off (Sandoval and Nosil 2005). However, the maintenance of polyphagy in these *Timema* species suggests that selection from predators for dietary specialization with specific camouflage is not the only factor operating on the evolution of herbivore diet breadth. Opposing selection from host-plant quality seems to be a critical factor as well, thus maintaining the use of both host-plant species. In caterpillars, this trade-off is seen in the lichenivorous larvae of the moth *Cleorodes lichenaria*, which feed on the genus *Ramalina*, whose shrubby appearance more closely matches the coloration of the larvae than other potential hosts (Pöykkö 2011). Even though other host plants provide greater food quality to the caterpillars, mortality from predation is significantly lower for larvae on *Ramalina* (Pöykkö 2011). Again, we see the role of enemy-free space via specific camouflage in expanding or maintaining polyphagy due to trade-offs between selection from predators and selection from plants.

Taken together, the studies discussed in this section provide provisional evidence supporting specific camouflage as a means by which predators select for dietary specialization in insect herbivores. This evidence is ample enough to warrant further study of this mechanism. Future research should investigate how selection on camouflage traits influences the evolution of dietary specialization. Caterpillar populations that vary in both diet breadth and camouflage traits would be ideal for such a study (see Plate 1).

Efficiency of Feeding Decisions

A more recent extension of the enemy-free space hypothesis relates to neural constraints on information processing, which may favor the evolution of dietary specialization due to improved accuracy and efficiency of feeding decisions by specialist herbivores (Bernays 2001). Indeed, dietary specialist arthropods can make more accurate and efficient feeding and oviposition decisions than dietary generalist arthropods (e.g., Bernays 1998; Bernays 2001; Janz 2003; Egan and Funk 2006). Our focus will be on the efficiency of feeding decisions, by which we mean the decision of individual herbivores to feed or not. The speed of making such decisions is predicted to mediate the enemy-free space advantages of dietary specialization.

In light of evidence that the act of feeding by caterpillars increases their risk of predation (Bernays 1997), more efficient feeding decisions of dietary specialists could limit predation risk (Bernays 1998; Bernays and Funk 1999). Generalist herbivores make decisions by processing a wider range of neurosensory inputs compared to specialist herbivores (Farris and Roberts 2005), yet arthropod brains have limited processing capacity (Bernays 2001). The use of sign stimuli by dietary specialists (Bernays and Chapman 1994) is expected to enable rapid decision-making, thus offering greater opportunity for attention to be dedicated to anti-predator vigilance and reduced time in a vulnerable position (Bernays 1998). Consequently, dietary generalists are expected to be less efficient at making feeding decisions and less vigilant to predation risk than specialists, giving the latter a fitness advantage (Bernays 2001).

Several experiments with well-designed comparisons have demonstrated greater efficiency of feeding decisions by specialist herbivores compared to their generalist counterparts. One of the most innovative of these conditioned individuals of the grasshopper *Schistocerca americana* to be either specialist or generalist phenotypes by feeding them artificial diets of either single or multiple flavors, respectively (Bernays 1998). The conditioned specialist phenotype made faster decisions than the generalist phenotype (Bernays 1998). Bernays and Funk (1999) found similar advantages in feeding behavioral efficiency in comparisons of specialist versus generalist populations of the aphid *Uroleucon ambrosiae*. Aphids from the specialist population were more efficient at finding, selecting, accepting, sampling, and settling on their host plants (Bernays and Funk 1999). That the specialist and generalist populations were genetically differentiated suggests the potential for selection to act on this trait and cause the evolution of dietary specialization (Funk and Bernays 2001). Turning to caterpillars, comparisons between the *Physalis* specialist, *Heliothis subflexa*, and its polyphagous close relative, *H. virescens*, strongly suggest that behavioral efficiency in host-plant use confers an enemy-free space advantage to the specialist (Oppenheim and Gould 2002). Mortality from parasitoids was over ten times higher and predation from ants and wasps was over twice as high for *H. virescens* versus *H. subflexa* on *Physalis* plants. The enemy-free space advantage stemmed from *H. subflexa*'s more efficient behavior in securing a refuge from enemies in the lantern-like calyx surrounding fruits on which it feeds. Additional study revealed that the specialist had greater taste and behavioral sensitivity to deterrent compounds, which may translate into increased efficiency in host-plant acceptance decisions (Bernays et al. 2000). While there is substantial evidence that dietary specialization entails increased feeding efficiency, further studies are needed to identify behavioral efficiency and associated neurosensory traits as a cause rather than a consequence of the evolution of dietary specialization.

Manipulation of Host-Plant Volatiles

Although Bernays and Graham (1988) did not postulate how dietary specialists may be able to manipulate their host plants, recent research has increasingly explored this idea. Since the 1990s, research on tri-trophic interactions has focused on herbivore-induced plant volatile compounds as a cue for parasitoids and predators. A recent offshoot of this research concerns the possibility that specialist herbivores might suppress herbivore-induced plant volatiles (HIPVs) better than generalist herbivores. However, we found limited evidence for this prediction, and there was a much larger body of research demonstrating no significant relationship between dietary specialization and the ability to suppress HIPVs.

HIPVs are one mechanism by which herbivore enemies are attracted to plants with feeding herbivores (Turlings and Erb 2018), and by suppressing them herbivores might protect themselves. Many contemporary studies focus on HIPVs attracting parasitoids (e.g., Williams III et al. 2008; Peñaflor et al. 2017; Gols et al. 2012; Rodriguez-Saona et al. 2005), which often—but not always—have high host specificity (e.g., Collatz and Dorn 2013; Lucas-Barbosa et al. 2014; Najar-Rodriguez et al. 2015), although predators may also be attracted to HIPVs (e.g., Kessler and Baldwin 2001). It is possible that HIPVs contribute to the increased predation risk caterpillars and other herbivores face while feeding (Bernays 1997). Moreover, chewing herbivores may have enzymes in their oral secretions that can disrupt the host plant's signaling pathways, and thus reduce HIPV emissions (Bede et al. 2006; Sarmento et al. 2011).

Evidence in support of dietary specialists being able to suppress volatiles more so than generalists is sparse. Oral secretions of the specialist velvetbean caterpillar (*Anticarsia gemmatalis*), when compared to those of the generalist fall armyworm (*Spodoptera frugiperda*), contain an amino acid substitution allowing them to diminish the induced response from their host plant *Vigna unguiculata*, resulting in lower volatile emissions (Schmelz et al. 2012). In another study, Pareja et al. (2012) used gas chromatography to measure the effects of phloem-feeding on floral volatile emissions and found that feeding by the specialist aphid *Lipaphis erysimi* on *Sinapis alba* resulted in a greater reduction of emitted floral volatiles than did feeding by the generalist aphid *Myzus persicae*. However, suppression of floral volatiles alone was not enough to prevent ladybirds from locating the aphids, as the predators could instead use the volatiles released from vegetative tissue feeding (Pareja et al. 2012). Although these studies demonstrated differences between dietary specialists and generalists, they do not provide a strong basis of support for the enemy-free space hypothesis.

The literature concluding that there is no clear relationship between dietary specialization and the ability to suppress volatiles to maintain enemy-free space is far more abundant. Glucose oxidase, an enzyme in caterpillar oral secretions responsible for suppressing induced plant defenses, has been a subject of interest for researchers measuring plant responses to chewing herbivores (Merkx-Jacques and Bede 2004). Contrary to the enemy-free space hypothesis, Eichenseer et al. (2010) measured glucose oxidase activity in 88 lepidopteran and hymenopteran species and found it to be higher in the polyphagous species than in the oligophagous species. One explanation for this result is that generalist herbivores may have a generalized counterdefense to the phylogenetically conserved jasmonic acid-regulated induced defense pathway in plants (Ali and Agrawal 2012).

There also appears to be context-dependency when it comes to an herbivore's ability to suppress HIPVs. The generalist fall armyworm caterpillar, although used as a basis of comparison for a specialist species that suppresses HIPVs (Schmelz et al. 2012), has the ability to suppress HIPVs in maize while other generalist caterpillar species do not (De Lange et al. 2020). However, it does not have the same ability on other host plant species, and this suppression does not have an effect on parasitoid attraction (De Lange et al. 2020).

From our analysis of the current literature, suppression of HIPVs does not seem to mediate the evolution of dietary specialization in insect herbivores. Supporting cases are few (e.g., Schmelz et al. 2012; Pareja et al. 2012), whereas the opposing evidence encompasses a greater variety of herbivore species (e.g., Eichenseer et al. 2010; Rowen and Kaplan 2016; Ali and Agrawal 2012). However, few of the studies measured enemy behavior and attraction to volatiles in the presence of herbivores of differing diet breadths. Future research that fills this gap could more clearly indicate if and how the enemy-free space hypothesis applies to HIPV suppression.

Appraisal and Refinement of the Hypothesis

Natural selection from generalist predators remains a plausible hypothesis for the evolution of dietary specialization of insect herbivores. We found quantitative support for a key prediction: a greater survival advantage of dietary specialist over generalist herbivores faced with generalist predators compared to parasitoids (Fig. 1). As caterpillars were the herbivores in most of these studies, the evidence mainly applies to this group. The small sample size of studies limits the strength of

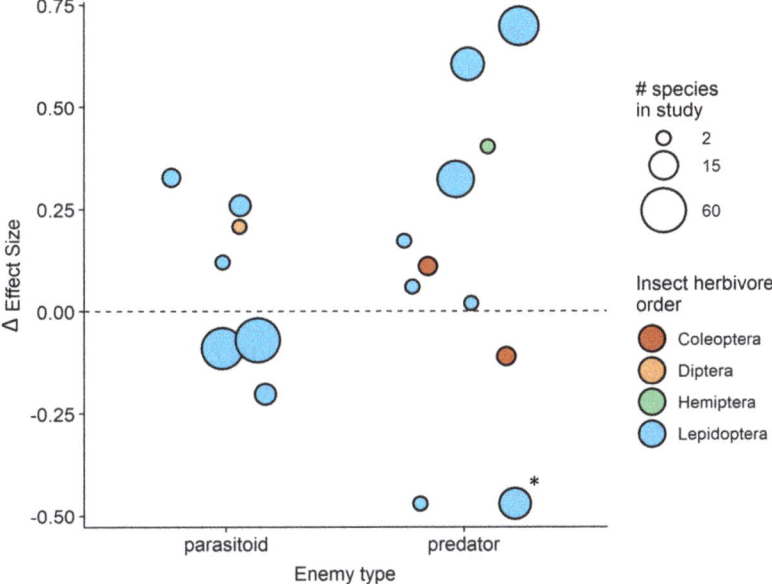

Fig. 1 The magnitude of the survival advantage (Δeffect size >1) or disadvantage (Δeffect size <1) of dietary specialist relative to generalist herbivores as a function of enemy type (parasitoid versus predator). Each data point is a survival comparison between specialist and generalist herbivores within a single study, with the size and color of the points denoting the number of herbivore species being compared and the taxonomic order of the herbivores being compared, respectively. Data points are jiggered for visibility. (*This data point represents Singer et al. (2019), which is discussed further (see Appraisal and refinement of the hypothesis))

our conclusion, especially for herbivores other than caterpillars. We also note that some studies did not support the enemy-free space hypothesis. This inconsistent support reinforces the current view that no single factor is likely to fully explain herbivore diet breadth. The importance of predation relative to other factors, such as host-plant defenses, is receiving increasing attention (e.g., Coley et al. 2006; Forister et al. 2011; Mooney et al. 2012; Katsanis et al. 2016). Enemies of herbivores, on the one hand, and plant traits, on the other, have similarly large effects on insect herbivore performance (Vidal and Murphy 2018a), suggesting that they jointly select on herbivore traits, such as diet breadth. Studies that directly measure selection from both enemies and plant traits on host-use traits by herbivores would be valuable additions to the field of tri-trophic interactions. Comparisons among conspecific heritable phenotypes varying in dietary specialization would be especially enlightening, and caterpillar species with intraspecific variation in diet breadth offer excellent opportunities for study.

Further exploration of the mechanisms by which dietary specialization confers anti-predator defense can generate new predictions and opportunities or testing. Our review shows that the anti-predator advantages of dietary specialization by insect herbivores stem mainly from sequestration of plant allelochemicals, specific camouflage, and perhaps behavioral efficiency, but not from suppression of herbivore-induced volatiles. What are the relative roles of these mechanisms in the evolution of dietary specialization? In theory, it is possible that they play equivalent roles as alternative evolutionary pathways, and the preponderance of dietary specialization is a consequence of multiple roads leading to the same destination. However, we think it more plausible that sequestration of plant allelochemicals is the more traveled road because it entails selection from both plant defenses and predators acting in concert (Singer 2008). For example, lineages of herbivores that have evolved some degree of physiological specialization to particular classes of phytochemicals may be under selection from plants and predators to maintain and even escalate the pertinent behavioral, physiological, and morphological traits that reinforce host or chemical specificity (Vencl et al. 2011). With respect to caterpillars, the especially well-studied case of milkweed butterflies supports this notion, as escalating concentrations of sequestered cardenolides (not concentrations in host-plants) are responsible for escalating molecular and physiological resistance to cardenolides in larval milkweed butterflies (Petschenka and Agrawal 2015). As the tri-trophic players coevolve, this evolutionary maintenance and escalation is likely to continue in tri-trophic modules of communities that persist over long periods of time, resulting in the phylogenetic and/or phytochemical conservatism of traits of host-plant use in herbivore lineages (Winkler and Mitter 2008) that captured the attention of Ehrlich and Raven (1964) and many others since.

Lineages of herbivores that have evolved alternatives to sequestration of allelochemicals may be more evolutionarily labile with respect to traits of host-plant use. The numerous cases of polyphenism of camouflage traits in polyphagous insect herbivores (Ruxton et al. 2018) as well as specific camouflage coupled with polyphagy in *Timema* stick insects (Crespi and Sandoval 2000) imply that specific camouflage does not necessarily constrain herbivores to highly restricted diets on a particular plant species. However, evidence for both strong selection on specifically camouflaged phenotypes (Sandoval and Nosil 2005) and superior camouflage of dietary specialists (Singer et al. 2014) suggests that selection from predators may still favor host specificity. Selection from predators acting in concert with selection from plant traits might cause dietary specialization characterized by escalating morphological, chemical, and behavioral specific camouflage (e.g., De Moraes and Mescher 2004; Sandoval and Crespi 2008; Whitehead et al. 2014). Opposing selection from plants and predators might create food quality versus enemy-free space trade-offs among host plants, thus favoring the use of multiple host-plant taxa and adaptive plasticity of morphological, behavioral, and physiological traits of polyphagous herbivores (e.g., polyphenism and masquerade [Higginson et al. 2012]). Several case studies show that benefits of enemy-free space and food-quality trade off among alternative host plants used by polyphagous populations of caterpillars (e.g., Mira and Bernays 2002; Singer et al. 2004; Pöykkö 2011; Rodrigues and Freitas 2013; Stoepler and Lill 2013; Murphy and Loewy 2015). However, the morphological, behavioral, and physiological plasticity of such polyphagous herbivores has received limited study.

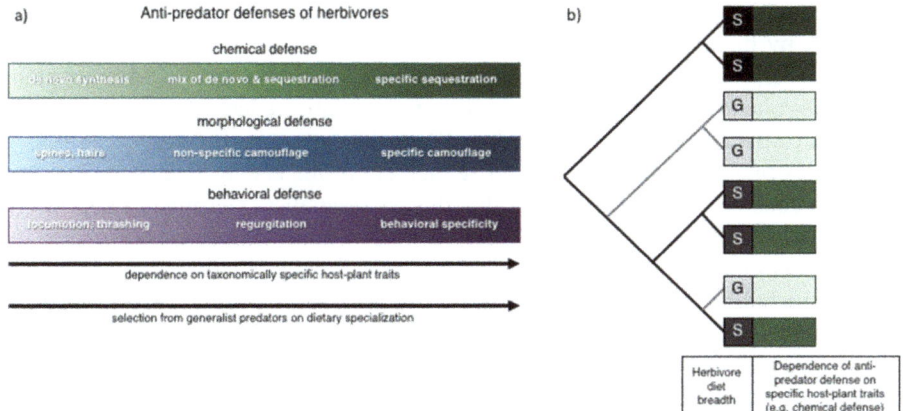

Fig. 2 (**a**) Hypothesized gradient of dependence of anti-predator defenses of insect herbivores on taxonomically specific host-plant traits for chemical (green), morphological (blue), and behavioral (purple) anti-predator traits. Selection from generalist predators is hypothesized to increase with increasing dependence of anti-predator traits on taxonomically specific host-plant traits. (**b**) Hypothetical example of the predicted macroevolutionary consequence of selection from predators. Note the visibly apparent association on the herbivore phylogeny between dietary specialization (S = specialists, G = generalists, with grayscale denoting a continuous measure of diet breadth) and dependence of anti-predator traits on taxonomically specific host-plant traits. The gradient in green shaded boxes refers to the gradient in host-plant dependence for chemical defense in **a**)

This mechanistic perspective inspired us to propose a refinement of the enemy-free space hypothesis: the greater the dependence of anti-predator defenses on taxonomically specific host plant traits, the stronger the selection for the evolution of dietary specialization in herbivores (Fig. 2). This refinement reframes the hypothesis in at least three important ways. First, predation is not exclusive to other factors, such as plant defenses, that are likely to select on dietary specialization (Dyer 1995, Singer and Stireman 2005, Vidal and Murphy 2018a). Second, by explicitly linking anti-predator defense traits to plant traits, this refinement can better guide research on evolutionary-genetic mechanisms (e.g., Gompert et al. 2014). Third, it makes a new phylogenetic prediction: repeated, independently evolved associations between dietary specialization and anti-predator traits that depend on taxonomically specific host-plant traits (e.g., sequestration of specific allelochemicals; Fig. 2). In contrast, polyphagy is predicted to be most strongly associated with key anti-predator traits that have limited dependence on taxonomically specific host-plant traits, such as de novo chemical defenses, many types of morphological defenses (e.g., spines, hairs, non-specific crypsis or masquerade), and many types of behavioral defenses (e.g., locomotion, evasion, thrashing, biting). In support of this prediction, polyphagy in British geometrid and drepanid caterpillars is associated with masquerade, but not crypsis or brightly coloration (Higginson et al. 2012). Sequestration of plant allelochemicals (e.g., Engler-Chaouat and Gilbert 2007), specific camouflage (e.g., De Moraes and Mescher 2004), and behavior (e.g., Lichter-Marck et al. 2015) all may be highly dependent on taxonomically specific plant traits. Dependence of anti-predator traits on taxonomically specific plant traits could be studied experimentally by manipulating plant traits (e.g., secondary chemistry for sequestering species, morphology for camouflaged species) and herbivore diet (if different plant species or phenotypes vary in traits contributing to anti-predator defense) and measuring the effect on anti-predator traits and/or predation of herbivores.

To illustrate the significance of our refinement of the enemy-free space hypothesis, we consider the recent finding that dietary specialist caterpillars suffer more ant predation than dietary generalist caterpillars in Connecticut forests (Singer et al. 2019). This finding clearly contradicts the conventional prediction of the hypothesis, and including this study in our quantitative analysis considerably weakens support for this prediction (Fig. 1). Importantly, this result does not contradict the prediction that dietary specialization is associated with anti-predator traits that depend on taxonomically specific host-plant traits. In this case, dietary specialists possess a syndrome of life history traits that confers protection from bird predation, but renders them susceptible to ant predation. Dietary generalists, while more susceptible to bird predation, have traits that protect them from ant predation, especially behavioral defenses (Singer et al. 2019). In accordance with the refined hypothesis, the anti-predator defenses of the dietary specialists are dependent on taxonomically specific traits of host plants, such as putative sequestration of host-plant allelochemicals, specific camouflage, and behavioral fidelity to particular structures on the host plant (Singer et al. 2014). The behavioral defenses that protect dietary generalists from ant predation, such as biting, thrashing, locomotion, and rappelling down from the plant on silk threads, do not derive their defensive efficacy from taxonomically specific host-plant traits (Singer et al. 2019). According to our refined hypothesis, birds select for dietary specialization in this community whereas ants do not because the caterpillar defenses against birds depend on taxonomically specific traits of host plants. More generally, we expect the degree to which predators select for taxonomically restricted herbivore diets will depend on the degree to which herbivores rely on taxonomically specific host plants for anti-predator defense.

Conclusions

Our review and others over the last several decades make a strong case for the centrality of tri-trophic interactions in driving the evolution of dietary specialization in caterpillars and other insect herbivores (e.g., Vidal and Murphy 2018a; Abdala-Roberts et al. 2019). That development alone is an important legacy of Bernays and Graham (1988) and other pioneering research of that era (reviewed in Abdala-Roberts et al. 2019). With respect to the hypothesis proposed by Bernays and Graham (1988), our review of the evidence suggests that generalist predators probably play an important role in the evolution of dietary specialization by insect herbivores. However, despite it being a highly cited idea in the literature, the enemy-free space hypothesis has received surprisingly little testing. At this time, empirical support for this hypothesis and its relative importance for the evolution of dietary specialization remains equivocal. In hopes of spurring more research on this topic, we reframe the hypothesis based on mechanisms by which specialist herbivores gain enemy-free space advantages over generalists. This refined hypothesis offers predictions that are more amenable to evolutionary analyses, which we encourage future research to explore.

Questions for Future Study

Among many possible research questions for future study, we pose a few that are most relevant to this volume. First, how well does the enemy-free space hypothesis apply to insect herbivores other than caterpillars? Theoretically, it should apply to arthropod herbivores in general, as originally proposed (Bernays and Graham 1988). Indirect evidence also suggests that it might be quite general, but the direct evidence mainly comes from studies of caterpillars, leaving the question open. Second, how do different types of predators and parasites affect the evolution of host-plant use by caterpillars? There are many types of predators of caterpillars, especially among arthropods (Montllor and Bernays 1993), and relatively few types have been addressed in the context of the enemy-free space hypothesis. Our refined hypothesis predicts that the type of enemy should matter less than how specifically caterpillars depend on particular host plant taxa for defense against each type of enemy. In other words, the biological details of tri-trophic interactions are key, and further study of physiological and behavioral mechanisms of tri-trophic interactions will continue to be valuable. Finally, there is the question as to how tri-trophic dynamics and caterpillar diet breadth will be influenced by rapid environmental change. Habitat fragmentation and loss, climate change, invasive species, pesticide use, and pollution all have potentially important influences, but little work has directly addressed this question (Stireman III and Singer 2018). We hypothesize that the complex dependence of dietary specialist herbivores on specific host plants for food and anti-predator defense would make them especially susceptible to changes in the plant and predator community. However, if they can track their specific host plants in the face of rapid environmental change, they might be able to maintain these dependencies. According to our refined hypothesis, the loss of specific groups of predators from a community could change selection on dietary specialization of caterpillars and other herbivores. We encourage future study of the enemy-free space hypothesis with explicit consideration of rapid environmental change.

Acknowledgments We thank Bob Marquis and Suzanne Koptur for the invitation to contribute to this volume, and for helpful feedback on the manuscript. We thank the anonymous reviewers for their critical feedback as well. Funding was provided by Wesleyan University's College of the Environment, College of Integrative Sciences, and the US National Science Foundation (NSF DEB: 1557086).

References

Abdala-Roberts L, Puentes A, Finke DL, Marquis RJ, Montserrat M, Poelman EH, Rasmann S, Sentis A, van Dam NM, Wimp G, Mooney K, Björkman C (2019) Tri-trophic interactions: bridging species, communities and ecosystems. Ecol Lett 22:2151–2167

Agrawal AA, Conner JK, Rasmann S (2010) Tradeoffs and negative correlations in evolutionary ecology. In: Bell MA (ed) Evolution since Darwin: the first 150 years. Sinauer, Sunderland, pp 243–268

Ali JG, Agrawal AA (2012) Specialist versus generalist insect herbivores and plant defense. Trends Plant Sci 17:293–302

Ali JG, Agrawal AA (2017) Trade-offs and tritrophic consequences of host shifts in specialized root herbivores. Funct Ecol 31:153–160

Anderson RM, Dallar NM, Pirtel NL, Connors CJ, Mickley J, Bagchi R, Singer MS (2019) Bottom-up and top-down effects of forest fragmentation differ between dietary generalist and specialist caterpillars. Front Ecol Evol 7:452

Bede JC, Musser RO, Felton GW, Korth KL (2006) Caterpillar herbivory and salivary enzymes decrease transcript levels of *Medicago truncatula* genes encoding early enzymes in terpenoid biosynthesis. Plant Mol Biol 60:519–531

Berdegue M, Trumble JT, Hare JD, Redak RA (1996) Is it enemy-free space? The evidence for terrestrial insects and freshwater arthropods. Ecol Entomol 21:203–217

Bernays EA (1988) Host specificity in phytophagous insects: selection pressure from generalist predators. Entomol Exp Appl 49:131–140

Bernays EA (1989) Host range in phytophagous insects: the potential role of generalist predators. Evol Ecol 3:299–311

Bernays EA (1997) Feeding by lepidopteran larvae is dangerous. Ecol Entomol 22:121–123

Bernays EA (1998) The value of being a resource specialist: behavioral support for a neural hypothesis. Am Nat 151:451–464

Bernays EA (2001) Neural limitations in phytophagous insects: implications for diet breadth and evolution of host affiliation. Annu Rev Entomol 46:703–727

Bernays EA, Chapman RF (1994) Host-plant selection by phytophagous insects. Chapman and Hall, New York

Bernays EA, Cornelius ML (1989) Generalist caterpillar prey are more palatable than specialists for the generalist predator *Iridomyrmex humilis*. Oecologia 79:427–430

Bernays EA, Funk DJ (1999) Specialists make faster decisions than generalists: experiments with aphids. Proc R Soc Lond Ser B Biol Sci 266:151–156

Bernays E, Graham M (1988) On the evolution of host specificity in phytophagous arthropods. Ecology 69:886–892

Bernays EA, Oppenheim S, Chapman RF, Kwon H, Gould F (2000) Taste sensitivity of insect herbivores to deterrents is greater in specialists than in generalists: a behavioral test of the hypothesis with two closely related caterpillars. J Chem Ecol 26:547–563

Bosc C, Roets F, Hui C, Pauw A (2018) Interactions among predators and plant specificity protect herbivores from top predators. Ecology 99:1602–1609

Brues CT (1924) The specificity of food-plants in the evolution of phytophagous insects. Am Nat 58:127–144

Camara MD (1997) Predator responses to sequestered plant toxins in buckeye caterpillars: are tritrophic interactions locally variable? J Chem Ecol 23:2093–2106

Coley PD, Bateman ML, Kursar TA (2006) The effects of plant quality on caterpillar growth and defense against natural enemies. Oikos 115:219–228

Collatz J, Dorn S (2013) Tritrophic consequences arising from a host shift between apple and walnut in an oligophagous herbivore. Biol Control 65:330–337

Core Team R (2018) R: a language and environment for statistical computing. R Foundation for Statistical Computing, Vienna

Cornell HV, Hawkins BA (2003) Herbivore responses to plant secondary compounds: a test of phytochemical coevolution theory. Am Nat 161:507–522

Crespi BJ, Sandoval CP (2000) Phylogenetic evidence for the evolution of ecological specialization in *Timema* walking-sticks. J Evol Biol 13:249–262

De Lange ES, Laplanche D, Guo HJ, Xu W, Vlimant M, Erb M, Ton J, Turlings TCJ (2020) *Spodoptera frugiperda* caterpillars suppress herbivore-induced volatile emissions in maize. J Chem Ecol 46:344–360

De Moraes CM, Mescher MC (2004) Biochemical crypsis in the avoidance of natural enemies by an insect herbivore. Proc Natl Acad Sci 101:8993–8997

Dethier VG (1954) Evolution of feeding preferences in phytophagous insects. Evolution 8:33–54

Dyer LA (1995) Tasty generalists and nasty specialists? Antipredator mechanisms in tropical lepidopteran larvae. Ecology 76:1483–1496

Dyer LA (1997) Effectiveness of caterpillar defenses against three species of invertebrate predators. J Res Lepidoptera 34:48–68

Dyer LA, Floyd T (1993) Determinants of predation on phytophagous insects: the importance of diet breadth. Oecologia 96:575–582

Egan SP, Funk DJ (2006) Individual advantages to ecological specialization: insights on cognitive constraints from three conspecific taxa. Proc R Soc B Biol Sci 273:843–848

Ehrlich PR, Raven PH (1964) Butterflies and plants: a study in coevolution. Evolution 18:586–608

Eichenseer H, Mathews MC, Powell JS, Felton GW (2010) Survey of a salivary effector in caterpillars: glucose oxidase variation and correlation with host range. J Chem Ecol 36:885–897

Engler-Chaouat HS, Gilbert LE (2007) De novo synthesis vs. sequestration: negatively correlated metabolic traits and the evolution of host plant specialization in cyanogenic butterflies. J Chem Ecol 33:25–42

Farris SM, Roberts NS (2005) Coevolution of generalist feeding ecologies and gyrencephalic mushroom bodies in insects. Proc Natl Acad Sci USA 102:17394–17399

Forister ML, Gompert Z, Nice CC, Forister GW, Fordyce JA (2011) Ant association facilitates the evolution of diet breadth in a lycaenid butterfly. Proc R Soc B Biol Sci 278:1539–1547

Forister ML, Dyer LA, Singer MS, Stireman JO III, Lill JT (2012) Revisiting the evolution of ecological specialization, with emphasis on insect–plant interactions. Ecology 93:981–991

Fraenkel GS (1959) The raison d'etre of secondary plant substances. Science 129:1466–1470

Fry JD (1996) The evolution of host specialization: are trade-offs overrated? Am Nat 148:S84–S107

Fry JD (2003) Detecting ecological trade-offs using selection experiments. Evolution 84:1672–1678

Funk DJ, Bernays EA (2001) Geographic variation in host specificity reveals host range evolution in *Uroleucon ambrosiae* aphids. Ecology 82:726–739

Futuyma DJ, Agrawal AA (2009) Macroevolution and the biological diversity of plants and herbivores. Proc Natl Acad Sci 106:18054–18061

Futuyma DJ, Moreno G (1988) The evolution of ecological specialization. Annu Rev Ecol Syst 19:207–233

Gentry GL, Dyer LA (2002) On the conditional nature of neotropical caterpillar defenses against their natural enemies. Ecology 83:3108–3119

Gols R, Veenemans C, Potting RPJ, Smid HM, Dicke M, Harvey JA, Bukovinszky T (2012) Variation in the specificity of plant volatiles and their use by a specialist and a generalist parasitoid. Anim Behav 83:1231–1242

Gompert Z, Messina FJ (2016) Genomic evidence that resource-based trade-offs limit host-range expansion in a seed beetle. Evolution 70:1249–1264

Gompert Z, Comeault AA, Farkas TE, Feder JL, Parchman TL, Beurkle CA, Nosil P (2014) Experimental evidence for ecological selection on genome variation in the wild. Ecol Lett 17:369–379

Hardy NB, Kaczvinsky C, Bird G, Normark BB (2020) What we don't know about diet-breadth evolution in herbivorous insects. Annu Rev Ecol Evol Syst 51:103–122

Higginson AD, De Wert L, Rowland HM, Speed MP, Ruxton GD (2012) Masquerade is associated with polyphagy and larval overwintering in Lepidoptera. Biol J Linn Soc 106:90–103

Jaenike J (1990) Host specialization in phytophagous insects. Annu Rev Ecol Syst 21:243–273
Janz N (2003) The cost of polyphagy: oviposition decision time vs error rate in a butterfly. Oikos 100:493–496
Janzen DH (1985) A host plant is more than its chemistry. Ill Nat Hist Surv Bull 33:141–174
Jeffries MJ, Lawton JH (1984) Enemy free space and the structure of ecological communities. Biol J Linn Soc 23:269–286
Jones CG, Whitman DW, Compton SJ, Silk PJ, Blum MS (1989) Reduction in diet breadth results in sequestration of plant chemicals and increases efficacy of chemical defense in a generalist grasshopper. J Chem Ecol 15:1811–1822
Joshi A, Thompson JN (1995) Trade-offs and the evolution of host specialization. Evol Ecol 9:82–92
Karban R, Agrawal AA (2002) Herbivore offense. Annu Rev Ecol Syst 33:641–664
Katsanis A, Rasmann S, Mooney KA (2016) Herbivore diet breadth and host plant defense mediate the tri-trophic effects of plant toxins on multiple coccinellid predators. PLoS One 11:e0155716
Kelly CA, Bowers MD (2018) Host plant iridoid glycosides mediate herbivore interactions with natural enemies. Oecologia 188:491–500
Kessler A, Baldwin IT (2001) Defensive function of herbivore-induced plant volatile emissions in nature. Science 291:2141–2144
Köpf A, Rank NE, Roininen H, Tahvanainen H (1997) Defensive larval secretions of leaf beetles attract a specialist predator *Parasyrphus nigritarsis*. Ecol Entomol 22:176–183
Koricheva J, Gurevitch J, Mengersen K (eds) (2013) Handbook of meta-analysis in ecology and evolution. Princeton University Press, Princeton
Krieger RI, Feeny PP, Wilkinson CF (1971) Detoxication enzymes in the guts of caterpillars: an evolutionary answer to plant defenses? Science 172:579–581
Kumar P, Rathi P, Schöttner M, Baldwin IT, Pandit S (2014) Differences in nicotine metabolism of two *Nicotiana attenuata* herbivores render them differentially susceptible to a common native predator. PLoS One 9:e95982
Lajeunesse MJ (2015) Bias and correction for the log response ratio in ecological meta-analysis. Ecology 96:2056–2063
Lampert EC, Dyer LA, Bowers MD (2014) Dietary specialization and the effects of plant species on potential multitrophic interactions of three species of nymphaline caterpillars. Entomol Exp Appl 153:207–216
Leuthardt FL, Glauser G, Baur B (2013) Composition of alkaloids in different box tree varieties and their uptake by the box tree moth *Cydalima perspectalis*. Chemoecology 23:203–212
Lichter-Marck IH, Wylde M, Aaron E, Oliver JO, Singer MS (2015) The struggle for safety: effectiveness of caterpillar defenses against bird predation. Oikos 124:525–533
Lucas-Barbosa D, Poelman EH, Aartsma Y, Snoeren TAL, van Loon JJA, Dicke M (2014) Caught between parasitoids and predators - survival of a specialist herbivore on leaves and flowers of mustard plants. J Chem Ecol 40:621–631
Merkx-Jacques M, Bede JC (2004) Caterpillar salivary enzymes: "eliciting" a response. Phytoprotection 85:33–37
Mira A, Bernays EA (2002) Trade-offs in host use by *Manduca sexta*: plant characters vs natural enemies. Oikos 97:387–397
Montllor CB, Bernays EA (1993) Invertebrate predators and caterpillar foraging. In: Stamp NE, Casey TM (eds) Caterpillars: ecological and evolutionary constraints on foraging. Chapman and Hall, New York, pp 170–202
Montllor CB, Bernays EA, Cornelius ML (1991) Responses of two hymenopteran predators to surface chemistry of their prey: significance for an alkaloid-sequestering caterpillar. J Chem Ecol 17:391–399
Mooney KA, Pratt RT, Singer MS (2012) The tri-trophic interactions hypothesis: interactive effects of host plant quality, diet breadth and natural enemies on herbivores. PLoS One 7:e34403
Murphy SM (2004) Enemy-free space maintains swallowtail butterfly host shift. Proc Natl Acad Sci 101:18048–18052

Murphy SM, Loewy KJ (2015) Trade-offs in host choice of an herbivorous insect based on parasitism and larval performance. Oecologia 179:741–751
Murphy SM, Lill JT, Bowers MD, Singer MS (2014) Enemy-free space for parasitoids. Environ Entomol 43:1465–1474
Najar-Rodriguez AJ, Friedli M, Klaiber J, Dorn S (2015) Aphid-deprivation from *Brassica* plants results in increased isothiocyanate release and parasitoid attraction. Chemoecology 25:303–311
Oppenheim SJ, Gould F (2002) Behavioral adaptations increase the value of enemy-free space for *Heliothis subflexa*, a specialist herbivore. Evolution 56:679–689
Pareja M, Qvarfordt E, Webster B, Mayon P, Pickett J, Birkett M, Glinwood R (2012) Herbivory by a phloem-feeding insect inhibits floral volatile production. PLoS One 7:e31971
Peñaflor MFGV, Goncalves FG, Colepicolo C, Sanches PA, Bento JMS (2017) Effects of single and multiple herbivory by host and non-host caterpillars on the attractiveness of herbivore-induced volatiles of sugarcane to the generalist parasitoid *Cotesia flavipes*. Entomol Exp Appl 165:83–93
Petschenka G, Agrawal AA (2015) Milkweed butterfly resistance to plant toxins is linked to sequestration, not coping with a toxic diet. Proc R Soc B Biol Sci 282:20151865
Pöykkö H (2011) Enemy-free space and the host range of a lichenivorous moth: a field experiment. Oikos 120:564–569
Price PW, Bouton CE, Gross P, McPheron BA, Thompson JN, Weis AE (1980) Interactions among three trophic levels: influence of plants on interactions between insect herbivores and natural enemies. Annu Rev Ecol Syst 11:41–65
Rausher MD (1988) Is coevolution dead? Ecology 69:898–901
Rodrigues D, Freitas AVL (2013) Contrasting egg and larval performances help explain polyphagy in a folivorous butterfly. Arthropod Plant Interact 7:159–167
Rodriguez-Saona C, Chalmers JA, Raj S, Thaler JS (2005) Induced plant responses to multiple damagers: differential effects on an herbivore and its parasitoid. Oecologia 143:566–577
Rothwell EM, Holeski LM (2020) Phytochemical defenses and performance of specialist and generalist herbivores: a meta-analysis. Ecological Entomology 45:396–405
Rowen E, Kaplan I (2016) Eco-evolutionary factors drive induced plant volatiles: a meta-analysis. New Phytol 210:284–294
Ruxton GD, Allen WL, Sherratt TN, Speed MP (2018) Avoiding attack: the evolutionary ecology of Crypsis, aposematism, and mimicry. Oxford University Press, Oxford
Sandoval CP (1994) Differential visual predation on morphs of *Timema cristinae* (Phasmatodeae: Timemidae) and its consequences for host range. Biol J Linn Soc 52:341–356
Sandoval CP, Crespi BJ (2008) Adaptive evolution of cryptic coloration: the shape of host plants and dorsal stripes in *Timema* walking-sticks. Biol J Linn Soc 94:1–5
Sandoval CP, Nosil P (2005) Counteracting selective regimes and host preference evolution in ecotypes of two species of walking-sticks. Evolution 59:2405–2413
Sarmento RA, Lemos F, Bleeker PM, Schuurink RC, Pallini A, Oliveira MGA, Lima ER, Kant M, Sabelis MW, Janssen A (2011) A herbivore that manipulates plant defence. Ecol Lett 14:229–236
Schmelz EA, Huffaker A, Carroll MJ, Alborn HT, Ali JG, Teal PEA (2012) An amino acid substitution inhibits specialist herbivore production of an antagonist effector and recovers insect-induced plant defenses. Plant Physiol 160:1468–1478
Singer MS (2008) Evolutionary ecology of polyphagy. In: Tilmon KJ (ed) Specialization, speciation, and radiation: the evolutionary biology of herbivorous insects. University of California Press, Berkeley, pp 29–42
Singer MS, Stireman JO III (2005) The tri-trophic niche concept and adaptive radiation of phytophagous insects. Ecol Lett 8:1247–1255
Singer MS, Rodrigues D, Stireman JO III, Carrière Y (2004) Roles of food quality and enemy-free space in host use by a generalist insect herbivore. Ecology 85:2747–2753
Singer MS, Lichter-Marck IH, Farkas TE, Aaron E, Whitney KD, Mooney KA (2014) Herbivore diet breadth mediates the cascading effects of carnivores in food webs. PNAS 111:9521–9526

Singer MS, Clark RE, Johnson ER, Lichter-Marck IH, Mooney KA, Whitney KD (2019) Dietary specialization is conditionally associated with increased ant predation risk in a temperate forest caterpillar community. Ecol Evol 9:12099–12112

Skelhorn J, Rowland HM, Speed MP, Ruxton GD (2010) Masquerade: camouflage without crypsis. Science 327:51

Smilanich AM, Dyer LA, Chambers JQ, Bowers MD (2009) Immunological cost of chemical defence and the evolution of herbivore diet breadth. Ecol Lett 12:612–621

Stamp N (2001) Enemy-free space via host plant chemistry and dispersion: assessing the influence of tri-trophic interactions. Oecologia 128:153–163

Stireman JO III, Singer MS (2003) Determinants of parasitoid–host associations: insights from a natural tachinid–lepidopteran community. Ecology 84:296–310

Stireman JO III, Singer MS (2018) Tritrophic niches of insect herbivores in an era of rapid environmental change. Current Opinion in Insect Science 29:117–125

Stoepler TM, Lill JT (2013) Direct and indirect effects of light environment generate ecological trade-offs in herbivore performance and parasitism. Ecology 94:2299–2310

Strong DR (1988) Special feature: insect host range. Ecology 69:885–915

Thompson JN (1988) Coevolution and alternative hypotheses on insect/plant interactions. Ecology 69:893–895

Traugott MS, Stamp NE (1997) Effects of chlorogenic acid-and tomatine-fed caterpillars on performance of an insect predator. Oecologia 109:265–272

Turlings TC, Erb M (2018) Tritrophic interactions mediated by herbivore-induced plant volatiles: mechanisms, ecological relevance, and application potential. Annu Rev Entomol 63:433–452

Vencl FV, Nogueira-de-Sá F, Allen BJ, Windsor DM, Futuyma DJ (2005) Dietary specialization influences the efficacy of larval tortoise beetle shield defenses. Oecologia 145:404–414

Vencl FV, Trillo PA, Geeta R (2011) Functional interactions among tortoise beetle larval defenses reveal trait suites and escalation. Behav Ecol Sociobiol 65:227–239

Vidal MC, Murphy SM (2018a) Bottom-up vs. top-down effects on terrestrial insect herbivores: a meta-analysis. Ecol Lett 21:138–150

Vidal MC, Murphy SM (2018b) Quantitative measure of fitness in tri-trophic interactions and its influence on diet breadth of insect herbivores. Ecology 99:2681–2691

Whitehead SR, Reid E, Sapp J, Poveda K, Royer AM, Posto AL, Kessler A (2014) A specialist herbivore uses chemical camouflage to overcome the defenses of an ant-plant mutualism. PLoS One 9:e102604

Wiens JJ, Lapoint RT, Whiteman NK (2015) Herbivory increases diversification across insect clades. Nature Communications 6(1) https://doi.org/10.1038/ncomms9370

Williams L III, Rodriguez-Saona C, Castle SC, Zhu S (2008) EAG-active herbivore-induced plant volatiles modify behavioral responses and host attack by an egg parasitoid. J Chem Ecol 34:1190–1201

Winkler IS, Mitter C (2008) The phylogenetic dimension of insect-plant interactions: a review of recent evidence. In: Tilmon KJ (ed) Specialization, speciation, and radiation: the evolutionary biology of herbivorous insects. University of California Press, Berkeley, pp 240–263

Zvereva EL, Kozlov MV (2016) The costs and effectiveness of chemical defenses in herbivorous insects: a meta-analysis. Ecol Monogr 86:107–124

Zvereva EL, Rank NE (2004) Fly parasitoid *Megaselia opacicornis* uses defensive secretions of the leaf beetle *Chrysomela lapponica* to locate its host. Oecologia 140:516–522

Zvereva EL, Kozlov MV, Rank NE (2016) Does ant predation favour leaf beetle specialization on toxic host plants? Biol J Linn Soc 119:201–212

Zvereva EL, Zverev V, Kruglova OY, Kozlov MV (2017) Strategies of chemical anti-predator defences in leaf beetles: is sequestration of plant toxins less costly than de novo synthesis? Oecologia 183:93–106

Zvereva EL, Doktorovova L, Svadova KH, Zverev V, Stys P, Adamova-Jezova D, Kozlov MV, Exnerova A (2018) Defence strategies of *Chrysomela lapponica* (Coleoptera: Chrysomelidae) larvae: relative efficacy of secreted and stored defences against insect and avian predators. Biol J Linn Soc 124:533–546

Caterpillar Responses to Ant Protectors of Plants

Suzanne Koptur, Jaeson Clayborn, Brittany Harris, Ian Jones, Maria Cleopatra Pimienta, Andrea Salas Primoli, and Paulo S. Oliveira

Sulphur butterfly caterpillar (*Phoebis* sp., Pieridae) on *Senna chapmannii* flower. (Photo by Brittany Harris)

S. Koptur (✉) · B. Harris · M. C. Pimienta · A. S. Primoli
Department of Biological Sciences, International Center for Tropical Botany,
Florida International University, Miami, FL, USA
e-mail: kopturs@fiu.edu

J. Clayborn
Department of Biological Sciences and International Center for Tropical Botany,
Florida International University, Miami, FL, USA

Miami-Dade College, Padrón Campus, Miami, FL, USA

© The Author(s), under exclusive license to Springer Nature Switzerland AG 2022
R. J. Marquis, S. Koptur (eds.), *Caterpillars in the Middle*, Fascinating Life Sciences, https://doi.org/10.1007/978-3-030-86688-4_10

Introduction

Adult butterflies emerge from their chrysalises to find mates and reproduce; however, during their caterpillar larval stage, time is spent feeding to gain enough mass and energy to form a chrysalis and successfully transform into an adult. During this comparatively sedentary larval stage, caterpillars are vulnerable to predation. Consequently, caterpillars have evolved a host of chemical, morphological, visual, and behavioral traits to defend themselves against diverse predators (Bernays 1997; Dyer 1997; Sugiura 2020).

When consuming plants, caterpillars are exposed to a variety of natural enemies, including predators and parasitoids (Singer et al. 2017; Sendoya and Oliveira 2017). A common defense of caterpillars against many natural enemies includes being poisonous or distasteful by sequestering plants' chemical defenses for their own defense (see Bowers, Chapter "Sequestered Caterpillar Chemical Defenses: From "Disgusting Morsels" to Model Systems"). The iconic monarch butterfly is well-known for feeding on milkweeds, making the caterpillars and adult butterflies distasteful (Brower and Glazier 1975; Calvert et al. 1979); many other Lepidoptera utilize other hostplants in this and other plant families to similar effects. Some of these defenses are carried over to the adult stage, providing defense in more than one phase of the life history.

In addition to the chemicals themselves, many Lepidoptera use aposematic coloration to warn predators that prey are distasteful (Skelhorn et al. 2016a). Other Lepidoptera have irritating hairs or spines to deter predators from consuming them or as a physical barrier to parasitoids' ovipositors (Murphy et al. 2010; Sugiura and Yamazaki 2014; Kageyama and Sugiura 2016) just as trichomes can protect plants from caterpillars (Kariyat et al. 2017).

Some caterpillars use colorful startle tactics, revealing eyespots or other warnings (osmeteria that look like snake tongues with a pungent odor) when harassed (Hossie and Sherratt 2012), and some have color patterns and behaviors that mimic larger predatory animals, a ruse for scaring away some predators. Many hide themselves within their host plant by spinning together plant parts and residing within or cutting leaves and silking them to construct shelters (Ito and Higashi 1991; Diniz et al. 2012; Lill et al. 2007; Marquis et al. 2019; Marquis et al. Chapter "The Impact Of Construct-Building By Caterpillars On Arthropod Colonists In A World Of Climate Change") to escape detection from certain enemies. Of those, some fling their frass (Weiss 2003, Moraes et al. 2012) or even make frass chains (Freitas and Oliveira 1992, 1996) to avoid detection when they climb out on the chains away from the leaf.

I. Jones
Department of Biological Sciences and International Center for Tropical Botany, Florida International University, Miami, FL, USA

Faculty of Forestry, University of Toronto, Toronto, ON, Canada

P. S. Oliveira
Departamento de Biologia Animal, Universidade Estadual de Campinas, UNICAMP, Campinas, Brazil

Larval Lepidoptera may pose as inedible objects, a strategy described as masquerade (Skelhorn and Ruxton 2010). Caterpillars of Geometridae may pose as twigs, holding on to branches with their back prolegs and keeping their body erect, at angles with the stem during the day. Others deploy a combination of crypsis with parts they can exert or flash to startle predators when disturbed (Stevens et al. 2008).

Early instars of the giant swallowtail butterfly look like shiny bird droppings on their citrus (Rutaceae) hostplants (Minno and Emmel 1992), while later instars may resemble small snakes (McAuslane 2009). Many tropical Sphingidae caterpillars are also snake mimics, a ruse to scare away predators (Janzen 1980) rather than to attract birds to eat snakes (Castellano and Cermelli 2015); the caterpillars can display behavior that furthers the ruse such as changing their posture (Hossie and Sherratt 2014), advantageous if the predators encounter the same prey repeatedly during their lives (Skelhorn et al. 2016b).

The least studied of the caterpillars may be those that are assumed to be palatable, and cryptically colored, avoiding detection by staying still on the plant surface (Bernays and Cornelius 1989; Dyer 1997; Henrique et al. 2005; Skelhorn and Ruxton 2010; Gaitonde et al. 2018). Animals may escape visual detection by blending into their surroundings, a phenomenon termed "background matching" (Skelhorn and Ruxton 2010; Ruxton et al. 2018). Young caterpillars of many Sphingidae align themselves with the midrib on the abaxial side of the leaf, not feeding during the day, to avoid detection. Many moth caterpillars demonstrate colors and patterns that easily camouflage against plant surfaces, including leaves and bark. Some employ "disruptive coloration" that makes their body harder to detect on patterned surfaces, or compound leaves, making it difficult for predators to grab them (Ruxton et al. 2018).

Birds and ants are considered the main enemies of caterpillars on foliage (Remmel et al. 2011; Singer et al. 2012), though parasitoids may also have a large negative effect (Wanner et al. 2007). Since foliage-dwelling ants detect prey at close range using mostly substrate-borne vibration and/or chemical cues (Cerdá and Dejean 2011; Wüst and Menzel 2017), and parasitoids can locate their prey by detection of kairomones (Rutledge 1996; Dutton et al. 2000; Afsheen et al. 2008; Wölfling and Rostás 2009), visual camouflage in caterpillars likely results from selective pressure exerted by visually hunting insectivorous birds that spot prey from greater distances (Edmunds 1974; Heinrich and Collins 1983).

Ant foraging on vegetation can be an important source of mortality to lepidopteran caterpillars on host plants (Floren et al. 2002). Since caterpillars usually move slowly on foliage, host plant selection by adult butterflies is one of the most important factors affecting caterpillar survival, which can be markedly lower on plants with high levels of ant visitation (Thompson and Pellmyr 1991; Sendoya and Oliveira 2015, 2017). Some butterflies are able to detect the presence of ants and avoid certain host plants in response (Mota and Oliveira 2016). The presence of ant attractants on plants (extrafloral nectaries, food/pearl bodies, fleshy fruits, insect trophobionts) has been demonstrated to induce ant presence on leaves and promote ant-caterpillar antagonism, which can negatively affect infestation levels by lepidopterans in numerous plant species (Koptur 1984, 1992, 2005; Heil and McKey

2003; Oliveira and Freitas 2004; Dutra et al. 2006; Rico-Gray and Oliveira 2007). Recent meta-analyses have concluded that ants attracted to and maintained by plant rewards are beneficial to plants, providing biotic protection in most cases (Chamberlain and Holland 2009; Rosumek et al. 2009).

Despite their pugnacious bodyguards (Bentley 1977), plants with ant attractants are still eaten by herbivores. Most well-known are some caterpillars that themselves secrete honeydew, appeasing the ant defenders to include them in their patrols, and protect them from their parasitoids (Pierce and Mead 1981; Pierce et al. 2002; Pierce & Dankowicz Chapter "Specializations for Ant Association in Caterpillars"), sometimes utilizing plants with extrafloral nectaries as "enemy free space" (Atsatt 1981). Some ant-tended Lycaenidae caterpillars have exocrine secretions that tantalize their ant attendants (Hojo et al. 2015). Caterpillars without such features may avoid ants by dropping from the plant, usually suspended by a silk strand up which they may return later (Sugiura and Yamazaki 2006). Some may even repel the ant guards by working in groups to fight back, such as the gregarious larvae of the noctuid moth *Dyops* on *Cecropia* plants protected by *Azteca* ants in Brazil (Ramos et al. 2018).

The Study System: Cryptic Caterpillars on Ant-Tended Plants

Many butterflies in the Pieridae family have green caterpillars, or otherwise blend into the color of plant parts on which they feed. Their eggs may be laid singly (e.g., *Phoebis* spp.) or in groups (e.g., *Ascia monuste*). Our experimental studies focused on pierid butterflies that oviposit on *Senna* spp. as hostplants. These butterflies have cryptically colored larvae that occur with a variety of patterns (Minno et al. 2005).

In southern Florida, native and ornamental species of *Senna* serve as host plants for *Phoebis philea*, *P. sennae*, and other Pieridae, including *Abaeis nicippe*. Ants are associated with all *Senna* spp., as these plants provide extrafloral nectar. Native *Senna* spp. associate with more species of native ants than do exotic species, though associations involve more species in urban areas than in natural areas (Koptur et al. 2017). Experimental exclusion of native ants from *Chamaecrista* (syn. *Cassia*) *fasciculata* in natural areas in northern Florida demonstrated that ants were effective in reducing numbers and damage to the plants by *P. sennae* larvae (Barton 1986); ant exclusion experiments in Iowa revealed that ants were important in reducing damage from another sulphur butterfly, *Pyrisitia* (syn. *Eurema*) *lisa* (Kelly 1986). In human-disturbed habitats, invasive species are more common than in natural areas, and fire ants, in particular, readily recruit to and occupy space near plants providing food rewards. The protective ability of imported fire ants was dramatically demonstrated by Fleet and Young (2000), who excluded red imported fire ants from *Senna occidentalis* in Texas and found that plants suffered much more herbivory from *P. sennae* and *A. nicippe* caterpillars in their absence, resulting in shorter plants that produced fewer and lighter fruits, and fewer seeds.

Extrafloral nectaries (EFNs) of *Senna mexicana* (Jacq.) H.S. Irwin & Barneby var. *chapmanii* (Isely) H.S. Irwin & Barneby (hereafter referred to as *Senna chapmanii*) attract ants and other predators that provide protection against herbivores (Koptur et al. 2015). When ants are excluded from plants, there are more caterpillars on the plants, and the nectaries attract numerous other predators and parasitoids. Artificial defoliation experiments showed that *S. chapmanii* plants produce more extrafloral nectar (EFN) in response to leaf damage (Jones and Koptur 2015a). Greenhouse experiments showed that plants produce more nectar at higher light intensities (Jones and Koptur 2015b), and field experiments showed that ant protection is most effective in sunny locations, where plants with ants received less herbivore damage, grew larger, and produced more flowers and fruits (Jones et al. 2017). More than eight species of ants have been observed at the foliar nectaries of *S. chapmanii* (Koptur et al. 2015; Jones et al. 2017), with two of the most common being the native Florida carpenter ant (*Camponotus floridanus* Buckley) and the invasive red imported fire ant (*Solenopsis invicta* Buren). Carpenter ants make their nests in fallen wood on soil or decaying wood in standing trees; conversely, fire ants nest in the ground, sometimes even at the base of *S. chapmanii* plants (Koptur et al. 2015). We have observed sulphur caterpillars (*Phoebis* spp.) on plants with both ant species in the field, sometimes being bothered or attacked by ants, other times ignored by the ants (Fig. 1).

Fig. 1 Typical interactions between carpenter ants, nectaries, and sulphur butterfly caterpillars. (**a**) *Camponotus floridanus* ants at extrafloral nectaries of *S. chapmanii* in an experimental colony (photo M.C. Pimienta); (**b**) *Solenopsis invicta* at extrafloral nectaries on a field-growing plant (photo I.M. Jones); (**c**) *C. floridanus* worker touching caterpillar with its antennae (photo M.C. Pimienta); (**d**) *S. invicta* workers attacking large sulphur caterpillar (photo M.C. Pimienta)

The efficacy of defense mechanisms by caterpillars may be crucial for their survival chances on ant-visited plants (Bernays 1997; Salazar and Whitman 2001; Sendoya and Oliveira 2017). On the other hand, ant-induced deterrence of caterpillars may also depend on the ant species' weaponry and aggressiveness (e.g., Sendoya et al. 2009). Consequently, ant-derived benefits to EFN-bearing plants may be conditioned by traits on either side of the ant-caterpillar interaction (Koptur 1992; Rico-Gray and Oliveira 2007, and references therein). Here, we sought to compare the interactions between sulphur butterfly caterpillars and the two main ant species visiting *Senna chapmanii* (*Camponotus floridanus* and *Solenopsis invicta*) to better understand the dynamics of the ant-herbivore interactions and gain insight into how so many caterpillars live to pupation on plants in nature.

Methods

To compare the response of these two ants to the caterpillars, we created captive colonies in the laboratory. In this way, we could control for environmental variables, allowing us to examine the effects of only caterpillar size and ant species in the interactions.

Senna chapmanii plants were grown from seeds collected from pine rockland natural areas in south Florida the previous year. Seeds were scarified with a razor blade, soaked overnight, and planted individually in soil-filled germination trays. After seedlings were large enough (and their roots protruded from the bottom of the seedling cell), they were potted up into 4″ pots where they grew for at least 4 months prior to being used in our experiments.

The plastic containers housing the ants were fashioned into mesocosms for the study. Six mesocosms were set up the same way before the ant colonies were introduced: one potted *Zanthoxylum coriaceum* plant was partially buried and centered in the container, two potted *Senna chapmanii* plants were partially buried and equidistant from the central plant, and six sticks were placed around each partially buried, potted plant as a pathway to ascend and descend. Periodically, plants were pruned to avoid touching edges, shelves, and lights above each box (so they would not serve as pathways for ant escapes). Grow-lights were suspended from the ceiling above the mesocosms to keep the plants healthy and to provide semi-natural light. Grow-lights were kept on a 12/12 light schedule.

In the spring of 2016, we collected native carpenter ants (*Camponotus floridanus*, Fig. 1a, c) from woodlands in city parks and gardens in Miami, Florida, by following workers to their nesting site, usually a rotting log lying on the ground surrounded by leaf litter. We collected the suspected colony into a large plastic tub, and in the laboratory opened it up enough to see if it contained a queen. We then put the colony in an open-topped plastic container filled with several inches of sandy loam soil, leaf litter, and pieces of decaying wood. The edges of the plastic container were coated with Fluon® Insect-a-Slip, a slippery surface that prevents ants from escaping. Three carpenter ant colonies were established and used throughout the duration of experiments.

Invasive red imported fire ants (*Solenopsis invicta*, Fig. 1b, d) were also collected in city parks and gardens in Miami, Florida. Fire ant nests are typically located in open areas with low vegetation and demonstrate a small, slightly raised, oval-shaped mound appearance. Once mounds were detected, a portion of it was dug up and transferred to a 5-gallon bucket. In the lab, fire ants were collected from the substrate by taking advantage of their rafting abilities against floods. A slow-drip technique was applied that inundates the substrate over a period of several days (Banks et al. 1981); afterwards, the floating raft of ants was transferred to a plastic container containing a sandy loam soil, leaf litter, and pieces of decaying wood. As for the carpenter ants, the edges of the plastic container were coated with Fluon® Insect-a-Slip. Three fire ant colonies were established and used throughout the duration of experiments.

Ant colonies were provided test tubes filled with sugar water and plain water using cotton wool to stop the liquids from spilling out. For protein and lipids, ant colonies were provided chunks of hard-boiled eggs and potted meat (Spam) every other day on small plastic dishes, which were cleaned every 2–3 days. Test tubes were replaced and cleaned every 3 days. Prior to experimentation with caterpillars, all food was removed for 2 days, so ants would be more likely to forage for solid food and collect foliar nectar produced by *Senna chapmanii*.

Caterpillars of the cloudless sulphur butterfly (*Phoebis sennae* (L.)) and the orange-barred sulphur butterfly (*P. philea* (L.)) were field-collected in Miami, Florida, and used for experimentation. The two caterpillar species are very similar in appearance and size in their earlier instars, and both occur in many different patterns, and are therefore difficult to distinguish prior to adult emergence. Therefore, we grouped both species together as "sulphur caterpillars." In our experiments we used a variety of patterns, though all the caterpillars we used were green and not yellow (green caterpillars are found eating foliage, yellow ones on flowers).

To measure the ants' responses to caterpillars, a single caterpillar was placed on the distal part of the plant's leaf with a small paintbrush or leaf fragment, observing it until ants discovered it. Ant colonies were randomly selected each day and rotated until approximately equal numbers of caterpillars (small and large, < 1.5 cm vs. > 1.5 cm) were tested against carpenter and fire ants. Small caterpillars represented 2nd instars, while large caterpillars included 4th and 5th instars.

For each trial, the caterpillar was introduced slowly onto the plant using a paintbrush or a leaf fragment. The placed caterpillar was observed for 15 min; if there was no encounter between caterpillar and ant then the trial was terminated, and no other data were recorded that day for that particular ant colony. If an encounter occurred, we recorded the time and observed all subsequent activity for 30 min or until the caterpillar was removed from or left the plant. Time to discovery was measured as the time at which an ant came into direct contact with the caterpillar after the caterpillar had settled on the leaf, irrespective of their behavioral response. The observer then recorded the behavioral responses of ants, categorizing them as ignore, inspect (Fig. 2), attack (Fig. 3), and/or removal from plant. Behavioral

Fig. 2 Carpenter ant touching then ignoring a sulphur caterpillar (photos J.T. Clayborn): (**a**) ant encounters caterpillar; (**b**) ant investigates with its mouthparts and antennae; (**c**) ant grooming antenna from material picked up from the caterpillar's surface; (**d**) ant investigates further; (**e**) ant departs

responses of caterpillars were also recorded such as twitch, flick, bite, bleed, thrash, or drop from the plant. We recorded time to discovery, time to attack, caterpillar behavior, and caterpillar fate for each trial. We summarized the behavioral interactions between the ants and caterpillars using the following categories:

1. Contact without violence (inspect and ignore/inspect and investigate).
2. Contact with agitation (ant inspects the caterpillar and the caterpillar moves—head flick or head butt)
3. Violence and agitation (ant bites the caterpillar and the caterpillar responds with a head butt or tail flick)
4. Caterpillar regurgitating after ant attack
5. Caterpillar bleeding after attack
6a. Caterpillar drops on silk strand from the leaf
6b. Ants remove and kill the caterpillar

We repeated a smaller number of placement experiments using a caterpillar that does not eat *Senna*, the long-tailed skipper (*Urbanus proteus* (L.); Hesperiidae). The goal was to see if these might be more easily detected than the sulphur caterpillars that normally eat the plants. These butterflies oviposit on butterfly pea (*Centrosema virginiana* (L.) Benth.) as well as on weedy species of *Desmodium* (*D. incanum*

Fig. 3 Carpenter ant attacking a caterpillar that moved when touched—a fatal flaw (photos J.T. Clayborn). (**a–f**) ant struggling with wriggling caterpillar, biting, wrestling; (**g**) ant attempts to carry caterpillar down plant; (**h**) help is on the way as another ant approaches

(Sw.) DC, *D. tortuosum* (Sw.) DC), and garden beans (*Phaseolus vulgaris* L.). Garden bean plants were the sole food source for the caterpillars used in these experiments. As in the other caterpillar trials, small caterpillars were less than 1.5 cm in length, and the large were greater than 1.5 cm in length.

We compared average time to discovery for each ant/caterpillar size combination using t-tests for independent samples, with Bonferroni probability corrections for multiple t-tests. Time to attack (post-discovery) was also compared using the same statistical tests.

We compared caterpillar fate in the presence of each ant species (proportions alive, dropped, or dead) using contingency table Chi-squared tests. We further explored caterpillar fate (dead or alive) using logistic regression of several variables (time to discovery, time to attack, and number of ants on plants).

We developed flow diagrams summarizing the sequences of behaviors and outcomes observed for caterpillars of each size in the presence of each ant species. These flow diagrams compare the proportions of caterpillars that conformed to each pathway.

Results

Many of the trials resulted in no encounters, and those data were not part of the behavioral responses. Of the 94 trials we attempted with carpenter ants, 33% of them were discarded, as the ants never contacted the caterpillars placed on the plant. Of the 63 trials attempted with fire ants, 14% of them were discarded as no shows.

The average time-to-discovery (i.e., the time when ants come into contact with the caterpillar), in those trials where an encounter was observed, showed no significant difference between caterpillar sizes for carpenter ants, but it took fire ants substantially longer to find the large caterpillars than the small ones (Fig. 4a). Time-to-attack (post-discovery) was nearly instantaneous for fire ants (within 5 s), whereas carpenter ants took much longer if they did attack the caterpillar (Fig. 4b). Carpenter ants did not differ in their attacks on small and large caterpillars, but fire ants attacked large caterpillars more quickly than small caterpillars (Fig. 4b). A logistic regression analysis revealed that attack time was the only variable correlated with caterpillar survival (p = 0.037); ants that attacked more quickly were more likely to kill and remove caterpillars from the plants.

The fate of sulphur caterpillars differed markedly between carpenter ant and fire ant encounters (Fig. 4c). In the presence of carpenter ants, more large caterpillars dropped from the plant than did small caterpillars (16% vs. 10%), and a similar proportion of caterpillars of both sizes were killed by the carpenter ants (10%). Most of the caterpillars detected remained alive and in place, with no significant differences in any of these responses between small and large caterpillars. In contrast, few caterpillars remained alive on the plants in the presence of fire ants (4%). As with carpenter ants, larger caterpillars were much more likely to drop off the plant than smaller caterpillars, but this difference was much larger in the presence of fire ants (72% vs. 19%). Most remaining ants that did not drop were killed by the fire ants (14% of the large caterpillars vs. 65% of the small caterpillars). The mortality rate of small sulphur caterpillars was significantly higher than that of large sulphur caterpillars from encounters with fire ants, but similar numbers of both sizes remained on the plants alive (15% and 14%). The ant responses to caterpillars of different sizes differed (Fig. 4d): carpenter ants were more likely to inspect, then poke or touch the caterpillars, and only sometimes attack, whereas fire ants were less likely to inspect before attacking the caterpillars.

Non-*Senna* eating skipper caterpillars placed on plants also sometimes went undetected by both ant species, although rates differed between ants. Half (50%) of these caterpillars were undetected by carpenter ants (n = 26), and 16% were undetected by fire ants (n = 38). The sample sizes of trials with these caterpillars with carpenter ants were therefore small, but we report them here for descriptive comparison.

Though half of the long-tailed skipper (*Urbanus proteus*) caterpillars placed on the plants were not detected by carpenter ants, those that were found took considerably longer to be detected (more than 10 min) than those found by fire ants (less than 5 min on average) (Fig. 4e). Once discovered by carpenter ants, however, the

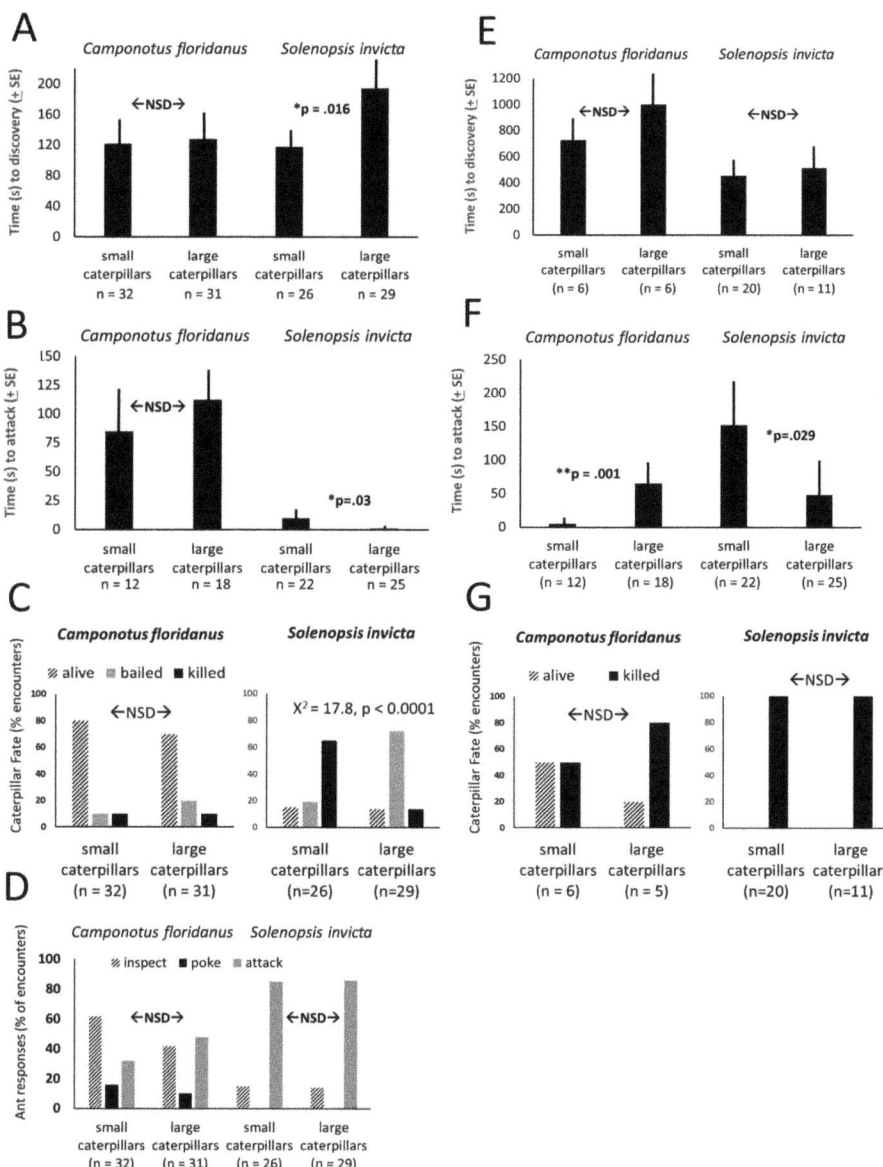

Fig. 4 Caterpillar placement experiment results: (**a**) Average time to discovery of small and large sulphur caterpillars (*Phoebis* spp.) placed on *Senna chapmanii* plants in mesocosms by carpenter ants (*Camponotus floridanus*) and fire ants (*Solenopsis invicta*). NSD indicates no significant difference between small and large caterpillars; (**b**) average time to attack of caterpillars, as in A; (**c**) fate of sulphur caterpillars placed on plants; (**d**) response of ants to sulphur caterpillars on encounter; (**e**) average time to discovery by carpenter ants and fire ants of small and large skipper caterpillars (*Urbanus proteus*) placed on *Senna chapmanii* plants in mesocosms; (**f**) average time to attack of caterpillars, as in E; (**g**) fate of skipper caterpillars placed on plants, as in E

caterpillars were attacked in less than 5 s. Fire ants took longer on average to attack a detected skipper caterpillar (Fig. 4f). The fate of skipper caterpillars encountered by fire ants was more dire than that of sulphur caterpillars. Some of them began to silk together leaves, in which they might have evaded detection, but none were observed to silk off or bail from plants. When compared with sulphur caterpillars, more of these skipper caterpillars were removed from the plants by carpenter ants, especially the smaller caterpillars (Fig. 4g), though the small sample sizes precluded statistical comparisons. With fire ants, the fate of every skipper caterpillar encountered was the same; all (small and large) were removed or killed by fire ants (Fig. 4g).

Comparing the sequence of events in the observed interactions reveals some differences in the ants' responses to small and large sulphur caterpillars (Fig. 5). A large proportion of the small caterpillars were ignored after encounter by carpenter ants (Fig. 5a), while those that were discovered were attacked and dispatched (a few dropped from the plant). Carpenter ants recognized most of the large caterpillars, but many of those resisted ant attack and a large proportion of them dropped from the plant (Fig. 5b). Fire ants found nearly all the caterpillars, both small (Fig. 5c) and large (Fig. 5d). Most small caterpillars succumbed to ant attack (Fig. 5c), while a substantial number of the larger caterpillars escaped by dropping from the plant (Fig. 5d).

Discussion

Size Matters

As in many other experimental studies (Tilman 1978; Smiley 1985, 1986; Koptur 1984; Freitas and Oliveira 1996; Fleet and Young 2000), we found that sulphur caterpillar mortality from ant predation is size-dependent: larger caterpillars were more likely to take evasive action by silking off and dropping from the plant and have higher survival. With carpenter ants, if the sulphur caterpillar did not move, it was not usually detected. It may be that individual caterpillars that hold still in the presence of ants are those likely to survive to larger sizes.

All Ants Are Not Created Equal

Ant species vary widely in size, behavior, and food preferences and exist as a mosaic throughout natural habitats interacting with the plant communities (Bluethgen et al. 2004; Leston 1978; Sendoya and Oliveira 2015). Experiments have revealed that some species are better than others in providing protection for plants (Horvitz and Schemske 1984; Letourneau 1983; Koptur 1984; Rico-Gray and Thien 1989; Mody and Linsenmair 2004; Sendoya and Oliveira 2015, 2017; Melati and Leal 2018).

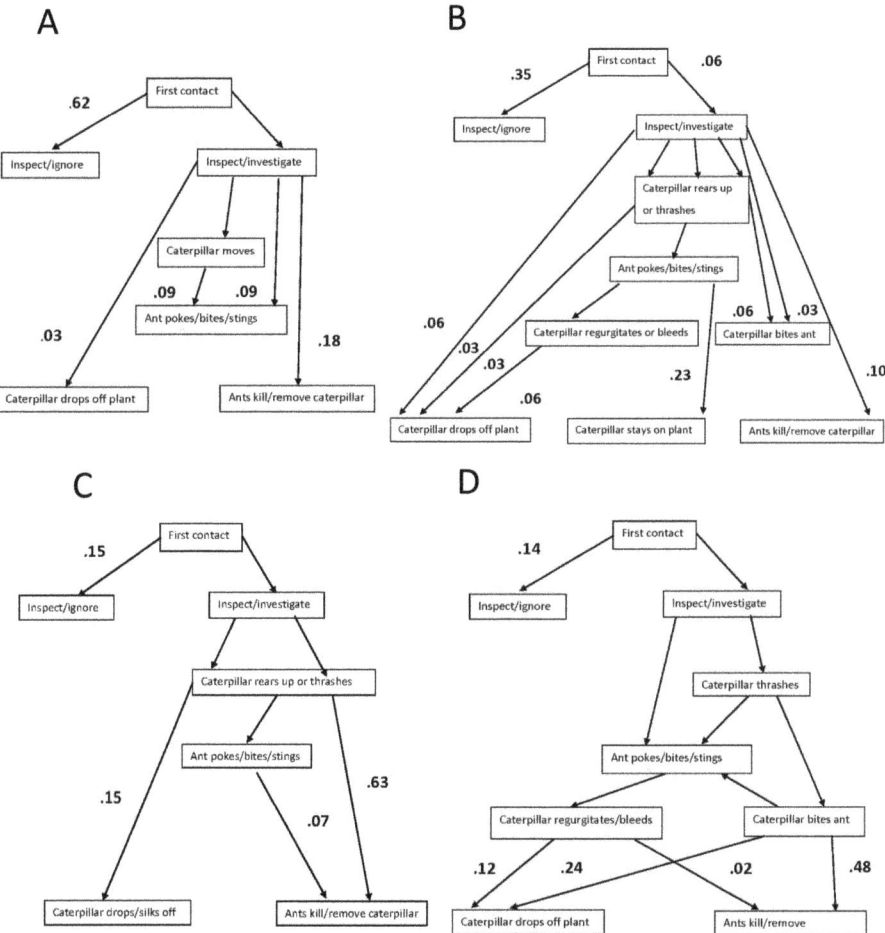

Fig. 5 Pathways of different ant/caterpillar interactions. (**a**) Carpenter ants and small sulphur caterpillars; (**b**) carpenter ants and large sulphur caterpillars; (**c**) fire ants and small sulphur caterpillars; (**d**) fire ants and large sulphur caterpillars. Numbers on lines are the proportion of interactions that follow that pathway

Camponotus spp. respond to both extrafloral nectar and homopteran produced honeydew (Sendoya et al. 2016) and are found more abundantly where those resources are available (Bluethgen and Feldhaar 2010). In the Brazilian cerrado, *Camponotus* spp. are attracted to these liquid plant rewards and are effective plant bodyguards (Del-Claro and Oliveira 2000; Oliveira and Freitas 2004), not only harassing caterpillars but also discouraging oviposition by butterflies (Sendoya et al. 2009). Several species of *Camponotus* associated with *Pseudocedrela* in Cote d'Ivoire differed in their ability to repel herbivores (Mody and Linsenmair 2004), as did those on *Inga* in Costa Rica (Koptur 1984).

Solenopsis invicta, an invasive species, can facilitate establishment of invasive plant species by protecting them against herbivores (Ackerman et al. 2014). However, sometimes the most aggressive ant species are not the most beneficial to the plant bearing extrafloral nectaries (Melati and Leal 2018). They may repel pollinators as well as herbivores (Ness 2006), thwarting a plant's reproduction. In addition, they can displace native ants and attack other beneficial insects, such as parasitoids (Ness 2003). Imported fire ants have been recognized as having some benefits in agriculture as their aggressive behavior has the potential to repel and eliminate many pest species (Reagan 1986), including the cotton boll weevil (Jones and Sterling 1979) and the sugarcane borer (Adams et al. 1981), though also potentially harmful to agricultural workers.

Fire ants were quick to find the caterpillars in our experiments, and poked and prodded them until they flinched and were mercilessly attacked with bites and stings. After 2 days with no food supplied, they were ready for more than just nectar, they needed caterpillar meat (the "tenderloin of the insect world," to quote David Wagner). Their ravenous attacks resulted in the dropping off or demise of nearly all the caterpillars in the experiments. In the field, we have occasionally observed these ants eating the seeds from the developing fruits of *Senna chapmanii*, another way in which an overly voracious bodyguard may harm plant fitness.

Appearance May Not Matter to Ants, But What Does?

It is thought that many lepidopteran species use cryptic coloration to evade visual predators, but few studies have explored the possible mechanisms by which such species avoid predation by insects (but see Henrique et al. 2005). Ants, for example, use largely tactile and olfactory senses to discover prey and orient themselves in the environment (Cerdá and Dejean 2011). What appears as visual camouflage may be effective against birds and other visual predators, but not be important to predators using tactile or olfactory cues to detect prey.

Phoebis sennae have been described as having "common, aposematic caterpillars" (Quicke 2017) and the US Forest Service states on its website (Cole 2017) that "Both *Senna* and *Cassia* are poisonous, which allows the caterpillars to accumulate a toxic deterrent to would-be predators." These reports are surprising, as we have often observed caterpillars of this species, and of *P. philea*, to be consumed by birds, lizards, spiders, and large wasps during our field studies. In addition, *Senna* is also consumed by humans as a laxative, so it is not very toxic to those primates. Another legume-consuming Pieridae tested for palatability with *Paraponera* ants (*Anteos clorinde*, hostplant *Cassia fruticosa*) was considered neither tasty nor nasty, but neutral (Dyer 1995). As the *Phoebis* spp. caterpillars in our experiments eat a similar legume hostplant, we assume they do not have chemicals in their bodies that make them unpalatable to ants. Indeed, those individuals that dropped into the captive ant colonies in our mesocosms were readily consumed.

Fig. 6 Sulphur caterpillars feeding on *Senna* flowers are yellow; here are two of several contrasting pattern morphs. (**a**) Yellow sulphur caterpillar with longitudinal black stripes; tiny droplets are visible along its body (Photo M.C. Pimienta). (**b**) Yellow sulphur caterpillar with black rings around its body in an inflorescence of *S. chapmanii* (photo S. Koptur). (**c**) Closer view of the tiny droplets on the sulphur caterpillar surface (photo M.C. Pimienta)

Given the lack of visual and chemical defenses exhibited by *Phoebis* spp. caterpillars, the question remains: how do so many beat the odds, surviving to maturity in the presence of motivated predatory ants? The answer to this question may lie in the sequestering of host-plant chemicals after all, not for the accrual of distasteful chemicals, but to develop a form of chemical camouflage. Photographers have documented tiny droplets at the ends of setae covering the bodies of *Phoebis* caterpillars (Fig. 6). It is likely that these droplets contain compounds similar to their hostplants, allowing the caterpillars to avoid detection by non-visual predators by blending in with the surrounding foliage. Chemical camouflage has been observed in Ithomiidae caterpillars, whose consumption of their hostplant led to their bodies expressing epicuticular waxes similar to those of the host plant, making them less likely to be found by *Camponotus* ants (Henrique et al. 2005). In our study, the more aggressive nature of the fire ants led them to prod anything they encountered, and once the caterpillars moved, they attacked. With carpenter ants, if the caterpillars remained motionless, the ants walked over them, occasionally grooming their antennae as if they had encountered some substance they needed to clean off, but not apparently detecting the caterpillars, and not attacking them. Future experiments using freeze-dried caterpillars and hexane extracts of these compounds are planned to determine if this phenomenon is a defense in these caterpillars against some of their ant predators (Henrique et al. 2005). In addition, repeating the present study, but placing the sulphur caterpillars on a non-host plant with extrafloral nectaries and ants, might reveal if the caterpillars can quickly discover caterpillars with cuticular chemicals that do not match the non-host plant.

Sulphur Butterfly Caterpillars and Ants in a Changing World

Temperature is apparently the most important variable predicting the diversity of ants in a given location (Jenkins et al. 2011), and ant ecologists agree there is much we still need to know about how changes in the warmest parts of the earth, both wet and dry, will be affected by climate change. These parts of the world are currently where the greatest ant diversity exists. The same is true for plant diversity, and species that are not tolerant of warmer temperatures may migrate higher in elevation or further north or south of the equator if connections exist and habitat is available for migration (Feeley 2012; Feeley and Rehm 2012).

Development encroaching on natural areas means less wildlife habitat and fewer native plants available for species that depend upon them. The host plant range of *Phoebis philea* and *P. sennae* butterflies includes a number of genera in the Fabaceae, and they utilize both native and introduced species (Koptur et al. 2017), making them one of the best ambassadors for butterfly gardening in urban areas (Minno and Minno 1999). Areas where people live often have more pest control (spraying for mosquitoes, fleas, ticks, etc.) that can reduce the numbers of butterflies as well as their natural enemies, particularly parasitoid wasps and flies. Depending on the timing of these chemical controls, the numbers of butterflies may increase when parasitoids and predators are suppressed, and at such times the butterflies can be numerous and a beautiful sight on city streets. When the timing is wrong, the urban landscape can be devoid of not only butterflies but other beautiful and beneficial insects, as few insecticides target only problem species.

While *P. sennae* is native to Florida and the southeastern USA., *P. philea* is naturalized to Florida, having become established around 1920 from its native range further south (Minno and Minno 1999). It is likely that as conditions warm up along the east coast of the USA, the range of tolerance of the hostplants will extend further north. Perhaps hostplants that require more tropical conditions may move northward and provide new alternative hosts for these butterflies; but likewise, a wider variety of Fabaceae-consuming Pieridae may also follow northward.

Camponotus floridanus is native but abundantly present in areas of human habitation, nesting in dead wood on trees and sometimes in man-made structures. *Solenopsis invicta* is an invasive species, outcompeting many native ants and occurring especially around the edges of many natural areas. They are frequent pests of developed areas, including home gardens and lawns, and people take many measures to limit their presence as their stings are painful and long-lasting. Wherever the sulphur butterfly caterpillars occur in southern Florida, they are likely to encounter both ant species; as the climate warms, *S. invicta* is spreading across the southern USA and may continue northward over the coming years (Fitzpatrick et al. 2007). Perhaps *C. floridanus* will do the same, as their populations are likely limited by freezing temperatures.

Conclusion

Caterpillars exhibit a broad spectrum of defensive traits, ranging from aposematic coloration and the sequestering of plant toxins to aggressive flicks and the flashing of osmeteria. Nevertheless, *Phoebis* spp. maintain large populations, seemingly in the absence of any such defenses. It is possible that evasion of ants is due to cuticular chemicals sequestered from their hostplants that function as chemical camouflage to the ants, but this suggestion requires further examination and experimentation.

Acknowledgments We thank Brianna Chin and Thomas Spencer for laboratory help. Ian Jones was supported by a Dissertation Year Fellowship from FIU; Andrea Salas Primoli and Brittany Harris by Dissertation Enhancement Awards from FIU; Paulo S. Oliveira was supported by CAPES – (Finance Code 001), and by research grants from the CNPq (302219/2017-0) and FAPESP (BIOTA Program, 2014/23141-1, 2014/15721-8, 2017/16645-1).

References

Ackerman JD, Falcon W, Molinari J, Vega C, Espino I, Cuevas AA (2014) Biotic resistance and invasional meltdown: consequences of acquired interspecific interactions for an invasive orchid, *Spathoglottis plicata* in Puerto Rico. Biol Invasions 16:2435–2447

Adams CT, Summers TE, Lofgren CS, Focks DA, Prewitt JC (1981) Interrelationship of ants and the sugarcane borer *Diatraea saccharalis* in Florida USA sugarcane fields. Environ Entomol 10:415–418

Afsheen S, Wang X, Li R, Zhu CS, Lou YG (2008) Differential attraction of parasitoids in relation to specificity of kairomones from herbivores and their byproducts. Insect Sci 15:381–397

Atsatt PR (1981) Lycaenid butterflies and ants: selection for enemy-free space. Am Nat 118:638–654

Banks WA, Lofgren CS, Jouvenaz DP, Stringer CE, Bishop PM, Williams DF, Wojcik PD, Glancey BM (1981) Techniques for collecting, rearing, and handling imported fire ants. US Department of Agriculture Technician AAT-S-21. 9 p

Barton AM (1986) Spatial variation in the effect of ants on an extrafloral nectary plant. Ecology 67:495–504

Bentley B (1977) Extrafloral nectaries and protection by pugnacious bodyguards. Annu Rev Ecol Syst 88:407–427

Bernays EA (1997) Feeding by lepidopteran larvae is dangerous. Ecol Entomol 22:121–123

Bernays EA, Cornelius ML (1989) Generalist caterpillar prey are more palatable than specialists for the generalist predator *Iridomyrmex humilis*. Oecologia 79:427–430

Bluethgen N, Feldhaar H (2010) Food and shelter: how resources influence ant ecology. In: Lach L, Parr CL, Abbott KL (eds) *Ant ecology*, pp 115–136

Bluethgen N, Stork NE, Fiedler K (2004) Bottom-up control and co-occurrence in complex communities: honeydew and nectar determine a rainforest ant mosaic. Oikos 106:344–358

Brower LP, Glazier SC (1975) Localization of heart poisons in the monarch butterfly. Science 188:19–25

Calvert WH, Hedrick LE, Brower LP (1979) Mortality of the monarch butterfly (*Danaus plexippus* L.) due to avian predation at five overwintering sites in Mexico. Science 204:847–851

Castellano S, Cermelli P (2015) Preys' exploitation of predators' fear: when the caterpillar plays the gruffalo. Proc R Soc Biol Sci Ser B 282:20151786

Cerda X, Dejean A (2011) Chapter 3: Predation by ants on arthropods and other animals. In: Polidori C (ed) Predation in the Hymenoptera: an evolutionary perspective, pp 39–78. ISBN: 978-81-7895-530-8

Chamberlain SA, Holland JN (2009) Quantitative synthesis of context dependency in ant-plant protection mutualisms. Ecology 90:2384–2392

Cole E (2017) U.S. Forest Service webpage pollinator of the month – cloudless Sulphur butterfly. https://www.fs.fed.us/wildflowers/pollinators/pollinator-of-the-month/cloudless-sulphur-butterfly.shtml

Del Claro K, Oliveira PS (2000) Conditional outcomes in a neotropical treehopper-ant association: temporal and species-specific variation in ant protection and homopteran fecundity. Oecologia 124:156–165

Diniz IR, Hay JD, Rico-Gray V, Greeney HF, Morais HC (2012) Shelter-building caterpillars in the cerrado: seasonal variation in relative abundance, parasitism, and the influence of extra-floral nectaries. Arthropod Plant Interact 6:583–589

Dutra HP, Freitas AVL, Oliveira PS (2006) Dual ant attraction in the Neotropical shrub *Urera baccifera* (Urticaceae): the role of ant visitation to pearl bodies and fruits in herbivore deterrence and leaf longevity. Funct Ecol 20:252–260

Dutton A, Mattiacci L, Dorn S (2000) Plant-derived semiochemicals as contact host location stimuli for a parasitoid of leafminers. J Chem Ecol 26(10):2259–2273

Dyer LA (1995) Tasty generalists and nasty specialists? Antipredator mechanisms in tropical lepidopteran larvae. Ecology 76:1483–1496

Dyer LA (1997) Effectiveness of caterpillar defenses against three species of invertebrate predators. J Res Lepidop 34:48–68

Edmunds M (1974) Defence in animals. Longman, Harlow

Feeley KJ (2012) Distributional migrations, expansions, and contractions of tropical plant species as revealed in dated herbarium records. Glob Chang Biol 18:1335–1341

Feeley KJ, Rehm EM (2012) Amazon's vulnerability to climate change heightened by deforestation and man-made dispersal barriers. Glob Chang Biol 18:3606–3614

Fitzpatrick MC, Weltzin JF, Sanders NJ, Dunn RR (2007) The biogeography of prediction error: why does the introduced range of the fire ant over-predict its native range? Glob Ecol Biogeogr 16:24–33

Fleet RR, Young BL (2000) Facultative mutualism between imported fire ants (*Solenopsis invicta*) and a legume (*Senna occidentalis*). Southwest Nat 45:289–298

Floren A, Biun A, Linsenmair KE (2002) Arboreal ants as key predators in tropical lowland rainforest trees. Oecologia (Berlin) 131:137–144

Freitas AVL, Oliveira PS (1992) Biology and behavior of the neotropical butterfly *Eunica bechina* (Nymphalidae) with special reference to larval defence against ant predation. J Res Lepidop 31:1–11

Freitas AVL, Oliveira PS (1996) Ants as selective agents on herbivore biology: effects on the behaviour of a non-myrmecophilous butterfly. J Anim Ecol 65:205–210

Gaitonde N, Joshi J, Kunte K (2018) Evolution of ontogenic change in color defenses of swallowtail butterflies. Ecol Evol 8:9751–9763

Heil M, McKey D (2003) Protective ant-plant interactions as model systems in ecological evolutionary research. Annu Rev Ecol Syst 34:425–453
Heinrich B, Collins S (1983) Caterpillar leaf damage, and the game of Hide-and-seek with birds. Ecology 64(3):592–602. https://doi.org/10.2307/1939978
Henrique A, Portugal A, Trigo JR (2005) Similarity of cuticular lipids between a caterpillar and its host plant: a way to make prey undetectable for predatory ants? J Chem Ecol 31:2551–2561
Hojo MK, Pierce NE, Tsuji K (2015) Lycaenid caterpillar secretions manipulate attendant ant behavior. Curr Biol 25:2260–2264
Horvitz CC, Schemske DW (1984) Effects of nectar-harvesting ants and an ant-tended herbivore on seed production of a neotropical herb. Ecology 65(5):1369–1378
Hossie TJ, Sherratt TN (2012) Eyespots interact with body colour to protect caterpillar-like prey from avian predators. Anim Behav 84:167–173
Hossie TJ, Sherratt TN (2014) Does defensive posture increase mimetic fidelity of caterpillars with eyespots to their putative snake models? Curr Zool 60:76–89
Ito F, Higashi S (1991) Variance of ant effects on the different life forms of moth caterpillars. J Anim Ecol 60:327–334
Jenkins CN, Sanders NJ, Andersen AN, Arnan X, Brühl CA, Cerda X, Ellison AM, Fisher BL, Fitzpatrick MC, Gotelli NJ, Gove AD, Guénard B, Lattke JE, Lessard J, McGlynn P, Menke SB, Parr CL, Philpott SM, Vasconcelos HL, Weiser MD, Dunn RR (2011) Global diversity in light of climate change: the case of ants. Divers Distrib 17:652–662
Jones D, Sterling WL (1979) Manipulation of red imported fire ants in a trap crop for boll weevil suppression. Environ Entomol 8:1073–1077
Janzen DH (1980) Two potential coral snake mimics in a tropical deciduous forest. Biotropica 12(1):77–78.
Jones IM, Koptur S (2015a) Dynamic Extrafloral nectar production: the timing of leaf damage affects the defensive response in *Senna mexicana* var. *chapmanii* (Fabaceae). Am J Bot 102:58–66
Jones IM, Koptur S (2015b) Quantity over quality: light intensity, but not red/far-red ratio, affects extrafloral nectar production in *Senna mexicana* var. *chapmanii*. Ecol Evol 5:4108–4114
Jones IM, Koptur S, Gallegos HR, Tardanico JP, Trainer PA, Peña J (2017) Changing light conditions in pine rockland habitats affect the intensity and outcome of ant-plant interactions. Biotropica 49:83–91
Kageyama A, Sugiura S (2016) Caterpillar hairs as an anti-parasitoid defence. Sci Nat 103:86
Kariyat RR, Smith JD, Stephenson AG, De Moraes CM, Mescher MC (2017) Non-glandular trichomes of *Solanum carolinense* deter feeding by *Manduca sexta* caterpillars and cause damage to the gut peritrophic matrix. Proc R Soc Biol Sci Ser B 284:20162323
Kelly CA (1986) Extrafloral nectaries: ants, herbivores and fecundity in *Cassia fasciculata*. Oecologia 69:600–605
Koptur S (1984) Experimental evidence for defense of *Inga* saplings (Mimosoideae) by ants. Ecology 65:1787–1793
Koptur S (1992) Extrafloral nectary-mediated interactions between insects and plants. In: Bernays E (ed) Insect–plant interactions, vol 4. CRC, Boca Raton, pp 81–129
Koptur S (2005) Chapter 3: Nectar as fuel for plant protectors. In: Wäckers FL, van Rijn PCJ, Bruin J (eds) Plant-provided food for carnivorous insects: a protective mutualism and its applications. Cambridge University Press, Cambridge, pp 75–108
Koptur S, Jones IM, Liu H, Diaz-Castelazo C (2017) Playing the system: the impacts of invasive ants and plants on facultative ant-plant interactions. In: Oliveira PS, Koptur S (eds) Ant-plant interactions - impacts of humans on terrestrial ecosystems. Cambridge University Press, Cambridge, pp 249–266
Koptur S, Jones IM, Peña J (2015) The influence of host plant extrafloral nectaries on multitrophic interactions: an experimental investigation. PLoS One 10:e0138157
Leston D (1978) A neotropical ant mosaic. Ann Entomol Soc Am 71:649–653

Letourneau DK (1983) Passive aggression an alternative hypothesis for the *Piper Pheidole* association. Oecologia (Berlin) 60:122–126

Lill JT, Marquis RJ, Walker MA, Peterson L (2007) Ecological consequences of shelter sharing by leaf-tying caterpillars. Entomol Exp Appl 124:45–53

Marquis RJ, Lill JT, Forkner RE, LeCorff J, Landosky JM, Whitfield JB (2019) Declines and resilience of communities of leaf chewing insects on Missouri oaks following spring frost and summer drought. Front Ecol Evol 7:396. https://doi.org/10.3389/fevo.2019.00396

McAuslane HJ (2009) Giant swallowtail, orangedog, *Papilio cresphontes* Cramer (Insecta: Lepidoptera: Papilionidae). EDIS 2009:4. Archival copy, update of EENY-008

Melati BG, Leal LC (2018) Aggressive bodyguards are not always the best: preferential interaction with more aggressive ant species reduces reproductive success of plant bearing extrafloral nectaries. PLoS One 13:e0199764

Minno MC, Butler JF, Hall DW (2005) Florida butterfly caterpillars and their host plants. University of Florida Press, Gainesville

Minno MC, Emmel TC (1992) Larval protective coloration in swallowtails from the Florida Keys (Lepidoptera: Papilionidae). Trop Lepidop Res 3:47–49

Minno MC, Minno M (1999) Florida butterfly gardening. A complete guide to attracting, identifying, and enjoying butterflies of the lower south. University Press of Florida, Gainesville

Mody K, Linsenmair KE (2004) Plant-attracted ants affect arthropod community structure but not necessarily herbivory. Ecol Entomol 29:217–225

Moraes AR, Greeney HF, Oliveira PS, Barbosa EP, Freitas AVL (2012) Morphology and behavior of the early stages of the skipper, *Urbanus esmeraldus*, on *Urera baccifera*, an ant-visited host plant. J Insect Sci 12:52

Mota LL, Oliveira PS (2016) Myrmecophilous butterflies utilize ant–treehopper associations as visual cues for oviposition. Ecol Entomol 41:338–343

Murphy SM, Leahy SM, Williams LS, Lill JT (2010) Stinging spines protect slug caterpillars (Limacodidae) from multiple generalist predators. Behav Ecol 21:153–160

Ness JH (2003) Contrasting exotic *Solenopsis invicta* and native *Forelius pruinosus* ants as mutualists with *Catalpa bignonioides*, a native plant. Ecol Entomol 28:247–251

Ness JH (2006) A mutualism's indirect costs: the most aggressive plant bodyguards also deter pollinators. Oikos 113:506–514

Oliveira PS, Freitas AVL (2004) Ant-plant-herbivore interactions in the neotropical cerrado savanna. Naturwissenschaften 91:557–570

Pierce NE, Braby MF, Heath A, Lohman DJ, Mathew J, Rand DB, Travassos MA (2002) The ecology and evolution of ant association in the Lycaenidae (Lepidoptera). Annu Rev Entomol 47:733–771

Pierce NE, Mead PS (1981) Parasitoids as selective agents in the symbiosis between lycaenid butterfly larvae and ants. Science (Washington D C) 211:1185–1187

Quicke DLJ (2017) Mimicry, crypsis, masquerade and other adaptive resemblances. Wiley Blackwell, Hoboken

Ramos RR, Freitas AVL, Francini RB (2018) Defensive strategies of a noctuid caterpillar in a myrmecophytic plant: are *Dyops* larvae immune to *Azteca* ants? Sociobiology 65(3):397–402

Reagan TE (1986) Beneficial aspects of the imported fire ant: a field ecology approach. In: Lofgren CS, VanderMeer RK (eds) Fire ants and leaf-cutting ants. Westview Press, Boulder, pp 58–71

Remmel T, Davison J, Tammaru T (2011) Quantifying predation on folivorous insect larvae: the perspective of life-history evolution. Biol J Linn Soc 104:1–18

Rico-Gray V, Oliveira PS (2007) The ecology and evolution of ant-plant interactions. The University of Chicago Press, Chicago, 331 pp

Rico-Gray V, Thien LB (1989) Effect of different ant species on reproductive fitness of *Schomburgkia tibicinis* (Orchidaceae). Oecologia 81:487–489

Rosumek FB, Silveira FAO, de Neves SF et al (2009) Ants on plants: a meta-analysis of the role of ants as plant biotic defenses. Oecologia 160:537–549. https://doi.org/10.1007/s00442-009-1309-x

Rutledge CE (1996) A survey of identified kairomones and synomones used by insect parasitoids to locate and accept their hosts. Chemoecology 7(3):121–131

Ruxton GD, Allen WL, Sherratt TN, Speed MP (2018) Avoiding attack: the evolutionary ecology of crypsis, aposematism, and mimicry, 2nd edn. Oxford University Press, Oxford

Salazar BA, Whitman DW (2001) Defensive tactics of caterpillars against predators and parasitoids. In: Ananthakrishnan TN (ed) Insect and plant defense dynamics. New Hampshire Science Publisher, pp 161–207

Sendoya SF, Blüthgen N, Tamashiro JY, Fernandez F, Oliveira PS (2016) Foliage-dwelling ants in a neotropical savanna: effects of plant and insect exudates on ant communities. Arthropod Plant Interact 10:183–195

Sendoya SF, Freitas AVL, Oliveira PS (2009) Egg-laying butterflies distinguish predaceous ants by sight. Am Nat 174:134–140

Sendoya SF, Oliveira PS (2015) Ant-caterpillar antagonism at the community level: Interhabitat variation of tritrophic interactions in a Neotropical savanna. J Anim Ecol 84:442–452

Sendoya SF, Oliveira PS (2017) Behavioural ecology of defence in a risky environment: caterpillars versus ants in a Neotropical savanna. Ecol Entomol 42:553–564

Singer MS, Clark RE, Lichter-Marck IH, Johnson ER, Mooney KA (2017) Predatory birds and ants partition caterpillar prey by body size and diet breadth. J Anim Ecol 86:1363–1371

Singer MS, Farkas TE, Skorik CM, Mooney KA (2012) Tritrophic interactions at a community level: effects of host plant species quality on bird predation of caterpillars. Am Nat 179:363–374

Skelhorn J, Halpin CG, Rowe C (2016a) Learning about aposematic prey. Behav Ecol 27:955–964

Skelhorn J, Holmes GG, Hossie TJ, Sherratt TN (2016b) Multicomponent deceptive signals reduce the speed at which predators learn that prey are profitable. Behav Ecol 27:141–147

Skelhorn J, Ruxton GD (2010) Predators are less likely to misclassify masquerading prey when their models are present. Biol Lett 6:597–599

Smiley J (1985) *Heliconius* caterpillar mortality during establishment on plants with and without attending ants. Ecology 66:845–849

Smiley J (1986) Ant constancy at *Passiflora* extrafloral nectaries: effects on caterpillar survival. Ecology 67:516–521

Stevens M, Stubbins CL, Hardman CJ (2008) The anti-predator function of 'eyespots' on camouflaged and conspicuous prey. Behav Ecol Sociobiol 62:1787–1793

Sugiura S (2020) Predators as drivers of insect defenses. Entomol Sci 23:316–337

Sugiura S, Yamazaki K (2006) The role of silk threads as lifelines for caterpillars: pattern and significance of lifeline-climbing behaviour. Ecol Entomol 31(1):52–57

Sugiura S, Yamazaki K (2014) Caterpillar hair as a physical barrier against invertebrate predators. Behav Ecol 25:975–983

Thompson JN, Pellmyr O (1991) Evolution of oviposition behavior and host preference in Lepidoptera. Annu Rev Entomol 36:65–90

Tilman D (1978) Cherries, ants, and tent caterpillars: timing of nectar production in relation to susceptibility of caterpillars to ant predation. Ecology 59:686–692

Trager MD, Bhotika S, Hostetler JA, Andrade GV, Rodriguez-Cabal MA, McKeon CS, Osenberg CW, Bolker BM (2010) Benefits for plants in ant-plant protective mutualisms: a meta-analysis. PLoS One 5(12):e14308. https://doi.org/10.1371/journal.pone.0014308

Wanner H, Gu H, Hattendorf B, Günther D, Dorn S (2007) Tracking parasitoids with the stable isotope ^{44}CA in agroecosystems. Agric Ecosyst Environ 118:143–148

Weiss MR (2003) Good housekeeping: why do shelter-dwelling caterpillars fling their frass? Ecol Lett 6:361–370

Wölfling M, Rostás M (2009) Parasitoids use chemical footprints to track down caterpillars. Commun Integr Biol 4:353–355

Wüst M, Menzel F (2017) I smell where you walked - how chemical cues influence movement decisions in ants. Oikos 126:149–160

The Natural History of Caterpillar-Ant Associations

Naomi E. Pierce and Even Dankowicz

A larva of *Nudina artaxidia* (Erebidae) steals honeydew from a monophlebid scale insect attended by *Lasius nipponensis*, as described in Komatsu and Itino (2014). (Photo by Takashi Komatsu)

The original version of this chapter was revised. The correction to this chapter is available at https://doi.org/10.1007/978-3-030-86688-4_21

N. E. Pierce (✉) · E. Dankowicz
Harvard University, Museum of Comparative Zoology, Cambridge, MA, USA
e-mail: npierce@oeb.harvard.edu; danko@brandeis.edu

© The Author(s) 2022, Corrected Publication 2022
R. J. Marquis, S. Koptur (eds.), *Caterpillars in the Middle*, Fascinating Life Sciences, https://doi.org/10.1007/978-3-030-86688-4_11

Introduction

Caterpillars have a fantastic array of chemical, physical, and behavioral defenses to protect themselves against ants (Borges et al. 2014; Darling et al. 2001; DeVries 1991a; Dyer 1995; Freitas 1999; Honda 1983; Peterson et al. 1987; Rostás 1657; Roux et al. 2011; Uemura et al. 2017). Larvae of diverse Lepidoptera are ignored by marauding ants foraging on their host plants, either due to chemical manipulation and camouflage (Akino et al. 2004; Eubanks et al. 1997; Portugal and Trigo 2005) or physical concealment (Bächtold and Alves-Silva 2013; Farquharson et al. 1922; Ito and Higashi 1991; Jones et al. 2002; Loeffler 1996; Sendoya and Oliveira 2017). Unharmed larvae of various butterfly and moth species are also occasionally known to live close to or within ant nests (Fiedler 1991; Kistner 1982; Lamborn et al. 1914, iNaturalist #65727498). Larvae that can survive encounters with ants and colonize ant territories, whether on host plants or inside structures built by ants, may enjoy a range of benefits including reduced competition, enemy-free space, and favorable microclimates (Atsatt 1981a; Hinton 1951; Koptur 1985; Saarinen and Daniels 2006). Passive coexistence of larvae and ants, through physical/chemical protection or signaling by larvae, may be an important prerequisite to the appearance of stable ant associations in caterpillars (DeVries 1991b; Fiedler 1991) much as in other arthropod groups (Cushing 1997; Cushing 2012; Hölldobler and Wilson 1990; Parker 2016; Stadler and Dixon 2005; Vantaux et al. 2012). Particularly in tropical tree canopies, mosaics of competing ant colonies and ant species play a major role in diversifying available host plant niches, structuring caterpillar communities and creating specialized niches for those able to coexist with them (Agassiz and Kallies 2018; Baker et al. 2016; Blüthgen and Stork 2007; Camarota et al. 2020; Dejean et al. 2017; Floren et al. 2002; Sendoya and Oliveira 2014; Seufert and Fiedler 1996; Wiens et al. 1993).

In this chapter, we provide an overview of caterpillar-ant associations. A number of recent reviews focus on ant associations in Lycaenidae and Riodinidae, including Pierce et al. (2002) and Casacci et al. (2019b). Other treatments such as Kistner (1982), Hölldobler and Wilson (1990) and Pierce (1995) have reviewed the caterpillars found in nests of social insects. However, ant associations have not been summarized and critically examined across all Lepidoptera since Hinton (1951). Many novel relationships have been uncovered in the intervening 70 years, and we discuss factors that may contribute to the phylogenetic distribution and biogeography of these unusual life histories at the end of the chapter. We have not used comparative methods to analyze potential correlates of different forms of ant association, although we plan to do so in a subsequent publication that will include additional phylogenetic and quantitative life history measurements. Our goal here is to describe the full range of natural histories exhibited by these taxa and to identify questions that require further study.

Over 70% of species in the large butterfly family Lycaenidae appear to be ant-associated, making them the largest single group of lepidopteran myrmecophiles (Tables 1 and 2). Two additional radiations of ant associates make up 20% of

Table 1 Based on life history records and recent phylogenies, myrmecophily appears to have arisen at least 30 times across the Lepidoptera as a whole in at least 17 families (Espeland et al. 2018; Kawahara et al. 2019; Léger et al. 2021; Mitter et al. 2017; Regier et al. 2015)

Family	Ant-associated group	Degree of association and type of relationship with ants	Number of ant-associated species	Distribution	References
Psychidae	*Iphierga, Ardiosteres* (may constitute more than one distinct group of ant associates)	Obligate. Larvae feed on debris or ants in *Iridomyrmex* or other nests	3 species	Australia	Hinton (1951), Kistner (1982)
Tineidae	Myrmecozelinae (in part may constitute more than one distinct group of ant associates)	Obligate. *Myrmecozela ochraceella* feed on *Formica* nest material and possibly also ants. *Ippa* are carnivorous and along with others occur with diverse ant groups	>8 species in >3 genera	Europe to New Guinea	Ahn et al. (2014), Gray (1974), Hinton (1951), Hölldobler and Kwapich (in review), Kistner (1982), Parmentier et al. (2014)
	Setomorpha melichrosta	Obligate (?). Larvae feed on plant materials in fungus gardens of *Atta* and *Acromyrmex* leaf-cutter ants	1 species	New World tropics/ subtropics	Kistner (1982), Robinson and Nielsen (1993)
	Amydria anceps	Obligate. Feed on fungal substrate accumulations outside of *Atta* nests	1 species	Mexico	Sanchez-Pena et al. (2003)
Tortricidae	*Hystrichophora* spp.	Obligate (?). Larvae feed within *Vachellia* ant-plant domatia	3 species	East Africa	Agassiz (2011), Baker et al. (2016)
	Semutophila saccharopa	Facultative (?). Trophobiotic relationship	1 species	Malaysian peninsula	Maschwitz et al. (1986)

(continued)

Table 1 (continued)

Family	Ant-associated group	Degree of association and type of relationship with ants	Number of ant-associated species	Distribution	References
Sesiidae	*Osmanthedon domaticola*	Facultative (?). Larvae feed on *Vachellia* ant-plant domatia within silk shelters	1 species	East Africa	Agassiz and Kallies (2018)
Cyclotornidae	*Cyclotorna* spp.	Obligate. Ant-attended and parasitic within ant nests	12 species	Australia	Dodd (1902), Dodd (1912), Pierce (1995)
Coleophoridae	*Batrachedra myrmecophila*	Obligate. Preys on ant brood	1 species	Java	Hinton (1951), Pierce (1995)
Oecophoridae	*Stathmopoda* sp.	Obligate (?). Known from *Oecophylla* nests	1 species	Australia	Downes and Edwards (2016)
Pyralidae	*Pachypodistes goeldii*	Obligate. Larvae feed on *Dolichoderus* ant nest cartons	1 species	Brazil	Hinton (1951), Pierce (1995)
	Stenachroia myrmecophila	Obligate. Larvae may feed on *Crematogaster* brood	1 species	Australia	Hinton (1951), Pierce (1995)
	Gen. sp.	Obligate (?). Found in *Dinomyrmex* nest	1 species	Borneo	Orr et al. (1996)
	Gen. sp.	Obligate (?). Found in *Oecophylla* nest	1 species	Cameroon	Dejean et al. (2017)
	Gen. sp.	Facultative (?). Found only on plants with *Crematogaster*	1 species	Cameroon	Dejean et al. (2017)
	Gen. sp.	Facultative (?). Found only on plants with *Oecophylla*	1 species	Cameroon	Dejean et al. (2017)
Crambidae	*Niphopyralis* and allies	Obligate. Feed on *Oecophylla* eggs and brood	4 species	Australia, Java, and Cameroon	Dejean et al. (2017), Hinton (1951), Pierce (1995)

(continued)

Table 1 (continued)

Family	Ant-associated group	Degree of association and type of relationship with ants	Number of ant-associated species	Distribution	References
Noctuidae	Dyops spp.	Facultative. Larvae feed on Cecropia ant-plants defended by Azteca ants	>10 species	Central and South America	Janzen and Hallwachs (2021, Ramos et al. 2018)
Erebidae	Coxina spp.	Facultative (?). Larvae feed on Acacia ant-plants	1 species	Central America	Janzen (1967), Janzen and Hallwachs (2021)
	Eublemma albifascia	Obligate (?). Larvae feed on Oecophylla regurgitations	1 species	Cameroon	Dejean et al. (2016, (2017)
	Homodes spp.	Obligate (?). Larvae feed on foliage around Oecophylla ants	>6 species	Tropical Asia and Australia	Entomological Network of Singapore (2017), Fiedler (1991), Holloway (2005), Leong and D'Rozario (2012), and additional references in text
	Nudina artaxidia	Obligate (?). Larvae feed from ant-attended scale insects	1 species	Japan	Komatsu and Itino (2014)
Notodontidae	Rosema dentifera	Facultative (?). Larvae feed only on Acacia ant-plants	1 species	Central America	Janzen (1967), Janzen and Hallwachs (2021)
	Gen. sp. (near Stauropus)	Obligate (?). May solicit trophallaxis from Oecophylla	1 species	Cameroon	Dejean et al. (2017)

(continued)

Table 1 (continued)

Family	Ant-associated group	Degree of association and type of relationship with ants	Number of ant-associated species	Distribution	References
Saturniidae	*Syssphinx mexicana*	Facultative (?). Larvae feed only on *Acacia* ant-plants	1 species	Central America	Janzen (1967), Janzen (1984), Janzen and Hallwachs (2021)
Hesperiidae	*Lotongus calathus*	Obligate (?). Larvae build nests always shared with ants	1 species	Malaysia	Igarashi and Fukuda (1997)
Pieridae	*Catopsilia* spp.	Facultative. Larvae regularly attract ants to excretions and leaf exudates	>3 species	Africa and tropical Asia	Williams (1995-2020) and additional references in text
Lycaenidae	Lycaenidae	See Table 2. Most form trophobiotic relationships with ants	>3830 species estimated	Widespread globally	See Table 2
Riodinidae	Eurybiina (Riodininae: Eurybiini)	See Table 2. All appear to form trophobiotic relationships with ants	>35 species estimated	Central and South America	See Table 2
	Nymphiidini (Riodininae)	See Table 2. Most form trophobiotic relationships with ants	> 273 species estimated	Central and South America	See Table 2

In the absence of detailed phylogenies, we base this estimate on the assumption that a myrmecophilous species observed in a clade of taxa whose larvae are not otherwise known to be ant-associated is likely to have independently evolved ant association, and for those families that show multiple cases of myrmecophily, each also appears embedded in a lineage with other species whose caterpillars are not ant-associated. Ant associations in which trophobiotic caterpillars consistently provide ants with food rewards are not as common and to date have only been well-documented in Tortricidae, Cyclotornidae, Pieridae, Lycaenidae, and twice in Riodinidae. Additional small radiations of caterpillars that appear obligately ant-associated are known from Psychidae; at least three groups of Tineidae, Tortricidae, Coleophoridae, Oecophoridae, Crambidae; at least four groups of Pyralidae; and three groups of Erebidae, Notodontidae, and Hesperiidae. Caterpillars specializing on ant-plants are often poorly described but include numerous additional ant-associated taxa as discussed in the text. Please refer to the text for explanation regarding criteria for inclusion as a myrmecophilous species

Table 2 Ant associations in Lycaenidae and Riodinidae

	Number of described species	Distribution	Ant association		Degree of association, if associated			Trophobiosis	
			Non-ant-associated	Ant-associated	Facultative	Obligate		Non-trophobiotic	Trophobiotic
LYCAENIDAE	5390	Global	16 (96)	881 (3830)	354 (1761)	344 (1281)		217 (1116)	687 (2096)
CURETINAE	18	AU, OR, PA	0 (0)	5 (18)	3 (18)	0 (0)		6 (18)	0 (0)
THECLINAE + POLYOMMATINAE	4019	Global	9 (62)	622 (2878)	341 (1654)	134 (442)		95 (339)	591 (1833)
LYCAENINAE	114	Global	1 (1)	10 (79)	6 (79)	0 (0)		25 (106)	0 (0)
MILETINAE	208	AT, AU, NA, OR, PA	0 (0)	51 (207)	4 (10)	32 (197)		42 (207)	0 (0)
APHNAEINAE	302	AT, OR, PA	0 (0)	122 (264)	0 (0)	115 (264)		0 (0)	96 (263)
PORITIINAE	729	AT, AU, OR	6 (33)	71 (384)	0 (0)	63 (378)		49 (446)	0 (0)
RIODINIDAE	1562	Global	0 (0)	68 (308)	3 (40)	22 (163)		145 (982)	62 (257)
NEMEOBIINAE	301	Global	0 (0)	0 (0)	0 (0)	0 (0)		32 (252)	0 (0)
RIODININAE	1261	NA, NT	0 (0)	68 (308)	3 (40)	22 (163)		113 (730)	62 (257)
Eurybiini	247	NT	0 (0)	7 (35)	0 (0)	3 (35)		37 (199)	7 (35)
Dianesiini	1	NT	0 (0)	0 (0)	0 (0)	0 (0)		0 (0)	0 (0)
Calydniini	27	NT	0 (0)	0 (0)	0 (0)	0 (0)		1 (21)	0 (0)
Nymphidiini	367	NT	0 (0)	61 (273)	3 (40)	19 (128)		3 (26)	55 (222)
Symmachiini, Emesidini, Riodinini, and Helicopini	615	NA, NT	0 (0)	0 (0)	0 (0)	0 (0)		72 (484)	0 (0)

Based on available life history information, **all described species** were classified as **ant-associated**, **non-ant-associated**, or data deficient. The number of species in this latter category is not listed. **Ant association** indicates whether larvae occur with ants, either based on direct observation or inferred from adult behavior—associations may range from mutualism to parasitism. We were unable to locate detailed descriptions that confirm the lack of ant associations in any non-trophobiotic riodinid caterpillars. Caution is definitely necessary: regular associates, such as *Stalachtis* (Riodinidae: Nymphidiini) and *Deloneura*

Table 2 (continued)

(Lycaenidae: Poritiinae), have been overlooked and scored as non-ant-associated in previous reports. Ant-associated species were further classified as **obligate** myrmecophiles, **facultative** myrmecophiles, or myrmecophiles whose degree of association could not be determined due to low sample size. Finally, we classified each species with available observations as **trophobiotic** or **non-trophobiotic**. Where field observations are lacking, a functional dorsal nectary organ is a good indicator that larvae are ant-attended, while long bristles on the thorax and abdomen reliably signal the absence of ant attendance (e.g., DeVries 1991c). Each cell in the table first reports the number of documented species records. For characters not known to vary among known species of a genus, the same character state was assigned to all other congeners. The number of these inferred species counts is given in **parentheses** within each cell.

Species lists were modified from G. Lamas (personal communication). For species whose life histories we could not document from published literature, we searched for relevant reports on Google Scholar, iNaturalist, and BugGuide. Altogether, ant association data of any kind remain unavailable for more than 75% of lycaenid and riodinid species. *AT* Afrotropical, *AU* Australasian, *NA* Nearctic, *NT* Neotropical, *OR* Oriental, *PA* Palearctic

References compiled for table: (Aibar-Abregú 2014; Albanese et al. 2007; Alves-Silva et al. 2018; Austin et al. 2008; Bálint and Benyamini 2001; Ballmer 2008; Ballmer and Pratt 1991; Ballmer and Wright 2008; Bascombe et al. 1999; Basu and Kunte 2020; Benyamini 1995; Benyamini 2013; Benyamini and Bálint 1995; Benyamini et al. 2018; Benyamini et al. 2019; Braby 2011; Braby 2012; Braby 2015; Braby and Douglas 2004; Bury and Savchuk 2015; Callaghan 1985; Callaghan 1986; Callaghan 1992a; Callaghan 1992b; Callaghan 1997; Callaghan 2003; Callaghan 2008; Casagrande et al. 2009; Castillo Guevara and Rico Gray 2002; Claassens 1996; Cock 2010; Comstock and Dammers 1932; Cottrell 1984; Dantchenko 1997; Dejean et al. 2017; DeVries 1984; DeVries 1988a; DeVries 1991c; DeVries 1991d; DeVries 1997; DeVries 1991c; DeVries et al. 2004; DeVries et al. 1986; Duarte and Robbins 2008; Duarte and Robbins 2010; Eastwood et al. 2005; Eastwood et al. 2008b; Faynel and González-Mercado 2019; Fiedler 1989a), supplemental table from (Fiedler 2001; Fiedler et al. 1995; Fukuda et al. 1984; Gibbs 1980; Hall 1998; Hall 2018; Hall et al. 2004; Harvey and Longino 1989; Harvey and Webb 1980; Hawkeswood et al. 2016; Heath 1997; Heath and Claassens 2003; Heredia and Robbins 2016; Hinton 1951; Hsu and Johnson 1998; Hsu et al. 2004; Igarashi and Fukuda 1997; Igarashi and Fukuda 2000; Itioka et al. 2009; Jackson 1937; Janzen 1967; Janzen and Hallwachs 2021; Jerathithikul et al. 2011; Johnson and Valentine 2001; Kaminski 2008a; Kaminski 2008b; Kaminski 2017; Kaminski and Carvalho-Filho 2012; Kaminski et al. 2016; Kaminski et al. 2012a; Kaminski et al. 2014; Kaminski et al. 2013; Kaminski et al. 2015; Kaminski et al. 2020a; Kaminski et al. 2010b; Kaminski et al. 2020b; Kim and Ho 2012; Kitching and Luke 1985; Kubik and Schorr 2018; Kumar et al. 2017; Lafranchis 2019; Lafranchis et al. 2007; Larsen 2005; Lo et al. 2017; Lohman and Samarita 2009; Martins et al. 2013; Maschwitz et al. 1988; Megens et al. 2005; Mota et al. 2014; Neild and Bálint 2014; New 1993; Nielsen and Kaminski 2018; Nishida 2010; Obregón et al. 2015; Okubo et al. 2009; Opler 1999; Pan and Morishita 1990; Parsons 1984; Parsons 1999; van der Poorten and van der Poorten 2016; Riva et al. 2017; Robbins and Aiello 1982; Robbins et al. 1996; Ross 1964; Saarinen 2005; Saarinen and Daniels 2006; Safian 2012; Sáfián 2015a; Sáfián 2015b; Sáfián and Collins 2014; Sáfián and Collins 2015; Sáfián and Larsen 2009; Sands 1986; Santos et al. 2014; Sariot and Ginés 2011), supplemental table from (Schär et al. 2018; Schmidt et al. 2010; Schmidt et al. 2014; Schurian and Eckweiler 2002; Schurian and Fiedler 1994; Schurian and Reif 1992; Schurian et al. 2005; Seufert and Fiedler 1996; Shapiro 2007; Shimizu-Kaya et al. 2015; Silva et al. 2014; Singh 2003; Stradomsky and Fomina 2009; Talavera et al. 2016; Tennent 1996; Torres and Pomerantz 2016; Tshikolovets 2011; Vargas and Duarte 2016; Williams 1995-2020; Yago et al. 2010; Youngsteadt and Devries 2005; Zanuncio et al. 2009; Zanuncio et al. 2013; Zhou and Zhuang 2018), BugGuide #1133247, BugGuide #1503993, BugGuide #1745213, BugGuide #1824230, BugGuide #329907, BugGuide #393584, BugGuide #405974, BugGuide #731179, BugGuide #803004, BugGuide #866019, BugGuide #1745213, iNaturalist #19541629, iNaturalist #19626729, iNaturalist #21747729, iNaturalist #2805952, iNaturalist #3270216, iNaturalist #3656419, iNaturalist #40848069, iNaturalist #4312429, iNaturalist #5888085, iNaturalist #9351793, iNaturalist #67275324, https://entomologytoday.org/2015/12/07/caterpillar-depends-on-parasitic-plants-and-nectar-drinking-ants, https://butterflycircle.blogspot.com/2011/11/life-history-of-singapore-four-line.html, https://butterflycircle.blogspot.com/2010/06/life-history-of-branded-imperial.html, https://butterflycircle.blogspot.com/2014/11/lycaenid-butterflies-and-ants.html, https://www.flickr.com/photos/142712970@N03/3332969114)

species in the closely related butterfly family Riodinidae (Table 2). While caterpillars in these two families are generally characterized as ant mutualists, we discuss evidence suggesting that interactions with negative consequences for ants are far more common than previously recognized, and that despite appearances, these associations might be better characterized as parasitic on the part of the lycaenids or, at best, reciprocally parasitic by both parties. Most other ant-associated groups, like the Australian moth family Cyclotornidae, are individually species-poor and rarely encountered but collectively span almost the entire lepidopteran tree of life and display great diversity, particularly in the tropics (Table 1). We show that myrmecophilous caterpillars that passively coexist with ants are far more diverse than previously recognized and suggest that many such caterpillar groups remain undiscovered.

Terminology and Overview

Myrmecophiles are "ant loving" organisms with adaptations that enable them to benefit from ant association, and we will refer to them interchangeably as **ant associates** (narrower definitions are also sometimes used: (Hölldobler and Wilson 1990; Kronauer and Pierce 2011; Nichols 1989). Specializations that help these species find or attract and subsequently stay in contact with ants are important and could be considered part of a basic signature of myrmecophily. Ants themselves, their pheromones, and even volatiles released by other organisms disturbed by ants are used as cues by adults or larvae to find ants, as discussed below. Within Lepidoptera, we consider caterpillars ant-associated if we can directly observe or infer from available evidence that caterpillars or ovipositing females use these cues to locate ants or that caterpillars themselves produce secretions or vibratory signals specialized to attract ants. Caterpillars may also qualify as ant-associated if they appear specialized to live in close proximity to ants on myrmecophytes, plants with a strong mutualistic relationship with ants and that typically provide ants with cavities for shelter.

Obligate ant associates are species that cannot complete their life cycle without ants. In cases where full life histories have been well documented, these species are easily identified. However, for cases where relationships must be inferred, a species is likely to be an obligate ant associate if the caterpillars are never found without ants nearby; if caterpillars rely on ants as a food source; if females hesitate or refuse to oviposit, even in captivity, without ants present; or if adults are typically only observed near the openings of ant nests. In contrast, **facultative** ant associates are sometimes found without ants. Facultative association of caterpillars with ants has only been well documented in Lycaenidae and Riodinidae, although it seems likely to occur in other groups that have not been so well characterized. Obligate ant associates usually associate with ants from only one genus or species, while most facultative myrmecophiles associate with multiple ant genera and subfamilies. A number of exceptions exist to these broad generalizations (Eastwood and Fraser 1999; Fiedler 2001; Glasier et al. 2018). For example, the obligately ant-associated

Australian lycaenid, *Jalmenus eichhorni*, is attended by ants from different genera during the day and night (Dunn 2007). Larvae of a congener, *J. evagoras*, are typically associated with only a few ant species in the genus *Iridomyrmex* but during "breakout" periods of high abundance can readily be found associating with other genera (Pierce and Nash 1999).

Like other conditional interactions with ants, caterpillar-ant associations vary spatially and temporally, ranging from **mutualisms**, where both parties derive net fitness benefits from their interaction, through to **parasitisms**, where one party (in this case usually the ants) pays a fitness cost due to the association. Many appear to be commensal or only mildly parasitic in the sense that caterpillars benefit while ant fitness seems largely unaffected.

Many insects produce secretions that serve as a food source to attract and maintain a standing guard of ants and are described as being **trophobiotic**. We refer to lycaenid and riodinid caterpillars that do this as **ant-attended**. We use the term **non-trophobiotic** to describe caterpillars that are not actively ant-attended. The term "myrmecoxenous" has been used as a substitute for "non-trophobiotic" in recent literature but confusingly describes either a symphile, an insect that is a guest in ant nests (Nichols 1989), or a non-myrmecophile, an insect that is simply not ant-associated (Kitching and Luke 1985; Paul 1977), so we have avoided using it here.

Parasites found in ant nests often belong to groups that prey on ant-attended hemipterans and thus already possess appropriate defensive and feeding-related adaptations to coexist with ants (Eisner et al. 1972; Malicky 1970; Pierce 1995). These include numerous genera within the subfamily Miletinae [Lycaenidae], *Shirozua* [Lycaenidae], a few riodinids, *Eublemma* [Erebidae], Cyclotornidae, and perhaps *Stathmopoda* [Tineidae] and *Baratrachedra* [Coleophoridae]. This pattern is not confined to Lepidoptera: ant brood and trophallaxis feeding have been reported in species from nearly every prominent hemipteran-associated arthropod group, including ladybug beetles (Orivel et al. 2004; Vantaux et al. 2010), flower flies (Hölldobler and Wilson 1990), green lacewings (Tauber and Winterton 2014; Tauber et al. 2020), and even certain aphids themselves (Salazar et al. 2015).

Many butterfly and moth larvae have ant associations that have been potentially overlooked because the relationship is defined largely by its absence: these are cases where ants cannot detect or appear indifferent to the caterpillars. These caterpillars typically only associate with ants near nests and food sources—habitats that are hotspots for lepidopteran ant associates more generally. For example, a veritable menagerie of potentially ant-associated Lepidoptera lives on the African ant-acacia *Vachellia drepanolobium*, the dominant tree species in the "black cotton" vertisols of East African savannas. Eighteen species of Lycaenidae, some attended by ants, were documented on these ant-plants at field sites in Kenya and Tanzania over a 5-year period (Fig. 1) (Baker et al. 2016; Martins et al. 2013; Whitaker et al. 2019). Numerous species of Tineidae, Tortricidae, Sesiidae, Blastobasidae, Gelechiidae, and Geometridae have been reared from the swollen thorn ant domatia of *V. drepanolobium*, and many others feed in the tree canopy (Adamski 2017; Agassiz 2011; Agassiz and Bidzilya 2016; Agassiz and Harper 2009; Agassiz and Kallies 2018; Baker et al. 2016; Hocking 1970). Some of these species are polyphagous and have

Fig. 1 Lycaenid larvae, almost certainly *Kipepeo kedonga* (formerly known as *Chilades kedonga* (Parmentier et al. 2014)) that were abundant in swollen thorns of *Vachellia drepanolobium* in Suyian, Kenya. (Photo by Dino Martins)

Fig. 2 The brown silk envelope on the left was built by a tortricid caterpillar feeding inside a thorn domatium of *Vachellia drepanolobium* occupied by *Crematogaster mimosae* in Kitengela, Kenya. (Photo by Naomi Pierce)

been described as having greater abundance in the absence of ants (Agassiz 2011), and we would not describe these ones as being ant-associated. The majority are not sufficiently well known to be able to characterize them as ant-associated or not.

A few specialist myrmecophiles have nonetheless been documented on ant-plants. For example, larvae of *Hystrichophora* (Tortricidae) build strong, membranous, dome-like shelters within hollowed-out *V. drepanolobium* domatia that are frequently shared with ants (Fig. 2) (Agassiz 2011). Caterpillars of *H. griseana* are common on trees inhabited by colonies of *Crematogaster mimosae* or *C. nigriceps*, but they are almost never found on trees inhabited by colonies of *Tetraponera penzigi* (Baker et al. 2016). Similarly, caterpillars of *Syssphinx mexicana* (Saturniidae), *Rosema dentifera* (Notodontidae), and *Coxina* spp. (Erebidae) specialize on Central American acacias, *Vachellia cornigera*, and its relatives, which are inhabited by aggressive *Pseudomyrmex* ants, whose defenses the caterpillars are able to overcome (Janzen 1967; Janzen 1984; Janzen and Hallwachs 2021). The larvae of *Dyops* spp. (Noctuidae) are essentially immune to ant attack and feed on various species of Urticaceae, including *Cecropia* ant-plants defended by *Azteca* ants (Janzen and

Hallwachs 2021; Ramos et al. 2018). Many other species reported from ant-plants may prove to be ant-associated upon further investigation. Tunnels and silk shelters built by *Stenoma charitarca* (Oecophoridae), and leaf rolls built by *Acrospila gastralis* (Crambidae), allow caterpillars to persist on *Maieta guianensis* plants occupied by *Pheidole* ants (Vasconcelos 1991), much as certain crambid larvae are protected from ants within leaf rolls on *Tococa* ant-plants (Michelangeli 2003). The database of macrocaterpillar food plants of the Area de Conservacion Guanacaste, Costa Rica (Janzen and Hallwachs 2021), does not indicate whether caterpillar host plants were actually occupied by ants but nonetheless includes dozens of butterfly and moth species that have been exclusively reared from ant-plant species, such as *Lygropia cernalis* (Crambidae) from *Triplaris melaenodendron*, *Conchylodes nolckenialis* (Crambidae) and *Munona robpuschendorfi* (Erebidae) from *Cordia alliodora*, and *Macalla* sp. (Pyralidae) from *Cecropia obtusifolia*. Many Lycaenidae and Riodinidae also prominently infiltrate ant-plants (e.g., DeVries and Baker 1989; Eastwood and Fraser 1999; Heredia and Robbins 2016; Heredia and Robbins 2016; Kaminski 2008b; Kaminski et al. 2010a; Kaminski et al. 2012b; Kaminski et al. 2020a; Maschwitz et al. 1984; Sands 1986; Shimizu-Kaya et al. 2015).

Many caterpillar species that do not directly interact with ants are polyphagous and occur on different host plants only as they become occupied by ants. For example, the obligate ant associations of many species in the butterfly tribe Liptenini (Lycaenidae) only became evident based on the observation that the large, attractive adults had only been observed around arboreal *Crematogaster* nests (see discussion below). Similarly, *Homodes* (Erebidae) are large and unusual caterpillars that occur on a wide variety of host plants but generally only when the plants are also patrolled by *Oecophylla* ants (see discussion below) (Fiedler 1991; Holloway 2005; Leong and D'Rozario 2012; Lokkers 1990). This kind of "cryptic" association probably exists even in less charismatic lepidopterans, such as leaf mining micromoths (compare Bily et al. (2008)).

Dejean et al. (2017) undertook the most extensive study to date of the extent of ant-caterpillar associations in tropical habitats. Defoliator and nectarivorous caterpillars were collected and reared from 50 to 100 m transects of the extrafloral nectary-bearing plant *Alchornea cordifolia* along forest edges in Cameroon, each transect exclusively dominated by one of five species of aggressive ants. Each of the tree-nesting species *Crematogaster striatula*, *Oecophylla longinoda*, *Tetramorium aculeatum*, and *Camponotus brutus* were represented by 30 transects, along with 10 transects dominated by the ground-nesting species *Myrmicaria opaciventris*. Of the 22 species of caterpillar found, only 1 was found with more than 1 ant species, although many were collected from numerous transects. All species showed distinct specializations to coexist with ants, including some parasites that could solicit trophallaxis or appeared to feed within ant nests. This study may be the first to systematically document the full spectrum of defoliator and nectarivorous caterpillars on a host plant dominated by specific ant species and shows that previously unknown ant associations across diverse lepidopteran families can be uncovered by careful observations in tropical habitats.

Synopsis of Caterpillar-Ant Associations

Tineidae and Psychidae

Diverse species of Tineidae and Psychidae are known to scavenge exclusively within ant nests, encased with debris or protected by silk webbing, and some of these probably feed on ant brood or food resources. Pending genus-level phylogenies that may reveal additional origins, ant associations appear to have originated independently in at least three tineid clades, represented respectively by the genera *Myrmecozela*, *Setomorpha*, and *Amydria*, as well as in the psychid genera *Iphierga* and *Ardiosteres*; see Regier et al. (2015) for a higher level molecular phylogeny of 62 representatives of the main lineages within Tineoidea) (Ahn et al. 2014; Gray 1974; Hinton 1951; Kistner 1982; Parmentier et al. 2014; Robinson and Nielsen 1993; Sanchez-Pena et al. 1993). Caterpillars in the Palearctic and Oriental genus *Ippa* (Tineidae) have been found in ant nests of *Crematogaster* (Myrmicinae), *Polyrhachis*, *Lasius*, *Dolichoderus*, and *Anoplolepis* (Formicinae) (Hinton 1951; Hölldobler and Kwapich in review). *Ippa* caterpillars build a flattened protective case, and while *I. dolichoderella* larvae in Java are only known to consume brood, *I. conspersa* larvae in Japan also feed on adult ants (Hinton 1951; Hölldobler and Kwapich in review). Although not obligately ant-associated, the free-living larvae of *Perisceptis carnivora* (Psychidae) in Panama build portable defensive cases and frequently feed on worker ants (Davis et al. 2008).

Tortricidae

Malaysian caterpillars of *Semutophila saccharopa* (Tortricidae) live in silk shelters constructed on bamboo and associate with ants from at least seven genera in a manner similar to aphids. Ants feed on the sugar-rich anal droplets provided by the caterpillars. The caterpillars prefer to excrete waste in the presence of ants, but the droplets can be withdrawn back into the anus and jettisoned several centimeters away from the larval shelter if ants remain unavailable (Maschwitz et al. 1986).

Cyclotornidae

In the Australian family Cyclotornidae, which comprises the single genus *Cyclotorna*, larvae start out as external parasites of ant-attended leafhoppers or scale insects (Fig. 3) (Dodd 1902, 1912; Pierce 1995). Second-instar larvae of *Cyclotorna monocentra* are flattened and produce an anal secretion that attracts ants. Workers of *Iridomyrmex purpureus* carry them into the nest, where they feed on brood until leaving to pupate under bark (Epstein et al. 1999; Pierce 1995). The *Cyclotorna* larvae will die if their anal secretions are not removed by ants (Hinton

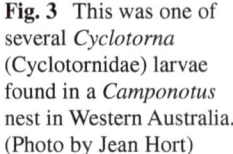

Fig. 3 This was one of several *Cyclotorna* (Cyclotornidae) larvae found in a *Camponotus* nest in Western Australia. (Photo by Jean Hort)

1951). Epipyropidae, the apparent sister group to Cyclotornidae (Hall et al. 2004; Heikkila et al. 2015), are ectoparasites of planthoppers and cicadas (Hemiptera) but are not known to interact with ants (Pierce 1995).

Coleophoridae and Oecophoridae

Many *Batrachedra* (Coleophoridae) prey on scale insects, but larvae of the Indonesian species *B. myrmecophila* feed on ant brood in nests of *Polyrhachis dives*, protected from ants by portable cases (Hinton 1951; Pierce 1995). While several *Stathmopoda* spp. (Oecophoridae) feed on scale insects (Pierce 1995), one Australian species builds webs in *Oecophylla* nests where it may feed on ants (Downes and Edwards 2016).

Pyralidae

Many Pyralidae are associated with ants. Larvae of the Brazilian *Pachypodistes goeldii* (Chrysauginae) chew *Dolichoderus gibbosoanalis* nest cartons, which they use to construct a protective case, and may also feed on the brood (Hinton 1951; Pierce 1995). Adults of this species are covered in long, loose setae that are likely to help freshly eclosed adults escape attack by ants (Kistner 1982). An Australian species, *Stenachroia myrmecophila* (Galleriinae), may feed on *Crematogaster* brood (Hinton 1951; Pierce 1995). Larvae of other unidentified pyralids have been found in *Dinomyrmex* nest debris in Borneo (Orr et al. 1996) and in *Oecophylla* nests in Cameroon (Dejean et al. 2017). Caterpillar silk weaving may also help herbivorous Pyralidae coexist with ants. Dejean et al. (2017) found an unidentified species of pyralid that uses silk to cordon off young leaves of *Alchornea cordifolia* inhabited by *Crematogaster striatula*. Caterpillars of another unidentified pyralid species were found only on *A. cordifolia* occupied by *Oecophylla longinoda*, in communal caterpillar nests resembling *Oecophylla* nests from which they emerge at night to feed when the ants are less active. *Crematogaster* ants were recently found nesting within a shelter built by larvae of *Triphassa* (Pyralinae) on an *Erica imbricata* heath in South Africa [iNaturalist #23039584]. More work will be needed to determine if this remarkable relationship is coincidental or occurs regularly.

Other than Lycaenidae and Riodinidae, species of Tineidae and Pyralidae are the most prominent caterpillar guests in ant nests (Table 1). These species are herbivores, detritivores, and parasites and include the only caterpillars found in colonies of ants, such as leaf-cutter ants [Attini], that do not harvest nectar from plants and hemipterans (Kistner 1982; Robinson and Nielsen 1993; Sanchez-Pena et al. 2003). Other species of Tineidae and Pyralidae feed within social wasp, bee, termite, and even communal spider nests (Ahn et al. 2014; Brandl et al. 1996; Davis and Davis 2007; Deyrup et al. 2004; Kistner 1982; Pierce 1995). Most Lepidoptera found within human dwellings also belong to these two families (Bertone et al. 2016; Linsley 1944). Flexible diets, along with defenses that help larvae avoid aggression, may be among the factors that help these families to thrive alongside diverse host ant associates, and more species will undoubtedly be found in association with ants as new life histories are uncovered.

Crambidae

In the family Crambidae, at least two lineages in the largely phytophagous subfamily Spilomelinae may be associated with ants. *Cirrhochrista saltusalis* (Spilomelinae: Margaroniini) caterpillars have been found alongside *Pheidole* ants and *Oboronia punctatus* caterpillars (Lycaenidae) within debris nests constructed by the ants on flowerheads, but this cohabitation may be an unusual occurrence (Lamborn 1911; Lamborn et al. 1914). Immature stages remain unknown from most Wurthiini, but several feed on brood of arboreal ants, in addition to a single phytophagous species (Mally et al. 2019). *Niphopyralis aurivillii* (Spilomelinae: Wurthiini), a possibly chemical mimic of host ants known from Java, feeds on the brood of *Polyrhachis bicolor* and may help maintain the silken nest structure (Hinton 1951; Pierce 1995). Another species found in Java, *N. myrmecophila*, feeds on *Oecophylla smaragdina* brood and has a flattened portable case for protection (Hinton 1951). *Niphopyralis chionesis* is suspected to prey on brood of *Oecophylla smaragdina* in Australia (Pierce 1995), and Dejean et al. (2017) found a related larva feeding on *Oecophylla longinoda* eggs in Cameroon (Fig. 4).

Erebidae

Larvae of lichen moths (Erebidae: Lithosiini) secrete toxins that protect them from ants (Chialvo et al. 2018; Palting 2020). Ayre (1958) observed hundreds of British Columbian *Crambidia casta* larvae that sheltered and pupated in *Formica* nests,

Fig. 4 (**a, b**) Caterpillar on an *Oecophylla* nest in Guinea, near Conakry. Larvae of this undescribed species near *Niphopyralis* (Crambidae) feed voraciously on weaver ant eggs (Dejean et al. 2017). (Photo by Piotr Naskrecki)

although this behavior has not been found in other populations of this species (Palting 2020). Larvae of another small lichen moth found in Japan, *Nudina artaxidia*, are obligate associates of *Lasius* ants and feed on honeydew from scale insects, along with lichen (chapter frontispiece) (Komatsu and Itino 2014).

Many *Eublemma* spp. (Erebidae: Boletobiinae) feed on scale insects, where they are concealed from attending ants by a portable protective casing (Dejean et al. 2016; Lamborn et al. 1914; Pierce 1995; Susilo and Susilo 2015). In Cameroon, Dejean et al. (2016) found that *Eublemma albifascia* lays eggs on ant nests, and first-instar caterpillars are carried into *Oecophylla longinoda* brood chambers by workers. Subsequent instars are fed by ants and steal from trophallaxis between workers, and ants groom their bodies and drink their anal secretions. The larvae acquire colony odors and do not require physical protection from host ants (Dejean et al. 2016). Dejean et al. (2017) found 359 caterpillars of *Eublemma albifascia* in only four colonies of *Oecophylla longinoda*. Due to their intense trophallaxis requirements, *Eublemma albifascia* parasites generally cause the death of the queen through neglect, though their numbers are regulated by some parasitoid wasps (Dejean et al. 2016). Eclosed adults are mostly ignored and, if occasionally attacked, are protected by long, dense scales (Dejean et al. 2016).

Fig. 5 (**a, b**) A weaver ant-mimicking *Homodes* larva (Erebidae) in Singapore. (Photo by Lionel Lim. Soh Kam Yung [K. Y. Soh] provides another full-habitus view at iNaturalist #37480826)

In a few wasmannian ant mimics, the same specialized tactile structures are used to integrate with ants and to scare off other predators (von Beeren et al. 2018; Kronauer and Pierce 2011). A few Oriental and Australasian species of the genus *Homodes* (Erebidae: Boletobiinae) occur on a wide range of host plants but never far from *Oecophylla smaragdina* weaver ants (Fiedler 1991; Holloway 2005; Leong and D'Rozario 2012, iNaturalist #65316827, iNaturalist #27728866). These caterpillars are excellent mimics of *Oecophylla* ants at both the front and the back, with a false head on the posterior abdomen and long clubbed setae resembling ant appendages (Fig. 5). Waving these setae not only deters visual predators but appears to placate *Oecophylla* workers (video at https://www.facebook.com/watch/?v=1938845709677099) (Entomological Network of Singapore 2017). Structurally similar, possibly glandular setae are found on the thorax and abdomen of related larvae documented on iNaturalist, which are not known to be ant-associated [e.g., iNaturalist #21087410, iNaturalist #38085822, iNaturalist #21414510]. Lokkers (1990) found ant-mimicking looper moth caterpillars in north Queensland exclusively on *Oecophylla*-occupied trees, which may have been larvae of *Homodes* or another group with a similar life history.

Fig. 6 *Stauropus* larva (Notodontidae) feeding on *Salix* in Italy. The elongated thoracic legs help early-instar larvae mimic ants and in some cases are used to communicate with ants. (Photo by Paolo Mazzei)

Notodontidae

Phytophagous larvae of *Stauropus* and *Neostauropus* (Notodontidae) have enlarged mesothoracic and metathoracic legs used to mimic ants in early instars, and spiders once larvae become larger, with a terrifying threat display (Fig. 6) (Poulton 1890; Pratt et al. 2016). In Britain, photographer Andy Newman experimentally brought together first-instar *Stauropus fagi* larvae and *Formica* ants and discovered that larvae were ignored after waving their mesothoracic legs and contacting the ants' antennae [http://www.andynewman.org/html/lobster_moth.html]. Dejean et al. (2017) discovered related larvae in Cameroon that use their enlarged mesothoracic legs to solicit trophallaxis from associated *Oecophylla longinoda* ants. The larvae also fed on young leaves and extrafloral nectaries. An unidentified larva of this species from southern Nigeria may have also been described by Farquharson et al. (1922). Larvae of Afrotropical *Amyops ingens* strongly resemble *Stauropus* larvae and have much shorter, but still notably elongated, thoracic legs of unknown function [(iNaturalist #11244196, iNaturalist #11446507]). Perhaps they are used to handle soft-bodied Hemiptera or honeydew as in some Lycaenidae and Riodinidae (DeVries and Penz 2000; Dejean et al. 2017). The biology of these fascinating Notodontidae remains largely undocumented; more research is needed to understand their ecology and diversity.

Papilionoidea (Hesperiidae, Nymphalidae, Pieridae)

With over 900 well-documented and more than 4000 inferred myrmecophilous species, the butterfly families Lycaenidae and Riodinidae account for an overwhelming proportion of caterpillar-ant associations (Table 2). At least a few butterfly species in other families are also ant-associated. Malaysian *Lotongus calathus* caterpillars

Fig. 7 Glistening droplets on spines of larva of *Phoebis philea* (Pieridae) feeding on *Senna mexicana* being inspected by an unidentified ant, with a second ant feeding on an extrafloral nectary nearby. The droplets are thought to be defensive but may in some cases (depending on the ant species, host plant, and location) be strikingly attractive to ants (e.g., photo of *Catopsilia pyranthe* surrounded by *Anoplolepis gracilipes* ants at http://pureoxygengenerators.blogspot.com/2017/10/some-nature-finds.html). (Photo by James Spencer, kindly provided by Nadia Spencer)

(Hesperiidae) build leaf shelters that are always shared with nesting *Dolichoderus* ants (Igarashi and Fukuda 1997). Chemically protected larvae of Neotropical *Vettius tertianus* (Hesperiidae) are usually found living with predatory ants in ant gardens, although not enough is known of their biology to conclude whether or not they are true myrmecophiles (Orivel and Dejean 2000).

Ants gathering to drink from leaf exudates generated by herbivores are not uncommon, although rarely analyzed, and result in facultative ant interaction with caterpillars of various butterfly and moth species (Fiedler 1991; Larsen 2005). For example, Young (1978) observed ants using their antennae to stroke a larva of the nymphalid butterfly *Mechanitis isthmia* in Costa Rica, whereupon the larva would withdraw from the leaf edge and allow the ants to drink exudates from the newly cut surface. Diverse ants commonly drink from the feeding sites of *Catopsilia* larvae (Pieridae), and some ant species appear to find the caterpillars themselves more attractive than the leaf exudates (Williams 1995-2020, iNaturalist #10726006 iNaturalist #15027508, http://pureoxygengenerators.blogspot.com/2017/10/some-nature-finds.html, https://www.flickr.com/photos/129254524@N06/16162943814/). Larvae of many Pieridae and Saturniidae produce potent secretions to deter ants, and occasional reports suggest that the secretions themselves are consumed by ants under rare circumstances (Fig. 7) (Fiedler 1991; Hinton 1951; Smedley et al. 2002).

Ant Association in the Lycaenidae and Riodinidae

Throughout Lepidoptera, only the families Lycaenidae and Riodinidae contain ant-associated taxa that number more than a few dozen species. The ability to actively attract ants with food rewards and sophisticated signaling may help account for their surprisingly massive radiation compared with other ant-associated larvae whose interactions are more limited and rarely involve food rewards. Non-trophobiotic myrmecophiles are limited to ant "hotspots," where enemy-free space is strongest and unique resources are available: either around ant-attended hemipterans, within ant nests, on ant-plants, or within the arboreal territories of highly aggressive ants like *Oecophylla*. Correspondingly, trophobiotic organs in Lycaenidae and Riodinidae that obligately occur around ant-tended hemipterans and ant nests are often lost or modified, most notably in the lycaenid subfamilies Miletinae and Poritiinae and in riodinids like *Aricoris arenarum* (Kaminski et al. 2020b; Shimizu-kaya et al. 2013).

Recent comparative analyses using a well-resolved tribal level phylogeny of butterflies indicate that ant association arose once in the ancestor of the Lycaenidae nearly 80 mya, twice more recently in its sister family, the Riodinidae, once in the subtribe Eurybiina, and once in the Nymphidiini (Espeland et al. 2018). Thus, similar traits used in ant-caterpillar associations appear to have arisen independently at least three times in these two butterfly families.

Adaptations of Adults

Ant-related visual and chemical cues are used during mate finding and oviposition by many ant-associated Lycaenidae and Riodinidae (e.g., Atsatt 1981b; Casacci et al. 2019b; Dejean et al. 2017; DeVries 1997; Elgar and Pierce 1988; Elgar et al. 2016; Fiedler and Maschwitz 1989a; Fiedler and Maschwitz 1989b; Fraser et al. 2002; Kaminski et al. 2013; Heath 1997; Henning 1983; Kaminski and Carvalho-Filho 2012; Martins et al. 2013; Pierce 1984; Pierce and Elgar 1985; Pierce and Nash 1999; van der Poorten and van der Poorten 2016; Pringle et al. 1994; Seufert and Fiedler 1996; Williams 1995-2020), even in species that are facultatively ant-attended (Mota and Oliveira 2016; Wagner and Kurina 1997) or non-trophobiotic (Bächtold et al. 2014; Fiedler and Maschwitz 1989b; Funk 1975; Sáfián and Collins 2014; Sáfián and Larsen 2009; Rodrigues et al. 2010). Many obligate ant associates will not oviposit unless ants are present (e.g., Heath 1997)). Chemical eavesdropping on ants is widespread among myrmecophiles, and lycaenid adults may detect ant pheromones as well as visual cues (e.g., Adams et al. 2020; Kaliszewska et al. 2015; Sáfián and Larsen 2009; Williams 1995-2020). Visual and chemical cues are also used by non-myrmecophiles to avoid ovipositing near ant territories (Freitas and Oliveira 1996; Van Mele et al. 2009; Sendoya et al. 2009).

Phengaris (=*Maculinea*) is one of two lycaenid genera with species whose larvae are obligately phyto-predaceous, with eggs laid on specific plant hosts that serve as food for the early instars and that later drop to the ground to be carried by workers into the ant nest, where they feed on the brood or solicit regurgitations to complete development. Recent research on ovipositing females of *Phengaris* species has started to resolve a longstanding puzzle regarding whether or not these parasitic butterflies use ants as cues to locate oviposition sites (Carleial et al. 2018; Casacci et al. 2019b; Czekes et al. 2014; van Dyck and Regniers 2010; Fürst and Nash 2010; Musche et al. 2006; Patricelli et al. 2011; Thomas and Elmes 2001; Wynhoff et al. 2008; Wynhoff et al. 2015). Apparently *Myrmica* ants nesting at the base of *Origanum vulgare* plants (Lamiaceae) damage the roots and thereby induce the plants to release defense-related volatile organic compounds, or VOCs, including the monoterpenoid carvacrol and its isomer thymol. Ovipositing females of *Phengaris arion* can detect these compounds and use them to identify plants with appropriate ant hosts located beneath them (Pech et al. 2007; Patricelli etal 2015). The larvae of other species of *Phengaris* also feed on host plants in the Gentianaceae and Rosaceae (Als et al. 2004), and it seems likely that a similar mechanism exists on other host plants whereby damage to plant roots caused by ant colonies nesting underground may induce the release of VOCs that attract ovipositing females. Cues from a number of different plant families may be used by ovipositing females in this way, but this remains to be tested.

Chemical signals seem to mediate ant interactions with adults of many lycaenid and riodinid butterflies, generally with ants that are also associated with caterpillars (Atsatt 1981a; Farquharson et al. 1922; Fiedler and Maschwitz 1989a; Pierce et al. 2002). These semiochemicals may be particularly important in species that pupate within ant nests (Elfferich 1998; Lohman 2004). Various adult Lycaenidae and Riodinidae are inspected or groomed by ants (DeVries 1984; Fiedler and Maschwitz 1989b; van der Poorten and van der Poorten 2016, iNaturalist #36616206, iNaturalist #5526494, iNaturalist #62627204, iNaturalist #56774612, iNaturalist #66838365). Adults of most Poritiinae and Miletinae (Lycaenidae) feed exclusively from extrafloral nectaries and carbohydrate-rich insect exudates, both frequently attended by workers of the same ant species that are associated with their own larvae (Figs. 8 and 9) (Atsatt 1981a; Callaghan 1992b; Cottrell 1984; Dejean et al. 2017; Farquharson et al. 1922; Fiedler and Maschwitz 1989b). Certain Riodinidae may have similar habits (Torres and Pomerantz 2016).

Adaptations of Caterpillars and Pupae

Before pupation, and in some species whenever not feeding, larvae of diverse Riodinidae (e.g., DeVries 1997; Kaminski and Carvalho-Filho 2012; Kaminski et al. 2020b; Ross 1966) and Lycaenidae enter special shelters built for them by ants (e.g., Eastwood et al. 2005; Eastwood et al. 2008a; Ekka and Rastogi 2019; Webster and Nielsen 1984) or the ants' nests themselves (e.g., Benyamini and Bálint 1995; Bury and Savchuk 2015; Mizuno et al. 2019; Wagner 1995). These cohabitation behaviors appear to co-opt existing ant behaviors widely used to shelter

Fig. 8 *Lachnocnema* butterflies (Lycaenidae: Miletinae) collecting honeydew from ant-attended scale insects in Gorongosa National Park, Mozambique. (Photo by Piotr Naskrecki)

Fig. 9 An adult *Miletus biggsii* (Lycaenidae: Miletinae) perches among aphid-tending dolichoderine ants in Thailand. (Photo by Henrik Petersen. A related *Logania malayica* perches similarly among myrmicine ants in another photo at iNaturalist #50360170)

hemipterans. Many caterpillars in seasonally arid and cold regions enter underground ant nests, likely to escape unfavorable conditions. The need to escape the increasingly dry conditions and the associated risk of fires that occurred during the aridification of Africa in the Miocene may have been an important driver leading to the relatively large number of obligately parasitic relationships found in the dry savanna habitats of southern Africa and Australia. These regions are also hotspots for myrmecochorous plants, those plants with seeds dispersed by ants (Lengyel et al. 2010), possibly for similar reasons, although the phosphorus-poor soils of these regions are also likely to have been important (see discussion below) (Westoby et al. 1982). Larvae of a number of species have been reported to follow ant trail pheromones, but only a few cases of this behavior have been experimentally confirmed (Dejean and Beugnon 1996; Fiedler et al. 1996).

Hinton (1951) noted that ant-attended larvae, even within ant nests, may be attacked if ants are sufficiently alarmed by an intruder. Most lycaenid larvae can retract their head beneath a sclerotized prothoracic plate and are ventrally flattened, shielding vulnerable body parts (Ballmer and Pratt 1988; Fiedler 1991; Malicky 1969; Malicky 1970; Pierce et al. 2002). Larvae that live in close proximity with ants may have a wrinkled cuticle up to 20 times thicker than that of other Lepidoptera to avoid harm from the occasional bite (Bächtold and Alves-Silva 2013; Fiedler 1991; Gnatzy et al. 2017; Malicky 1969; Malicky 1970). In general, those with facultative associations with ants have thicker cuticles than those with obligate associations, although this depends in part on the mandible size of the ant associates (Dupont 2012). Lycaenid caterpillars also generally lack the thrash reflex to disturbance found in other Lepidoptera, which can elicit enhanced attack from ants (Bächtold and Alves-Silva 2013; Fiedler 1991).

Ant-attended Lycaenidae and Riodinidae possess a variety of multimodal "ant organs" to attract and signal to ants via chemicals or stridulation. Cuticular hydrocarbons and similar substances protect lycaenid larvae from most ant aggression, as described in a later section. In addition, many ant-associated lycaenid and riodinid caterpillars are attractive to ants, which groom and antennate various parts of their bodies. Ants are often drawn to specific parts of lycaenid larvae bearing dense single-celled epidermal glands that Malicky (Malicky 1970) described in English as "perforated cupola organs" (PCOs). Kitching (g 1983) translated Malicky's original *"porenkuppeln"* (Malicky 1969) as "pore cupola organs" (PCOs), and this term has been adopted generally. PCOs are also found in many pupae (e.g., Duarte et al. 2001; Fiedler 1989b; Fiedler and Seufert 1995; Hinton 1951; Malicky 1970; Pierce and Nash 1999). PCOs or putative homologs have been found in the larvae of all Lycaenidae and Riodinidae that have been examined (Dupont et al. 2016; Fiedler 1991; Mota et al. 2014; Nielsen and Kaminski 2018; Pierce et al. 2002; Santos et al. 2014). As a result, Pierce et al. (2002) suggested that PCOs may represent a key preadaptation for the radiations of myrmecophilous Lycaenidae and Riodinidae. The ant-associated functions of these organs are likely to be convergent given what we now know about the phylogeny of these groups. The function of PCOs in non-myrmecophilous caterpillars has not been carefully explored: PCOs are widespread among caterpillars of non-myrmecophilous Riodinidae as well as the non-myrmecophilous family Hesperiidae, where they were originally called "lenticles" (DeVries 1991c; Franzl et al. 1984).

Larval PCOs are often concentrated around spiracles and secretory organs (e.g., Downey and Allyn 1979; Fiedler 1991; Kitching and Luke 1985; Mota et al. 2014; Mota et al. 2020; Pierce and Nash 1999). Many Lycaenidae also have a higher density of PCOs on thoracic segments that are attractive to ants (Pierce and Nash 1999). Comparing related species or different populations of a single species, PCOs may be more numerous or productive in larvae that are more closely ant-associated (e.g., Ballmer and Pratt 1991; Kaminski et al. 2013).

In addition, a large number of wedge-shaped, dendritic, mushroom, and other highly modified setae appear important to ant interactions of various larvae and pupae (DeVries et al. 1986; Downey and Allyn 1979; Duarte et al. 2001; Dupont et al. 2016; Fiedler 1989a; Fiedler 1991; Hall and Harvey 2001; Hall et al. 2004; Kaminski and Carvalho-Filho 2012; Kaminski et al. 2013; Kaminski et al. 2020b;

Fig. 10 *Plebejus idas* larva (Lycaenidae: Polyommatini) in Italy with everted tentacle organs (on left), attended by *Lasius emarginatus* ants. (Photo by Paolo Mazzei)

Pierce et al. 2002). The presence of dendritic setae appears to be strongly correlated with the ants' interest in larvae (Ballmer and Pratt 1991). These specialized setae are generally concentrated near PCOs and other secretory organs and may help disperse secretions to arouse ants (Ballmer and Pratt 1991). Others are mechanoreceptors that respond to attending ants (Tautz and Fiedler 1992).

Tentacle organs (TOs) are paired, typically eversible structures on the eighth abdominal segment of many riodinid and lycaenid larvae that are operated hydrostatically by specialized muscles (Fig. 10) (Basu and Kunte 2020; Gnatzy et al. 2017; Hinton 1951; Vegliante and Hasenfuss 2012). While TOs are potentially part of the lycaenid and riodinid ground-plan, they are absent in the riodinid subfamily Nemeobiinae, the lycaenid subfamilies Poritiinae and Lycaeninae, all of the Miletinae except the genus *Aslauga,* and a few other genera (Campbell and Pierce 2003; Fiedler 1991; Pierce et al. 2002). Their function is usually defensive and often specialized to signal to ants as discussed below.

Vibratory Signaling

Larvae of various Lepidoptera produce vibratory signals to deter predators, defend larval territories, or attract additional larvae (see Yack, Ch. 7) (e.g., Bura et al. 2009; Bura et al. 2011; Dookie et al. 2017; Fletcher et al. 2006; Sanetra and Fiedler 1996; Yack et al. 2001; Yadav et al. 2017). Stridulations are a widespread method for ants to recruit nestmates for foraging or defense and have correspondingly been adapted by some larvae to attract attention (Schönrogge et al. 2017). One of the earliest reports of larval stridulation came from naturalist Charles O. Farquharson, who noted a sensation like an electric shock from touching different lycaenid caterpillars (Farquharson et al. 1922). Substrate-borne acoustic signals produced by numerous lycaenid and riodinid larvae encourage ant attendance and are similar to those made by attending ants (e.g., Fiedler et al. 1996; Lin et al. 2019; Riva et al. 2017; Schurian and Fiedler 1994; Travassos and Pierce 2000). Larval sounds or sound-producing organs have been observed in all examined ant-attended lycaenid and riodinid larvae and are only known to be absent in some non-myrmecophilous Riodinidae and New World Lycaenidae of the tribe Eumaeini (DeVries 1990; DeVries 1991d). Some non-trophobiotic larvae are able to produce sounds, but all belong to genera that facultatively associate with ants (Elfferich 1998; Pierce et al. 2002; Riva et al. 2017).

The few described sound production mechanisms in lycaenid larvae are all stridulatory (Hill 1993; Schönrogge et al. 2017; Schurian and Fiedler 1994). The stridulatory organ of both the larva and pupa of *Arhopala madytus* is located between the fifth and sixth abdominal segments (Hill 1993), as is the stridulatory organ of most lycaenid pupae (Downey 1966). However, in the pupa, the file (sixth segment) is posterior to the stridulatory plate (fifth segment), whereas in the larva of *A. madytus,* their placements are reversed. The discrete organs giving rise to these substrate-borne vibrations have proved difficult to identify in many species. In the Australian lycaenid, *Jalmenus evagoras*, they seem likely to consist of rings of tiny, serially repeating teeth and scrapers occurring between each pair of larval abdominal segments. When the larva is calling, these areas can be seen to vibrate using high speed video (Pierce et al. 2002; Travassos and Pierce 2000).

Pupae of Lycaenidae and Riodinidae also produce several types of vibrations, including "chirping" noises audible to humans, using plate-and-file stridulatory mechanisms located on membranes between abdominal segments 4 and 7 (Downey and Allyn 1973; Downey and Allyn 1978). In addition, "tooth-cast" systems, in which one opposing structure of the sound-producing organ is an imprint of the other, are found in diverse Lycaenidae (Downey and Allyn 1973), as in pupae of Nymphalidae and Papilionidae (Dolle et al. 2018). Acoustic signals play an important role in ant recruitment and appeasement by myrmecophilous Lycaenidae and Riodinidae but are also widespread in non-myrmecophilous pupae, presumably serving as deimatic displays to startle predators as in other Lepidoptera (Dodd 1916; Dolle et al. 2018; Downey and Allyn 1973; Elfferich 1998; Lin et al. 2019; Pierce et al. 2002; Travassos and Pierce 2000).

Lycaenidae

The Lycaenidae contain over 5000 species in more than 400 genera distributed worldwide (Eliot 1973; Espeland et al. 2018; Pierce et al. 2002). Although different species vary in the relative strength and context of ant association, all lycaenid subfamilies have species that are either ant-attended or form some kind of regular association with ants (Table 2).

Curetinae

The lycaenid subfamily Curetinae consists of a single genus (*Curetis*) of 18 species and is distributed from India to the Solomon Islands (Eliot 1990). The genus is significant inasmuch as it is sister to all other Lycaenidae and may illustrate plesiomorphic traits shared with riodinids but lost in other lycaenids (Espeland et al. 2018). *Curetis* larvae can produce loud substrate-borne vibrations (Fiedler et al. 1995). *Curetis* TOs are housed in large, sclerotized cylinders, which evert long filamentous processes when the larva is disturbed, exciting nearby ants (videos at https://www.youtube.com/watch?v=2AAg26XDtgM, https://www.youtube.com/watch?v=zhSX_7edW44) (DeVries 1984). Much like those of some non-trophobiotic riodinids described below (Nielsen and Kaminski 2018), *Curetis* TOs evert and appear to emit repulsive chemicals, in response to ants and other attackers including parasitoid flies and wasps (video at https://www.youtube.com/watch?v=LUKxmq3_6MU) (Ballmer 2015; DeVries et al. 1986; Fiedler et al. 1995; de Niceville 1890; van der Poorten and van der Poorten 2016). Ants usually show little interest in *Curetis* larvae but often accompany them to drink from leaf exudates where larvae have been feeding (Fig. 11) (DeVries 1984; Fiedler et al. 1995).

The remaining Lycaenidae form a clade that is ancestrally ant-attended (Espeland et al. 2018). Most species of the subfamilies Aphnaeinae, Theclinae, and Polyommatinae have a dorsal nectary organ [DNO], a unique slit-like glandular invagination on the 7th abdominal segment that produces attractive secretions for ants and appears in the 2nd or 3rd instar (Daniels et al. 2005; Fiedler 1991; Hinton 1951; Pierce et al. 2002). A superficially similar abdominal invagination found in Curetinae may be a vestigial DNO or perhaps simply a muscle attachment site (DeVries et al. 1986). The DNO contains 2–4 individual glands, which structurally and developmentally resemble modified setae (Hinton 1951; Malicky 1970; Newcomer 1912; Pierce and Nash 1999; Vegliante and Hasenfuss 2012). Muscles around the DNO usually allow it to push upward and extrude liquid droplets or retract and suck back these secretions (video at https://www.youtube.com/

Fig. 11 *Curetis thetis* (Curetinae) larva with ants in Sri Lanka. (Photo by Nuwan Chathuranga)

watch?v=fCho3Vrt2bU) (Basu and Kunte 2020; Pierce and Nash 1999). Larvae of many obligately ant-attended species have been reported to die in captivity from mold and/or infection without ants to remove built-up secretions around the opening of the DNO (Cottrell 1984; Hinton 1951; Williams 1995-2020).

Caterpillars of some species have been shown experimentally to deploy their DNO secretions strategically, increasing the rate of droplets provided when they are vulnerable or under perceived attack and decreasing per capita secretions in larger larval aggregations (Agrawal and Fordyce 2000; Axen and Pierce 1998; Axén et al. 1996; Leimar and Axén 1993). Caterpillars may also increase secretion rates when more ants are present; this might allow them to retain a larger retinue of ants (Axén 2000; Fiedler and Hagemann 1992). Curiously, the dorsal nectary organ remains functional in many parasites that enter the ant nest such as *Niphanda fusca* and species of *Phengaris*, suggesting that secretions from the DNO in these species may contain essential substances enabling them to manipulate attendant ants.

The TOs of species in the Aphnaeinae and the Theclinae-Polyommatinae assemblage appear to secrete volatile chemicals that excite ants to defend the larva (Casacci et al. 2019b; Fiedler 1991; Fiedler et al. 1996; Henning 1983; Pierce et al. 2002). Lycaenid TOs are most frequently everted to attract ants when caterpillars are disturbed or are traveling to a new location or when ant-caterpillar interactions first begin (Axén et al. 1996; Fiedler et al. 1996; Fiedler and Hagemann 1992; Leimar and Axén 1993). Secretions from the tentacle organs of lycaenids have been difficult to detect and/or characterize chemically (Gnatzy et al. 2017; Pierce and Nash 1999). The TOs of the Japanese species *Shirozua jonasi* (Theclinae: Theclini) were described to contain dendrolasin (Yamagushi and Shirozu 1988), a compound found in some ant alarm pheromones (Hölldobler and Wilson 1990). Although the chemicals involved are unknown, extracts from the TOs of *Aleiodes dentatis* (Aphnaeinae) were shown to elicit an alarm response from workers of the attendant ant species (Henning 1983). Alarm pheromones are also mimicked by many myrmecophilous rove beetles and wasps (Stoeffler et al. 2007; Thomas et al. 2002).

In terms of delivery, some authors have speculated that the tentacle organs of Lycaenidae might disperse chemical signals that are coated on their long, finely branched apical setae when the tentacle is withdrawn into an evagination formed by the cuticle (Fiedler et al. 1996; Fiedler et al. 1995; Hinton 1951; Kitching and Luke 1985; Pierce and Nash 1999; Sanetra and Fiedler 1996). Additional research is warranted, as Gnatzy et al. (2017) carefully examined the histology of these setae and found no evidence that they were glandular in nature.

The Theclinae-Polyommatinae Assemblage

Theclinae and Polyommatinae are both polyphyletic as traditionally defined but together form a well-supported monophyletic group (Espeland et al. 2018). The Theclinae-Polyommatinae assemblage is widespread, including over 4000 species in nearly 350 genera. Larvae are mostly phytophagous and ant-associated, but

Fig. 12 The larvae of several species of *Hypolycaena* (Theclinae: Hypolycaenini) are attended by *Oecophylla* ants, such as this *H. erylus* in Malaysia being accompanied as it travels. (Photo by Masatoshi Sone)

several lineages are non-myrmecophilous (Table 2, Fig. 12, videos at https://www.youtube.com/watch?v=GsSlcA0WXnk, https://www.youtube.com/watch?v=43vmltWoSdo].

With over 1300 species that typically only form facultative ant associations, the tribe Polyommatini is the largest tribe of Lycaenidae. Only a few obligately ant-associated taxa are known in this tribe outside of the two unique genera, *Lepidochrysops* and *Phengaris* (Polyommatini). Larvae of the some 130 species of Afrotropical *Lepidochrysops* typically feed on flowers until the 3rd instar, when they begin to mimic ant brood and are carried by workers of species of *Camponotus* (subfamily Formicinae) into the nest to feed on brood and/or engage in trophallaxis (Heath and Claassens 2003; Henning 1983).

Like *Lepidochrysops,* the approximately ten species of Palearctic *Phengaris* (= *Maculinea*) are also phyto-predaceous. The larvae of different species of *Phengaris* initially feed on flowers and in the fourth instar are carried by *Myrmica* workers (subfamily Myrmicinae) into the nest, where different larvae, even those derived from eggs laid in the same year, will remain parasitic for either 1 or 2 years (Elmes et al. 2019; Thomas et al. 1998; Witek et al. 2006). Acceptance of *Phengaris* by host ants is mediated by specialized chemical mimicry of ant hosts (Akino et al. 1999; Casacci et al. 2019b; Casacci et al. 2019a; Nash et al. 2008; Schönrogge et al. 2004; Solazzo et al. 2013). Although most *Phengaris* feed directly on ant brood, a group of "cuckoo" species have larvae that specialize on trophallaxis (Als et al. 2004; Thomas and Elmes 1998). Both predatory *Phengaris arion* and cuckoo *Phengaris rebeli* are nest parasites whose larvae have been reported to produce acoustic signals resembling those of their host ant queens and giving them extreme priority in feeding and protection (Barbero et al. 2009a; Barbero et al. 2009b; Barbero et al. 2012; Sala et al. 2014; Thomas et al. 2013). Most *Phengaris* species can parasitize nests of multiple ant species, although local populations are often strongly specialized on different hosts (Pech et al. 2007; Tartally et al. 2019; Ueda et al. 2016; Witek et al. 2011; Witek et al. 2008; Sielezniew et al. 2010; Thomas et al. 2013). *Phengaris arion* has become a classic conservation success story, after recognition of its obligate relationship with a single ecologically restricted *Myrmica* species in the UK facilitated the reintroduction of the caterpillar species (Thomas et al. 2009).

First-instar larvae of East Asian *Niphanda fusca* (Niphandini) feed on aphid honeydew, but later-instar larvae enter *Camponotus* nests, where they chemically mimic male ants and are fed by workers (Hojo et al. 2014a; Hojo et al. 2009). Larvae of *Phengaris*, *Lepidochrysops*, and *Niphanda fusca* that enter the ant nest in later instars have an unusual growth pattern, growing more than ten times as much once in the ant nest as would be predicted from their earlier stages (Elmes et al. 2001). Two Afrotropical species of *Anthene* (Lycaenesthini) are parasites in nests of species of *Crematogaster* (Williams 1995-2020). A few related larvae—Tropical Asian *Chilades lajus* (Polyommatini) and Afrotropical *Triclema lamias* (Lycaenesthini)—may prey on aphids and scale insects (Farquharson et al. 1922; Pierce 1995). Many other plant-feeding species supplement their larval diet with hemipteran honeydew under certain conditions (Fig. 13) (e.g., Pierce and Elgar 1985).

Only a few other parasitic species can be found within the remaining tribes that are currently non-monophyletically grouped as Theclinae. All 11 species of the Australian genus *Acrodipsas* (Eastwood and Hughes 2003; Miller and Lane 2004; Sands and Sands 2015) and a few species within the mostly phytophagous and highly ant-associated genera *Ogyris* and *Arhopala* are brood predators in ant nests (Braby 2000; Fiedler 2012; Pierce 1995). Palearctic *Shirozua* larvae mostly feed on hemipterans and their excretions but also sometimes on *Lasius* or *Camponotus* ant trophallaxis (Fiedler 2012; Pierce 1995; Zhou and Zhuang 2018). *Shirozua jonasi* may enter ant nests to pupate, and adults are protected by dense cotton-like hairs (Cottrell 1984).

Although widely distributed, over 90% of the approximately 1096 species in the tribe Eumaeini are found in the Neotropical region, and all are either non-myrmecophilous or facultatively so, usually only sporadically ant-attended. The Old World taxa are clustered in a single clade consisting largely of the species-rich sections *Callophrys*, *Erora*, and *Satyrium*. Their huge radiation appears to be associated with intense sexual selection, as males have a great diversity of secondary sexual traits such as brush organs associated with the genitalia and androconial

Fig. 13 A fourth-instar larva of *Jalmenus daemeli* (Theclinae: Zesiini) feeds on secretions from a margarodid scale, while both are tended by workers of *Iridomyrmex rufoniger*. These Australian caterpillars are usually herbivorous but may facultatively feed on honeydew secretions. (Photo by Naomi Pierce)

Fig. 14 Cycad-feeding *Eumaeus* larvae (Theclinae: Eumaeini), such as these *E. toxea* in Nayarit, Mexico, are toxic and not ant-associated. (Photo by Juan Cruzado Cortés)

wing scent pads and patches that waft pheromones (Valencia-Montoya et al. In review). Caterpillars of several genera are aposematically colored or bear defensive tubercles and scoli, resembling Limacodidae (Fig. 14) (e.g., Kaminski et al. 2010b; Silva et al. 2014). Some respond to disturbance by curling their body or hanging off the substrate on a silk thread, behaviors otherwise unknown in Lycaenidae (Fiedler 1991; Silva et al. 2014). The approximately 175 species in the detritivorous Neotropical subtribe Calycopidina have never been reported with ants, but limited evidence suggest that some species might be facultatively ant-associated (Duarte and Robbins 2010; Nishida and Robbins 2020, supplemental table from Schär et al. 2018; Silva et al. 2014).

A number of studies have looked at the developmental effects of ant attendance on caterpillars of the Theclinae-Polyommatinae assemblage. Different attendant ant species differ in their impact on survival and development (Fraser et al. 2001; Kaminski and Rodrigues 2011; Mizuno et al. 2019; Trager and Daniels 2009; Saarinen and Daniels 2006; Wagner 1993). The costs and benefits of ant attendance are also borne differently by males and females, probably based on differing physiological demands on adults of each sex to ensure reproductive success (Mizuno et al. 2019; Pierce et al. 1987). Measured effects of ant attendance on developmental times and adult sizes vary extensively between different species (Baylis and Pierce 1992; Cushman et al. 1994; Fiedler and Hölldobler 1992; Fiedler and Hummel 1995; Fiedler and Saam 1994; Fraser et al. 2001; Kaminski and Rodrigues 2011; Mizuno et al. 2019; Pierce and Nash 1999; Pierce et al. 1987; Robbins 1991; Saarinen and Daniels 2006; Trager et al. 2013; Wagner 1993). The methods employed in quite a few of these studies involve placing ants and larvae together in disturbed laboratory environments in order to create an "ant-attended" treatment. Controlled experiments using intact ant colonies containing queens and with naturally foraging workers tending caterpillars feeding on live host plants are difficult to carry out, but they seem likely to yield different results from treatments in which individual workers are simply enclosed with caterpillars feeding on cuttings to simulate natural tending. For example, field versus laboratory experiments found different effects on developmental times of facultatively ant-associated larvae of *Glaucopsyche lygdamus* (Fraser et al. 2001; Pierce and Easteal 1986).

Leguminous host plant use is broadly correlated with ant attendance within the Theclinae-Polyommatinae assemblage (Fiedler 1995; Pellissier et al. 2012a; Pierce 1985). The relationship may not be causal, but protein-rich foods could help caterpillars produce nitrogen-rich secretions for ants. For example, individual larvae of *Jalmenus evagoras* larvae are tended by more ants per capita when they are fed higher-quality host plants that have been treated with nitrogenous fertilizer than when feeding on lower-quality control plants (Baylis and Pierce 1991). Similarly, feeding on flowers may lead to greater larval growth and in some cases has been shown to increase the volume of DNO secretions (Burghardt and Fiedler 1996; Collier 2007; Pierce and Easteal 1986; Wagner and Kurina 1997). The distribution of legumes and their symbiotic bacteria might also exert indirect effects on lycaenid biogeography (Steidinger et al. 2019). Feeding on Fabaceae appears to be an ancestral state of all phytophagous lycaenid subfamilies with the exception of the Lycaeninae (Boyle et al. 2015; Espeland et al. 2018; Fiedler 1991). Thus, the correlation between ant attendance and legume feeding might be more appropriately viewed as one where species that switch to less nutritious food sources are unlikely to remain ant-attended (Fiedler 1995).

Lycaeninae

The approximately 110 species of Lycaeninae, which form a sister group to the Theclinae-Polyommatinae assemblage, have an unusually wide, disjunct distribution that includes all major zoogeographic regions. All described species of Lycaeninae lack a dorsal nectary organ and tentacle organs, but larvae and pupae possess stridulatory organs and sometimes enter ant nests (Bascombe et al. 1999; DeVries 1991d; Downey and Allyn 1973; Fiedler 1991; Gibbs 1980; Heath and Claassens 2003; Yago et al. 2010). Furthermore, a few species have been reported possibly to rely on ants for oviposition, and these caterpillars may also be somewhat attractive to ants (Ballmer and Pratt 1991; Fiedler 1989a; Funk 1975; Oliver 2007).

Miletinae

The lycaenid subfamily Miletinae is notably missing from the Neotropics and western Palearctic (and has only one species in the Nearctic). All 190 species in 13 genera are thought to be entomophagous, eating either ants, their regurgitations, or ant-associated hemipterans and their secretions (Fig. 15, video at https://www.youtube.com/watch?v=ZmCz2UxKaHA) (Cottrell 1984; Eliot 1986; Kaliszewska et al. 2015; Pierce 1995). Many adult Miletinae have an especially long, sclerotized abdomen and legs, possibly to protect against the occasional ant bite while alighting and/

Fig. 15 This ant-associated *Lachnocnema laches* larva (Miletinae) feeding on treehopper nymphs was reared in South Africa. (Photo by Suncana Bradley)

or ovipositing among hemipteran prey that are being tended by ants (Cottrell 1984; Pierce 1995).

Ants intensively palpate and display interest in the larvae of many Miletinae, but larvae lack dorsal nectary organs, and only species in the Afrotropical genus *Aslauga* possess tentacle organs (Bascombe et al. 1999; Claassens and Heath 1997; Cottrell 1984; Dejean et al. 2017; Lohman and Samarita 2009; Pierce et al. 2002). Most appear obligately ant-associated (Table 2). Larvae of Oriental and Palearctic *Spalgis* and *Taraka* spp. and Nearctic *Feniseca tarquinius* appear facultatively ant-associated and are protected by silk shelters or cuticular hydrocarbons of ant-attended prey (photos at iNaturalist #57006925 and iNaturalist #14834663) (Cottrell 1984; Lohman et al. 2006; Youngsteadt and Devries 2005). Larvae of *F. tarquinius* produce vibratory signals that may be ant-related (Mathew et al. 2008).

Kaliszewska et al. (Kaliszewska et al. 2015) found that the subfamily of hemipteran-attending ant is strongly conserved phylogenetically within Miletinae, whereas hemipteran host preference can be quite broad (Fiedler and Maschwitz 1989b; Lohman and Samarita 2009). For example, lycaenids in the genus *Miletus* appear to associate only with species of ants in the genus *Dolichoderus,* which adults use to find their hemipteran prey. All 27 species in the southern African genus *Thestor* are thought to parasitize ants in the genus *Anoplolepis* (Formicinae), particularly *A. custodiens (*Claassens and Dickson 1980*;* Clark and Dickson 1971*;* Pringle et al. 1994*).* While the larvae of the majority of Miletinae feed on Hemiptera, later instars may occasionally be carried into the ant nest, where they feed on ant regurgitations and sometimes also ant eggs and detritus (Clark and Dickson 1960; Clark and Dickson 1971; Heath and Claassens 2000; Heath and Claassens 2003; Heath and Pringle 2004; Williams and Joannou 1996).

Caterpillars of the sister genera *Liphyra* and *Euliphyra* inhabit the nests of weaver ants in the genus *Oecophylla* (Fig. 16). Oriental *Liphyra brassolis* and *Liphyra grandis* feed voraciously on ant brood and are protected from occasional

Fig. 16 *Liphyra brassolis* larva (Miletinae) displaying its highly sclerotized, tank-like dorsum on the outside of an *Oecophylla* nest in Queensland, Australia. (Photo by Martin Lagerwey)

attack by a thick, bulky "chain link" integument derived from modified setae (Braby 2000; Dupont et al. 2016; Pierce 1995). African species of *Euliphyra* also coax trophallaxis, intercept trophallaxis between workers, and steal brood within *Oecophylla* nests (Dejean et al. 2017; Fiedler 2012). *Liphyra* pupae are entirely enclosed within the hardened exuviae of the last larva instar, or puparium, and *Euliphyra* pupae only partially emerge during larval-pupal ecdysis (Eltringham 1913). *Euliphyra* larvae have been shown experimentally to find *Oecophylla* nests by following ant trail pheromones (Dejean and Beugnon 1996), and *Liphyra* probably do this as well (Common and Waterhouse 1981; Pierce et al. 2002). Larvae of *Liphyra*, along with those of *Thestor*, have short, stubby antennae used to seek and manipulate prey (Dupont et al. 2016). Adult *Liphyra* are protected upon eclosion by a thick vestiture of greasy, loose scales that slip off in the mandibles of vicious attacking ants, similar to several other species that pupate in ant nests (Atsatt 1981a; Cottrell 1984; Dodd 1902; Hinton 1951; Pierce 1995).

The Miletinae likely constitute the largest radiation of entomophagous lepidopterans (Cottrell 1984; Pierce 1995), and ant association may be linked to the success of this dramatic dietary shift. Body plan constraints may limit the success of predatory caterpillars, except around concentrated food resources or in situations where there are few competing predators (Pierce 1995)—ant brood and ant-attended hemipterans meet both of these conditions. The reverse dietary shift in spiders follows the same principle: the only two spider species with known specializations for plant-feeding are found on well-defended ant-plants with few other herbivores (Meehan et al. 2009; Nyffeler et al. 2016; Painting et al. 2017).

Aphnaeinae

The Aphnaeinae, which along with the Poritiinae are sister to the Miletinae (Espeland et al. 2018), are a largely African subfamily that seems to have been ancestrally associated with *Crematogaster* ants and legume feeding (Boyle et al.

Fig. 17 *Cigaritis takanonis* larvae (Aphnaeinae) tended by *Crematogaster* ants in South Korea. (Photo by iNaturalist user clurarit)

2015). All species whose life histories are known appear obligately ant-associated (Fig. 17, Table 2). Moreover, at least one species in each of nine genera is aphytophagous, feeding on hemipterans or the eggs, brood, or regurgitations of ants (Boyle et al. 2015; Pierce 1995; Sanetra and Fiedler 1996). One species, *Aleiodes pallida*, is known to feed in early instars on species of *Aspalathus* (Fabaceae), but Heath and Claassens (Heath and Claassens 2000) were able to rear final-instar caterpillars of this species in observation nests of the formicine ant, *Lepisiota capensis*, where caterpillars selectively ate only the ant eggs and not the brood. Additional evidence suggests that several other species of *Aloiedes* may share this ability to shift from eating plants to eating ant eggs in the final instar. Other species, with exclusively parasitic larval habits, appear in several otherwise phytophagous genera (Basu and Kunte 2020; Fiedler 2012; Heath and Claassens 2003; Pierce 1995). Many Aphnaeinae depend on the presence of a specific species of ant to oviposit (Heath 1997). Dish organs or dew patches are dish-like depressions found on the anterior abdomen in several ant-attended genera of Aphnaeinae that appear to produce reward secretions (Basu and Kunte 2020; Clark and Dickson 1971; Cottrell 1984; Vegliante and Hasenfuss 2012). Several authors note that caterpillars of different species of Aphnaeinae will die if ants are not present to remove secretions from the dew patches and the DNO to prevent them from growing moldy (e.g., Heath 1997; Williams 1995-2020). Tentacle organs of aphnaeine larvae are often housed in protruding cylindrical bases and can be deployed almost like a cat-o'-nine-tails to shoo away overly persistent ants from the DNO (video at https://www.youtube.com/watch?v=Qkd23Pmucmk) (Fiedler 1991).

Poritiinae

The Poritiinae are a subfamily of lycaenid butterflies with non-trophobiotic caterpillars. The approximately 729 species are divided into two clades: the small Asian tribe Poritiini and the large African tribe Liptenini (sometimes split further). Among these, the Liptenini are notable for their lichenivorous diet, although larvae of *Deloneura* have also been recorded feeding on honeydew near ants (Heath and Claassens 2003; Williams 2006). Larvae of some species feed on lichens growing on bark, rocks, or sticks along the ground and may not be found around ants (Larsen 2005; Williams 2006). While poritiine larvae in the generally open habitats of southern Africa are generally facultatively ant-associated (with the exception of *Deloneura*), most species in the wetter forests of West Africa seem to be obligately ant-associated (Bampton 1995). Adults of many species are only found around individual colonies of arboreal *Crematogaster* ants (Larsen 2005; Sáfián 2015b), and caterpillars of several genera have been reared from ant-infested trees (Callaghan 1992a; Dejean et al. 2017; Jackson 1937; Sáfián 2015a; Sáfián and Collins 2014; Sáfián and Larsen 2009). Over 50% of Poritiinae belong to genera that appear to contain obligate ant associates (Table 2). Obligately ant-associated poritiines tend to be rare, and some species have only ever been found in association with a single arboreal ant colony, raising considerable conservation concern (Larsen 2005; Williams 1995-2020). The lack of obligate ant association in some genera of Poritiinae is perhaps a secondary loss—except for the Poritiini that remain understudied, all major lineages of Poritiinae include species apparently only found on trees along with their associated ant species. Together, the subfamilies Miletinae, Aphnaeinae, and Poritiinae probably constitute the largest single radiation of obligately ant-associated Lepidoptera.

All caterpillars of Poritiinae are covered in long bristles that appear to repel ants (Callaghan 1992b; Dejean et al. 2017). They are probably also chemically defended, as larvae of many species interact with ants with no sign of overt conflict (Farquharson et al. 1922; Sáfián and Collins 2014; Sáfián and Larsen 2009). Ants are repelled from many liptenine caterpillars, perhaps because they secrete toxic chemicals. Some species form large larval aggregations, and others appear to be aposematic (Sáfián 2015a; Sáfián and Larsen 2009). Tussock moth caterpillars (Erebidae: Lymantriinae) protected by defensive glands are sometimes found near poritiine caterpillars in Africa and may similarly be associated with arboreal ant colonies (Farquharson et al. 1922; Hinton 1951). These lymantriine and poritiine caterpillars are visually similar and possibly form a Müllerian mimicry complex (Farquharson et al. 1922).

Riodinidae

Riodinidae are sister to Lycaenidae, and while the 153 genera of Riodinidae are distributed worldwide, more than 1400 species are found in Central and South America. The *ca.* 120 Old World species are concentrated in Southeast Asia (Espeland et al. 2015; Seraphim et al. 2018). Most Riodinidae are not known to be ant-associated and possess long setae and chemical defenses that prevent ants from getting too close (e.g., Ballmer and Pratt 1988; DeVries 1988a; Fiedler 1991; Kaminski 2008a; Mota et al. 2014; Nishida 2010; Vélez-Arango et al. 2010). Larval

Fig. 18 Aposematic *Emesis aurimna* larva (Riodininae: Emesidini) in Costa Rica. (Photo by Karl Kroeker)

aggregation and aposematism are also widespread among riodinids (Fig. 18) (Allen 2010; Callaghan 1986; Janzen and Hallwachs 2021; Nishida 2010). Recorded ant associations are limited to the tribe Nymphidiini and subtribe Eurybiina of the tribe Eurybiini, both of which are in the strictly Neotropical subfamily Riodininae (Espeland et al. 2015). Almost a thousand riodinid species belong to genera that are non-trophobiotic and generally not known to be ant-associated (Table 2).

Eurybiini

In ant-attended larvae of the subtribe Eurybiina of the riodinid tribe Eurybiini, modified TOs, called tentacle nectary organs (TNOs), evert to release a drop of fluid that ants eagerly drink (Horvitz et al. 1987). However, larvae of the subtribe Mesosemiina, sister to the subtribe Eurybiina (Espeland et al. 2018; Seraphim et al. 2018), have never been found with ants, and their TOs are protected by defensive bristles (Nielsen and Kaminski 2018; Vélez-Arango et al. 2010). Nielsen and Kaminski (2018) found that TOs of these larvae evert and extrude a droplet of liquid when attacked by various predators including wasps, biting midges, and lacewing larvae. Ants that came into contact with this liquid cleaned themselves and shunned the larva (Nielsen and Kaminski 2018). Larvae of Symmachiini (Riodininae), which are not ant-associated, also possess tentacle organ openings that may prove to have a similar function (Seraphim et al. 2018).

Nymphidiini

In the riodinid tribe Nymphidiini, all known larvae are ant-associated and typically secrete liquid droplets from glandular tissue within the tentacle organs for ants to imbibe (Fig. 19) (Callaghan 1986; DeVries 1988b; DeVries 1997; DeVries and

Fig. 19 Ant-attended *Synargis calyce* larva (Nymphidiini) in Brazil. (Photo by Kel Silva)

Penz 2000; Hall and Harvey 2001; Kaminski 2008b; Kaminski and Carvalho-Filho 2012; Kaminski et al. 2016; Kaminski et al. 2013; Mota et al. 2020; Ross 1964; Torres and Pomerantz 2016). The TNOs may be everted most often when the larva is vulnerable or attending ants have only started to arrive (DeVries 1988b).

In a handful of related Nymphiidini, a pair of metathoracic anterior tentacle organs (ATOs) induce alarm in attending ants, sensitizing them to future threats much like the TOs of Lycaenidae (DeVries 1988b, 1997; Kaminski and Carvalho-Filho 2012; Kaminski et al. 2016; Penz and DeVries 2006). Brush-like setae at the apex of the ATOs likely help disperse volatile chemicals (DeVries 1997; Ross 1964). DeVries (1988b) found that the ATOs are important for these larvae to maintain the attention of attending ants and activate most often when the larva is initiating contact or vulnerable. Some phylogenetically earlier-branching Nymphidiini have thoracic PCO clusters that appear homologous in position to the anterior tentacle organs and similarly excite ants (Kaminski et al. 2013).

Balloon setae, swollen structures on the prothorax, may play a role in myrmecophilous interactions in some Nymphidiini that lack ATOs (Kaminski 2008a; Penz and DeVries 2006). However, balloon setae appear to serve a largely defensive function and are shared by many non-myrmecophilous caterpillars (Fig. 20) (Hall et al. 2004; Kaminski et al. 2013; Mota et al. 2014). While *Zabuella paucipuncta* (Nymphidiini) lacks ATOs, a unique cervical gland that is exposed when ants antennate the balloon setae causes the ants to react in alarm (DeVries et al. 2004).

Adaptations for ant attendance have largely been lost in the riodinid genus *Stalachtis* (Nymphidiini), but caterpillars remain facultatively ant-associated, and their cuticle appears attractive to diverse ants, much as in some non-trophobiotic lycaenids (Espeland et al. 2015; Seraphim et al. 2018, https://www.flickr.com/photos/142712970@N03/33322969114, https://www.flickr.com/photos/142712970@N03/40459961724, https://www.flickr.com/photos/142712970@N03/38713147222/, https://www.flickr.com/photos/142712970@N03/27298752638/, https://www.flickr.com/photos/142712970@N03/34660951953/, https://www.flickr.com/photos/142712970@N03/48374845847/).

Fig. 20 Non-myrmecophilous *Caria ino* larva (Riodininae: Riodinini) in Texas, displaying orange balloon setae. (Photo by Joseph Connors IV)

Vibratory signals in ant-attended riodinid larvae are produced through several different mechanisms. Larvae of ant-attended Eurybiini produce sound by rubbing small teeth on the cervical membrane against granulations on the head (DeVries and Penz 2000; Travassos et al. 2008). Larvae of ant-attended Nymphidiini produce sounds using vibratory papillae, small rodlike structures on the prothorax that rub against granulations on the head. Larvae can adjust the beat frequency of the vibratory papillae, with higher rates attracting more ants. Vibratory papillae of *Thisbe irenea* beat fastest when the larva is stressed, traveling, or during initial contact with ants (DeVries 1988b).

Riodinidae in several genera have independently evolved hemipteran diets (DeVries 1997; Mota et al. 2020). Many species in both Eurybiini and Nymphidiini cohabit with ants, including a single species, *Aricoris arenarum*, in which the first two instars steal honeydew from ant-attended hemipterans and solicit trophallaxis, and later instars feed by trophallaxis within *Camponotus* nests (DeVries 1997; Kaminski et al. 2020b; Robbins et al. 1996). Another riodinid caterpillar was recently found preying on ant brood in arboreal nests of *Neoponera villosa* (Rocha et al. 2020). As in the Lycaenidae, adults of aphytophagous Riodinidae frequently have greasy wings that may help them to escape ants (DeVries 1997; Espeland et al. 2015; Hall and Harvey 2002), and the greasiness of wings has been used to successfully predict larval diet in at least one instance (Hall 2007; Mota et al. 2020). The TNOs no longer secrete rewards in ant-associated hemipterophagous riodinids, but still signal to ants (Kaminski et al. 2020b; Mota et al. 2020), much as nectary organs have been lost in the predatory lycaenid subfamily Miletinae. The predatory larvae of Neotropical *Pachythone* spp. (Mota et al. 2020) are also remarkably convergent in appearance and adaptive morphology to the ecologically similar larvae of Afrotropical *Aslauga* spp. (Lycaenidae: Miletinae) (Dejean et al. 2017).

Mutualism and Manipulation: Caterpillar-Ant Trophobiosis in Lycaenidae and Riodinidae

All ant species known to tend trophobiotic caterpillars are agricultural in the sense that they also harvest plant extrafloral nectar and the honeydew produced by Hemiptera (DeVries 1991b; Eastwood and Fraser 1999; Fiedler 2001, 2006; Pierce and Elgar 1985). They include genera such as *Iridomyrmex*, *Oecophylla*, *Camponotus*, and *Crematogaster* that are among the most dominant ants in the regions where they occur, with wide distributions and large colony sizes that are often polydomous in structure. Caterpillars may take advantage of ant preadaptations to harvest carbohydrate rewards, which are essential resources for ants in many environments (Blüthgen and Fiedler 2002, 2004; Blüthgen et al. 2003; Davidson et al. 2003; Dejean et al. 2007; Grover et al. 2007; Kaspari et al. 2020; Kaspari et al. 2012; Pohl et al. 2016; Ribas and Schoereder 2004). Although biochemically modified to attract ants in many species, hemipteran honeydew is an excrement, produced whether attendant ants are present to collect it or not (Stadler and Dixon 2005). *Semutophila saccharopa* (Tortricidae) are the only caterpillars that produce sugar-rich excrement for ants in a manner similar to aphids. In contrast, lycaenid and riodinid larvae produce secretions tailored specifically for their ant associates and released from specialized exocrine glands. This distinction is an important one because exocrine glands provide opportunities for lycaenids and riodinids to fine-tune their secretions to manipulate ant behavior, without necessarily providing nutritious rewards.

Ants, although they confer substantial overall benefit to aphid populations, sometimes consume honeydew-producing hemipteran mutualists, particularly when alternative carbohydrate sources are available (Offenberg 2001; Shibao et al. 2009; Silveira et al. 2010; Stadler and Dixon 2005). In contrast, ants have been rarely reported to attack lycaenid larvae except under unnatural circumstances in captivity. Ants suffer a serious opportunity cost when they invest in protecting caterpillars rather than preying on them, especially facultatively ant-attended lycaenids whose secretions may provide poor-quality rewards. The striking absence of overt ant predation also suggests that lycaenid caterpillars must be able to manipulate ants, at least sufficiently to avoid aggression (Fiedler 1998a; Fiedler et al. 1996).

A number of lycaenid larvae mimic the cuticular hydrocarbon (CHC) profiles of host plants or ants or conceal themselves entirely by lacking recognizable molecules (Barbero 2016; Inui et al. 2015; Lima et al. 2020; Lohman 2004; Morozumi et al. 2019), much as reported for different honeydew-producing hemipterans (Endo and Itino 2013; Silveira et al. 2010). This "cloak of invisibility" can ensure that a caterpillar is not attacked by the ants, even if it is not actively tended. CHC mimicry of ants plays an intimate role in the adoption of parasitic species like *Phengaris* by host ants, as reviewed by Barbero (2016) and Casacci et al. (2019b). Chemical disguise may be observed in other groups when more lycaenid species are studied, but it is clearly not universal (Hojo et al. 2014a; Omura et al. 2009). Other mechanisms might also exist to avoid ant predation. Pupae of facultatively ant-attended *Lycaeides argyrognomon*, which sometimes cohabit with *Camponotus* or *Formica* host ants, have been found to subdue ant aggression through the presence in their cuticle of several long-chained aldehydes not seen in larvae (Mizuno et al. 2018).

DNO and TNO secretions contain amino acids and carbohydrates and have been studied in about ten species (Cushman et al. 1994; Daniels et al. 2005; DeVries 1988b; Pierce and Nash 1999; Pierce et al. 2002; Wada et al. 2001). Secretions of obligate or steadily ant-attended larvae have a higher nutritive content than those of less myrmecophilous species (Daniels et al. 2005).

Aphid mimicry may be one way that facultatively attended larvae attract ants. Melezitose is an aphid gut compound that serves as an attractant for aphid-attending ant workers and is a major component of the nectary secretions in *Polyommatus icarus* and *Zizeeria knysna* [Polyommatini], one of the few well-studied species whose caterpillars are weakly attended (Daniels et al. 2005; Depa et al. 2020; Detrain et al. 2010; Vantaux et al. 2011).

Hojo et al. (2015) determined that the facultatively ant-attended Japanese lycaenid *Arhopala* (=*Narathura*) *japonica* produces DNO secretions that manipulate the dopaminergic pathway in the brains of their attendant ants, workers of *Pristomyrmex punctatus*. Reduced levels of dopamine are correlated with a reduction in worker activity levels (thereby increasing their fidelity to the caterpillar) and heightening aggression toward intruders. Specialized cuticular hydrocarbons (CHCs) of *A. japonica* act as a signal to host ants that they learn to associate with reward after attending larvae (Hojo et al. 2014a).

The parasitic species *Niphanda fusca* has larvae that secrete primarily only trehalose and glycine for host *Camponotus japonicus* ants (Hojo et al. 2008; Wada et al. 2001). Glycine alone is ignored by ants at low concentrations but acts as a manipulative "umami" taste enhancer substance when added to trehalose, increasing host ant interest in this specific sugar (Hojo et al. 2008; Wada et al. 2001). The relative simplicity of this "umami" mechanism for taste enhancement (i.e., the coupling of an amino acid or small peptide with a sugar reward) makes it an attractive candidate for further research into how and why lycaenid caterpillars can be so extremely attractive to their associated ants. However, these same taste preferences are not shared by the closely related *Camponotus obscuripes*, indicating that lycaenid secretion components may be specialized to individual ant species (Hojo et al. 2008).

By comparing the foraging behavior of colonies of the attendant ant species, *Iridomyrmex mayri,* fed on high-protein, high-carbohydrate, or mixed diets, Pohl et al. (2016) showed experimentally that the nutritional state of the attendant ant colony influenced the number of attendant workers foraging on larval secretions from the Australian lycaenid, *Jalmenus evagoras* (Fig. 21). Workers from colonies fed on either carbohydrate- or protein-restricted diets were inconsistent in their compensatory behavior. Those on low-carbohydrate diets compensated by foraging more on sugars, but those on low-protein diets did not show compensatory behavior by foraging more on amino acids. However, workers from colonies that were diet restricted were significantly more interested in foraging on secretions from the larvae than those from well-fed colonies. Workers were not strongly attracted to the

Fig. 21 *Jalmenus evagoras* larval aggregation (Theclinae: Zesiini) in Victoria, Australia. (Photo by Ron I. Greer)

amino acid serine, which had been thought to be the primary amino acid in *Jalmenus evagoras* larval secretions (Pierce 1984) nor did they show an "umami" response when serine was coupled with sugar. More recent analysis suggests that glutamine rather than serine is the primary amino acid in *J. evagoras* larval secretions (Zemeitat 2017), and further work will be necessary to explore the relationship between larval secretions and ant attendance. The chemical composition of liquid secretions produced by different species of ant-attended larvae varies depending on the species and seems likely to be shaped at least in part by the feeding preferences of the ants that attend each caterpillar species (Daniels et al. 2005; Pierce 1984).

Certain obligate lycaenid-ant associations may provide sufficient fitness benefits to both partners under some conditions to warrant being classified as mutualists (Cushman et al. 1994; Fiedler and Maschwitz 1988; Fiedler et al. 1996). For example, caterpillar secretions from the Australian *Jalmenus evagoras* confer a net benefit in terms of positive growth rates (potentially resulting in a greater production of alates) for colonies of its most common ant associate, *Iridomyrmex mayri* (Pierce et al. 1987).

However, even associations with larvae of *Jalmenus evagoras* may sometimes be detrimental to ants. Under experimental conditions, small ant colonies grew faster when allowed to collect secretions from *Jalmenus evagoras* larvae, but colonies provided with only one larva grew significantly faster than colonies given access to five larvae, although this may have been because larvae were allowed to aggregate on the host plants (Nash 1990; Pierce and Nash 1999). Subsequent experiments showed that once small groups of workers and brood passed below a minimum ratio of workers to brood, workers would consistently chose to tend *J. evagoras* larvae and neglect their own brood, allowing them to perish (Merrill 1997; Pierce and Nash 1999). As has been demonstrated for symbioses more generally, whether the relationship is mutualistic or parasitic varies spatially and temporally and is influenced by a number of factors (Madeiros et al. 2018; Thompson 2005; Warren et al. 2019). In the case of caterpillar-ant interactions, this context dependency can include the size of the ant colony, the availability of alternative resources, the number of caterpillars, and the relative cost of their defense.

In certain contexts, manipulation can stabilize mutualisms (Heil et al. 2014; Sachs 2006). However, the manipulation and asymmetry seen in typical lycaenid-ant associations suggest that despite their superficial similarity to honeydew-secreting insects such as aphids, the majority of phytophagous lycaenids might best be viewed as only rarely mutualistically associated with ants and more often mildly parasitic upon them. This association could also help to explain why lycaenids are unusually prone to shifts to overt parasitism and obligate aphytophagy (Sachs and Simms 2006). A number of different ant species are obligately dependent upon associated plants, fungi, and even hemipterans, but it is perhaps significant that none are known to be obligately associated with caterpillars (Chomicki and Renner 2015; Eastwood and Fraser 1999; Ivens 2015).

Because many non-trophobiotic and trophobiotic caterpillars associate with the same species of ants, the primary advantage for trophobiotic caterpillars seems to be that they can attract ants with additional food rewards, not simply that they are able to appease them. There are many reasons why myrmecophilous lycaenids may utilize ants for defense rather than relying on toxic secondary compounds or other means of caterpillar protection. Different circumstances in combination with factors such as larval size or feeding activity can influence predation risk for non-myrmecophilous caterpillars (Berger et al. 2006; Bernays 1997; Dmitriew 2011; Gotthard 2000; Mänd et al. 2007). Selection will favor conditional ant association for protection if maintaining an ant guard is both possible and metabolically less expensive than other means of defense (see discussion below) (Mizuno et al. 2019; Wagner 1993). Different attending ant species can confer significantly different levels of protection (Fraser et al. 2001), but predation or parasitism rates are typically many-fold higher in the absence of ants, both for facultative and obligate ant associates (Atsatt 1981a; Forister et al. 2017; Kaminski et al. 2010a; Rodrigues et al. 2010; Peterson 1993; Pierce and Easteal 1986; Pierce and Mead 1981; Thomas et al. 2020; Weeks 2003). For example, the obligately ant-associated juveniles of the Australian lycaenid, *Jalmenus evagoras*, suffered nearly 100% mortality from parasites and predators when ants were experimentally removed (Pierce et al. 1987), and even the facultatively attended larvae of the North American lycaenid, *Glaucopsyche lygdamus*, were shown experimentally in one field season to suffer up to a 12-fold difference in mortality without ants (Pierce and Easteal 1986).

Highly specialized parasitoids and predators may seek out myrmecophilous caterpillars by using chemical or vibrational cues from their associated ants to locate their prey (e.g., Dejean et al. 2016; DeVries 1991b; Elgar et al. 2016; Fiedler et al. 1992; Pierce and Nash 1999; Thomas and Elmes 1993; Thomas et al. 2002). In situations like this where specialized enemies use attendant ants to find their prey, selection might favor the loss of ant association. Ant associations in plants and insects are prone to frequent loss or modification, and Lepidoptera are no exception (Chomicki and Renner 2015; Sachs and Simms 2006; Stadler and Dixon 2005; Weber and Keeler 2013; Yao 2014). For example, several Australian species in the obligately ant-associated lycaenid genus *Hypochrysops* have likely either lost ant association (*Hypochrysops byzos*, *H. pythias*) or become facultatively associated with them (*H. polycletus*) (Braby 2000). In another Australian genus, *Ogyris*, while most species have caterpillars that are never found without ants, often with only a single genus or species, the clade containing *Ogyris amaryllis*, *O. oroetes*, *O. olane*, and *O. barnardi* has lost or greatly reduced facultative ant associations (Eastwood and Fraser 1999; Schmidt and Rice 2002). The *Taraka-Spalgis-Feniseca* clade (Lycaenidae: Miletinae) and several genus-groups of Poritiinae represent other cases where phylogenetic studies will probably reveal extensive secondary loss of obligate ant associations. Ant associations appear lost most frequently in facultatively ant-associated groups of the Theclinae-Polyommatinae assemblage, a pattern also observed in facultative associations more generally (Chomicki et al. 2020).

Like many classic symbioses, ant-caterpillar associations typically have additional hidden partners. Hemipteran-ant mutualisms often benefit host plants (Campbell et al. 2013; Pringle et al. 2011; Styrsky and Eubanks 2007), and host plants may similarly benefit when lycaenid and riodinid caterpillars attract ants that drive away other herbivores and deposit nutrients, as recently documented for *Euchrysops cnejus* caterpillars attended by *Camponotus* ants on *Vigna* plants in India (Ekka et al. 2020).

Abiotic Effects, Obligate Associations, and Biogeography

One of the more significant insights gained in recent years from a worldwide consideration of the drivers of interspecies symbiosis (e.g., Kaspari 2020; Steidinger et al. 2019) is the importance of abiotic factors in determining the distribution of species interactions such as those seen between caterpillars and ants. Pierce (1987) pointed out a striking pattern in the biogeographic distribution of lycaenid-ant interactions: obligate interactions are considerably more common in the Southern Hemisphere, particularly Australia and Southern Africa, compared to those in the Northern Hemisphere, including the Nearctic and Palearctic. These patterns appear to extend into wet tropical Africa and Southeast Asia, where life histories of Lycaenidae are comparatively less well-documented. All but one tribe of lycaends has representatives in both hemispheres, and thus this pattern cannot be explained by a single vicariance event involving ant-associated and non-associated lineages. Rather, it is due to a heterogenous distribution of tribes with different levels of ant association, with Poritiinae, Aphnaeinae, Miletinae, and strongly ant-associated genera of Theclinae-Polyommatinae generally limited to the Afrotropical, Oriental, and Australasian regions. In the same way, ant-associated Riodinidae and other Lepidoptera are almost entirely limited to the Neotropical, Afrotropical, Oriental, and Australasian regions (Table 1).

Two nonexclusive explanations for this pattern include (1) climate differences and (2) bottom-up effects of soil micronutrients and precipitation that affect plants, microbes, and species that interact with them (e.g., Steidinger et al. 2019). For example, the phosphorus-poor soils of southern Africa and Australia have been evoked as potentially playing a role in the high percentage of ant-dispersed myrmecochorous plants in these areas (Westoby et al. 1982). Research has accumulated over the past 25 years in what Kaspari has called "ionic ecology" (Kaspari 2020), demonstrating the importance of the stoichiometry of essential elements like Na, P, Cl, K, Mg, and Ca that flux across membranes of organisms at different trophic levels.

Functional mechanisms involving different species are complex and undoubtedly vary depending upon the circumstances, but numerous experiments at a variety of spatial scales focusing on invertebrates ranging from termites (with a rich gut microbiome) to caterpillars (largely devoid of a gut microbiome) have shown that levels of accessible environmental sodium either from direct access through soil or "mud puddles" or indirectly through plant tissues consumed by herbivores can have an enormous effect not only on the abundance and distribution of invertebrates but also on the entire network of parasites and predators interacting with them (Baker et al. 2020; Kaspari 2020). The significance of these results in considering caterpillar-ant interactions seems especially clear given the trophobiotic nature of most of their associations. Coping mechanisms are required in habitats with soils that are poor in phosphorus: an essential, rare, and limited nutrient needed for ribosomes, ATP, and nucleic acids. For insects with gut microbiota, movement into the cell can be facilitated by bacteria with surface proteins that can cotransport Na-P across membranes (Werner and Kinne 2001). This means that sodium can be at a premium for organisms with microbial associations like ants because of its role in facilitating cotransport of phosphates into cells of their symbionts (Kaspari 2020). A growing body of research has shown that the availability of sodium and phosphorus can place constraints on ant growth (Bujan et al. 2016; Goitía and Jaffé 2009; Kaspari et al. 2008, 2009, 2020). Any mechanism that could enhance sodium acquisition and/or facilitate sodium ion transport might be especially favored in regions where soils have low phosphorus or sodium.

For plants, this could be achieved through extrafloral or floral nectars or through seeds with attractive eliasomes. For caterpillars, this could perhaps be achieved with secretions and could explain the appearance of obligate, intense ant associations in arid habitats with low phosphorus soils such as those found in central Australia and Southern Africa. The same ability to attract and manipulate ant partners would not exist in habitats with well-fertilized soils because ants might not be limited by essential micronutrients in the same way. This difference in soil fertility could also help to explain in part why ant plants are restricted to the tropics. For example, although the genus *Macaranga* is widely distributed in the Old World tropics, the clade containing ant plants occurs in West Melanesian rainforest, a region also characterized by phosphorus-poor soils (Davies et al. 2001). Similarly, natural variation as well as experimental manipulations in nutrient exposure of obligately associated ant plants ranging from *Cordia* and *Cecropia* in the Neotropics to *Macaranga* and *Vachellia* in the Old World tropics have resulted in differences in plant growth rates and turnover of ant inhabitants (Folgarait and Davidson 1995; Heil et al. 2001; Pringle et al. 2013).

Aridification has played an additional role in affecting ant associations, both by driving caterpillars to seek shelter and possibly food in ant nests (Espeland et al. in review) and by generating extremes in the distribution of soil micronutrients over a large spatial scale (Bui et al. 2014). It seems no coincidence that relatively high levels of ant association are observed in caterpillar-ant interactions in Australia, Southern Africa, and the Cerrado of Brazil.

Finally, differences in soil composition across large spatial scales may have also played a more important role in shaping ant association across landscapes of the Southern Hemisphere including Australia and Southern Africa because weathering and erosion processes have taken place in the absence of the kind of severe, cyclical history of glaciation observed in the Northern Hemisphere (Hopper 2009). This explanation has the advantage of accounting for different levels of obligate association in similar temperate climates of the Western Palearctic (Fiedler 1998b) versus in Australia (Pierce 1987).

Experimental evidence suggests that seasonal temperature fluctuations can break up established partnerships by disrupting ant ecological partitioning. Ant territorialism on caterpillar host plants seems to facilitate obligate ant-caterpillar association, as exemplified by species of aggressive, tropical, arboreal *Oecophylla* and *Crematogaster* found with diverse specialized Lepidoptera. Ecologically dominant ant species typically have the most abundant and aggressive workers, and their large colonies create stable, high-quality habitats for myrmecophiles (Eastwood and Fraser 1999; Fiedler 1991; Fiedler 2001; Fiedler 2006; Hölldobler and Wilson 1990). The combined suitability of a host plant to feed on and enemy free space afforded by the ability to appease otherwise threatening ants can create "ecological islands" of opportunity for obligately specialized myrmecophilous Lepidoptera, and this can have strong effects on their subsequent diversity through restriction of population size and/or structuring of populations (Eastwood et al. 2006; Pellissier et al. 2012; Pierce et al. 2002; Schär et al. 2018). This helps explain why New World *Eciton* and Old World *Dorylus* army ant species, although the former have the most diverse myrmecophile communities known, do not have caterpillar associates—their lepidopteran interactions are restricted to a few species of Papilionidae, Hesperiidae, and Nymphalidae whose adults use ant columns to find nitrogen-rich bird droppings (Ivens et al. 2016; Kistner 1982; Rettenmeyer et al. 2011). The foraging strategies of army ants afford little opportunity for caterpillars, which are relatively sedentary and typically herbivores, to form stable associations (Pierce 1995; Powell et al. 1998). Ant partners in temperate regions might also be less desirable due to the various documented effects of climate on colony traits and community structure (Dunn et al. 2010; Dunn et al. 2009; Kaspari and Vargo 1995; Kaspari et al. 2000).

Besides making suitable ant partners difficult to find, climate fluctuations in temperate areas might increase caterpillar developmental times and reduce access to nutritious host plants (Fiedler 2006; Pellissier et al. 2012a). These hypotheses are supported by the observation that obligate ant associates in temperate areas generally spend most of their life cycle within ant nests, where nutritious food sources, favorable microclimates, and attendant ants are always available (Fiedler 2006). Caterpillar life histories are generally more specialized and diverse in the tropics (Dyer et al. 2007; Forister et al. 2015). Further studies are needed to elucidate the underlying ecological causes.

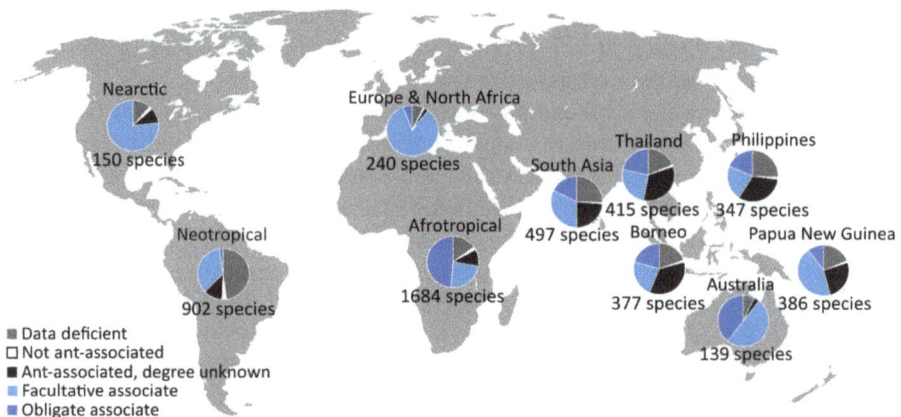

Fig. 22 Geographical distribution of ant associations in Lycaenidae. Known and inferred lycaenid life histories reported in Table 2, or the lack thereof, were cross-referenced with regional catalogs in (Braby 2000; Ek-Amnuay 2012; van Gasse 2013; Hardy and Lawrence 2017; Lamas et al. 2004; Opler 1992, 1999; Parsons 1999; Seki et al. 1991; Tshikolovets 2011; Williams 1995-2020). Facultative associates are often more widely distributed than obligate associates, and they may comprise an increasingly large fraction of total species when looking at small, disturbed, or isolated regions

Whether based on climate, soils, the availability of required ants, or other factors, only non-myrmecophilous or flexible, facultative ant associates in the families Lycaenidae and Riodinidae are found extensively beyond the Old World and New World southern zoogeographic regions (Fig. 22). The Lycaenidae are thought to have originated in the Old World tropics, where extant phylogenetic diversity remains heavily concentrated (Espeland et al. 2018). Only five lycaenid groups occur substantially beyond the Old World tropics and subtropics into the Palearctic, and all five have also entered the New World through Beringia or across the Atlantic (Fric et al. 2019; Gompert et al. 2008; Vila et al. 2011): the *Spalgis-Taraka-Feniseca* clade [Miletinae], the subfamily Lycaeninae, and the tribes Theclini, Eumaeini, and Polyommatini of the Theclinae-Polyommatinae assemblage. Except for a handful of obligately associated species of Polyommatini found within otherwise facultatively associated genus groups, species in these groups are all largely facultatively ant-associated or not ant-associated (Table 2, Fig. 23). The Riodinidae have diversified primarily in the Neotropics. A single lineage within the subfamily Nemeobiinae has colonized the Old World, most likely via Beringia, with a secondary return to the Neotropics of the genera *Styx* and *Corrachia* (Espeland et al. 2015). None of the over 300 species in this subfamily are known to be ant-associated (Table 2). It is tempting to conclude that strongly ant-associated butterflies have been unable to disperse between the Old World and the New World due to the challenge of finding suitable ant partners, especially when confronted with climatic conditions found in temperate regions.

Obligate myrmecophiles' double reliance on associated ants and food sources may make them especially sensitive to disturbance and vulnerable to extinction

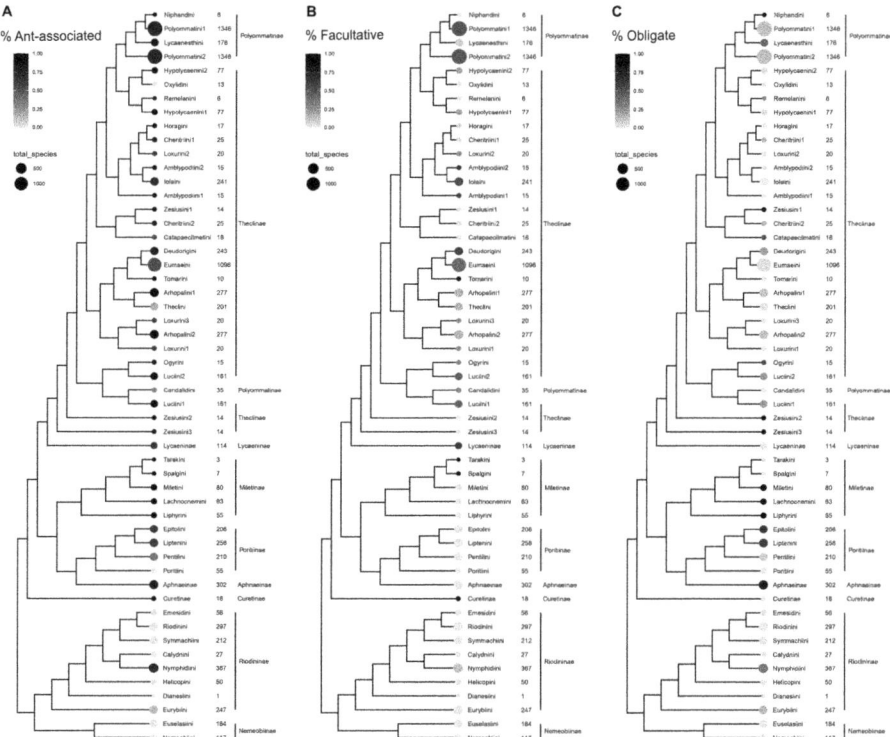

Fig. 23 Phylogenetic tree of extant Lycaenidae and Riodinidae based on data from Espeland et al. (Espeland et al. 2018), showing (**a**) distribution of species with larval ant association overall, (**b**) distribution of facultative larval ant associations, and (**c**) distribution of obligate larval ant associations (from data summarized in Table 2). The size of each circle is proportional to the number of species represented. Percentages correspond to the number of records known or inferred from congeners divided by the total number of species, with groups where life history information such as degree of obligacy is unavailable scored as 0%. Tribes that were rendered non-monophyletic appear multiple times in the tree (e.g., the Polyommatini). We have illustrated the same species counts and ant association proportions next to each appearance as current phylogenies do not allow us to break down these data further. (Figure prepared by João Tonini)

(Chomicki et al. 2020; Geyle et al. 2021; Koh et al. 2004; Pierce 1995; Pierce et al. 2002). Invasive ant species, habitat disturbance and destruction, and climate change are particular concerns (e.g., Braby et al. 2021; Geyle et al. 2021). Over 60% of the threatened Lycaenidae and Riodinidae on the IUCN Red List are recorded as obligate ant associates or predicted to be obligate ant associates based on congeners (Table 2) (IUCN 2020). Among the obligately ant-associated species, those that are "aphytophagous," having at least one life stage obligately dependent on animal rather than plant tissue for nutrition, are particularly vulnerable. Species of Lepidoptera with this rare life history trait comprise only 400 species at most, representing only about 0.25% of the estimated 160,000 species (Pierce 1995). However, aphytophagous species are greatly overrepresented on the IUCN Red List of Threatened Species of butterflies, appearing almost two orders of magnitude more likely to be included in the categories of "Extinct," "Critically Endangered," and "Endangered" than are herbivorous species (IUCN 2020). Most obligate ant associations remain poorly studied. Moreover, countless groups of ant-associated caterpillars likely remain undiscovered, as nearly two-thirds of currently known myrmecophile groups are known from only a single species, almost entirely in the tropics (Table 1). Surprising and unique forms of ant association continue to be described regularly (e.g., Agassiz and Kallies 2018; Dejean et al. 2017; Komatsu and Itino 2014; Ramos et al. 2018; Rocha et al. 2020). Future experiments will need to measure the abundance and distribution of myrmecophiles in different regions and habitats to let us estimate their true global diversity, document their often-unbelievable biology, and ensure their future.

Acknowledgments We thank Robert Marquis and Suzanne Koptur for their encouragement and patience in organizing and editing this volume. Bert Hölldobler and Roger L. Kitching provided inspiration and guidance in improving our understanding of myrmecophiles. Chris Baker, Colin E. Congdon, Alan Heath, Masaru Hojo, Daniel H. Janzen and Winnie Hallwachs, Daniel Kronauer, Dino Martins, David Nash, Piotr Naskrecki, Lucas Kaminski, and Joe Parker helped resolve some key concepts and clarify critical ambiguities in published information. We also thank Marianne Espeland, Michael Braby, and Jaret Daniels for their insightful comments and criticisms of the manuscript. David J. Lohman gave expert advice and assisted in numerous ways, especially in helping to consider how to structure the data for Table 2. João Tonini gave advice about data gathering for Table 2 and helped prepare Fig. 23. Andrew Berry gave advice and support throughout. Numerous naturalists, scientists, and photographers generously gave permission to reproduce their images (their respective names accompany each figure). Takashi Komatsu and Takao Itino kindly provided photographs from their study of the Japanese arctiid moth, *Nudina artaxidia*, associating with *Lasius* ants and homopterans, for our chapter frontispiece. Support for this research came from NSF DEB-1541560 to NEP.

References

Adams RM, Wells RL, Yanoviak SP, Frost CJ, Fox EG (2020) Interspecific eavesdropping on ant chemical communication. Front Ecol Evol 8:24
Adamski D (2017) A New species of *Calosima* Dietz, 1910 from Kenya (Lepidoptera: Gelechioidea: Blastobasidae) reared from the Domatia of *Vachellia drepanolobium* (Fabaceae). Proc Entomol Soc Wash 119(sp1):697–702
Agassiz DJ (2011) The Lepidoptera of acacia domatia in Kenya, with description of two new genera and six new species. J Nat Hist 45(29–30):1867–1893
Agassiz DJ, Bidzilya OV (2016) Gelechiidae (Lepidoptera) bred from acacia in Kenya with description of eight new species. Ann Ditsong Nat Mus Nat Hist 6(07):116–145
Agassiz DJ, Harper DM (2009) The Macrolepidoptera fauna of acacia in the Kenyan Rift Valley (part 1). Trop Lepidop Res 1:4–8
Agassiz DA, Kallies AX (2018) A new genus and species of myrmecophile clearwing moth (Lepidoptera: Sesiidae) from East Africa. Zootaxa 4392(3):588–594
Agrawal AA, Fordyce JA (2000) Induced indirect defence in a lycaenid-ant association: the regulation of a resource in a mutualism. Proc R Soc Lond Ser B Biol Sci 267:1857–1861
Ahn N, Bae Y, Kang E (2014) A first record of *Gaphara conspersa* (Matsumura, 1931) (Lepidoptera, Tineidae) in Korea. Reg Gen Meet Korean Soc Appl Insects Spring Confer. Accessed at https://www.semanticscholar.org in February 2021

Aibar-Abregú P (2014) Scientific note: hostplant records for the myrmecophilous butterfly *Harveyope densemaculata* (Hewitson, 1870)(Lepidoptera: Riodinidae). Trop Lepidop Res 24(2):121

Akino T, Knapp JJ, Thomas JA, Elmes GW (1999) Chemical mimicry and host specificity in the butterfly *Maculinea rebeli*, a social parasite of *Myrmica* ant colonies. Proc R Soc Lond Ser B Biol Sci 266:1419–1426

Akino T, Nakamura KI, Wakamura S (2004) Diet-induced chemical phytomimesis by twig-like caterpillars of *Biston robustum* Butler (Lepidoptera: Geometridae). Chemoecology 14(3–4):165–174

Albanese GE, Nelson MW, Vickery PD, Sievert PR (2007) Larval feeding behavior and ant association in frosted elfin, *Callophrys irus* (Lycaenidae). J Lepidop Soc 61(2):61–66

Allen PE (2010) Group size effects on survivorship and adult development in the gregarious larvae of *Euselasia chrysippe* (Lepidoptera, Riodinidae). Insect Soc 57(2):199–204

Als TD, Vila R, Kandul NP, Nash DR, Yen SH, Hsu YF, Mignault AA, Boomsma JJ, Pierce NE (2004) The evolution of alternative parasitic life histories in large blue butterflies. Nature 432(7015):386–390

Alves-Silva E, Bächtold A, Del-Claro K (2018) Florivorous myrmecophilous caterpillars exploit an ant–plant mutualism and distract ants from extrafloral nectaries. Austral Ecol 43(6):643–650

Atsatt PR (1981a) Lycaenid butterflies and ants: selection for enemy-free space. Am Nat 118(5):638–654

Atsatt PR (1981b) Ant-dependent food plant selection by the mistletoe butterfly *Ogyris amaryllis* (Lycaenidae). Oecologia 48:60–63

Austin GT, Boyd BM, Murphy DD (2008) *Euphilotes ancilla* (Lycaenidae) in the Spring Mountains, Nevada: more than one species. J Lepidop Soc 62(3):148–160

Axén AH (2000) Variation in behavior of lycaenid larvae when attended by different ant species. Evol Ecol 14(7):611–625

Axen AH, Pierce NE (1998) Aggregation as a cost-reducing strategy for lycaenid larvae. Behav Ecol 9(2):109–115

Axén AH, Leimar O, Hoffman V (1996) Signalling in a mutualistic interaction. Anim Behav 52(2):321–333

Ayre GL (1958) Notes on insects found in or near nests of *Formica subnitens* Creighton (Hymenoptera: Formicidae) in British Columbia. Insect Soc 5(1):1–7

Bächtold A, Alves-Silva E (2013) Behavioral strategy of a lycaenid (Lepidoptera) caterpillar against aggressive ants in a Brazilian savanna. Acta Ethologica 16(2):83–90

Bächtold A, Alves-Silva E, Kaminski LA, Del-Claro K (2014) The role of tending ants in host plant selection and egg parasitism of two facultative myrmecophilous butterflies. Naturwissenschaften 101(11):913–919

Baker CC, Bittleston LS, Sanders JG, Pierce NE (2016) Dissecting host-associated communities with DNA barcodes. Philos Trans R Soc B Biol Sci 371(1702):2015.0328

Baker CC, Castillo Vardaro JA, Doak DF, Pansu J, Puissant J, Pringle RM, Tarnita CE (2020) Spatial patterning of soil microbial communities created by fungus-farming termites. Mol Ecol 29(22):4487–4501

Bálint Z, Benyamini D (2001) Taxonomic notes, faunistics and species descriptions of the austral south American polyommatine lycaenid genus *Pseudolucia* (Lepidoptera: Lycaenidae): the *chilensis* and *collina* species-groups. Annls Hist Nat Mus Nat Hung 93:107–149

Ballmer GR (2008) Life history of *Purlisa gigantea* (Lepidoptera: Lycaenida, Theclini) in South Thailand. Trop Lepidop Res 1:32–39

Ballmer GR (2015) Pursued by adrenalin. In: Pursuit of dopamine. The lives of lepidopterists. Springer, pp 41–48

Ballmer GR, Pratt GF (1988) A survey of the last instar larvae of the Lycaenidae (Lepidoptera) of California. J Res Lepidop 27(1):1–81

Ballmer GR, Pratt GF (1991) Quantification of ant attendance (myrmecophily) of lycaenid larvae. J Res Lepidop 30(1–2):95–112

Ballmer GR, Wright DM (2008) Life history and larval chaetotaxy of *Ahmetia achaja* (Lepidoptera, Lycaenidae, Lycaeninae, Theclini, Cheritrina). Zootaxa 1845(1):47–59

Bampton I (1995) A discussion on the larval food of the subfamily Lipteninae (Lepidoptera, Lycaenidae). Metamorphosis 6(4):162–166

Barbero F (2016) Cuticular lipids as a cross-talk among ants, plants and butterflies. Int J Mol Sci 17(12):1966

Barbero F, Bonelli S, Thomas JA, Balletto E, Schönrogge K (2009a) Acoustical mimicry in a predatory social parasite of ants. J Exp Biol 212(24):4084–4090

Barbero F, Thomas JA, Bonelli S, Balletto E, Schönrogge K (2009b) Queen ants make distinctive sounds that are mimicked by a butterfly social parasite. Science 323(5915):782–785

Barbero F, Patricelli DA, Witek MA, Balletto E, Casacci LP, Sala MA, Bonelli S (2012) *Myrmica* ants and their butterfly parasites with special focus on the acoustic communication. Psyche 2012. https://doi.org/10.1155/2012/725237

Bascombe MJ, Johnston G, Bascombe FS (1999) The butterflies of Hong Kong. Academic, San Diego

Basu DN, Kunte K (2020) Tools of the trade: MicroCT reveals native structure and functional morphology of organs that drive caterpillar–ant interactions. Sci Rep 10(1):1–7

Basu DN, Churi P, Soman A, Sengupta A, Bhakare M, Lokhande S, Bhoite S, Huertas B, Kunte K (2019) The genus *Tarucus* Moore, [1881] (Lepidoptera: Lycaenidae) in the Indian subcontinent. Trop Lepidop Res 29(2):87–110

Baylis M, Pierce NE (1991) The effect of host-plant quality on the survival of larvae and oviposition by adults of an ant-tended lycaenid butterfly, *Jalmenus evagoras*. Ecol Entomol 16:1–9

Baylis M, Pierce NE (1992) Lack of compensation by final instar larvae of the myrmecophilous lycaenid butterfly, *Jalmenus evagoras*, for the loss of nutrients to ants. Physiol Entomol 17:107–114

Benyamini D (1995) Synopsis of biological studies of the Chilean Polyommatini (Lepidoptera, Lycaenidae). Reports of the Museum of Natural History, vol 52. University of Wisconsin, Stevens Point, pp 1–51

Benyamini D (2013) *Pseudolucia balinti* sp. n. of the *plumbea-sibylla* species group in central-West Argentina (Lepidoptera: Lycaenidae: Polyommatinae). Fol Entomol Hung 74:157–174

Benyamini D, Bálint Z (1995) Studies of life history and myrmecophily in certain Chilean *Pseudolucia* Nabokov (Lepidoptera, Lycaenidae). Reports of the Museum of Natural History, vol 51. University of Wisconsin, Stevens Point, pp 1–7

Benyamini D, Aristophanous M, Aristophanous A, John E (2018) The biology of the Cyprus endemic blue *Glaucopsyche paphos* Chapman, 1920 (Lepidoptera: Lycaenidae, Polyommatinae). Entomol Gaz 69:151–165

Benyamini D, Mega NO, Romanowski HP, Moser A, Vila R, Bálint Z (2019) Distribution, life history and conservation assessment of the critically endangered butterfly *Pseudolucia parana* (Lepidoptera: Lycaenidae). Fol Entomol Hung 80:303–325

Berger D, Walters R, Gotthard K (2006) What keeps insects small?—size dependent predation on two species of butterfly larvae. Evol Ecol 20:575–589

Bernays EA (1997) Feeding by lepidopteran larvae is dangerous. Ecol Entomol 22:121–123

Bertone MA, Leong M, Bayless KM, Malow TL, Dunn RR, Trautwein MD (2016) Arthropods of the great indoors: characterizing diversity inside urban and suburban homes. PeerJ 4:e1582

Bily S, Fikacek M, Sipek P (2008) First record of myrmecophily in buprestid beetles: immature stages of *Habroloma myrmecophila* sp. nov.(Coleoptera: Buprestidae) associated with *Oecophylla* ants (Hymenoptera: Formicidae). Insect Syst Evol 39:1399–1560

Blüthgen N, Fiedler K (2002) Interactions between weaver ants *Oecophylla smaragdina*, homopterans, trees and lianas in an Australian rain forest canopy. J Anim Ecol 71(5):793–801

Blüthgen N, Fiedler K (2004) Competition for composition: lessons from nectar-feeding ant communities. Ecology 85(6):1479–1485

Blüthgen N, Stork NE (2007) Ant mosaics in a tropical rainforest in Australia and elsewhere: a critical review. Austral Ecol 32(1):93–104

Blüthgen N, Gebauer G, Fiedler K (2003) Disentangling a rainforest food web using stable isotopes: dietary diversity in a species-rich ant community. Oecologia 137(3):426–435

Borges ED, Borges ME, Zarbin PH (2014) Defensive behavior associated with secretions from the prosternal paired glands of the larvae of *Heliconius erato phyllis* Fabricius (Lepidoptera, Nymphalidae). Rev Bras Entomol 58(2):161–167

Boyle JH, Kaliszewska ZA, Espeland M, Suderman TR, Fleming J, Heath A, Pierce NE (2015) Phylogeny of the Aphnaeinae: myrmecophilous African butterflies with carnivorous and herbivorous life histories. Syst Entomol 40(1):169–182

Braby MF (2000) Butterflies of Australia: their identification, biology and distribution. CSIRO Publishing, Melbourne

Braby MF (2011) New larval food plant associations for some butterflies and diurnal moths (Lepidoptera) from the Northern Territory and eastern Kimberley, Australia. Beagle Rec Mus Art Gall North Territory 27:85–105

Braby MF (2012) New larval food plants and biological notes for some butterflies (Lepidoptera: Papilionoidea) from eastern Australia. Aust Entomol 39(2):65–68

Braby M (2015) New larval food plant associations for some butterflies and diurnal moths (Lepidoptera) from the Northern Territory and Kimberley, Australia. Part II. Rec West Aust Mus 30:73–97. https://doi.org/10.18195/issn.0312-3162.30(2).2015.073-097

Braby MF, Douglas F (2004) The taxonomy, ecology and conservation status of the Golden-rayed blue: a threatened butterfly endemic to western Victoria, Australia. Biol J Linn Soc 81(2):275–299

Braby MF, Williams MR, Douglas F, Beardsell C, Crosby DF (2021) Changes in a peri-urban butterfly assemblage over 80 years near Melbourne, Australia. Aust Entomol 60:27–51

Brandl R, Bagine RN, Kaib M (1996) The distribution of *Schedorhinotermes lamanianus* (Isoptera: Rhinotermitidae) and its termitophile *Paraclystis* (Lepidoptera: Tineidae) in Kenya: its importance for understanding east African biogeography. Glob Ecol Biogeogr Lett 5(3):143–148

Bui EN, González-Orozco CE, Miller JT (2014) Acacia, climate, and geochemistry in Australia. Plant Soil 381(1–2):161–175

Bujan J, Wright SJ, Kaspari M (2016) Biogeochemical drivers of Neotropical ant activity and diversity. Ecosphere 7(12):e01597

Bura VL, Fleming AJ, Yack JE (2009) What's the buzz? Ultrasonic and sonic warning signals in caterpillars of the great peacock moth (*Saturnia pyri*). Naturwissenschaften 96(6):713–718

Bura VL, Rohwer VG, Martin PR, Yack JE (2011) Whistling in caterpillars (*Amorpha juglandis*, Bombycoidea): sound-producing mechanism and function. J Exp Biol 214(1):30–37

Burghardt F, Fiedler K (1996) The influence of diet on growth and secretion behaviour of myrmecophilous *Polyommatus icarus* caterpillars (Lepidoptera: Lycaenidae). Ecol Entomol 21(1):1–8

Bury JA, Savchuk VL (2015) New data on the biology of ten lycaenid butterflies (Lepidoptera: Lycaenidae) of the genera *Tomares* Rambur, 1840, *Pseudophilotes* Beuret, 1958, *Polyommatus* Latreille, 1804, and *Plebejus* Kluk, 1780 from the Crimea and their attending ants (Hymenoptera: Formicidae). Act Entomol Siles 23:1–6

Callaghan CJ (1985) Notes on the biology of *Stalachtis susanna* (Lycaenidae: Riodininae) with a discussion of riodinine larval strategies. J Res Lepidop 24(3):258–263

Callaghan CJ (1986) Restinga butterflies: biology of *Synargis brennus* (Stichel) (Riodinidae). J Lepidop Soc 40(2):93–96

Callaghan CJ (1992a) Notes on the biology of a myrmecophilous African lycaenid, *Aphnaeus adamsi* Stempffer (Lepidoptera, Lycaenidae). Bull Soc Entomol France 97(4):339–342

Callaghan CJ (1992b) Biology of epiphyll feeding butterflies in a Nigerian cola forest (Lycaenidae: Liptenidae). J Lepidop Soc 46(3):203–214

Callaghan CJ (1997) The biology of *Abisara neophron neophron* (Hewitson, 1860) from Nepal (Lepidoptera, Riodinidae). Bull Soc Entomol France 2:129–132

Callaghan CJ (2003) The biology of *Melanis leucophlegma* (Stichel, 1910) (Riodinidae) in western Peru. J Lepidop Soc 57(3):193–196

Callaghan CJ (2008) The biology of *Rhamma arria* in Colombia (Lepidoptera: Lycaenidae). Trop Lepidop Res 18(1):60–61

Camarota F, Vasconcelos HL, Marquis RJ, Powell S (2020) Revisiting ecological dominance in arboreal ants: how dominant usage of nesting resources shapes community assembly. Oecologia 194(1):151–163

Campbell DL, Pierce NE (2003) Phylogenetic relationships of the Riodinidae: implications for the evolution of ant association. In: Butterflies: ecology and evolution taking flight. University of Chicago Press, Chicago, pp 395–408

Campbell H, Townsend IR, Fellowes MD, Cook JM (2013) Thorn-dwelling ants provide antiherbivore defence for camelthorn trees, *Vachellia erioloba*, in Namibia. Afr J Ecol 51(4):590–598

Carleial S, Maurel N, van Kleunen M, Stift M (2018) Oviposition by the mountain Alcon blue butterfly increases with host plant flower number and host ant abundance. Bas Appl Ecol 28:87–96

Casacci LP, Schönrogge K, Thomas JA, Balletto E, Bonelli S, Barbero F (2019a) Host specificity pattern and chemical deception in a social parasite of ants. Sci Rep 9(1):1–10

Casacci LP, Bonelli S, Balletto E, Barbero F (2019b) Multimodal signaling in myrmecophilous butterflies. Front Ecol Evol 7:454

Casagrande M, Penz C, DeVries PJ (2009) Description of early stages of *Chorinea licursis* (Fabricius) (Riodinidae). Trop Lepidop Res 19(2):89–93

Castillo Guevara C, Rico Gray V (2002) Is cycasin in *Eumaeus minyas* (Lepidoptera: Lycaenidae) a predator deterrent? Interciencia 27(9):465–470

Chialvo CH, Chialvo P, Holland JD, Anderson TJ, Breinholt JW, Kawahara AY, Zhou X, Liu S, Zaspel JM (2018) A phylogenomic analysis of lichen-feeding tiger moths uncovers evolutionary origins of host chemical sequestration. Mol Phylogenet Evol 121:23–34

Chomicki G, Renner SS (2015) Phylogenetics and molecular clocks reveal the repeated evolution of ant-plants after the late Miocene in Africa and the early Miocene in Australasia and the Neotropics. New Phytol 207(2):411–424

Chomicki G, Kiers ET, Renner SS (2020) The evolution of mutualistic dependence. Annu Rev Ecol Evol Syst 51:409–432

Claassens AJ (1996) Notes on the feeding habits and protective measures of the larvae of Durbaniopsis saga van Son (Lepidoptera: Lycaenidae). Metamorphosis 7(3):127–128

Claassens AJM, Dickson CGC (1980) Butterflies of the Table Mountain range. C. Struik, Cape Town

Claassens AJ, Heath A (1997) Notes on the myrmecophilous early stages of two species of *Thestor* Hübner (Lepidoptera: Lycaenidae) from South Africa. Metamorphosis 8(2):56–60

Clark GC, Dickson CGC (1960) The life- histories of two species of *Thestor* (Lepidoptera: Lycaenidae). J Entomol Soc S Afr 23:278–283

Clark GC, Dickson CGC (1971) Life histories of the south African lycaenid butterflies. Purnell, Cape Town

Cock MJ (2010) A note on the biology of *Pirascca sagaris sagaris* (Cramer) (Lepidoptera: Riodinidae) in Trinidad, West Indies. Living World J Trinidad Tobago Field Nat Club 2010:85–86

Collier N (2007) Identifying potential evolutionary relationships within a facultative lycaenid-ant system: ant association, oviposition, and butterfly-ant conflict. Insect Sci 14(5):401–409

Common IFB, Waterhouse DF (1981) Butterflies of Australia, 2nd edn. Angus and Robertson, Sydney

Comstock JA, Dammers CM (1932) Early stages of *Melitaea leanira wrightii* Edw. and *Calephelis nemesis* Edw. (Lepidoptera). Bull S Cal Acad Sci 31:9–15

Cottrell CB (1984) Aphytophagy in butterflies: its relationship to myrmecophily. Zool J Linnean Soc 80(1):1–57

Cushing PE (1997) Myrmecomorphy and myrmecophily in spiders: a review. Fl Entomol 80(2):165–193

Cushing PE (2012) Spider-ant associations: an updated review of myrmecomorphy, myrmecophily, and myrmecophagy in spiders. Psyche 2012:151989

Cushman JH, Rashbrook VK, Beattie AJ (1994) Assessing benefits to both participants in a lycaenid-ant association. Ecology 75(4):1031–1041

Czekes Z, Markó B, Nash DR, Ferencz M, Lázár B, Rákosy L (2014) Differences in oviposition strategies between two ecotypes of the endangered myrmecophilous butterfly *Maculinea alcon* (Lepidoptera: Lycaenidae) under unique syntopic conditions. Insect Conserv Divers 7:122–131

Daniels H, Gottsberger G, Fiedler K (2005) Nutrient composition of larval nectar secretions from three species of myrmecophilous butterflies. J Chem Ecol 31(12):2805–2821

Dantchenko AV (1997) Notes on the biology and distribution of the damone and damocles species-complexes of the subgenus *Polyommatus* (*Agrodiaetus*) (Lepidoptera: Lycaenidae). Nachr Entomol Ver Apollo Suppl 16:23–42

Darling CD, Schroeder FC, Meinwald J, Eisner M, Eisner T (2001) Production of a cyanogenic secretion by a thyridid caterpillar (*Calindoea trifascialis*, Thyrididae, Lepidoptera). Naturwissenschaften 88(7):306–309

Davidson DW, Cook SC, Snelling RR, Chua TH (2003) Explaining the abundance of ants in lowland tropical rainforest canopies. Science 300(5621):969–972

Davies SJ, Lum SKY, Chan R, Wang LK (2001) Evolution of myrmecophytism in West Malesian *Macaranga* (Euphorbiaceae). Evolution 55:1542–1559

Davis DR, Davis MM (2007) Neotropical Tineidae, V: a new genus and species of Tineidae associated with social hymenoptera and re-examination of two poorly known genera with similar biology (Lepidoptera: Tineidae, Lyonetiidae). Proc Entomol Soc Wash 109(4):741–764

Davis DR, Quintero DA, Cambra RA, Aiello A (2008) Biology of a new Panamanian bagworm moth (Lepidoptera: Psychidae) with predatory larvae, and eggs individually wrapped in setal cases. Ann Entomol Soc Am 101(4):689–702

de Niceville L (1890) The butterflies of India. Calcutta Central Press Company, Burmah/Ceylon

Dejean A, Beugnon G (1996) Host-ant trail following by myrmecophilous larvae of Liphyrinae (Lepidoptera, Lycaenidae). Oecologia 106(1):57–62

Dejean A, Corbara B, Orivel J, Leponce M (2007) Rainforest canopy ants: the implications of territoriality and predatory behavior. Func Ecosyst Commun 1(2):105–120

Dejean A, Orivel J, Azémar F, Hérault B, Corbara B (2016) A cuckoo-like parasitic moth leads African weaver ant colonies to their ruin. Sci Rep 6(1):1–9

Dejean A, Azémar F, Libert M, Compin A, Herault B, Orivel J, Bouyer T, Corbara B (2017) Ant-lepidopteran associations along African forest edges. Sci Nat 104(1–2):7

Depa Ł, Kaszyca-Taszakowska N, Taszakowski A, Kanturski M (2020) Ant-induced evolutionary patterns in aphids. Biol Rev 95(6):1574–1589

Detrain C, Verheggen FJ, Diez L, Wathelet B, Haubruge E (2010) Aphid–ant mutualism: how honeydew sugars influence the behaviour of ant scouts. Physiol Entomol 35(2):168–174

DeVries PJ (1984) Of crazy-ants and Curetinae: are Curetis butterflies tended by ants? Zool J Linnean Soc 80(1):59–66

DeVries PJ (1988a) The use of epiphylls as larval hostplants by the neotropical riodinid butterfly, *Sarota gyas*. J Nat Hist 22:1447–1450

DeVries PJ (1988b) The larval ant-organs of *Thisbe irenea* (Lepidoptera: Riodinidae) and their effects upon attending ants. Zool J Linnean Soc 94(4):379–393

DeVries PJ (1990) Enhancement of symbioses between butterfly caterpillars and ants by vibrational communication. Science 248(4959):1104–1106

DeVries PJ (1991a) Foam barriers, a new defense against ants for milkweed butterfly caterpillars (Nymphalidae: Danainae). J Res Lepidop 30(3–4):261–266

DeVries PJ (1991b) Mutualism between *Thisbe irenea* butterflies and ants, and the role of ant ecology in the evolution of larval-ant associations. Biol J Linn Soc 43(3):179–195

DeVries PJ (1991c) Ecological and evolutionary patterns in myrmecophilous riodinid butterflies. In: Ant-plant interactions. Oxford University Press, Oxford, pp 143–156

DeVries PJ (1991d) Call production by myrmecophilous riodinid and lycaenid butterfly caterpillars (Lepidoptera): morphological, acoustical, functional, and evolutionary patterns. Am Mus Novit 3025:1–24

DeVries PJ (1997) The butterflies of Costa Rica and their natural history, volume II. Princeton University Press, Riodinidae
DeVries PJ, Baker I (1989) Butterfly exploitation of an ant-plant mutualism: adding insult to herbivory. J N Y Entomol Soc 97(3):332–340
DeVries PJ, Penz CM (2000) Entomophagy, behavior, and elongated thoracic legs in the Myrmecophilous Neotropical butterfly *Alesa amesis* (Riodinidae). Biotropica 32(4a):712–721
DeVries PJ, Harvey DJ, Kitching IJ (1986) The ant associated epidermal organs on the larva of the lycaenid butterfly *Curetis regula* Evans. J Nat Hist 20(3):621–633
DeVries PJ, Cabral BC, Penz CM (2004) The early stages of *Apodemia paucipuncta* (Riodinidae): myrmecophily, a new caterpillar ant-organ and consequences for classification. Milwaukee Pub Mus Contrib 1002:1–13
Deyrup M, Kraus J, Eisner T (2004) A Florida caterpillar and other arthropods inhabiting the webs of a subsocial spider (Lepidoptera: Pyralidae; Araneida: Theridiidae). Fl Entomol 87(4):554–558
Dmitriew CM (2011) The evolution of growth trajectories: what limits growth rate? Biol Rev 86(1):97–116
Dodd FP (1902) Contribution to the life-history of *Liphyra brassolis*, Westw. Entomologiste 35:153–188
Dodd FP (1912) Some remarkable ant-friend Lepidoptera of Queensland. Trans R Entomol Soc Lond 59(3–4):577–590, plate 48
Dodd FP (1916) Noise-producing Lepidoptera. Etud Lépid compar, Fascicule 11 bis, pp. 13–14
Dolle P, Klein P, Fischer OW, Schnitzler HU, Gilbert LE, Boppré M (2018) Twittering pupae of papilionid and nymphalid butterflies (Lepidoptera): novel structures and sounds. Ann Entomol Soc Am 111(6):341–354
Dookie AL, Young CA, Lamothe G, Schoenle LA, Yack JE (2017) Why do caterpillars whistle at birds? Insect defence sounds startle avian predators. Behav Process 138:58–66
Downes MF, Edwards T (2016) An undescribed concealer moth, *'Stathmopoda'* sp. (Lepidoptera: Oecophoridae) in nests of the weaver ant *'Polyrhachis australis'* Mayr (Hymenoptera: Formicidae). Aust Entomol 43(3):161
Downey JC (1966) Sound production in pupae of Lycaenidae. J Lepidop Soc 20:129–155
Downey JC, Allyn AC (1973) Butterfly ultrastructure. 1. Sound production and associated abdominal structures in pupae of Lycaenidae and Riodinidae. Bull Allyn Mus 14:1–47
Downey JC, Allyn AC (1978) Sounds produced in pupae of Lycaenidae. Bull Allyn Mus 48:1–14
Downey JC, Allyn AC (1979) Morphology and biology of the immature stages of *Leptotes cassius theonus* (Lucas) (Lepid: Lycaenidae). Bull Allyn Mus 55:1–28
Duarte M, Robbins RK (2008) Immature stages of *Calycopis bellera* (Hewitson) and *C. janeirica* (Felder) (Lepidoptera, Lycaenidae, Theclinae, Eumaeini): taxonomic significance and new evidence for detritivory. Zootaxa 2325:39–61
Duarte M, Robbins RK (2010) Description and phylogenetic analysis of the *Calycopidina* (Lepidoptera, Lycaenidae, Theclinae, Eumaeini): a subtribe of detritivores. Rev Bras Entomol 54(1):45–65
Duarte M, Almeida GL, Casagrande MM, Mielke OH (2001) Notes on the last instar larva and pupa of *Hemiargus hanno* (Stoll) (Lepidoptera, Lycaenidae, Polyommatinae). Rev Bras Zool 18(4):1097–1105. https://doi.org/10.1590/S0101-81752001000400008
Dunn KL (2007) Nocturnal ant attendance of *Jalmenus eichhorni* Staudinger (Lepidoptera: Lycaenidae) on Cape York Peninsula, Queensland. Calodema 10:35–40
Dunn RR, Agosti D, Andersen AN, Arnan X, Bruhl CA, Cerdá X, Ellison AM, Fisher BL, Fitzpatrick MC, Gibb H, Gotelli NJ (2009) Climatic drivers of hemispheric asymmetry in global patterns of ant species richness. Ecol Lett 12(4):324–333
Dunn RR, Guenard B, Weiser MD, Sanders NJ (2010) Geographic gradients. Ant ecology. Oxford University Press, New York
Dupont S (2012) Structural evolution and diversity of the caterpillar trunk. PhD thesis, University of Copenhagen

Dupont ST, Zemeitat DS, Lohman DJ, Pierce NE (2016) The setae of parasitic *Liphyra brassolis* butterfly larvae form a flexible armour for resisting attack by their ant hosts (Lycaenidae: Lepidoptera). Biol J Linn Soc 117(3):607–619

Dyer LA (1995) Tasty generalists and nasty specialists? Antipredator mechanisms in tropical lepidopteran larvae. Ecology 76(5):1483–1496

Dyer LA, Singer MS, Lill JT, Stireman JO, Gentry GL, Marquis RJ, Ricklefs RE, Greeney HF, Wagner DL, Morais HC, Diniz IR (2007) Host specificity of Lepidoptera in tropical and temperate forests. Nature 448(7154):696–699

Eastwood R, Fraser AM (1999) Associations between lycaenid butterflies and ants in Australia. Aust J Ecol 24(5):503–537

Eastwood R, Hughes JM (2003) Molecular phylogeny and evolutionary biology of *Acrodipsas* (Lepidoptera: Lycaenidae). Mol Phylogent Evol 27(1):93–102

Eastwood R, Kitching RL, Manh HB (2005) Behavioral observations on the early stages of *Jamides celeno* (Cramer) (Lycaenidae) at Cat Tien National Park, Vietnam: an obligate myrmecophile? J Lepidop Soc 59(4):219

Eastwood R, Pierce NE, Kitching RL, Hughes JM (2006) Do ants enhance diversification in lycaenid butterflies? Phylogeographic evidence from a model myrmecophile, *Jalmenus evagoras*. Ecology 60(2):315–327

Eastwood R, Braby MF, Lohman DJ, King A (2008a) New ant-lycaenid associations and biological data for some Australian butterflies (Lepidoptera: Lycaenidae). Aust Entomol 35(1):47–56

Eastwood R, Braby MF, Schmidt DJ, Hughes JM (2008b) Taxonomy, ecology, genetics and conservation status of the pale imperial hairstreak (*Jalmenus eubulus*) (Lepidoptera: Lycaenidae): a threatened butterfly from the Brigalow Belt, Australia. Invertebr Syst 22:407–423

Eisner T, Jutro P, Aneshansley DJ, Niedhauk R (1972) Defense against ants in a caterpillar that feeds on ant-guarded scale insects. Ann Entomol Soc Am 65(4):987–988

Ek-Amnuay P (2012) Butterflies of Thailand, 2nd edn. Amarin, Thailand

Ekka PA, Rastogi N (2019) A single lycaenid caterpillar gets an ant-constructed shelter and uninterrupted ant attendance. Entomol Exp Appl 167(12):1012–1019. https://doi.org/10.1111/eea.12859

Ekka PA, Rastogi N, Singh H, Singh HB, Ray S (2020) Lycaenid-tending ants can contribute to fitness gain of the infested host plants by providing nutrients. Arthropod Plant Interact 14(6):745–757

Elfferich NW (1998) Is the larval and imaginal signalling of Lycaenidae and other Lepidoptera related to communication with ants. Deinsea 4:91–95

Elgar MA, Pierce NE (1988) Mating success and fecundity in an ant-tended lycaenid butterfly. Reproductive success. University of Chicago Press, Chicago, pp 59–75

Elgar MA, Nash DR, Pierce NE (2016) Eavesdropping on cooperative communication within an ant-butterfly mutualism. Sci Nat 103(9–10):84

Eliot JN (1973) The higher classification of the Lycaenidae (Lepidoptera): a tentative arrangement. Bull Brit Mus Nat Hist (Entomol) 28:371–505

Eliot JN (1986) A review of the Miletini (Lepidoptera: Lycaenidae). Bull Br Mus Nat Hist 53:1–105

Eliot JN (1990) Notes on the genus *Curetis* Hubner (Lepidoptera, Lycaenidae). Tyo to Ga 41(4):201–225

Elmes GW, Thomas JA, Munguira ML, Fiedler K (2001) Larvae of lycaenid butterflies that parasitize ant colonies provide exceptions to normal insect growth rules. Biol J Linn Soc 73(3):259–278

Elmes GW, Wardlaw JC, Schönrogge K, Thomas JA (2019) Evidence of a fixed polymorphism of one-year and two-year larval growth in the myrmecophilous butterfly *Maculinea rebeli*. Insect Conserv Div 12(6):501–510

Eltringham H (1913) The larva of *Euliphyra mirifica*. Trans Entomol Soc Lond 1913:509–515

Endo S, Itino T (2013) Myrmecophilous aphids produce cuticular hydrocarbons that resemble those of their tending ants. Popul Ecol 55(1):27–34

Entomological Network of Singapore (2017) *Homodes* larva mimics *Oecophylla smaragdina*. Available via Facebook. https://www.facebook.com/watch/?v=1938845709677099. Accessed 28 Dec 2020

Epstein ME, Geertsema H, Naumann CM, Tarmann GM (1999) The Zygaenoidea. In: Lepidoptera, moths and butterflies, volume 1: evolution, systematics, and biogeography. Walter de Gruyter, Berlin, pp 159–180

Espeland M, Hall JP, DeVries PJ, Lees DC, Cornwall M, Hsu YF, Wu LW, Campbell DL, Talavera G, Vila R, Salzman S, Ruehr S, Lohman DJ, Pierce NE (2015) Ancient Neotropical origin and recent recolonisation: phylogeny, biogeography and diversification of the Riodinidae (Lepidoptera: Papilionoidea). Mol Phylogent Evol 93:296–306

Espeland M, Breinholt J, Willmott KR, Warren AD, Vila R, Toussaint EF, Maunsell SC, Aduse-Poku K, Talavera G, Eastwood R, Jarzyna MA, Guralnick R, Lohman DJ, Pierce NE, Kawahara AY (2018) A comprehensive and dated phylogenomic analysis of butterflies. Curr Biol 28(5):770–778

Espeland M, Chazot N, Condamine FL, Lemmon A, Lemmon EM, Pringle E, Heath A, Collins S, Tiren W, Nzisa M, Lees D, Fisher S, Woodhall S, Tropek R, Ahlborn SS, Cockburn K, Webb P, Dobson J, Bouyer T, Kaliszewska Z, Baker CM, Talavera G, Vila R, Gardiner A, Williams M, Martins DJ, Safian S, Edge D, Pierce NE (In review) Explosive radiation of ant parasitic butterflies in the genus *Lepidochrysops* (Lycaenidae) following extinction events during the Miocene aridification of Africa

Eubanks MD, Nesci KA, Petersen MK, Liu Z, Sanchez HB (1997) The exploitation of an ant-defended host plant by a shelter-building herbivore. Oecologia 109(3):454–460

Farquharson CO, Poulton EB, Bagnall RS, Bethune-Baker GT, Chappman TA, Collin JE, Durrant JH, Edwards FW, Eltringham H, Gatenby JB, Newstead R (1922) Five Years' observations (1914–1918) on the bionomics of southern Nigerian insects, chiefly directed to the investigation of Lycaenid life-histories and to the relation of Lycaenidae, Diptera, and other insects to ants. Trans R Entomol Soc Lond 69(3–4):319–531

Faynel C, González-Mercado JH (2019) Crianza de *Michaelus phoenissa* (Lepidoptera: Lycaenidae) sobre *Senna alata* (Fabaceae) en Perú. Rev Peru Biol 26(2):265–270. [publication in Spanish]

Fiedler K (1989a) Differences in the behaviour of ants towards two larval instars of *Lycaena tityrus* (Lep., Lycaenidae). Deut Entomol Zeits 36(4–5):267–271

Fiedler K (1989b) The relationships between lycaenid pupae (Lepidoptera: Lycaenidae) and ants (Hymenoptera: Formicidae). Nachr Ent Ver Apollo Frankfurt 9(1):33–58. [publication in German]

Fiedler K (1991) Systematic, evolutionary, and ecological implications of myrmecophily within the Lycaenidae (Insecta: Lepidoptera: Papilionoidea). Bonn Zool Monogr 31:1–210

Fiedler K (1995) Lycaenid butterflies and plants: is myrmecophily associated with particular host-plant preferences? Ethol Ecol Evol 7(2):107–132

Fiedler K (1998a) Lycaenid-ant interactions of the *Maculinea* type: tracing their historical roots in a comparative framework. J Insect Conserv 2(1):3–14

Fiedler K (1998b) Geographical patterns in life-history traits of Lycaenidae butterflies—ecological and evolutionary patterns. Zoology 100:336–347

Fiedler K (2001) Ants that associate with Lycaeninae butterfly larvae: diversity, ecology and biogeography. Divers Distrib 7(1–2):45–60

Fiedler K (2006) Ant-associates of Palaearctic lycaenid butterfly larvae (Hymenoptera: Formicidae; Lepidoptera: Lycaenidae)–a review. Myrmecol Nachr 9:77–87

Fiedler K (2012) The host genera of ant-parasitic Lycaenidae butterflies: a review. Psyche A J Entomol 2012:153975

Fiedler K, Hagemann D (1992) The influence of larval age and ant number on myrmecophilous interactions of the African Grass Blue butterfly, *Zizeeria knysna* (Lepidoptera: Lycaenidae). J Res Lepidop 31:213–232

Fiedler K, Hölldobler B (1992) Ants and *Polyommatus icarus* immatures (Lycaenidae)—sex-related developmental benefits and costs of ant attendance. Oecologia 91(4):468–473

Fiedler K, Hummel V (1995) Myrmecophily in the brown argus butterfly, *Aricia agestis* (Lepidoptera: Lycaenidae): effects of larval age, ant number and persistence of contact with ants. Zoology 99:128–137

Fiedler K, Maschwitz U (1988) Functional analysis of the myrmecophilous relationships between ants (Hymenoptera: Formicidae) and lycaenids (Lepidoptera: Lycaenidae). Oecologia 75(2):204–206

Fiedler K, Maschwitz U (1989a) The symbiosis between the weaver ant, *Oecophylla smaragdina*, and *Anthene emolus*, an obligate myrmecophilous lycaenid butterfly. J Nat Hist 23(4):833–846

Fiedler K, Maschwitz U (1989b) Adult myrmecophily in butterflies: the role of the ant *Anoplolepis longipes* in the feeding and oviposition behaviour of *Allotinus unicolor* (Lepidoptera, Lycaenidae). Tyo to Ga 40(4):241–251

Fiedler K, Saam C (1994) Does ant-attendance influence development in 5 European Lycaenidae butterfly species?(Lepidoptera). Nota Lepidop 17(1/2):5–24

Fiedler K, Seufert P (1995) The mature larva and pupa of *Semanga superba* (Lepidoptera: Lycaenidae). Nachr Entoml Ver Frankfurt (NF) 16:1–2

Fiedler K, Seufert P, Pierce NE, Pearson JG, Baumgarten HT (1992) Exploitation of lycaenid-ant mutualisms by braconid parasitoids. J Res Lepidop 31:153–168

Fiedler K, Seufert P, Maschwitz U, Azarae I (1995) Notes on larval biology and pupal morphology of Malaysian *Curetis* butterflies (Lepidoptera: Lycaenidae). Tyo to Ga 45(4):287–299

Fiedler K, Hölldobler B, Seufert P (1996) Butterflies and ants: the communicative domain. Experientia 52(1):14–24

Fletcher LE, Yack JE, Fitzgerald TD, Hoy RR (2006) Vibrational communication in the cherry leaf roller caterpillar *Caloptilia serotinella* (Gracillarioidea: Gracillariidae). J Insect Behav 19(1):1–8

Floren A, Biun A, Linsenmair EK (2002) Arboreal ants as key predators in tropical lowland rainforest trees. Oecologia 131(1):137–144

Folgarait PJ, Davidson DW (1995) Myrmecophytic *Cecropia*: antiherbivore defenses under different nutrient treatments. Oecologia 104(2):189–206

Forister ML, Gompert Z, Nice CC, Forister GW, Fordyce JA (2011) Ant association facilitates the evolution of diet breadth in a lycaenid butterfly. Proc R Soc B Biol Sci 278(1711):1539–1547

Forister ML, Novotny V, Panorska AK, Baje L, Basset Y, Butterill PT, Cizek L, Coley PD, Dem F, Diniz IR, Drozd P, Fox M, Glassmire A, Hazen R, Hrcek J, Jahner JP, Kaman O, Kozubowski TJ, Kursar TA, Lewis OT, Lill J, Marquis RJ, Miller SE, Morais HC, Murakami M, Nickel H, Pardikes NA, Ricklefs RE, Singer MS, Smilanich AM, Stireman JO, Villamarin-Cortez S, Vodka S, Volf M, Wagner DL, Walla T, Weiblen GD, Dyer LA (2015) The global distribution of diet breadth in insect herbivores. Proc Natl Acad Sci 112(2):442–447

Franzl S, Locke M, Huie P (1984) Lenticles: innervated secretory structures that are expressed at every other larval moult. Tissue Cell 16:251–268

Fraser AM, Axén AH, Pierce NE (2001) Assessing the quality of different ant species as partners of a myrmecophilous butterfly. Oecologia 129(3):452–460

Fraser AM, Tregenza T, Wedell N, Elgar MA, Pierce NE (2002) Oviposition tests of ant preference in a myrmecophilous butterfly. J Evol Biol 15(5):861–870

Freitas AV (1999) An anti-predator behavior in larvae of *Libytheana carinenta* (Nymphalidae, Libytheinae). J Lepidop Soc 53:130–131

Freitas AV, Oliveira PS (1996) Ants as selective agents on herbivore biology: effects on the behaviour of a non-myrmecophilous butterfly. J Anim Ecol 65(2):205–210

Fric ZF, Maresova J, Kadlec T, Tropek R, Pyrcz TW, Wiemers M (2019) World travellers: phylogeny and biogeography of the butterfly genus *Leptotes* (Lepidoptera: Lycaenidae). Syst Entomol 44(3):652–665

Fukuda H, Hama E, Kuzuya T, Taka-hashi A, Takahashi M et al (1984) The life histories of butterflies in Japan, vol III. Hoikusha, Osaka

Funk RS (1975) Association of ants with ovipositing *Lycaena rubidus* (Lycaenidae). J Lepidop Soc 29:261–262
Fürst MA, Nash DR (2010) Host ant independent oviposition in the parasitic butterfly *Maculinea alcon*. Biol Lett 6(2):174–176
Geyle HM, Braby MF, Andren M, Beaver EP, Bell P, Bryne C, Castles M, Douglas F, Glatz RV, Haywood B, Hendry P, Kitching RL, Lambkin TA, Meyer CE, Moore MD, Moss JT, Nally S, Rew TR, Palmer CM, Petrie E, Potter-Craven J, Richards K, Sanderson C, Stolarski A, Taylor GS, Williams MR, Woinarski JCZ, Garnett ST (2021) Butterflies on the brink: identifying the Australian butterflies (Lepidoptera) at greatest risk of extinction. Aust Entomol 60:98–110
Gibbs GW (1980) New Zealand butterflies. William Collins Publishers, Auckland
Glasier JR, Poore AG, Eldridge DJ (2018) Do mutualistic associations have broader host ranges than neutral or antagonistic associations? A test using myrmecophiles as model organisms. Insect Soc 65(4):639–648
Gnatzy W, Jatho M, Kleinteich T, Gorb SN, Hustert R (2017) The eversible tentacle organs of *Polyommatus* caterpillars (Lepidoptera, Lycaenidae): morphology, fine structure, sensory supply and functional aspects. Earth Struct Dev 46(6):788–804
Goitía W, Jaffé K (2009) Ant-plant associations in different forests in Venezuela. Neotrop Entomol 38(1):7–31
Gompert Z, Fordyce JA, Forister ML, Nice CC (2008) Recent colonization and radiation of north American *Lycaeides (Plebejus)* inferred from mtDNA. Mol Phylogent Evol 48:481–490
Gotthard K (2000) Increased risk of predation as a cost of high growth rate: an experimental test in a butterfly. J Anim Ecol 69(5):896–902
Gray B (1974) Associated fauna found in nests of *Myrmecia* (hymenoptera: Formicidae). Insect Soc 21(3):289–299
Grover CD, Kay AD, Monson JA, Marsh TC, Holway DA (2007) Linking nutrition and behavioural dominance: carbohydrate scarcity limits aggression and activity in argentine ants. Proc R Soc B Biol Sci 274(1628):2951–2957
Hall JP (1998) A review of the genus *Sarota* (Lepidoptera: Riodinidae). Trop Lepidop 9(supplement 1):1–21
Hall JP (2007) Phylogenetic revision of the new neotropical riodinid genus *Minstrellus* (Lepidoptera: Riodinidae). Ann Entomol Soc Am 100(6):773–786
Hall JP (2018) A monograph of the Nymphidiina (Lepidoptera: Riodinidae: Nymphidiini): phylo-geny, taxonomy, biology, and biogeography. Entomological Society of Washington, Washington, DC
Hall JP, Harvey DJ (2001) A phylogenetic analysis of the Neotropical riodinid butterfly genera *Juditha, Lemonias, Thisbe* and *Uraneis*, with a revision of *Juditha* (Lepidoptera: Riodinidae: Nymphidiini). Syst Entomol 26(4):453–490
Hall JP, Harvey DJ (2002) Basal subtribes of the Nymphidiini (Lepidoptera: Riodinidae): phylogeny and myrmecophily. Cladistics 18(6):539–569
Hall JP, Harvey DJ, Janzen DH (2004) Life history of *Calydna sturnula* with a review of larval and pupal balloon setae in the Riodinidae (Lepidoptera). Ann Entomol Soc Am 97(2):310–321
Hardy PB, Lawrence JB (2017) Field guide to butterflies of the Philippines. Siri Scientific Press
Harvey DJ, Longino J (1989) Myrmecophily and larval food plants of *Brephidium isophthalma pseudofea* (Lycaenidae) in the Florida keys. J Lepidop Soc 43(4):332–333
Harvey DJ, Webb TA (1980) Ants associated with *Harkenclenus titus, Glaucopsyche lygdamus*, and *Celastrina argiolus* (Lycaenidae). J Lepidop Soc 34(4):371–372
Hawkeswood TJ, Dunn KL, Sommung B (2016) *Acanthus ilicifolius* L.(Acanthaceae), a new larval host plant for *Hypolycaena erylus himavantus* Fruhstorfer, 1912 (Lepidoptera: Lycaenidae) from Thailand, with a review on larval host plants of *H. erylus* and its weaver ant association. Calodema 416:1–8
Heath A (1997) A review of African genera of the tribe Aphnaeini (Lepidoptera: Lycaenidae). Metamorphosis Occ Suppl 2:1–60

Heath A, Claassens AJ (2000) New observations of ant associations and life history adaptations (Lepidoptera: Lycaenidae) in South Africa. Metamorphosis 11:3–19

Heath A, Claassens AJ (2003) Ant-association among southern African Lycaenidae. J Lepidop Soc 57(1):1–16

Heath A, Pringle EL (2004) A review of the southern African genus *Thestor* Hübner (Lepidoptera: Lycaenidae: Miletinae). Metamorphosis 15:89–151

Heikkilä M, Mutanen M, Wahlberg N, Sihvonen P, Kaila L (2015) Elusive ditrysian phylogeny: an account of combining systematized morphology with molecular data (Lepidoptera). BMC Evol Biol 15(1):260

Heil M, Hilpert A, Fiala B, Linsenmair K (2001) Nutrient availability and indirect (biotic) defence in a Malaysian ant-plant. Oecologia 126(3):404–408

Heil M, Barajas-Barron A, Orona-Tamayo D, Wielsch N, Svatos A (2014) Partner manipulation stabilises a horizontally transmitted mutualism. Ecol Lett 17(2):185–192

Henning SF (1983) Chemical communication between lycaenid larvae (Lepidoptera: Lycaenidae) and ants (Hymenoptera: Formicidae). J Entomol Soc S Afr 46(2):341–366

Heredia MD, Robbins RK (2016) Natural history of the mistletoe-feeding *Thereus lomalarga* (Lepidoptera, Lycaenidae, Eumaeini) in Colombia. Zootaxa 4117(3):301–320

Hill CJ (1993) The myrmecophilous organs of *Arhopala madytus* Fruhstorfer (Lepidoptera: Lycaenidae). J Aust Entomol Soc 32:283–288

Hinton HE (1951) Myrmecophilous Lycaenidae and other Lepidoptera—a summary. Proc Lond Entomol Nat Hist Soc 1949–1950:111–175

Hocking B (1970) Insect associations with the swollen thorn acacias. Trans R Entomol Soc Lond 122(pt. 7):211–255

Hojo MK, Wada-Katsumata A, Akino T, Yamaguchi S, Ozaki M, Yamaoka R (2009) Chemical disguise as particular caste of host ants in the ant inquiline parasite *Niphanda fusca* (Lepidoptera: Lycaenidae). Proc R Soc B Biol Sci 276(1656):551–558

Hojo MK, Wada-Katsumata A, Ozaki M, Yamaguchi S, Yamaoka R (2008) Gustatory synergism in ants mediates a species-specific symbiosis with lycaenid butterflies. J Comput Physiol A Neuroethol Sens Neural Behav Physiol 194(12):1043–1052

Hojo MK, Yamaguchi S, Akino T, Yamaoka R (2014a) Adoption of lycaenid *Niphanda fusca* (Lepidoptera: Lycaenidae) caterpillars by the host ant *Camponotus japonicus* (Hymenoptera: Formicidae). Entomol Sci 17(1):59–65

Hojo MK, Yamamoto A, Akino T, Tsuji K, Yamaoka R (2014b) Ants use partner specific odors to learn to recognize a mutualistic partner. PLoS One 9(1):e86054

Hojo MK, Pierce NE, Tsuji K (2015) Lycaenid caterpillar secretions manipulate attendant ant behavior. Curr Biol 25(17):2260–2264

Hölldobler B, Kwapich C (in review) The guests of ants: how myrmecophiles interact with their hosts

Hölldobler B, Wilson EO (1990) The ants. Harvard University Press, Cambridge. https://www.antwiki.org/wiki/The_Ants

Holloway JD (2005) The moths of Borneo parts 15 & 16: Family Noctuidae, Subfamily Catocalinae. Southdene Sdn Bhd. Available at https://www.mothsofborneo.com/part-15-16/index.htm. Accessed 3 Jan 2020

Honda K (1983) Defensive potential of components of the larval osmeterial secretion of papilionid butterflies against ants. Physiol Entomol 8(2):173–179

Hopper SD (2009) OCBIL theory: towards an integrated under- standing of the evolution, ecology and conservation of biodiversity on old, climatically buffered, infertile landscapes. Plant Soil 322:49–86

Horvitz CC, Turnbull C, Harvey DJ (1987) Biology of immature *Eurybia elvina* (Lepidoptera: Riodinidae), a myrmecophilous metalmark butterfly. Ann Entomol Soc Am 80(4):513–519

Hsu YF, Johnson KU (1998) A New species of *Zinaspa* from China (Lepidoptera: Lycaenidae: Theclinae). Holarct Lepidop 5(2):39–42

Hsu YF, Ding D, Yen SH, Qian ZQ (2004) Systematic problems surrounding *Howarthia melli* (Forster)(Lepidoptera: Lycaenidae: Theclinae), with description of a new species and a review of Rhododendron-association in lycaenid butterflies. Ann Entomol Soc Am 97(4):653–666

Igarashi S, Fukuda H (1997) The life histories of Asian butterflies, vol 1. Tokai University Press, Tokyo

Igarashi S, Fukuda H (2000) The life histories of Asian butterflies, vol 2. Tokai University Press, Tokyo

Inui Y, Shimizu-kaya U, Okubo T, Yamsaki E, Itioka T (2015) Various chemical strategies to deceive ants in three *Arhopala* species (Lepidoptera: Lycaenidae) exploiting *Macaranga* myrmecophytes. PLoS One 10(4):e0120652

Itioka T, Yamamoto T, Tzuchiya T, Okubo T, Yago M, Seki Y, Ohshima Y, Katsuyama R, Chiba H, Yata O (2009) Butterflies collected in and around Lambir Hills National Park, Sarawak, Malaysia in Borneo. Contrib Biol Lab Kyoto Univ 30(1):25–68

Ito F, Higashi S (1991) Variance of ant effects on the different life forms of moth caterpillars. J Anim Ecol 1:327–334

IUCN (2020) The IUCN Red list of threatened species. https://www.iucnredlist.org

Ivens AB (2015) Cooperation and conflict in ant (Hymenoptera: Formicidae) farming mutualisms: a review. Myrmecol News 21:19–36

Ivens AB, von Beeren C, Blüthgen N, Kronauer DJ (2016) Studying the complex communities of ants and their symbionts using ecological network analysis. Annu Rev Entomol 61:353–371

Jackson TH (1937) The early stages of some African Lycaenidae (Lepidoptera), with an account of the larval habits. Trans R Entomol Soc Lond 86(12):201–237

Jackson TH (1957) Protective mimicry in the pupa of *Epitola urania* (Lep.: Lycaenidae). J Entomol Soc S Afr 20(2):234

Janzen DH (1967) Interaction of the bull's-horn acacia (*Acacia cornigera* L.) with an ant inhabitant (*Pseudomyrmex ferrugineus* F. Smith) in eastern Mexico. Kansas Univ Sci Bull 47:315–558

Janzen DH (1984) Two ways to be a tropical big moth: Santa Rosa saturniids and sphingids. Oxf Surv Evol Biol 1:85–146

Janzen DH, Hallwachs W (2021) Dynamic database for an inventory of the macrocaterpillar fauna, and its food plants and parasitoids, of Area de Conservacion Guanacaste (ACG), northwestern Costa Rica. http://janzen.sas.upenn.edu. Accessed March 2021

Jeratthitikul E, Yago M, Shizuya H, Yokoyama J, Hikida T (2011) Life history and morphology of the black cupid butterfly, *Tongeia kala* (De Nicéville)(Lycaenidae), from Myanmar. J Lepidop Soc 65(3):167–174

Johnson SJ, Valentine PS (2001) The life history of *Nesolycaena medicea* Braby (Lepidoptera: Lycaenidae). Aust Entomol 27(4):109–112

Jones MT, Castellanos I, Weiss MR (2002) Do leaf shelters always protect caterpillars from invertebrate predators? Ecol Entomol 27(6):753–757

Kaliszewska ZA, Lohman DJ, Sommer K, Adelson G, Rand DB, Mathew J, Talavera G, Pierce NE (2015) When caterpillars attack: biogeography and life history evolution of the Miletinae (Lepidoptera: Lycaenidae). Evolution 69(3):571–588

Kaminski LA (2008a) Immature stages of *Caria plutargus* (Lepidoptera: Riodinidae), with discussion on the behavioral and morphological defensive traits in nonmyrmecophilous riodinid butterflies. Ann Entomol Soc Am 101(5):906–914

Kaminski LA (2008b) Polyphagy and obligate myrmecophily in the butterfly *Hallonympha paucipuncta* (Lepidoptera: Riodinidae) in the Neotropical Cerrado savanna. Biotropica 40(3):390–394

Kaminski LA (2017) In: Diehl E (ed) Formigas, besouros e lepidópteros. Interações das formigas com outros organismos: diversidade ecológica e evolutiva. Oikos, São Leopoldo, pp 51–65

Kaminski LA, Carvalho-Filho FS (2012) Life history of *Aricoris propitia* (Lepidoptera: Riodinidae)—a myrmecophilous butterfly obligately associated with fire ants. Psyche 2012:1–10

Kaminski LA, Rodrigues D (2011) Species-specific levels of ant attendance mediate performance costs in a facultative myrmecophilous butterfly. Physiol Entomol 36(3):208–214

Kaminski LA, Freitas AV, Oliveira PS (2010a) Interaction between mutualisms: ant-tended butterflies exploit enemy-free space provided by ant-treehopper associations. Am Nat 176(3):322–334

Kaminski LA, Thiele SC, Iserhard CA, Romanowski HP, Moser A (2010b) Natural history, new records, and notes on the conservation status of *Cyanophrys bertha* (Jones) (Lepidoptera: Lycaenidae). Proc Entomol Soc Wash 112(1):54–60

Kaminski LA, Mota LL, Freitas AV (2012a) Life history of *Porthecla ravus* (Druce) (Lepidoptera: Lycaenidae), with discussion on the use of Annonaceae by Eumaeini butterflies. Annales de la Société entomologique de France 48(3–4):309–312

Kaminski LA, Rodrigues D, Freitas AV (2012b) Immature stages of *Parrhasius polibetes* (Lepidoptera: Lycaenidae): host plants, tending ants, natural enemies and morphology. J Nat Hist 46(11–12):645–667

Kaminski LA, Mota LL, Freitas AV, Moreira GR (2013) Two ways to be a myrmecophilous butterfly: natural history and comparative immature-stage morphology of two species of *Theope* (Lepidoptera: Riodinidae). Biol J Linn Soc 108(4):844–870

Kaminski LA, Mota LL, Freitas AV (2014) Larval cryptic coloration and mistletoe use in the metalmark butterfly *Dachetola azora* (Lepidoptera: Riodinidae). Entomol Am 120(1):18–23

Kaminski LA, Volkmann L, Ríos S, Vila R (2015) Ecology and evolution of *Aricoris chilensis* complex (Riodinidae): integrative taxonomy reveal cryptic species. En Lepid Neotrop

Kaminski LA, Iserhard CA, Freitas AV (2016) *Thisbe silvestre* sp. nov. (Lepidoptera: R iodinidae): a new myrmecophilous butterfly from the Brazilian Atlantic Forest. Aust Entomol 55(2):138–146

Kaminski LA, Carneiro E, Dolibaina DR, Casagrande MM, Mielke OH (2020a) Oviposition of *Minstrellus grandis* (Lepidoptera: Riodinidae) in a harmful ant-plant symbiosis. Acta Amazon 50(3):256–259

Kaminski LA, Volkmann L, Callaghan CJ, DeVries PJ, Vila R (2020b) The first known riodinid 'cuckoo' butterfly reveals deep-time convergence and parallelism in ant social parasites. Zool J Linnean Soc 192:1–20

Kaspari M (2020) The seventh macronutrient: how sodium shortfall ramifies through populations, food webs and ecosystems. Ecol Lett 23:1153–1168. https://doi.org/10.1111/ele.13517

Kaspari M, Vargo EL (1995) Colony size as a buffer against seasonality: Bergmann's rule in social insects. Am Nat 145(4):610–632

Kaspari M, Alonso L, O'Donnell S (2000) Three energy variables predict ant abundance at a geographical scale. Proc R Soc Lond Ser B Biol Sci 267(1442):485–489

Kaspari M, Yanoviak SP, Dudley R (2008) On the biogeography of salt limitation: a study of ant communities. Proc Natl Acad Sci 105(46):17848–17851

Kaspari M, Yanoviak SP, Dudley R, Yuan N, Clay NA (2009) Sodium shortage as a constraint on the carbon cycle in an inland tropical rainforest. Proc Natl Acad Sci 106(46):19405–19409

Kaspari M, Donoso D, Lucas JA, Zumbusch T, Kay AD (2012) Using nutritional ecology to predict community structure: a field test in Neotropical ants. Ecosphere 3(11):1–5

Kaspari M, Welti EA, de Beurs KM (2020) The nutritional geography of ants: gradients of sodium and sugar limitation across North American grasslands. J Anim Ecol 89(2):276–284

Kawahara AY, Plotkin D, Espeland M, Meusemann K, Toussaint EF, Donath A, Gimnich F, Frandsen PB, Zwick A, dos Reis M, Barber JR, Peters RS, Liu S, Zhou X, Mayer C, Podsiadlowski L, Storer C, Yack JE, Misof B, Breinholt JW (2019) Phylogenomics reveals the evolutionary timing and pattern of butterflies and moths. Proc Natl Acad Sci 116(45):22657–22663

Kim SS, Ho YH (2012) Life histories of Korean butterflies. Sagyejel, Paju [publication in Korean]

Kistner DH (1982) The social insects' bestiary. In: Social insects. Academic, New York, pp 1–244

Kitching RL (1983) Myrmecophilous organs of the larvae and pupa of the lycaenid butterfly *Jalmenus evagoras* (Donovan). J Nat Hist 17(3):471–481

Kitching RL, Luke B (1985) The myrmecophilous organs of the larvae of some British Lycaenidae (Lepidoptera): a comparative study. J Nat Hist 19(2):259–276

Koh LP, Dunn RR, Sodhi NS, Colwell RK, Proctor HC, Smith VS (2004) Species coextinctions and the biodiversity crisis. Science 305(5690):1632–1634

Komatsu T, Itino T (2014) Moth caterpillar solicits for homopteran honeydew. Nat Sci Rep 4(1):1–3

Koptur S (1985) Alternative defenses against herbivores in *Inga* (Fabaceae: Mimosoideae) over an elevational gradient. Ecology 66(5):1639–1650

Kronauer DJ, Pierce NE (2011) Myrmecophiles. Curr Biol 21(6):R208–R209

Kubik TD, Schorr RA (2018) Facultative myrmecophily (Hymenoptera: Formicidae) in the hops blue butterfly, *Celastrina humulus* (Lepidoptera: Lycaenidae). Entomol News 127(5):490–498

Kumar KP, Kamala Jayanthi PD, Verghese A, Chakravarthy AK (2017) Facultative myrmecophily in *Deudorix isocrates* (Fabricius) (Lepidoptera: Lycaenidae). J Entomol Zool Stud 5(5):870–875

Lafranchis T (2019) Notes on the biology of some butterflies in Greece (Lepidoptera: Papilionoidea). Entomol Gaz 70(2):113–134

Lafranchis T, Gil-T FE, Lafranchis A (2007) New data on the ecology of 8 taxa of *Agrodiaetus* Hubner, 1822 from Greece and Spain: hostplants, associated ants and parasitoids. Atalanta 38(1/2):189–197

Lamas G, Callaghan C, Casagrande MM, Mielke O, Pyrcz T, Robbins R, and Viloria A (2004) Hesperioidea – Papilionoidea. Association for Tropical Lepidoptera/Scientific Publishers, Gainesville. Digitized by Symons F at https://www.ucl.ac.uk/taxome/gbn/

Lamborn WA (1911) Some ant-tended lycaenid larvae observed by Mr. W. A. Lamborn in the Lagos District. Proc Entomol Soc Lond 1911:xcix–cvii

Lamborn WA, Bethune-Baker GT, Distant WL, Eltringham H, Poulton EB, Durrant JH, Newstead R (1914) XX. On the relationship between certain west African insects, especially ants, Lycaenidae and Homoptera. Trans R Entomol Soc Lond 61(3):436–498

Larsen TB (2005) Butterflies of West Africa. Apollo Books, Stenstrup

Léger T, Mally R, Neinhuis C, Nuss M (2021) Refining the phylogeny of Crambidae with complete sampling of subfamilies (Lepidoptera, Pyraloidea). Zool Scr 50(1):84–99

Leimar O, Axén AH (1993) Strategic behaviour in an interspecific mutualism: interactions between lycaenid larvae and ants. Anim Behav 46(6):1177–1182

Lengyel S, Gove AD, Latimer AM, Majer JD, Dunn RR (2010) Convergent evolution of seed dispersal by ants, and phylogeny and biogeography in flowering plants: a global survey. Perspect Plant Ecol Evol Syst 12(1):43–55

Leong TM, D'Rozario V (2012) Mimicry of the weaver ant, *Oecophylla smaragdina* by the moth caterpillar, *Homodes bracteigutta*, the crab spider, *Amyciaea lineatipes*, and the jumping spider, *Myrmarachne plataleoides*. Nat Singapore 5:39–56

Lima LD, Trigo JR, Kaminski LA (2020) Chemical convergence between a guild of facultative myrmecophilous caterpillars and host plants. Ecol Entomol. https://doi.org/10.1111/een.12941

Lin YH, Liao YC, Yang CC, Billen J, Yang MM, Hsu YF (2019) Vibrational communication between a myrmecophilous butterfly *Spindasis lohita* (Lepidoptera: Lycaenidae) and its host ant *Crematogaster rogenhoferi* (Hymenoptera: Formicidae). Sci Rep 9(1):1

Linsley E (1944) Natural sources, habitats, and reservoirs of insects associated with stored food products. Hilgardia 16(4):185–224

Lo YF, Li F, Ding L (2017) Description of a new subspecies of *Talicada nyseus* (Guérin-Méneville, 1843) from Hainan, China (Lepidoptera: Lycaenidae), with notes on the genus *Talicada* Moore, 1881. Zootaxa 4269(4):586–592

Loeffler CC (1996) Caterpillar leaf folding as a defense against predation and dislodgment: staged encounters using *Dichomeris* (Gelechiidae) larvae on goldenrods. J Lepidop Soc 50(3):245–260

Lohman DJ (2004) Cuticular hydrocarbons and chemical camouflage in butterfly-ant symbioses (Lepidoptera: Lycaenidae; Hymenoptera: Formicidae). PhD Thesis, Harvard University, Cambridge

Lohman DJ, Samarita VU (2009) The biology of carnivorous butterfly larvae (Lepidoptera: Lycaenidae: Miletinae: Miletini) and their ant-tended hemipteran prey in Thailand and the Philippines. J Nat Hist 43(9–10):569–581
Lohman DJ, Liao Q, Pierce NE (2006) Convergence of chemical mimicry in a guild of aphid predators. Ecol Entomol 31(1):41–51
Lokkers C (1990) Colony dynamics of the green tree ant (*Oecophylla smaragdina* Fab.) in a seasonal tropical climate. PhD Thesis, James Cook University, Queensland
Madeiros LP, Garcia G, Thompson JN, Guimarães PR (2018) The geographic mosaic of coevolution in mutualistic networks. Proc Natl Acad Sci U S A 115:12017–12022
Malicky H (1969) Versuch einer Analyse der okologischen Beziehungen zwischen Lycaeniden (Lepidoptera) und Formiciden (Hymenoptera). Tijd Entomol 112:213–298. (publication in German)
Malicky H (1970) New aspects of the association between lycaenid larvae (Lycaenidae) and ants (Formicidae, Hymenoptera). J Lepidop Soc 24(3):190–202
Mally R, Hayden JE, Neinhuis C, Jordal BH, Nuss M (2019) The phylogenetic systematics of Spilomelinae and Pyraustinae (Lepidoptera: Pyraloidea: Crambidae) inferred from DNA and morphology. Earth Syst Phylogent 77(1):141–204
Mänd T, Tammaru T, Mappes J (2007) Size dependent predation risk in cryptic and conspicuous insects. Evol Ecol 21(4):485–498
Martins DJ, Collins SC, Congdon C, Pierce NE (2013) Association between the African lycaenid, *Anthene usamba*, and an obligate acacia ant, *Crematogaster mimosae*. Biol J Linn Soc 109(2):302–312
Maschwitz U, Schroth M, Hänel H, Pong TY (1984) Lycaenids parasitizing symbiotic plant-ant partnerships. Oecologia 64(1):78–80
Maschwitz U, Dumpert K, Tuck KR (1986) Ants feeding on anal exudate from tortricid larvae: a new type of trophobiosis. J Nat Hist 20(5):1041–1050
Maschwitz U, Nassig WA, Dumpert K, Fiedler K (1988) Larval carnivory and myrmecoxeny, and imaginal myrmecophily in miletine lycaenids (Lepidoptera, Lycaenidae) on the Malay peninsula. Tyo to Ga 39(3):167–181
Mathew J, Travassos MA, Canfield MR, Murawski DA, Kitching RL, Pierce NE (2008) The singing reaper: diet, morphology and vibrational signaling in the nearctic species *Feniseca tarquinius* (Lepidoptera: Lycaenidae, Miletinae). Trop Lepidop Res 1:24–29
Meehan CJ, Olson EJ, Reudink MW, Kyser TK, Curry RL (2009) Herbivory in a spider through exploitation of an ant–plant mutualism. Curr Biol 19(19):R892–R893
Megens HJ, De Jong R, Fiedler K (2005) Phylogenetic patterns in larval host plant and ant association of indo-Australian Arhopalini butterflies (Lycaenidae: Theclinae). Biol J Linn Soc 84(2):225–241
Merrill, DN (1997) Deception in lycaenid-ant mutualism? Senior thesis, Harvard University, Cambridge
Michelangeli FA (2003) Ant protection against herbivory in three species of *Tococa* (Melastomataceae) occupying different environments 1. Biotropica 35(2):181–188
Miller CG, Lane DA (2004) A new species of *Acrodipsas* Sands (Lepidoptera: Lycaenidae) from the Northern Territory. Aust Entomol 31:141–146
Mitter C, Davis DR, Cummings MP (2017) Phylogeny and evolution of Lepidoptera. Annu Rev Entomol 62:265–283
Mizuno T, Hagiwara Y, Akino T (2018) Chemical tactic of facultative myrmecophilous lycaenid pupa to suppress ant aggression. Chemoecology 28(6):173–182
Mizuno T, Hagiwara Y, Akino T (2019) Varied effects of tending ant species on the development of facultatively myrmecophilous lycaenid butterfly larvae. Insects 10(8):234
Morozumi Y, Murakami T, Watanabe M, Ohta S, Ômura H (2019) Absence of cuticular alkenes allows lycaenid larvae to avoid predation by *Formica japonica* ants. Entomol Sci 22(2):126–136
Mota LL, Oliveira PS (2016) Myrmecophilous butterflies utilise ant–treehopper associations as visual cues for oviposition. Ecol Entomol 41:338–343

Mota LL, Kaminski LA, Freitas AV (2014) Last instar larvae and pupae of *Ourocnemis archytas* and *Anteros formosus* (Lepidoptera: Riodinidae), with a summary of known host plants for the tribe Helicopini. Zootaxa 3838(4):435–444

Mota LL, Kaminski LA, Freitas AV (2020) The tortoise caterpillar: carnivory and armoured larval morphology of the metalmark butterfly *Pachythone xanthe* (Lepidoptera: Riodinidae). J Nat Hist 54(5–6):309–319

Musche M, Anton C, Worgan A, Settele J (2006) No experimental evidence for host ant related oviposition in a parasitic butterfly. J Insect Behav 19(5):631–643

Nash DR (1990) Cost-benefit analysis of a mutualism between lycaenid butterflies and ants. PhD thesis, Oxford University

Nash DR, Als TD, Maile R, Jones GR, Boomsma JJ (2008) A mosaic of chemical coevolution in a large blue butterfly. Science 319(5859):88–90

Neild AF, Bálint Z (2014) Notes on the identity of *Evenus coronata* (Hewitson, 1865) (Lepidoptera: Lycaenidae: Theclinae: Eumaeini) with the description of a remarkably overlooked sibling species. Trop Lepidop Res 24(2):105–120

New TR (ed) (1993) Conservation biology of Lycaenidae (butterflies). IUCN, Gland

Newcomer EJ (1912) Some observations on the relations of ants and lycaenid caterpillars, and a description of the relational organs of the latter. J N Y Entomol Soc 20(1):31–36

Nichols SW (1989) Torre-Bueno glossary of entomology. Entomological Society, New York

Nielsen GJ, Kaminski LA (2018) Immature stages of the Rubiaceae-feeding metalmark butterflies (Lepidoptera: Riodinidae), and a new function for the tentacle nectary organs. Zootaxa 4524(1):1–32

Nishida K (2010) Description of the immature stages and life history of *Euselasia* (Lepidoptera: Riodinidae) on *Miconia* (Melastomataceae) in Costa Rica. Zootaxa 2466:1–74

Nishida K, Robbins RK (2020) One side makes you taller: a mushroom–eating butterfly caterpillar (Lycaenidae) in Costa Rica. Neotrop Biol Conserv 15(4):463–470

Nyffeler M, Olson EJ, Symondson WO (2016) Plant-eating by spiders. J Arachnol 44:15–27

Obregón R, Shaw MR, Fernández-Haeger J, Jordano D (2015) Parasitoid and ant interactions of some Iberian butterflies (Insecta: Lepidoptera). SHILAP Rev Lepidopterol 43(171):439–454

Offenberg J (2001) Balancing between mutualism and exploitation: the symbiotic interaction between *Lasius* ants and aphids. Behav Ecol Sociobol 49(4):304–310

Okubo T, Yago M, Itioka T (2009) Immature stages and biology of Bornean *Arhopala* butterflies (Lepidoptera, Lycaenidae) feeding on myrmecophytic *Macaranga*. Trans Lepidop Soc Jpn 60(1):37–51

Oliver JC (2007) Population, phylogenetic, and coalescent analyses of character evolution in gossamer-winged butterflies (Lepidoptera: Lycaenidae). Dissertation, University of Arizona

Omura H, Watanabe M, Honda K (2009) Cuticular hydrocarbons of larva and pupa of Reverdin's blue, *Lycaeides argyrognomon* (Lycaenidae) and its tending ants. Trans Lepidop Soc Jpn 60(3):203–210

Opler PA (1992) A field guide to eastern butterflies. Houghton Mifflin Co, New York

Opler PA (1999) A field guide to western butterflies. Houghton Mifflin Co, New York

Orivel J, Dejean A (2000) Myrmecophily in Hesperiidae. The case of *Vettius tertianus* in ant gardens. Comptes Rend l'Académ Sci-Ser III-Sci Vie 323(8):705–715

Orivel J, Servigne P, Cerdan P, Dejean A, Corbara B (2004) The ladybird *Thalassa saginata*, an obligatory myrmecophile of *Dolichoderus bidens* ant colonies. Naturwissenschaften 91(2):97–100

Orr AG, Charles JK, Yahya HH, Sharebini NH (1996) Nesting and colony structure in the giant forest ant, *Camponotus gigas* (Latreille) (Hymenoptera: Formicidae). Raffles Bull Zool 44:247–252

Painting C, Nicholson CC, Bulbert MW, Norma-Rashid Y, Li D (2017) Nectary feeding and guarding behavior by a tropical jumping spider. Front Ecol Environ 15(8):469–470

Palting JD (2020) A Molecular Phylogenetic Assessment of the North American Lichen Tiger Moths (Lepidoptera; Erebidae; Arctiinae; Lithosiini) with Life History Observations and a Description of a New Species from Central Arizona. Dissertation, University of Arizona

Pan TS, Morishita K (1990) Notes on *Nacaduba normani* Eliot in Sabah, North Borneo (Lepidoptera, Lycaenidae). Tyo to Ga 41(3):149–154

Parker J (2016) Myrmecophily in beetles (Coleoptera): evolutionary patterns and biological mechanisms. Myrmecol News 22:65–108

Parmentier T, Dekoninck W, Wenseleers T (2014) A highly diverse microcosm in a hostile world: a review on the associates of red wood ants (*Formica rufa* group). Insect Soc 61(3):229–237

Parsons M (1984) Life histories of four species of *Philiris* Roeber (Lepidoptera: Lycaenidae) from Papua New Guinea. J Lepidop Soc 38(1):15–22

Parsons M (1999) The butterflies of Papua New Guinea: their systematics and biology. Academic, Cambridge

Patricelli D, Barbero F, Occhipinti A, Bertea CM, Bonelli S, Casacci LP, Zebelo SA, Crocoll C, Gershenzon J, Maffei ME, Thomas JA (2015) Plant defences against ants provide a pathway to social parasitism in butterflies. Proc R Soc B Biol Sci 282(1811):20151111

Patricelli D, Barbero F, La Morgia V, Casacci LP, Witek M, Balletto E, Bonelli S (2011) To lay or not to lay: oviposition of *Maculinea arion* in relation to *Myrmica* ant presence and host plant phenology. Anim Behav 82(4):791–799

Paul RG (1977) Aspects of the biology and taxonomy of British myrmecophilous root aphids. PhD thesis, University of London, London

Pech P, Fric Z, Konvicka M (2007) Species-specificity of the *Phengaris* (*Maculinea*)–*Myrmica* host system: fact or myth?(Lepidoptera: Lycaenidae; Hymenoptera: Formicidae). Sociobiology 50(3):983–1004

Pellissier L, Litsios G, Guisan A, Alvarez N (2012) Molecular substitution rate increases in myrmecophilous lycaenid butterflies (Lepidoptera). Zool Scr 41(6):651–658

Pellissier L, Litsios G, Fiedler K, Pottier J, Dubuis A, Pradervand JN, Salamin N, Guisan A (2012a) Loss of interactions with ants under cold climate in a regional myrmecophilous butterfly fauna. J Biogeogr 39(10):1782–1790

Pellissier L, Rasmann S, Litsios G, Fiedler K, Dubuis A, Pottier J, Guisan A (2012b) High host-plant nitrogen content: a prerequisite for the evolution of ant–caterpillar mutualism? J Evol Biol 25(8):1658–1666

Penz CM, DeVries PJ (2006) Systematic position of *Apodemia paucipuncta* (Riodinidae), and a critical evaluation of the nymphidiine transtilla. Zootaxa 1190(1):1–50

Peterson MA (1993) The nature of ant attendance and the survival of larval *Icaricia acmon* (Lycaenidae). J Lepidop Soc 47(1):8–16

Peterson SC, Johnson ND, LeGuyader JL (1987) Defensive regurgitation of allelochemicals derived from host cyanogenesis by eastern tent caterpillars. Ecology 68(5):1268–1272

Pierce NE (1984) Amplified species diversity: a case study of an Australian lycaenid butterfly and its attendant ants. In: The biology of butterflies. Academic, London, pp 197–200

Pierce NE (1985) Lycaenid butterflies and ants: selection for nitrogen fixing and other protein rich food plants. Am Nat 125:888–895

Pierce NE (1987) The evolution and biogeography of associations between lycaenid butterflies and ants. Oxf Surv Evol Biol 4:89–116

Pierce NE (1995) Predatory and parasitic Lepidoptera: carnivores living on plants. J Lepidop Soc 49(4):412–453

Pierce NE, Easteal S (1986) The selective advantage of attendant ants for the larvae of a lycaenid butterfly, *Glaucopsyche lygdamus*. J Anim Ecol 55(2):451–462

Pierce NE, Elgar MA (1985) The influence of ants on host plant selection by *Jalmenus evagoras*, a myrmecophilous lycaenid butterfly. Behav Ecol Sociobiol 3:209–222

Pierce NE, Mead PS (1981) Parasitoids as selective agents in the symbiosis between lycaenid butterfly larvae and ants. Science 211(4487):1185–1187

Pierce NE, Nash DR (1999) The imperial blue, *Jalmenus evagoras* (Lycaenidae). Mon Aust Lepidop 6:277–313

Pierce NE, Kitching RL, Buckley RC, Taylor MF, Benbow KF (1987) The costs and benefits of cooperation between the Australian lycaenid butterfly, *Jalmenus evagoras*, and its attendant ants. Behav Ecol Sociobiol 21(4):237–248

Pierce NE, Braby MF, Heath A, Lohman DJ, Mathew J, Rand DB, Travassos MA (2002) The ecology and evolution of ant association in the Lycaenidae (Lepidoptera). Annu Rev Entomol 47(1):733–771

Pohl S, Frederickson ME, Elgar MA, Pierce NE (2016) Colony diet influences ant worker foraging and attendance of myrmecophilous lycaenid caterpillars. Front Ecol Evol 4:114

Portugal AHA, Trigo JR (2005) Similarity of cuticular lipids between a caterpillar and its host plant: a way to make prey undetectable for predatory ants? J Chem Ecol 31(11):2551–2561

Poulton EB (1890) The colours of animals, their meaning and use, especially considered in the case of insects. Appleton and Company, New York. https://www.biodiversitylibrary.org/bibliography/11353

Powell JA, Mitter C, Farrell B (1998) 20. Evolution of larval food preferences in Lepidoptera. Evol Syst Biogeogr 1:403–422

Pratt PD, Herdocia K, Valentin V, Makinson J, Purcell MF, Mattison E, Rayamajhi MB, Moran P, Raghu S (2016) Development rate, consumption, and host fidelity of *Neostauropus alternus* (Walker, 1855) (Lepidoptera: Notodontidae). Pan-Pacific Entomol 92(4):200–209

Pringle EL, Henning GA, Ball JB (1994) Pennington's butterflies of southern Africa, 2nd edn. Struik Winchester, Cape Town

Pringle EG, Dirzo R, Gordon DM (2011) Indirect benefits of symbiotic coccoids for an ant-defended myrmecophytic tree. Ecology 92(1):37–46

Pringle EG, Akçay E, Raab TK, Dirzo R, Gordon DM (2013) Water stress strengthens mutualism among ants, trees, and scale insects. PLoS Biol 11(11):e1001705

Ramos RR, Freitas AV, Francini RB (2018) Defensive strategies of a noctuid Caterpillar in a Myrmecophytic plant: are *Dyops* larvae immune to *Azteca* ants? Sociobiology 65(3):397–402

Regier JC, Mitter C, Davis DR, Harrison TL, Sohn JC, Cummings MP, Zwick A, Mitter KT (2015) A molecular phylogeny and revised classification for the oldest ditrysian moth lineages (Lepidoptera: Tineoidea), with implications for ancestral feeding habits of the mega-diverse Ditrysia. Syst Entomol 40(2):409–432

Rettenmeyer CW, Rettenmeyer ME, Joseph J, Berghoff SM (2011) The largest animal association centered on one species: the army ant *Eciton burchellii* and its more than 300 associates. Insect Soc 58(3):281–292

Ribas CR, Schoereder JH (2004) Determining factors of arboreal ant mosaics in cerrado vegetation (Hymenoptera: Formicidae). Sociobiology 44(1):49–68

Riva F, Barbero F, Bonelli S, Balletto E, Casacci LP (2017) The acoustic repertoire of lycaenid butterfly larvae. Bioacoustics 26(1):77–90

Robbins RK (1991) Cost and evolution of a facultative mutualism between ants and lycaenid larvae (Lepidoptera). Oikos 62(3):363–369

Robbins RK, Aiello A (1982) Foodplant and oviposition records for Panamanian Lycaenidae and Riodinidae. J Lepidop Soc 36(2):65–75

Robbins RK, Lamas G, Mielke OH, Harvey DJ, Casagrande MM (1996) Taxonomic composition and ecological structure of the species-rich butterfly community at Pakitza, Parque Nacional del Manu, Peru. In: Manu: the biodiversity of southeastern Peru. The biodiversity of southeastern Peru. Smithsonian Institution Press, Washington, DC, pp 217–252

Robinson GF, Nielsen ES (1993) Tineid genera of Australia (Lepidoptera). CSIRO Publishing, Melbourne

Rocha FH, Lachaud JP, Pérez-Lachaud G (2020) Myrmecophilous organisms associated with colonies of the ponerine ant *Neoponera villosa* (Hymenoptera: Formicidae) nesting in *Aechmea bracteata* bromeliads: a biodiversity hotspot. Myrmecol News 30:73–92

Rodrigues D, Kaminski LA, Freitas AV, Oliveira PS (2010) Trade-offs underlying polyphagy in a facultative ant-tended florivorous butterfly: the role of host plant quality and enemy-free space. Oecologia 163(3):719–728

Rosi-Denadai CA, Scallion ML, Merrett CG, Yack JE (2018) Vocalization in caterpillars: a novel sound-producing mechanism for insects. J Exp Biol 221(4):jeb169466

Ross GN (1964) Life history studies on Mexican butterflies. III. Early stages of *Anatole rossi*, a new myrmecophilous metalmark. J Res Lepidop 3(2):81–94

Ross GN (1966) Life-history studies on Mexican butterflies. IV. The ecology and ethology of *Anatole rossi*, a myrmecophilous metalmark (Lepidoptera: Riodinidae). Ann Entomol Soc Am 59(5):985–1004

Rostás M (1657) Blassmann K (2009) insects had it first: surfactants as a defence against predators. Proc R Soc B Biol Sci 276:633–638

Roux O, Céréghino R, Solano PJ, Dejean A (2011) Caterpillars and fungal pathogens: two co-occurring parasites of an ant-plant mutualism. PLoS One 6(5):e20538

Saarinen EV (2005) Life history and myrmecophily of *Neomyrina nivea periculosa* (Lycaenidae: Theclinae). J Lepidop Soc 59(2):112–115

Saarinen EV (2006) Differences in worker caste behaviour of *Oecophylla smaragdina* (Hymenoptera: Formicidae) in response to larvae of *Anthene emolus* (Lepidoptera: Lycaenidae). Biol J Linn Soc 88(3):391–395

Saarinen EV, Daniels JC (2006) Miami blue butterfly larvae (Lepidoptera: Lycaenidae) and ants (Hymeoptera: Formicidae): new information on the symbionts of an endangered taxon. Fl Entomol 899(1):69–74

Sachs JL (2006) Cooperation within and among species. J Evol Biol 19(5):1415–1418

Sachs JL, Simms EL (2006) Pathways to mutualism breakdown. Trends Ecol Evol 21(10):585–592

Safian SZ (2012) Butterflies across the river. Report on the rapid butterfly surveys for the 'Across The River Project' in Sierra Leone and Liberia in 2011. Report for birdlife international and the society for the conservation of nature, Liberia. https://doi.org/10.13140/RG.2.1.1956.3607

Sáfián S (2015a) Observations on the little-known *Geritola frankdaveyi* Libert, 1999 (Lepidoptera: Lycaenidae: Poritiinae). Metamorphosis 26:19–23

Sáfián S (2015b) Two new Epitolini from Liberia in the genera *Stempfferia* Jackson, 1962 and *Cephetola* Libert, 1999 (Lepidoptera: Lycaenidae). Metamorphosis 26:12–19

Sáfián S, Collins SC (2014) A new *Iridana* Aurivillius, 1920 and a new *Teratoneura* Dudgeon, 1909 (Lepidoptera: Lycaenidae) from tropical Africa. Metamorphosis 25:90–96

Safian S, Collins SC (2015) Establishment of a new genus for *Eresiomera paradoxa* (Schultze, 1917) and related taxa (Lepidoptera: Lycaenidae) with description of two new species. Zootaxa 4018(1):124–136

Sáfián S, Larsen TB (2009) On the ecology and behavior of *Cerautola crowleyi* (Sharpe, 1890), *Cerautola ceraunia* (Hewitson, 1873) and *Cerautola miranda* (Staudinger, 1889) with descriptions of early stages (Lepidoptera: Lycaenidae, Epitolini). Trop Lepidop Res 1:22–28

Sala M, Casacci LP, Balletto E, Bonelli S, Barbero F (2014) Variation in butterfly larval acoustics as a strategy to infiltrate and exploit host ant colony resources. PLoS One 9(4):e94341

Salazar A, Fürstenau B, Quero C, Pérez-Hidalgo N, Carazo P, Font E, Martínez-Torres D (2015) Aggressive mimicry coexists with mutualism in an aphid. Proc Natl Acad Sci 112(4):1101–1106

Sanchez-Pena SR, Davis DR, Mueller UG (2003) A gregarious, mycophagous, myrmecophilous moth, *Amydria anceps* Walsingham (Lepidoptera: Acrolophidae), living in *Atta mexicana* (F. Smith) (Hymenoptera: Formicidae) spent fungal culture accumulations. Proc Entomol Soc Wash 105(1):186–194

Sands DP (1986) A revision of the genus *Hypochrysops* C. and R. Felder (Lepidoptera: Lycaenidae). Brill, Leiden

Sands DPA, Sands MC (2015) Review of variation in *Acrodipsas cuprea* (Sands, 1965) and *A. aurata* Sands, 1997 (Lepidoptera: Lycaenidae), with descriptions of a new subspecies of *A. cuprea* and a new species of *Acrodipsas* Sands from inland southern Queensland. Aust Entomol 42:197–218

Sanetra M, Fiedler K (1996) Behaviour and morphology of an aphytophagous lycaenid caterpillar: *Cigaritis* (*Apharitis*) *acamas* Klug, 1834 (Lepidoptera: Lycaenidae). Nota Lepidop 18(1):57–76

Santos FL, Dias FM, Leite LA, Dolibaina DR, Casagrande MM, Mielke OH (2014) Taxonomic notes on some species of *Euselasia* Hübner, [1819] from the "Uriiformes" group, with the description of the immature stages of *Euselasia satyroides* Lathy, 1926, stat. rev. (Lepidoptera: Riodinidae: Euselasiinae). Zootaxa 3869(5):501–522

Sariot M, Ginés M (2011) Biología y Ecología de los Licénidos Espanõles. Granada

Schär S, Eastwood R, Arnaldi KG, Talavera G, Kaliszewska ZA, Boyle JH, Espeland M, Nash DR, Vila R, Pierce NE (2018) Ecological specialization is associated with genetic structure in the ant-associated butterfly family Lycaenidae. Proc R Soc B Biol Sci 285:20181158

Schmid S, Schmid VS, Kamke RA, Steiner JO, Zillikens AN (2010) Association of three species of *Strymon* Hübner (Lycaenidae: Theclinae: Eumaeini) with bromeliads in southern Brazil. J Res Lepidop 42:50–55

Schmidt DJ, Rice SJ (2002) Association of ants with juvenile *Ogyris amaryllis* Hewitson (Lepidoptera: Lycaenidae) in South-Eastern Queensland. Aust J Entomol 41(2):164–169

Schmidt DJ, Grund R, Williams MR, Hughes JM (2014) Australian parasitic *Ogyris* butterflies: east–west divergence of highly-specialized relicts. Biol J Linn Soc 111(2):473–484

Schönrogge K, Wardlaw JC, Peters AJ, Everett S, Thomas JA, Elmes GW (2004) Changes in chemical signature and host specificity from larval retrieval to full social integration in the myrmecophilous butterfly *Maculinea rebeli*. J Chem Ecol 30(1):91–107

Schönrogge K, Barbero F, Casacci LP, Settele J, Thomas JA (2017) Acoustic communication within ant societies and its mimicry by mutualistic and socially parasitic myrmecophiles. Anim Behav 134:249–256

Schurian KG, Eckweiler W (2002) Beitrag zur Biologie und Ökologie von *Polyommatus* (*Aricia*) *teberdina nahizericus* (Eckweiler, 1978) aus der Nordostürkei (Lepidoptera: Lycaenidae). Nachr Entomol Ver Apollo 22(4):211–214. [publication in German]

Schurian KG, Fiedler K (1994) Zur Biologie von *Polyommatus* (*Lysandra*) *dezinus* (De Freina & Witt) (Lepidoptera: Lycaenidae). Nachr Entomol Ver Apollo 14(4):339–353. [publication in German]

Schurian KG, Reif A (1992) Beitrag zur Biologie von *Polyommatus* (*Aricia*) *isaurica* (Staudinger) (Lepidoptera: Lycaenidae). Nachr Entomol Ver Apollo 12(4):255–261. [publication in German]

Schurian KG, ten Hagen W, Eckweiler W (2005) Beitrag zur Biologie und Ökologie von *Polyommatus* (*Agrodiaetus*) *peilei* Bethune-Baker, 1921 (Lepidoptera, Lycaenidae). Nachr Entomol Ver Apollo 26(4):197–206. [publication in German]

Seki Y, Takanami Y, Otsuka K (1991) Butterflies of Borneo (vol 2, no. 1). Tobishima, Tokyo

Sendoya SF, Oliveira PS (2015) Ant–caterpillar antagonism at the community level: interhabitat variation of tritrophic interactions in a neotropical savanna. J Anim Ecol 84(2):442–452

Sendoya SF, Oliveira PS (2017) Behavioural ecology of defence in a risky environment: caterpillars versus ants in a Neotropical savanna. Ecol Entomol 42(5):553–564

Sendoya SF, Freitas AV, Oliveira PS (2009) Egg-laying butterflies distinguish predaceous ants by sight. Am Nat 174(1):134–140

Seraphim N, Kaminski LA, Devries PJ, Penz C, Callaghan C, Wahlberg N, Silva-Brandao KL, Freitas AV (2018) Molecular phylogeny and higher systematics of the metalmark butterflies (Lepidoptera: Riodinidae). Syst Entomol 43(2):407–425

Seufert P, Fiedler K (1996) The influence of ants on patterns of colonization and establishment within a set of coexisting lycaenid butterflies in a south-east Asian tropical rain forest. Oecologia 106(1):127–136

Shapiro A (2007) Field guide to butterflies of the San Francisco Bay and Sacramento Valley regions. University of California Press, Berkeley

Shibao H, Morimoto M, Okumura Y, Shimada M (2009) Fitness costs and benefits of ant attendance and soldier production for the social aphid *Pseudoregma bambucicola* (Homoptera: Aphididae: Hormaphidinae). Sociobiology 54(3):673

Shimizu-kaya U, Okubo T, Yago M, Inui Y, Itioka T (2013) Myrmecoxeny in *Arhopala zylda* (Lepidoptera, Lycaenidae) larvae feeding on *Macaranga* myrmecophytes. Entomol News 123(1):63–70

Shimizu-Kaya U, Kishimoto-Yamada K, Itioka T (2015) Biological notes on herbivorous insects feeding on myrmecophytic *Macaranga* trees in the Lambir Hills National Park, Borneo. Contrib Biol Lab Kyoto Univ 30(2):85–125

Sielezniew M, Wlostowski M, Dziekanska I (2010) *Myrmica schencki* (Hymenoptera: Formicidae) as the primary host of *Phengaris (Maculinea) arion* (Lepidoptera: Lycaenidae) at heathlands in eastern Poland. Sociobiology 55(1):95–106

Silva NA, Duarte M, Araújo EB, Morais HC (2014) Larval biology of anthophagous Eumaeini (Lepidoptera: Lycaenidae, Theclinae) in the Cerrado of Central Brazil. J Insect Sci 14(1):184

Silveira HC, Oliveira PS, Trigo JR (2010) Attracting predators without falling prey: chemical camouflage protects honeydew-producing treehoppers from ant predation. Am Nat 175(2):261–268

Singh AP (2003) New records on the distribution and ecology of common gem butterfly, *Poritia hewitsoni hewitsoni* Moore from the lower Western Himalayas: a lesser known taxa. J Lepidop Soc 57(4):295–298

Smedley SR, Schroeder FC, Weibel DB, Meinwald J, Lafleur KA, Renwick JA, Rutowski R, Eisner T (2002) Mayolenes: labile defensive lipids from the glandular hairs of a caterpillar (*Pieris rapae*). Proc Natl Acad Sci 99(10):6822–6827

Solazzo G, Moritz RF, Settele J (2013) Choice behaviour of *Myrmica rubra* workers between ant larvae and larvae of their *Phengaris (Maculinea) nausithous* nest parasites. Insect Soc 60(1):57–64

Stadler B, Dixon AF (2005) Ecology and evolution of aphid-ant interactions. Annu Rev Ecol Evol Syst 36:345–372

Steidinger BS, Crowther TW, Liang J, Van Nuland ME, Werner GD, Reich PB, Nabuurs GJ, de Miguel S, Zhou M, Picard N, Hérault B, Zhao X, Zhang C, Routh D, GFBI consortium, Peay KG (2019) Climatic controls of decomposition drive the global biogeography of forest-tree symbioses. Nature 569(7756):404–408

Stoeffler M, Maier TS, Tolasch T, Steidle JL (2007) Foreign-language skills in rove-beetles? Evidence for chemical mimicry of ant alarm pheromones in myrmecophilous *Pella* beetles (Coleoptera: Staphylinidae). J Chem Ecol 33(7):1382–1392

Stradomsky BV, Fomina EA (2009) The developmental stages of some blue butterflies (Lepidoptera: Lycaenidae) of Russian south. Caucas Entomol Bull 5(2):269–272. [publication in Russian]

Styrsky JD, Eubanks MD (2007) Ecological consequences of interactions between ants and honeydew-producing insects. Proc R Soc B Biol Sci 274(1607):151–164

Susilo I, Susilo FX (2015) Preliminary study on *Eublemma* sp. (Eublemminae): a lepidopteran predator of *Coccus viridis* (Hemiptera: Coccidae) on coffee plants in Bandarlampung, Indonesia. Jurnal Hama dan Penyakit Tumbuhan Tropika 15(1):10–16

Talavera G, Kaminski LA, Freitas AV, Vila R (2016) One-note samba: the biogeographical history of the relict Brazilian butterfly *Elkalyce cogina*. J Biogeogr 43(4):727–737

Talavera G, Lukhtanov V, Pierce NE, Vila R (2021) DNA barcodes combined with multi-locus data of representative taxa can generate reliable higher-level phylogenies. Syst Biol. [in press]

Tartally A, Thomas JA, Anton C, Balletto E, Barbero F, Bonelli S, Bräu M, Casacci LP, Csősz S, Czekes Z, Dolek M et al (2019) Patterns of host use by brood parasitic *Maculinea* butterflies across Europe. Philos Trans R Soc B 374(1769):20180202

Tauber CA, Winterton SL (2014) Third instar of the myrmecophilous *Italochrysa insignis* (Walker) from Australia (Neuroptera: Chrysopidae: Belonopterygini). Zootaxa 3811(1):95–106

Tauber CA, Kilpatrick SK, Oswald JD (2020) Larvae of *Abachrysa eureka* (Banks) (Neuroptera: Chrysopidae: Belonopterygini): descriptions and a discussion of the evolution of myrmecophily in Chrysopidae. Zootaxa 4789(2):481–507

Tautz J, Fiedler K (1992) Mechanoreceptive properties of caterpillar hairs involved in mediation of butterfly-ant symbioses. Naturwissenschaften 79(12):561–563

Tennent J (1996) The butterflies of Morocco, Algeria and Tunisia. Gem Publishing Company, Wallingford

Thomas JA, Elmes GW (1993) Specialized searching and the hostile use of allomones by a parasitoid whose host, the butterfly *Maculinea rebeli*, inhabits ant nests. Anim Behav 45(3):593–602

Thomas JA, Elmes GW (1998) Higher productivity at the cost of increased host-specificity when *Maculinea* butterfly larvae exploit ant colonies through trophallaxis rather than by predation. Ecol Entomol 23(4):457–464

Thomas JA, Elmes GW (2001) Food–plant niche selection rather than the presence of ant nests explains oviposition patterns in the myrmecophilous butterfly genus *Maculinea*. Proc R Soc Lond Ser B Biol Sci 268(1466):471–477

Thomas JA, Elmes GW, Wardlaw JC (1998) Polymorphic growth in larvae of the butterfly *Maculinea rebeli*, a social parasite of *Myrmica* ant colonies. Proc R Soc Lond Ser B Biol Sci 265(1408):1895–1901

Thomas JA, Knapp JJ, Akino T, Gerty S, Wakamura S, Simcox DJ, Wardlaw JC, Elmes GW (2002) Parasitoid secretions provoke ant warfare. Nature 417(6888):505–506

Thomas JA, Simcox DJ, Clarke RT (2009) Successful conservation of a threatened *Maculinea* butterfly. Science 325(5936):80–83

Thomas JA, Elmes GW, Sielezniew M, Stankiewicz-Fiedurek A, Simcox DJ, Settele J, Schönrogge K (2013) Mimetic host shifts in an endangered social parasite of ants. Proc R Soc B Biol Sci 280(1751):20122336

Thomas CC, Tillberg CV, Schultz CB (2020) Facultative mutualism increases survival of an endangered ant-tended butterfly. J Insect Conserv 24(2):385–395

Thompson JN (2005) The geographic mosaic of coevolution. University of Chicago Press, Chicago

Torres PJ, Pomerantz AF (2016) Butterfly kleptoparasitism and first account of immature stages, myrmecophily, and bamboo host plant of the metalmark *Adelotypa annulifera* (Riodinidae). J Lepidop Soc 70(2):130–138

Trager MD, Daniels JC (2009) Ant tending of Miami blue butterfly larvae (Lepidoptera: Lycaenidae): partner diversity and effects on larval performance. Fl Entomol 92(3):474–482

Trager MD, Thom MD, Daniels JC (2013) Ant-related oviposition and larval performance in a myrmecophilous lycaenid. Int J Ecol 2013:152139

Travassos MA, Pierce NE (2000) Acoustics, context and function of vibrational signalling in a lycaenid butterfly–ant mutualism. Anim Behav 60(1):13–26

Travassos MA, DeVries PJ, Pierce NE (2008) A novel organ and mechanism for larval sound production in butterfly caterpillars: *Eurybia elvina* (Lepidoptera: Riodinidae). Trop Lepidop Res 18(1):20–23

Tshikolovets VV (2011) Butterflies of Europe and the Mediterranean area. Tshikolovets Publications, Pardubice

Ueda S, Komatsu T, Itino T, Arai R, Sakamoto H (2016) Host-ant specificity of endangered large blue butterflies (*Phengaris* spp., Lepidoptera: Lycaenidae) in Japan. Sci Rep 6(1):1–5

Uemura M, Perkins LE, Zalucki MP, Cribb BW (2017) Predator-prey interaction between greenhead ants and processionary caterpillars is mediated by chemical defence. Anim Behav 129:213–222

Valencia-Montoya WA, Quental TB, Tonini JFR, Talavera G, Crall JD, Lamas G, Busby RC, Carvalho APS, Morais AB, Mega NO, Romanowski HP, Liénard MA, Salzman S, Whitaker MRL, Kawahara AY, Lohman DJ, Robbins RK, Pierce NE (In review) Evolutionary tradeoffs between male secondary sexual traits revealed by a phylogeny of the hyperdiverse tribe Eumaeini (Lepidoptera: Lycaenidae). Proc R Soc B

van der Poorten GM, van der Poorten NE (2016) The butterfly fauna of Sri Lanka. Lepodon Books

van Dyck H, Regniers S (2010) Egg spreading in the ant-parasitic butterfly, *Maculinea alcon*: from individual behaviour to egg distribution pattern. Anim Behav 80(4):621–627

van Gasse P (2013) Butterflies of India – annotated checklist. Butterflies of India, Kruibeke

Van Mele P, Vayssieres JF, Adandonon A, Sinzogan A (2009) Ant cues affect the oviposition behaviour of fruit flies (Diptera: Tephritidae) in Africa. Physiol Entomol 34(3):256–261

Vantaux A, Roux O, Magro A, Ghomsi NT, Gordon RD, Dejean A, Orivel J (2010) Host-specific myrmecophily and myrmecophagy in the tropical coccinellid *Diomus thoracicus* in French Guiana. Biotropica 42(5):622–629

Vantaux A, van den Ende W, Billen J, Wenseleers T (2011) Large interclone differences in melezitose secretion in the facultatively ant-tended black bean aphid *Aphis fabae*. J Insect Physiol 57(12):1614–1621

Vantaux A, Roux O, Magro A, Orivel J (2012) Evolutionary perspectives on myrmecophily in ladybirds. Psyche 2012:591570

Vargas HA, Duarte M (2016) First host plant record for *Strymon davara* (Hewitson) (Lepidoptera, Lycaenidae) in the highly human-modified coastal valleys of the Atacama desert. Rev Bras de Entomol 60(4):352–355

Vasconcelos HL (1991) Mutualism between *Maieta guianensis* Aubl., a myrmecophytic melastome, and one of its ant inhabitants: ant protection against insect herbivores. Oecologia 87(2):295–298

Vegliante F, Hasenfuss I (2012) Morphology and diversity of exocrine glands in lepidopteran larvae. Annu Rev Entomol 57:187–204

Vélez-Arango AM, Vélez PD, Wolff M (2010) The life cycle of *Mesosemia mevania* (Hewitson 1857) (Riodinidae) in a lower montane humid forest in Colombia. J Lepidop Soc 64(1):23–28

Vila R, Bell CD, Macniven R, Goldman-Huertas B, Ree RH, Marshall CR, Balint Z, Johnson K, Benyamini D, Pierce NE (2011) Phylogeny and palaeoecology of *Polyommatus* blue butterflies show Beringia was a climate-regulated gateway to the New World. Proc R Soc B Biol Sci 278(1719):2737–2744

von Beeren C, Brückner A, Maruyama M, Burke G, Wieschollek J, Kronauer DJ (2018) Chemical and behavioral integration of army ant-associated rove beetles–a comparison between specialists and generalists. Front Zool 15(1):1–6

Wada A, Isobe Y, Yamaguchi S, Yamaoka R, Ozaki M (2001) Taste-enhancing effects of glycine on the sweetness of glucose: a gustatory aspect of symbiosis between the ant, *Camponotus japonicus*, and the larvae of the lycaenid butterfly, *Niphanda fusca*. Chem Senses 26(8):983–992

Wagner D (1993) Species-specific effects of tending ants on the development of lycaenid butterfly larvae. Oecologia 96(2):276–281

Wagner D (1995) Pupation site choice of a North American lycaenid butterfly: the benefits of entering ant nests. Ecol Entomol 20:384–392

Wagner D, Kurina L (1997) The influence of ants and water availability on oviposition behaviour and survivorship of a facultatively ant-tended herbivore. Ecol Entomol 22(3):352–360

Warren RJ, Elliott KJ, Giladi I, King JR, Bradford MA (2019) Field experiments show contradictory short-and long-term myrmecochorous plant impacts on seed-dispersing ants. Ecol Entomol 44(1):30–39

Weber MG, Keeler KH (2013) The phylogenetic distribution of extrafloral nectaries in plants. Ann Bot 111(6):1251–1261

Webster RP, Nielsen MC (1984) Myrmecophily in the Edward's hairstreak butterfly *Satyrium edwardsii* (Lycaenidae). J Lepidop Soc 38(2):124–133

Weeks JA (2003) Parasitism and ant protection alter the survival of the lycaenid *Hemiargus isola*. Ecol Entomol 28(2):228–232

Werner A, Kinne RK (2001) Evolution of the Na-pi cotransport systems. Am J Physiol Regul Int Comparat Physiol 280:R301–R312

Westoby M, Rice B, Shelley JM, Haig D, Kohen JL (1982) Plants' use of ant dispersal at west head, New South Wales. Ant-plant interactions. Junk Publishers, The Hague, pp 7–9

Whitaker MR, Baker CC, Salzman SM, Martins DJ, Pierce NE (2019) Combining stable isotope analysis with DNA metabarcoding improves inferences of trophic ecology. PLoS One 14(7):e0219070

Wiens JA, Chr N, Van Horne B, Ims RA (1993) Ecological mechanisms and landscape ecology. Oikos:369–380

Williams MC (1995-2020) Afrotropical butterflies. Lepid Soc Africa. https://www.lepsocafrica. org/?p=publications&s=atb

Williams MC (2006) What do the larvae of *Alaena amazoula* (Boisduval, 1847) (Lepidoptera: Lycaenidae: Poritiinae) feed on? Metamorphosis 17(4):140–150

Williams MC, Joannou JG (1996) Observations on the oviposition behaviour and caterpillars of *Thestor basutus capeneri* Dickson (Lepidoptera: Lycaenidae) in South Africa. Metamorphosis 7:12–16

Witek M, Sliwinska EB, Skorka P, Nowicki P, Settele J, Woyciechowski M (2006) Polymorphic growth in larvae of *Maculinea* butterflies, as an example of biennialism in myrmecophilous insects. Oecologia 148(4):729–733

Witek M, Śliwińska E, Skorka P, Nowicki P, Wantuch M, Vrabec V, Settele J, Woyciechowski M (2008) Host ant specificity of large blue butterflies *Phengaris* (*Maculinea*) (Lepidoptera: Lycaenidae) inhabiting humid grasslands in east-Central Europe. Eur J Entomol 105(5):871–877

Witek M, Skórka P, Śliwińska EB, Nowicki P, Moroń D, Settele J, Woyciechowski M (2011) Development of parasitic *Maculinea teleius* (Lepidoptera, Lycaenidae) larvae in laboratory nests of four *Myrmica* ant host species. Insect Soc 58(3):403–411

Wynhoff I, Grutters M, Van Langevelde F (2008) Looking for the ants: selection of oviposition sites by two myrmecophilous butterfly species. Anim Biol 58(4):371–388

Wynhoff I, Bakker RB, Oteman B, Arnaldo PS, van Langevelde F (2015) *Phengaris* (*Maculinea*) *alcon* butterflies deposit their eggs on tall plants with many large buds in the vicinity of *Myrmica* ants. Insect Conserv Divers 8(2):177–188

Yack JE, Smith ML, Weatherhead PJ (2001) Caterpillar talk: acoustically mediated territoriality in larval Lepidoptera. Proc Natl Acad Sci 98(20):11371–11375

Yadav C, Guedes RN, Matheson SM, Timbers TA, Yack JE (2017) Invitation by vibration: recruitment to feeding shelters in social caterpillars. Behav Ecol Sociobiol 71(3):51. https://doi.org/10.1007/s00265-017-2280-x

Yago M, Miyagawa T, Yokoyama J, Williams M (2010) Life history of the Guatemalan copper, *Iophanus pyrrhias* (Godman & Salvin) (Lepidoptera, Lycaenidae). Trans Lepidop Soc Jpn 60(4):259–267. https://doi.org/10.18984/lepid.60.4_259

Yamagushi S, Shirozu T (1988) The life histories of five Myrmecophilous Lycaenid butterflies of Japan. Kodansha, Tokyo

Yao I (2014) Costs and constraints in aphid-ant mutualism. Ecol Res 229(3):383–391. https://doi.org/10.1007/s11284-014-1151-4

Young AM (1978) Possible evolution of mutualism between *Mechanitis* caterpillars and an ant in northeastern Costa Rica. Biotropica 10:77–78. https://doi.org/10.2307/2388114

Youngsteadt E, Devries PJ (2005) The effects of ants on the entomophagous butterfly caterpillar *Feniseca tarquinius*, and the putative role of chemical camouflage in the *Feniseca*–ant interaction. J Chem Ecol 31(9):2091–2109. https://doi.org/10.1007/s10886-005-6079-2

Zanuncio JC, Torres JB, Sediyama CA, Pereira FF, Pastori PL, Wermelinger ED, Ramalho FS (2009) Mortality of the defoliator *Euselasia eucerus* (Lepidoptera: Riodinidae) by biotic factors in an *Eucalyptus urophylla* plantation in Minas Gerais State, Brazil. Ann Acad Bras Ciê 81(1):61–66. https://doi.org/10.1590/S0001-37652009000100008

Zanuncio JC, Soares MA, Zanuncio TV, Mielke OH, Ramalho FD, Júnior SL, Wilcken CF (2013) *Euselasia hygenius occulta* (Riodininae): first report of feeding on *Psidium guajava* (Myrtaceae) in Minas Gerais State, Brazil. J Lepidop Soc 67(3):221–224. https://doi.org/10.18473/lepi.v67i3.a8

Zemeitat DS (2017) Evolution of cooperative behaviour in Australian lycaenid butterflies and ants. PhD Thesis, University of Melbourne

Zhou L, Zhuang H (eds) (2018) *Zephyrus* hairstreaks from China I. Hong Kong Lepidopterists' Society Limited, Hong Kong

Open Access This chapter is licensed under the terms of the Creative Commons Attribution 4.0 International License (http://creativecommons.org/licenses/by/4.0/), which permits use, sharing, adaptation, distribution and reproduction in any medium or format, as long as you give appropriate credit to the original author(s) and the source, provide a link to the Creative Commons license and indicate if changes were made.

The images or other third party material in this chapter are included in the chapter's Creative Commons license, unless indicated otherwise in a credit line to the material. If material is not included in the chapter's Creative Commons license and your intended use is not permitted by statutory regulation or exceeds the permitted use, you will need to obtain permission directly from the copyright holder.

Part IV
Multiple Interactive Effects Among All Three Trophic Levels

Caterpillars, Plant Chemistry, and Parasitoids in Natural vs. Agroecosystems

Paul J. Ode

Pieris rapae (Pieridae) caterpillars (cabbage white butterfly) on *Brassica oleracea* (collard). Photo by Dhaval Vyas

Introduction

Caterpillars (larval Lepidoptera) are among the most widespread and important herbivores in both natural and agricultural ecosystems (Turcotte et al. 2014a). In natural ecosystems, insect herbivory has selected for a diverse array of chemical and physical defenses in plants (Speed et al. 2015); in turn, plant defense traits select for a suite of physiological and behavioral traits in insect herbivores to excrete, metabolize, avoid, or tolerate plant defenses (Schoonhoven et al. 2005). Such plant-herbivore interactions (particularly those based on plant defense chemistry) are

P. J. Ode (✉)
Department of Agricultural Biology, Graduate Degree Program in Ecology, Colorado State University, Fort Collins, CO, USA
e-mail: paul.ode@colostate.edu

widely thought to have driven high rates of speciation within both groups (Mitter et al. 1988; Futuyma and Agrawal 2009; Richards et al. 2015). Indeed, some of the classic studies of plant-insect herbivore coevolution have focused on caterpillars (e.g., Ehrlich and Raven 1964; Berenbaum 1983).

In addition to the bottom-up effects of plant defense chemistry on caterpillars, most species of caterpillars are food themselves for at least one, and often multiple, species of parasitoids, predators, and pathogens. Caterpillars have evolved elaborate behavioral and physiological mechanisms to avoid consumption by their natural enemies (e.g., crypsis, production of repellent regurgitants, sequestration of plant toxins, encapsulation of parasitoid eggs) (Zvereva and Kozlov 2016; Kaplan et al. 2016). Furthermore, recent studies have shown that parasitoids can modify plant chemistry-caterpillar interactions (Fatouros et al. 2005; Poelman et al. 2011, 2012; Ode et al. 2016; Tan et al. 2020). In these ways, tritrophic interactions have contributed to species diversification across all three trophic levels (Singer and Stireman III 2005). That top-down forces on caterpillars are important has been supported by a recent meta-analysis (Vidal and Murphy 2018). Clearly, the success and diversification of Lepidoptera reflect their ability to adapt to bottom-up as well as top-down selective pressures.

Caterpillars in agroecosystems are expected to experience trophic interactions that differ, both quantitatively and qualitatively, from their counterparts in natural ecosystems. The effects of domestication on plant traits and their consequences for trophic interactions involving herbivores and natural enemies have been reviewed elsewhere several times (e.g., Evans 1993; Meyer et al. 2012; Meyer and Purugganan 2013; Chen et al. 2015a, b, 2018; Chen 2016; Whitehead et al. 2017) including a review specifically focused on caterpillars (Barbosa 1993). Plants in agroecosystems differ from those in natural ecosystems in three ways. First, most domesticated crops have been introduced widely throughout the world, often away from their region of origin (Meyer et al. 2012; Meyer and Purugganan 2013). Such introductions result in ecological and evolutionary mismatches between plants, insect herbivores, and their natural enemies compared to related plants in natural ecosystems (Chen 2016). Mismatches are most extreme in cases where domesticated plants have been introduced to geographic regions without their coevolved herbivores or their natural enemies. Indeed, the underlying premise of importation (classical) biological control programs is to reestablish historic trophic assemblages such that top-down control of herbivore pests is renewed (Risch et al. 1983; Landis et al. 2000; Cohen and Crowder 2017; Heimpel and Mills 2017). At the other extreme, plants domesticated in their region of origin often retain many of the same herbivores and natural enemy complexes as found in their wild progenitors (Chen et al. 2015a). Second, plant community structure is generally far less complex in agroecosystems than in natural ecosystems. Planting practices typically result in monocultures or highly simplified plant communities with uniform phenologies compared to the more species diverse natural ecosystems from which a given crop originated (Root 1973; Landis et al. 2000). Such simplifications undoubtedly affect the predator/parasitoid communities attacking caterpillars (Gurr et al. 2017; González et al. 2020) with often highly system-specific effects on the regulation of herbivore populations by their natural enemies (Chaplin-Kramer et al. 2011; Bianchi et al. 2006; Tscharntke et al. 2016; Karp et al. 2018). Third, plant breeding efforts to improve crop products during and post domestication have also resulted in changes in plant traits that have been generally shown to decrease resistance to herbivory, benefitting insect herbivores (Whitehead et al. 2017). One important experimental study of 29 crop species and their wild relatives found that resistance to the generalist caterpillar, beet armyworm *Spodoptera exigua*, was generally reduced on domesticated crops (Turcotte et al. 2014b). Domestication has typically increased the size of plant organs (Evans 1993) and increased palatability, which is often the result of decreased production of plant defensive chemistry (Meyer et al. 2012).

Domesticated crops grown in their region of origin are generally presumed to retain relatively rich herbivore and predator/parasitoid fauna compared to where the crop has been introduced, yet this notion has been infrequently explored with any rigor (Chen et al. 2015a), e.g., maize *Zea mays* (Rosenthal and Dirzo 1997; Moya-Raygoza et al. 2019), lima bean *Phaseolus lunatus* (Bustos-Segura et al. 2020), and rice *Oryza sativa* (Chen et al. 2013). A classic example of this is sunflower *Helianthus annuus*, which is native to North America. In many areas of North America, cultivated sunflower is grown sympatrically with wild populations. Consequently, the herbivore and parasitoid fauna associated with both wild and cultivated sunflower are very similar (Rogers 1988; Ode et al. 2011). One of the most important herbivores of wild and cultivated sunflower is the sunflower moth *Homoeosoma electellum* (Lepidoptera: Pyralidae) (Chen and Welter 2002). Larvae feed inside the developing seeds where they are sometimes attacked by many native parasitoids including the braconid wasp *Dolichogenidea homoeosomae*. Female moths showed a strong preference for ovipositing on cultivated sunflower, which have larger flower heads and larger seeds (Chen and Welter 2003). Interestingly, *D. homoeosomae* females spent more time foraging on wild sunflowers resulting in parasitism rates that were 19 times higher on wild sunflower compared to cultivated sunflower. In part, these differences were explained by the observation that moth larvae were more likely to be found feeding on the fewer and smaller florets of wild sunflower compared to cultivated sunflower where a higher proportion of caterpillars feed within the seeds. This made it easier for the parasitoid female to locate and parasitize the moth larvae on wild sunflowers compared to cultivated sunflower (Chen and Welter 2007). Seeds of wild sunflowers develop a phytomelanin-containing seed coat before most moth larvae are large enough to penetrate the seed coat (Rogers and Kreitner 1983). Furthermore, wild sunflowers have higher concentrations of diterpenoids and sesquiterpene lactones that are both feeding deterrents and toxins to many sunflower herbivores (Gershenzon et al. 1985; Rogers et al. 1987). The combination of quick-forming, phytomelanin-containing seed coats and the production of higher levels of terpenes presumably force *H. electellum* larvae to remain feeding on pollen and florets where they are more exposed to parasitism (Chen and Welter 2007).

While there is strong evidence that plant domestication has resulted in plant traits that are generally beneficial to insect herbivores (e.g., larger plant organs/tissues, reduced chemical defenses, higher nutritional content), the effects of plant domestication on parasitoids and predators are far less studied (Chen et al. 2015b). In this chapter, I largely restrict my focus to consideration of the role of domestication in altering plant secondary metabolites that are involved in direct and indirect defense against caterpillar herbivores (see examples in Plate 1) and their interactions with natural enemies. Surprisingly few studies have directly compared trophic interactions involving plant secondary metabolites in both agroecosystems and natural ecosystems, and I highlight those that do. Elsewhere, I draw reasonable inferences from studies that have either focused on natural systems or agroecosystems.

Plate 1 Ten "bad cats": (**a**) Native to southeast Asia, the beet armyworm *Spodoptera exigua* (Noctuidae) has a cosmopolitan distribution. This species is broadly polyphagous, having been recorded from 34 plant families ("Spodoptera exigua" by Macreando is licensed under CC BY-NC-ND 2.0); (**b**) adult *S. exigua*. ("[2385] Small Mottled Willow (Spodoptera exigua)" by

Bennyboymothman is licensed with CC BY 2.0). (**c**) Fall armyworm *Spodoptera frugiperda* (Frank Peairs, Colorado State University, Bugwood.org). While the fall armyworm has a broad reported host range, it primarily feeds on grasses, especially maize. Native to the Americas it has recently spread in Africa, South Asia, Australia, and Europe where it causes significant damage to maize; (**d**) Adult fall armyworm ("File:- 9666 – Spodoptera frugiperda – Fall Armyworm Moth (male) (22653159377).jpg" by Andy Reago & Chrissy McClarren is licensed under CC BY 2.0). (**e**) A highly polyphagous herbivore of deciduous and coniferous trees, the gypsy moth *Lymantria dispar* is one of the most damaging caterpillar pests of natural and commercial forests. This species has been the focus of many biocontrol introductions, including the generalist tachinid *Compsilura concinnata* that have had disastrous non-target effects on native Lepidoptera. ("Gypsy Moth - Lymantria dispar" by Björn S... is licensed under CC BY-SA 2.0); (**f**) Adult gypsy moth ("Lymantridae - Lymantria dispar-1" by Ettore Balocchi is licensed under CC BY 2.0). (**g**) The diamondback moth *Plutella xylostella* has a cosmopolitan distribution and feeds almost exclusively on plants within the Brassicaceae. This species is notable for its ability to detoxify glucosinolates as well as its resistance to a broad range of insecticides including Bt toxins. Several biocontrol agents provide some natural control including *Cotesia vestalis* and *Diadegma insulare* ("110203 新竹 關西 小菜蛾 Plutella xylostella") by Bettaman is licensed under CC BY-NC-SA 2.0); (**h**) Adult *P. xylostella* ("091213 苗栗 公館 小菜蛾 Plutella xylostella" by Bettaman is licensed under CC BY-NC-SA 2.0). (**i**) *Spodoptera littoralis* (African cotton leafworm) is native to the Middle East, the Mediterranean Europe, and Africa. This species has been intercepted at US border controls numerous times. It is highly polyphagous, causing serious economic damage to cotton, tomato, maize, cabbage, and potato, among others ("Spodoptera littoralis" by Macreando is licensed under CC BY-NC-ND 2.0); (**j**) Adult *S. littoralis* ("Mediterranean Brocade .Spodoptera littoralis. Noctuidae" by gailhampshire is licensed under CC BY 2.0). (**k**) The Asiatic rice borer *Chilo suppressalis* (Crambidae) is found from South to East Asia and has been introduced to Spain and Hawaii and is a serious pest of rice. Its relative, the spotted stalk borer *Chilo partellus*, attacks a wide range of grass species including maize, sugarcane, sorghum, and rice. The spotted stalk borer is likely native to South Asia and has spread throughout Africa in the twentieth century. ("Stem borer" by IRRI Images is licensed under CC BY-NC-SA 2.0); (**l**) Adult spotted stalk borer ("File:Chilo suppressalis (41866967681).jpg" by LiCheng Shih is licensed under CC BY 2.0). (**m**) The cotton bollworm *Helicoverpa armigera* (Noctuidae) is native to Asia and Africa and is a polyphagous pest of many crop plants including cotton, tomato, chickpea, rice, and maize among many others ("Helicoverpa armigera" by fturmog is licensed under CC BY-NC-SA 2.0). The related *Helicoverpa zea* (corn earworm, aka cotton bollworm, tomato fruitworm) is also polyphagous but is native to the Americas; (**n**) Adult *H. armigera* ("Helicoverpa armigera" by xulescu_g is licensed under CC BY-SA 2.0). (**o**) The codling moth *Cydia pomonella* (Tortricidae) has a cosmopolitan distribution and is one of the most destructive pests of a wide variety of fruit crops; (**p**) Adult *C. pomonella* ("[1261] Cydia pomonella" by Bennyboymothman is licensed under CC BY 2.0). (**q**) The large white *Pieris brassicae* (Pieridae) and the related small white *P. rapae* are both pests of a wide variety of plants in the Brassicaceae. While both are native to Europe (and the large white also found in north Africa and to the Himalaya), they have different invasion ranges. The small white is widespread throughout North America, eastern Asia, and New Zealand/Australia. The large white has established in South Africa and temporarily in New Zealand ("'Cap a l'aplec amb mes germanes' (Pieris brassicae)" by AntoniGarciaLlorca is licensed under CC BY-NC-SA 2.0); (**r**) Adult *P. brassicae* ("Large White - Pieris brassicae" by Björn S... is licensed under CC BY-SA 2.0). (**s**) The cabbage moth *Mamestra brassicae* is another polyphagous noctuid that is distributed from Europe to Asia. While it feeds on several crops in the Brassicaceae, it also attacks a wide range of other crops including tobacco, tomato, and sunflower ("73.274 BF2154 Cabbage Moth, Mamestra brassicae" by Patrick Clement. is licensed under CC BY 2.0); (**t**) Adult *M. brassicae* ("[2154] Cabbage Moth (Mamestra brassicae)" by Bennyboymothman is licensed under CC BY 2.0)

Domestication and Plant Breeding Effects on Plant Secondary Metabolites

Plants produce a wide array of secondary metabolites that directly act as toxins or repellants to herbivores ("direct defenses") as well as volatiles that are attractive to natural enemies of herbivores ("indirect defenses") (Gols 2014). Direct defenses include protease inhibitors (PIs) that inhibit the functioning of herbivore proteases (Zhu-Salzman and Zeng 2015), anti-nutritive enzymes such as polyphenol oxidases (PPOs) (González-Santoyo and Córdoba-Aguilar 2012), and an astounding array of often plant-specific mixtures of toxins (e.g., alkaloids, furanocoumarins, iridoid glycosides, cardiac glycosides, glucosinolates to name a few classes) (Kessler and Baldwin 2002; De Geyter et al. 2012; Heckel 2014). Indirect defenses are generally thought of as plant volatiles (many of which are inducible by herbivory) that attract parasitoids and predators of insect herbivores. Other plant traits such as extrafloral nectaries, domatia, and other structures that can house natural enemies of herbivores, and traits that enhance natural enemy foraging success, are also considered indirect defenses (Pearse et al. 2020). Yet, the distinction between so-called direct and indirect defenses is somewhat artificial as the metabolites involved in both categories of defenses often influence the behavior and physiology of both herbivore and natural enemy (Gols 2014).

Plant Breeding, Plant Toxins, and Caterpillars

Plant domestication and breeding for traits beneficial for agriculture are widely presumed to result in reduced production of secondary metabolites associated with resistance against herbivory, a concept sometimes referred to as the "plant domestication-reduced defense" hypothesis (Evans 1993, Meyer et al. 2012, Whitehead et al. 2017). Indeed, reduction of plant chemical defenses has been posed as one of the main reasons why cultivated crops often experience higher damage from herbivores compared to their wild progenitors (Chen et al. 2015a, b; Fernandez et al. 2021). In many cases, reduced secondary plant chemistry is the direct result of breeding crops for increased palatability of plant organs to be consumed as many plant compounds in wild progenitors are bitter tasting or outright toxic to humans. In other cases, selection on other agriculturally favorable plant traits such as increased yield may indirectly select for reduced plant secondary chemistry if investment in yield trades off with investment in defense (Benrey et al. 1998). In still other cases, cultivated plants may be selected to produce increased levels of defensive chemicals in cases where taste is positively correlated with plant chemistry (e.g., condiment mustard), where there may be medical or nutritional benefits (e.g., anticancer properties of glucosinolates; Fahey et al. 2001), or there is intentional selection for increased resistance to herbivory (e.g., Degenhardt et al. 2009; Tamiru et al. 2011; Bleeker et al. 2012).

Selection on plant chemistry (whether intentional or unintentional) can influence direct (via changes to plant toxins) and indirect (e.g., via herbivore-induced plant volatiles, HIPVs; see following sections) plant defenses. Yet, whether domestication results in increased resistance against herbivores depends on the identity of the herbivore. Whereas more generalist herbivores enjoy faster growth rates and reduced mortality on domesticated plants compared to their wild progenitors, this is not always the case for co-evolved specialist herbivores. For example, while the generalist *Lymantria dispar* fed more and enjoyed reduced mortality on domesticated blueberries (*Vaccinium corymbosum*), the native specialist blueberry leafroller *Sparganothis sulfureana* performed equally well on domesticated and wild blueberries (Hernandez-Cumplido et al. 2018). In a study of eight specialist and generalist herbivores of maize and its wild ancestor teosinte, domestication most strongly benefitted the more generalist herbivores (Gaillard et al. 2018). In other cases, domestication may result in increased resistance against specialist herbivores. Domestication of murtilla (strawberry myrtle, *Ugni molinae*) has resulted in the reduction of four flavonols, which are important feeding stimulants for the native erebid moth, *Paracles* (= *Chelisia*) *rudis* (Chacón-Fuentes et al. 2015).

Crops in the genus *Brassica* (Brassicaceae) are excellent examples of how plants can either be intentionally or unintentionally bred for either increased or decreased chemical levels depending on the plant part used for consumption (Hopkins et al. 2009; Gols and Harvey 2009). In the Brassicaceae, glucosinolates are plant toxins that confer resistance to many insect herbivores as well as pathogens. Glucosinolates also give many brassicaceous crop plants (e.g., cabbage, Brussels sprouts, cauliflower, mustard) their characteristic bitter taste (van Doorn et al. 1998). In some cases, breeding programs have selected for increased glucosinolate concentrations. For instance, the discovery that some glucosinolates have anticancer properties has prompted efforts to increase the glucosinolate content in some crops such as broccoli (Farnham et al. 2004). In other cases, breeding programs have attempted to select for decreased glucosinolate content. The widespread use of oilseed meal as a source of protein in animal feeds prompted efforts to select for seeds low in glucosinolates (with limited success) (Hopkins et al. 2009). Interestingly, selection for decreased glucosinolate content in the seeds of *B. napus* did not translate into reduced glucosinolate content elsewhere in the plant. In other oilseed species (*B. juncea*, *Sinapis alba*), selection for decrease glucosinolate concentrations in the seeds also resulted in decreased glucosinolate concentrations in the leaves. Lines of *B. juncea* containing low levels of glucosinolates were more susceptible to lepidopteran herbivores, although this depended on whether the herbivore was a relative generalist or specialist (Bodnaryk 1997). Whereas the crucifer specialist diamondback moth *Plutella xylostella* fed equally on low and high glucosinolate lines of *B. juncea*, the generalist bertha armyworm *Mamestra configurata* fed five times more on low glucosinolate lines compared to high glucosinolate lines (Bodnaryk 1997). The levels of constitutive and inducible glucosinolates were higher in wild *B. oleracea* populations than in cultivated populations (Gols et al. 2008a, b; Ode et al. 2016). While development time increased and adult body size decreased of both the specialists (*P. xylostella* and small cabbage white *Pieris rapae*) and the generalist (*M. brassicae*) when feeding on wild plants, the survivorship of *M. brassicae* was also significantly reduced (Gols et al. 2008a, b), indicating that domestication/cultivation may benefit generalist herbivores more than specialist herbivores.

Plant Volatile-Mediated Interactions with Caterpillars and Parasitoids

Ovipositing females of most herbivore species are attracted to volatile chemicals released by their host plants. Adults of specialist herbivores are typically far more sensitive to specific blends of herbivore-induced plant volatiles (HIPVs) than are generalists, and slight changes to these blends may make plants unapparent to ovipositing specialists. Therefore, while generalist herbivores may be expected to respond more strongly to changes in plant defense toxins (Rothwell and Holeski 2020), specialist herbivores may be expected to respond more to changes in specific HIPVs that result from domestication and plant breeding. In turn, caterpillars often alter plant volatile phenotypes by inducing different volatile blends in herbivore-specific ways (e.g., Poelman et al. 2011).

Parasitoids (including hyperparasitoids) and predators are well known to respond to HIPVs to locate plants attacked by their host caterpillars (Vet and Dicke 1992; De Moraes et al. 1998; Mumm and Dicke 2010; Kessler and Heil 2011; Poelman et al. 2012; Stam et al. 2014; Rowen and Kaplan 2016). Parasitoids can alter the HIPV profiles of plants by altering the quality of the elicitors in the oral secretions and regurgitants of their host caterpillars (e.g., Poelman et al. 2011; Tan et al. 2020). Indeed, the HIPV blend induced by caterpillars can depend on the species of parasitoid developing within the caterpillar (Poelman et al. 2011). In a study of three parasitoid species that attack *Pieris rapae* and *P. brassicae*, application of regurgitant from caterpillars parasitized by one of the three parasitoid species differentially induced genes in the jasmonic acid pathway responsible for HIPV production, protease inhibitors, and toxin production (Poelman et al. 2011). Changes in the HIPV profiles from plants fed upon by caterpillars attacked by different species of parasitoids not only have been shown to influence oviposition behavior by subsequent herbivores (Poelman et al. 2011) but attack by parasitoids in the fourth trophic level as well (Poelman et al. 2012). In a further twist, obligate symbiotic polydnaviruses (which are found within the Ichneumonoidea and injected along with the parasitoid egg into their hosts) have been shown to alter the quality of elicitors in the caterpillars' regurgitant (Zhu et al. 2018). Studies such as these highlight the sometimes highly species-specific nature of the multitrophic interactions based on plant volatiles. Therefore, changes to plant volatile production resulting from crop domestication may alter trophic interactions in unexpected ways.

While domestication tends to result in reduced production of many toxic secondary metabolites, this is not necessarily the case for plant volatile production. A meta-analysis of 236 experiments found that domesticated species in general produced higher overall levels green leaf volatiles and sesquiterpenes (albeit in less complex blends) than did their wild relatives, whereas other HIPVs (e.g., benzenoid, homo- and monoterpenes) were unaffected by breeding (Rowen and Kaplan 2016), although exceptions exist (e.g., Loughrin et al. 1995). Nevertheless, many domesticated crops are generally less attractive to natural enemies of caterpillars and other herbivores (Chen et al. 2015a) suggesting that specific volatiles, and not total HIPV production, make damaged plants attractive to specific natural enemies of caterpillars (Clavijo McCormick et al. 2012; Li et al. 2018). This is especially the case for specialist parasitoids that respond to minor compounds rather than the most abundant HIPVs (e.g., D'Alessandro et al. 2009). The production of less diverse volatile profiles may be a consequence of domestication whereby cultivated crops have lost the ability to synthesize certain volatiles present in their wild relatives. As a case in point, wild cabbage populations produce isothiocyanates (volatile breakdown products of glucosinolates), whereas cultivated varieties do not when fed upon by the small cabbage white *Pieris rapae* (Gols et al. 2011). These volatiles are highly attractive to the specialist parasitoid *Cotesia rubecula*, which are more likely to parasitize *P. rapae* feeding on wild cabbage plants compared to caterpillars feeding on cultivated cabbage plants (Gols et al. 2011). However, whether crop domestication results in decreased attractiveness to generalist vs. specialist natural enemies to their host caterpillars depends on which specific volatiles are reduced; an overall reduction in volatile production would likely decrease attractiveness to both generalist and specialist natural enemies (Steidle and van Loon 2003). Such observations may help to explain the increased leaf damage seen in agroecosystems compared to natural ecosystems (Chen et al. 2015a).

Several studies have compared HIPV production in commercial hybrid cultivars of maize (*Zea mays*) with those from traditional landraces or wild teosinte species and their consequences for trophic interactions involving caterpillars and their parasitoids. Commercial cultivars produce lower levels of HIPVs relative to landrace varieties, which can significantly compromise attractiveness to parasitoids. Oviposition by the spotted stalk borer *Chilo partellus*, a serious pest of maize in southern and eastern Africa and South Asia, results in HIPV emission in three landrace varieties (from Brazil, Cuba, and Haiti), but not in two commercial maize varieties (Tamiru et al. 2011). Consequently, both *Trichogramma bournieri* (an egg parasitoid) and *Cotesia sesamiae* (a larval parasitoid) were attracted to *C. partellus* on the landrace varieties, but not on the commercial hybrids, suggesting that the ability to produce HIPVs in response to oviposition was lost during plant breeding (Tamiru et al. 2011). Several studies have shown significant variation in volatile profiles across different maize cultivars, suggesting that HIPVs can change as the result of plant breeding. In a study of the genetic variability of HIPVs produced by 31 inbred lines of maize, significant variation was found in 23 volatile compounds (Degen et al. 2004) including several that are induced by the elicitor "volicitin" found in the regurgitant of the beet armyworm *Spodoptera exigua* (Alborn et al. 1997; Turlings et al. 2000). These volatile compounds are known to be attractive to the parasitoid *Cotesia marginiventris* (Turlings et al. 1990). This parasitoid showed a strong preference for maize cultivars that produced higher quantities of HIPVs when fed upon by the African cotton leafworm *S. littoralis*, although the specific composition of these HIPVs also influenced preference (Fritzsche Hoballah et al. 2002). Interestingly, a comparison of the total HIPVs produced by 11 maize cultivars and five wild *Zea* (teosinte) species when attacked by *S. littoralis* found that while HIPV amounts were highly variable across cultivars and teosinte populations, the amounts were similarly variable across the maize cultivars and the wild teosinte species, suggesting that domestication did not significantly alter total HIPV production in maize (Gouinguené et al. 2001). Nevertheless, cultivars and teosinte species often differed in amounts of sesquiterpenes produced (Gouinguené et al. 2001), highlighting the importance of specific, not total, amounts of HIPVs. In another set of studies, different cultivars of maize may vary substantially in the production of (*E*)-ß-caryophyllene, a sesquiterpene that has been shown to be produced in response to herbivory by both the caterpillar *S. littoralis* and the corn rootworm *Diabrotica virgifera virgifera* (Turlings et al. 1998; Rasmann et al. 2005). Teosinte and European maize cultivars have maintained the ability to produce (*E*)-ß-caryophyllene in response to herbivory, whereas North American maize lines have lost this ability and, consequently, are less attractive to *C. marginiventris* (Köllner et al. 2008). Efforts to transform North American maize lines with a (*E*)-ß-caryophyllene synthase gene have demonstrated that it is possible to restore indirect defense by creating the ability of these maize lines to produce this sesquiterpene (Degenhardt et al. 2009). Finally, it is important to recognize that herbivore species may differ in their ability to suppress HIPV production. In a study of four noctuid herbivores of maize, herbivory by the fall armyworm *Spodoptera frugiperda* suppressed HIPV emissions in maize compared to the other three, more generalist herbivores *S. littoralis*, *S. exigua*, and the corn earworm *Helicoverpa armigera*, and that these results were likely due to differences in the regurgitants of these four caterpillar species (De Lange et al. 2020). Somewhat surprisingly, the suppressed production of HIPVs when attacked by *S. frugiperda* had little apparent effect on the attractiveness to *C. marginiventris*, although it may be that other parasitoids and predators in this system, not studied here, may be less attracted.

Most plants, in both natural and agroecosystems, are attacked by multiple species of herbivores, which in turn are attacked by often, complex suites of natural enemies (Stam et al. 2014). HIPVs play a central role in mediating interactions between the plant and its herbivore and natural enemy communities (Turlings and Erb 2018). In many parts of Asia, cultivated (*Oryza sativa*) and wild rice (*O. rufipogon*) are attacked by two important insect pests, the rice striped stem borer *Chilo suppressalis* and the brown planthopper *Nilaparvata lugens*. The brown planthopper is strongly attracted to HIPVs released by rice plants (both wild and cultivated) already attacked by the stem borer in large part because such plants represent enemy-free space from their own egg parasitoids *Anagrus nilaparvatae* (Hu et al. 2020). Whereas the parasitoid is strongly attracted to the HIPVs produced by plants infested with their host, the brown planthopper, this attraction disappears (and becomes somewhat repellent) when the *C. suppressalis* caterpillars are present. Identification of several volatiles produced by rice only when fed upon by *C. suppressalis* also showed that these compounds were repellent to *A. nilaparvatae* (Hu et al. 2020). Hu et al. (2020) argue that ovipositing *N. lugens* use HIPVs released in response to *C. suppressalis* damage as "enemy-free space" for their offspring. Studies such as this illustrate the importance of considering the community complexities and interactions among species in addition to focusing on more simplistic trophic interactions that involve only a single member at each level.

Despite the examples discussed above, much remains to be learned regarding how natural enemies decipher information about the status of their hosts from the complex volatile mixtures emitted by the plant (Dicke 2016). Much of what we know about domestication effects on plant volatile production comes from a handful of crop systems (e.g., maize, rice – Poaceae). Yet, many crops are dicots, and more research needs to be conducted before we can say how universally plant attractiveness to parasitoids is reduced by domestication. Furthermore, parasitoids are well known to learn to associate novel HIPV blends with successful foraging for hosts (e.g., Meiners et al. 2003; De Rijk et al. 2018). This makes it difficult to predict how plant attractiveness and parasitoid foraging will be affected by plant domestication.

Plant Toxin-Mediated Interactions with Caterpillars and Parasitoids

The direction and magnitude of the effects of changes in plant toxins resulting from domestication and cultivation on higher trophic levels depend on the degree of host plant or host specificity exhibited by herbivores and their natural enemies (Ali and Agrawal 2012; Ode 2019). Herbivores with narrow host ranges often have highly specific detoxification enzymes allowing them to specialize on otherwise highly toxic plants (e.g., Ratzka et al. 2002; Wittstock et al. 2004; Mao et al. 2006; Calla et al. 2020). Specialized herbivores are often relatively insensitive to changes in concentrations of plant toxins encountered as they feed (Rothwell and Holeski 2020). Rather, these herbivores either metabolize or excrete plant toxins in their frass. Therefore, changes in plant defensive chemistry resulting from crop breeding may not have a substantial effect on the larval performance of specialist herbivores or their natural enemies. Important exceptions to this prediction are cases where specialist herbivores sequester plant toxins in their tissues as defense against their own natural enemies (Nishida 2002; Ode 2006; Ali and Agrawal 2012). While this idea remains to be tested, specialist herbivores may be harmed by plant breeding efforts that result in lower concentrations of plant toxins as these sequestering herbivores may be more susceptible to their own natural enemies (although see the discussion on interactions between plant toxins and caterpillar immune responses, below). Furthermore, many specialist caterpillars use plant metabolites as ovipositional and feeding stimulants (e.g., Reddy et al. 2002; Renwick et al. 2006; Hopkins et al. 2009). In such cases, reductions in plant secondary metabolites in cultivated plants may result in lower oviposition and feeding by some specialist herbivores compared to wild plants (e.g., Altesor et al. 2014). The net effect of reduced plant toxins on specialist caterpillars depends on the balance between changes in larval performance, the risk of being consumed by predators and parasitoids, and oviposition rates, making predictions of how specialist caterpillars respond to agroecosystems challenging. Generalist herbivores, on the other hand, are often highly sensitive to changes in the levels of encountered plant toxins (Li et al. 2004; Wen et al. 2009; Lampert et al. 2011a). Domestication and plant breeding efforts that result in decreased plant toxin concentrations are generally expected to benefit generalist herbivores.

While few studies have explicitly compared plant toxin effects on the third trophic level in cultivated vs. wild plant systems (e.g., Harvey et al. 2007; Harvey and Gols 2011; Garvey et al. 2020), breeding changes to plant secondary metabolite production undoubtedly influence higher trophic levels. Plant toxins may influence parasitoid developmental success in at least one of three ways (Kaplan et al. 2016):

(a) Plant toxins reduce caterpillar size/quality: Toxic plant metabolites may indirectly affect parasitoid performance if host caterpillar size is diminished even if parasitoids do not directly encounter plant toxins. Smaller hosts may support fewer or smaller parasitoid larvae – an issue particularly germane for gregarious parasitoids with large clutch sizes. That plant chemistry often results in smaller, poorer-quality host caterpillars may reflect the cost to the caterpillar of metabolizing plant toxins (e.g., Whittaker and Feeny 1971; Brattsten et al. 1977; Appel and Martin 1992; Cresswell et al. 1992). For instance, metabolism of the furanocoumarin xanthotoxin by the parsnip webworm *Depressaria pastinacella* under conditions of dietary protein limitation resulted in decreased growth rates. Detoxification ability in this species is maintained at the expense of growth rate (Berenbaum and Zangerl 1994). On the other hand, ingestion of plant toxins may slow host development resulting in an increased window of vulnerability for the caterpillar for which it can be attacked by parasitoids – the so-called slow-growth, high-mortality hypothesis (Benrey and Denno 1997; Williams 1999; Chen and Chen 2018).

(b) Accumulation/sequestration of plant toxins in the hemolymph of the caterpillar: Plant toxins may be ingested by caterpillars and subsequently passed from the midgut either unmetabolized or as toxic breakdown products into the hemolymph where they would be encountered directly by feeding parasitoids or predators. To a large degree, the strength of these effects is mediated by where the herbivore falls on the generalist–specialist continuum. Compared to generalist caterpillars, relatively specialized caterpillars (apart from those species that sequester plant toxins, see below) are less likely to pass ingested plant toxins to their hemolymph (Lampert et al. 2011a). Therefore, parasitoids and predators attacking generalist caterpillars are expected to directly suffer the negative effects of plant toxins more so than specialists (Ode 2006, 2013). In one of the few studies to explicitly test these predictions, we found that hemolymph of the generalist cabbage looper *Trichoplusia ni* contained nearly twice the concentration of the plant toxin xanthotoxin as that found in the highly specialized parsnip webworm *Depressaria pastinacella* despite feeding on identical artificial diets (Lampert et al. 2011a). Differences in the levels of xanthotoxin found in the hemolymph of these two caterpillars can be explained by the significantly lower detoxification abilities of the generalist *T. ni* (Lee and Berenbaum 1989). Correspondingly, brood sizes and survivorship of *Copidosoma floridanum* that parasitize *T. ni* were significantly lower when the caterpillar fed on xanthotoxin-containing diet compared to a diet free of xanthotoxin; brood sizes and survivorship of *C. sosares* that parasitize parsnip webworm were unaffected by the presence or absence of xanthotoxin in the diet of their webworm host (Lampert et al. 2008, 2011a). Studies of nicotine-containing *Nicotiana* and iridoid glycoside-containing *Plantago* also show that generalist herbivores and their parasitoids are more negatively affected by plant toxins than are specialists (Barbosa et al. 1986, 1991; Harvey et al. 2005). In a common garden study of 18 wild and domesticated plants in the Solanaceae, the tobacco hornworm *Manduca sexta* generally preferred to oviposit on plants in the genera *Nicotiana* and *Datura* (both of which contain nicotine) rather than other genera such as *Capsicum* and *Physalis* that do not contain nicotine (Garvey et al. 2020). While *M. sexta* larvae generally perform more poorly on *Nicotiana* and *Datura*, they also experience reduced parasitism rates by the parasitoid *Cotesia congregata* (Garvey et al. 2020). A final example of how parasitoids can be the beneficiaries of the detoxification abilities of their hosts comes from a study of the diamondback moth *Plutella xylostella* and its parasitoid *Diadegma semiclausum* (Sun et al. 2020). Normally, *P. xylostella* can feed on brassicaceous plants due to glucosinolate sulphatases that prevent the formation of toxic isothiocyanates (Ratzka et al. 2002). Using plant-mediated RNA interference, Sun et al. (2019, 2020) silenced the *Pxgss* genes responsible for the glucosinolate sulphatase production in *P. xylostella* showing that toxic isothiocyanates accumulated within the caterpillar hemolymph; this, in turn, resulted in delayed development and reduced survivorship of *D. semiclausum* (Sun et al. 2020).

Very few studies have examined whether parasitoids are able to detoxify plant metabolites encountered in their caterpillar hosts (Ode 2006, 2013). Those few that have, including studies on *C. floridanum* and *C. sosares*, found no evidence that parasitoid larvae were able to metabolize plant toxins found in the hemolymph of their caterpillar hosts (McGovern et al. 2006; Lampert et al. 2008, 2011a). This suggests that the developmental success of many parasitoids is dependent on the amount of plant toxins ingested by the host caterpillar and whether that caterpillar can detoxify, excrete, or sequester those plant toxins.

As noted above, many caterpillars sequester plant toxins providing "enemy-free space" from parasitoids and predators (Jeffries and Lawton 1984; Dyer 1995; Stamp 2001; Nishida 2002; Ode 2006). Many of these sequestering species are specialists that, when they feed on toxic plants, become "nasty hosts" to their parasitoids and predators (Gauld et al. 1992; Dyer 1995; Lampert et al. 2010; Lampert et al. 2014; Ali and Agrawal 2012; Petschenka and Agrawal 2016). Sequestration often incurs a cost for the caterpillar in terms of reduced growth rates (Singer et al. 2004), suggesting that the benefits of "enemy-free space" outweigh costs associated with selective uptake, storage, and deployment of the sequestered plant toxins (Heckel 2014). To the extent that the benefits provided by sequestration outweigh their costs, plant breeding efforts that reduce plant toxin concentrations may result in higher predation and parasitism rates in agroecosystems. Yet, little is known about the magnitude of such trade-offs (Strauss and Zangerl 2002; Hunter 2003) let alone how these may be shifted in agroecosystems compared to natural ecosystems.

(c) Plant toxins interfere with the caterpillar's immune response system: Plant secondary metabolites are known to mediate the expression of caterpillar immune responses against parasitoids as well as a variety of protozoan, bacterial, and viral pathogens (Klemola et al. 2007). In general, increasing plant toxin concentrations suppress immune function of specialist caterpillars against parasitoids and pathogens, a phenomenon that has been termed the "immunocompromised host hypothesis" (Reudler et al. 2011) or the "vulnerable host hypothesis" (Smilanich et al. 2009a). This pattern contradicts the "nasty host hypothesis" (e.g., Gauld et al. 1992), especially in situations involving generalist caterpillars that accumulate unmetabolized toxins in their hemolymph or specialist species that sequester plant toxins as defense against natural enemies (see previous section). Most studies of plant defensive chemistry effects on caterpillar immune responses have measured encapsulation and melanization of injected beads or nylon microfilaments, a technique that has been widely taken as a proxy for defense against parasitoids and pathogens (Smilanich et al. 2009b; Lampert and Bowers 2015).

Several studies have shown that host plant species (and, by extension, host plant chemistry; Smilanich et al. 2009a) can affect the level of immunity expressed by lepidopteran herbivores against entomoviruses (Wang et al. 2020), bacteria, protists, and parasitoids (Lampert and Bowers 2015; Hansen et al. 2017). For example, the strength of the immune system of specialist caterpillars in the genus *Eois* (Geometridae) is strongly dependent on the host plant species on which they feed. Immune system strength (measured by phenoloxidase activity) of both *Eois apyraria* and *E. nympha* was greatly weakened when these caterpillars fed on *Piper cenocladum* (Piperaceae), a host plant that produces a high diversity of secondary metabolites compared to its congener *P. imperiale* (Hansen et al. 2017). Correspondingly, successful parasitism rates in the field were significantly higher on *P. cenocladum* than on *P. imperiale* (Hansen et al. 2017). That plant secondary chemistry is likely involved in suppression of immune responses of these caterpillars has been supported by studies using experimental diets to which known quantities of *Piper* defensive compounds (several amides) were added (Richards et al. 2010). Furthermore, while both the specialist *Eois* caterpillars and the generalist *Spodoptera frugiperda* (an important crop pest native throughout the Americas and a recent invader in much of the rest of the world) were negatively affected by *Piper* amides, the mechanisms were different (Richards et al. 2010). *Eois* caterpillars experienced increased parasitism, likely because their immune system was compromised by investing in metabolically expensive sequestration. *Spodoptera frugiperda* experienced increased larval mortality when fed a diet high in these *Piper* amides. Yet, not all studies have shown that increasing plant toxin concentrations result in decreased immune effectiveness in herbivores. For instance, the polyphagous beet armyworm *Spodoptera exigua* experiences high viral loads and low rates of melanization (with low expression of genes responsible for three key enzymes involved in the melanization process) when feeding on soybean *Glycine max*, intermediate viral loads and melanization rates when feeding on collards *Brassica oleracea*, and low viral loads and high melanization capabilities when feeding on *Ipomoea aquatica* (Wang et al. 2020). *Ipomoea aquatica*, followed by *B. oleracea*, have higher levels of foliar total phenolics than *G. max* (Wan et al. 2018). Foliar phenolics have been shown to inhibit the infectivity of entomoviruses (Ali et al. 2002; Felton et al. 1987).

Very few studies have compared the effects of plant defense chemistry variation in wild vs. cultivated plants on the ability of caterpillars to mount a successful encapsulation response. In one such study, Bukovinszky et al. (2009) showed that small cabbage white (*Pieris rapae*) caterpillars grew larger on cultivated cabbage (Brussels sprouts; *Brassica oleracea* gemmifera cv. Cyrus) than on wild populations and that encapsulation rates against the parasitoid *Cotesia glomerata* were higher in larger caterpillars. *Pieris rapae* caterpillars that fed on wild cabbage plants were smaller and had weaker immune responses against parasitism by *C. glomerata* (Bukovinszky et al. 2009). That caterpillars were smaller when developing on wild cabbage likely reflects a cost of being able to efficiently metabolize and detoxify glucosinolates. Cabbage plants from wild populations produced significantly higher glucosinolate levels, both constitutive and herbivore-induced, in their leaves (Harvey et al. 2007; Gols et al. 2008a, b; Bukovinszky et al. 2009; Ode et al. 2016). *Pieris rapae* is highly adapted to feeding on glucosinolate-containing plants because it possesses a nitrile-specifier protein that results in the production of less toxic nitriles, which are excreted, instead of the production of the more toxic isothiocyanates (Wittstock et al. 2004). Neither *P. rapae* nor the closely related *P. brassicae* sequester glucosinolates (Müller et al. 2003). Because of this, *C. glomerata* larvae do not encounter significant quantities of glucosinolates or their breakdown products when developing inside *P. rapae* larvae.

As many of the examples presented above suggest, changes in secondary plant chemistry may have variable effects on parasitoids depending on the mechanism involved. While chemical defense via sequestered or otherwise accumulated plant toxins may directly harm predators and parasitoids (see section b, above), these same toxins may suppress the immune system of the caterpillar resulting in an increased likelihood of successful parasitism or pathogen infection (see section c, above). Studies supporting the "vulnerable host" hypothesis stand in contrast to the large number of studies showing that parasitoid success is lower when attacking caterpillars that feed on toxic plants (Campbell and Duffey 1979; Barbosa et al. 1991; Singer and Stireman III 2003; Ode 2006, 2013; Kaplan et al. 2016). This includes several studies documenting self-medication by herbivores to "cure" themselves of parasitoids/parasites (pharmacophagy, feeding on toxic diets to rid oneself of parasites and pathogens; Singer and Stireman III 2003; de Roode et al. 2013).

In general, use of chemical defense by caterpillars against predators and parasites in the form of sequestration appears to come at the cost of compromised cellular immunity (encapsulation responses, melanization, and phagocytosis) against parasitoids and pathogens. Therefore, trade-offs likely exist between the ability to sequester and the maintenance of an effective immune system (Smilanich et al. 2009a; Quintero et al. 2014; Slinn et al. 2018). That such trade-offs occur is suggested in a study of two closely related sphingid caterpillars, one of which sequesters the iridoid glycoside catalpol (*Ceratomia catalpae*), while the other does not (*C. undulosa*) (Lampert and Bowers 2015). The melanization response (to injected beads) of only *C. catalpae* was negatively correlated with the amount of catalpol sequestered. The authors suggest that while sequestration of catalpol may enhance protection against predators, it likely comes at the cost of high rates of parasitism (Lampert and Bowers 2015). Indeed, field populations of *C. catalpae* are often heavily attacked by the endoparasitoid *Cotesia congregata* (Lampert et al. 2010, 2011b), supporting the hypothesis that sequestering caterpillars may be "safe havens" for endoparasitoids that develop in hosts whose immune systems have been compromised (Dyer 1995). In caterpillar species that sequester, it is unclear whether chemical defense or immune responses provide more effective control against parasitoids. Both detoxification and mounting an immune response are metabolically costly, strongly suggesting trade-offs between these two activities. That immune responses are energetically costly to maintain has been shown in a study correlating encapsulation response with increased standard metabolic rate in the large cabbage white butterfly *Pieris brassicae* (Freitak et al. 2003). The magnitude of this trade-off is expected to be exacerbated in wild cabbage plants compared to domesticated varieties.

Of course, which type of defense is effective also depends on the degree of adaption of different parasitoid populations to the immune response of their hosts in terms of resisting encapsulation and melanization, a phenomenon that has been documented in the parasitoid *Asobara tabida* that attacks drosophilid flies (Kraaijeveld and van Alphen 1995; Kraaijeveld and Godfray 2001). Differential adaptation across parasitoid populations highlights one of the challenges of using beads and monofilaments as proxies for parasitoids. Many parasitoids, particularly those in the superfamily Ichneumonoidea, have effective mechanisms to counter encapsulation and melanization responses of their host caterpillars including the injection of venoms (Poirié et al. 2014) and polydnaviruses (both ichnoviruses and bracoviruses; Drezen et al. 2014). This parasitoid superfamily, comprising the Ichneumonidae and Braconidae, includes many species that attack caterpillars in both natural and agroecosystems and are widely used as biocontrol agents. The ichneumonid wasp *Pimpla hypochondriaca* injects a venom at oviposition that inhibits hemocytes from forming capsules in its host caterpillar, the bright-line brown eye (*Lacanobia oleracea*; a.k.a. tomato moth) (Richards and Edwards 1999; Richards and Parkinson 2000). The braconid wasp *Microplitis demolitor* injects a bracovirus (MdBV) at oviposition into its hosts, the soybean looper *Pseudoplusia* (=*Chrysodeixis*) *includens* and several other agronomically important noctuids. The MdBV inhibits the cellular immune response of these caterpillars by preventing the hemocytes, granulocytes, and plasmatocytes from adhering to parasitoid eggs, thereby preventing encapsulation as well as phagocytosis by the caterpillars (Strand et al. 2006). Assessment of whether chemical defense or immune response is a more effective defense against parasitoid attack depends not only on understanding host immune responses but also parasitoid countermeasures. Understanding the mechanism by which plant secondary chemistry affects parasitoid success is crucial for predicting how caterpillar – parasitoid interactions, especially those involving biocontrol efforts – respond to changes in plant defenses that arise from agricultural practices.

Climate Change, Plant Secondary Metabolites, and Phenological Match/Mismatch Between Trophic Levels

Human activities such as burning of fossil fuels, deforestation, and raising ruminant livestock result in the production of greenhouse gases (e.g., carbon dioxide, nitrous oxide, nitrogen oxides, methane, ozone) that absorb solar radiation and drive anthropogenic climate change (IPCC 2014; Blunden and Arndt 2015). Agricultural practices themselves are estimated to contribute about 15% of all greenhouse gas emissions and land conversion to agriculture contributes an additional, similar amount (Godfray and Garnett 2014).

Increased production of greenhouse gases can alter the structure of trophic interactions in two broad ways (Ode 2013). First, increased levels of greenhouse gases increase global temperatures and result in increasingly variable weather patterns. Herbivores can respond to increasing temperatures by expanding their ranges, altering the timing of life history events including the increasing the number of generations per year, displaying increased fluctuations in year-to-year population sizes, and experiencing changes to trophic interactions (Lehmann et al. 2020). Altered temperature and precipitation patterns result in changes in the life histories and developmental rates of plants, insect herbivores, and their natural enemies creating phenological mismatches among trophic levels (Parmesan 2006; Parmesan and Yohe 2003). Such trophic matches and mismatches have consequences for reciprocal selection between trophic levels and the resulting expression of plant secondary metabolites. Second, greenhouse gases themselves (in particular, CO_2 and O_3) can modify the expression of plant defense signaling pathways, notably the jasmonic acid and salicylic acid pathways, which are implicated in defenses against chewing herbivores (e.g., caterpillars) and microbial pathogens/piercing-sucking insect herbivores, respectively (Ode et al. 2014). While a case can be made that agroecosystems generate more greenhouse gases than natural systems, it is not clear that trophic interactions in these two ecosystems are differentially affected by climate change.

Temperature and phenological matches/mismatches: As ectothermic organisms, the timing of key life history events of plants and their insect associates is strongly dependent on temperature (Bale et al. 2002; van der Putten et al. 2010). Many trophic interactions have narrow temporal windows in which the phenologies of plants, herbivores, and natural enemies must be aligned if they are to persist (van Nouhuys and Lei 2004; van Asch and Visser 2007; Chidawanyika et al. 2019; Yang and Cenzer 2020). Where different trophic members differentially respond to changes in temperature, old trophic associations can be broken and new ones formed under changing temperature regimes. There are numerous examples of plants and insects that have undergone range expansions in response to rising temperatures, becoming pests as they leave their natural enemies behind (Menéndez et al. 2008, van der Putten et al. 2010, Battisti and Larsson 2015; Lindström and Lehmann 2015). Even in the classic example of the importance of synchrony between ovipositing winter moth *Operophtera brumata* and budburst in the oak *Quercus robur* (van Dongen et al. 1997), warming climate can disrupt the phenological match between herbivore and plant (Visser and Holleman 2001; Hagen et al. 2007). Of 17 herbivore-parasitoid pairs, in 11 cases the critical thermal limits and optimal development temperatures of the hosts were higher than the parasitoids (Hance et al. 2007; Furlong and Zalucki 2017), suggesting that climate warming may likely result in many herbivores escaping control by their natural enemies.

Warming climates can differentially affect the population dynamics of herbivores and their parasitoids. For instance, the invasive larch casebearer *Coleophora laricella* experienced population outbreaks every 2–6 years in the 20 years before two introduced parasitoids (*Agathis pumila* and *Chrysocharis laricinellae*) began to control larch casebearer populations at low, stable densities (Ward et al. 2020). Since the 1990s, however, warmer temperatures have disrupted control by these two parasitoids, leading to resumed casebearer outbreaks (Ward et al. 2020). Warmer temperatures may alter host immune responses in caterpillars, particularly when mounting a successful immune response trades off with other life history traits (Karl et al. 2011; Laughton et al. 2017). While rarely explicitly studied, such changes in phenological matches and mismatches between trophic levels almost certainly drive changes in the underlying plant secondary chemistry and their effects on herbivores and their natural enemies. Little is known about how temperature-driven phenological mismatches will play out in natural vs. agroecosystems. Within agricultural systems, phenological matches underlying trophic relationships in longer-term systems such as orchards or agroforestry systems may be more susceptible to changing climates (e.g., Castex et al. 2018; Hamann et al. 2021).

Greenhouse gases and plant secondary metabolites: As a key substrate for photosynthesis, elevated CO_2 levels generally result in increased net primary productivity (Ainsworth and Long 2005). Elevated CO_2 may alter plant nutritive quality for herbivores, particularly by increasing C:N ratios. In addition to driving increasing global temperatures and the potential for phenological mismatches as well as the formation of novel associations, anthropogenic changes in the concentrations of atmospheric gases, especially carbon dioxide (CO_2) and ozone (O_3), are recognized to directly influence the production of plant secondary metabolites (Ode et al. 2014).

Increasing levels of CO_2 are known to suppress the jasmonic acid signaling pathway (largely responsible for induced defenses against chewing herbivores such as caterpillars) (e.g., Hall et al. 2020) and increase activity of the salicylic acid pathway (associated with defense against plant pathogens) via crosstalk between these two pathways (Zavala et al. 2013). These patterns notwithstanding, the effects of elevated CO_2 on the expression of specific plant volatiles and toxins are highly idiosyncratic, depending strongly on the plant species and the specific plant compound in question (Karowe et al. 1997; DeLucia et al. 2012; Ode et al. 2014).

Ozone induces the production of reactive oxygen species that upregulate the salicylic acid pathway. On the one hand, ozone has been shown to increase the production of several terpenoids involved in the attraction of herbivore natural enemies, but ozone may also interact with other plant volatiles disrupting their roles in indirect plant defense (Ode et al. 2014). It is tempting to predict that increasing atmospheric CO_2 levels will suppress jasmonic acid-based defenses against caterpillars, but how general this proves to be will depend on much more research in a broader range of plant-herbivore-natural enemy systems than has been conducted to date.

Conclusions/Future Directions

This chapter has focused on how crop domestication alters plant secondary chemistry and the consequences for herbivore-natural enemy interactions. Our understanding of the effects of domestication, breeding, and agricultural practices on the interactions between plant chemical defenses (both direct and indirect), caterpillar herbivores, and their parasitoid and predator natural enemies is based largely on a few well-studied systems: *Brassica* species, tomatoes, maize, rice, *Phaseolus* beans, and *Helianthus* sunflowers, among others. In some of these systems, notably Brassicaceae and Solanaceae, there are excellent genomic resources via *Arabidopsis*, *Lycopersicon*, and *Nicotiana* allowing for detailed, mechanistic understandings of how selection for agronomically desirable traits may affect plant chemical defenses. Yet, these systems represent a fraction of all crop plants, and understanding how caterpillars (and other insect herbivores) and their natural enemies respond to domestication will undoubtedly be enhanced by further studies. In other systems, we must make reasonable, semi-informed guesses about how domestication has altered trophic relationships, but undoubtedly other novel changes remain to be discovered. Such an understanding is even more pressing given the relentless onslaught of climate change. If agriculture is to adapt to changing climate in all its facets, we must have a clearer understanding of the complex relationships between increasing temperatures and changes in precipitation patterns and what this means for trophic matches and mismatches and whether caterpillar species will likely increase the number of generations per year.

We also need a better understanding of how greenhouse gases interact with plant signaling pathways in order to understand and predict how plant relationships with their herbivores will be altered, both in terms of direct defenses and indirect defenses. Further studies are needed that incorporate more complex trophic interactions allowing for emergent phenomena such as apparent competition, apparent predation, and other more complex indirect effects that cannot be studied with more unidimensional trophic relationships. Finally, this chapter has focused on how crop domestication alters plant secondary chemistry and the consequences for herbivore-natural enemy interactions. Many of the studies covered here have focused on individual plants and their interactions with caterpillars and their natural enemies. Yet, the importance of scale and landscape characteristics must be taken into consideration if we are to understand how plant-caterpillar-natural enemy interactions differ between natural and agroecosystems. Patch size and plant neighborhood characteristics at different scales can differentially influence the attractiveness of HIPVs to parasitoids (Aartsma et al. 2020). Proximity and the size of natural areas to agricultural fields can determine the spillover of both herbivores and natural enemies between these two habitat types (Tscharntke et al. 2005; González et al. 2020). Our understanding of the effects of domestication on plant secondary chemistry and interactions with caterpillars and their natural enemies will be greatly improved as we explore these issues at multiple spatial scales.

Acknowledgments I am grateful for the constructive comments on earlier drafts of this chapter from members of my research group as well as those from anonymous reviewers. I also thank Bob Marquis and Suzanne Koptur for the invitation to contribute to this book.

References

Aartsma Y, Pappagallo S, van der Werf W, Dicke M, Bianchi FJJA, Poelman EH (2020) Spatial scale, neighboring plants and variation in plant volatiles interactively determine the strength of host-parasitoid relationships. Oikos 129:1429–1439

Ainsworth EA, Long SP (2005) What have we learned from 15 years of free-air CO_2 enrichment (FACE)? A meta-analytic review of the responses of photosynthesis, canopy properties and plant production to rising CO_2. New Phytol 165:351–371

Alborn HT, Turlings TCJ, Jones TH, Stenhagen G, Loughrin JH, Tumlinson JH (1997) An elicitor of plant volatiles from beet armyworm oral secretion. Science 276:945–949

Ali JG, Agrawal AA (2012) Specialist versus generalist insect herbivores and plant defense. Trends Plant Sci 17:293–302

Ali MI, Young SY, Felton GW, McNew RW (2002) Influence of the host plant on occluded virus production and lethal infectivity of a baculovirus. J Invertebr Pathol 81:158–165

Altesor P, García Á, Font E, Rodríguez-Haralambides A, Vilaró OM, Soler R, González A (2014) Glycoalkaloids of wild and cultivated *Solanum*: effects on specialist and generalist insect herbivores. J Chem Ecol 40:599–608

Appel HM, Martin MM (1992) Significance of metabolic load in the evolution of host specificity of *Manduca sexta*. Ecology 73:216–228

Bale JS, Masters GJ, Hodkinson ID, Awmack C, Bezemer TM, Brown VK, Butterfield J, Buse A, Coulson JC, Farrar J, Good JEG, Harrington R, Hartley S, Jones TH, Lindroth RL, Press MC, Symrnioudis I, Watt AD, Whittaker JB (2002) Herbivory in global climate change research: direct effects of rising temperature on insect herbivores. Glob Chang Biol 8:1–16

Barbosa P (1993) Lepidopteran foraging on plants in agroecosystems: constraints and consequences. In: Stamp NE, Casey TM (eds) Caterpillars: ecological and evolutionary constraints on foraging. Routledge, Chapman & Hall, New York, pp 523–566

Barbosa P, Saunders JA, Kemper J, Trumbule R, Olechno J, Martinat P (1986) Plant allelochemicals and insect parasitoids: effects on nicotine on *Cotesia congregata* (Say) (Hymenoptera: Braconidae) and *Hyposoter annulipes* (Cresson) (Hymenoptera: Ichneumonidae). J Chem Ecol:12–1319

Barbosa P, Gross P, Kemper J (1991) Influence of plant allelochemicals on the tobacco hornworm and its parasitoid, *Cotesia congregata*. Ecology 72:1567–1575

Battisti A, Larsson S (2015) Climate change and insect pest distribution range. In: Björkman C, Niemelä P (eds) Climate change and insect pests. CAB International Press, Oxfordshire, UK, pp 1–15

Benrey B, Denno RF (1997) The slow-growth-high-mortality hypothesis: a test using the cabbage butterfly. Ecology 78:987–999

Benrey B, Callejas A, Rios L, Oyama K, Denno RF (1998) The effects of domestication on *Brassica* and *Phaseolus* on the interaction between phytophagous insects and parasitoids. Biol Control 11:130–140

Berenbaum MR (1983) Coumarins and caterpillars: a case for coevolution. Evolution 37:163–179

Berenbaum MR, Zangerl AR (1994) Costs of inducible defense: protein limitation, growth, and detoxification in parsnip webworms. Ecology 75:2311–2317

Bianchi FJJA, Booij CJH, Tscharntke T (2006) Suistainable pest regulation in agricultural landscapes: a review on landscape composition, biodiversity and natural pest control. Proc R Soc B 273:1715–1727

Bleeker PM, Mirabella R, Diergaarde PJ, VanDoorn A, Tissier A, Kant MR, Prins M, de Vos M, Haring MA, Schuurink RC (2012) Improved herbivore resistance in cultivated tomato with the sesquiterpene biosynthetic pathway from a wild relative. Proc Natl Acad Sci 109:20124–20129

Blunden J, Arndt DS (eds) (2015) State of the climate in 2014. Bull Am Meteorol Soc 96:S1–S267

Bodnaryk RP (1997) Will low-glucosinolate cultivars of the mustards *Brassica juncea* and *Sinapis alba* be vulnerable to insect pests? Can J Plant Sci 77:283–287

Brattsten LB, Wilkinson CF, Eisner T (1977) Herbivore-plant interactions: mixed-function oxidases and secondary plant substances. Science 196:1349–1352

Bukovinszky T, Poelman EH, Gols R, Prekatsakis G, Vet LEM, Harvey JA, Dicke M (2009) Consequences of constitutive and induced variation in plant nutritional quality for immune defence of a herbivore against parasitism. Oecologia 160:299–308

Bustos-Segura C, Cuny MAC, Benrey B (2020) Parasitoids of leaf herbivores enhance plant fitness and do not alter caterpillar-induced resistance against seed beetles. Funct Ecol 34:586–596

Calla B, Wu W-Y, Dean CAE, Schuler MA, Berenbaum MR (2020) Substrate-specificity of cytochrome P450-mediated detoxification as an evolutionary strategy for specialization on furanocoumarin-containing hostplants: CYP6AE89 in parsnip webworms. Insect Mol Biol 29:112–123

Campbell BC, Duffey SS (1979) Tomatine and parasitic wasps: potential incompatibility of plant antibiosis with biological control. Science 205:700–702

Castex V, Beniston M, Calanca P, Fleury D, Moreau J (2018) Pest management under climate change: the importance of understanding trophic relations. Sci Total Environ 616-617:397–407

Chacón-Fuentes M, Parra L, Rodriguez-Saona C, Sequel I, Ceballos R, Quiroz A (2015) Domestication in murtilla (*Ugni molinae*) reduced defensive flavonol levels but increased resistance against a native herbivorous insect. Environ Entomol 44:627–637

Chaplin-Kramer R, O'Rourke ME, Blitzer EJ, Kremen C (2011) A meta-analysis of crop pest and natural enemy response to landscape complexity. Ecol Lett 14:922–932

Chen YH (2016) Crop domestication, global human-mediated migration, and the unresolved role of geography in pest control. Elementa Sci Anthropocene 4:000106

Chen K-W, Chen Y (2018) Slow-growth high-mortality: a meta-analysis for insects. Insect Sci 25:337–351

Chen YH, Welter SC (2002) Abundance of a native moth *Homoeosoma electellum* (Lepidoptera: Pyralidae) and activity of indigenous parasitoids in native and agricultural sunflower habitats. Environ Entomol 31:626–636

Chen YH, Welter SC (2003) Confused by domestication: incongruent behavioral responses of the sunflower moth, *Homoeosoma electellum* (Lepidoptera: Pyralidae) and its parasitoid, *Dolichogenidea homoeosomae* (Hymenoptera: Braconidae), towards wild and domesticated sunflowers. Biol Control 2003:180–190

Chen YH, Welter SC (2007) Crop domestication creates a refuge from parasitism for a native moth. J Appl Ecol 44:238–245

Chen YH, Langellotto GA, Barrion AT, Cuong NL (2013) Cultivation of domesticated rice alters arthropod biodiversity and community composition. Ann Entomol Soc Am 106:100–110

Chen YH, Gols R, Benrey B (2015a) Crop domestication and its impact on naturally selected trophic interactions. Annu Rev Entomol 60:35–58

Chen YH, Gols R, Stratton CA, Brevik KA, Benrey B (2015b) Complex tritrophic interactions in response to crop domestication: predictions from the wild. Entomol Exp Appl 157:40–59

Chen YH, Ruiz-Arocho J, von Wettberg EJB (2018) Crop domestication: anthropogenic effects on insect-plant interactions in agroecosystems. Curr Opin Insect Sci 29:56–63

Chidawanyika F, Mudavanhu P, Nyamukondiwa C (2019) Global climate change as a driver of bottom-up and top-down factors in agricultural landscapes and the fate of host-parasitoid interactions. Front Ecol Evol 7:article 80

Clavijo McCormick A, Unsicker SB, Gershenzon J (2012) The specificity of herbivore-induced plant volatiles in attracting herbivore enemies. Trends Plant Sci 17:303–310

Cohen AL, Crowder DW (2017) The impacts of spatial and temporal complexity across landscapes on biological control: a review. Curr Opin Insect Sci 20:13–18

Cresswell JE, Merritt SZ, Martin MM (1992) The effect of dietary nicotine on the allocation of assimilated food to energy metabolism and growth in fourth-instar larvae of the southern armyworm, *Spodoptera eridania* (Lepidoptera: Noctuidae). Oecologia 89:449–453

D'Alessandro M, Brunner V, von Mérey G, Turlings TCJ (2009) Strong attraction of the parasitoid *Cotesia marginiventris* towards minor volatile compounds of maize. J Chem Ecol 35:999–1008

De Geyter N, Gholami A, Goormachtig S, Goossens A (2012) Transcriptional machineries in jasmonate-elicited plant secondary metabolism. Trends Plant Sci 17:349–359

De Lange ES, Laplanche D, Guo H, Xu W, Vlimant M, Erb M, Ton J, Turlings TCJ (2020) *Spodoptera frugiperda* caterpillars suppress herbivore-induced volatile emissions in maize. J Chem Ecol 46:344–360.

De Moraes CM, Lewis WJ, Paré PW, Alborn HT, Tumlinson JH (1998) Herbivore-infested plants selectively attract parasitoids. Nature 393:570–573

De Rijk M, Sánchez VC, Smid HM, Engel B, Vet LEM, Poelman EH (2018) Associative learning of host presence in non-host environments influences parasitoid foraging. Ecol Entomol 43:318–325

de Roode JC, Lefèvre T, Hunter MD (2013) Self-medication in animals. Science 340:150–151

Degen T, Dillmann C, Marion-Poll F, Turlings TCJ (2004) High genetic variability of herbivore-induced volatile emission within a broad range of maize inbred lines. Plant Physiol 135:1928–1938

Degenhardt J, Hiltpold I, Köllner TG, Frey M, Gierl A, Gershenzon J, Hibbard BE, Ellersieck MR, Turlings TCJ (2009) Restoring a maize root signal that attracts insect-killing nematodes to control a major pest. Proc Natl Acad Sci 106:13213–13218

DeLucia EH, Nabity PD, Zavala JA, Berenbaum MR (2012) Climate change: resetting plant-insect interactions. Plant Physiol 160:1677–1685

Dicke M (2016) Induced plant volatiles: plant body odours structuring ecological networks. New Phytol 210:10–12

Drezen J-M, Chevignon G, Louis F, Huguet E (2014) Origin and evolution of symbiotic viruses associated with parasitoid wasps. Curr Opin Insect Sci 6:35–43

Dyer LA (1995) Tasty generalists and nasty specialists? Antipredator mechanisms in tropical lepidopteran larvae. Ecology 76:1483–1496

Ehrlich PR, Raven PH (1964) Butterflies and plants: a study in coevolution. Evolution 18:586–608

Evans LT (1993) Crop Evolution, Adaptation, and Yield. Cambridge University Press, Cambridge, UK

Fahey JW, Zalcmann AT, Talalay P (2001) The chemical diversity and distribution of glucosinolates and isothiocyanates among plants. Phytochemistry 56:5–51

Farnham MW, Wilson PE, Stephenson KK, Fahey JW (2004) Genetic and environmental effects on glucosinolate content and chemoprotective potency of broccoli. Plant Breed 123:60–65

Fatouros NE, van Loon JJA, Hordijk KA, Smid HM, Dicke M (2005) Herbivore-induced plant volatiles mediate in-flight host discrimination by parasitoids. J Chem Ecol 31:2033–2047

Felton GW, Duffey SS, Vail PV, Kaya HK, Manning J (1987) Interaction of nuclear polyhedrosis virus with catechols: potential incompatibility for host-plant resistance against noctuid larvae. J Chem Ecol 13:947–957

Fernandez AR, Sáez A, Quintero C, Gleiser G, Aizen MA (2021) Intentional and unintentional selection during plant domestication: herbivore damage, plant defensive traits and nutritional quality of fruit and seed crops. New Phytol. https://doi.org/10.1111/nph.17452

Freitak D, Ots I, Vanatoa A, Hõrak P (2003) Immune response is energetically costly in white cabbage butterfly pupae. Proc R Soc Lond B 270(Suppl):S220–S222

Fritzsche Hoballah ME, Tamò C, Turlings TCJ (2002) Differential attractiveness of induced odors emitted by eight maize varieties for the parasitoid *Cotesia marginiventris*: is quality or quantity important? J Chem Ecol 28:951–968

Furlong MJ, Zalucki MP (2017) Climate change and biological control: the consequences of increasing temperatures on host-parasitoid interactions. Curr Opin Insect Sci 20:39–44

Futuyma DJ, Agrawal AA (2009) Macroevolution and the biological diversity of plants and herbivores. Proc Natl Acad Sci 106:18054–18061

Gaillard MDP, Glauser G, Robert CAM, Turlings TCJ (2018) Fine-tuning the 'plant domestication-reduced defense' hypothesis: specialist vs generalist herbivores. New Phytol 217:355–366

Garvey MA, Creighton JC, Kaplan I (2020) Tritrophic interactions reinforce a negative preference-performance relationship in the tobacco hornworm (*Manduca sexta*). Ecol Entomol 45:783–794

Gauld ID, Gaston KJ, Janzen DH (1992) Plant allelochemicals, tritrophic interactions and the anomalous diversity of tropical parasitoids: the "nasty" host hypothesis. Oikos 65:353–357

Gershenzon J, Rossiter MC, Mabry TJ, Rogers CE, Blust MH, Hopkins TL. 1985. Insect antifeedant terpenoids in wild sunflower: a possible source of resistance to the sunflower moth. American Chemical Society Symposium Series, The Society 276: 433–446. Washington, DC

Godfray HCJ, Garnett T (2014) Food security and sustainable intensification. Philos Trans R Soc B 369:20120273

Gols R (2014) Direct and indirect chemical defences against insects in a multitrophic framework. Plant Cell Environ 37:1741–1752

Gols R, Harvey JA (2009) Plant-mediated effects in the Brassicaceae on the performance and behaviour of parasitoids. Phytochem Rev 8:187–206

Gols R, Bukovinszky T, van Dam NM, Dicke M, Bullock JM, Harvey JA (2008a) Performance of generalist and specialist herbivores and their endoparasitoids differs on cultivated and wild *Brassica* populations. J Chem Ecol 34:132–143

Gols R, Wagenaar R, Bukovinszky T, van Dam NM, Dicke M, Bullock JM, Harvey JA (2008b) Genetic variation in defense chemistry in wild cabbages affects herbivores and their endoparasitoids. Ecology 89:1616–1626

Gols R, Bullock JM, Dicke M, Bukovinszky T, Harvey JA (2011) Smelling the wood from the trees: non-linear parasitoid responses to volatile attractants produced by wild and cultivated cabbage. J Chem Ecol 37:795–807

González E, Landis DA, Knapp M, Valladares G (2020) Forest cover and proximity decrease herbivory and increase crop yield via enhanced natural enemies in soybean fields. J Appl Ecol 57:2296–2306

González-Santoyo I, Córdoba-Aguilar A (2012) Phenoloxidase: a key component of the insect immune system. Entomol Exp Appl 142:1–16

Gouinguené S, Degen T, Turlings TCJ (2001) Variability in herbivore-induced odour emissions among maize cultivars and their wild ancestors (teosinte). Chemoecology 11:9–16

Gurr GM, Wratten SD, Landis DA, You M (2017) Habitat management to suppress pest populations: progress and prospects. Annu Rev Entomol 62:91–109

Hagen SB, Jepsen JU, Ims RA, Yoccoz NG (2007) Shifting altitudinal distribution of outbreak zones of winter moth *Operophtera brumata* in sub-arctic birch forest: a response to recent climate warming? Ecography 30:299–307

Hall CR, Mikhael M, Hartley SE, Johnson SN (2020) Elevated atmospheric CO_2 suppresses jasmonate and silicon-based defences without affecting herbivores. Funct Ecol 34:993–1002

Hamann E, Blevins C, Franks SJ, Jameel MI, Anderson JT (2021) Climate change alters plant–herbivore interactions. New Phytol 229:1894–1910

Hance T, van Baaren J, Vernon P, Boivin G (2007) Impact of extreme temperatures on parasitoids in a climate change perspective. Annu Rev Entomol 52:107–126

Hansen AC, Glassmire AE, Dyer LA, Smilanich AM (2017) Patterns in parasitism frequency explained by diet and immunity. Ecography 40:803–805

Harvey JA, Gols (2011) Population-related variation in plant defense more strongly affects survival of an herbivore than its solitary parasitoid wasp. J Chem Ecol 37:1081–1090

Harvey JA, van Nouhuys S, Biere A (2005) Effects of quantitative variation in allelochemicals in *Plantago lanceolate* on development of a generalist and a specialist herbivore and their endoparasitoids. J Chem Ecol 31:287–302

Harvey JA, Gols R, Wagenaar R, Bezemer TM (2007) Development of an insect herbivore and its pupal parasitoid reflect differences in direct plant defense. J Chem Ecol 33:1556–1569

Heckel DG (2014) Insect detoxification and sequestration strategies. Annu Plant Rev 47:77–114

Heimpel GE, Mills NJ (2017) Biological Control: ecology and applications. Cambridge University Press, Cambridge

Hernandez-Cumplido J, Giusti MM, Zhou Y, Kyryczenko-Roth V, Chen YH, Rodriguez-Saona (2018) Testing the 'plant domestication-reduced defense' hypothesis in blueberries: the role of herbivore identity. Arthropod Plant Interact 12:483–493

Hopkins RJ, van Dam NM, van Loon JJA (2009) Role of glucosinolates in insect-plant relationships and multitrophic interactions. Annu Rev Entomol 54:57–83

Hu X, Su S, Liu Q, Jiao Y, Peng Y, Li Y, Turlings TCJ (2020) Caterpillar-induced rice volatiles provide enemy-free space for the offspring of the brown planthopper. elife 9:e55421

Hunter MD (2003) Effects of plant quality on the population ecology of parasitoids. Agric For Entomol 5:1–8

IPCC (2014) Climate Change 2014: Synthesis Report. Contribution of Working Groups I, II and III to the Fifth Assessment Report of the Intergovernmental Panel on Climate Change [Core Writing Team, R.K. Pachauri and L.A. Meyer (eds)]. IPCC, Geneva, Switzerland, 151 pp.

Jeffries MJ, Lawton JH (1984) Enemy free space and the structure of ecological communities. Biol J Linn Soc 23:269–286

Kaplan I, Carrillo J, Garvey M, Ode PJ (2016) Indirect plant-parasitoid interactions mediated by changes in herbivore physiology. Curr Opin Insect Sci 14:112–119

Karl I, Stoks R, De Block M, Janowitz SA, Fischer K (2011) Temperature extremes and butterfly fitness: conflicting evidence from life history and immune function. Glob Chang Biol 17:676–687

Karowe DN, Seimens DH, Mitchell-Olds T (1997) Species-specific response of glucosinolate content to elevated atmospheric CO_2. J Chem Ecol 23:2569–2582

Karp DS, 156 additional authors (2018) Crop pests and predators exhibit inconsistent responses to surrounding landscape composition. Proc Natl Acad Sci 115:E7863–E7870

Kessler A, Baldwin IT (2002) Plant responses to insect herbivory: the emerging molecular analysis. Annu Rev Plant Biol 53:299–328

Kessler A, Heil M (2011) The multiple faces of indirect defences and their agents of natural selection. Funct Ecol 25:348–357

Klemola N, Klemola T, Rantala MJ, Ruuhola T (2007) Natural host-plant quality affects immune defence of an insect herbivore. Entomol Exp Appl 123:167–176

Köllner TG, Held M, Lenk C, Hiltpold I, Turlings TCJ, Gershenzon J, Degenardt J (2008) A maize (E)-ß-caryophyllene synthase implicated in indirect defense responses against herbivores is not expressed in most American maize varieties. Plant Cell 20:482–494

Kraaijeveld AR, Godfray HCJ (2001) Is there local adaptation in Drosophila–parasitoid interactions? Evol Ecol Res 3:107–116

Kraaijeveld AR, van Alphen JJM (1995) Geographical variation in encapsulation ability of Drosophila melanogaster larvae and evidence for parasitoid-specific components. Evol Ecol 9:10–17

Lampert EC, Bowers MD (2015) Incompatibility between plant-derived defensive chemistry and immune response of two sphingid herbivores. J Chem Ecol 41:85–92

Lampert EC, Zangerl AR, Berenbaum MR, Ode PJ (2008) Tritrophic effects of xanthotoxin on the polyembryonic parasitoid Copidisoma sosares (Hymenoptera: Encyrtidae). J Chem Ecol 34:783–790

Lampert EC, Dyer LA, Bowers MD (2010) Caterpillar chemical defense and parasitoid success: Cotesia congregata parasitism of Ceratomia catalpa. J Chem Ecol 36:992–998

Lampert EC, Zangerl AR, Berenbaum MR, Ode PJ (2011a) Generalist and specialist host-parasitoid associations respond differently to wild parsnip (Pastinaca sativa) defensive chemistry. Ecol Entomol 36:52–61

Lampert EC, Dyer LA, Bowers MD (2011b) Chemical defense across three trophic levels: Catalpa bignonioides, the caterpillar Ceratomia catalpa, and its endoparasitoid Cotesia congregata. J Chem Ecol 37:1063–1070

Lampert EC, Dyer LA, Bowers MD (2014) Dietary specialization and the effects of plant species on potential multitrophic interactions of three species of nymphaline caterpillars. Entomol Exp Appl 153:207–216

Landis DA, Wratten SD, Gurr WM (2000) Habitat management to conserve natural enemies of arthropod pests in agriculture. Annu Rev Entomol 45:175–201

Laughton AM, O'Connor CO, Knell RJ (2017) Responses to a warming world: integrating life history, immune investment, and pathogen resistance in a model insect species. Ecol Evol 7:9699–9710

Lee K, Berenbaum MR (1989) Action of antioxidant enzymes and cytochrome P-450 monooxygenases in the cabbage looper in response to plant phototoxins. Arch Insect Biochem Physiol 10:151–162

Lehmann P, Ammunét T, Barton M, Battisti A, Eigenbrode SD, Jepsen JU, Kalinkat G, Neuvonen S, Niemelä TJS, Økland B, Björkman (2020) Complex responses of global insect pests to climate warming. Front Ecol Environ 18:141–150

Li X, Baudry J, Berenbaum MR, Schuler MA (2004) Structural and functional divergence of insect CYP6B proteins: from specialist to generalist cytochrome P450. Proc Natl Acad Sci 101:2939–2944

Li X, Garvey M, Kaplan I, Li B, Carrillo J (2018) Domestication of tomato has reduced the attraction of herbivore natural enemies to pest-damaged plants. Agric For Entomol 20:390–401

Lindström L, Lehmann P (2015) Climate change effects on agricultural insect pests in Europe. In: Björkman C, Niemelä P (eds) Climate change and insect pests. CAB International Press, Oxfordshire, pp 136–153

Loughrin JH, Manukian A, Heath RR, Tumlinson JH (1995) Volatiles emitted by different cotton varieties damaged by feeding beet armyworm larvae. J Chem Ecol 21:1217–1227

Mao W, Rupasinghe S, Zangerl AR, Schuler MA, Berenbaum MR (2006) Remarkable substrate-specificity of CYP6AB3 in Depressaria pastinacella, a highly specialized caterpillar. Insect Mol Biol 15:169–179

McGovern JL, Zangerl AR, Ode PJ, Berenbaum MR (2006) Furanocoumarins and their detoxification in a tri-trophic interaction. Chemoecology 16:45–50

Meiners T, Wäckers F, Lewis WJ (2003) Associative learning of complex odours in parasitoid host location. Chem Senses 28:231–236

Menéndez R, González-Megías A, Lewis OT, Shaw MR, Thomas CD (2008) Escape from natural enemies during climate-driven range expansion: a case study. Ecol Entomol 33:413–421

Meyer RS, Purugganan MD (2013) Evolution of crop species: genetics of domestication and diversification. Nat Rev Genet 14:840–852

Meyer RS, DuVal AE, Jensen HR (2012) Patterns and processes in crop domestication: an historical review and quantitative analysis of 203 global food crops. New Phytol 196:29–48

Mitter C, Farrell B, Wiegman B (1988) The phylogenetic study of adaptive zones: has phytophagy promoted insect diversification? Am Nat 132:107–128

Moya-Raygoza G, Cuevas-Guzmán R, Pinedo-Escatel JA, Morales-Arias JG (2019) Comparison of leafhopper (Hemiptera: Cicadellidae) diversity in maize and its wild ancestor teosinte, and plant diversity in the teosinte habitat. Ann Entomol Soc Am 112:99–106

Müller C, Agerbirk N, Olsen CE (2003) Lack of sequestration of host plant glucosinolates in *Pieris rapae* and *P. brassicae*. Chemoecology 13:47–54

Mumm R, Dicke M (2010) Variation in natural plant products and the attraction of bodyguards involved in indirect plant defense. Can J Zool 88:628–667

Nishida R (2002) Sequestration of defensive substances from plants by Lepidoptera. Annu Rev Entomol 47:57–92

Ode PJ (2006) Plant chemistry and natural enemy fitness: effects on herbivore and natural enemy interactions. Annu Rev Entomol 51:163–185

Ode PJ (2013) Plant defences and parasitoid chemical ecology. In: Wajnberg E, Colazza S (eds) Chemical Ecology of Insect Parasitoids. Wiley-Blackwell, Oxford, pp 11–36

Ode PJ (2019) Plant toxins and parasitoid trophic ecology. Curr Opin Insect Sci 32:118–123

Ode PJ, Charlet LD, Seiler GJ (2011) Sunflower stem weevil and its larval parasitoids in native sunflowers: is parasitoid abundance and diversity greater in the US Southwest? Environ Entomol 40:15–22

Ode PJ, Johnson SN, Moore BD (2014) Atmospheric change and induced plant secondary metabolites – are we reshaping the building blocks of multi-trophic interactions? Curr Opin Insect Sci 5:57–65

Ode PJ, Harvey JA, Reichelt M, Gershenzon J, Gols R (2016) Differential induction of plant chemical defenses by parasitized and unparasitized herbivores: consequences for reciprocal, multitrophic interactions. Oikos 125:1398–1407

Parmesan C (2006) Ecological and evolutionary responses to recent climate change. Annu Rev Ecol Syst 37:637–669

Parmesan C, Yohe G (2003) A globally coherent fingerprint of climate change impacts across natural systems. Nature 399:579–583

Pearse IS, LoPresti E, Schaeffer RN, Wetzel WC, Mooney KA, Ali JG, Ode PJ, Eubanks MD, Bronstein JL, Weber MG (2020) Generalising indirect defence and resistance of plants. Ecol Lett 23:1137–1152

Petschenka G, Agrawal AA (2016) How herbivores coopt plant defenses: natural selection, specialization, and sequestration. Curr Opin Insect Sci 14:17–24

Poelman EH, Zheng S-J, Zhang Z, Heemskerk NM, Cortesero A-M, Dicke M (2011) Parasitoid-specific induction of plant responses to parasitized herbivores affects colonization by subsequent herbivores. Proc Natl Acad Sci 108:19647–19652

Poelman EH, Bruinsma M, Zhu F, Weldegergis BT, Boursault AE, Jongema Y, van Loon JJA, Vet LEM, Harvey JA, Dicke M (2012) Hyperparasitoids use herbivore-induced plant volatiles to locate their parasitoid host. PLoS Biol 10:e1001435

Poirié M, Colinet D, Gatti J-L (2014) Insights into the function and evolution of parasitoid wasp venoms. Curr Opin Insect Sci 6:52–60

Quintero C, Lampert EC, Bowers MD (2014) Time is of the essence: direct and indirect effects of plant ontogenetic trajectories on higher trophic levels. Ecology 95:2589–2602

Rasmann S, Köllner TJ, Degenhardt J, Hiltpold I, Toepfer S, Kulhmann U, Gershenzon J, Turlings TCJ (2005) Recruitment of entomopathogenic nematodes by insect-damaged maize roots. Nature 434:732–737

Ratzka A, Vogel H, Kliebenstein DJ, Mitchell-Olds T, Kroymann J (2002) Disarming the mustard oil bomb. Proc Natl Acad Sci 99:11233–11228

Reddy GVP, Holopainen JK, Guerrero A (2002) Olfactory responses of *Plutella xylostella* natural enemies to host pheromone, larval frass, and green leaf cabbage volatiles. J Chem Ecol 28:131–143

Renwick JAA, Haribal M, Gouinguené SE (2006) Isothiocyanates stimulating oviposition by the diamondback moth, *Plutella xylostella*. J Chem Ecol 32:755–766

Reudler JH, Biere A, Harvey JA, van Nouhuys S (2011) Differential performance of a specialist and two generalist herbivores and their parasitoids on *Plantago lanceolata*. J Chem Ecol 37:765–778

Richards EH, Edwards JP (1999) Parasitization of *Lacanobia oleracea* (Lepidoptera: Noctuidae) by the ectoparasite wasp, *Eulophus pennicornis*: effects of parasitization, venom and starvation on host haemocytes. J Insect Physiol 45:1073–1083

Richards EH, Parkinson NM (2000) Venom from the endoparasitis wasp *Pimpla hypochondriaca* adversely affects the morphology, viability, and immune function of hemocytes from larvae of the tomato moth, *Lacanobia oleracea*. J Invertebr Pathol 76:33–42

Richards LA, Dyer LA, Smilanich AM, Dodson CD (2010) Synergistic effects of amides from two *Piper* species on generalist and specialist herbivores. J Chem Ecol 36:1105–1113

Richards LA, Dyer LA, Forister ML, Smilanich AM, Dodson CD, Leonard MD, Jeffrey CS (2015) Phytochemical diversity drives plant-insect community diversity. Proc Natl Acad Sci 112:10973–10978

Risch SJ, Andow DA, Altieri MA (1983) Agroecosystem diversity and pest control: data, tentative conclusions, and new research directions. Environ Entomol 12:625–629

Rogers CE (1988) Insects from native and cultivated sunflowers (*Helianthus*) in the southern latitudes of the United States. J Agric Entomol 5:267–287

Rogers CE, Kreitner GL (1983) Phytomelanin of sunflower achenes: a mechanism for pericarp resistance to abrasion by larvae of the sunflower moth (Lepidoptera: Pyralidae). Environ Entomol 12:277–285

Rogers CE, Gershenzon J, Ohno N, Mabry TJ, Stipanovic RD, Kreitner GL (1987) Terpenes of wild sunflowers (*Helianthus*): an effective mechanism against seed predation by larvae of the sunflower moth, *Homoeosoma electellum* (Lepidoptera: Pyralidae). Environ Entomol 16:586–592

Rosenthal JP, Dirzo R (1997) Effects of life history, domestication and agronomic selection on plant defence against insects: evidence from maizes and wild relativels. Evol Ecol 11:337–355

Root RB (1973) Organization of a plant-arthropod association in simple and diverse habitats: the fauna of collards (*Brassica oleracea*). Ecol Monogr 43:95–124

Rothwell EM, Holeski LM (2020) Phytochemical defences and performance of specialist and generalist herbivores: a meta-analysis. Ecol Entomol 45:396–405

Rowen E, Kaplan I (2016) Eco-evolutionary factors drive induced plant volatiles: a meta-analysis. New Phytol 210:284–294

Schoonhoven LM, van Loon JJA, Dicke M (2005) Insect-plant biology, 2nd edn. Oxford University Press, Oxford

Singer MS, Stireman JO III (2003) Does anti-parasitoid defense explain host-plant selection by a polyphagous caterpillar? Oikos 100:554–562

Singer MS, Stireman JO III (2005) The tri-trophic niche concept and adaptive radiation of phytophagous insects. Ecol Lett 8:1247–1255

Singer MS, Carrière Y, Theuring C, Hartmann T (2004) Disentangling food quality from resistance against parasitoids: diet choice by a generalist caterpillar. Am Nat 164:423–429

Slinn HL, Richards LA, Dyer LA, Hurtado PJ, Smilanich AM (2018) Across multiple species, phytochemical diversity and herbivore diet breadth have cascading effects on herbivore immunity and parasitism in a tropical model system. Front Plant Sci 9:article 656

Smilanich AM, Dyer LA, Chambers JQ, Bowers MD (2009a) Immunological cost of chemical defence and the evolution of herbivore diet breadth. Ecol Lett 12:612–621

Smilanich AM, Dyer LA, Gentry GL (2009b) The insect immune response and other putative defenses as effective predictors of parasitism. Ecology 90:1434–1440

Speed MP, Fenton A, Jones MG, Ruxton GD, Brockhurst MA (2015) Coevolution can explain defensive secondary metabolite diversity in plants. New Phytol 208:1251–1263

Stam JM, Kroes A, Li Y, Gols R, van Loon JJA, Poelman EH, Dicke M (2014) Plant interactions with multiple insect herbivores: from community to genes. Annu Rev Plant Biol 65:689–713

Stamp N (2001) Enemy-free space via host plant chemistry and dispersion: assessing the influence of tri-trophic interactions. Oecologia 128:153–163

Steidle JLM, van Loon JJA (2003) Dietary specialization and infochemical use in carnivorous arthropods: testing a concept. Entomol Exp Appl 108:133–148

Strand MR, Beck MH, Lavine MD, Clark KD (2006) *Microplitis demolitor* bracovirus inhibits phagocytosis by hemocytes from *Pseudoplusia includens*. Arch Insect Biochem Physiol 61:134–145

Strauss SY, Zangerl AR (2002) Plant-insect interactions in terrestrial ecosystems. In: Herrera CM, Pellmyr O (eds) Plant-animal Interactions: an evolutionary approach. Blackwell Science, Oxford, pp 77–106

Sun R, Jiang X, Reichelt M, Gershenzon J, Pandit SS, Vassão DG (2019) Tritrophic metabolism of plant chemical defenses and its effects on herbivore and predator performance. elife 8:e51029. https://doi.org/10.7554/eLife.51029

Sun R, Gols R, Harvey JA, Reichelt M, Gershenzon J, Pandit SS, Vassão DG (2020) Detoxification of plant defensive glucosinolates by an herbivorous caterpillar is beneficial to its endoparasitic wasp. Mol Ecol 29:4014–4031

Tamiru A, Bruce TJA, Woodcock CM, Caulfield JC, Midega CAO, Ogol CKPO, Mayon P, Birkett MA, Pickett JA, Khan ZR (2011) Maize landraces recruit egg and larval parasitoids in response to egg deposition by a herbivore. Ecol Lett 14:1075–1083

Tan C-W, Peiffer ML, Ali JG, Luthe DS, Felton GW (2020) Top-down effects from parasitoids may mediate plant defence and plant fitness. Funct Ecol 34:1767–1778. https://doi.org/10.1111/1365-2435.13617

Tscharntke T, Rand TA, Bianchi FJJA (2005) The landscape context of trophic interactions: insect spillover across the crop–noncrop interface. Ann Zool Fenn 42:421–432

Tscharntke T, Karp DS, Chaplin-Kramer R, Batáry P, DeClerck F, Gratton C, Hunt L, Ives A, Jonsson M, Larsen A, Martin EA, Martínez-Salinas A, Meehan TD, O'Rourke M, Poveda K, Rosenheim JA, Rusch A, Schellhorn N, Wanger TC, Wratten S, Zhang W (2016) When natural habitat fails to enhance biological pest control – five hypotheses. Biol Conserv 204:449–458

Turcotte MM, Thomsen CJM, Broadhead GT, Fine PVA, Godfrey RM, Lamarre GPA, Meyer ST, Richards LA, Johnson MTJ (2014a) Percentage leaf herbivory across vascular plant species. Ecology 95:788

Turcotte MM, Turley NE, Johnson MTJ (2014b) The impact of domestication on resistance to two generalist herbivores across 29 independent domestication events. New Phytol 204:671–681

Turlings TCJ, Erb (2018) Tritrophic interactions mediated by herbivore-induced plant volatiles: mechanisms, ecological relevance, and application potential. Annu Rev Entomol 63:433–452

Turlings TCJ, Tumlinson JH, Lewis WJ (1990) Exploitation of herbivore-induced plant odors by host-seeking parasitic wasps. Science 250:1251–1253

Turlings TCJ, Urs B, Lengwiler UB, Bernasconi ML, Wechsler D (1998) Timing of induced volatile emissions in maize seedlings. Planta 207:146–152

Turlings TCJ, Alborn HT, Loughrin JH, Tumlinson JH (2000) Volicitin, an elicitor of maize volatiles in oral secretions of *Spodoptera exigua*: isolation and bioactivity. J Chem Ecol 26:189–202 van Asch M, Visser ME (2007) Phenology of forest caterpillars and their host trees: the importance of synchrony. Annu Rev Entomol 52:37–55

van der Putten WH, Macel M, Visser ME (2010) Predicting species distribution and abundance responses to climate change: why it is essential to include biotic interactions across trophic levels. Philos Trans R Soc B 365:2025–2034

van Dongen S, Backeljau T, Matthysen E, Dhondt AA (1997) Synchronization of hatching date with budburst of individual host trees (*Quercus robur*) in the winter moth (*Operophtera brumata*) and its fitness consequences. J Anim Ecol 66:113–121

van Doorn HE, van der Kruk GC, van Holst G-J, Raaijmaker-Ruijs NCME, Postma E, Groeneweg B, Jongen WHF (1998) The glucosinolates sinigrin and progoitrin are important determinants for taste preference and bitterness of Brussels sprouts. J Sci Food Agric 78:30–38

van Nouhuys S, Lei G (2004) Parasitoid-host metapopulation dynamics: the causes and consequences of phenological asynchrony. J Anim Ecol 73:526–535

Vet LEM, Dicke M (1992) Ecology of infochemical use by natural enemies in a tritrophic context. Annu Rev Entomol 37:141–172

Vidal MC, Murphy SM (2018) Bottom-up vs. top-down effects on terrestrial insect herbivores: a meta-analysis. Ecol Lett 21:138–150

Visser ME, Holleman LJM (2001) Warmer springs disrupt the synchrony of oak and winter moth phenology. Proc R Soc Lond Ser B 268:289–294

Wan N-F, Li X, Guo L, Ji X-Y, Zhang H, Chen Y-J, Jiang J-X (2018) Phytochemical variation mediates the susceptibility of insect herbivores to entomoviruses. J Appl Entomol 142:705–715

Wang J-Y, Zhang H, Siemann E, Ji X-Y, Chen Y-J, Wang Y, Jiang J-X, Wan N-F (2020) Immunity of an insect herbivore to an entomovirus is affected by different host plants. Pest Manag Sci 76:1004–1010

Ward SF, Aukema BH, Fei S, Liebhold AM (2020) Warm temperatures increase population growth of a non-native defoliator and inhibit demographic responses by parasitoids. Ecology 101:e03156. https://doi.org/10.1002/ecy.3156

Wen Z, Zeng RS, Niu G, Berenbaum MR, Schuler MA (2009) Ecological significance of induction of broad-substrate cytochrome P450s by natural and synthetic inducers in *Helicoverpa zea*. J Chem Ecol 35:183–189

Whitehead SR, Turcotte MM, Poveda K (2017) Domestication impacts on plant-herbivore interactions: a meta-analysis. Philos Trans R Soc B 372:20160034

Whittaker RH, Feeny PP (1971) Allelochemicals: chemical interactions between species. Science 171:757–770

Williams IS (1999) Slow-growth, high-mortality – a general hypothesis, or is it? Ecol Entomol 24:490–495

Wittstock U, Agerbirk N, Stauber EJ, Olsen CE, Hippler M, Mitchell-Olds T, Gershenzon J, Vogel H (2004) Successful herbivore attack due to metabolic diversion of a plant chemical defense. Proc Natl Acad Sci 101:4859–4864

Yang LH, Cenzer ML (2020) Seasonal windows of opportunity in milkweed–monarch interactions. Ecology 101:e02880

Zavala JA, Nabity PD, DeLucia EH (2013) An emerging understanding of mechanisms governing insect herbivory under elevated CO_2. Annu Rev Entomol 58:79–97

Zhu F, Cusumano A, Bloem J, Weldegergis BT, Villela A, Fatouros NE, van Loon JJA, Dicke M, Harvey JA, Vogel H, Poelman EH (2018) Symbiotic polydnavirus and venom reveal parasitoid to its hyperparasitoids. Proc Natl Acad Sci 115:5205–5210

Zhu-Salzman K, Zeng R (2015) Insect response to plant defense protease inhibitors. Annu Rev Entomol 60:233–252

Zvereva EL, Kozlov MV (2016) The costs and effectiveness of chemical defenses in herbivorous insects: a meta-analysis. Ecol Monogr 86:107–124

Host Plants as Mediators of Caterpillar-Natural Enemy Interactions

John T. Lill and Martha R. Weiss

Aposematic, sequestering catalpa sphinx *Ceratomia catalpae* caterpillar with parasitoid cocoons. Photo: John T. Lill

J. T. Lill (✉)
Department of Biological Sciences, George Washington University, Washington, DC, USA
e-mail: lillj@gwu.edu

M. R. Weiss
Georgetown University, Washington, DC, USA

Introduction

The preceding chapters in this book have summarized the large and growing body of work focusing on adjacent trophic level interactions (e.g., those between plants and caterpillars and between caterpillars and their various natural enemies). The chapters in this section consider some of the complex ways that three (or more) trophic levels can interact to shape caterpillar ecology and evolution. Beginning in the early 1980s, as Price et al. (1980) and Bernays and Graham (1988) began the initially contentious task of integrating across multiple trophic levels, and as evidence of terrestrial trophic cascades involving plants, herbivores, and natural enemies began to accumulate (Marquis and Whelan 1994; Schmitz et al. 2000; Mooney et al. 2010; Mooney and Singer 2012; Abdala-Roberts and Mooney 2013), we gained new insights into the myriad ways that members of these three trophic levels could interact with one another, both directly and indirectly (Abdala-Roberts et al. 2019). Additionally, the evolutionary implications of these tritrophic interactions began to become increasingly apparent. Singer and Stireman (2005) pushed these ideas forward with a new conceptual framework, outlining how tritrophic niches are both formed and occupied, and empirical studies from a variety of systems are now validating these ideas (e.g., Condon et al. 2014).

While it is now widely recognized that both bottom-up and top-down forces often interact to shape the phenology, morphology, behavior, physiology, and fitness of caterpillars (Hunter and Price 1992; Kaplan et al. 2016), in this chapter we focus on describing the accumulated evidence supporting and/or refuting proposed mechanisms for host plant mediation of interactions between caterpillars and their natural enemies. The evidence that genetically based intraspecific trait variation mediates herbivore-natural enemy interactions has been reviewed elsewhere (e.g., Hare 1992) and is beyond the scope of this chapter. We instead focus here on the much wider trait variation that can be found among host plant species utilized by oligophagous or polyphagous caterpillars (e.g., *Epargyreus clarus*; Plate 1). Interspecific variation in plant traits can alter the risk of mortality from natural enemies directly or can function indirectly, via effects on caterpillar behavioral and/or physiological responses to their host plants, which in turn affect larval risk of mortality from natural enemies (Agrawal 2000). Because many traits can influence both the 2nd and the 3rd trophic levels simultaneously, we have organized our discussion around the major themes of plant traits implicated in tritrophic interactions (rather than by interaction pathway) and indicate the relevant interaction pathway(s) within each section (Plate 2).

While many of these traits and their hypothesized effects on herbivore-natural enemy interactions were initially described in Price et al. (1980), a considerable amount of empirical evidence has accumulated in the ensuing 40 years, and additional interaction pathways have emerged that require further exploration. For example, there is now widespread appreciation for the rapid and sometimes incredibly specific ways in which plant damage by caterpillars can elicit the release of herbivore-induced plant volatiles (HIPVs; Turlings and Erb 2018) that have

Plate 1 Silver-spotted skipper (*Epargyreus clarus*, Hesperiidae) caterpillars interact with their leguminous host plants in several ways that may influence their vulnerability to predators and parasitoids: all silver-spotted skipper caterpillars cut and silk their host leaves to produce shelters, a time-consuming activity that increases their exposure to natural enemies (**a**); caterpillars often must spend time removing trichomes (i.e., "mowing the lawn") on densely pubescent leaves prior to initiating shelter-building (**b**); sharp trichomes, like those on the surface of *Desmodium* leaves, can impale a young caterpillar, making it vulnerable to enemies and/or desiccation and impeding its forward motion, resulting in a thick white crescent of silk on leaves with impaled caterpillars (**c**); and dense trichomes found on the edges and surfaces of young kudzu leaves slow the forward progress of an early instar larva (**d**) (all photos by M. Weiss)

important impacts on herbivore mortality. The myriad ways in which plants directly or indirectly provision a broad set of natural enemies that engage in mutualistic relationships with plants have also become increasingly apparent (Bronstein 1998; Stapley 1998; Wäckers et al. 2005). Recent research has also highlighted the importance of plant architecture and foliar pubescence as two important drivers of caterpillar and natural enemy foraging and patch use. Finally, because plant traits and their influence on both herbivores and natural enemies have been shown to change over ontogeny (Boege and Marquis 2005; Boege et al. 2019), this axis of variation must be explicitly considered in designing experimental tests to examine the direct and indirect effects of these traits.

We have worked for many years on the ecology and behavior of the silver-spotted skipper, *Epargyreus clarus* (Hesperiidae), a widespread oligophagous butterfly, the biology of which provides useful insights into the many ways in which plant traits can influence the outcomes of herbivore-natural enemy interactions. We highlight some of our most recent findings in pertinent sections below.

Plate 2 Examples of ways in which plant traits can mediate caterpillar-natural enemy interactions. Caterpillars like this monarch larva feeding on lactiferous plants often spend considerable time trenching prior to feeding, which can increase exposure to natural enemies (**a**; photo by A. Agrawal); host plant variation in provisioning, including the extrafloral nectaries of peonies that strongly recruit ants, can pose differential threats to caterpillars (**b**; photo by M. Weiss); host plant variation in concentrations of plant secondary chemicals like catalpol in *Catalpa* spp. can influence levels of sequestration by specialist herbivores like these catalpa sphinx caterpillars, which in turn can impact larval defenses against parasitoids and predators (**c**; photo by J. Lill); plant architectural traits like the growth form and branching pattern of Menzie's goldenbush have been shown to influence access to these plants by insectivorous birds, which in turn impacts the strength of trophic cascades in the California chaparral (**d**; Nell and Mooney 2019; photo by D. Valov); foliar water content of alternative host plants can influence caterpillar defensive behaviors, as in this silver-spotted skipper caterpillar (**e**; photo by C. Block); and the chemical composition of floral nectar can influence parasitoid foraging, potentially increasing encounters with host caterpillars (**f**; photo by A. Zemenick)

Chemical Defenses

We begin by briefly discussing plant chemical defenses, which can indirectly affect natural enemies by altering caterpillar behavioral and physiological responses. While we follow the traditional division of plant secondary metabolites into lethal vs. sublethal "defensive" compounds, we note that this narrative has been recently challenged (Smilanich et al. 2016). As such, we encourage future researchers to assess multitrophic interactions without presumption and focus on elucidating the specific modes of action of these compounds within their natural ecological contexts.

Lethal Compounds Host plants that defend their tissues with toxins (e.g., furanocoumarins, cardenolides, glucosinolates, pyrrolizidine alkaloids, cyanogenic and iridoid glycosides, and aristolochic acids) are typically attacked by highly specialized caterpillar taxa that have evolved physiological and/or behavioral detoxification mechanisms that enable larval feeding (Berenbaum 1983; Boppré 1990;

Nishida 2002; Wittstock et al. 2004; Dobler et al. 2011; Agrawal et al. 2012). Because these toxins often deter or kill most herbivores that attempt to feed on them, their role in modulating caterpillar-natural enemy interactions is somewhat limited, as noted by Price et al. (1980). However, the various host plant species imbued with these chemicals can vary considerably in both toxin concentration and composition (e.g., Agrawal et al. 2015). For caterpillars that have evolved to subvert, tolerate, or detoxify these compounds (rather than sequester them), the interspecific variation in plant chemical defenses encountered on alternative hosts can theoretically translate into variation in caterpillar susceptibility to natural enemies. However, beyond demonstrating host plant variation in caterpillar performance measures on alternative toxin-containing plants (e.g., non-sequestering milkweed-feeding caterpillars assayed in Petschenka and Agrawal 2015), we are unaware of any studies that clearly link the toxin concentrations of alternative host plants and differential natural enemy attack for non-sequestering specialist caterpillars.

For the subset of specialist caterpillars that has evolved the ability not only to tolerate, but to *sequester* plant toxins (e.g., monarchs, queens, pipevine swallowtails, buckeyes, catalpa sphinx, and woolly bears (discussed in detail in Bowers, Chapter "Sequestered Caterpillar Chemical Defenses: From "Disgusting Morsels" to Model Systems"), reduced caterpillar palatability and/or rates of mortality by a variety of natural enemies often result (Järvi et al. 1981; Bowers and Farley 1990; Dyer and Floyd 1993; Dyer 1995; Theodoratus and Bowers 1999; Sime 2002). A diverse set of caterpillar taxa sequestering different secondary compounds has also been found to deposit these compounds in their integuments (reviewed in Sime 2002), presumably as an honest signal of their unpalatability that may prevent lethal injury by naive predators. In systems where sequestration levels reflect dietary toxin concentrations, alternative host plant associations could result in differential attack and consumption of these variably defended caterpillars by natural enemies (Dyer and Bowers 1996; Camara 1997; Theodoratus and Bowers 1999). The natural enemies themselves could respond either positively (e.g., some specialist parasitoids; Richards et al. 2012) or negatively (e.g., many generalist predators) to sequestered toxin concentrations of their hosts and prey, respectively. In their study involving specialist caterpillars feeding on two different toxin-containing host plants, Richards et al. (2012) found that iridoid glycoside (IG) sequestration levels of buckeye caterpillars (*Junonia coenia*, Nymphalidae) were positively correlated with growth but negatively correlated with host immune responses to parasitism, suggesting possible ecological trade-offs in how these toxins may affect different measures of caterpillar fitness.

For predators that use olfaction as the proximate cue in assessing caterpillar prey quality/palatability (e.g., many ant species, most small mammals), there may be a greater potential for host plant-determined levels of sequestration to modulate prey selection behaviors. Moreover, for specialist parasitoids that often use host plant-produced volatile organic compounds (VOCs) as coarse-grained host location cues (Godfray 1994), such non-toxin cues could be honest signals of potential host sequestration levels that result in differential parasitism. As a hypothetical example, parasitoids of catalpa sphinx (*Ceratomia catalpae*) caterpillars feeding on species of *Catalpa* differing in levels of IGs may prefer VOC signals of the tree species hosting caterpillars with the highest IG sequestration levels, which consequently have the lowest immune responses (following from Richards et al. 2012, cited above). We propose that additional research on the tritrophic consequences of a wider array of specialist caterpillars reared on alternative host plants in natural settings or common gardens could potentially shed light on how different natural enemy complexes may be shaping local patterns of host plant preference and use.

Sublethal Compounds A second major category of plant traits hypothesized to influence caterpillar-natural enemy interactions includes sublethal plant secondary chemicals, which while not typically toxic to most caterpillars have been shown to have significant impacts on a variety of caterpillar performance measures. Historically referred to as "quantitative" defenses (Feeny 1976), these include a wide array of mostly higher molecular weight, nonvolatile secondary compounds such as phenolics, lectins, some terpenoids, and proteinase inhibitors as well as structural compounds such as lignins, cellulose, etc. (Rosenthal and Berenbaum 1991). While these "dose-dependent" compounds can theoretically defend plants by directly impeding caterpillar feeding and biomass gain (reviewed in Ayres et al. 1997), the primary mode of action is thought to be indirect; by slowing down caterpillar growth, particularly during highly vulnerable early instars, these plant compounds could potentially increase natural enemy-mediated mortality and decrease plant damage (Benrey and Denno 1997). This slow-growth high-mortality hypothesis (hereafter abbreviated as SG-HM hypothesis; Clancy and Price 1987) highlights the importance of the third trophic level as a critical component of the arsenal of plant defenses. Historically, an additional appeal of the SG-HM hypothesis has been that it neatly resolves a paradox of sublethal plant defenses: in the absence of natural enemies, compounds that reduce the ability of herbivores to assimilate nutrients from plant food (i.e., that reduce digestibility or limit access to limiting nutrients) increase herbivore consumption and plant damage (Moran and Hamilton 1980; Price et al. 1980; but see Neuvonen and Haukioja 1984). For example, when oxidized, some phenolic compounds found in the foliage of many broad-leaved trees produce quinones, which bind to leaf proteins, inhibiting their digestion in caterpillar guts (Taranto et al. 2017). Presumably, caterpillars feeding on plants containing these compounds would grow more slowly, thus exposing them to increased risk of predation and parasitism.

While the basic tenets of the SG-HM hypothesis – that caterpillars exhibit differential growth rates and/or "windows of vulnerability" when reared on host plants differing in nutritional quality – have been demonstrated in multiple laboratory/ greenhouse studies (e.g., Isenhour et al. 1989; Benrey and Denno 1997; Medina et al. 2005; Coley et al. 2006; Kursar et al. 2006; Shikano et al. 2018), evidence from field studies has mostly failed to find empirical support for the SG-HM hypothesis in caterpillars (Lill and Marquis 2001; Murphy 2004; Cornelissen and Stiling 2006; Kursar et al. 2006) or has demonstrated the opposite, i.e., that natural enemy pressure was *greater* for faster-growing caterpillars on higher-quality plants

(Damman 1987; Leather and Walsh 1993; Idris and Grafius 1996; Benrey and Denno 1997; Medina et al. 2005; Singer et al. 2012). It is worth noting that the only manipulative field experiments involving caterpillars that have supported the predictions of the SG-HM hypothesis have altered caterpillar growth using intraspecific "treatments" (as noted in Farkas and Singer 2013). These have included adding fertilizer to produce plants with different levels of foliar nitrogen (Loader and Damman 1991), rearing caterpillars on plants with different leaf ages (Parry et al. 1998), or contrasting caterpillar performance on resistant vs. susceptible cultivars (Johnson and Gould 1992). In contrast, we are not aware of any field experiments in which caterpillar growth was manipulated by rearing them on alternative host plant species that have supported the hypothesis.

Two reviews of the SG-HM hypothesis (Williams 1999; Chen and Chen 2018) have pointed out the need for rigorous experiments that track herbivore development and survival under natural field conditions where they are exposed to the full complement of potential natural enemies, for the duration of larval development. In our recent work on the evolutionary ecology of host use in the silver-spotted skippers (*E. clarus*), we tracked the individual fates of hundreds of bagged and exposed caterpillars from egg hatch to the end of larval development on six different leguminous host plants which varied markedly in nutritional content. Despite strong and consistent differences in development time on these alternative host plants, we found little to no support for the SG-HM hypothesis across 3 years and eight generations of *E. clarus* caterpillars, in what we believe is the most comprehensive field-based test of the hypothesis to date (Weiss, Lill & Lind, in preparation). Four decades after it was first proposed to explain the functioning of sublethal plant chemical defenses, the SG-HM hypothesis has been little supported in tritrophic systems involving caterpillars.

Physical Defenses

Physical features of the plant, including trichomes, surface waxes, and leaf toughness, can directly affect the movements and behaviors of herbivores (Kaur et al. Chapter "Surface Warfare: How Plant Structural Defenses Battle Caterpillars") and can also impact those of natural enemies both directly and indirectly (Kennedy 2003; Cortesero et al. 2000; Peterson et al. 2016). Plant physical defenses are often integrated into biological control strategies, but the dynamics can be complicated, as plant surface traits that impede herbivores may also help or hinder the abilities of natural enemies to attack targeted pests (Eigenbrode et al. 1995; Cortesero et al. 2000; Peterson et al. 2016).

Trichomes, both nonglandular and glandular, can directly affect caterpillar behaviors in ways that impact natural enemies. For example, trichomes can impede feeding, movement, and other larval "establishment" activities (shelter-building, burrowing into a leaf to produce a mine), slowing them down and thereby increasing visibility, exposure time, and/or vulnerability to natural enemies. These impacts can be particularly consequential for first instar larvae, which suffer very high mortality, the specific causes of which are difficult to untangle (Zaluki et al. 2002).

In a field investigation of the success of first instar silver-spotted skipper (*Epargyreus clarus*) hatchlings at establishment on six different leguminous host plants, Weiss et al. (in preparation) found a consistent sixfold range in larval failure to construct a leaf shelter within 24 h, depending on plant species. We note that all *E. clarus* larvae obligately construct leaf shelters on their host plants, starting soon after hatching, and making four or five different shelters across their larval lifespan (Weiss et al. 2003); first instar larvae that cannot make a shelter do not survive. Trichome densities on the plants ranged from smooth to densely hairy, but this factor alone was not predictive of hatchling establishment; the host plant with the highest trichome density had the lowest hatchling failure rate, and the plant with the lowest trichome density had the highest failure rate. The plants also differed in leaf toughness and chemistry, traits that are likely to contribute to success at establishment. Although trichome density in the field did not correlate with failure to construct a shelter, follow-up laboratory studies demonstrated that *time* to complete a shelter was in fact correlated with trichome density and that larval behavior varied consistently across the three host plants with the densest trichome coverage, adding another layer of complexity to the plant-herbivore interface (Weiss et al., in prep). As Bernays (2003) demonstrated in her oft-cited paper, "Feeding by lepidopteran larvae is dangerous," caterpillars may face as much a 100-fold higher risk of predation when feeding than when resting, presumably because head movements (and perhaps chewing sounds) attract the attention of predators. Shelter-building by *E. clarus* is even more visually apparent than feeding, and so it is likely that an extended shelter-building time exposes larvae to increased predation risk.

Consistent with predictions by Boege and Marquis (2005), plant ontogeny can affect the course of plant-herbivore interactions, as can the age or ontogenetic stage of the herbivore. On kudzu (*Pueraria montana*), first instar *E. clarus* larvae were unable to initiate shelter construction on densely haired young leaves and were able to do so only after the leaf had expanded to the point that the larvae were able to contact the leaf surface. Third and fourth instar larvae, however, were able to initiate feeding on younger leaves (Weiss et al., in prep.). Kariyat et al. (2018) reported, similarly, that first and second instar *Manduca sexta* (Sphingidae) larvae were effectively deterred from initiating feeding by the presence of nonglandular trichomes on a number of different solanaceous species, while third instar larvae were not similarly inhibited.

In addition to directly impacting feeding, movement, and establishment by caterpillars, leaf surface characteristics can indirectly affect herbivores by impeding or, rarely, facilitating movement by predators or parasitoids. In many cases, trichome coverage causes more harm than good to natural enemies (Riddick and Simmons 2014). The interaction between trichomes and the third trophic level has been particularly well studied in tomatoes and their relatives (*Solanum* and *Lycopersicon* spp.), the leaves of which bear many types of glandular and nonglandular trichomes (Kennedy 2003, and refs therein). It has been demonstrated, for example, that on *Lycopersicon* spp., nonglandular trichomes can hamper searching behavior of predators and parasitoids, glandular trichomes can entrap small hymenopterous parasitoids in sticky exudates and reduce predator mobility, and exudates of glandular trichomes can be directly toxic to natural enemies of pests. Similarly, on tobacco leaves, tiny *Trichogramma minutum* wasps get trapped in the sticky exudate of glandular trichomes and are unable to parasitize *Manduca sexta* eggs, though the wasps readily parasitize the eggs on other substrates (Keller 1987). Lacewings, predatory beetles, parasitoids, and true bugs can die when they get stuck in exudates from glandular trichomes or are impaled on sharp nonglandular trichomes, though a consequent negative impact on plant fitness has not been demonstrated (Riddick and Simmons 2014). Conversely, some "sticky plants" enjoy higher densities of predators that are attracted by the accumulated insect carrion found on plants bearing glandular trichomes. This easily accessible prey can be particularly important during the juvenile stages of these predaceous insects, which in later developmental stages shift from feeding on immobile carrion to mobile prey, including caterpillars (Krimmel and Pearse 2013). In an interesting twist on the defensive role of trichomes, Weinhold and Baldwin (2011) reported that consumption of glandular trichomes of *Nicotiana attenuata* by neonate *Manduca sexta* larvae imparts a distinctive volatile profile to both their body and frass and that ground-foraging *Pogonomyrmex rugosus* ants use these distinctive odors to locate their larval prey.

Plant exudates, including latex and resins, are another widespread category of defense, with both physical and chemical components. These sticky exudates can gum up the mouthparts of chewing herbivores and can also dose them with a range of toxic chemicals (Doussord 2017 and refs therein). Caterpillars in many families have evolved behavioral counteradaptations to these defenses, including vein cutting, trenching, girdling, and leaf clipping, which drain the latex or resin and allow the caterpillar to safely feed distal to the cut (Doussourd and Eisner 1987; Doussord 2017). While the exudates themselves are a direct plant defense against herbivory, the caterpillars' time-consuming and visible pre-feeding behaviors, like the shelter-building behaviors described above for *E. clarus*, are likely to increase caterpillar exposure to natural enemies (Bergelson and Lawton 1988; Greeney et al. 2012) and thus could constitute an indirect defense, the effectiveness of which may vary among host plants.

Epicuticular waxes, another common leaf surface feature, can positively or negatively affect the movement of herbivores and their enemies, such that the net effect of three-dimensional waxy blooms on herbivory will vary from system to system (Eigenbrode 2004). Commonly, on surfaces covered with dense 3D waxes, insects are able to attach only weakly, and are manifold less able to hold on, relative to non-waxy surfaces (Gorb and Gorb 2017 and refs. therein). Eigenbrode and Espelie (1995) reported that three different generalist predators were more effective at capturing *Plutella xylostella* larvae on wax-deficient glossy-leaved than on normal-wax cabbage plants, due to the improved mobility of the walking predators on glossy leaf surfaces. They also reported reduced mining by *P. xylostella* on glossy leaves, potentially increasing larval exposure to predation.

The protective nature of leaf toughness is generally considered in a bitrophic context, with recent work exploring the role of various leaf structural traits in defense against different kinds of herbivores and with attention to ontogenetic changes in both insect feeding behavior and plant structural defenses (Hanley et al. 2007; Malishev and Sanson 2015; Caldwell et al. 2016). Such ontogenetic changes in both plant and insect may interact in ways that are affected by the third trophic level. For example, Damman (1987) described an interesting situation in which *Omphalocera munroei* (Pyralidae) larvae collectively built leaf ties on older, tougher leaves, which, though nutritionally inferior to tender young leaves, were more resistant to predator attack. The toughness of the older leaves necessitated that larvae feed in groups of at least 20 individuals, which together were able to manipulate the relatively inflexible leaf. Given the ubiquity of shelter-building and use by lepidopteran larvae (Marquis et al. Chapter "The Impact of Construct-Building by Caterpillars on Arthropod Colonists in a World of Climate Change"), additional studies relating leaf toughness to the defensive properties of their shelters against natural enemies, particularly for caterpillars that feed on multiple host plant taxa, are needed.

Plant Architecture

It is now widely recognized that plant architecture, which is also referred to as plant structural complexity (McCoy and Bell 1991), is an important plant trait that can mediate predator-prey dynamics broadly (Grof-Tisza et al. 2017) and herbivore-natural enemy interactions specifically (reviewed in Marquis and Whelan 1996). Here we treat plant architecture as a trait that acts directly on the third trophic level, by facilitating or impeding natural enemy foraging activities and/or access to caterpillar prey. While several studies have found that leaf shape, size, arrangement, and surface features can influence arthropod predator and parasitoid foraging success (e.g., Kauffman and Kennedy 1989; Grevstad and Kleptetka 1992; Clark and Messina 1998; Reynolds and Cuddington 2012), most of these studies have not involved caterpillars. There is some evidence that plant architecture can influence the dynamics of caterpillar-vespid wasp (e.g., Geizenauer and Bernays 1996) and caterpillar-parasitoid (e.g., Pimentel 1961) interactions, but the overwhelming majority of studies involving caterpillars have focused on caterpillar-insectivorous bird interactions on woody plants with variable architectures. The physical size, arrangement, and density of structural elements (leaves and stems) on woody plants have been shown to affect multiple aspects of avian foraging, including host plant preference, searching time, and energy expenditures (Holmes and Robinson 1981; Robinson and Homes 1984; Whelan 1989, 2001), with resulting impacts on caterpillar densities, folivory, and plant biomass (e.g., Marquis and Whelan 1994; Van Bael et al. 2008; Whelan et al. 2008; Mäntylä et al. 2011). More recent work has focused on how woody plant architecture, specifically the density and heterogeneity of structural elements, can serve as a size-selective filter that could limit the access to potential predators (Grof-Tisza et al. 2017). A recent study found a strong negative relationship between the structural complexity of shrubs in the California coastal sage scrub ecosystem and the ability of resident birds to reduce insect herbivore abundance on those plants (Nell and Mooney 2019); plants with more open, less dense branching patterns offered birds greater access to the interior of these shrubs, thus enhancing their ability to glean herbivores from their foliage. Moreover, because avian-mediated indirect defenses traded off with direct, anti-herbivore plant defenses in this set of co-occurring shrubs, the earlier suggestion by Marquis and Whelan (1996) that natural enemies can exert selective pressures on plant architectural traits was strongly supported (Nell and Mooney 2019). We expect that advances in three-dimensional imaging technologies may facilitate the speedy and precise quantification of plant architectural traits (e.g., the fractal index and plant complexity index described in Halley et al. 2004 and Grof-Tisza et al. 2017, respectively), spurring additional research in this fruitful area.

Resources

We briefly discuss two categories of plant resources that are used by the third trophic level: plant provisioning of tangible resources (e.g., food and shelter) and plant provisioning of information (i.e., volatiles that can indicate location and in some cases identity of the caterpillar).

Tangible Resources In a botanical illustration of the adage, "the enemy of my enemy is my friend," plants from a broad taxonomic range provide physical resources, including food, water, and shelter, to natural enemies of their herbivores, potentially resulting in a reduction of herbivore damage (Koptur et al. 2015; Heil 2015; Wackers et al. 2005). Such relationships can be obligate or facultative.

The most common category of obligate relationship occurs between ants and their partner "ant plants," or myrmecophytes. These relationships, which are often very specialized, involve tropical angiosperms in at least 20 families (Davidson and McKey 1993) and ants in 5 subfamilies of Formicidae (Davidson 1997; Bronstein et al. 2006). The plants commonly provide nourishment to the ants in the form of carbohydrate from extrafloral nectaries and protein from specialized food bodies (pearl bodies, Beltian bodies, etc.), as well as specialized structures in which the ants can make a nest. The ants, for their part, patrol the plant and chase off or kill potential herbivores (e.g., Rico-Gray and Oliveira 2007). Exclusion of ants from myrmecophytes results in increased levels of herbivory (Heil and McKey 2003; Heil 2008). Several relatively recent reviews document these fascinating relationships (Bronstein et al. 2006; Rico-Gray and Oliveira 2007; Heil 2015).

More taxonomically and geographically widespread than obligate ant-plant relationships, many plants have facultative relationships in which they provide food resources to natural enemies, including ants as well as mites, spiders, parasitic hymenopterans, and predatory hymenopterans, hemipterans, and beetles (Wäckers et al. 2005; Koptur et al. 2015). Plants in over 108 plant families (Weber and Keeler 2013) secrete extrafloral nectar (EFN) from specialized secretory cells that can be located on many different plant tissues (e.g., leaf bases and blades, flower buds, petioles, and developing fruits). The number, distribution, and nectar production rate of EFNs can be induced by herbivory, either directly or through exposure of undamaged plants to herbivore-induced volatiles (Heil 2015 and references therein; Yamawo and Suzuki 2018).

In a number of experimental trials in which ants were excluded from EFNs, herbivory levels on the test plants were increased over those on unmanipulated controls; some studies explicitly note attacks on or removal of lepidopteran herbivores (Koptur et al. 2015; Yamawo et al. 2012; Yamawo and Suzuki 2018). Hymenopteran parasitoids were reported to remain longer and attack more larvae of *Helicoverpa zea*, in the presence of extrafloral nectar (Stapel et al. 1997). Patrolling ants may or may not reduce oviposition by butterflies (Koptur et al. 2015), and their effectiveness varies depending on the growing conditions of the plant with nectaries (Jones et al. 2017). Rudgers and Strauss (2004), in one of the few studies that integrates trait expression, plant fitness, and genetic variation, demonstrated that reduction of nectar production by EFNs on wild cotton resulted in reduced ant recruitment, increased herbivory, and decreased plant fitness.

While EFNs provide resources for predators and parasitoids, and flowers often provide nectar for pollinators, natural enemies can also benefit from access to floral nectar (Heimpel and Jervis 2005; Zemenick et al. 2019; Patt et al. 1997; Tooker et al. 2006). A range of hymenopteran and dipteran predators and parasitoids (including syrphids, tachinids, and eulophids) have been recorded nectaring on a diversity of flowering plants, particularly on flowers with either exposed or partially exposed nectaries, such as many plants in the Apiaceae (Patt et al. 1997). As with extrafloral nectar, acquisition of carbohydrates from floral nectar can contribute to the longevity, fecundity, and, potentially, prey encounter rate of natural enemies (Russell 2015 and references therein).

Guttation droplets are also a potentially important plant resource that may play a role in multi-trophic interactions. Though droplets released from the margins of leaves through guttation have long been considered a water source for insects, Urbaneja-Bernat et al. (2020) recently reported that the liquid in fact contains proteins and carbohydrates and is a reliable source of nutrients for herbivores, parasitoids, and predators throughout the growing season. Insects fed on guttation droplets had higher fecundity and longevity relative to insects fed on water, and furthermore, in field trials, the presence of guttation droplets increased the number of predators and parasitic wasps visiting the plants. Further studies should investigate the generality of this finding and the potential impacts of guttation droplets on multi-trophic plant-insect interactions.

Leaves of many plants bear domatia, very small invaginations in the leaf epidermis, or dense tufts of hair at vein junctions that provide refuge for small predatory arthropods, most commonly predatory or fungivorous mites (Walter and O'Dowd 1992; Norton et al. 2001; Romero and Benson 2005) but also predatory minute pirate bugs (Agrawal et al. 2000). This mutualistic association between arthropods and plants, which is particularly common in temperate deciduous forest trees (as well as a range of agricultural crops), has been shown to decrease herbivore damage to leaves, particularly through reduction of herbivorous mites by predatory mites (Romero and Benson 2005). Although predatory mites rarely attack caterpillars, small predaceous arthropods such as the minute pirate bug do, so it is likely that the presence of leaf domatia may indirectly impact lepidopteran herbivores.

Information Resources Over the last several decades an explosion of research in the field of chemical ecology has revolutionized our understanding of how the complex blends of herbivore-induced plant volatiles (HIPVs) produced by plants in response to herbivore interactions with plants convey information to at least four trophic levels: plants, caterpillars (and adult moths and butterflies), a variety of natural enemies, and even hyperparasitoids (Kessler and Heil 2011; Peterson et al. 2016; Turlings and Erb 2018; Cusumano et al. 2019). The HIPVs of a wide array of plants, including some well-studied model systems (e.g., *Arabidopsis thaliana*, *Nicotiana tabacum*, *Zea mays*, *Solanum lycopersicum*, various *Brassica* spp.), have been characterized using gas chromatography/mass spectroscopy (GC/MS) approaches. By capturing and analyzing the headspace around plants exposed to lepidopterans, researchers have detected HIPV-based responses to caterpillar feeding (Karban and Baldwin 1997; Howe and Jander 2008; Body et al. 2019), oviposition (e.g., Cusumano et al. 2015; Hilker and Fatouros 2015), and even arthropod movement on leaf surfaces (e.g., Hilker and Meiners 2010; Tooker et al. 2010). Three major types of HIPVs that respond to caterpillar activities, green leaf volatiles (GLVs), terpenoids, and the sulfur- and nitrogen-containing volatiles associated with the myrosinase/glucosinolate defenses of brassicaceous plants (Wittstock et al. 2003; Muller and Wittstock 2005; Mumm and Dicke 2010; Turlings and Erb 2018) have each been shown to be the most responsive to actual caterpillar feeding, with caterpillar oral secretions playing a pivotal role (Alborn et al. 1997). In addition, both GLV and terpenoid blends have been shown to convey specific information about the species of caterpillar causing the induced response (Allmann and Baldwin 2010), likely due to differences in the elicitors found in caterpillar oral secretions (Turlings and Erb 2018). Moreover, the HIPV signature for a given plant species can vary with the amount and type of damage (Delphia et al. 2007; Peterson et al. 2016), ontogenetic stage (Takabayashi et al. 1995), and even parasitization status (e.g., Poelman et al. 2011; Zhu et al. 2015; Cusumano et al. 2019) of the lepidopteran, providing a wealth of ecological and taxonomic information to "receivers" from any of the four trophic levels mentioned above.

While initial studies of caterpillar-based HIPVs considered parasitoid wasps to be the primary receivers of this chemically communicated information (Turlings and Erb 2018), the taxonomic makeup of documented receivers (and responders) has broadened considerably (Mumm and Dicke 2010) to include social wasps (Vespidae; McPheron and Mills 2012, Saraiva et al. 2017), ladybird beetles (James 2005), predatory bugs (e.g., *Geocoris* sp. and *Orius tristicolor*; James 2005; Allmann and Baldwin 2010), entomopathogenic nematodes (Ali et al. 2012), multiple fly families (Tachinidae, Chloropidae, Sarcophagidae, Syrphidae, and Agromyzidae; James 2005), and even insectivorous birds (Amo et al. 2013; Mäntylä et al. 2008; Mrazova and Sam 2017, 2019). Both innate and learned responses to

HIPVs have been documented in a variety of predators and parasitoids (Vet et al. 1995; Steidle and van Loon 2003; De Boer and Dicke 2006; Glinwood et al. 2011), and natural enemy abilities to detect subtle differences in HIPV blends appear to exceed that of available instrumentation (Mumm and Dicke 2010).

Acknowledging that the use of chemical information by the natural enemies of caterpillars is now well established, its functioning as a "defense trait" in plants is much less certain (Peñuelas and Llusiá 2004; Kessler and Heil 2011). Plant de novo synthesis of novel volatiles in response to herbivory appears likely to have evolved primarily as a means of rapid intraplant communication and/or a means of deterring additional herbivores, raising the possibility that most natural enemy responses to HIPVs can be attributed to "eavesdropping" rather than plant-derived "SOS" signals (Dicke 2009; Dicke and Baldwin 2010). Because the koinobiont strategy used by many caterpillar parasitoids (Godfray 1994) allows their hosts to continue feeding and developing, functional responses of parasitoids to this information (i.e., increased parasitism rates) are unlikely to decrease plant damage in the short term (but see van Loon et al. 2000); however, for long-lived plants, numerical declines in herbivore populations resulting from increased parasitism may hold greater promise for reducing damage. By contrast, when plant signaling attracts generalist predators that effectively remove caterpillars, thus preventing further damage, these signaling traits have a greater likelihood of serving as indirect plant defenses. Agricultural field tests with lures containing synthetic HIPVs to augment biocontrol efforts of lepidopteran pests are likely to provide valuable insights into this debate, although there are a number of mitigating factors that may limit its effectiveness (reviewed in Kaplan 2012).

Future Directions

Complementing the well-studied mechanisms detailed above, additional types of plant trait-mediated caterpillar-natural enemy interactions warrant further study based on some intriguing initial findings. These include investigations of how host plant affiliations affect the expression and/or effectiveness of antipredator behaviors such as regurgitation (a common caterpillar response to natural enemy attack) and caterpillar dropping behaviors (reviewed in Greeney et al. 2012). Because the foliage of alternative host plant species can differ in water content and allelochemical composition, both the amount of regurgitant produced (Peiffer and Felton 2009; Block et al. in review) and the effectiveness of the regurgitant in deterring predators (Theodoratus and Bowers 1999) may vary among host plants. Additional experiments testing the expression and effectiveness of regurgitation as an antipredator behavior in a wider range of oligophagous and polyphagous caterpillar taxa, and against a wider range of natural enemies (both arthropod predators and parasitoids), are needed. A second common caterpillar antipredator behavior that is likely to be impacted by host plant identity is their dropping/escape behavior. A variety of caterpillars respond to perceived predation threats by vigorously wriggling and/or simply dropping off of their host plant (Gross 1993; Greeney et al. 2012) to avoid predation. When these behaviors are elicited by substrate-borne vibrations produced by natural enemies, as was demonstrated for the geometrid *Macaria aemulataria* (Castellanos and Barbosa 2006), physical and structural features of their host plant (e.g., wood density, leaf stiffness) could facilitate or impede signal transmission (McNett and Cocroft 2008; Hill 2009), potentially altering the effectiveness of these escape strategies.

Lastly, because some important natural enemies of caterpillars are themselves spatially restricted (e.g., many ant species), traits that determine a plant's microhabitat preferences (e.g., physiological tolerances for light, moisture, and various edaphic features) are likely to affect the abundance and/or community composition of natural enemies patrolling particular host plant species. In one of the most convincing demonstrations to date of how host plant shifts can be promoted and/or maintained by enemy-free space, Murphy (2004) found that the novel host plants (Asteraceae) used by the Alaskan swallowtail butterfly (*Papilio machaon aliaska*) grew at higher elevations than their ancestral host plants (Apiaceae); as a consequence, caterpillars feeding on the novel host plants were not exposed to their most important predator (*Formica podzolica*), which co-occurs only with the ancestral host plant. Plant species restricted to serpentine soils, many of which also hyperaccumulate metals that can be toxic to a variety of herbivores (Brady et al. 2005), might also expose their caterpillars to different natural enemy pressures than their close relatives growing on non-serpentine soils (Robinson 2017). In short, by expanding the suite of "plant traits" hypothesized to influence caterpillar-natural enemy interactions to include interspecific differences in plant physiological tolerances for environmental variables, we can begin to assess their hypothesized role in environmental filtering (e.g., Bazzaz 1991; Swenson et al. 2012). Alternative host plants may also differ in their biotic associations (e.g., mycorrhizal fungi, nitrogen-fixing bacteria, and foliar endophytes), all of which have been shown to affect the expression of antiherbivore chemical defenses in their plant hosts (Kempel et al. 2009; Saikkonen et al. 2010; Jung et al. 2012), which in turn can modulate caterpillar-natural enemy interactions by any of the mechanisms described above. By extending the concept of plant phenotypes to include both their microhabitat associations and their many symbionts, a greater array of ecologically relevant modes of interaction can be investigated.

References

Abdala-Roberts L, Mooney KA (2013) Environmental and plant genetic effects on tri-trophic interactions. Oikos 122:1157–1166

Abdala-Roberts L, Puentes A, Finke DL, Marquis RJ, Montserrat M, Poelman EH, Rasmann S, Sentis A, van Dam NM, Wimp G, Mooney K, Björkman NSC (2019) Tri-trophic interactions: bridging species, communities, and ecosystems. Ecol Lett 22:2151–2167

Agrawal AA (2000) Mechanisms, ecological consequences and agricultural implications of tri-trophic interactions. Curr Opin Plant Biol 3:329–395

Agrawal AA, Karban R, Colfer RG (2000) How leaf domatia and induced plant resistance affect herbivores, natural enemies and plant performance. Oikos 89:70–80

Agrawal AA, Petschenka G, Bingham RA, Weber MG, Rasmann S (2012) Toxic cardenolides: chemical ecology and coevolution of specialized plant–herbivore interactions. New Phytol 194:28–45

Agrawal AA, Ali JG, Rasmann S, Fishbein M (2015) Macroevolutionary trends in the defense of milkweeds against monarchs: latex, cardenolides, and tolerance of herbivory. In: Oberhauser KS, Nail KR, Altizer S (eds) Monarchs in a changing world: biology and conservation of an iconic insect. Cornell University Press, Ithaca

Alborn HT, Turlings TCJ, Jones TH, Stenhagen G, Loughrin JH, Tumlinson JH (1997) An elicitor of plant volatiles from beet armyworm oral secretions. Science 276:945–949

Ali JG, Alborn HT, Campos-Herrera R, Kaplan F, Duncan LW, Rodriguez-Saona C, Koppenhöfer AM, Stelinski LL (2012) Subterranean, herbivore-induced plant volatile increases biological control activity of multiple beneficial nematode species in distinct habitats. PLoS ONE 7:e38146

Allmann S, Baldwin IT (2010) Insects betray themselves in nature to predators by rapid isomerization of green leaf volatiles. Science 329:1075–1078

Amo L, Jansen JJ, van Dam NM, Dicke M, Visser ME, Turlings T (2013) Birds exploit herbivore-induced plant volatiles to locate herbivorous insects. Ecol Lett 16:1348–1355

Ayres MP, Clausen TP, MacLean SF Jr, Redman AM, Reichardt PB (1997) Diversity of structure and antiherbivore activity in condensed tannins. Ecology 78:1696–1712

Bazzaz FA (1991) Habitat selection in plants. Am Nat 137:S116–S130

Benrey B, Denno RF (1997) The slow-growth–high-mortality hypothesis: a test using the cabbage butterfly. Ecology 78:987–999

Berenbaum MR (1983) Coumarins and caterpillars: a case for coevolution. Evolution 37:163–179

Bergelson JM, Lawton JH (1988) Does foliage damage influence predation on the insect herbivores of birch? Ecology 69:434–445

Bernays EA (2003) Feeding by lepidopteran larvae is dangerous. Ecol Entomol 22:121–123

Bernays EA, Graham M (1988) On the evolution of host specificity in phytophagous arthropods. Ecology 69:886–892

Block C, Weiss MR, Lind E, Lill JT (in review). Diet and ontogeny affect anti-predator behavior and predation risk in an insect herbivore. Ecol Entomol

Body MJA, Neer WC, Vore C, Lin C-H, Vu DC, Schultz JC, Cocroft RB, Appel HM (2019) Caterpillar chewing vibrations cause changes in plant hormones and volatile emissions in *Arabidopsis thaliana*. Front Plant Sci 10:810

Boege K, Marquis RJ (2005) Facing herbivory as you grow up: the ontogeny of resistance in plants. Trends Ecol Evol 20:4410448

Boege K, Agrawal AA, Thaler JS (2019) Ontogenetic strategies in insect herbivores and their impact on tri-trophic interactions. Curr Opin Insect Sci 32:61–67

Boppré M (1990) Lepidoptera and pyrrolizidine alkaloids: exemplification of complexity in chemical ecology. J Chem Ecol 16:165–185

Bowers MD, Farley S (1990) The behavior of grey jays, *Perisoreus canadensis*, towards palatable and unpalatable Lepidoptera. Anim Behav 39:699–705

Brady JU, Kruckeberg AR, Bradshaw HD Jr (2005) Evolutionary ecology of plant adaptation to serpentine soils. Annu Rev Ecol Evol Syst 36:243–266

Bronstein JL (1998) The contribution of ant-plant protection studies to our understanding of mutualism. Biotropica 30:151–160

Bronstein JL, Alarcón R, Geber M (2006) The evolution of plant–insect mutualisms. New Phytol 172:412–428

Caldwell E, Read J, Sanson GD (2016) Which leaf mechanical traits correlate with insect herbivory among feeding guilds? Ann Bot 117:349–361

Camara MD (1997) Predator responses to sequestered plant toxins in buckeye caterpillars: are tritrophic interactions locally variable? J Chem Ecol 23:2093–2106

Castellanos I, Barbosa P (2006) Evaluation of predation risk by a caterpillar using substrate-borne vibrations. Anim Behav 72:461–469

Chen K-W, Chen Y (2018) Slow-growth high-mortality: a meta-analysis for insects. Insect Sci 25:337–351

Clancy KM, Price PW (1987) Rapid herbivore growth enhances enemy attack: sublethal plant defenses remain a paradox. Ecology 68:736–738

Clark TL, Messina FJ (1998) Foraging behavior of lacewing larvae (Neuroptera: Chrysopidae) on plants with divergent architectures. J Insect Behav 11:303–317

Coley PD, Bateman ML, Kursar TA (2006) The effects of plant quality on caterpillar growth and defense against natural enemies. Oikos 115:219–228

Condon MA, Scheffer SJ, Lewis ML, Wharton R, Adams DC, Forbes AA (2014) Lethal interactions between parasites and prey increase niche diversity in a tropical community. Science 343:1240–1244

Cornelissen T, Stiling P (2006) Does low nutritional quality act as a plant defence? An experimental test of the slow-growth, high-mortality hypothesis. Ecol Entomol 31:32–40

Cortesero AM, Stapel JO, Lewis WJ (2000) Understanding and manipulating plant attributes to enhance biological control. Biol Control 17:35–49

Cusumano A, Weldergergis BT, Colazza S, Dicke M, Fatouros NE (2015) Attraction of egg-killing parasitoids toward induced plant volatiles in a multi-herbivore context. Oecologia 179:163–174

Cusumano A, Harvey JA, Dicke N, M. (2019) Hyperparasitoids exploit herbivore-induced plant volatiles during host location to assess host quality and non-host identity. Oecologia 189:699–709

Damman H (1987) Leaf quality and enemy avoidance by the larvae of a pyralid moth. Ecology 68:88–97

Davidson DW (1997) The role of resource imbalances in the evolutionary ecology of tropical arboreal ants. Biol J Linn Soc 61:153–181

Davidson DW, McKey D (1993) Ant-plant symbioses: stalking the chuyachaqui. Trends Ecol Evol 8:326–332

De Boer JG, Dicke M (2006) Olfactory learning by predatory arthropods. Anim Biol 56:143–155

Delphia CM, Mescher MC, De Moraes CM (2007) Induction of plant volatiles by herbivores with different feeding habits and the effects of induced defenses on host-plant selection by thrips. J Chem Ecol 33:997–1012

Dicke M (2009) Behavioural and community ecology of plants that cry for help. Plant Cell Environ 32:654–665

Dicke M, Baldwin IT (2010) The evolutionary context for herbivore-induced plant volatiles: Beyond the 'cry for help'. Trends Plant Sci 15(3):167–175

Dobler S, Petschenka G, Pankoke H (2011) Coping with toxic plant compounds: the insect's perspective on iridoid glycosides and cardenolides. Phytochemistry 72:1593–1604

Doussourd DE, Eisner T (1987) Vein-cutting behavior insect counterploy to the latex defense of plants. Science 237:898–901

Dussourd DE (2017) Behavioral sabotage of plant defenses by insect folivores. Annu Rev Entomol 62:15–34

Dyer LA (1995) Tasty generalists and nasty specialists? Antipredator mechanisms in tropical lepidopteran larvae. Ecology 76:1483–1496

Dyer LA, Bowers MD (1996) The importance of sequestered iridoid glycosides as a defense against an ant predator. J Chem Ecol 22:1527–1539

Dyer LA, Floyd T (1993) Determinants of predation on phytophagous insects: the importance of diet breadth. Oecologia 96:575–582

Eigenbrode SD (2004) The effects of plant epicuticular waxy blooms on attachment and effectiveness of predatory insects. Arthropod Struct Dev 33(1):91–102

Eigenbrode SD, Espelie (1995) Effects of plant epicuticular lipids on insect herbivores. Annu Rev Entomol 40:171–194

Eigenbrode SD, Moodie S, Castagnola T (1995) Predators mediate host plant resistance to a phytophagous pest in cabbage with glossy leaf wax. Entomol Exp Appl 77:335–342

Farkas TE, Singer MS (2013) Can caterpillar density or host-plant quality explain host-plant-related parasitism of a generalist forest caterpillar assemblage? Oecologia 173:971–983

Feeny P (1976) Plant apparency and chemical defense. In: Wallace JW, Mansell RL (eds) Biochemical interactions between plants and insects. Plenum, New York, pp 1–40

Geitzenauer HL, Bernays EA (1996) Plant effects on prey choice by a vespid wasp, *Polistes arizonensis*. Ecol Entomol 21:227–234

Glinwood R, Ahmed E, Qvarfordt E, Ninkovic V (2011) Olfactory learning of plant genotypes by a polyphagous insect predator. Oecologia 166:637–647

Godfray HCJ (1994) Parasitoids: behavioral and evolutionary ecology. Princeton University Press, New Jersey. 473 pp

Gorb EV, Gorb SN (2017) Anti-adhesive effects of plant wax coverage on insect attachment. J Exp Bot 68:5323–5337

Greeney HF, Dyer LA, Smilanich AM (2012) Feeding by lepidopteran larvae is dangerous: a review of caterpillars' chemical, physiological, morphological, and behavioral defenses against natural enemies. Invertebr Surviv J 9:7–34

Grevstad FS, Klepetka BW (1992) The influence of plant architecture on the foraging efficiencies of a suite of ladybird beetles feeding on aphids. Oecologia 92:399–404

Grof-Tisza P, LoPresti E, Heath SK, Karban R (2017) Plant structural complexity and mechanical defense mediate predator-prey interactions in an odonate-bird system. Ecol Evol 7:1650–1659

Gross P (1993) Insect behavioral and morphological defenses. Annu Rev Entomol 38:251–273

Halley J, Hartley S, Kallimanis A, Kunin W, Lennon J, Sgardelis S (2004) Uses and abuses of fractal methodology in ecology. Ecol Lett 7:254–271

Hanley ME, Lamont BB, Fairbanks MM, Rafferty CM (2007) Plant structural traits and their role in anti-herbivore defence. Persp Plant Ecol Evol Syst 8:157–178

Hare JD (1992) Effects of plant variation on herbivore-natural enemy interactions. In: Fritz RS, Simms EL (eds) Plant resistance to herbivores and pathogens. The University of Chicago Press, Chicago, pp 278–298

Heil M (2008) Indirect defence via tritrophic interactions (Tansley Review). New Phytol 178:41–61

Heil M (2015) Extrafloral nectar at the plant-insect interface: a spotlight on chemical ecology, phenotypic plasticity, and food webs. Annu Rev Entomol 60:213–232

Heil M, McKey D (2003) Protective ant-plant interactions as model systems in ecological and evolutionary research. Annu Rev Ecol Evol Syst 34:425–553

Heimpel GE, Jervis MA (2005) Does floral nectar improve biological control by parasitoids? In: Plant-provided food for carnivorous insects: a protective mutualism and its applications. Cambridge University Press, pp 267–304

Hilker M, Fatouros NE (2015) Plant responses to egg deposition. Annu Rev Entomol 60:493–515

Hilker M, Meiners T (2010) How do plants "notice" attack by herbivorous arthropods? Biol Rev 85:267–280

Hill PSM (2009) How do animals use substrate-borne vibrations as an information source? Naturwissenschaften 96:1355–1371

Holmes RT, Robinson SK (1981) Tree species preferences of foraging insectivorous birds in a northern hardwoods forest. Oecologia 48:31–35

Howe GA, Jander G (2008) Plant immunity to insect herbivores. Annu Rev Plant Biol 59:41–66

Hunter MD, Price PW (1992) Playing chutes and ladders: heterogeneity and the relative roles of bottom-up and top-down forces in natural communities. Ecology 73:724–732

Idris AB, Grafius E (1996) Effects of wild and cultivated host plants on oviposition, survival, and development of diamondback moth (Lepidoptera: Plutellidae) and its parasitoid *Diadegma insulare* (Hymenoptera: Ichneumonidae). Environ Entomol 25:825–833

Isenhour DJ, Wiseman BR, Layton RC (1989) Enhanced predation by *Orius insidiosus* (Hemiptera: Anthocoridae) in larvae of *Heliothis zea* and *Spodoptera frugiperda* (Lepidoptera: Noctuidae) caused by prey feeding on resistant corn genotypes. Environ Entomol 18:418–422

James DG (2005) Further field evaluation of synthetic herbivore-induced plan volatiles as attractants for beneficial insects. J Chem Ecol 31:481–495

Järvi T, Sillen-Tullberg B, Wiklund C (1981) The cost of being aposematic. An experimental study of predation on larvae of *Papilio machaon* by the great tit *Varus major*. Oikos 36:267–272

Johnson MT, Gould F (1992) Interaction of genetically engineered host plant resistance and natural enemies of Heliothis virescens (Lepidoptera: Noctuidae) in tobacco. Environ Entomol 21:586–597

Jones IM, Koptur S, Gallegos HR, Tardanico JP, Trainer PA, Pena J (2017) Changing light conditions in pine rockland habitats affect the intensity and outcome of ant–plant interactions. Biotropica 49:83–91

Jung SC, Martinez-Medina A, Lopez-Raez JA, Pozo MJ (2012) Mycorrhiza-induced resistance and priming of plant defenses. J Chem Ecol 38:651–664

Kaplan I (2012) Attracting carnivorous arthropods with plant volatiles: the future of biocontrol or playing with fire? Biol Control 60:77–89

Kaplan I, Carrillo J, Garvey M, Ode PJ (2016) Indirect plant-parasitoid interactions mediated by changes in herbivore physiology. Curr Opin Insect Sci 14:112–119

Karban R, Baldwin IT (1997) Induced responses to herbivory. University of Chicago Press, Chicago

Kariyat RR, Hardison SB, Ryan AB, Stephenson AG, De Moraes CM, Mescher MC (2018) Leaf trichomes affect caterpillar feeding in an instar-specific manner. Commun Integr Biol 11:1–6

Kaufmann WC, Kennedy GC (1989) Relationship between trichome density in tomato and parasitism of *Heliothis* sp. (Lepidoptera: Noctuidae) eggs by *Trichogramma* spp. (Hymenoptera: Trichogrammatidae). Environ Entomol 18:698–704

Keller MA (1987) Influence of leaf surfaces on movements by the hymenopterous parasitoid *Trichogramma exiguum*. Entomol Exp Appl 43:55–59

Kempel A, Brandl R, Schädler M (2009) Symbiotic soil microorganisms as players in aboveground plant–herbivore interactions – the role of rhizobia. Oikos 118:634–660

Kennedy GC (2003) Tomato, Pests, parasitoids, and predators: Tritrophic interactions involving the genus *Lycopersicon*. Annu Rev Entomol 48:51–72

Kessler A, Heil M (2011) The multiple faces of indirect defences and their agents of natural selection. Funct Ecol 25:348–357

Koptur S, Jones IM, Peña JE (2015) The influence of host plant extrafloral nectaries on multitrophic interactions: an experimental investigation. PLoS ONE 10:e0138157. https://doi.org/10.1371/journal.pone.0138157

Krimmel BA, Pearse IS (2013) Sticky plant traps insects to enhance indirect defence. Ecol Lett 16:219–224

Kursar TA, Wolfe BT, Epps MJ, Coley PD (2006) Food quality, competition, and parasitism influence feeding preference in a neotropical lepidopteran. Ecology 87:3058–3069

Leather SR, Walsh PJ (1993) Sub-lethal plant defences: the paradox remains. Oecologia 93:153–155

Lill JT, Marquis RJ (2001) The effects of leaf quality on herbivore performance and attack from natural enemies. Oecologia 126:418–428

Loader C, Damman H (1991) Nitrogen content of food plants and vulnerability of Pieris rapae to natural enemies. Ecology 75:1586–1590

Malishev M, Sanson GD (2015) Leaf mechanics and herbivory defence: how tough tissue along the leaf body deters growing insect herbivores. Aust Ecol 40:300–308

Mäntylä E, Alessio GA, Blande JD, Heijari J, Holopainen JK, Laaksonen T, Piirtola P, Klemola T (2008) From plant to birds: higher avian predation rates in trees responding to insect herbivory. PLoS ONE 3:e2832

Mäntylä E, Klemola T, Laaksonen T (2011) Birds help plants: a meta-analysis of top-down trophic cascades caused by avian predators. Oecologia 165:143–151

Marquis RJ, Whelan CJ (1994) Insectivorous birds increase growth of white oak by consuming its herbivores. Ecology 75:2007–2014

Marquis RJ, Whelan CJ (1996) Plant morphology and recruitment of the third trophic level: subtle, unrecognized defenses? Oikos 75:330–334

McCoy ED, Bell SS (1991) Habitat structure: the evolution and diversification of a complex topic. In: Bell SS, McCoy ED, Mushinsky HR (eds) Habitat structure: the physical arrangement of objects in space. Chapman & Hall, London, pp 3–27

McNett GD, Cocroft RB (2008) Host shifts favor vibrational signal divergence in *Enchenopa binotata* treehoppers. Behav Ecol 19:650–656

McPheron LJ, Mills NJ (2012) Influence of visual and olfactory cues on the foraging behavior of the paper wasp *Mischocyttarus favitarsis* (Hymenoptera: Vespidae). Entomol General 30:105–118

Medina RF, Barbosa P, Waddell K (2005) Parasitism levels in *Orgyia leucostigma* feeding on two tree species: implications for the slow-growth-high-mortality hypothesis. Entomol Exp Appl 115:193–197

Mooney KA, Singer MS (2012) Plant effects on herbivore–enemy interactions in natural systems. In: Ohgushi T, Schmitz OJ, Holt RD (eds) Trait-mediated indirect interactions: ecological and evolutionary perspectives. Cambridge University Press, pp 107–130

Mooney KA, Gruner DS, Barber NA, Van Bael SA, Philpott SM, Greenberg R (2010) Interactions among predators and the cascading effects of vertebrate insectivores on arthropod communities and plants. PNAS 107:7335–7340

Moran N, Hamilton WD (1980) Low nutritive quality as defense against herbivores. J Theor Biol 86:247–254

Mrazova A, Sam K (2017) Application of methyl jasmonate to grey willow (*Salix cinerea*) attracts insectivorous birds in nature. Arthropod Plant Interact 12:1–8

Mrazova A, Sam K (2019) Exogenous application of methyl jasmonate to *Ficus hahliana* attracts predators of insects along an altitudinal gradient in Papua New Guinea. J Trop Ecol 35:157–164

Müller C, Wittstock U (2005) Uptake and turnover of glucosinolates sequestered in the sawfly Athalia rosae. Insect Biochem Mol Biol 10:1189–1198

Mumm R, Dicke M (2010) Variation in natural plant products and the attraction of bodyguards involved in indirect plant defense. Can J Zool 88:628–667

Murphy SM (2004) Enemy-free space maintains swallowtail butterfly host shift. Proc Natl Acad Sci U S A 101:18048–18052

Nell CS, Mooney KA (2019) Plant structural complexity mediates trade-off in direct and indirect plant defense by birds. Ecology 100:e02853

Neuvonen S, Haukioja E (1984) Low nutritive quality as a defence against herbivores: induced responses in birch. Oecologia 63:71–74

Nishida R (2002) Sequestration of defensive substances from plants by Lepidoptera. Annu Rev Entomol 47:57–92

Norton AP, English-Loeb G, Belden E (2001) Host plant manipulation of natural enemies: leaf domatia protect beneficial mites from insect predators. Oecologia 126:535–542

Parry D, Spence JR, Volney WJA (1998) Budbreak phenology and natural enemies mediate survival of first-instar forest tent caterpillar (Lepidoptera: Lasiocampidae). Environ Entomol 27:1368–1374

Patt JM, Hamilton GC, Lashomb JH (1997) Foraging success of parasitoid wasps on flowers: interplay of insect morphology, floral architecture and searching behavior. Entomol Exp Appl 83:21–30

Peiffer M, Felton GW (2009) Do caterpillars secrete "oral secretions"? J Chem Ecol 35:326–335. https://doi.org/10.1007/s10886-009-9604-x

Peñuelas J, Llusiá J (2004) Plant VOC emissions: making use of the unavoidable. Trends Ecol Evol 19:402–404

Peterson JA, Ode PJ, Oliveira-Hofman C, Harwood JD (2016) Integration of plant defense traits with biological control of arthropod pests: challenges and opportunities. Front Plant Sci 7:1794

Petschenka G, Agrawal AA (2015) Milkweed butterfly resistance to plant toxins is linked to sequestration, not coping with a toxic diet. Proc R Soc B 282:20151865

Pimintel D (1961) An evaluation of insect resistance in broccoli, brussel sprouts, cabbage, collards, and kale. J Econ Entomol 54:156–158

Poelman EH, Zheng SJ, Zhang Z, Heemskerk NM, Cortesero AM, Dicke M (2011) Parasitoid-specific induction of plant responses to parasitized herbivores affects colonization by subsequent herbivores. Proc Natl Acad Sci 108:19647–19652

Price P, Bouton C, Gross P, McPheron B, Thompson J, Weiss A (1980) Interactions among three trophic levels: influence of plants on interactions between insect herbivores and natural enemies. Annu Rev Ecol Syst 11:41–65. https://doi.org/10.1146/annurev.es.11.110180.000353

Reynolds PG, Cuddington K (2012) Effects of plant gross morphology on predator searching behaviour. Environ Entomol 41:516–522

Richards LA, Lampert EC, Bowers MD, Dodson CD, Smilanich AM, Dyer LA (2012) Synergistic effects of iridoid glycosides on the survival, development and immune response of a specialist caterpillar, *Junonia coenia* (Nymphalidae). J Chem Ecol 38:1276–1284

Rico-Gray V, Oliveira PS (2007) The ecology and evolution of ant-plant interactions. University of Chicago Press

Riddick EW, Simmons AM (2014) Do plant trichomes cause more harm than good to predatory insects? Pest Manag Sci 70(11):1655–1665

Robinson M (2017) When soils cascade: effects of soil resource availability on interactions between plants, herbivores, and their natural enemies. Dissertation, University of California-Davis

Robinson SK, Holmes RT (1984) Effects of plant species and foliage structure on the foraging behavior of forest birds. Auk 101:672–684

Romero GQ, Benson WW (2005) Biotic interactions of mites, plants and leaf domatia. Curr Opin Plant Biol 8:436–440

Rosenthal GA, Berenbaum MR (1991) Herbivores: their interactions with secondary metabolites. Academic, New York

Rudgers JA, Strauss S (2004) A selection mosaic in the facultative mutualism between ants and wild cotton. Proc R Soc Lond B 271:2481–2488

Russell M (2015) A meta-analysis of physiological and behavioral responses of parasitoid wasps to flowers of individual plant species. Biol Control 82:96–103

Saikkonen K, Saari S, Helander M (2010) Defensive mutualism between plants and endophytic fungi? Fungal Divers 41:101–113

Saraiva NB, Prezoto F, Fonseca M d G, Blassioli-Moraes MC, Borges M, Laumann RA, Auad AM (2017) The social wasp *Polybia fastidiosuscula* Saussure (Hymenoptera: Vespidae) uses herbivore-induced maize plant volatiles to locate its prey. J Appl Entomol 141:620–629

Schmitz OJ, Hamback PA, Beckerman AP (2000) Trophic cascades in terrestrial systems: a review of the effects of carnivore removals on plants. Am Nat 155:141–153

Shikano I, McCarthy E, Hayes-Plazolles N, Slavicek JM, Hoover K (2018) Jasmonic acid-induced plant defenses delay caterpillar developmental resistance to a baculovirus: slow-growth, high-mortality hypothesis in plant-insect-pathogen interactions. J Invertebr Pathol 158:16–23

Sime K (2002) Chemical defence of *Battus philenor* larvae against attack by the parasitoid *Trogus pennator*. Ecol Entomol 27:337–345

Singer MS, Stireman JO III (2005) The tri-trophic niche concept and adaptive radiation of phytophagous insects. Ecol Lett 8:1247–1255

Singer MS, Farkas TE, Skorik CM, Mooney KA (2012) Tritrophic interactions at a community level: effects of host plant species quality on bird predation of caterpillars. Am Nat 179:363–374

Smilanich AM, Fincher RM, Dyer LA (2016) Does plant apparency matter? Thirty years of data provide limited support by reveal clear patterns of the effects of plant chemistry on herbivores. New Phytol 210:1044–1057

Stapel JO, Cortesero AM, De Moraes CM, Tumlinson JH, Lewis WJ (1997) Effects of extrafloral nectar, honeydew and sucrose on searching behavior and efficiency of *Microplitis croceipes* (Hymenoptera: Braconidae) in cotton. *Environ Entomol* 26:617–623

Stapley L (1998) The interaction of thorns and symbiotic ants as an effective defence mechanism of swollen-thorn acacias. Oecologia 115:401–405

Steidle JLM, van Loon JJA (2003) Dietary specialization and infochemical use in carnivorous arthropods: testing a concept. Entomol Exp Appl 108:133–148
Swenson NG, Enquist BJ, Pither J, Kerkhoff AJ, Boyle B, Weiser MD et al (2012) The biogeography and filtering of woody plant functional diversity in North and South America. Glob Ecol Biogeogr 21:798–808
Takabayashi J, Takahashi S, Dicke M, Posthumus MA (1995) Developmental stage of herbivorePseudaletia separata affects production of herbivore-induced synomone by corn plants. J Chem Ecol 21:273–287
Taranto F, Pasqualone A, Mangini G, Tripodi P, Miazzi MM, Pavan S, C. Montemurro. C. (2017) Polyphenol oxidases in crops: biochemical, physiological and genetic aspects. Int J Mol Sci 18:377
Theodoratus DH, Bowers MD (1999) Effects of sequestered iridoid glycosides on prey choice of the prairie wolf spider, *Lycosa carolinensis*. J Chem Ecol 25:283–295
Tooker JF, Hauser M, Hanks LM (2006) Floral host plants of Syrphidae and Tachinidae (Diptera) of central Illinois. Ann Entomol Soc Am 99:96–112
Tooker J, Peiffer M, Luthe DS, Felton GW (2010) Trichomes as sensors: detecting activity on the leaf surface. Plant Signal Behav 5:73–75
Turlings TCJ, Erb M (2018) Tritrophic interactions mediated by herbivore-induced plant volatiles: mechanisms, ecological relevance, and application potential. Annu Rev Entomol 63:433–453
Urbaneja-Bernat P, Tena A, González-Cabrera J, Rodriguez-Saona C (2020) Plant guttation provides nutrient-rich food for insects. Proc R Soc B287:20201080
Van Bael SA, Philpott SM, Greenberg RS, Bichier P, Barber NA, Mooney KA, Gruner DS (2008) Birds as predators in tropical agroforestry systems. Ecology 89:928–934
van Loon JJA, de Boer JG, Dicke M (2000) Parasitoid–plant mutualism: parasitoid attack of herbivore increases plant reproduction. Entomol Exp Appl 97:219–227
Vet LE, Lewis WJ, Carde RT (1995) Parasitoid foraging and learning. In: Chemical ecology of insects 2. Springer, Boston, pp 65–101
Wäckers FL, van Rijn PCJ, Bruin J (2005) Plant-provided food for carnivorous insects: a protective mutualism and its applications. Cambridge University Press, Cambridge
Walter DE, O'Dowd DJ (1992) Leaf morphology and predators: effect of leaf domatia on the abundance of predatory mites (Acari: Phytoseiidae). Environ Entomol 21:478–484
Weber MG, Keeler KH (2013) The phylogenetic distribution of extrafloral nectaries in plants. Ann Bot 111:1251–1261
Weinhold A, Baldwin IT (2011) Trichome-derived O-acyl sugars are a first meal for caterpillars that tags them for predation. Proc Natl Acad Sci 108:7855–7859
Weiss MR, Lind EM, Jones MT, Long JD, Maupin JL (2003) Uniformity of leaf shelter construction by larvae of *Epargyreus clarus* (Hesperiidae), the silver-spotted skipper. J Insect Behav 16:465–480
Whelan CJ (1989) Avian foliage structure preferences for foraging and the effect of prey biomass. Anim Behav 38:839–846
Whelan CJ (2001) Foliage structure influences foraging of insectivorous forest birds: an experimental study. Ecology 82:219–231
Whelan CJ, Wenny D, Marquis RJ (2008) Ecosystem services provided by birds. Ann N Y Acad Sci 1134:25–60
Williams IS (1999) Slow-growth, high-mortality – a general hypothesis, or is it? Ecol Entomol 24:490–295
Wittstock U, Kliebenstein DJ, Lambrix V, Reichelt M, Gershenzon J (2003) Glucosinolate hydrolysis and its impact on generalist and specialist insect herbivores. In: Romeo JT (ed) Recent advances in phytochemistry. Pergamon, Amsterdam, pp 101–125
Wittstock U, Agerbirk N, Stauber EJ, Olsen CE, Hippler M, Mitchell-Olds T, Gershenzon J, Vogel H (2004) Successful herbivore attack due to metabolic diversion of a plant chemical defense. Proc Natl Acad Sci U S A 101:4859–4864

Yamawo A, Suzuki N (2018) Induction and relaxation of extrafloral nectaries in response to simulated herbivory in young *Mallotus japonicus* plants. J Plant Res 131:255–260

Yamawo A, Suzuki N, Tagawa J, Hada Y (2012) Leaf ageing promotes the shift in defence tactics in *Mallotus japonicus* from direct to indirect defence. J Ecol 100:802–809

Zalucki MP, Clarke AR, Malcolm SB (2002) Ecology and behavior of first instar larval Lepidoptera. Annu Rev Entomol 47:361–393

Zemenick AT, Kula RR, Russo L, Tooker J (2019) A network approach reveals parasitoid wasps to be generalized nectar foragers. Arthropod Plant Interact 13:239–251

Zhu F, Broekgaarden C, Weldegergis BT, Harvey JA, Vosman B, Dicke M, Poelman EH (2015) Parasitism overrides herbivore identity allowing hyperparasitoids to locate their parasitoid host by using herbivore induced plant volatiles. Mol Ecol 24:2886–2899

Host Plant Effects on the Caterpillar Immune Response

Angela M. Smilanich and Nadya D. Muchoney

Tarchon felderi (Apatelodidae) in Costa Rica. (Photo by Angela Smilanich)

Introduction

In this chapter, we review the literature on host plant-derived effects on lepidopteran immunity to natural enemies. Much of the work reviewed here falls under the umbrella of ecological immunology or ecoimmunology, a growing area of research aimed at understanding the ecological basis of variation in the immune response of animals (Sheldon and Verhulst 1996; Schulenburg et al. 2009; Brock et al. 2014).

A. M. Smilanich (✉) · N. D. Muchoney
Program in Ecology, Evolution, and Conservation Biology, University of Nevada, Reno, NV, USA

Department of Biology, University of Nevada, Reno, NV, USA
e-mail: asmilanich@unr.edu; nmuchoney@nevada.unr.edu

Given the topic of this book, we focus exclusively on studies with Lepidoptera, which are many since they are a prime subject for ecological studies (Price et al. 2011). The majority of Lepidoptera are herbivorous in the larval stage; thus there are ample opportunities for consequential interactions between host plants and the caterpillar immune response. Plants primarily affect lepidopteran immune responses through nutrient content and secondary metabolite composition (Fig. 1; Singer et al. 2014), and here we summarize the work and results that have been discovered to date. We also highlight studies that include outcomes of caterpillar immune variation for the third trophic level (i.e., parasites, parasitoids, and pathogens) since these consequences are most important for understanding the evolution of the immune response (Schmid-Hempel and Ebert 2003), as well as life history traits such as diet breadth (Smilanich et al. 2009a). We end the chapter with a look at areas of the literature that would benefit from more investigation and suggest avenues for future research.

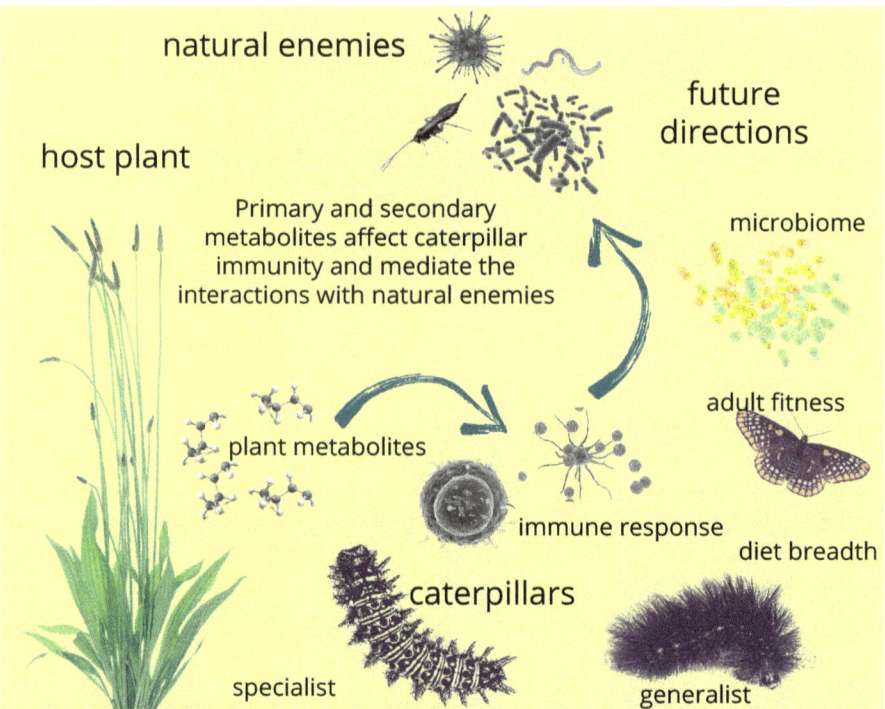

Fig. 1 Overview of the information that we cover in this chapter. Host plant chemistry can have a variety of effects on the immune response of caterpillars, which we review in this chapter. We also consider and highlight research that links host plant chemistry to natural enemy performance via the immune response. We end the chapter with future directions that are beginning to be explored and would benefit from increased research attention

Insect Immunity

Since many excellent reviews, chapters, and books have been written describing insect immunity (Gillespie et al. 1997; Lavine and Strand 2002; Cerenius and Söderhäll 2004; Kanost et al. 2004; Siva-Jothy et al. 2005; Beckage 2008; Carton et al. 2008; Strand 2008; Rolff and Reynolds 2009; Jiang et al. 2010; González-Santoyo and Córdoba-Aguilar 2012), we provide a brief overview covering information and details needed for effectively reading this chapter.

Insects have an innate immune response capable of responding to foreign invaders that breech the primary (e.g., shelter-building, warning coloration) and secondary (thrashing/biting, spines, regurgitating) lines of defense (Gross 1993; Smilanich et al. 2009b). Parasitoids may be deterred by the first two lines of defense, but bacteria, fungi, protozoans, and viruses will mostly be targeted by the immune response alone. Once inside the hemocoel, the invader will be attacked by both cellular and humoral components of the immune system. These two branches do not operate in isolation but work in concert to suppress infection. The cellular response encompasses encapsulation, nodule formation, clotting, and phagocytosis. Large eukaryotic enemies such as parasitoids or protozoans are likely to be attacked by encapsulation, wherein hemocytes such as plasmatocytes and granulocytes adhere to the surface of the parasitoid egg or parasite and build layers of cells around the invader, isolating and asphyxiating it (Nappi and Christensen 2005). As noted, the humoral response accompanies the cellular response and in this example would be followed by melanogenesis initiated by the phenoloxidase (PO) cascade in the hemolymph. Once activated, the PO cascade produces melanin, which is deposited on the encapsulated invader. The process of melanogenesis creates free radicals that are locally cytotoxic and contribute to suppressing the invader (Kanost et al. 2004; Nappi and Christensen 2005; Nakhleh et al. 2017). The humoral response is also characterized by antimicrobial peptides (AMPs) and pattern recognition receptors (PRRs) that bind to molecular components on the surface of microbial pathogens, including lipopolysaccharide, peptidoglycan, and β-1,3-glucan. These immune components have been well characterized from model systems including lepidopterans (i.e., *Manduca sexta*, Sphingidae) (Kanost et al. 2004; Beckage 2008; Ragan et al. 2009).

While not as complex as the vertebrate immune system, the insect immune system is still sophisticated. Accurate measures of immunocompetence remain challenging (Adamo 2004b). Immune assays take many forms, but common measures include encapsulation and melanization of foreign objects inserted into the caterpillar's hemocoel (Fig. 2; Lavine and Beckage 1996; Rantala and Roff 2007; Smilanich et al. 2009a) and counting and differentiating immune cells from samples of hemolymph (Ribeiro and Brehélin 2006; Strand 2008; Triggs and Knell 2012b). In addition, measurement of humoral responses like PO enzyme activity and bacterial-killing assays are common and relatively simple (Siva-Jothy et al. 2001; Adamo 2004a; Freitak et al. 2007). As discussed by Adamo (2004b), immune assays are meant to act as a metric of immune strength, which should then translate into disease

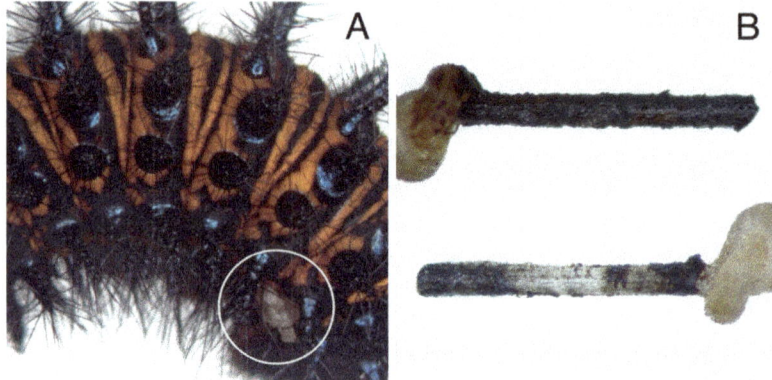

Fig. 2 Filament encapsulation assay used to measure the strength of the encapsulation and melanization responses in caterpillars. (**a**) Nylon monofilament implanted into a sixth instar Baltimore checkerspot (*Euphydryas phaeton*) caterpillar. Filament was knotted at one end to facilitate implantation and removal; knot is visible resting outside of the abdominal cuticle (circled in white). (**b**) Examples of extracted filaments (length: 2 mm) after 24 h within the *E. phaeton* hemocoel, showing varying degrees of melanization and cellular encapsulation

resistance. However, the line between immune assay and disease outcome is not straightforward, and, in many instances, investigators find that the immune strength does not predict disease outcome (further discussed in "Linking Immunity to Fitness via Enemy Attack," below). Adamo (2004b) points out that immune assays may be specific to certain enemies (e.g., encapsulation assays targeting parasitoid enemies). Furthermore, it may also be that immune assays do not always reflect disease resistance as there could be threshold barriers of detection for the assay or the fact that a statistically significant assay result does not mean a significant biological outcome (Adamo 2004b). To help remedy these discrepancies, Adamo (2004b) suggests measuring multiple immune parameters (e.g., PO activity, encapsulation, hemocyte counts, antibacterial activity), as well as including a measure of disease resistance (e.g., pathogen load). Measures of multiple immune parameters have largely been adopted in the lepidopteran literature (Barthel et al. 2014; see Tables 1 and 2), while including measures of disease resistance and host fitness is less common (Graham et al. 2011). The specificity and memory of the insect immune response are certainly greater than once thought, and although it does not have antibodies like the vertebrate immune system, studies show that antimicrobial peptides can be specific to parasite isolates (Riddell et al. 2009), and memory in the form of immune priming can arm insects for a second attack (Rodrigues et al. 2010). Thus, in some situations, it is imperative that the correct immune parameters are measured since immune specificity to natural enemies is an important consideration.

Transcript and transcriptomic analyses represent a relatively new tool for measuring the immune response of insects (Freitak et al. 2007; Vogel et al. 2011; Pascual et al. 2012; Gunaratna and Jiang 2013; Cotter et al. 2019; Tan et al. 2019). These studies have produced rich and detailed datasets that comprise the expression of immune genes in a variety of insects and lepidopterans, in particular (e.g., *Galleria*

Host Plant Effects on the Caterpillar Immune Response 453

Table 1 Effects of dietary macronutrients on caterpillar functional immune responses and interactions with natural enemies (as measured by survival following pathogen challenge). Positive associations between nutritional variables and immune parameters or survival are indicated by "(+)"; nonsignificant associations are indicated by "(N.S.)"

Nutritional parameter	Species	Diet type	Immune effect(s)	Survival effect	Reference
Protein-to-carbohydrate ratio	Spodoptera littoralis	Artificial	Antibacterial activity (+) Implant encapsulation (+) Hemocyte density (N.S.) Phenoloxidase activity (+[a])	Survival of S. littoralis nucleopolyhedrovirus infection (+)	Lee et al. (2006)
Protein-to-carbohydrate ratio	Spodoptera exempta	Artificial	Antibacterial activity (+) Phenoloxidase activity (+)	Survival of Bacillus subtilis infection (+)	Povey et al. (2009)
Protein-to-carbohydrate ratio	Spodoptera exempta	Artificial	Antibacterial activity (+) Hemocyte density (+) Phenoloxidase activity (+)	Survival of S. exempta nucleopolyhedrovirus infection (+)	Povey et al. (2013)
Protein-to-carbohydrate ratio	Manduca sexta	Artificial	Hemocyte density (N.S.) Implant encapsulation (+[a]) Phenoloxidase activity (+) Prophenoloxidase activity (N.S.)		Wilson et al. (2019)[c]
Protein content	Spodoptera littoralis	Semi-artificial	Antibacterial activity (+) Phenoloxidase activity (+) Prophenoloxidase activity (+)	Survival of Xenorhabdus nematophila infection (+)	Cotter et al. (2019)
Protein content	Lycaeides melissa	Plant	Implant encapsulation (N.S.) Phenoloxidase activity (+[b])		Yoon et al. (2019)
Protein quality	Spodoptera littoralis	Semi-artificial	Antibacterial activity (+) Phenoloxidase activity (N.S.)		Lee et al. (2008)

(continued)

Table 1 (continued)

Nutritional parameter	Species	Diet type	Immune effect(s)	Survival effect	Reference
Protein consumption	*Spodoptera littoralis*	Artificial	Antibacterial activity (+) Phenoloxidase activity (N.S.)		Cotter et al. (2011)[c]
Carbohydrate consumption	*Spodoptera littoralis*	Artificial	Antibacterial activity (+) Phenoloxidase activity (N.S.)		Cotter et al. (2011)[c]
Protein consumption	*Grammia incorrupta*	Artificial	Implant encapsulation (+[a])		Mason et al. (2014)
Carbohydrate consumption	*Grammia incorrupta*	Artificial	Implant encapsulation (+)		Mason et al. (2014)

[a]Reported as marginally significant ($P = 0.05$–0.08)
[b]Indirect effect mediated by weight gain on host plant *Medicago sativa*
[c]Effects summarized; see paper for detailed response surfaces

mellonella, Trichoplusia ni, Bombyx mori, Manduca sexta). While patterns are still being explored, several studies have documented increased expression of extracellular immune-related proteins in response to bacterial infection (Freitak et al. 2007; Bel et al. 2013; Gunaratna and Jiang 2013; Cotter et al. 2019). Lysozyme activity increased in *Trichoplusia ni* (Noctuidae) larvae reared on diets laced with nonpathogenic bacteria, while PO activity decreased (Freitak et al. 2007). Accordingly, transcript expression of hemolymph antimicrobial peptides (including lysozymes) and a PO inhibiting enzyme increased on bacterial diets, showing congruence between the functional assays and RNA transcripts. Cotter et al. (2019) found that functional immune assays (PO and lysozyme activity) were increased in bacterially challenged *Spodoptera littoralis* (Noctuidae) larvae, and associated transcripts were also upregulated; however, the congruence was dependent upon the amount of protein that was consumed in the diet. In studies with transcriptome data, upregulation of immune genes in response to infection appears to produce mixed results, where some immune genes are upregulated, but some are downregulated (Liu et al. 2014; Tan et al. 2019; Lin et al. 2020). These kinds of studies are rarely paired with functional assays, such as measuring encapsulation of an artificial implant (Fig. 2) or quantifying PO or antibacterial activity in the hemolymph. Interestingly, Tan et al. (2019) found that no immune genes were differentially expressed between monarch larvae (*Danaus plexippus*, Nymphalidae) infected with a protozoan pathogen and uninfected larvae; however, they did find differences in immune gene expression according to host plant diet, echoing results from Cotter et al. (2019). With such an enormous amount of genetic data generated by this type of analysis, it is perhaps not too surprising that results would be complex and not straightforward, especially in non-model species (Peterson et al. 2019). Overall, transcript and transcriptome data

Table 2 Effects of plant secondary chemistry on caterpillar functional immune responses and interactions with natural enemies (as measured by survival or infection loads following pathogen challenge). Positive associations between phytochemicals and immune or survival/resistance parameters are indicated by "(+)," negative associations are indicated by "(-)," and nonsignificant associations are indicated by "(N.S.)." The ability of each herbivore species to sequester the focal phytochemical ("Sequest.") is indicated by "Yes," "No," or unknown ("Unk.")

Secondary metabolite	Species	Sequest.	Diet type(s)	Immune effect(s)	Survival/resistance effect	Reference
Cardenolides	*Danaus plexippus*	Yes	Plant	Implant encapsulation (-)		Decker et al. (2020)
Glucosinolates	*Pieris rapae*	No	Plant	Encapsulation of *Cotesia glomerata* eggs (-ᵃ)		Bukovinszky et al. (2009)
Glucosinolates	*Plutella xylostella*	Unk.	Plant	Phenoloxidase activity (N.S.)	Emergence of *Diadegma semiclausum* parasitoids (N.S.)	Sun et al. (2020)
Iridoid glycosides	*Anartia jatrophae*	Yes	Plant	Implant encapsulation (+)		Lampert et al. (2014)
Iridoid glycosides	*Anartia jatrophae*	Yes	Plant	Hemocyte density (N.S.) Implant encapsulation (N.S.)	Survival of Junonia coenia densovirus infection (+)	Muchoney et al. unpublished data
Iridoid glycosides	*Ceratomia catalpae*	Yes	Plant, foliar application	Implant encapsulation (-)		Lampert and Bowers (2014)
Iridoid glycosides	*Ceratomia undulosa*	No	Plant, foliar application	Implant encapsulation (-)		Lampert and Bowers (2014)
Iridoid glycosides	*Euphydryas phaeton*	Yes	Plant	Hemocyte density (-) Implant encapsulation (N.S.) Phenoloxidase activity (N.S.)	Junonia coenia densovirus infection load (-)	Muchoney et al. (2021)
Iridoid glycosides	*Grammia incorrupta*	Yes	Artificial, plant	Implant encapsulation (N.S.)		Smilanich et al. (2011b)
Iridoid glycosides	*Junonia coenia*	Yes	Artificial, plant, foliar application	Implant encapsulation (-)		Smilanich et al. (2009)

(continued)

Table 2 (continued)

Secondary metabolite	Species	Sequest.	Diet type(s)	Immune effect(s)	Survival/resistance effect	Reference
Iridoid glycosides	*Junonia coenia*	Yes	Artificial	Implant encapsulation (−)[b]		Richards et al. (2012)
Iridoid glycosides	*Junonia coenia*	Yes	Plant	Implant encapsulation (−)		Quintero et al. (2014)
Iridoid glycosides	*Junonia coenia*	Yes	Plant	Implant encapsulation (−)		Lampert et al. (2014)
Iridoid glycosides	*Junonia coenia*	Yes	Plant	Hemocyte density (N.S.) Phenoloxidase activity (N.S.) Phenoloxidase resp. (Y/N) (+)	Survival of Junonia coenia densovirus infection (+)	Smilanich et al. (2018)
Iridoid glycosides	*Junonia coenia*	Yes	Plant	Hemocyte density (N.S.) Implant encapsulation (N.S.)		Carper et al. (2019)
Iridoid glycosides	*Melitaea cinxia*	Yes	Plant	Implant encapsulation (+)	Survival of *Serratia marcescens* infection (N.S.)	Laurentz et al. (2012)
Iridoid glycosides	*Vanessa cardui*	Yes	Plant	Implant encapsulation (+)		Lampert et al. (2014)
Monoterpenes	*Lophocampa ingens*	No	Artificial	Implant encapsulation (−)		Trowbridge et al. (2016)
Phenolics	*Epirrita autumnata*	Unk.	Plant	Implant encapsulation (−/N.S.)[c]		Haviola et al. (2007)
Phenolics	*Manduca sexta*	Yes	Artificial	Hemocyte density (+) Phenoloxidase activity (N.S.)	Survival of *Enterococcus faecalis* infection (+) Survival of *Pseudomonas aeruginosa* infection (N.S.)	del Campo et al. (2013)

Pyrrolizidine alkaloids	Grammia incorrupta	Yes	Artificial	Implant encapsulation (N.S.)		Smilanich et al. (2011a)
Withanolides	Heliothis subflexa	Unk.	Artificial	Antibacterial activity (N.S.) Phenoloxidase activity (+)	Survival of *Bacillus thuringiensis* infection (+)	Barthel et al. (2016)
Withanolides	Heliothis virescens	Unk.	Artificial	Antibacterial activity (N.S.) Phenoloxidase activity (N.S.)	Survival of *Bacillus thuringiensis* infection (N.S.)	Barthel et al. (2016)

[a] Reduced encapsulation in plants induced by previous herbivory
[b] Effect summarized; see paper for detailed path analyses
[c] Negative effects of hydrolyzable tannins, but no effects of condensed tannins or flavonoid glycosides

appear to be useful for understanding immune regulation but should be paired with functional assays and pathogen loads to better understand the outcome for infection (Cotter et al. 2019). A final cautionary note for immune assays in the lab concerns the possibility that organisms may already be infected with a parasite/pathogen, in which case any manipulation that is applied will have a background infection that can add noise to the data. In some cases, it is impossible to assess background infections, but in cases where an infection is applied, then control groups should always be checked to ensure they are infection-free.

Effects of Nutritional Content

Nutritional Content and the Caterpillar Immune Response

As immune responses are costly to maintain and activate (Moret and Schmid-Hempel 2000; Freitak et al. 2003; Ardia et al. 2012), the quantity and quality of nutritional resources available to insects are expected to impact immunocompetence. In this section, we use diet "quantity" to refer to the availability of energy, which can be manipulated through food limitation treatments or diets varying in caloric density. Diet "quality" may refer to the relative ability of a certain diet to support herbivore development, as indicated by metrics such as growth rate or body mass, or more specifically to variation in the composition of the diet, including macronutrient, micronutrient, or secondary metabolite content. We begin with an overview of the effects of diet quantity and intrinsic quality on caterpillar immunity, followed by a review of studies that examine the impacts of specific nutrients or nutrient blends on immune responses.

Nutritional Quantity and Quality A number of studies have demonstrated that quantitative resource limitation, in the form of food restriction or starvation, can be detrimental to insect immune responses (Suwanchaichinda and Paskewitz 1998; Siva-Jothy and Thompson 2002; Rantala et al. 2003; Ayres and Schneider 2009), while others have documented positive effects (e.g., Becker et al. 2010). In caterpillars, immunological responses to food restriction have been mixed, indicating that varying degrees of resource limitation may trigger reconfiguration, rather than wholesale downregulation, of immune processes (see Adamo et al. 2016). Periodic food restriction reduced both PO activity and hemocyte concentrations in monarch caterpillars (*Danaus plexippus*), an effect which appeared to be mediated by reduced body size, but did not affect the probability of infection by a protozoan parasite, *Ophyrocystis elektroscirrha*, following inoculation (McKay et al. 2016). Similarly, food limitation reduced PO activity in the western tent caterpillar, *Malacosoma californicum pluviale* (Lasiocampidae), but did not impact hemocyte density or susceptibility to nucleopolyhedrovirus infection (Myers et al. 2011). In another outbreaking herbivore, the autumnal moth *Epirrita autumnata* (Geometridae), however, gradual starvation impaired the encapsulation response but increased PO activity

(Yang et al. 2007). This enhancement of PO activity was mirrored in severely food-limited *Manduca sexta*, possibly due to a reduced threshold for enzyme activation, while lysozyme-like antimicrobial activity was unaffected by resource limitation (Adamo et al. 2016). Notably, food-deprived *M. sexta* also exhibited higher constitutive expression of several immune genes, compared to better-fed controls, but did not upregulate expression of these genes following immune challenge. These results suggest that when resources are limited, caterpillars may prioritize constitutive defenses, which are maintained in the absence of infection and available to act immediately upon challenge, over inducible defenses that are activated in response to infection.

Consumption of host plants or artificial diets that support high herbivore performance, often referred to as "high-quality," has been associated with immune enhancement in caterpillars (e.g., Klemola et al. 2008; Diamond and Kingsolver 2011; Triggs and Knell 2012a, b; but see Krams et al. 2015). In studies comparing the effects of high- and low-quality host plants, it is important to note that intrinsic quality may be influenced by nutritional composition (see below), secondary chemistry (see section on "Effects of Plant Secondary Chemistry"), or traits such as leaf toughness and water content (Klemola et al. 2008). Though it may not be possible to identify a single plant trait that is responsible for differences in herbivore performance and immunity in these cases, these studies provide valuable insight into the relationship between developmental performance (i.e., growth rate, body condition) and immune performance. For example, *Manduca sexta* reared on a high-quality host plant species (*Nicotiana tabacum*) exhibited stronger melanization and encapsulation responses than those reared on a low-quality host plant (*Proboscidea louisianica*) (Diamond and Kingsolver 2011). Using structural equation models, the authors revealed that the indirect effect of host plant quality, mediated by body condition, was substantial for encapsulation but negligible for melanization. Such effects may also arise due to intraspecific variation in host plant quality: wild *Epirrita autumnata* confined to high-quality mountain birch (*Betula pubescens* ssp. *czerepanovii*) trees (defined by relative growth rate and pupal masses on trees in previous years) demonstrated stronger encapsulation responses during the pupal stage than those reared on low-quality trees (Klemola et al. 2008; but see Klemola et al. 2007 for contrasting results under laboratory conditions). However, this increased ability to encapsulate an artificial antigen did not appear to correspond to improved defense against wild hymenopteran parasitoids, as parasitism levels did not differ on high- and low-quality trees. Importantly, variation in larval diet quality may also give rise to trans-generational effects on the immune responses of offspring. Indian meal moths, *Plodia interpunctella* (Pyralidae), which were reared on poor-quality artificial diets (lower in protein, lipid, and micronutrients) not only exhibited reduced PO activity and hemocyte densities but also produced offspring that exhibited 11–26% reductions in PO activity, depending on whether the mother, father, or both parents experienced a poor diet (Triggs and Knell 2012b). The authors hypothesized that these trans-generational effects may be the result of imprinting mechanisms (e.g., epigenetic markers that affect offspring immune gene expression) rather than differential parental resource allocation.

Macronutritional Composition Substantial progress in understanding the roles of specific dietary nutrients in regulating caterpillar immunocompetence has been gleaned through experiments that manipulate the composition of macronutritional resources (typically protein and carbohydrate), rather than overall diet quality or caloric density. In particular, positive effects of dietary protein on immune responses have been documented in numerous studies with Lepidoptera (summarized in Table 1). In *Spodoptera littoralis*, consumption of artificial diets with high protein-to-carbohydrate ratios (i.e., protein-skewed diets) increased antibacterial activity of the hemolymph and encapsulation of an artificial implant, compared to more carbohydrate-skewed diets (Lee et al. 2006). A similar pattern was documented for *Spodoptera exempta*: consumption of protein-biased diets resulted in elevated antibacterial activity and PO activity, though caterpillars challenged with bacteria-derived immune elicitors invested relatively more in antibacterial activity and less in PO activity than control larvae (Povey et al. 2009). In *S. exempta* caterpillars infected with a low dose of nucleopolyhedrovirus, multiple immune parameters were enhanced on diets with high protein-to-carbohydrate ratios, although certain relationships were nonlinear (e.g., antibacterial activity peaked on a marginally protein-skewed diet) (Povey et al. 2013).

Few studies have investigated whether the benefits of protein extend to caterpillars consuming plant-based diets; however, Yoon et al. (2019) found that foliar protein content of the host plant *Medicago sativa* had an indirect positive effect on PO activity in Melissa blue butterfly caterpillars (*Lycaeides melissa*, Lycaenidae), which was mediated by weight gain. Importantly, consumption of high-protein diets is often associated with increased levels of protein in the hemolymph (Lee et al. 2006; Povey et al. 2009, 2013), which were shown by Povey et al. (2009) to be positively correlated with antibacterial activity and PO activity in *S. exempta*. These results provide strong evidence that constitutive immune responses of caterpillars are protein-intensive and may be constrained on low-protein diets due to depletion of the "hemolymph protein pool." By extension, experiments manipulating protein-to-carbohydrate ratios have also demonstrated that carbohydrate may be less important than protein in regulating immunity; however, Mason et al. (2014) found that the encapsulation response of the woolly bear *Grammia incorrupta* (Erebidae) was positively correlated with carbohydrate intake but not protein intake, suggesting that the relative importance of carbohydrate consumption may vary across species.

Geometric Framework for Nutrition Further insight into the macronutritional requirements of caterpillar immune processes has been provided by studies employing the Geometric Framework for Nutrition (Simpson and Raubenheimer 1995), which have focused on determining nutritional optima or "intake targets" for different performance parameters within multidimensional nutritional space (Behmer 2009; Ponton et al. 2011). This approach addresses the "qualitative resource constraints hypothesis," which proposes that variation in the specific blend of nutrients ingested, rather than their overall quantity, will determine the relative performance of physiological traits (including immune responses) that may differ in their nutritional requirements (Cotter et al. 2011). Cotter et al. (2011) manipulated both the macronutritional composition and caloric density of artificial diets consumed by *Spodoptera littoralis*, demonstrating that different immune parameters peaked in distinct regions of nutritional space and were more responsive to variation in nutrient composition than quantity. In particular, hemolymph protein levels and lysozyme-like activity exhibited strong positive responses to increased protein consumption, with peaks indicating optimal ranges of protein intake. In contrast, PO activity was not significantly influenced by diet variables and peaked in a more carbohydrate-rich region of nutrient space than the other immune-related parameters. In a specialist herbivore, *Manduca sexta*, immune parameters again varied in their responses to dietary macronutrients, with protein generally exerting a stronger effect on immune performance than carbohydrate, though patterns were complex (Wilson et al. 2019). In a second study with *S. littoralis*, Cotter et al. (2019) found that while functional immune responses were negatively impacted by low-protein diets (Table 1), the expression of immune genes exhibited variable patterns. Interestingly, the relationship between immune gene expression and its functional outcomes (enzyme activity) was dependent upon the protein content of the diet; for example, both pro-PO activity and PPO gene expression were low on protein-poor diets but showed an increasingly linear relationship as dietary protein increased, indicating that low-protein diets may limit the translation of immune genes.

Together, these experiments have provided a wealth of data demonstrating that while no single diet can optimize all components of the caterpillar immune response, protein consumption is often an important determinant of immunocompetence. As the immunological effects of macronutrients likely vary across caterpillar species differing in their dietary preferences and nutritional requirements (e.g., specialists compared to generalists; Wilson et al. 2019), research investigating the influence of protein and carbohydrate intake in a wider variety of caterpillar species will be most fruitful for seeking general patterns. In addition, while experiments utilizing artificial diets have facilitated the precise manipulation of macronutrient composition, extending this work to understand how natural variation in the protein content of host plants affects herbivore immunity (see Yoon et al. 2019) represents an important next step in this field.

Nutritional Content and Caterpillar-Natural Enemy Interactions

No-Choice Studies Effects of dietary macronutrients on caterpillar immune responses may be expected to produce bottom-up effects on their interactions with endoparasites and pathogens, with enhanced immune performance favoring reduced enemy fitness and increased host survival. The importance of dietary protein in determining host survival following viral and bacterial challenge has been provided by multiple studies (Table 1). Lee et al. (2006) and Povey et al. (2013) found that consumption of diets with higher protein-to-carbohydrate ratios increased both immune responses (see above) and survival following nucleopolyhedrovirus infection in *Spodoptera littoralis* and *Spodoptera exempta*, respectively. In both cases, differences in mortality were quite large, for example, ranging from 77% survival of NPV-infected *S. littoralis* on the most protein-biased diet to just 25% survival on the most carbohydrate-biased diet, while uninfected caterpillars exhibited high survival on all diets (Lee et al. 2006). A similar pattern was reported for interactions between *S. exempta* and a bacterial pathogen: caterpillars that consumed protein-rich diets exhibited strengthened immune responses (antibacterial activity and PO activity) and increased survival following infection with Gram-positive *Bacillus subtilis* (Povey et al. 2009). In *S. littoralis* injected with a live LD_{50} dose of the Gram-negative bacterium *Xenorhabdus nematophila*, the amount of protein consumed by caterpillars increased immune responses and strongly influenced time to death, with low-protein diets resulting in rapid mortality and high-protein diets facilitating longer survival (Cotter et al. 2019). Notably, Cotter et al. (2019) also found that expression of a gene that produces an antimicrobial peptide, Moricin, was both upregulated in response to infection and highly correlated with survival; however, infected larvae with especially high expression of Moricin were less likely to survive than those with low expression. This pattern was noted by the authors as an indication that high Moricin expression may be a signature of high bacterial loads, rather than an indicator of immunocompetence. Overall, the high consistency between measured immune parameters and survival documented in these studies indicates that protein-mediated immune enhancement likely plays a role in determining the outcomes of pathogen infection, possibly by augmenting hemolymph protein reserves required to mount an immune response.

Self-Selection Studies While the examples above utilized "no-choice" diet treatments, these experiments have often been paired with dietary "self-selection" experiments, in which pathogen- or parasitoid-challenged caterpillars are able to compose their own diets. Such experiments provide an important source of insight into the impacts of macronutritional composition on caterpillar resistance to natural enemies, revealing that some caterpillars can modify their feeding behavior to preferentially ingest nutrients that are required to fight off infection. This behavior can constitute a form of "self-medication," provided that the behavioral change induced by infection improves the fitness of infected hosts but reduces host fitness in the absence of infection (Singer et al. 2009). While self-medication has been the focus of several studies involving the antiparasitic effects of plant secondary compounds (see "Self-Medication with Secondary Chemicals," below), insects may also self-medicate by altering their intake of nutritive substances, including protein and carbohydrate (de Roode and Hunter 2019).

The majority of examples of macronutrient self-selection in infected caterpillars have documented an increased preference for consuming protein. For example, NPV-challenged *S. littoralis* that went on to survive infection selected diets that were more protein-biased than those chosen by caterpillars that eventually succumbed to infection and uninfected controls from day three post-infection onward (Lee et al. 2006). In *S. exempta*, which typically prefer a more carbohydrate-biased diet, caterpillars selected a higher protein diet than controls when injected with bacteria (Povey et al. 2009) and when challenged with either a lethal (LD_{50}) or sublethal (LD_{10}) dose of NPV (Povey et al. 2013). Caterpillars infected with a high dose of NPV that went on to survive infection also demonstrated an earlier shift toward a high protein-to-carbohydrate ratio (on day one post-inoculation) and an overall decrease in food consumption over time (indicative of illness-induced anorexia; Adamo et al. 2007) than those that later died from infection. This behavior provides a clear example of self-medication with macronutrients, as high levels of protein intake enhanced survival of infected caterpillars but reduced performance of uninfected caterpillars (Povey et al. 2013). In contrast, Cotter et al. (2011) found no change in the amount of protein or carbohydrate consumed by *S. littoralis* challenged with lyophilized bacteria; however, lysozyme activity on the chosen diet was higher in challenged larvae than in controls, suggesting an increased allocation of available nutrients to this immune response.

Changes in caterpillar feeding behavior in response to attack by different types of natural enemies (e.g., parasitoids, entomopathogenic fungi) have received less attention; however, Mason et al. (2014) found that *Grammia incorrupta* caterpillars parasitized by tachinid fly parasitoids (*Chetogena* spp.) reduced their consumption of high-protein food, compared to unparasitized controls. Importantly, parasitoid development is closely linked to host physiology, and parasitized hosts may alter their feeding behavior in ways that positively affect the fitness of the parasitoid, as has been shown in *Manduca sexta* parasitized by *Cotesia congregata* wasps (Thompson et al. 2005). Thus, while the survival benefits of increased protein selection in *Spodoptera* caterpillars challenged with bacteria or baculoviruses have been well documented, additional research will be necessary to determine how common this behavior is in response to different types of natural enemy attack (e.g., parasitoids versus pathogens with an oral route of infection). In addition, future studies that quantify pathogen loads of infected caterpillars reared on diets varying in macronutrient composition will be very constructive for elucidating the extent to which protein consumption facilitates survival through suppression of pathogen replication (i.e., resistance), as opposed to increasing survival at a given pathogen burden (i.e., tolerance).

Effects of Plant Secondary Chemistry

Plant Secondary Chemistry and the Caterpillar Immune Response

Plant secondary metabolites are highly diverse and can serve a variety of functions in plants, including defense against herbivores, protection from pathogen infection and abiotic stressors, and mediation of interactions with competitors and mutualists (Iason et al. 2012). Secondary metabolite profiles can vary substantially in both concentration and composition across plant species, populations, and individuals, potentially giving rise to differential impacts on the physiology and fitness of the herbivores that consume these plants (Rosenthal and Berenbaum 1991) and shaping the overall composition and diversity of herbivore communities (Richards et al. 2015; Glassmire et al. 2016). Investigations of the impacts of plant secondary metabolites on lepidopteran immune responses have employed a variety of experimental approaches, documenting effects that range from negative to neutral to positive (summarized in Table 2; Lampert 2012), and have focused relatively more on assays of the melanization and encapsulation response (e.g., PO activity and implant encapsulation; Fig. 2) than have studies focusing on macronutritional effects (Table 1).

A particular area of interest has been determining the immunological consequences of phytochemical sequestration, the process by which certain herbivores co-opt plant defensive compounds and employ them in their own defense (Nishida 2002). While this adaptation may increase herbivore resistance to parasitism in cases where endoparasites are unable to cope with toxins ingested by their hosts (i.e., the "nasty host hypothesis"; Gauld et al. 1992), consumption of high levels of secondary metabolites may also inflict negative effects on herbivore immune responses, rendering hosts immunologically vulnerable to parasitism (the "vulnerable host hypothesis"; Smilanich et al. 2009a). These hypotheses provide a useful framework for considering both the direct and indirect impacts of plant secondary chemistry on caterpillar-natural enemy interactions, the consequences of which may be challenging to disentangle. In this section, we review examples from the literature of phytochemically mediated variation in caterpillar immune responses. We particularly highlight examples involving iridoid glycosides and phenolics, as compounds within these groups have been the focus of multiple studies (Table 2), as well as research evaluating the influences of induced plant defenses, phytochemical mixtures, and overall phytochemical diversity, which illustrate the complexity and context dependence of interactions between plant secondary chemistry and caterpillar immunity.

Iridoid Glycosides Evidence supporting the "vulnerable host hypothesis" has primarily been provided by studies focusing on iridoid glycosides (IGs), a class of monoterpenoid secondary metabolites that can be toxic or deterrent to non-adapted herbivores (Puttick and Bowers 1988). Certain specialized herbivores can sequester these compounds in high concentrations, thereby obtaining chemical defense against vertebrate (Bowers 1980) and invertebrate predators (Dyer and Bowers 1996). In one such specialist, the buckeye (*Junonia coenia*, Nymphalidae), Smilanich et al. (2009a) found that consumption of high concentrations of IGs was associated with suppression of the larval encapsulation response against an artificial antigen (injected beads). This effect was consistent across experiments: the encapsulation response was weaker in *J. coenia* caterpillars consuming a high-IG host plant (*Plantago lanceolata*) compared to a low-IG host plant (*Plantago major*) and in larvae consuming *P. major* leaves with high concentrations of IGs added to the foliar surface, compared to control leaves. Sequestration of IGs played a role in mediating this response: the encapsulation response of caterpillars reared on artificial diets supplemented with IGs (aucubin and catalpol) was negatively associated with the amount of catalpol that was sequestered from diets. Interestingly, this immunological cost of sequestration did not appear to be explained by increased metabolic demands of processing IGs, as respiration rates were lower on the diet containing high IGs, compared to low IGs (Smilanich et al. 2009a).

This negative effect of IG sequestration on the *J. coenia* encapsulation response has been corroborated by three additional studies (Richards et al. 2012; Lampert et al. 2014; Quintero et al. 2014), one of which revealed negative synergistic effects of aucubin and catalpol sequestration on encapsulation (Richards et al. 2012; see "Synergistic and Combinatorial Phytochemistry," below). Quintero et al. (2014) examined the impact of host plant ontogeny on this relationship, finding that IG concentrations in *P. lanceolata* increased with age and, consequently, that *J. coenia* caterpillars feeding on older plants exhibited increased sequestration and reduced encapsulation compared to larvae consuming younger plants. In addition, Carper et al. (2019) found that while encapsulation tended to decrease with increasing sequestration of IGs, this relationship varied considerably across larval instars. Together, experiments in the buckeye-IG system have revealed a strong connection between sequestration of plant secondary metabolites and the strength of the encapsulation-melanization response (but see Smilanich et al. 2018 for an exploration of other immune parameters in *J. coenia*), indicating that heightened investment in chemical defense may impair components of herbivore immune defense.

Immunosuppressive effects of iridoid glycoside sequestration have been documented in two additional caterpillar species, both of which sequester IGs in high concentrations. Lampert and Bowers (2015) found that the encapsulation response of the catalpa sphinx, *Ceratomia catalpae* (Sphingidae), was negatively correlated with the amount of catalpol sequestered by caterpillars consuming a typical host plant, *Catalpa bignonioides*. Baltimore checkerspots, *Euphydryas phaeton* (Nymphalidae), exhibited reduced hemocyte densities when sequestering high concentrations of IGs (aucubin and catalpol) from wild host plants, while PO activity and encapsulation were not correlated with sequestration (Muchoney et al. 2021). In a third IG-sequestering species, the Glanville fritillary (*Melitaea cinxia*, Nymphalidae), however, the larval encapsulation response was stronger on artificially selected *Plantago lanceolata* lines containing high concentrations of catalpol, though the quantity of IGs sequestered (measured in adults) was not correlated with encapsulation (Laurentz et al. 2012). This result is somewhat surprising in the context of well-documented negative effects of catalpol on encapsulation in other specialists (e.g., *J. coenia*). While it is certainly possible that relationships between IG sequestration and encapsulation vary across species, the range of IG concentrations used in different experiments may also be an important factor to consider. For example, Laurentz et al. (2012) utilized five *P. lanceolata* diets, ranging from 1.62% to 3.42% IGs by dry weight, whereas Smilanich et al. (2009a) used artificial diets containing either 1% or 5% IGs and *P. major* diets supplemented with either 2% or 10% IGs. It is therefore possible that negative immunological effects of sequestered

IGs only become apparent in herbivores consuming very high concentrations of these compounds.

Complementary studies focusing on species that do not sequester IGs, or sequester IGs in low concentrations, have produced varying results. The encapsulation response of the generalist *Grammia incorrupta*, which can sequester low levels of IGs, did not differ when caterpillars were reared on artificial diets or host plants containing varying concentrations of IGs (Smilanich et al. 2011b). In white peacocks (*Anartia jatrophae*, Nymphalidae), which can also sequester IGs at low concentrations, one study documented a positive relationship between IG sequestration and melanization (Lampert et al. 2014), while another found that hemocyte concentrations and implant encapsulation did not differ when caterpillars were reared on a plant containing high IGs (*Plantago lanceolata*) or a plant containing no IGs (*Bacopa monnieri*) (Muchoney et al. 2021). In contrast, the encapsulation response of the waved sphinx (*Ceratomia undulosa*, Sphingidae), which consumes plants containing seco-iridoid glycosides but does not appear to sequester IGs in significant amounts, was reduced when caterpillars consumed a host plant supplemented with catalpol (Lampert and Bowers 2015). These results indicate that the extent to which caterpillars sequester secondary metabolites may play an important role in mediating interactions between host plants and immunity; however, sequestration may not be a prerequisite for immune effects in all cases (see section on "Specialist and Generalist Immunity," below).

Phenolics In addition to iridoid glycosides, multiple studies have investigated the impacts of phenolic compounds on caterpillar immune defense, documenting negative, neutral, or positive effects (Table 2). Phenolics are a large group of secondary compounds that can contribute to plant defense against herbivores and may also possess antioxidant and antimicrobial properties. Haviola et al. (2007) characterized the phenolic composition of mountain birch foliage (*Betula pubescens* ssp. *czerepanovii*), which can contain over 20% phenolics by dry weight. This study found that the encapsulation response of an important defoliator of mountain birch, *Epirrita autumnata*, exhibited variable interactions with different types of phenolic compounds when measured during the pupal stage. Encapsulation was negatively correlated with 8 out of 24 hydrolyzable tannins, which exhibited generally positive relationships with larval performance, while condensed tannins and flavonoid glycosides tended to be negatively associated with larval survival but uncorrelated with encapsulation. In contrast, del Campo et al. (2013) examined the effects of a single, widespread phenolic compound, chlorogenic acid (CA), on immune responses of *Manduca sexta*, finding that artificial diets supplemented with CA increased the number of hemocytes in the larval hemolymph but did not impact PO activity. Phenolic compounds were also investigated by Ojala et al. (2005), who found that the encapsulation response of the generalist caterpillar *Parasemia plantaginis* (Erebidae) was highest on host plants containing high levels of antioxidants, consisting of phenolics (flavonoids and condensed tannins) and carotenoids, and lowest on an artificial diet containing trace amounts of secondary metabolites. As consumption of high concentrations of foliar phenolics has been shown to correspond to high levels of antioxidant activity in caterpillar hemolymph (Johnson and Felton 2001), these studies raise the intriguing question of whether the antioxidant properties of certain phenolics (e.g., flavonoids, CA) may play a role in enhancing or stimulating caterpillar immune responses.

Induced Phytochemistry Intraspecific variation in plant secondary chemistry triggered by previous herbivory (induced defenses) may additionally impact the strength of lepidopteran immune responses. Bukovinszky et al. (2009) compared the effects of wild and cultivated populations of cabbage (*Brassica oleracea*), which differ substantially in both constitutive and inducible levels of defensive glucosinolates (Gols et al. 2008), on the encapsulation response of the small cabbage white, *Pieris rapae* (Pieridae). Caterpillars reared on wild host plants, which contain higher constitutive levels of glucosinolates, exhibited a decreased ability to encapsulate eggs of the parasitoid *Cotesia glomerata* (Hymenoptera, Braconidae) than those reared on a cultivated variety with lower glucosinolate concentrations. Furthermore, previous herbivory by congeneric larvae (*Pieris brassicae*) resulted in a reduction of the *P. rapae* encapsulation response on both cultivated and wild plants, indicating that herbivore immunity was indirectly impacted by plant defenses induced by previous herbivores, along with constitutive defenses. In a contrasting example, caterpillar damage to mountain birch trees (*Betula pubescens* ssp. *czerepanovii*) was associated with an increase in the encapsulation response of *Epirrita autumnata* pupae in the subsequent year (Kapari et al. 2006). Which aspects of plant nutritional or secondary chemistry were responsible for this upregulation is unknown; however, certain phenolics in mountain birch have been associated with reduced encapsulation in *E. autumnata* (Haviola et al. 2007), while another phenolic (CA) was associated with increased hemocyte density in *Manduca sexta* (del Campo et al. 2013; see above). Importantly, previous herbivory does not exclusively result in increased concentrations of plant secondary metabolites; for example, Trowbridge et al. (2016) reported that herbivore-damaged piñon pine (*Pinus edulis*) trees contain approximately 30% lower concentrations of foliar monoterpenes than undamaged trees, likely due to volatile emissions that offset increased synthesis. These authors conducted an experiment in which Southwestern tiger moth caterpillars (*Lophocampa ingens*, Erebidae) were reared on artificial diets mimicking the monoterpene concentrations of damaged and undamaged needles. The results revealed that damaged diets (i.e., lower monoterpene concentrations) were associated with a 40% increase in the encapsulation response relative to undamaged diets, which may have been facilitated by increased consumption and energy allocation toward immunity.

Synergistic and Combinatorial Phytochemistry While most research to date has addressed the effects of a small number of individual compounds, an emerging area of research focuses on the impacts of phytochemical mixtures, as well as overall phytochemical diversity, on herbivore immunity. Examination of biologically relevant mixtures of plant secondary metabolites is key to understanding their effects on caterpillar immune responses, as compounds may exhibit additive or synergistic impacts on herbivore physiology (Berenbaum et al. 1991; Calcagno et al. 2002; Dyer et al. 2003; Richards et al. 2010, 2012, 2016), and all plant species contain more than one and sometimes hundreds of secondary compounds. Several experiments have utilized artificial diets supplemented with crude extracts of secondary chemicals from host plants (Smilanich et al. 2009a, 2011b; Richards et al. 2012; Barthel et al. 2016) or chemical mixtures mimicking those found in wild host plants (e.g., Trowbridge et al. 2016) to investigate effects on caterpillar immune parameters. Richards et al. (2012) directly evaluated the synergistic impacts of two iridoid glycosides, aucubin and catalpol, on *Junonia coenia* performance and immunity, finding that caterpillars that consumed mixtures of IGs exhibited higher growth rates and survival, greater IG sequestration, and weaker encapsulation responses than those reared on diets containing single compounds. This result is compatible with the "vulnerable host hypothesis" (Smilanich et al. 2009a), as positive synergistic effects of IGs on chemical defense (sequestration) were accompanied by negative synergistic effects on the encapsulation response. In the non-sequestering specialist *Lophocampa ingens*, diets containing higher concentrations of monoterpene mixtures (consisting of four compounds) were associated with reduced encapsulation, while diets containing single compounds varied considerably in their effects on encapsulation (Trowbridge et al. 2016). Such mixture-dependent effects of secondary chemicals on herbivore immune responses may provide explanatory power in systems in which concentration effects alone fail to explain patterns.

The impacts of total phytochemical diversity on caterpillar immunity have received less attention in the literature; however, Slinn et al. (2018) examined the effects of interspecific phytochemical diversity of *Piper* plants on the PO activity of five neotropical caterpillar species. This study revealed that phytochemical diversity was negatively correlated with PO activity in specialist herbivores in the genus *Eois* (Geometridae) at a sampling site in Costa Rica, while the opposite pattern was documented at a site in Ecuador. Though an explanation for these contrasting patterns is unclear, the authors highlight substantial differences in site elevation (2100 m in Ecuador, 35–140 m in Costa Rica) and parasitoid pressure on *Eois* (three times higher in Costa Rica) as potential factors to consider. In addition, there was no correlation between phytochemical diversity and PO activity of the focal generalist herbivore in the study, *Quadrus cerealis* (Hesperiidae). Another study found that PO activity of *Eois nympha* and *E. apyraria* caterpillars was reduced by 46% when consuming a *Piper* species with high phytochemical diversity (*P. cencocladum*) compared to a congeneric species with lower phytochemical diversity (*P. imperiale*) (Hansen et al. 2017). Yoon et al. (2019) characterized intraspecific phytochemical diversity of wild alfalfa plants (*Medicago sativa*; mean, 33.7 effective compounds per sample), finding that phytochemical diversity had a direct negative effect on PO activity, but no effect on encapsulation, in Melissa blue caterpillars (*Lycaeides melissa*). In this study, no individual saponin or phenolic compound was significantly associated with the strength of the PO response, providing additional evidence of the ability of natural chemical mixtures to inhibit immune responses. These results indicate that in some cases, consumption of plants with high levels of phytochemical diversity may suppress the PO enzyme, possibly due to an increased potential for negative synergistic effects (Richards et al. 2012) or the elevated resource costs of detoxifying a wide array of defensive chemicals. However, it remains to be determined whether this effect is generalizable to other immune components and across additional plant-herbivore pairings.

To conclude, recent research has demonstrated that caterpillar immune responses can be impacted by both the composition and concentration of secondary metabolites ingested from host plants. Though the diversity of secondary chemicals consumed by lepidopterans is vast and only a small subset of compounds have been studied, it is clear that multiple types of metabolites can strongly influence caterpillar immunity, including iridoid glycosides (Table 2), phenolics (Haviola et al. 2007; del Campo et al. 2013), glucosinolates (Bukovinszky et al. 2009; but see Sun et al. 2020), monoterpenes (Trowbridge et al. 2016), and withanolides (Barthel et al. 2016; see below). Extending these findings to understand how these phytochemically mediated effects cascade to impact natural enemies represents a relatively new but exciting avenue of research.

Plant Secondary Chemistry and Caterpillar-Natural Enemy Interactions

A substantial body of literature has demonstrated that variation in host plant secondary chemistry can impact interactions between caterpillars and their natural enemies, including parasitoids (e.g., Campbell and Duffey 1979; Barbosa et al. 1991; Ode et al. 2004; Ode 2006 and references therein; Lampert et al. 2008; Singer et al. 2009; Richards et al. 2010, Slinn et al. 2018) and pathogens (Cory and Hoover 2006 and references therein; De Roode et al. 2008; del Campo et al. 2013; Barthel et al. 2016; Smilanich et al. 2018; Muchoney et al. 2021). In many examples, the fitness of internal parasites and pathogens is negatively impacted by host consumption of secondary metabolites, which may or may not be accompanied by sequestration. It is important to note that in many cases, effects of plant secondary metabolites on natural enemies may be mediated through non-immunological routes including direct toxicity (e.g., α-tomatine and parasitoids; Campbell and Duffey 1981), interference with pathogen infection via the caterpillar midgut (e.g., phenolics and baculoviruses; Felton and Duffey 1990), or indirect effects mediated by negative impacts on caterpillar development (e.g., Bloem and Duffey 1990). As the impacts of plant chemistry on herbivore-parasitoid (Ode 2006) and herbivore-pathogen (Cory and Hoover 2006) interactions have been reviewed elsewhere, here we focus on research that explicitly includes the immune response in investigations of plant-caterpillar-natural enemy interactions, asking whether phytochemically mediated variation in caterpillar immunity produces tangible effects on the outcomes of infection. We also highlight the special case of caterpillars that self-medicate with toxic plant compounds against parasite infection, exploring the repercussions of this behavior for the caterpillar immune response.

Studies that explicitly link phytochemically mediated changes in caterpillar immunity to interactions with natural enemies are relatively rare and have produced complex results (Table 2). However, a number of examples have been provided in which positive effects of plant secondary metabolites on caterpillar immunity are accompanied by increased survival following pathogen infection. For example, *Manduca sexta* caterpillars reared on diets supplemented with chlorogenic acid (CA) exhibited increased hemocyte densities and greater survival following infection by the Gram-positive bacterium *Enterococcus faecalis* (del Campo et al. 2013). This survival benefit did not appear to be mediated by improved attenuation of bacterial proliferation, as infection loads did not differ between diet treatments at 24 h postinfection; however, it is possible that hemocyte-mediated defenses played a role in controlling *E. faecalis* at a later stage in the infection. Notably, CA supplementation did not increase *M. sexta's* survival of a more virulent, Gram-negative bacterium (*Pseudomonas aeruginosa*), indicating that the defensive benefits of CA may be pathogen-specific (del Campo et al. 2013). In *Heliothis subflexa* (Noctuidae), which specializes on host plants containing herbivore-deterrent withanolides (*Physalis* spp.), artificial diets supplemented with withanolides had a positive effect on PO activity, accompanied by upregulation of certain immune genes, and corresponded to increased survival following infection by the Gram-positive bacterium *Bacillus thuringiensis* (Bt) (Barthel et al. 2016). As withanolides were found to exhibit potent antibacterial activity, it is possible that Bt-infected *H. subflexa* benefitted from both the direct antibacterial action of withanolides and from their immunostimulatory effects. In contrast, withanolides did not impact PO activity or survival of Bt infection in the generalist congener *Heliothis virescens*, suggesting that the immunostimulatory role of withanolides may be confined to species that are adapted to ingesting these compounds. Laurentz et al. (2012) found that diets high in an IG, catalpol, enhanced the encapsulation response of *Melitaea cinxia* but that dietary IG content was not correlated with survival following infection by the Gram-negative bacterium *Serratia marcescens*. However, the strength of the encapsulation response was positively correlated with survival time following infection, providing support for the functional relevance of this assay in evaluating *M. cinxia* resistance to *S. marcescens*. These examples demonstrate that positive effects of phytochemicals on immune responses may be associated with increased survival of infection in some cases; however, determining the role of assayed immune parameters in mediating resistance to specific natural enemies (e.g., through suppression of pathogen replication) remains an important goal.

Case Study: Iridoid Glycosides and JcDV Plant secondary metabolites that suppress, rather than enhance, components of the caterpillar immune response are predicted to have indirect positive effects on natural enemy fitness (i.e., the "vulnerable host hypothesis"). However, recent examples involving iridoid glycosides, which have exhibited suppressive effects on immunity in specialist caterpillars (Table 2), illustrate that this may not be the case for all types of enemies. This research focuses on elucidating the role of IGs in mediating interactions between nymphalid butterflies and a viral pathogen, Junonia coenia densovirus (JcDV). Smilanich et al. (2018) evaluated variation in immune responses and resistance to JcDV infection in buckeyes (*Junonia coenia*) feeding on a plant high in IGs (*Plantago lanceolata*) or low in IGs (*P. major*). Virus-infected caterpillars consuming the high-IG host plant were more likely to exhibit an activated (nonzero) PO response and were also more likely to survive to the adult stage, though PO activity and hemocyte density did not differ based on host plant species. A similar effect was documented in white peacocks (*Anartia jatrophae*): consumption of the same high-IG host plant (*P. lanceolata*) increased survival of JcDV infection compared to a host plant containing no IGs, but this survival benefit was not accompanied by enhancement of the encapsulation response or hemocyte densities (Muchoney et al. 2021). However, increased survival on *P. lanceolata* was associated with significantly lower postmortem viral loads, highlighting the possibility that the high-IG plant conferred greater resistance to JcDV infection through processes external to measured immune responses, including direct interference with virus infectivity or replication in host tissues. Evidence supporting this hypothesis was documented in wild-collected individuals of a third host species, the Baltimore checkerspot (*Euphydryas phaeton*). In this species, naturally JcDV-infected individuals that sequestered higher IG concentrations harbored lower in situ viral loads, once again suggesting IG-mediated suppression of infection, but also exhibited reduced hemocyte densities (Muchoney et al. 2021). In this case, larval immune parameters did appear to interact with infection: in particular, granulocyte concentrations were negatively correlated with JcDV loads, but this effect was inhibited at high levels of IG sequestration. This example illustrates that even phytochemicals that typically reduce herbivore immune responses may positively impact resistance against natural enemies through alternative pathways, providing a cautionary tale against the assumption that reduced immunocompetence necessarily correlates with reduced fitness of infected individuals (see "Linking Immunity to Fitness via Enemy Attack," below).

Self-Medication with Secondary Chemicals Woolly bears (*Grammia incorrupta*) and monarchs (*Danaus plexippus*) are notable examples of caterpillars that "self-medicate" through preferential consumption of toxic plant secondary metabolites when parasitized (Singer et al. 2009; Lefèvre et al. 2010). While this behavioral change effectively increases herbivore resistance to key parasites (De Roode et al. 2008; Singer et al. 2009), putatively through direct toxicity of sequestered compounds, its compatibility with caterpillar immune defenses is important to consider (Smilanich et al. 2009a). *Grammia incorrupta* caterpillars increase their consumption of pyrrolizidine alkaloids (PAs) when parasitized, which increases survival following parasitoid attack but reduces caterpillar fitness in the absence of parasitism (Singer et al. 2009). Smilanich et al. (2011a) examined the effects of PAs on the *G. incorrupta* immune response, finding that the strength of the melanization response did not differ across caterpillars reared on high-PA, low-PA, or control diets. In this example, consumption of PAs was not detrimental to the melanization response, indicating that plant-derived antiparasitic compounds and certain immune defenses may act in concert against parasitoid infection.

In *D. plexippus*, consumption of host plant species containing high concentrations of toxic cardenolides suppresses infection by a protozoan parasite, *Ophryocystis elektroscirrha* (OE) (De Roode et al. 2008; Gowler et al. 2015). In an example of trans-generational self-medication, female monarchs infected with OE preferentially oviposit on a milkweed species containing high levels of cardenolides, effectively reducing the parasite loads of their infected offspring (Lefèvre et al. 2010). Recently, Decker et al. (2020) found that consumption of the high-cardenolide species, *Asclepias curassavica*, reduced caterpillar melanization by 13%, compared to a low-cardenolide species (*A. incarnata*). This finding mirrored the results of a transcriptomic study by Tan et al. (2019), which documented downregulation of five canonical immune genes in *D. plexippus* caterpillars consuming the more toxic, medicinal milkweed species, accompanied by upregulation of several detoxification genes. These studies provide a contrasting example of potential incompatibility between exogenous sources of antiparasitic defense (i.e., sequestered toxins) and immune defenses, raising the compelling questions of whether selection may favor investment in effective forms of chemical defense, at the expense of the immune response, in certain species (Tan et al. 2019; see also Muchoney et al. 2021) and whether such investment may increase host vulnerability to other types of pathogens.

While evidence for indirect effects of plant secondary metabolites on caterpillar-natural enemy interactions, mediated by the immune response, has been provided in a handful of systems, there is still much to learn about these relationships. Secondary metabolites that improve host survival or resistance to infection (as indicated by reduced pathogen loads) have often been associated with enhancement of at least one immune parameter (but see Muchoney et al. 2021), but typically not all measured immune parameters (Table 2). In contrast, evidence for indirect *positive* effects of secondary chemistry on natural enemy performance, explicitly mediated by suppression of host immune responses, has been elusive (Table 2). Moving forward, research addressing the relationship between plant secondary chemistry, caterpillar immunity, and natural enemy performance (directly assessed via pathogen load or parasite development) in a greater variety of tritrophic systems will be essential for evaluating hypotheses regarding patterns of immunocompetence and putative fitness outcomes in caterpillars varying in their dietary preferences (e.g., the "Vulnerable Host Hypothesis"; Smilanich et al. 2009a).

Knowledge Gaps and Avenues of Future Research

In this section, we elaborate on interactions between host plants and caterpillar immunity that are just beginning to be discovered and would benefit from more research attention. We enumerate these areas below and suggest ways in which future research can address these knowledge gaps.

Linking Immunity to Fitness via Enemy Attack

A first area where persistent knowledge gaps remain is linking the measured immune parameter to fitness of the studied organism. As mentioned previously, immune parameters do not always respond in the same direction or with the same magnitude, thus measuring multiple parameters is ideal (Adamo 2004b). Beyond this methodological issue, how can researchers be certain that the strength of the immune response actually reflects the fitness outcome for an organism? The obvious answer is to measure the fitness of the parasite/pathogen challenged organisms after immune parameters have been measured and compared to the control group. Graham et al. (2011) called upon ecoimmunologists to include measures of fitness (i.e., reproductive success) and parasite/pathogen load along with measurements of immune parameters. In their paper, they give an especially potent example of potential pitfalls from ignoring fitness. The authors found that genotypes of the freshwater crustacean, *Daphnia magna*, which were resistant to the bacterial pathogen, *Pasteuria ramosa*, had lower plasmatocyte counts compared to the susceptible genotypes, which showed higher plasmatocyte counts. Since *P. ramosa* renders its host sterile, the high plasmatocyte count did not translate into improved reproductive success for *D. magna*. Without knowing the susceptibility of the differing genotypes and measuring parasite load, one might conclude that higher plasmatocyte counts indicate that individuals are better protected and thus have a higher fitness, when in fact the opposite was true. Adult reproductive success, parasite/pathogen loads, and immune response are rarely recorded concurrently in studies investigating the insect immune response (but see for recent examples Parker et al. 2017; Merrill et al. 2019) and even more rarely in lepidopteran studies. Most lepidopteran studies capture the caterpillar's immunity in response to an exogenous variable such as diet (Muller et al. 2015), and many do include a measure of survival against parasites/pathogens (see Tables 1 and 2; Adamo et al. 2007; Silva and Elliot 2016), but few studies include fecundity (but see Myers et al. 2011), or all three metrics (immunity, reproductive success, and parasite or pathogen load).

There is nothing inherently wrong with studies that focus on the effect of exogenous variables on the immune response. In fact, they have helped immensely to establish patterns of what causes variation in the immune response (see sections on nutrition and phytochemistry above). However, the obvious next step is to understand the repercussions of immunological variation for lepidopteran study species with regard to their interactions with natural enemies; such studies are essential if we are to reveal the importance of immunocompetence. As with the *D. magna* example, several studies with lepidopterans show that immunocompetence does not always translate into resistance against natural enemies (Karimzadeh and Wright 2008; Saejeng et al. 2010; Scholefield et al. 2019; Tan et al. 2019; Resnik and Smilanich 2020), which is not to conclude that the immune response is unimportant, but that understanding its importance requires putting the immune response into context. For example, Resnik and Smilanich (2020) found that virus-infected painted lady caterpillars (*Vanessa cardui*) had suppressed PO activity, high viral loads, and low survivorship when reared on an exotic host plant. However, when caterpillars fed on a native host plant, PO activity was unchanged by viral infection, viral loads were low, and survival was highest in the infected group. Including reproductive success can be challenging. Some immune assays require destructive sampling (e.g., injected beads, transcripts), and in many cases with lepidopterans it can be difficult or impossible to successfully mate adults in captivity. Possible ways around these challenges are the use of nondestructive methods when possible and proxies for fecundity such as pupal mass.

Microbiome and the Immune Response

Studies on the insect microbiome are a relatively new area of research, but they have already proven to be a fascinating topic unearthing many novel interactions between host and microbe (reviewed in Douglas 2015; Kucuk 2020). Many of the elucidated functions of resident microbes in the insect gut include facilitation of digestion and nutrition (Anand et al. 2010; Scully et al. 2014), detoxification of plant defenses (Ceja-Navarro et al. 2015), and also a defense function by inducing immune system priming (Broderick et al. 2006; Freitak et al. 2007; Brownlie and Johnson 2009). Of all the insect taxa studied to date, Lepidoptera represent the group with perhaps the lowest diversity of microbes, with few to no known obligate microbes (Hammer et al. 2017, 2019; Voirol et al. 2018). Hammer et al. (2017) showed the depauperate gut microbial community of temperate and neotropical caterpillars and pointed to the high alkalinity and lack of specialized microbe-harboring structures of the caterpillar gut as possible explanations for the low diversity. In addition, several studies have eliminated the gut microbiome of caterpillars and found no detrimental effect on growth or survival (Hammer et al. 2017; Phalnikar et al. 2019). Nonetheless, the evidence falls short of sweeping generalizations for the lack of importance of the lepidopteran microbiome (Voirol et al. 2018; Hammer et al. 2019; Duplouy et al. 2020). With regard to immune function, midgut bacteria of *Spodoptera exigua* have been directly linked to immune priming that then provides protection against *Bacillus thuringiensis* (Hernández-Martínez et al. 2010). In *S. littoralis*, the symbiont bacteria, *Enterococcus mundtii*, secrete antimicrobial peptides in the midgut that selectively suppresses pathogenic microbes and maintains healthy gut microbiota (Johnston and Rolff 2015; Shao et al. 2017). Similarly, Duplouy et al. (2020) showed that the gut microbiome of the specialist caterpillar, *Melitaea cinxia*, was associated with an upregulation of antimicrobial peptides. Since research shows that the gut microbiome in caterpillars can be host plant dependent (Hammer et al. 2017; Jones et al. 2019; but see Whitaker et al. 2016; Minard et al. 2019), studies exploring the relationship between microbiome and immunity should investigate it within the context of host plants. Jones et al. (2019) showed that the host plant is a primary factor explaining variation in the microbiome of two caterpillars, *Spodoptera frugiperda* and *Helicoverpa zea*. Further, Yoon et al. (2019) showed that microbial diversity in the midgut of Melissa blue butterfly larvae (*Lycaeides melissa*) varied by host plant. This same study also found that phenoloxidase activity increased with microbial diversity. This effect was host plant dependent: there was a positive relationship on the native host plant (*Astragalus canadensis*) and a negative relationship on the exotic host plant (*Medicago sativa*). Aside from these and a handful of other studies (Smilanich et al. 2018; Duplouy et al. 2020), there are few that explicitly test for an association between the microbiome and immune function in caterpillars, especially in non-model systems, making this a worthy pursuit for future studies.

Specialist and Generalist Immunity

Given the prominent use of caterpillars in studying the evolution of diet breadth (Dyer et al. 2007; Forister et al. 2012, 2015; Nylin et al. 2014; Hardy 2017), the question arises whether there are differences in immune strength between specialist and generalist caterpillars. In a recent review detailing plant-mediated indirect effects on parasitoid success, Kaplan et al. (2016) pointed to important differences between specialist and generalist herbivores that could lead to a pattern of specialists having lower immune responses compared to generalists (Lampert 2012). First, specialists are masters of sequestration (Nishida 2002; Opitz and Müller 2009; Erb and Robert 2016; Petschenka and Agrawal 2016), which is not to disregard generalist sequestration (Hartmann et al. 2004; Smilanich et al. 2011a; Lampert et al. 2014), but specialists are very good at concentrating sequestered compounds at high concentrations (Bowers and Puttick 1986; Bowers and Collinge 1992). The link between sequestration and depressed immunity has been demonstrated with the plant secondary metabolites, iridoid glycosides (IGs), and three specialist caterpillars, the buckeye (*Junonia coenia*), the Baltimore checkerspot (*Euphydryas phaeton*), and the catalpa sphinx (*Ceratomia catalpae*). These species are capable of sequestering IGs at up to 25% dry weight in their hemolymph (Bowers and Puttick 1986; Bowers and Collinge 1992; Lampert and Bowers 2015). Smilanich et al. (2009a) found that as sequestration increases in the buckeye caterpillar, the melanization response decreases (see also Quintero et al. 2014). Similarly, Muchoney et al. (2021) found that increased sequestration of IGs in the Baltimore checkerspot caterpillar led to decreased total hemocyte counts. In all cases, it was only at high concentrations that the immune response was suppressed (but see Laurentz et al. 2012). In fact, with Baltimore checkerspot larvae, low sequestration did not negatively affect total hemocyte count, and these individuals were able to keep viral loads low. At least in the case of iridoid glycosides, the effect does appear to be sequestration specific and not due to simply ingesting IGs from the diet (Smilanich et al. 2011b). The generalist caterpillar, *Grammia incorrupta*, which does not sequester IGs at high concentrations, reared on host plants containing IGs and a host plant without IGs had a similar melanization response regardless of diet (see also Lampert and Bowers 2015 for similar results). Thus, for specialists and generalists consuming plants containing iridoid glycosides, specialists appear to suffer more due to high sequestration (Lampert et al. 2014). As previously mentioned, Tan et al. (2019) demonstrated that monarch caterpillars (famous for their sequestration of cardenolides) downregulated immune-related genes when reared on host plants with high cardenolide concentrations. However, caterpillars reared on the low cardenolide host plant did not downregulate immune genes. Further, they found that there was no difference in immune gene regulation in response to infection from the protozoan *Ophryocystis elektroscirrha* (OE). These results with the specialist monarch caterpillars point again to sequestration being an important component of immune malaise in specialist lepidopterans. Whether this pattern holds in other well-known lepidopteran sequestration systems (e.g., cyanogenic glycosides, pyrrolizidine alkaloids) has yet to be tested.

A second difference pointed out by Kaplan et al. (2016) is the obvious host range difference between specialists and generalists. This difference could be important in terms of pathogen exposure. Barthel et al. (2014) found that the generalist caterpillar *Heliothis virescens* had a more robust immune response compared to the specialist, *H. subflexa*, and point out that the specialist in this system spends most of its larval stage enveloped within the flower calyx, possibly exposing it to fewer pathogens. This hypothesis of pathogen exposure differences in the context of immunity and diet breadth has not been well tested and deserves further study. Another possible benefit to generalist immunity is the option to use alternative host plants to optimize the immune response. Muller et al. (2015) found that this was the case with the polyphagous caterpillar *Lobesia botrana* (Tortricidae), a pest in wine vineyards that is capable of feeding on plants from 30 different families. When compared to individuals feeding on *Vitus vinifera* (grape cultivar), the hemocyte count, standing PO activity, and total PO activity of *L. botrana* were greater on eight alternative host plants. When surveyed across geographic locations and host plants, Vogelweith et al. (2011) found that these immune parameters varied by location and that there was a positive correlation between immunity and parasitism rates, showing that populations with high parasitism were investing in immunity. Overall patterns between generalists' host plant range and immunocompetence are still in the nascent stages of investigation, as some studies have shown no difference in generalist immune response across alternative host plants (Yang et al. 2008; Smilanich et al. 2011a), and many generalists are not browsers and are confined to the plant species where they were oviposited (Renwick and Chew 1994; Rosenwald et al. 2017).

Conclusion

The caterpillar immune response is strongly influenced by host plant diet. In this chapter, we have shown that plant primary and secondary metabolites have a pivotal place in determining the strength of the caterpillar immune response. While we have gained much knowledge within the last 25 years about the bitrophic relationship between host plant and caterpillar immunity, there is still much to be learned about how these interactions cascade to the third trophic level. Empirical studies that quantify damage to natural enemies, as well as survival and reproductive success of the host, will add much to our knowledge of how the immune response fits into plant-caterpillar-enemy interactions. There may be general trends that are beginning to form within caterpillar ecoimmunology. Consumption of dietary protein enhances a variety of immune parameters, and sequestration of secondary metabolites appears to be harmful to the immune response at high concentrations, which may be particularly detrimental for specialist caterpillars. Finally, there are pieces of this interaction that are completely unknown (e.g., latitudinal trends in immune strength) and worthy of pursuit for future studies.

References

Adamo SA (2004a) Estimating disease resistance in insects: phenoloxidase and lysozyme-like activity and disease resistance in the cricket *Gryllus texensis*. J Insect Physiol 50:209–216
Adamo SA (2004b) How should behavioural ecologists interpret measurements of immunity? Anim Behav 68:1443–1449
Adamo SA, Fidler TL, Forestell CA (2007) Illness-induced anorexia and its possible function in the caterpillar, *Manduca sexta*. Brain Behav Immun 21:292–300
Adamo SA, Davies G, Easy R, Kovalko I, Turnbull KF (2016) Reconfiguration of the immune system network during food limitation in the caterpillar *Manduca sexta*. J Exp Biol 219:706–718
Anand AAP, Vennison SJ, Sankar SG, Prabhu DIG, Vasan PT, Raghuraman T et al (2010) Isolation and characterization of bacteria from the gut of *Bombyx mori* that degrade cellulose, xylan, pectin and starch and their impact on digestion. J Insect Sci 10:1–20
Ardia DR, Gantz JE, Schneider BC, Strebel S (2012) Costs of immunity in insects: an induced immune response increases metabolic rate and decreases antimicrobial activity. Funct Ecol 26:732–739
Ayres JS, Schneider DS (2009) The role of anorexia in resistance and tolerance to infections in *Drosophila*. PLoS Biol 7:e1000150
Barbosa P, Gross P, Kemper J (1991) Influence of plant allelochemicals on the tobacco hornworm and its parasitoid, Cotesia congregata. Ecology 72:1567–1575
Barthel A, Kopka I, Vogel H, Zipfel P, Heckel DG, Groot AT (2014) Immune defence strategies of generalist and specialist insect herbivores. Proc R Soc B Biol Sci 281:1–9
Barthel A, Vogel H, Pauchet Y, Pauls G, Kunert G, Groot AT et al (2016) Immune modulation enables a specialist insect to benefit from antibacterial withanolides in its host plant. Nat Commun 7:1–11
Beckage NE (ed) (2008) Insect immunology. Elsevier Academic Press, San Diego
Becker T, Loch G, Beyer M, Zinke I, Aschenbrenner AC, Carrera P et al (2010) FOXO-dependent regulation of innate immune homeostasis. Nature 463:369–373
Behmer ST (2009) Insect herbivore nutrient regulation. Annu Rev Entomol 54:165–187
Bel Y, Jakubowska AK, Costa J, Herrero S, Escriche B (2013) Comprehensive analysis of gene expression profiles of the beet armyworm *Spodoptera exigua* larvae challenged with *Bacillus thuringiensis* Vip3Aa toxin. PLoS One 8:1–23
Berenbaum MR, Nitao JK, Zangerl AR (1991) Adaptive significance of furanocoumarin diversity in *Pastinaca sativa* (Apiaceae). J Chem Ecol 17:207–215
Benjamin J, Parker Seth M, Barribeau Alice M, Laughton Lynn H, Griffin Nicole M, Gerardo Sheena, C (2017) Life-history strategy determines constraints on immune function. Journal of Animal Ecology 86(3) 473–483. https://doi.org/10.1111/1365-2656.12657
Bloem KA, Duffey SS (1990) Interactive effect of protein and rutin on larval *Heliothis zea* and the endoparasitoid *Hyposoter exiguae*. Entomol Exp Appl 54:149–160

Bowers MD (1980) Unpalatability as a defense strategy of *Euphydryas phaeton* (Lepidoptera: Nymphalidae). Evolution (N. Y) 34:586–600

Bowers MD, Collinge SK (1992) Fate of iridoid glycosides in different life stages of the buckeye, *Junonia coenia* (Lepidoptera: Nymphalidae). J Chem Ecol 18:817–831

Bowers MD, Puttick GM (1986) Fate of ingested iridoid glycosides in lepidopteran herbivores. J Chem Ecol 12:169–178

Brock PM, Murdock CC, Martin LB (2014) The history of ecoimmunology and its integration with disease ecology. Integr Comp Biol 54:353–362

Broderick NA, Raffa KF, Handelsman J (2006) Midgut bacteria required for *Bacillus thuringiensis* insecticidal activity. Proc Natl Acad Sci U S A 103:15196–15199

Brownlie JC, Johnson KN (2009) Symbiont-mediated protection in insect hosts. Trends Microbiol 17:348–354

Bukovinszky T, Poelman EH, Gols R, Prekatsakis G, Vet LEM, Harvey JA et al (2009) Consequences of constitutive and induced variation in plant nutritional quality for immune defence of a herbivore against parasitism. Oecologia 160:299–308

Calcagno MP, Coll J, Lloria J, Faini F, Alonso-Amelot ME (2002) Evaluation of synergism in the feeding deterrence of some furanocoumarins on *Spodoptera littoralis*. J Chem Ecol 28:175–191

Campbell BC, Duffey SS (1979) Tomatine and parasitic wasps: Potential incompatibility of plant antibiosis with biological control. Science 205:700–702

Campbell BC, Duffey SS (1981) Alleviation of α-tomatine-induced toxicity to the parasitoid, *Hyposoter exiguae*, by phytosterols in the diet of the host, *Heliothis zea*. J Chem Ecol 7:927–946

Carper AL, Enger M, Deane Bowers M (2019) Host plant effects on immune response across development of a specialist caterpillar. Front Ecol Evol 7:1–11

Carton Y, Poirié M, Nappi AJ (2008) Insect immune resistance to parasitoids. Insect Sci 15:67–87

Ceja-Navarro JA, Vega FE, Karaoz U, Hao Z, Jenkins S, Lim HC et al (2015) Gut microbiota mediate caffeine detoxification in the primary insect pest of coffee. Nat Commun 6:7618

Cerenius L, Söderhäll K (2004) The prophenoloxidase-activating system in invertebrates. Immunol Rev 198:116–126

Cory JS, Hoover K (2006) Plant-mediated effects in insect-pathogen interactions. Trends Ecol Evol 21:278–286

Cotter SC, Simpson SJ, Raubenheimer D, Wilson K (2011) Macronutrient balance mediates trade-offs between immune function and life history traits. Funct Ecol 25:186–198

Cotter SC, Reavey CE, Tummala Y, Randall JL, Holdbrook R, Ponton F et al (2019) Diet modulates the relationship between immune gene expression and functional immune responses. Insect Biochem Mol Biol 109:128–141

de Roode JC, Hunter MD (2019) Self-medication in insects: when altered behaviors of infected insects are a defense instead of a parasite manipulation. Curr Opin Insect Sci 33:1–6

De Roode JC, Pedersen AB, Hunter MD, Altizer S (2008) Host plant species affects virulence in monarch butterfly parasites. J Anim Ecol 77:120–126

Decker LE, Jeffrey CS, Ochsenrider KM, Potts AS, De Roode JC, Smilanich AM et al (2020) Elevated atmospheric concentrations of CO_2 increase endogenous immune function in a specialist herbivore. J Anim Ecol 90:628–640

del Campo ML, Halitschke R, Short SM, Lazzaro BP, Kessler A (2013) Dietary plant phenolic improves survival of bacterial infection in *Manduca sexta* caterpillars. Entomol Exp Appl 146:321–331

Diamond SE, Kingsolver JG (2011) Host plant quality, selection history and trade-offs shape the immune responses of *Manduca sexta*. Proc R Soc B Biol Sci 278:289–297

Douglas AE (2015) Multiorganismal insects: diversity and function of resident microorganisms. Annu Rev Entomol 60:17–34

Duplouy A, Minard G, Saastamoinen M (2020) The gut bacterial community affects immunity but not metabolism in a specialist herbivorous butterfly. Ecol Evol 10:8755–8769

Dyer LA, Bowers MD (1996) The importance of sequestered iridoid glycosides as a defense against an ant predator. J Chem Ecol 22:1527–1539

Dyer LA, Dodson CD, Stireman JO, Tobler MA, Smilanich AM, Fincher RM et al (2003) Synergistic effects of three *Piper* amides on generalist and specialist herbivores. J Chem Ecol 29:2499–2514

Dyer LA, Singer MS, Lill JT, Stireman JO, Gentry GL, Marquis RJ et al (2007) Host specificity of Lepidoptera in tropical and temperate forests. Nature 448:696–699

Erb M, Robert CAM (2016) Sequestration of plant secondary metabolites by insect herbivores: molecular mechanisms and ecological consequences. Curr Opin Insect Sci 14:8–11

Felton GW, Duffey SS (1990) Inactivation of baculovirus by quinones formed in insect-damaged plant tissues. J Chem Ecol 16:1221–1236

Forister ML, Dyer LA, Singer MS, Stireman JO, Lill JT (2012) Revisiting the evolution of ecological specialization, with emphasis on insect–plant interactions. Ecology 93:981–991

Forister ML, Novotny V, Panorska AK, Baje L, Basset Y, Butterill PT et al (2015) The global distribution of diet breadth in insect herbivores. Proc Natl Acad Sci U S A 112:442–447

Freitak D, Ots I, Vanatoa A, Hõrak P (2003) Immune response is energetically costly in white cabbage butterfly pupae. Proc R Soc B Biol Sci 270:220–222

Freitak D, Wheat CW, Heckel DG, Vogel H (2007) Immune system responses and fitness costs associated with consumption of bacteria in larvae of *Trichoplusia ni*. BMC Biol 5:1–13

Gauld ID, Gaston KJ, Janzen DH (1992) Plant allelochemicals, tritrophic interactions and the anomalous diversity of tropical parasitoids: the "nasty" host hypothesis. Oikos 65:353–357

Gillespie JP, Kanost MR, Trenczek T (1997) Biological mediators of insect immunity. Annu Rev Entomol 42:611–643

Glassmire AE, Jeffrey CS, Forister ML, Parchman TL, Nice CC, Jahner JP et al (2016) Intraspecific phytochemical variation shapes community and population structure for specialist caterpillars. New Phytol 212:208–219

Gols R, Bukovinszky T, Van Dam NM, Dicke M, Bullock JM, Harvey JA (2008) Performance of generalist and specialist herbivores and their endoparasitoids differs on cultivated and wild *Brassica* populations. J Chem Ecol 34:132–143

González-Santoyo I, Córdoba-Aguilar A (2012) Phenoloxidase: a key component of the insect immune system. Entomol Exp Appl 142:1–16

Gowler CD, Leon KE, Hunter MD, de Roode JC (2015) Secondary defense chemicals in milkweed reduce parasite infection in monarch butterflies, *Danaus plexippus*. J Chem Ecol 41:520–523

Graham AL, Shuker DM, Pollitt LC, Auld SKJR, Wilson AJ, Little TJ (2011) Fitness consequences of immune responses: strengthening the empirical framework for ecoimmunology. Funct Ecol 25:5–17

Gross P (1993) Insect behavioral and morphological defenses against parasitoids. Annu Rev Entomol 38:251–273

Gunaratna RT, Jiang H (2013) A comprehensive analysis of the *Manduca sexta* immunotranscriptome. Dev Comp Immunol 39:388–398

Hammer TJ, Janzen DH, Hallwachs W, Jaffe SP, Fierer N (2017) Caterpillars lack a resident gut microbiome. Proc Natl Acad Sci U S A 114:9641–9646

Hammer TJ, Sanders JG, Fierer N (2019) Not all animals need a microbiome. FEMS Microbiol Lett 366:1–11

Hansen AC, Glassmire AE, Dyer LA, Smilanich AM (2017) Patterns in parasitism frequency explained by diet and immunity. Ecography (Cop) 40:803–805

Hardy NB (2017) Do plant-eating insect lineages pass through phases of host-use generalism during speciation and host switching? Phylogenetic evidence. Evolution (N Y) 71:2100–2109

Hartmann T, Theuring C, Beuerle T, Bernays EA (2004) Phenological fate of plant-acquired pyrrolizidine alkaloids in the polyphagous arctiid *Estigmene acrea*. Chemoecology 14:207–216

Haviola S, Kapari L, Ossipov V, Rantala MJ, Ruuhola T, Haukioja E (2007) Foliar phenolics are differently associated with *Epirrita autumnata* growth and immunocompetence. J Chem Ecol 33:1013–1023

Hernández-Martínez P, Naseri B, Navarro-Cerrillo G, Escriche B, Ferré J, Herrero S (2010) Increase in midgut microbiota load induces an apparent immune priming and increases tolerance to *Bacillus thuringiensis*. Environ Microbiol 12:2730–2737

Iason GR, Dicke M, Hartley SE (eds) (2012) The ecology of plant secondary metabolites: from genes to global processes. Oxford University Press, New York

Jiang H, Vilcinskas A, Kanost MR (2010) Immunity in lepidopteran insects. In: Söderhäll K (ed) *Invertebrate immunity*. Landes Bioscience/Springer/LLC, New York, pp 181–194

Johnson KS, Felton GW (2001) Plant phenolics as dietary antioxidants for herbivorous insects: a test with genetically modified tobacco. J Chem Ecol 27:2579–2597

Johnston PR, Rolff J (2015) Host and symbiont jointly control gut microbiota during complete metamorphosis. PLoS Pathog 11:e1005246

Jones AG, Mason CJ, Felton GW, Hoover K (2019) Host plant and population source drive diversity of microbial gut communities in two polyphagous insects. Sci Rep 9:1–11

Kanost MR, Jiang H, Yu X-Q (2004) Innate immune responses of a lepidopteran insect, *Manduca sexta*. Immunol Rev 198:97–105

Kapari L, Haukioja E, Rantala MJ, Ruuhola T (2006) Defoliating insect immune defense interacts with induced plant defense during a population outbreak. Ecology 87:291–296

Kaplan I, Carrillo J, Garvey M, Ode PJ (2016) Indirect plant-parasitoid interactions mediated by changes in herbivore physiology. Curr Opin Insect Sci 14:112–119

Karimzadeh J, Wright DJ (2008) Bottom-up cascading effects in a tritrophic system: interactions between plant quality and host-parasitoid immune responses. Ecol Entomol 33:45–52

Klemola N, Klemola T, Rantala MJ, Ruuhola T (2007) Natural host-plant quality affects immune defence of an insect herbivore. Entomol Exp Appl 123:167–176

Klemola N, Kapari L, Klemola T (2008) Host plant quality and defence against parasitoids: no relationship between levels of parasitism and a geometrid defoliator immunoassay. Oikos 117:926–934

Krams I, Kecko S, Kangassalo K, Moore FR, Jankevics E, Inashkina I et al (2015) Effects of food quality on trade-offs among growth, immunity and survival in the greater wax moth *Galleria mellonella*. Insect Sci 22:431–439

Kucuk RA (2020) Gut bacteria in the Holometabola: a review of obligate and facultative symbionts. J Insect Sci 20:22

Lampert EC (2012) Influences of plant traits on immune responses of specialist and generalist herbivores. Insects 3:573–592

Lampert EC, Bowers MD (2015) Incompatibility between plant-derived defensive chemistry and immune response of two sphingid herbivores. J Chem Ecol 41:85–92

Lampert EC, Zangerl AR, Berenbaum MR, Ode PJ (2008) Tritrophic effects of xanthotoxin on the polyembryonic parasitoid *Copidosoma sosares* (Hymenoptera: Encyrtidae). J Chem Ecol 34:783–790

Lampert EC, Dyer LA, Bowers MD (2014) Dietary specialization and the effects of plant species on potential multitrophic interactions of three species of nymphaline caterpillars. Entomol Exp Appl 153:207–216

Laurentz M, Reudler JH, Mappes J, Friman V, Ikonen S, Lindstedt C (2012) Diet quality can play a critical role in defense efficacy against parasitoids and pathogens in the Glanville fritillary (*Melitaea cinxia*). J Chem Ecol 38:116–125

Lavine MD, Beckage NE (1996) Temporal pattern of parasitism-induced immunosuppression in *Manduca sexta* larvae parasitized by *Cotesia congregata*. J Insect Physiol 42:41–51

Lavine MD, Strand MR (2002) Insect hemocytes and their role in immunity. Insect Biochem Mol Biol 32:1295–1309

Lee KP, Cory JS, Wilson K, Raubenheimer D, Simpson SJ (2006) Flexible diet choice offsets protein costs of pathogen resistance in a caterpillar. Proc R Soc B Biol Sci 273:823–829

Lee KP, Simpson SJ, Wilson K (2008) Dietary protein-quality influences melanization and immune function in an insect. Funct Ecol 22:1052–1061

Lefèvre T, Oliver L, Hunter MD, De Roode JC (2010) Evidence for trans-generational medication in nature. Ecol Lett 13:1485–1493

Lin J, Yu XQ, Wang Q, Tao X, Li J, Zhang S et al (2020) Immune responses to *Bacillus thuringiensis* in the midgut of the diamondback moth, *Plutella xylostella*. Dev Comp Immunol 107:103661

Liu Y, Shen D, Zhou F, Wang G, An C (2014) Identification of immunity-related genes in *Ostrinia furnacalis* against entomopathogenic fungi by RNA-seq analysis. PLoS One 9:e86436

Mason PA, Smilanich AM, Singer MS (2014) Reduced consumption of protein-rich foods follows immune challenge in a polyphagous caterpillar. J Exp Biol 217:2250–2260

McKay AF, Ezenwa VO, Altizer S (2016) Consequences of food restriction for immune defense, parasite infection, and fitness in monarch butterflies. Physiol Biochem Zool 89:389–401

Minard G, Tikhonov G, Ovaskainen O, Saastamoinen M (2019) The microbiome of the *Melitaea cinxia* butterfly shows marked variation but is only little explained by the traits of the butterfly or its host plant. Environ Microbiol 21:4253–4269

Moret Y, Schmid-Hempel P (2000) Survival for immunity: The price of immune system activation for bumblebee workers. Science 290:1166–1168

Muchoney ND, Bowers MD, Carper AL, Mason PA, Teglas MB, Smilanich AM (2021) Use of an exotic host plant shifts immunity, chemical defense, and viral burden in wild populations of a specialist insect herbivore. *In press.*

Muller K, Vogelweith F, Thiéry D, Moret Y, Moreau J (2015) Immune benefits from alternative host plants could maintain polyphagy in a phytophagous insect. Oecologia 177:467–475

Myers JH, Cory JS, Ericsson JD, Tseng ML (2011) The effect of food limitation on immunity factors and disease resistance in the western tent caterpillar. Oecologia 167:647–655

Nakhleh J, Christophides GK, Osta MA, O'Neill L (2017) The serine protease homolog CLIPA14 modulates the intensity of the immune response in the mosquito *Anopheles gambiae*. J Biol Chem 292:18217–18226

Nappi AJ, Christensen BM (2005) Melanogenesis and associated cytotoxic reactions: applications to insect innate immunity. Insect Biochem Mol Biol 35:443–459

Nishida R (2002) Sequestration of defensive substances from plants by Lepidoptera. Annu Rev Entomol 47:57–92

Nylin S, Slove J, Janz N (2014) Host plant utilization, host range oscillations and diversification in nymphalid butterflies: a phylogenetic investigation. Evolution (N. Y) 68:105–124

Ode PJ (2006) Plant chemistry and natural enemy fitness: effects on herbivore and natural enemy interactions. Annu Rev Entomol 51:163–185

Ode PJ, Berenbaum MR, Zangerl AR, Hardy ICW (2004) Host plant, host plant chemistry and the polyembryonic parasitoid Copidosoma sosares: indirect effects in a tritrophic interaction. Oikos 104:388–400

Ojala K, Julkunen-Tiitto R, Lindström L, Mappes J (2005) Diet affects the immune defence and life-history traits of an Arctiid moth *Parasemia plantaginis*. Evol Ecol Res 7:1153–1170

Opitz SEW, Müller C (2009) Plant chemistry and insect sequestration. Chemoecology 19:117–154

Pascual L, Jakubowska AK, Blanca JM, Cañizares J, Ferré J, Gloeckner G et al (2012) The transcriptome of *Spodoptera exigua* larvae exposed to different types of microbes. Insect Biochem Mol Biol 42:557–570

Peterson B, Sanko TJ, Bezuidenhout CC, van den Berg J (2019) Transcriptome and differentially expressed genes of *Busseola fusca* (Lepidoptera: Noctuidae) larvae challenged with Cry1Ab toxin. Gene 710:387–398

Petschenka G, Agrawal AA (2016) How herbivores coopt plant defenses: natural selection, specialization, and sequestration. Curr Opin Insect Sci 14:17–24

Phalnikar K, Kunte K, Agashe D (2019) Disrupting butterfly caterpillar microbiomes does not impact their survival and development. Proc R Soc B Biol Sci 286:20192438

Ponton F, Wilson K, Cotter SC, Raubenheimer D, Simpson SJ (2011) Nutritional immunology: a multi-dimensional approach. PLoS Pathog 7:1–4

Povey S, Cotter SC, Simpson SJ, Lee KP, Wilson K (2009) Can the protein costs of bacterial resistance be offset by altered feeding behaviour? J Anim Ecol 78:437–446

Povey S, Cotter SC, Simpson SJ, Wilson K (2013) Dynamics of macronutrient self-medication and illness-induced anorexia in virally infected insects. J Anim Ecol 83:245–255

Price PW, Denno RF, Eubanks MD, Finke DL, Kaplan I (2011) Insect ecology. Cambridge University Press, New York

Puttick GM, Bowers MD (1988) Effect of qualitative and quantitative variation in allelochemicals on a generalist insect: iridoid glycosides and the southern armyworm. J Chem Ecol 14:335–351

Quintero C, Lampert EC, Bowers MD (2014) Time is of the essence: direct and indirect effects of plant ontogenetic trajectories on higher trophic levels. Ecology 95:2589–2602

Ragan EJ, An C, Jiang H, Kanost MR (2009) Roles of haemolymph proteins in antimicrobial defences of *Manduca sexta*. In: Reynolds S, Rolff J (eds) *Insect infection and immunity: evolution, ecology, and mechanisms*. Oxford University Press, New York, pp 34–48

Rantala MJ, Roff DA (2007) Inbreeding and extreme outbreeding cause sex differences in immune defence and life history traits in *Epirrita autumnata*. Heredity (Edinb) 98:329–336

Rantala MJ, Kortet R, Kotiaho JS, Vainikka A, Suhonen J (2003) Condition dependence of pheromones and immune function in the grain beetle *Tenebrio molitor*. Funct Ecol 17:534–540

Renwick JAA, Chew FS (1994) Oviposition behavior in Lepidoptera. Annu Rev Entomol 39:377–400

Resnik JL, Smilanich AM (2020) The effect of phenoloxidase activity on survival is host plant dependent in virus-infected caterpillars. J Insect Sci 20:1–4

Ribeiro C, Brehélin M (2006) Insect haemocytes: what type of cell is that? J Insect Physiol 52:417–429

Richards LA, Dyer LA, Smilanich AM, Dodson CD (2010) Synergistic effects of amides from two *Piper species* on generalist and specialist herbivores. J Chem Ecol 36:1105–1113

Richards LA, Lampert EC, Bowers MD, Dodson CD, Smilanich AM, Dyer LA (2012) Synergistic effects of iridoid glycosides on the survival, development and immune response of a specialist caterpillar, *Junonia coenia* (Nymphalidae). J Chem Ecol 38:1276–1284

Richards LA, Dyer LA, Forister ML, Smilanich AM, Dodson CD, Leonard MD et al (2015) Phytochemical diversity drives plant-insect community diversity. Proc Natl Acad Sci U S A 112:10973–10978

Richards LA, Glassmire AE, Ochsenrider KM, Smilanich AM, Dodson CD, Jeffrey CS et al (2016) Phytochemical diversity and synergistic effects on herbivores. Phytochem Rev 15:1153–1166

Riddell C, Adams S, Schmid-Hempel P, Mallon EB (2009) Differential expression of immune defences is associated with specific host-parasite interactions in insects. PLoS One 4:2–5

Rodrigues J, Brayner FA, Alves LC, Dixit R, Barillas-Mury C (2010) Hemocyte differentiation mediates innate immune memory in *Anopheles gambiae* mosquitoes. Science 329:1353–1355

Rolff J, Reynolds SE (eds) (2009) Insect infection and immunity: evolution, ecology, and mechanisms. Oxford University Press, New York

Rosenthal G, Berenbaum M (eds) (1991) Herbivores: their interactions with secondary plant metabolites, volume I: the chemical participants, 2nd edn. Academic, San Diego

Rosenwald LC, Lill JT, Lind EM, Weiss MR (2017) Dynamics of host plant selection and host-switching by silver-spotted skipper caterpillars. Arthropod Plant Interact 11:833–842

Saejeng A, Tidbury H, Siva-Jothy MT, Boots M (2010) Examining the relationship between hemolymph phenoloxidase and resistance to a DNA virus, *Plodia interpunctella* granulosis virus (PiGV). J Insect Physiol 56:1232–1236

Schmid-Hempel P, Ebert D (2003) On the evolutionary ecology of specific immune defence. Trends Ecol Evol 18:27–32

Scholefield JA, Shikano I, Lowenberger CA, Cory JS (2019) The impact of baculovirus challenge on immunity: the effect of dose and time after infection. J Invertebr Pathol 167:107232

Schulenburg H, Kurtz J, Moret Y, Siva-Jothy MT (2009) Introduction. Ecological immunology. Philos Trans R Soc B Biol Sci 364:3–14

Scully ED, Geib SM, Carlson JE, Tien M, McKenna D, Hoover K (2014) Functional genomics and microbiome profiling of the Asian longhorned beetle (*Anoplophora glabripennis*) reveal insights into the digestive physiology and nutritional ecology of wood feeding beetles. BMC Genomics 15:1–21

Shao Y, Chen B, Sun C, Ishida K, Hertweck C, Boland W (2017) Symbiont-derived antimicrobials contribute to the control of the Lepidopteran gut microbiota. Cell Chem Biol 24:66–75

Sheldon BC, Verhulst S (1996) Ecological immunology: costly parasite defences and trade-offs in evolutionary ecology. Trends Ecol Evol 11:317–321

Silva FWS, Elliot SL (2016) Temperature and population density: interactional effects of environmental factors on phenotypic plasticity, immune defenses, and disease resistance in an insect pest. Ecol Evol 6:3672–3683

Simpson SJ, Raubenheimer D (1995) The geometric analysis of feeding and nutrition: a user's guide. J Insect Physiol 41:545–553

Singer MS, Mace KC, Bernays EA (2009) Self-medication as adaptive plasticity: increased ingestion of plant toxins by parasitized caterpillars. PLoS One 4:e4796

Singer MS, Mason PA, Smilanich AM (2014) Ecological immunology mediated by diet in herbivorous insects. Integr Comp Biol 54:913–921

Siva-Jothy MT, Thompson JJW (2002) Short-term nutrient deprivation affects immune function. Physiol Entomol 27:206–212

Siva-Jothy MT, Tsubaki Y, Hooper RE, Plaistow SJ (2001) Investment in immune function under chronic and acute immune challenge in an insect. Physiol Entomol 26:1–5

Siva-Jothy MT, Moret Y, Rolff J (2005) Insect immunity: an evolutionary ecology perspective, *Adv. Insect Physiol. Vol. 32*. Elsevier Academic Press, San Diego

Slinn HL, Richards LA, Dyer LA, Hurtado PJ, Smilanich AM (2018) Across multiple species, phytochemical diversity and herbivore diet breadth have cascading effects on herbivore immunity and parasitism in a tropical model system. Front Plant Sci 9:1–12

Smilanich AM, Dyer LA, Chambers JQ, Bowers MD (2009a) Immunological cost of chemical defence and the evolution of herbivore diet breadth. Ecol Lett 12:612–621

Smilanich AM, Dyer LA, Gentry GL (2009b) The insect immune response and other putative defenses as effective predictors of parasitism. Ecology 90:1434–1440

Smilanich AM, Mason PA, Sprung L, Chase TR, Singer MS (2011a) Complex effects of parasitoids on pharmacophagy and diet choice of a polyphagous caterpillar. Oecologia 165:995–1005

Smilanich AM, Vargas J, Dyer LA, Bowers MD (2011b) Effects of ingested secondary metabolites on the immune response of a polyphagous caterpillar *Grammia incorrupta*. J Chem Ecol 37:239–245

Smilanich AM, Langus TC, Doan L, Dyer LA, Harrison JG, Hsueh J et al (2018) Host plant associated enhancement of immunity and survival in virus infected caterpillars. J Invertebr Pathol 151:102–112

Strand MR (2008) The insect cellular immune response. Insect Sci 15:1–14

Sun R, Gols R, Harvey JA, Reichelt M, Gershenzon J, Pandit SS et al (2020) Detoxification of plant defensive glucosinolates by an herbivorous caterpillar is beneficial to its endoparasitic wasp. Mol Ecol 29:4014–4031

Suwanchaichinda C, Paskewitz SM (1998) Effects of larval nutrition, adult body size, and adult temperature on the ability of *Anopheles gambiae* (Diptera: Culicidae) to melanize Sephadex beads. J Med Entomol 35:157–161

Tan WH, Acevedo T, Harris EV, Alcaide TY, Walters JR, Hunter MD et al (2019) Transcriptomics of monarch butterflies (*Danaus plexippus*) reveals that toxic host plants alter expression of detoxification genes and down-regulate a small number of immune genes. Mol Ecol 28:4845–4863

Tara E, Stewart Merrill Spencer R, Hall Loren, Merrill Carla E, Cáceres (2019) Variation in Immune Defense Shapes Disease Outcomes in Laboratory and Wild Daphnia. Integrative and Comparative Biology 59(5):1203–1219. https://doi.org/10.1093/icb/icz079

Thompson SN, Redak RA, Wang LW (2005) Host nutrition determines blood nutrient composition and mediates parasite developmental success: *Manduca sexta* L. parasitized by *Cotesia congregata* (Say). J Exp Biol 208:625–635

Triggs A, Knell RJ (2012a) Interactions between environmental variables determine immunity in the Indian meal moth *Plodia interpunctella*. J Anim Ecol 81:386–394

Triggs AM, Knell RJ (2012b) Parental diet has strong transgenerational effects on offspring immunity. Funct Ecol 26:1409–1417

Trowbridge AM, Bowers MD, Monson RK (2016) Conifer monoterpene chemistry during an outbreak enhances consumption and immune response of an eruptive folivore. J Chem Ecol 42:1281–1292

Vogel H, Altincicek B, Glöckner G, Vilcinskas A (2011) A comprehensive transcriptome and immune-gene repertoire of the lepidopteran model host *Galleria mellonella*. BMC Genomics 12:1–19

Vogelweith F, Thiéry D, Quaglietti B, Moret Y, Moreau J (2011) Host plant variation plastically impacts different traits of the immune system of a phytophagous insect. Funct Ecol 25:1241–1247

Voirol LRP, Frago E, Kaltenpoth M, Hilker M, Fatouros NE (2018) Bacterial symbionts in Lepidoptera: their diversity, transmission, and impact on the host. Front Microbiol 9:1–14

Whitaker MRL, Salzman S, Sanders J, Kaltenpoth M, Pierce NE (2016) Microbial communities of lycaenid butterflies do not correlate with larval diet. Front Microbiol 7:1–13

Wilson JK, Ruiz L, Davidowitz G (2019) Dietary protein and carbohydrates affect immune function and performance in a specialist herbivore insect (*Manduca sexta*). Physiol Biochem Zool 92:58–70

Yang S, Ruuhola T, Rantala MJ (2007) Impact of starvation on immune defense and other life-history traits of an outbreaking geometrid, *Epirrita autumnata*: a possible causal trigger for the crash phase of population cycle. Ann Zool Fennici 44:89–96

Yang S, Ruuhola T, Haviola S, Rantala MJ (2008) Effects of host-plant shift on immune and other key life-history traits of an eruptive Geometrid, *Epirrita autumnata* (Borkhausen). Ecol Entomol 33:510–516

Yoon SA, Harrison JG, Philbin CS, Dodson CD, Jones DM, Wallace IS et al (2019) Host plant-dependent effects of microbes and phytochemistry on the insect immune response. Oecologia 191:141–152

Trophic Interactions of Caterpillars in the Seasonal Environment of the Brazilian Cerrado and Their Importance in the Face of Climate Change

Laura Braga and Ivone R. Diniz

Caterpillar of *Megalopyge lanata* (Stoll 1780). Photo by Geraldo Freire

L. Braga (✉) · I. R. Diniz
Departamento de Zoologia, Instituto de Ciências Biológicas, Universidade de Brasília, Brasília, Brazil

© The Author(s), under exclusive license to Springer Nature Switzerland AG 2022
R. J. Marquis, S. Koptur (eds.), *Caterpillars in the Middle*, Fascinating Life Sciences, https://doi.org/10.1007/978-3-030-86688-4_15

Introduction

In tropical ecosystems, plant leafing patterns may be relatively constant across the year or may vary greatly according to the season. In seasonally tropical ecosystems, in particular, plant phenology and food availability for caterpillars can differ greatly across the year. The seasonality of rainfall strongly influences, directly and indirectly, plant phenology and herbivorous insect abundance (Essens et al. 2014) and consequently affects all trophic interactions. Apparently the pattern of caterpillar seasonality in tropical dry forests is driven by the increase in food availability in the rainy season (Connahs et al. 2011), since most plants are deciduous during the dry season, producing new leaves at the beginning of each rainy season (Janzen 1988a, b, 1993).

Although the Cerrado is a seasonally dry ecosystem with deciduous and semi-deciduous woody vegetation (Oliveira 2008), as is the tropical dry forest of Santa Rosa National Park, Costa Rica, plant and caterpillar phenologies differ greatly between these two locations. In Santa Rosa, leaf production occurs at the beginning of the rainy season, and most plants lose their leaves in the dry season (Janzen 1988b, 1993). Caterpillars are extremely abundant only at the beginning of the rainy season (Janzen 1993), with highest damage rates on young leaves (Janzen 1988a, b). On the other hand, in the Cerrado (Brazilian savanna), some deciduous and semi-deciduous plants can retain their mature leaves during the dry season (Morais et al. 1995). This happens because leaf production and loss are asynchronous events (Morais et al. 1999). Leaf production occurs in the dry-rainy season transition, and the expansion of leaves occurs just before the peak of the herbivore abundance (Marquis et al. 2002; Pinheiro et al. 2002). As a result, in the Cerrado, the seasonal pattern of folivorous caterpillars appears to be distinct and unique. Caterpillars peak in abundance and species richness at the beginning of the dry season (Morais et al. 1999), decreasing from the middle to the end of that period.

What makes Cerrado caterpillars (Plate 1) exhibit this unique caterpillar seasonality? The answer may be due to the high availability of mature leaves throughout the dry season and the ability of many caterpillars to feed on this resource (Morais et al. 1995), whereas in the dry forest of Santa Rosa, caterpillars apparently do not use mature leaves as a resource. Because of that, in Santa Rosa, the density of folivorous caterpillars comes close to zero during the dry season (Janzen 1988a). It can also be argued that this seasonality is, probably, also reinforced due to the occurrence, during the Cerrado dry season, of enemy-free space (Morais et al. 1999) as well as the high incidence of sheltered caterpillars. Shelters seem to protect the shelter builder against severe drought conditions of the dry season (Diniz et al. 2012; Velasque and Del-Claro 2016). Our data show that the majority of the Cerrado caterpillars that occur in the dry season are shelter-builders, while in Costa Rican tropical dry forest, the caterpillars recorded in the dry season are those that are internal feeders, such as seed- and fruit-mining and wood-boring, and therefore are protected by the host plant (Janzen 1988a).

Plate 1. Cerrado caterpillars. (**a**) *Eacles* sp., Saturniidae; (**b**) *Chlamydastis platyspora*, Elachistidae; (**c**) *Platynota rostrana*, Tortricidae; (**d**) *Phobetron hipparchia*, Limacodidae; (**e**) *Tolype* prop. *innocens*, Lasiocampidae; (**f**) *Podalia annulipes*, Megalopygidae; (**g**) *Dalcerina tijucana*, Dalceridae; (**h**) *Chioides catillus catillus*, Hesperiidae and (**i**) *Kolana ergina*, Lycaenidae. (Photographs by Laura Braga, Neuza Silva, and Rosevaldo Pessoa-Queiroz)

Ecology of the Cerrado

The Cerrado is a phytogeographic domain, part of the South American "Corridor of Savannas" (Schmidt and Inger 1951). It is bordered by the Amazon and the Atlantic Forests, the main Brazilian rainforests, and connects the Caatinga and Pantanal biomes in the northeast-southwest direction (Fig. 1). A prototypical form of the Cerrado already existed in the Cretaceous, before the separation of the American and African continents (Ratter and Ribeiro 1996). It is now considered a phytogeographic unit with a unique evolutionary history, a high degree of endemism, and unique adaptations (Marinho-Filho et al. 2010).

The Cerrado currently occupies the central region of Brazil but also occurs in disjunct areas within the Caatinga, Amazon, Atlantic Forest, Pantanal, and pine forest biomes (Rizzini 1979). The occurrence of these disjunct areas can be explained

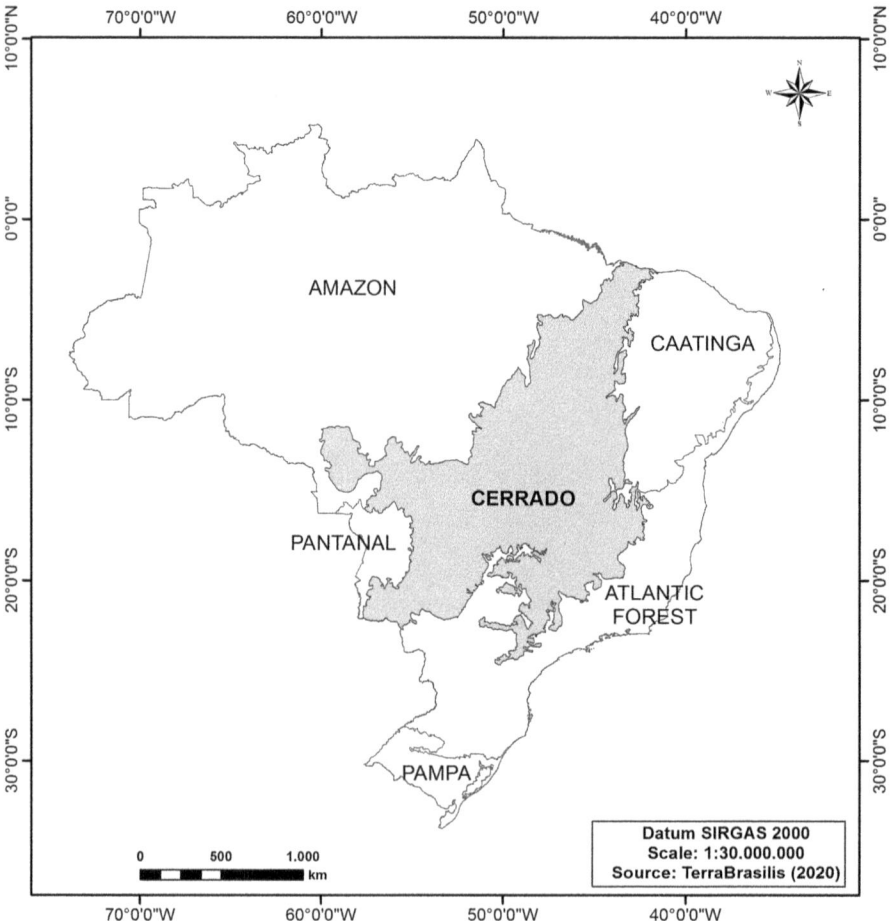

Fig. 1 Location map of the Cerrado and other biomes in Brazil

by their ancient dynamics in the Quaternary (Ledru 2002), when there were great climatic pulses, with long, cold and dry periods of glaciation interspersed with short, hot and humid interglacial periods, causing expansions and retractions of humid and dry forests in South America (Salgado-Laboriau 1994).

The Cerrado biome (edaphic woodland savanna) represents the largest extent of Brazilian savanna, composed of a heterogeneous mosaic of habitats including open grassland, savanna, and forest formations (Oliveira-Filho and Ratter 2002). Unfortunately, half of the original 1,783,200 km^2 of primary vegetation (Myers et al. 2000) has been lost to grazing and agriculture (Ministério do Meio Ambiente 2015; Noojipady et al. 2017). Even though these habitats occur under similar climatic conditions, they differ in structure and floristic composition, due to local factors, such as edaphic characteristics, topography, drainage, and fire dynamics (Werneck et al. 2012). The flora of the Cerrado is extremely rich (Machado et al. 2008), with around 12,350 angiosperm species (Mendonça et al. 2008).

The climate of the Cerrado displays the marked seasonality typical of savannas (Ribeiro and Walter 1998; Cardoso et al. 2014). Based on average monthly precipitation, the rainy season (103.8–264.5 mm per month) lasts from October to April, and the dry season, with only 7% of the annual rainfall (13.8–65.7 mm per month), lasts from May to September. However, there is some interannual variation: in some years, April can be drier, and September can be rainier than usual (Costa et al. 2012; Marcuzzo et al. 2012). The average monthly relative humidity decreases during the dry season, mainly in August and September. Unlike rainfall, relative humidity shows little interannual variation. The average annual temperature shows more stability, around 22–23 °C (Fig. 2) (Marcuzzo et al. 2012). Therefore, climate conditions in the Cerrado are quite predictable. Because of this environmental predictability, plant species employ a variety of phenological strategies, with phenophase selected to take advantage of ideal conditions (particularly the rainy season) for the establishment of seedlings (Oliveira 2008; Silva et al. 2011). While not all species share this pattern, the majority of woody plants in the Cerrado are deciduous or semideciduous with peak leaf production between September and October, just before the first rain (Morais et al. 1995; Oliveira 2008). Therefore, leaf expansion may occur before herbivorous insects reach peak numbers in the Cerrado (Marquis et al. 2002; Pinheiro et al. 2002). Among deciduous plants, leaves are produced and lost asynchronously, even within the same species (Morais et al. 1999). The Cerrado is therefore characterized by a marked seasonality. Having a diverse flora, it provides an excellent system to study the seasonality of insect-plant interactions.

Seasonal fluctuations in insect abundance are common in the tropics, and many insect populations tend to follow rainfall (Wolda 1988; Abarca 2019) and plant phenology patterns (Asch and Visser 2007). Insect abundance increases from the dry to the rainy season (Wolda 1988; Boinski and Scott 1998; Frith and Frith 1990; Pinheiro et al. 2002), with few exceptions, such as adult lepidopterans, bees, and Thysanoptera species (Boinski and Scott 1998; Tanaka and Tanaka 1982). While there is strong evidence for fluctuations in species richness and abundance, our knowledge is limited, as it is based on studies of only a few insect

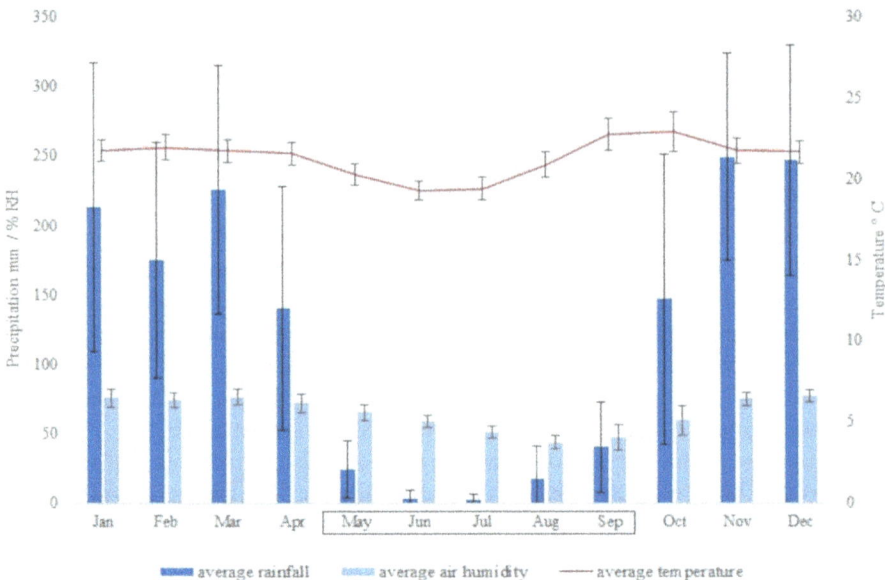

Fig. 2 Monthly averages of rainfall, relative humidity, and temperature from historical series 1991–2016 of climatic data from the Instituto Nacional de Metereologia (INMET) registered in Brasília, Distrito Federal, Brazil. Dry season: May to September

species. However, caterpillars are known to be seasonal in environments with a well-defined dry season, such as tropical dry forests of America (Janzen 1988a, b, 1993; Connahs et al. 2011; Essens et al. 2014) and the central Brazilian Cerrado (Morais et al. 1999).

Methods

Studies of interactions of folivorous caterpillars and their host plants and parasitoids in savanna formations of central Brazil were initiated in 1991. In the following 25 years, systematic surveys were conducted in the Cerrado sensu stricto in Distrito Federal, especially in Brasília, and in Serra dos Pireneus and Chapada dos Veadeiros the state of Goiás, which represent typical Cerrado. Each year, particular plant species (20–46 total) were selected for caterpillar surveys, and 15–20 plants per species were searched for caterpillars once a week, for at least 1 year, resulting in repeated sampling of 700–900 plants per year. The plants examined were 0.50–2.50 m tall. Censuses were conducted between 8:00 and 12:00 am. All externally feeding caterpillars (regardless of size or instar) found in the field were collected and reared in the laboratory of the Zoology Department of the University of Brasília. Caterpillars were supplied with leaves of the plant species on which they were found. In the laboratory, caterpillars were photographed and their characteristics recorded. These characteristics included the duration of the pupal stage, pupal parasitoid emergence (Diptera or Hymenoptera), and other interesting behaviors, such as shelter types, change of shelters throughout development, feeding period, morphological differences between instars, etc. Adult insects were identified by Vitor O. Becker and Keith Brown Jr. Vouchers and specimens were deposited in the Entomological Collection at the University of Brasília. Two researchers, Ivone Rezende Diniz and Helena Castanheira de Morais, and over a hundred University of Brasília students contributed to this long-term dataset, which resulted in over 50 publications on the ecology and natural history of caterpillars in the Cerrado. The dataset includes over 9,000 caterpillars collected during the surveys and 8,000 successfully reared in the laboratory, belonging to 594 species, 260 genera of 44 families of Lepidoptera (some species are listed in Diniz et al. 2001 and Oliveira 2010). The caterpillars were recorded on 145 species of woody plants belonging to 91 genera and 51 families.

Caterpillar abundance (i.e., the proportion of plants with caterpillars) increased in the Cerrado in mid-April and remained high until mid-July, the beginning of the dry season. However, more parasitoids emerged in the laboratory from caterpillars collected in October (Morais et al. 1999). Here we review the current knowledge about the seasonality of Cerrado caterpillars and their interactions based on 25 years of collecting and rearing caterpillars in the laboratory. Caterpillar abundance in the Brazilian Cerrado is known to be seasonal (Morais et al. 1999; Marquis et al. 2002). Species richness (Andrade et al. 1999; Bendicho-Lopez et al. 2006) and interactions with plants and parasitoids (Bendicho-Lopez et al. 2006; Scherrer et al. 2016) also vary according to the season, at least for some cohorts in our data. Therefore, our hypothesis is that the abundance, species richness, and the interactions of the caterpillars represent a repetitive phenomenon. We review here how the diversity of interactions and the specialization of the networks between caterpillars-plants and parasitoids-caterpillars vary over the months and seasons in the Cerrado across the 25-year sampling period. We also infer the potential effects of climate change on these interactions in the future.

The Seasonality of Caterpillar-Plant Interactions in the Cerrado

Among the approximately 600 woody plant species that occur in the Cerrado sensu stricto (Ratter et al. 2001; Gomes et al. 2020), 145 serve as principal host plants for caterpillars in our study areas, because these host plants were common, widely distributed, and easy to survey. Among these host plants, *Roupala montana* Aubl. (Proteaceae) had the highest richness of associated caterpillars (112 spp.; Table 1).

The peak abundance of caterpillars on plants in the Cerrado showed a consistent pattern over 10 years (Morais et al. 1995). Caterpillars occurred at low abundance per plant individual, possibly due to low air humidity, low leaf nutritional quality,

Table 1 Woody host plant species with high species richness of associated caterpillars in the Cerrado of Distrito Federal and Goiás, Brazil

Host plants		Species richness
Proteaceae	*Roupala montana* Aubl.	112
Erythroxylaceae	*Erythroxylum tortuosum* Mart.	56
	Erythroxylum deciduum A.St.-Hill.	50
Malpighiaceae	*Byrsonima coccolobifolia* Kunth.	53
Vochysiaceae	*Qualea parviflora* Mart.	51
	Qualea multiflora Mart.	46
	Qualea grandiflora Mart.	40
Caryocaraceae	*Caryocar brasiliense* Cambess.	48
Ochnaceae	*Ouratea hexasperma* (A.St.-Hil.) Baill.	37

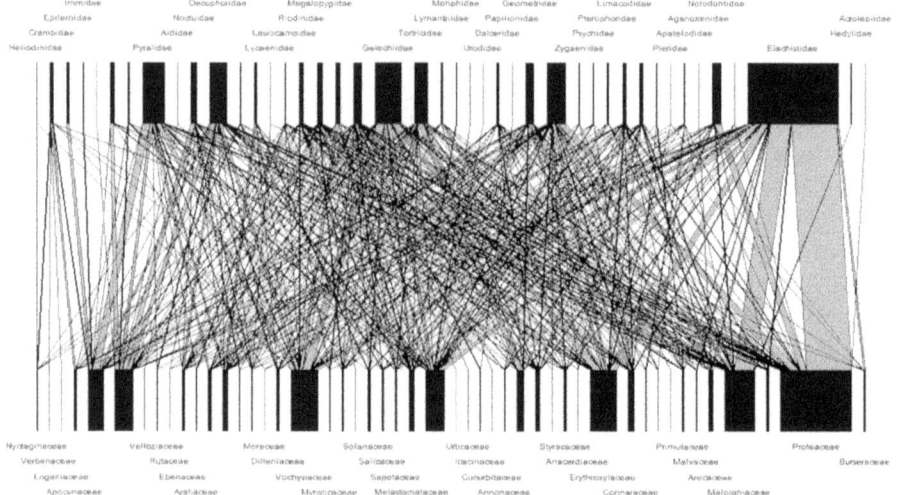

Fig. 3 Trophic networks illustrate caterpillars and host plants interactions. Nodes represent families within each trophic level, then ranked by node degree. Edge thickness represents relative link weights

and pressure from natural enemies in the Cerrado (Marquis et al. 2001, 2002). Five Lepidoptera families represented 63% of the abundance and 47.5% of the species richness: Elachistidae (with 31% of individuals and 19% of species richness), Gelechiidae (10% and 12%), Pyralidae (9% and 8%), Erebidae (Arctiinae: woolly bears; 6.5% and 5.5%), and Oecophoridae (6.5% and 3%). Elachistidae was most often associated with the host plant family Proteaceae, while Gelechiidae interacted mainly with Vochysiaceae, Pyralidae with Rubiaceae, Erebidae (Arctiinae) with Proteaceae, and Oecophoridae with Caryocaraceae (Fig. 3).

The shelter-building species with the highest number of records has yet to be described, likely belonging to an undescribed genus of Elachistidae (Vitor Becker, personal communication). This new species represents 9.5% of the total number of

Fig. 4 Undescribed genus of Elachistidae. (**a–b**) Shelter made with mature leaves of *Roupala montana* (Proteaceae) connected by silk. (**b**) The place where the caterpillars are feeding on the leaf are protected by feces and silk. (**c**) Open shelter with tunnels made with silk and feces, where the caterpillars hide. (**d**) Individual of an undescribed genus of Elachistidae

caterpillars sampled (Fig. 4). It was recorded feeding only on mature leaves of *R. montana*, building shelters in the form of rigid tunnels, using feces and silk to attach leaves to each other (Fig. 4c). About two to five fresh leaves are added to the shelter over the time of development (Fig. 4a). These caterpillars grow inside the tunnels and only leave these tunnels to feed on the leaf's epidermis. A thin layer made of silk and feces protects the feeding area (Fig. 4b). These caterpillars peak in abundance in May (dry season) ($z = 157{,}752$; $p < 0.05$) (Oliveira 2010), and adults emerge in the laboratory in October (transition to the rainy season). Below we describe other five most abundant species, together representing 16% of the total abundance. They exhibit a variety of behaviors, diet breadths, and patterns of seasonality (Fig. 5).

Fig. 5 The most abundant caterpillar species in the database. (**a**) *Stenoma cathosiota* (Elachistidae), (**b**) *Cerconota achatina* (Elachistidae), (**c**) *Eurata semiluna* (Erebidae), (**d**) *Idalus lineosus* (Erebidae), and (**e**) *Eustema opaca* (Notodontidae)

Stenoma cathosiota Meyrick, 1925, (Elachistidae: Stenomatinae) (Fig. 5a) are monophagous caterpillars that feed on mature leaves of *Roupala montana*. They exhibit shelter-building behavior, gathering two to four mature leaves from the host plant, and securing them to the stem with silk. The caterpillars remain in these shelters throughout their larval and pupal development. Over a year of sampling, the abundance of the species was highest at the beginning of the dry season ($z = 4.563$; $p < 0.05$) (Oliveira 2010); however, interannual variations in abundance and frequency have been reported by Morais and collaborators (2007). This species is subject to a high degree of parasitism by four genera of Braconidae (Hymenoptera), present in about 30% of caterpillars raised in the laboratory: *Apanteles* (Microgastrinae), *Orgilus* (Orgilinae), *Bracon* (Braconinae), and *Hyposoter* (Campopleginae). Parasitism was highest in July–August (late dry season) and November (Morais et al. 2007).

Cerconota achatina (Zeller, 1855) (Elachistidae: Stenomatinae) (Fig. 5b) are oligophagous caterpillars that feed on *Byrsonima* spp., particularly on *B. coccolobifolia* (78% of occurrences). They are primarily leaf feeders but are also present on the flowers and fruits of the same plant species (Diniz and Morais 2002). This species also shows a shelter-building behavior, joining leaves of the host plant to each other with silk and connecting them by silk tunnels embedded in their feces. The abundance of this multivoltine species is higher in the dry season, from May to August (Andrade et al. 1995). While the timing of their peak abundance varies from year to year (e.g. in June 1994 and in August 2006), it always occurs during the dry season. The rate of parasitism is low, around 3% of the total caterpillars reared in laboratory, with two recorded genera of Hymenoptera: *Meniscomorpha* (Ichneumonidae: Banchinae) and *Goniozus* (Bethylidae: Bethylinae).

Eurata semiluna Walker, 1854, (Erebidae: Arctiinae) (Fig. 5c) has the most extensive diet breadth among all Cerrado caterpillars, feeding on 41 species of host plants belonging to 28 families (Diniz et al. 2000, 2013). The highest numbers of records were for *Erythroxylum tortuosum*, *Byrsonima pachyphylla*, and *Ouratea hexasperma* St. Hil. (Ochnaceae). The caterpillars are solitary, external folivores and occur practically throughout the year, except for August and September, with peak abundance from March to June (Diniz et al. 2013). It is an irruptive species, apparently taking advantage of recently burned areas (Lepesqueur et al. 2012). They are parasitized by Diptera (Tachinidae: Tachininae and Exoristinae), and Hymenoptera (Ichneumonidae: *Casinaria*; Braconidae: *Aleiodes*); however, the rate of parasitism is low, present in about 4% of caterpillars reared in the laboratory.

Idalus lineosus Walker, 1869, (Erebidae: Arctiinae) (Fig. 5d) are solitary, monophagous, external folivores, feeding on mature leaves of *Roupala montana*. This caterpillar is active mainly in the dry season ($z = 25.88$, $p < 0.05$), with a peak abundance in May (Braga et al. 2014). The rate of parasitism is low, present in about 5% of caterpillars raised in the laboratory. They are parasitized by species of Diptera, Tachinidae (Exoristinae), and four genera of Hymenoptera: *Apanteles* and *Protopanteles* (Braconidae) and *Cidaphus* and *Pristomerus* (Ichneumonidae). In some cases, multiparasitism occurs with interactions between *Apanteles* and *Protopanteles* species and between *Cidaphus* and *Protopanteles* species (Braga et al. 2014).

Eustema opaca Schaus, 1922, (Notodontidae) (Fig. 5e) has only been recorded feeding on new leaves of *Roupala montana*; thus, it is considered locally monophagous. Caterpillars are gregarious and external feeders, with groups of over 60 individuals (Diniz et al. 2013). They occur only in November and December, at the beginning of the rainy season, during the period of leaf expansion. They can be parasitized by Diptera and Tachinidae (Exoristinae and Tachininae), with very high rate (~80%) of parasitism in caterpillars raised in laboratory. They also suffer strong predation pressure from an unidentified species of Hemiptera of the genus *Zelus* (Reduviidae) (Pessoa-Queiroz 2008).

Caterpillar species exhibit a variety of strategies and behaviors. External folivores can feed on the leaf blade or surrounded by a shelter made of silk, plant parts, bristles, or feces (Diniz et al. 2013). The Cerrado caterpillar community is characterized by a high proportion of shelter building, with around 70% of species and individuals showing this behavior (Pessoa-Queiroz 2008; Diniz et al. 2012). These shelter-builders are mostly microlepidopterans. During the dry season, shelter-building caterpillars are significantly more abundant than exposed caterpillars ($x^2 = 46.319$, $p = 0.0001$), indicating that the shelter acts as a protection against desiccation (Diniz et al. 2012), providing a suitable microclimate for the caterpillar's development (Loeffler 1996; Velasque and Del Claro 2016).

Synchrony with the flushing of new leaves of host plant is crucial for many insect herbivores (Asch and Visser 2007). In general, young leaves are more nutritious than mature leaves, causing herbivores to synchronize with leaf production (Niemelä et al. 1982; Aide and Londoño 1989; Aide 1993; Alonso and Herrera 2000; Marquis et al. 2001; Ivashov et al. 2002). Most woody species in the Cerrado are deciduous, losing their leaves in the dry season and having the highest production before the beginning of the rainfall (Franco 1998). However, in some individuals of the same species, leaves can be produced all year round. This characteristic can be found in several woody species in the Cerrado (Morais et al. 1995). Therefore, the availability of the leaves varies throughout the year and declines at the end of the dry season. Consequently, new leaves provide a less predictable resource for caterpillars compared to mature leaves (Morais et al. 1995). As a result, caterpillar peak abundance does not coincide with the peak of leaf flush in the dry-rainy transition, and the majority of caterpillar species feed on mature leaves (Morais et al. 1995, 1999; Price et al. 1995; Scherrer et al. 2010).

The relative abundance and species richness of caterpillars vary monthly in the Cerrado, which are concentrated at the beginning of the dry season (Rayleigh test: abundance: $z = 945.056$; $p < 0.001$; richness: $z = 49.568$; $p < 0.01$) (circular analysis: Fig. 6a–b), confirming our hypothesis based on previous data. This pattern was common in most families including Cosmopterigidae, Elachistidae, Erebidae (Arctiinae), Gelechiidae, Geometridae, Hesperiidae, Limacodidae, Noctuidae, Oecophoridae, Pyralidae, Riodinidae, Sphingidae, and Thyrididae. On the other hand, some families, such as Megalopygidae, Mimallonidae, Saturniidae, and Tortricidae, showed two abundance peaks throughout the year, while others, such as

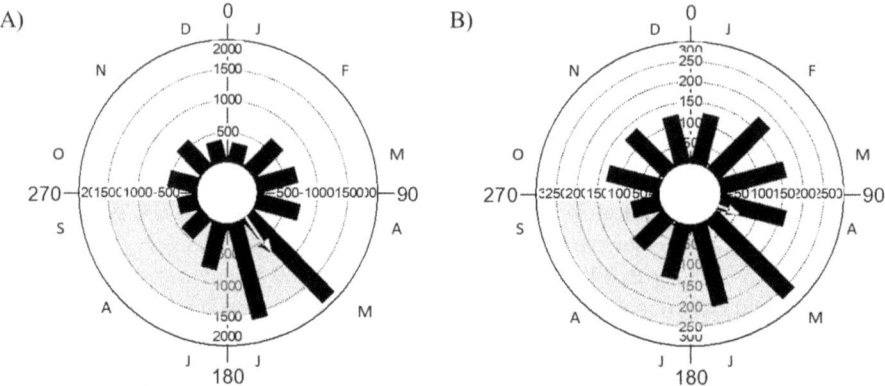

Fig. 6 Circular analysis of monthly (**a**) abundance and (**b**) species richness of Cerrado caterpillars (the number of individuals and species sampled in the field). 15° corresponds to January, 45° to February, etc. and 345° to December (represented by the initials of the months). Arrow indicates the mean vector, and the length of the arrow reflects the mean vector length (r) (the strength of the clustering among samples). The gray area corresponds to the dry season

Notodontidae, had only one peak during the rainy season. The seasonality of caterpillar occurrence in the central Brazilian Cerrado shows a unique pattern, not described in other tropical ecosystems (Janzen 1993; Connahs et al. 2011), i.e., their occurrence is highest in the early dry season (Marquis et al. 2002). This pattern may be associated with disadvantages that the rainy season brings for some species, like increased pressure from natural enemies, such as birds (Macedo 2002) and parasitoids (Morais et al. 1999; Pinheiro et al. 2002).

In the dry and rainy seasons, caterpillars have specific species composition and biotic interactions. Remarkably, only about 35% of species occur in both seasons. These "aseasonal" species are the most abundant species. The dissimilarity in species composition between consecutive months was 0.37–0.73 (Bray-Curtis index). Lower turnover occurs in May–June (0.37), at the beginning of the dry season, and in January–February (0.44), in the middle of the rainy season. The high turnover occurs in August–September (0.73) and September–October (0.73), evidencing how much the caterpillar community changes during the dry-rainy transition. With regard to the other months, dissimilarity was around 0.60, demonstrating a high temporal turnover in species composition. These patterns suggest a strong association between caterpillar species and the phenological and climatic variations of the Cerrado.

The Cerrado caterpillar community is characterized by high specialization, with monophagous species and oligophagous family specialists (Marquis et al. 2002; Diniz and Morais 2005; Morais et al. 2011; Mitchell et al. 2011), which is typical for tropical communities (Dyer et al. 2007). Among the species that occur in the two climatic seasons, 56% have restricted diet (monophagous or oligophagous), and 44% are polyphagous. It is difficult to fully understand the diet breadth of caterpillar species that are restricted to one season, since 60% of these species are rare (singletons). However, when we exclude singletons, the caterpillar community that occurs in a single season is also composed mainly of restricted diet species (78% are monophagous or oligophagous). The high-diet specialization of Cerrado caterpillars implies a higher similarity in the composition of associated caterpillar species among phylogenetically closest plants (Morais et al. 2011). In this context, the phenological availability of leaves of the host plant genus or families will determine the composition of the folivorous caterpillar fauna throughout the year.

We observed 8,022 interactions between caterpillar and plant species in the 25 years of study. During 1 year of observation, around 500 different trophic interactions were recorded between caterpillars and their host plants, with few interactions found in more than one season, considering four seasons (dry, rainy, and the two transitions) (Lepesqueur et al. 2018). Diet breadth can vary over time, as the number of host plants used by the same caterpillar species can increase, particularly during the dry season (Scherrer et al. 2016). This increase is possibly caused by the lower availability of leaves from deciduous host plant species (Lenza 2005; Lenza and Klink 2006). Thus, the high temporal β-diversity of trophic interactions between caterpillars and plants is mainly due to changes in interactions among species that co-occur in different seasons (Lepesqueur et al. 2018).

The diversity of trophic interactions between caterpillars and host plants remains high during the second half of the rainy season and the beginning of the dry season

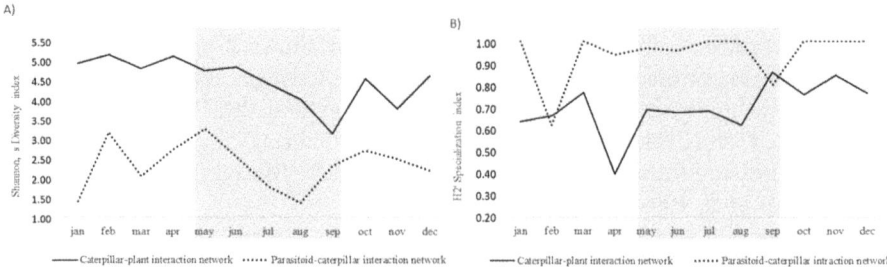

Fig. 7 Annual variation in the diversity of interactions (Shannon's index) (**a**) and degree of specialization (index H2; Blüthgen et al. 2006) (**b**) of the networks of interactions between caterpillars-plants and parasitoids-caterpillars in the Cerrado. Gray area corresponds to the dry season

(network metrics – Shannon's diversity index: Fig. 7a). This pattern of interaction diversity during the rainy-dry transition reflects the higher richness and abundance of caterpillars, as well as the higher number of interactions between caterpillars and plants (Lepesqueur et al. 2018), which is expected since the diversity of interactions is related to the diversity of herbivore species (Dyer et al. 2010). After this, the interaction diversity declines and reaches its lowest in September, at the end of the dry season, when the entire network is the most specialized during the year (network metrics – specialization index H2: Fig. 7b). At the end of the dry season, 53% of caterpillars in the network are singletons, and the most frequent species are monophagous. In this way, the diversity of interactions and the specialization of the network reflect the lower availability of resources and lower caterpillar activity characteristic of this time of the year. The late dry season has severe climatic conditions for both caterpillars and plants, resulting from the low air humidity and higher temperatures (Fig. 2), making interaction networks less diverse and more specialized.

The dry season and its effects on plant phenology and leaf quality interfere with the larval development of some species that have two generations per year. For instance, caterpillars of *Cerconota achatina* (oligophagous on *Byrsonima* spp.) and *Chlamydastis platyspora* (monophagous on *Roupala montana*) develop more slowly in the dry season than in the rainy season (Morais et al. 1999; Bendicho-Lopez et al. 2006). Thus, the dry season is a critical period for caterpillars, because in addition to the lower availability of food resources, the quality of the leaves is also inferior compared to the rainy season.

The Seasonality of Parasitoid-Caterpillar Interactions in the Cerrado

Parasitoids and their hosts have complex and multifaceted interactions. Here, we only discuss variation in abundance and species richness of parasitoids. We focus on two interaction networks between parasitoids and caterpillars over time and

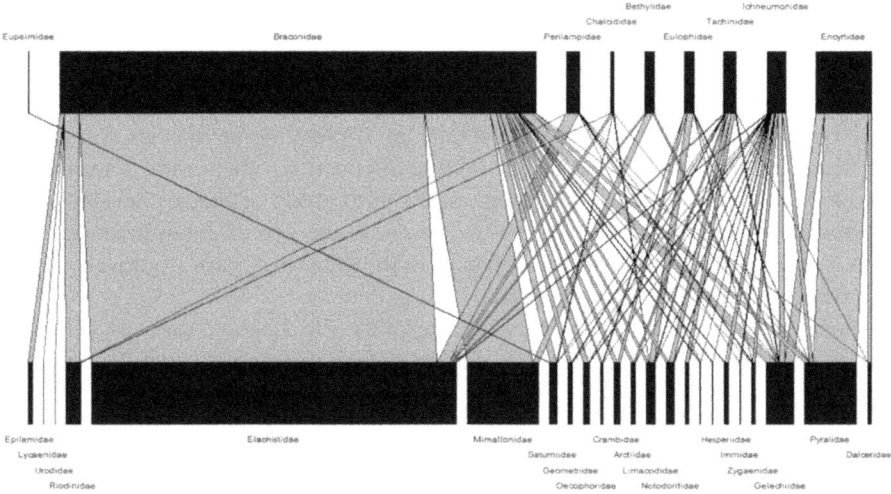

Fig. 8 Trophic networks illustrating interactions between parasitoids and caterpillars. Nodes represent families within each trophic level ranked by node degree. Edge thickness represents relative link weights

compare the two major taxa: Hymenoptera and Diptera. We have little data to discuss the factors that affect the relative proportion of parasitoid taxa attacking our caterpillar species.

Parasitoids are a main cause of mortality in caterpillars in our system. In our dataset, hymenopterans were the principal parasitoids of caterpillars, responsible for 80% of the registered parasitism events. They were represented by 250 species in 86 genera belonging to 13 families, with Braconidae and Ichneumonidae being the most common. Dipteran parasitism occurred less frequently than hymenopteran parasitism, with 20% of the registered parasitism events. Most dipterans were Tachinidae species, with a few records of Conopidae and Syrphidae species. Braconidae interacted most frequently with Elachistidae, while Encyrtidae interacted mainly with Pyralidae, Ichneumonidae with Gelechiidae, Perilampidae with Elachistidae, and Tachinidae with Notodontidae (Fig. 8).

Apatelodidae, Dalceridae, Limacodidae, Lymantriidae, Megalopygidae, Notodontidae, and Sphingidae were the lepidopteran families most heavily parasitized by Diptera. All of these families have morphological defenses, such as hairs, spines, or glands, used as protection against predators (Pessoa-Queiroz 2008). There are two possible explanations for the success of Tachinidae is these caterpillar families. They could be less susceptible to predators as hypothesized by Stireman and Singer (2003) in their "enemy-free space" hypothesis. The other explanation could be oviposition of tiny parasitoid eggs on the leaves of host plants that are eaten by caterpillars (Belshaw 1994). Our results are in line with the idea that the oviposition strategies used by many tachinid flies render morphological defenses useless (Gentry and Dyer 2002). However, these possible explanations for parasitoid-caterpillar interactions in the Cerrado need to be tested. The relatively low number of caterpillars parasitized by Diptera in our dataset could also be an artifact of our collection method. In some studies, researchers generally only collected caterpillars in the final instars. A review of dipteran parasitoid biology shows that larval development times vary widely in Diptera: some species delay their development and only kill the host at or close to pupation, others enter diapause and develop only after a change of seasons, and a third group of species develops after oviposition (Feener and Brown 1997). Therefore, the age of caterpillars (i.e., the time caterpillars are available for dipterans in the field) may have negatively biased our perception of dipteran parasitism.

Some ecological characteristics of caterpillar life history, such as behavior, dietary habits, and defense strategies, influence their interactions with the third trophic level (Gentry and Dyer 2002; Greeney et al. 2012). Thus, the incidence of parasitism in caterpillars varies according to their morphological and behavioral characteristics. Data from parasitoid species richness and the proportion of caterpillars parasitized showed that parasitoids responded to the abundance and life strategy (i.e., exposed or sheltered) of the caterpillars that use the host plant *Caryocar brasiliense* (Caryocaraceae) (Rodovalho et al. 2007). With regard to shelter, we see a different pattern in Cerrado caterpillars, as shelter-building behavior seems to be protecting them more efficiently from parasitoids compared to other ecosystems (Dyer and Gentry 1999; Gentry and Dyer 2002; Connahs et al. 2011). In the Cerrado, shelters seem to provide protection from parasitoids, because the rate of parasitism is higher in exposed caterpillars than in shelter caterpillars (Chi-square = 13.77; $p = 0.0002$) (Rodovalho et al. 2007; Pessoa-Queiroz 2008; Diniz et al. 2012). Some shelter-builder caterpillars in the Cerrado keep adding leaves to the shelter during their development, which can make it more difficult for parasitoids to reach the caterpillar. In these cases, shelter building can serve as an effective defense strategy against parasitoids. Therefore, we hypothesize that the combination of a long dry season with low relative humidity and high parasitism pressure can explain the predominance of shelter-building caterpillars in the Cerrado (Diniz et al. 2012). With regard to caterpillar diet breadth in the Cerrado, specialist caterpillars suffer the highest levels of parasitism (chi-square = 29.52; $p < 0.0001$) (Pessoa-Queiroz 2008), similar to other natural systems (Dyer and Gentry 1999). This may be related to the ability of parasitoids to use morphological and chemical characteristics of the host plants as possible cues (Weseloh 1993).

Parasitism of Cerrado caterpillars varies from 6% in May (early dry season) to 29% in November (early rainy season), being more pronounced in the rainy season (Rayleigh test: $z = 7.5$; $p < 0.05$) (Fig. 9a). Parasitoid wasps peak in November (Rayleigh test: $z = 353.684$; $p < 0.001$) (Fig. 9b), while parasitoid flies show two abundance peaks, one in December, during the rainy season, and the other in June, during the dry season, with uniform temporal distribution (Rayleigh test: $z = 2.837$; $p > 0.05$). The temporal distribution of parasitoid species richness is more even over the months (Rayleigh test: $z = 4.131$; $p = 0.016$) (Fig. 9c), with a higher number of species in February, May, June, and November and lower in August. Thus, it is possible that Cerrado caterpillar species use the temporal "enemy-free space" strategy

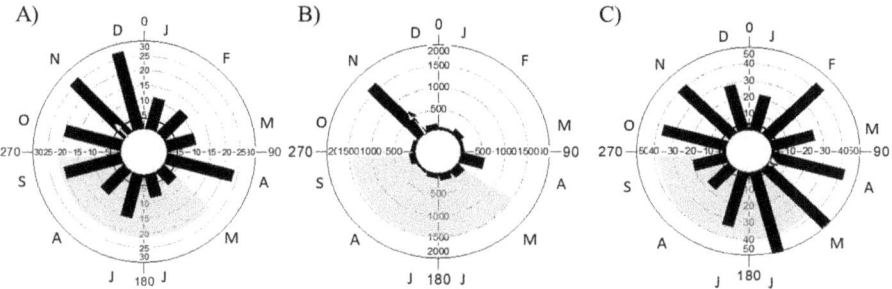

Fig. 9 Circular analysis of (a) the rate of parasitism, (b) abundance, and (c) species richness of parasitoids over time (the number of individuals and species sampled in the caterpillars sampled in the field). 15° corresponds to January, 45° to February, etc. and 345° to December (represented by the initials of the months). An arrow indicates the mean vector, and the length of the arrow reflects the mean vector length (r) (the strength of the clustering among samples). The gray area corresponds to the dry season

(Morais et al. 1999), as their abundance is higher in the early dry season, escaping from the seasonal peak in parasite prevalence in the early rainy season. In the early dry season, the abundance of Hymenoptera is low in Cerrado (Pinheiro et al. 2002). Nevertheless, the reason why a parasitoid peak does not follow caterpillar abundance is not yet fully understood (Connahs et al. 2011). However, the dry season in the Cerrado can be a limiting factor for most parasitoid species, since low humidity limits the parasitoid activity given the risk of dehydration (Shapiro and Pickering 2000).

The diversity of interactions between caterpillars and their parasitoids is more accentuated in the late rainy season and the early dry season, with peaks in February and May, respectively, and a less pronounced peak in October at the beginning of the rainy season. The diversity of interactions is lowest in August, during the late dry season, and in January, in the middle of the rainy season (network metrics – Shannon's diversity index: Fig. 7a). In August and January, the interaction networks are composed of very few species at each trophic level with entirely specialized interactions. These temporal variations in the diversity of interactions partly reflect parasitoid and caterpillar species richness and abundance over the months. The highest diversity of interactions between parasitoids and caterpillars occurred in February, May, and October, reflecting a higher parasitoid species richness with some generalist interactions in these networks (few caterpillar species are attacked by few parasitoid species, and few parasitoid species attack a variety of caterpillars).

The networks of parasitoid-caterpillar interactions had a higher degree of specialization than caterpillar-plant networks, as expected. However, in February and September, these networks were slightly less specialized (network metrics – specialization index H2: Fig. 7b). In the parasitoid-caterpillar interaction networks for each month, 90–100% of parasitoid species were registered with only one host species, while few (up to four) species of caterpillars were parasitized by more than one species of parasitoid. The dry season acme remains poor in interactions, as observed in caterpillar-plant networks, due to the more severe climatic conditions for all three trophic levels, especially for parasitoids, which are disadvantaged under dry conditions (Shapiro and Pickering 2000). Nevertheless, we need to understand better the climatic requirements and life histories of both hymenopteran and dipteran parasitoids in the Cerrado, to better assess the factors that influence the annual variation in host-parasitoid interactions.

The Effects of Climate Change on Caterpillar Interactions in the Cerrado

A discussion of the effects of climate change on the caterpillar community and their biotic interactions in the Cerrado is somewhat hampered by the lack of consistent experimental and other data on the subject. Thus, the discussion remains speculative, with several questions unanswered. However, knowledge about the seasonality of interactions between caterpillars-plants and parasitoids-caterpillars in the Cerrado in these 25 years allows us to infer the periods of the year when the interactions are most vulnerable to climate change.

Based on past climatic fluctuations in the Quaternary, along with current habitat loss and fragmentation in the Cerrado, we should expect dramatic changes in the near future, especially in species distributions and even extinctions. Models predict a 2–6 °C warming for Brazil by 2100 along with severe droughts (Salazar et al. 2007), which will cause a decrease in soil moisture in most regions of the biome. The dry season is expected to last longer in the north and northeast, with less precipitation and higher temperatures, while the southeastern region is predicted to receive higher rainfall and a decrease in the number of consecutive days of drought (Marengo et al. 2009a, b). For central Brazil, where the core Cerrado and our study areas are located, there is a tendency for increased occurrence of extreme climatic events, with both wetter and drier years (Bombardi and Carvalho 2008), higher temperatures, and possibly up to 30% reduction in annual precipitation (Marengo et al. 2009a). The duration, regularity, and distribution of rainfall are crucial to maintain the biodiversity of the Cerrado (Bombardi and Carvalho 2008). Plant phenology will likely respond to these climatic changes, and prolonged drought may alter species dominance and the ratio of different woody plants with different phenological strategies. Low water availability, soil depletion, and intense fires can favor herbaceous and annual species (Bustamante et al. 2012), while hydraulic failure and carbon starvation can increase the mortality of woody plants, tree seedlings, and shallow-rooted herbs (Franco et al. 2014). The balance between forest expansion and forest retreat into savanna might also be strongly affected by extreme drought events and warming, since forest species are more sensitive to drought than savanna species (Franco et al. 2014). Most of the 162 woody species in the Cerrado biome are predicted to suffer serious declines, losing over 50% of their potential distribution area. Depending on the scenario, 10–30% of these species may become locally extinct (Siqueira and Peterson 2003). This would result in a decrease or local extinction of Lepidoptera species, as higher local plant density can account for higher caterpillar abundance (Scherrer et al. 2010).

The accumulation of atmospheric CO_2 and associated higher temperatures will affect the physiological processes, development, reproduction, and phenology of plants, changing the nutritional quality and phytochemical defenses that directly affect the metabolic rate, growth, feeding rates, fecundity, longevity, and life-history traits of herbivorous insects, such as diet breadth and food preference (Jamieson et al. 2012; Pincebourde et al. 2017). With global environmental change, generalist herbivores would be more adapted to face an increase in quantitative defenses commonly present in host plants and lower nutritional quality (Massad and Dyer 2010). Generalist herbivores may be more able to change host plants due to phenological changes and the decreased availability of food resources (DeLucia et al. 2012). Changes in the development time at different trophic levels can cause phenological asynchrony (DeLucia et al. 2012; Dyer et al. 2013). Thus, the Cerrado caterpillar community, characterized by its high-diet specialization, may suffer greater selection pressure for asynchrony and face an even higher risk of local extinction. The survival of these caterpillars with extremely restricted diets is limited by the quality and availability of the host plant species, placing them in extreme danger of disappearance. The loss of woody plants and other phenological changes in the Cerrado flora (Siqueira and Peterson 2003; Bustamante et al. 2012; Franco et al. 2014) will determine the composition and temporal distribution of the caterpillar community in the future. Better leaf digestibility can explain the higher abundance of caterpillars on certain plant species in the Cerrado (Scherrer et al. 2010), so phenological leaf changes can affect the local lepidopteran community. Thus, some caterpillar species will be favored, while others will become extinct or decrease in response to global changes (Salcido et al. 2020). Furthermore, as the interactions between parasitoids and caterpillars in the Cerrado are even more specialized than caterpillar-plant interactions, there is an even higher risk of unlinking. The extreme climatic events predicted for the central Cerrado will exacerbate these disconnections between parasitoids and caterpillars (Stireman et al. 2005). Thus, the differences in the physiological responses of plants, herbivores, and parasitoids to varying climatic cues may, positively or negatively, affect their synchrony (Tylianakis et al. 2008). The variety of phenological responses in relation to climate change at each trophic level can increase the number of ecologically mismatched trophic interactions over time.

Due to current climatic conditions in the late dry season in the Cerrado, the diversity of caterpillar-plant and parasitoid-caterpillar interactions is lower than in other periods of the year. Future climate scenarios (i.e., more severe drought events) are predicted to worsen the conditions for these tritrophic interactions. Thus, the few interactions that currently occur during the dry season acme are more likely to decline or disappear. As these interaction networks are less diverse, more specialized, and without redundant interactions, they form networks with less stability. Moreover, the intensification of fires in the Cerrado further threatens these species and their interactions, affecting plant and caterpillar community structure, and consequently parasitoids, particularly in the late dry season, when the effects of fire on caterpillars are the strongest (Diniz et al. 2011). Climatic events that alter the distribution and duration of rainfall and droughts may negatively affect parasitoids, allowing irruptions of currently rare Cerrado caterpillars. As several species that feed on alternative native plants are considered pests, outbreaks of these species could affect the agriculture in this biome. Future climatic conditions, such as increased temperature and reduced humidity, are expected to cause changes in the community structure and interactions of caterpillar species. We hypothesize that future conditions will possibly further select for shelter-building caterpillars in the Cerrado. Finally, we can infer that the lower diversity of interactions at the late dry season indicates that more severe future climatic conditions may lead to a loss of interactions in the Cerrado and in that way affect community structure.

Acknowledgments We thank CNPq, CAPES, and FAPDF for research grants and scholarships to students and researchers during the construction of our database. Among others, we would like to highlight the dedicated work of Helena Castanheira de Morais, Cíntia Lepesqueur, Rosevaldo Pessoa-Queiroz, Scheila Scherrer, and Neuza Silva. We also thank Andre Rangel Nascimento and Roberto Pujol Luz for the parasitoid database. We are very grateful to Vitor Becker and Keith Brown Jr., who have identified most Lepidoptera species over the years. All of these researchers along with over a hundred students made it possible to build our knowledge about caterpillars and their biological interactions in the Cerrado.

References

Abarca M (2019) Herbivore seasonality responds to conflicting cues: untangling the effects of host, temperature, and photoperiod. PLoS One 14(9):e0222227

Aide TM (1993) Patterns of leaf development and herbivory in a tropical understory community. Ecology 74(2):455–466

Aide TM, Londono EC (1989) The effects of rapid leaf expansion on the growth and survivorship of a lepidopteran herbivore. Oikos 55(1):66–70

Alonso C, Herrera CM (2000) Seasonal variation in leaf characteristics and food selection by larval noctuids on an evergreen Mediterranean shrub. Acta Oecol 21:257–265

Andrade I, Diniz IR, Morais HC (1995) A lagarta de *Cerconota achatina* (Zeller) (Lepidoptera, Oecophoridae, Stenomatinae): biologia e ocorrência em plantas hospedeiras do gênero *Byrsonima* Rich (Malpighiaceae). Rev Bras Zool 12(4):735–741

Andrade I, Morais HC, Diniz IR, van der Berg C (1999) Richness and abundance of caterpillars on *Byrsonima* (Malpighiaceae) species in an area of cerrado vegetation in Central Brazil. Rev Biol Trop 47(4):691–669

Asch MV, Visser ME (2007) Phenology of forest caterpillars and their host trees: the importance of synchrony. Annu Rev Entomol 52:37–55

Belshaw R (1994) Life history characteristics of Tachinidae (Diptera) and their effect on polyphagy. In: Hawkins BA, Sheehan W (eds) Parasitoid community ecology. Oxford University Press, Oxford, pp 145–162

Bendicho-López A, Morais HC, Hay JD et al (2006) Folivore caterpillars on *Roupala montana* Aubl.(Proteaceae) in cerrado sensu stricto. Neotrop Entomol 35(2):182–191

Blüthgen N, Menzel F, Blüthgen N (2006) Measuring specialization in species interactions networks. BMC Ecol 6:9

Boinski S, Scott PE (1998) Association of birds with monkeys in Costa Rica. Biotropica 20:136–143

Bombardi RJ, Carvalho LMV (2008) Variabilidade do regime de monções sobre o Brasil: o clima presente e projeções para um cenário com 2xCO2 usando o modelo MIROC. Rev Bras Meteorol 23(1):58–72

Braga L, Lepesqueur C, Silva NA et al (2014) Immature stages and ecological characteristics of *Idalus lineosus* Walker (Erebidae: Arctiinae). J Lepid Soc 68(1):45–53

Bustamante MMC, Nardoto GB, Pinto AS et al (2012) Potential impacts of climate change on biogeochemical functioning of Cerrado ecosystems. Braz J Biol 72:655–671

Cardoso MRD, Marcuzzo FFN, Barros JR (2014) Classificação climática de Köppen–Geiger para o estado de Goiás e o Distrito Federal. Acta Geog 8:40–55

Connahs H, Aiello A, Bael SV et al (2011) Caterpillar abundance and parasitism in a seasonally dry versus wet tropical forest of Panama. J Trop Ecol 27:51–58

Costa HC, Marcuzzo FFN, Ferreira OM et al (2012) Espacialização e sazonalidade da precipitação pluviométrica no estado de Goiás e Distrito Federal. Rev Bras Geogr Fís 1:87–100

DeLucia EH, Nabity PD, Zavala JA et al (2012) Climate change: resetting plant–insect interactions. Plant Physiol 160(4):1677–1685

Diniz IR, Morais HC (2002) Local pattern of host plant utilization by lepidoptern larvae in the cerrado vegetation. Entomotropica 17(2):115–119

Diniz IR, Morais HC (2005) Abundância e amplitude de dieta de lagartas (Lepidoptera) no cerrado de Brasília (DF). In: Scariot A, Sousa-Silva JC, Felfili JM (eds) Cerrado: Ecologia, Biodiversidade e Conservação. MMA, Brasília, pp 305–318

Diniz IR, Morais HC, Scherrer S et al (2000) The polyphagous caterpillar *Fregela semiluna* (Lepidoptera: Arctiidae): occurrence on plants in the central Brazilian cerrado. Bol Herb Ezechias Paulo Heringer 5:103–112

Diniz IR, Morais HC, Camargo AJA (2001) Host plants of lepidopteran caterpillars in the cerrado of the Distrito Federal, Brazil. Rev Bras Entomol 45(2):107–122

Diniz IR, Higgins B, Morais HC (2011) How do frequent fires in the Cerrado alter the lepidopteran community? Biodivers Conserv 20(7):1415–1426

Diniz IR, Hay JD, Rico-Gray V et al (2012) Shelter–building caterpillars in the cerrado: seasonal variation in relative abundance, parasitism, and the influence of extra–floral nectaries. Arthropod–Plant Interact 6(4):583–589

Diniz IR, Braga L, Lepesqueur C et al (2013) Lagartas do Cerrado: guia de campo. Technical Books Editora, Rio de Janeiro

Dyer LA, Gentry GL (1999) Predicting natural-enemy responses to herbivores in natural and managed systems. Ecol Appl 9(2):402–408

Dyer LA, Singer MS, Lill JT et al (2007) Host specificity of Lepidoptera in tropical and temperate forests. Nature 448:696–700

Dyer LA, Walla TR, Greeney HF et al (2010) Interactions diversity: a metric for studies of biodiversity. Biotropica 42(3):281–289

Dyer LA, Richards LA, Short SA et al (2013) Effects of CO2 and temperature on tritrophic interactions. PLoS One. https://doi.org/10.1371/journal.pone.0062528

Essens T, Leyequién E, Pozo C et al (2014) Effects of climate and forest age on plant and caterpillar diversity in the Yucatan, Mexico. J Trop Ecol 30(5):419–434

Feener DH Jr, Brown VB (1997) Diptera as parasitoids. Annu Rev Entomol 42:73–97

Franco AC (1998) Seasonal patterns of gas exchange, water relations and growth of *Roupala montana*, an evergreen savanna species. Plant Ecol 136:69–76

Franco AC, Rossatto DR, Silva LDCR et al (2014) Cerrado vegetation and global change: the role of functional types, resource availability and disturbance in regulating plant community responses to rising CO 2 levels and climate warming. Theor Exp Plant Physiol 26(1):19–38

Frith D, Frith C (1990) Seasonality of litter invertebrate populations in an Australian upland tropical rain forest. Biotropica 22:181–190

Gentry GL, Dyer LA (2002) On the conditional nature of Neotropical caterpillar defenses against their natural enemies. Ecology 83(11):3108–3119

Gomes WB, Corrêa RS, Balduíno APC (2020) Richness of Cerrado Woody species engaged in ecological restoration in the Brazilian Federal District. Forest Ambient 27(4):1–23

Greeney HF, Dyer LA, Smilanich AM (2012) Feeding by lepidopteran larvae is dangerous: a review of caterpillars' chemical, physiological, morphological, and behavioral defenses against natural enemies. Invertebr Surviv J 9:7–34

Instituto Nacional de Meteorologia (2020) Banco de Dados Meteorológicos do INMET, Brasil. https://bdmep.inmet.gov.br/. Accessed 10 Oct 2020

Ivashov AV, Boyko GE, Simchuk AP (2002) The role of host plant phenology in the development of the oak leafroller moth, *Tortrix viridana* L. (Lepidoptera: Tortricidae). For Ecol Manag 157:7–14

Jamieson MA, Trowbridge AM, Raffa KF et al (2012) Consequences of climate warming and altered precipitation patterns for plant–insect and multitrophic interactions. Plant Physiol 160(4):1719–1727

Janzen DH (1988a) Ecological characterization of a Costa Rican Dry Forest Caterpillar Fauna. Biotropica 20(2):120–135

Janzen DH (1988b) Tropical dry forests: the most endangered major tropical ecosystem. In: Wilson EO, Peter FM (eds) Biodiversity. National Academy Press, Washington, pp 140–137

Janzen DH (1993) Caterpillar seasonality in a Costa Rican dry forest. In: Stamp NE, Casey TM (eds) Caterpillars: ecological and evolutionary constraints on foraging. Chapman and Hall, New York, pp 448–477

Ledru MP (2002) Late quaternary history and evolution of the cerrados as revealed by palynological records. In: Oliveira PS, Marquis RJ (eds) The cerrados of Brazil: ecology and natural history of a Neotropical savanna. Columbia University Press, New York, pp 33–50

Lenza E (2005) Fenologia, demografia foliar e características foliares de espécies lenhosas em um cerrado sentido restrito no Distrito Federal e suas relações com as condições climáticas. Thesis, University of Brasília

Lenza E, Klink CA (2006) Comportamento fenológico de espécies lenhosas em um cerrado sentido restrito de Brasília, DF. Rev Bras Bot 29(4):627–638

Lepesqueur C, Morais HC, Diniz IR (2012) Accidental fire in the Cerrado: its impact on communities of caterpillars on two species of *Erythroxylum*. Pysche. https://doi.org/10.1155/2012/101767

Lepesqueur C, Scherrer S, Vieira MC et al (2018) Changing interactions among persistent species as the major driver of seasonal turnover in plant–caterpillar interactions. PLoS One 13(9):e0203164

Loeffler CC (1996) Caterpillar leaf folding as a defense against predation and dislodgment: staged encounters using *Dichomeris* (Gelechiidae) larvae on goldenrods. J Lepid Soc 50(3):245–260

Macedo RH (2002) The avifauna: ecology, biogeography, and behavior. In: Oliveira PS, Marquis RJ (eds) The Cerrados of Brazil: ecology and natural history of a Neotropical savanna. Columbia University Press, New York, pp 242–265

Machado RB, Aguiar LMS, Castro AAJF et al (2008) Caracterização da Fauna e Flora do Cerrado. In: Bastos FMD, Faleiro FG, Araújo GP (eds) Palestras do XI Simpósio Nacional sobre o Cerrado e II Simpósio Internacional sobre Savanas Tropicais. Embrapa Cerrados, Brasília, p 12

Marcuzzo F, Faria TG, Pinto Filho RF (2012) Chuvas no estado de Goiás: análise histórica e tendência futura. Acta Geotech 6(12):125–137

Marengo JA, Jones R, Alves LM et al (2009a) Future change of temperature and precipitation extremes in South America as derived from the PRECIS regional climate modeling system. Int J Climatol 29(15):2241–2255

Marengo JA, Ambrizzi T, Rocha DA et al (2009b) Future change of climate in South America in the late twenty–first century: intercomparison of scenarios from three regional climate models. Clim Dyn 35(6):1073–1097

Marinho-Filho J, Machado RB, Henriques RPB (2010) Evolução do conhecimento e da conservação do Cerrado. In: Diniz IR, Marinho-Filho J, Machado RB, Cavalcanti RB (eds) Cerrado: conhecimento científico quantitativo como subsídio para ações de conservação. UnB Thesaurus Editora, Brasília, pp 13–32

Marquis RJ, Diniz IR, Morais HC (2001) Patterns and correlates of interspecific variation in foliar insect herbivory and pathogen attack in Brazilian cerrado. J Trop Ecol 17:127–148

Marquis RJ, Morais HC, Diniz IR (2002) Interactions among cerrado plants and their herbivores: unique or typical. In: Oliveira PS, Marquis RJ (eds) The cerrados of Brazil: ecology and natural history of a Neotropical savanna. Columbia University Press, New York, pp 306–328

Massad TJ, Dyer LA (2010) A meta-analysis of the effects of global environmental change on plant–herbivore interactions. Arthropod–Plant Interact 4(3):181–188

Mendonça RC, Felfili JM, Walter BMT et al (2008) Flora vascular do bioma Cerrado: checklist com 12.356 espécies. In: Sano SM, Almeida SDP, Ribeiro JF (eds) *Cerrado*: ecologia e flora. Embrapa Informação Tecnológica, Brasília, pp 421–442

Ministério do Meio Ambiente (2015) Mapeamento do uso e cobertura do Cerrado: projeto TerraClass Cerrado. MMA, Brasília

Mitchell RJ, Campbell CD, Chapman SJ et al (2011) The cascading effects of birch on heather moorland: a test for the top–down control of an ecosystem engineer. J Ecol 95:540–554

Morais HC, Diniz IR, Baumgarten LC (1995) Padrões de produção de folhas e sua utilização por larvas de Lepidoptera em um cerrado de Brasília. Rev Bras Bot 18(2):163–170

Morais HC, Diniz IR, Silva D (1999) Caterpillar seasonality in a central Brazilian cerrado. Rev Biol Trop 47(4):1025–1033

Morais HC, Cabral BC, Mangabeira JA et al (2007) Temporal and spatial variation of *Stenoma cathosiota* Meyrick (Lepidoptera: Elachistidae) Caterpillar abundance in the Cerrado of Brasília, Brazil. Neotrop Entomol 36(6):843–847

Morais HC, Sujii ER, Almeida-Neto M et al (2011) Host plant specialization and species turnover of caterpillars among hosts in the Brazilian Cerrado. Biotropica 43(4):467–472

Myers N, Mittermeier RA, Mittermeier CG et al (2000) Biodiversity hotspots for conservation priorities. Nature 403(6772):853–858

Niemelä P, Tahvanainen J, Sorjonen J et al (1982) The influence of host plant growth form and phenology on the life strategies of Finnish macrolepidopterous larvae. Oikos 39(2):164–170

Noojipady P, Morton DC, Macedo DC et al (2017) Forest carbon emissions from cropland expansion in the Brazilian Cerrado biome. Environ Res Lett 12:1–11

Oliveira PEAM (2008) Fenologia e biologia reprodutiva de espécies do Cerrado. In: Sano SM, Almeida SP, Ribeiro JF (eds) Cerrado: ecologia e flora. Embrapa, Planaltina, pp 273–290

Oliveira LB (2010) Diversidade e fenologia de lagartas folívoras em *Roupala montana* (Proteaceae) no Cerrado do Brasil Central. Dissertation, University of Brasília

Oliveira-Filho AT, Ratter JA (2002) Vegetation physiognomies and woody flora of the cerrado biome. In: Oliveira PS, Marquis RJ (eds) The cerrados of Brazil: ecology and natural history of a Neotropical savanna. Columbia University Press, New York, pp 91–120

Pessoa-Queiroz R (2008) Padrões de parasitismo em lagartas folívoras externas no Cerrado. Thesis, University of Brasília

Pincebourde S, Van Baaren J, Rasmann S et al (2017) Plant–insect interactions in a changing world. Adv Bot Res 81:289–332

Pinheiro F, Diniz IR, Coelho D, Bandeira MPS (2002) Seasonal pattern of insect abundance in the Brazilian cerrado. Austral Ecol 27(2):136

Price PW, Diniz IR, Morais HC et al (1995) The abundance of insect herbivore species in the tropics: the high local richness of rare species. Biotropica 27(4):468–478

Ratter JA, Ribeiro JF (1996) Biodiversity of the flora of the cerrado. In: Pereira RC, Nasser LCB (eds) Proceedings of the 1st international symposium on tropical savannas. Embrapa CPAC, Planaltina, pp 3–6

Ratter JA, Bridgewater S, Ribeiro JF (2001) Espécies lenhosas da fitofisionomia cerrado sentido amplo em 170 localidades do Bioma Cerrado. B. Herb. Ezechias Paulo Heringer 7:1–138

Ribeiro JF, Walter BMT (1998) Fitofisionomias do bioma Cerrado. In: Sano SM, Almeida SP (eds) Cerrado: ambiente e flora. Embrapa-CPAC, Planaltina, pp 89–166

Rizzini CT (1979) Tratado de Fitogeografia do Brasil, vol 2. Edusp, São Paulo, pp 212–223

Rodovalho SR, Laumann RA, Diniz IR (2007) Ecological aspects of lepidopteran caterpillar parasitoids from *Caryocar brasiliense* Camb. (Caryocaraceae) in a cerrado sensu stricto of Central Brazil. Biota Neotrop 7(3):239–243

Salazar LF, Nobre CA, Oyama MD (2007) Climate change consequences on the biome distribution in tropical South America. Geophys Res Lett. https://doi.org/10.1029/2007GL029695

Salcido DM, Forister ML, Lopez HG et al (2020) Loss of dominant caterpillar genera in a protected tropical forest. Sci Rep 10:422

Salgado-Laboriau ML (1994) História ecológica da Terra. Blucher, São Paulo, pp 255–291

Scherrer S, Diniz IR, Morais HC (2010) Climate and host plant characteristics effects on lepidopteran caterpillar abundance on *Miconia ferruginata* DC. and *Miconia pohliana* Cogn (Melastomataceae). Braz J Biol 70(1):103–109

Scherrer S, Lepesqueur C, Vieira MC et al (2016) Seasonal variation in diet breadth of folivorous Lepidoptera in the Brazilian cerrado. Biotropica 48(4):491–498

Schmidt KP, Inger RF (1951) Amphibians and reptiles of the Hopkins–Branner expedition to Brazil. Natural History Museum, Chicago

Shapiro BA, Pickering J (2000) Rainfall and parasitic wasp (Hymenoptera: Ichneumonoidea) activity in successional forest stages at Barro Colorado Nature Monument, Panama, and La Selva Biological Station, Costa Rica. Agric For Entomol 2(1):39–47

Silva IA, Silva DM, Carvalho GH et al (2011) Reproductive phenology of Brazilian savannas and riparian forests: environmental and phylogenetic issues. Ann For Sci 68:1207–1215

Siqueira MF, Peterson AT (2003) Consequences of global climate change for geographic distributions of cerrado tree species. Biota Neotrop 3:1–14

Stireman JO, Singer MS (2003) Determinants of parasitoid–host associations: insights from a natural tachinid–lepidopteran community. Ecology 84(2):296–310

Stireman JO, Dyer LA, Janzen DH et al (2005) Climatic unpredictability and parasitism of caterpillars: implications of global warming. Proc Natl Acad Sci 102(48):17384–17387

Tanaka IK, Tanaka SK (1982) Rainfall and seasonal changes in Arthropod abundance on a tropical oceanic island. Biotropica 14:114–123

Tylianakis JM, Didham RK, Bascompte J et al (2008) Global change and species interactions in terrestrial ecosystems. Ecol Lett 11:1351–1363

Velasque M, Del-Claro K (2016) Host plant phenology may determine the abundance of an ecosystem engineering herbivore in a tropical savanna. Ecol Entomol 41:421–430

Werneck FP, Nogueira C, Colli GR et al (2012) Climatic stability in the Brazilian Cerrado: implications for biogeographical connections of South American savannas, species richness and conservation in a biodiversity hotspot. J Biogeogr 39(9):1695–1706

Weseloh RM (1993) Potential effects of parasitoids on the evolution of caterpillar foraging behavior. In: Stamp NE, Casey TM (eds) Caterpillars: ecological and evolutionary constraints on foraging. Chapman Hall Press, New York, pp 203–223

Wolda H (1988) Insect seasonality: why? Annu Rev Ecol Syst 19:1–18

The Impact of Construct Building by Caterpillars on Arthropod Colonists in a World of Climate Change

Robert J. Marquis, Christina S. Baer, John T. Lill, and H. George Wang

Chionodes fuscomaculella (Chambers) (Gelechiidae), a common leaf-tying caterpillar on *Quercus* in Missouri, USA. (Photo: Robert J. Marquis)

R. J. Marquis (✉)
Department of Biology, Whitney R. Harris World Ecology Center,
University of Missouri–St. Louis, St. Louis, MO, USA
e-mail: robert_marquis@umsl.edu

Introduction

Caterpillars build an enormous array of structures on plants. In building these structures, they use their host plant as a container, platform, or scaffold, often incorporating plant parts. Materials used include silk and frass produced by the caterpillar and whole or partial leaves, flowers, fruits, stems, twigs, and branches from the plant (Packard 1877; Fukui 2001; Lill and Marquis 2007; Cornelissen et al. 2016; Calderón-Cortés 2020; Sane et al. 2020). In some cases these structures ("constructs") (see Box 1) are camouflaged with fruits, lichens, or sand (Jones and Parks 1928). The resulting constructs can include silk webs (e.g., Rota and Wagner 2008; Baer 2018), silk and frass enclosures (Fitzgerald 1995), silk scaffolds (Rathcke and Poole 1975; Cardoso 2008), frass chains (DeVries 1987; Freitas and Oliveira 1996), leaf folds (e.g., *Ancylis divisana*, MacKay 1959) (including leaf "tents": Lind et al. 2001), leaf ties (Marquis et al. 2019b), leaf rolls (e.g., Vieira and Romero 2013), hollowed herbaceous (e.g., Sidhu et al. 2013) and woody stems (e.g., Cory 1918; Yule and Burns 2017), leaf bags (Davis 1964; Rhainds et al. 2009; Yoshioka et al. 2019), leaf mines (e.g., Opler 1974; Faeth et al. 1981), galls (Riley 1869; Leiby 1922), frass and silk tubes (Neunzig 1972), nets (Dyar 1900), and cases (Bucheli et al. 2002) (Plates 1 and 2). While construct building is not unique to lepidopteran larvae among insects, caterpillars are diverse in the structures they build, producing the full gamut of types seen in terrestrial insects outside of the Lepidoptera, with the exception of nests built by Hymenoptera (Packard 1877; Sane et al. 2020) and termites (Korb 2003).

C. S. Baer
First-year Research Immersion Program, Binghamton University, Binghamton, NY, USA

J. T. Lill
Department of Biological Sciences, George Washington University, Washington, DC, USA

H. G. Wang
Department of Biological and Environmental Sciences, East Central University, Ada, OK, USA

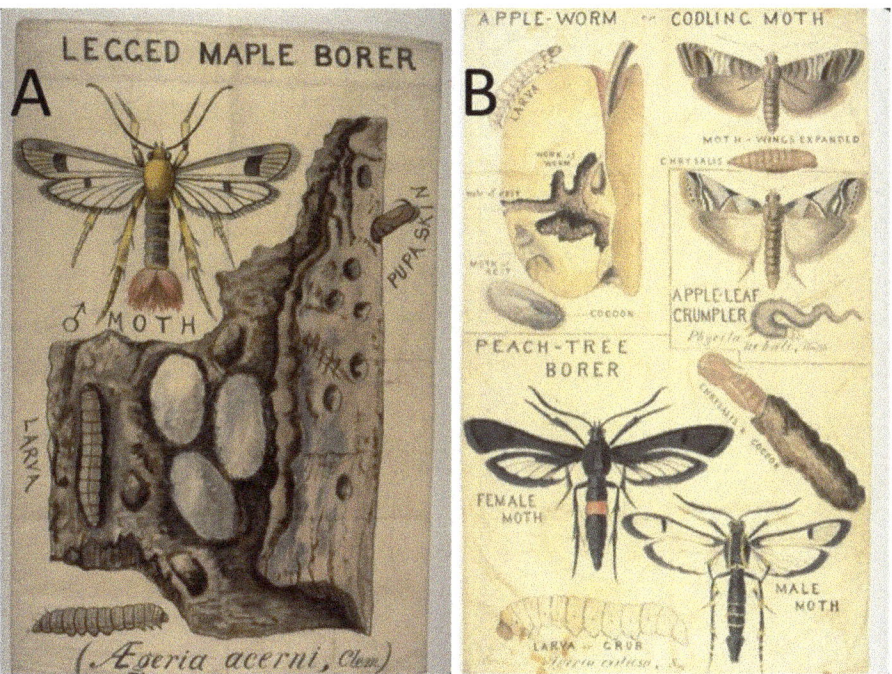
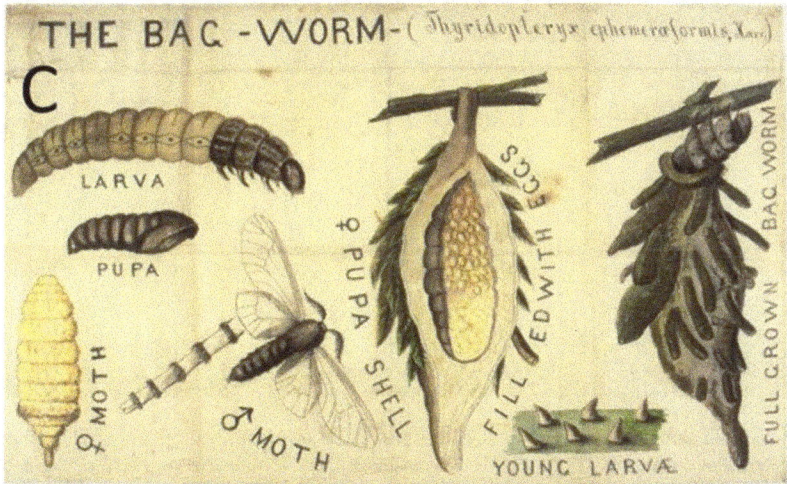

Plate 1 Examples of constructs and life cycles of the Lepidoptera larvae who make them. (**a**) Wood-boring *Aegeria acerni* (now *Synanthedon acerni*) (Sessidae), the legged maple borer. (**b**) Codling moth (*Cydia pomonella*, Tortricidae), whose larvae bore fruits of the Rosaceae; the apple-leaf crumpler *Phycita nebulella* (now *Acrobasis indigninella* group) (Pyralidae), which make frass tubes on leaves; and the peach tree borer, *Synanthedon exitiosa* (Sessidae), also a wood-boring species. (**c**) The bagworm, *Thyridopteryx ephemeraeformis*, often found on *Juniperus*, but is known from 125 species throughout its range (Moore and Hanks 2004). Drawings by C. V. Riley, images from the US National Agricultural Library (https://entomology.k-state.edu/department-info/links/national-ag.-library.html)

Plate 2 Sample showing the range of size, closure, and frass accumulation among different species of caterpillar constructs. (Photos by R. J. Marquis, except *A. cerasivorana*, Whitney Cranshaw, Bugwood). (**a**) Single web of *Hyphantria cunea* (Erebidae, Arctiinae) on *Juglans nigra*: open web, containing frass, 1.5 m in length. (**b**) Sheet web of *Acleris* sp. (Tortricidae) on *Quercus macrocarpa*: closed web, no frass (the hole at the bottom of the web allows frass pellets to drop out), 4 cm in length (River Falls, WI, USA). (**c**) Leaf pieces/silk construct of an unidentified Gelechiidae on *Quercus macrocarpa*: open, no frass, small (1 cm in length) (River Falls, WI, USA). (**d**) Ugly nest of the ugly nest caterpillar, *Archips cerasivorana* (Tortridae): partially open, frass, large (50–75 cm in length). (**e**) Multiple *H. cunea* webs on *J. nigra* (Union Cemetery, Brazil, Indiana, USA). (**f**) Silk and frass nest of *Malacosoma americana* on *Prunus* (University City, MO, USA)

Plate 2 (continued) (20 cm in length). (**g**) Open silk and frass construct on *Pastinaca sativa* by *Depressaria radiella* (previously *D. pastinacella*) (Depressariidae) (River Falls, WI, USA). (**h**) Open leaf roll on *Quercus imbricaria* (Cuivre River State Park, Troy, MO, USA) (8 cm in length). (**i**) Leaf tie on *Quercus alba*, showing silk strands holding leaves together and frass pellets on the leaves (Tyson Research Center, Eureka, MO, USA) (10 cm in length). (**j**) Sheet web on *Q. alba*, open at both ends, no frass, made by *Machimia tentoriferella* (Depressariidae) (Cuivre River State Park, Troy, MO, USA) (1.5 cm in length). (**k**) Bagworm on *Quercus stellata*, open at one end (Plainville, IN, USA) (1.5 cm in length)

> **Box 1**
> We use the term "construct" here ("physical structure" in Jones et al. 1997), instead of the more commonly used term "shelter" to avoid conflating terminology with any discussion regarding the adaptive value of these structures. Specifically, the world "shelter" has the connotation of providing protection against some external threat, but there may be additional or alternative adaptive reasons for building the construct and for colonizing the construct. In building constructs, caterpillars are acting as physical ecosystem engineers, "organisms that directly or indirectly control the availability of resources to other organisms by causing physical state changes in abiotic or biotic materials" (Jones et al. 1997). We use the term "colonists" to label the organisms using the caterpillar construct. These colonists might be organisms that have little impact on the original engineer, but they can also include predators and parasitoids that attack the engineering caterpillar itself.

Many constructs built by Lepidoptera are overlooked or ignored because most species of caterpillars that build them are small (Microlepidoptera: see Lill and Marquis 2007) or nondescript, at least to the naked eye. As a result, many construct-building caterpillars are omitted from caterpillar field guides (e.g., Wagner 2005; but see Ives and Wong 1988; Marquis et al. 2019b). Even less attention has been paid to the many other arthropods that colonize caterpillar constructs, either during or after the builder inhabits its construct. And yet there is increasing evidence that the presence of these constructs increases the diversity of arthropods on plants, encourages the initial invasion and abundance of exotic species, and influences the amount of damage inflicted on the plant (e.g., Lill and Marquis 2003; Marquis and Lill 2010; Baer and Marquis 2014; Henriques et al. 2019; see meta-analysis: Romero et al. 2015).

In building constructs, caterpillars potentially influence the resources available to other arthropods. This physical manipulation of the environment can change resource availability via creation of, maintenance of, or change in a new habitat. Various arthropods colonize the constructs (Plate 3), migrating from nearby leaves of the same plant or from other plants. As a result, both the local and regional abundance of other species can be changed. The term physical ecosystem engineering (Jones et al. 1997; Wright and Jones 2006) is used to describe this process: the organism causing the change is the physical engineer, and the portion of the physical environment that has been changed or built de novo is the construct. We refer here to species that use the construct but did not build it (although some may be able to do so) as colonists. Colonists of galls are often referred to as inquilines; the name is different but the process is similar. Generally, physical ecosystem engineering is thought to have a positive impact on diversity at the landscape scale (one that includes both engineered and non-engineered habitat) via an increase in habitat heterogeneity, an increase in the number of resources, or an increase in overall resource abundance. The concept of physical ecosystem engineering is different from that of

Plate 3 Construct, construct maker, and sample of colonists found in leaf rolls on *Tilia americana* (Malvaceae), Mound Park, River Falls, WI, USA (all photos are by R. J. Marquis). (**a**) Construct maker, *Pantographa limata* (Crambidae), with edge of leaf roll containing accumulated frass in the upper left corner. (**b**) Ladybug beetle (Coleoptera: Coccinellidae). (**c**) Wolf spider (Lycosidae). (**d**) Plant louse (Psocidae). (**e**) Construct (leaf roll) formed by a cut perpendicular to the axis of the leaf and then rolling the distal cut portion to the axis of the leaf. This particular leaf has two such constructs. (**f**) Thrips (Thysanoptera). (**g**) Assassin bug (Hemiptera: Reduviidae). (**H**) Lace bug (Hempitera: Tingidae)

mutualism in that there is no required reciprocal benefit between the engineer and species affected by the engineering. Negative or positive feedbacks may occur from the colonists to the constructor, which if they do occur, would make the relationship different from traditional commensalism.

All organisms have some physical influence on their local environment, and the same is true for caterpillars that build constructs. The challenge for the ecologist is to identify which engineering species have nontrivial impacts on associated species (Jones et al. 1997). In turn, the challenge for the caterpillar biologist is to identify which engineering caterpillar species have major impacts on associated arthropod species found on their shared host plants. We do not expect all engineering caterpillars to have equal impacts, just as we do not expect all species to be equally competitive or equally impactful on community structure as predators or mutualists. These impacts will include effects on the abundance and behavior of associated species, community attributes such as species richness and diversity, and ecosystem processes, such as nutrient cycling. A reasonable hypothesis is that the strength of these engineering impacts is a function of characteristics of the construct itself. The first step toward testing this hypothesis is to determine which traits of the construct are relevant for the arthropods that might colonize the construct.

One place to look to identify these traits is the list of proposed adaptive reasons caterpillars build constructs in the first place, as those benefits might also accrue to colonists. The main proposed adaptive reasons for building a construct on a plant are amelioration of a harsh environment (Connor and Taverner 1997, Fukui 2001), including protection against UV radiation (Connor and Taverner 1997; Vieira and Romero 2013), improved host plant quality (Sagers 1992), reduced competition (Green et al. 1998), and escape from natural enemies (e.g., Ito and Higashi 1991, Sendoya and Oliveira 2017; Baer and Marquis 2020) and pathogens (Green et al. 1998), including easier access to leaves than when hiding off plants to escape predation (Minno 1994). Each might be relevant to colonists. Benefits specific to the colonists might be that constructs serve as a source of food (e.g., frass, cast skins, prey, and hosts) other than plant tissue (Marquis and Lill 2006) or a site for locating potential mates (Fukui 2001). Minno (1994) also proposed that leaf tents made by nonpigmented skipper caterpillars (Hesperiini) might prevent chlorophyll in their guts from reacting with sunlight to form oxygen radicals.

One ultimate goal is to understand how construct building by caterpillars influences the abundance, diversity, and community structure of arthropods associated with the plants on which the constructs occur (Marquis and Lill 2006). We would like to know whether functional types of caterpillar constructs exist, allowing us to predict their impact by measuring their specific traits. We would also like to learn the scale at which these effects occur (local branch, whole plant, or multiple plants). These effects may ripple or diffuse out to neighboring plants depending on the engineered structure and the colonizing species (Priest et al. 2021). Historically, competition and predation have been considered the main structuring forces of arthropod communities (Strong et al. 1984; Price et al. 2011), and they still dominate the conversation (e.g., Bird et al. 2019). The mounting evidence for the impacts of physical ecosystem engineering by insects on plants, however, indicates that competition, predation, and mutualism are not the only biotic forces structuring arthropod communities on plants. Lill and Marquis (2007) reviewed the natural history of insects in general as ecosystem engineers, and Marquis and Lill (2006) discussed the possible pathways of interactions, both direct and indirect among engineers, colonists, and their host plants.

Many, if not all, benefits derived from building constructs or colonizing them are likely to be modified by the rising impact of humans on the global ecosystem. For example, climate change may influence the ability of caterpillar engineers to colonize the plant initially via changes in plant architecture (e.g., Marquis et al. 2002; Pangga et al. 2013) and in plant quality (e.g., DeLucia et al. 2012; Trebicki et al. 2017; Eisenach 2019). In turn, as climate changes, the value of a construct as shelter from desiccating temperatures is expected to increase. Similarly, the value of the construct as a source of food is likely to change because of climate impacts on leaf nutrient quality and secondary defense (DeLucia et al. 2012). Finally, as the relative abundance of natural enemies changes or shifts in seasonality, the value of a construct as refuge from those enemies may also shift. In particular, climate-driven changes in ant communities are likely to modify the success of construct-building caterpillars versus free-feeders when the plant species has extrafloral nectaries (Sendoya and Oliveira 2015, 2017). In sum, the population biology and evolution of the host plant, the constructor, and the colonists are all likely to be influenced.

Our goal with this chapter is threefold. First, we describe the results of natural history observations and experiments, mostly in Missouri (USA) oak forests, which show that constructs impact arthropod composition on host plants and the factors that influence that impact. Second, we define the traits of shelters that are likely to influence their colonization. To do so, we consider the full range of caterpillar construct builders. Specifically, we include both construct (shelter) builders and concealed feeders (leaf miners and gall makers), as defined by Lill and Marquis (2007). Third, we predict how climate change is likely to influence the impact that caterpillars have as ecosystem engineers. Climate change is happening rapidly, and predictions of carbon emissions on future temperature change have become more certain (Sherwood et al. 2020). Given that amelioration of the abiotic environment is one predicted benefit conferred for both constructors and colonists, climate change is directly relevant to our understanding of the impacts of caterpillars as ecosystem engineers, and in turn, the impact of climate change on arthropod community structure.

Data Sources: Natural History and Experiments

We begin by describing the two main data sources for clues to answer the questions posed above. Natural history provides essential data for determining the range of construct-building and construct-using behaviors exhibited by caterpillar engineers and colonists that determine the types of constructs they build and use. Natural history also provides the data needed for correlations between variables that might be relevant for determining underlying processes. Experiments in turn allow the researcher to manipulate the relevant variables that appear to be important based on observations and correlations.

Natural History and Correlations

Natural history observations are scattered throughout the literature describing constructs (e.g., Minno 1994), sometimes how they are built (Weiss et al. 2003), how structures change with caterpillar ontogeny (e.g., Kendall 1965; Minno 1981; Lind et al. 2001), and very rarely their colonists (e.g., Fukui 2001; Lill and Marquis 2007; Cornelissen et al. 2016). Early attempts have been made for a general classification scheme for Hesperiidae (Greeney 2009) and for caterpillars from multiple families found at a single location (Baer and Marquis 2020). However, we are still in the early stages of accumulating descriptions of construct types and, indeed, learning how to best describe them. Ecological studies of caterpillar traits that influence predation and parasitism most often lump all construct-builders together but treat them as a group separate from leaf miners. As a result, we know little of the individual construct traits that determine the success of the caterpillar and colonization levels by colonists. A database of construct descriptions, common colonists, host plants, and seasonal abundance would be most useful. Community surveys (e.g., Diniz et al. 2012), conducted at numerous locations across the globe (e.g., Dyer et al. 2007), would allow us to correlate local climate with the proportion of a particular caterpillar fauna making constructs, the distribution and relative abundance of construct types, and the role of host plant traits in structuring the community of construct builders.

Even seasonal surveys at local sites seem to be rare. Censuses of leaf ties by Lill (2004) on white oak in Missouri showed two peaks, one in June–July and the other in August–September, associated with the two generations found in most species of leaf-tying caterpillars at the study site. Sigmon and Lill (2013) found that the composition of colonists in artificial leaf ties placed on *Quercus alba* and *Fagus grandifolia* (both Fagaceae) changed seasonally but was also strongly affected by host plant species. Ernst et al. (unpublished data) censused all constructs found on all woody plants < 3 m in height along three 50-m-long transects, three times, May, July, and September 2018 (3302 ± 367 total leaves (mean ± SE) sampled per transect per census) (Fig. 1). They found that the absolute and relative abundance of leaf ties increased across the three censuses, whereas leaf folds, leaf rolls, and silk webs stayed approximately constant. The percentage of leaves with constructs was 2.5% on average across all censuses and all transects, which is within the range that can influence arthropod community composition on oaks (Reinhart and Marquis, in review). Colonists were mostly spiders, thrips, and ants, with the greatest abundance in the first census, declining thereafter. These results together suggest that, from the colonist's point of view, constructs are "moving targets" across the season:

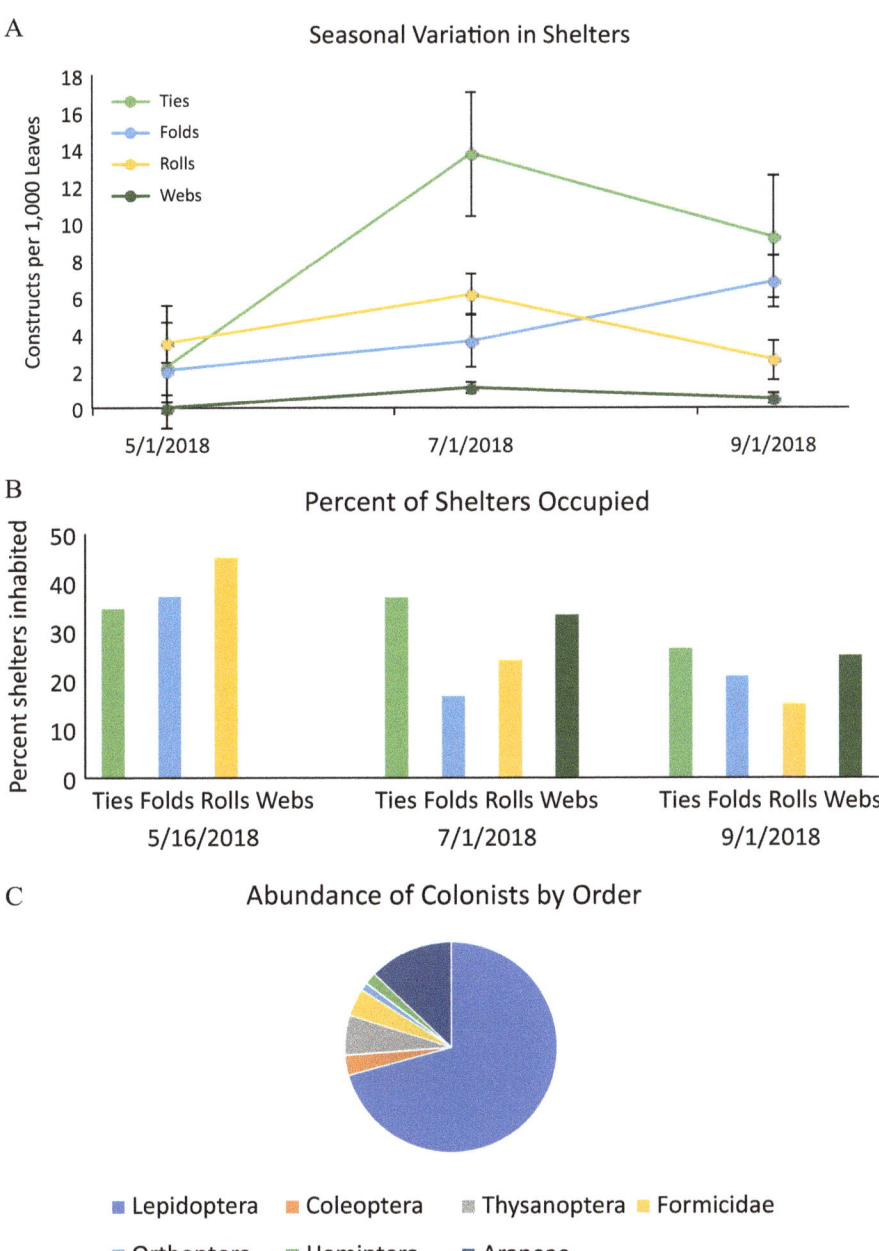

Fig. 1 Summary of sampling leaf ties, leaf rolls, leaf folds, and silk webs on woody plants at Cuivre River State Park, Troy, MO, USA, three times during the growing season of 2018. (**a**) Relative abundance (means ± SD) of each of the three types by census. (**b**) Percent of constructs inhabited by colonists or the original construct maker. (**c**) Relative abundance by order of construct inhabitants

they vary in number by season (see also Vieira and Romero 2013), host plant species, and construct type.

Part of this seasonal variation may be due to an impact of leaf age. *Urbanus esmeraldus* (Hesperiidae) preferentially build constructs on mature rather than young leaves, which might lead to their uneven distribution within the canopy of their host plants, *Urera baccifera* (Euphorbiaceae) (Moraes et al. 2012). In Missouri *Quercus*, leaf rolls are more often constructed on developing leaves, while leaf ties are overwhelming made on mature leaves (RJM, unpublished data). This seasonal variation in construct type is likely to influence the overall engineering impact of the community of caterpillar engineers on the community of colonizers. The choice of leaf age by *Omphalocera* (Pyralidae) caterpillars on *Asimina* (Annonaceae) for constructs reflects a tradeoff between the high nutrition of young leaves versus a greater protective effect against predators by constructs made of mature leaves (Damman 1987).

In designing field surveys of construct colonization, investigators should carefully consider both the temporal and spatial scales at which sampling will occur. Because constructs will attract arthropods already present on the plant (e.g., Marquis and Lill 2010; Romero et al. in review), they can redistribute arthropods within a plant without changing the plant-level abundance. For arthropod species that are highly mobile or transient, including active wanderers as well as passively dispersing individuals arriving at the plant via ballooning or parachuting, the presence of constructs could also serve to retain individuals whose presence would otherwise be fleeting. This is a phenomenon that requires further evaluation. Alternatively, constructs can actively attract flying gravid females to oviposit directly into the construct (Lill and Marquis 2004). In these latter two cases, arthropod abundance, biomass, and species richness and diversity at the whole plant level can all be increased due to the presence of the constructs. The clues that ovipositing females use to find constructs are currently unknown (Lill and Marquis 2004), but volatiles released following leaf damage, odors from frass and construct inhabitants, and visual damage are all likely candidates.

Sampling should also consider the influence of various microhabitats on these interactions within a particular location, including sun vs. shade (Barber and Marquis 2011), high-nutrient vs. low-nutrient soils, riparian vs. upland soils (Pugnaire et al. 2019), lower vs. upper canopy (Seifert et al. 2020), and north- vs. south-facing slope exposure. There are remarkably few studies of spatial variation in construct abundance or how it affects the responses of the associated colonists. In a study of the distribution of caterpillars with respect to canopy position in 15 deciduous tree species sampled in Virginia, USA, Seifert et al. (2020) found that the density of constructs (leaf ties, leaf rolls, and webs) was 75–100% greater in the upper tree canopy versus mid-story and lower canopy leaves, respectively. In a study of the effect of slope aspect (north- vs. south-facing slopes) and slope position (top, mid, and bottom along a 75 m transect) in an oak-hickory forest Missouri, USA, Gamui and Marquis (unpublished data) found that colonization of artificial leaf ties was greater on south-facing (drier and warmer) than north-facing slopes (65% vs. 40%) and that colonization by slope position varied with colonist identity.

Together, these findings suggest that both the production of constructs and their occupancy may be maximized on plant parts and plant individuals subject to warm, dry conditions that might lead to desiccation.

Experiments

There are two main types of experiments that can be used to test for the impact of the presence of engineering caterpillars on associated arthropods. First, artificial (supplementary) constructs can be added to the plants and the response of colonists measured. Second, the abundance of the engineer itself can be manipulated. Depending on the hypotheses being tested, these experiments can be conducted at the level of the construct (e.g., Marquis and Lill 2010; Wang et al. 2012), at the branch level (Marquis et al. 2002), at the whole plant level (Lill and Marquis 2003), or at the level of multiple plants (e.g., stands). They can also be replicated along ecological gradients and across biomes. Shelter occupancy can easily be included as a treatment by adding arthropod models made of plasticine (Tvardíková and Novotný 2012; Pereira et al. 2020).

In the first type of experiment, artificial constructs (along with appropriate controls), designed to carefully mimic natural constructs, can be added to the host plants or sections of host plants (e.g., Martinsen et al. 2000), after first removing naturally occurring constructs. Leaves can be rolled or clipped together, and artificial stem chambers (Powell et al. 2011) or galls (e.g., 3D printed) may all be added to plants. Intrinsic properties of the constructs can be modified to further delineate the nature of impact, for example, by selecting certain leaf sizes for leaf rolls and leaf ties, and by manipulating the size of the entrance hole and cavity in the case of artificial stem chambers and galls. The presence (Lill and Marquis 2003) and identity (Wang et al. 2012; Wang et al., in review) of the occupant in the experimental construct can also be manipulated in some cases to determine the independent effect of the presence of the engineer. Truly artificial leaf rolls and leaf ties, composed of cardboard, can be added to plants to determine the effect of the construct on the environment, excluding any effect of the leaf as a food source (Martinsen et al. 2000).

The second type of experiment involves manipulating the abundance of the engineer. Individual eggs, egg clusters, or caterpillars can be moved to plants of an "experimental" treatment to control for abundance and individual plant effects. Galls can be destroyed early in their development to reduce their impact on plant physiology. Constructs made of silk or leaves are readily removed from plants. These experimental plants can then be compared with unmanipulated controls, including controls for any manipulation, to test for the effects of background levels of engineering on colonization. The specific mechanisms by which the construct influences colonization, for example, by the production of volatiles, will require additional experiments.

The host plant is the starting point for all these interactions. The effect of host plant identity can be determined explicitly by including plant identity in the experimental design. Plant species (Wang et al. 2012; Baer and Marquis 2014) or plant genotype can be included as a treatment, leaf quality can be manipulated (e.g., via fertilization, artificial damage, or exogenous jasmonic acid application) or chosen prior to the experiment to incorporate extremes, and plant architecture can be manipulated (Marquis et al. 2002). These plant treatments can reveal the importance of plant traits for the establishment of the engineering caterpillar and the relative importance of plant traits versus physical manipulation of the host plant for arthropod community structure.

Experimental Evidence

We summarize here our understanding of the impacts of caterpillars as ecosystem engineers based on studies of interactions on Missouri, USA, oaks. We have spent the last 25 years studying the impacts of leaf-tying and leaf-rolling caterpillars on the associated arthropod fauna of eight *Quercus* species. The leaf-rolling species (17 total) are active in the spring (April–May) before the leaves harden, while the leaf-tying species (15 total), which tie two or more leaves together, one surface to the other, are active in June–October. There are 30 or so relatively common species of construct-building caterpillars at our Missouri study sites (Marquis et al. 2019a). Results of pertinent studies from other systems were previously reviewed (see Marquis and Lill 2006; Lill and Marquis 2007).

Colonists of Leaf Constructs on Missouri Quercus

Marquis and Lill (2010) tested the role of five *Quercus* species (and by proxy, leaf quality) on the local abundance of colonists of leaf ties. In that study, they compared differences in arthropod composition between artificial leaf ties and neighboring control leaves. They found that the addition of experimental leaf ties to saplings of these five tree species increased diversity of non-leaf-tying colonists, secondarily colonizing leaf-tying caterpillars, and leaf damage. The results also showed that interspecific variation in patterns of colonization by leaf-tying caterpillars was driven primarily by the leaf tissue quality of the five oak species rather than by plant architecture.

Reinhart and Marquis (in review) performed a "press" experiment (an experiment in which the treatments are continued over time) in a second-growth forest of Missouri, USA. The goal of this experiment was to determine the impact of leaf ties on arthropod composition at the whole plant level. In one treatment, they located and pulled apart all leaf ties (and removed the leaf-tying caterpillars) weekly from the crowns of 35 *Quercus alba* (white oak) saplings for an entire season. They compared the species composition of colonists from experimental plants with 35 control saplings containing an intact leaf tie assemblage. They then censused all arthropods on each plant 1, 2, and 3 months from the beginning of the experiment. For control plants, they opened leaf ties, censused their contents, and then closed them with a hair clip, removing the hair clip after a week. Finally, they measured the total N and total phenolic content of the foliage of each plant at the beginning of the experiment. They found that arthropod species richness was higher on control plants at each census, and community structure was significantly different between control and experimental plants combined across all censuses. In their analysis, they categorized control and experimental treatment plants as either "low quality" (low nitrogen and high phenolics) or "high quality" (high nitrogen and low phenolics), with approximately half of the plants of each treatment falling into each category. Importantly, the impact of the press treatment on species richness was greater for high quality plants. The observed differences in community structure and species richness occurred despite the fact that only 2–4% of the leaves on control plants were in leaf ties.

Lill and Marquis (2003) and Marquis and Lill (in prep.) also looked at the impact of leaf ties on community structure at the whole plant level in a "pulse" experiment, one in which the experimental treatment is applied for a short time period only. They either removed all natural ties (30 plants), removed all natural ties and constructed artificial ties on 10% of the leaves (30 plants), or removed all natural ties and added caterpillars of *Pseudotelphusa quercinigracella* (Gelechiidae), a common tie-making species in this system, into artificial ties, also comprising 10% of the leaves (30 plants). There were two goals: first, to determine the impact of the leaf ties on arthropod composition at the whole plant level and, second, to determine whether the presence of the leaf-tying caterpillars themselves (larvae of *P. quercinigracella*) had an additional impact on community composition, over and above the effect of its engineered constructs. They found a significant and lasting impact of a pulsed removal of early season leaf ties on herbivore richness, herbivore community composition (Lill and Marquis 2003), and overall arthropod community composition (Marquis and Lill, in prep.). There was no additional influence of the presence of the tie maker on any of these measures suggesting that the impact of this ecosystem engineer on associated arthropods is entirely attributable to the presence of its constructs. The results of this type of experiment might be dependent on both host plant species and the identity of the leaf tie maker (Wang et al. 2012).

There seem to be only two studies that have compared the relative impacts of two different construct types. Marquis and Khaja (unpublished data) compared the effect of leaf rolls versus leaf ties on colonization by arthropods in a pulse experiment on *Quercus alba* in Missouri. They found equal effects of the two construct types on species richness, but 30% more arthropods, and as a result greater biomass, in leaf rolls than in leaf ties. The difference may be due to the greater cavity area available in leaf rolls, as the size of the leaves used was similar between treatments. Vieira and Romero (2013) found no difference in the colonization of artificial funnel- versus cylinder-shaped leaf rolls in a Brazilian Atlantic rainforest.

Interactions with a Colonist: The Asiatic Oak Weevil

The Asiatic oak weevil, *Cyrtepistomus castaneus* (Coleoptera: Curculionidae), is of particular interest as a colonist of leaf constructs on oaks. *C. castaneus* is native to Japan, Korea, eastern Russia, and China. Introduced to the USA in 1957 in Delaware, it quickly spread across the eastern half of the USA and is now in Kansas and Texas (USDA Forest Service 2019). It is invasive, potentially reducing growth of tree seedlings when feeding as a larva (Marquis and Catano, unpublished data). Adults congregate in leaf ties and leaf rolls made by caterpillars in Missouri, particularly those on *Quercus*, causing substantial damage to the leaves of the construct and to the surrounding leaves.

Baer and Marquis (2014) quantified the impact of leaf ties on the abundance and feeding of this insect on six deciduous tree species. They made artificial leaf ties on two species of oak and four other common tree species at their study site to test the importance of leaf ties on the relative preference of *C. castaneus* for each of the hosts. They found that when increasing the availability of leaf ties, abundances of the weevil increased on each host species, damage by the weevils increased on all but one tree species, and host preference ranking based on weevil damage changed. This study suggests that host preference for a non-engineer can change due to the presence of constructs made by the caterpillar engineer(s). This is the first study to show that the relative abundance of engineered constructs can influence host plant use (and potentially diet breadth) by a colonist.

Why do adult *C. castaneus* use leaf constructs? Adults readily fly or drop to the ground when leaf rolls or ties are opened by human observers (RJM, CSB, pers. obs.). This behavior suggests that use of these constructs might be mostly predator-driven. This weevil species contributes to a significant portion of the diet of at least two bat species (*Eptesicus fuscus* (Brack and Duffey 2006) and *Myotis sodalis* (Tuttle et al. 2006)). Improvement in leaf quality (decreased total phenolics and higher nitrogen content) as a result of construct formation, at least in leaf ties, may also be relevant, but these changes occurred in only two out of eight Missouri oak species in which artificial ties were constructed (Wang et al., in review). Although not quantified, differences in temperature and humidity between inside and outside the leaf rolls and ties are likely to be very small in the warm, humid period of late June through August when adults are most active. Competition does not seem to be an issue, as adults readily congregate, with up to 17 individuals in one leaf tie (Baer and Marquis 2014). Because these weevils are herbivores, their food resources are not limited to leaf shelters. They are also parthenogenetic, so they are not attracted to leaf constructs to find mates. Their bodies are highly pigmented and sclerotized; therefore, any chlorophyll they consume is not likely to be strongly activated by UV radiation. Finally, abundances of tie and roll makers are driven in part by climate (Marquis et al. 2019a), and abundance in any given year of adult weevils is positively correlated with the abundance of leaf rolls, but not leaf ties in previous years (O'Brien and Marquis, in prep.).

Together, these results suggest that the climate influences the population dynamics of *C. castaneus* via the abundance of an assemblage of leaf-rolling caterpillars. Our working hypothesis is that leaf rolls and leaf ties both provide protection against predation, but leaf rolls have a greater impact on the population dynamics of the weevil. Leaf rolls are produced in the spring before adult weevils first emerge from the soil in late June, and thus have a greater impact on weevil survival. Leaf ties also contribute to weevil population dynamics, but not as much as leaf rolls because the caterpillars that make leaf ties are only beginning to do so when adults emerge in late June. Plant traits are important in determining whether leaf rolls and leaf ties are made at all. Both plant architecture (Baer and Marquis 2014) and leaf quality (Marquis and Lill 2010) appear to contribute to colonization by leaf-tying caterpillars in Missouri forests. These findings are summarized in the interaction diagram presented in Fig. 2.

Predictions: Effect of Construct Traits

We have begun documenting the phylogenetic distribution of construct-building species within the Lepidoptera (Baer and Marquis 2020), impacts of constructs on arthropod community composition at the sub-plant and whole-plant level, and the importance of plant traits on initial colonization by engineers (Marquis and Lill 2010). These initial studies have allowed us to develop a predictive framework for determining a priori the characteristics that determine the level of colonization. Here we present a set of predictions based on our observations and experiments (Fig. 3). It should be emphasized that these traits do not necessarily vary independently. Community, regional, and perhaps global surveys, coupled with experiments, will be required to tease apart the utility of each of these traits in predicting colonization.

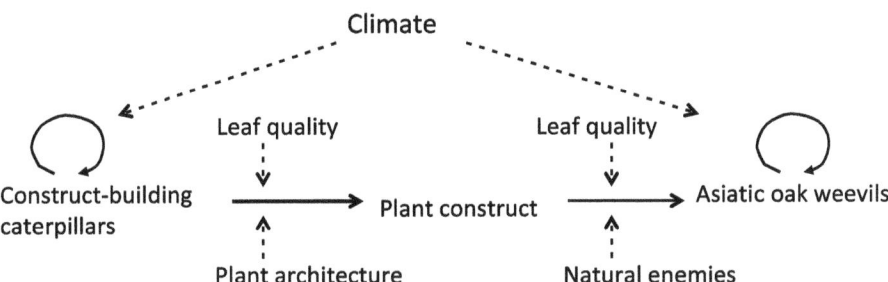

Fig. 2 Processes, indicated by solid arrows, and modifiers of processes, indicated by dashed arrows, which influence the impact of shelter-building caterpillars on the abundance and population dynamics of the Asiatic oak weevil (*Cyrtepistomus castaneus*) in the Missouri (USA) Ozark region via construction and maintenance of leaf ties and leaf rolls. The population growth for both the caterpillars and weevils is represented by a circular arrow

Construct size is a critical element in two ways. The larger the construct, the more volume there is to be colonized (X axis, Fig. 3). Fall webworm (*Hyphantria cunea*) webs (Plate 2), for example, can reach 1 m in length or more and may support tens of colonizing predatory arthropods (Morris 1972). At the other extreme are 4-mm-long cases decorated with pappus, enclosing larvae that eat developing seeds in the single-seeded fruits of plants in the Asteraceae (Steeves et al. 2008). Besides the mere size of the construct, the interior volume available for colonization depends on the size of the engineer relative to the interior volume. Leaf tents built by skipper caterpillars (Hesperiidae) are often just large enough to contain the body of the caterpillar building the tent. Thus we rarely see these leaf tents colonized by other arthropods unless they are abandoned (RJM, CSB, pers. obs.).

Whether or not the engineer is present at the time of colonization is also critical. The presence of the engineer ensures that the construct will be maintained. Continued maintenance would likely increase colonization. Thus, on average, constructs with the engineer present will have more colonists than constructs with the engineer absent (indicated by downward pointing arrows, Fig. 3). However,

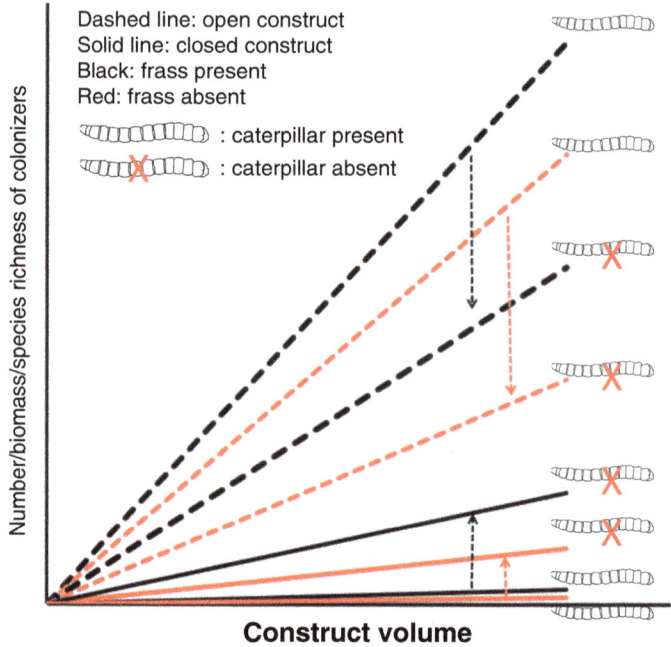

Fig. 3 Hypothesized effect of construct volume, presence of the engineering caterpillar, openness of construct, and presence of frass on colonization by secondary inhabitants, as measured by their biomass, numbers, and/or species richness. Arrows show the effect of the loss of the original engineer on colonization of the construct. In open constructs, this loss results in a decrease in colonization because the construct is no longer maintained, but in closed constructs, the loss of the engineer leads to increased colonization as entrances to the previously closed construct arise. The presence of frass increases colonization except when the construct is closed and the caterpillar constructor is present. The more durable and long-lasting the construct (leaves to hard galls to cavities in wood), the less impact there will be on colonization when the engineer departs from the construct. Not all possible combinations of the listed variables are shown

antagonistic behaviors of the engineer toward potential colonists will reduce the likelihood of colonization (e.g., Green et al. 1998). We suspect these antagonistic interactions are more likely to occur between caterpillars (e.g., Sigmon 2015), reducing the difference in colonization by other non-caterpillar species between constructs with and without the original engineer.

A fourth critical element is the degree to which the interior volume of the construct is accessible to potential colonists (dashed versus solid lines, Fig. 3). Constructs, particularly those built with silk and leaves, vary greatly in the degree to which they sealed. Some have no openings before the engineer completes its development, e.g., galls and leaf mines. Other examples include leaf rolls consisting of many leaf layers in the rolls, but with a closed chamber in the interior (e.g., fern frond balls made by *Herpetogramma theseusalis* caterpillars (Crambidae) (Lo Presti and Morse 2013) and leaf folds in which the folded over edge of the leaf is tightly sealed with silk all along the edges. The ugly nest caterpillar (*Archips cerasivorana*; Plate 2) encloses its "nest" with leaves sewed to form a tube around a branch, adding more leaves as the gregarious caterpillars grow (Powell 1975). Apparently, gaps in silk allow both parasitoids and predators to enter (Balduf 1965). In contrast, leaf ties and leaf tents often have numerous entrance points to the internal chamber that permit colonization. Ito and Higashi (1991) and Sendoya and Oliveira (2017) also provide examples of how construct openness increases predation by ants on the constructor. Leaf mines tend to be completely closed while the miner is present, but the emergence hole of the leaf miner can allow subsequent colonization (Kagata and Ohgushi 2004), or the mine will break apart and become colonizable once the miner has completed its development (RJM pers. obs.).

A fifth element is the presence/absence of frass. Depending on the identity of the receiver, frass can potentially serve as a resource and therefore, an attractant (Weiss 2003) or, if impregnated with noxious compounds, as a deterrent (Green et al. 1998). Caterpillar species vary in their defecation behavior: many let fecal pellets drop wherever they happen to be, while others shoot pellets via a specialized structure, the anal comb (e.g., Minno 1994; Weiss 2003), or fling the pellets with a shake of the head after they have grabbed a pellet with their mandibles. Leaf ties on oaks, in particular, tend to have either much frass or none at all, depending on the species of tie maker. Leaf ties with much frass have many colonizers, while ties with little or no frass have few or no colonizers (Marquis and Lill, unpublished data). Ejection of frass pellets reduces predation by ants in *Urbanus esmeraldus* (Moraes et al. 2012) and predation by wasps in *Epargyreus clarus* (both Hesperiidae) (Weiss 2003). The relative importance of frass as a cue versus as a food source is not known. We do know that many species of arthropods actively seek out leaf ties in which to oviposit, but we do not know the cues that are used (Lill and Marquis 2004).

Once the engineer has abandoned the construct, its persistence time (not shown in Fig. 3) will determine its availability for colonization and thus the builder's ecosystem engineering effect (Jones et al. 1997). At the short extreme, some caterpillars remove shelters from the plant as soon as they vacate them (e.g., larvae of *Epargyreus clarus* that actively cut the silken strands of their leaf tents when they leave to construct a new one; Weiss et al. 2003). At the other extreme, bored holes in twigs or tree trunks will persist as long as the bored portion of the plant persists, which can be many years. The persistence of constructs involving leaves will be a function of the stability of the silk used in the construct, the developmental stage of the leaves used, and the longevity of the leaves (months for most temperate zone plants; years for some tropical trees). Developing leaves are more likely to retain their construct shape. Thus, leaf rolls on oaks, formed in the spring using developing leaves, tend to persist, while leaf ties, formed after leaves have completed their development, fall apart unless subsequent caterpillars colonize and add silk. Colonists themselves, such as spiders and additional caterpillars, often actively contribute to the maintenance of constructs, which greatly extends their persistence and increases their engineering impacts. Because leaf constructs in temperate systems are often made on deciduous trees, their persistence is generally limited to a single growing season, however.

Climate Change

To be able to predict the role that construct-building caterpillars may have in mitigating the impacts of climate change on arthropod communities, we need to know both how the frequency of construct building varies with climate and how colonization and the use of those constructs varies with climate. The first determines whether constructs are available for colonization, and the second how the colonizing arthropod community responds to the availability of those constructs. Both are likely affected by current climatic conditions and will be affected by future climates. To the extent that similar factors may drive the frequency of construct building and their use (see above), these two may be correlated. As of yet, there have been no single studies or meta-analyses of how construct building by caterpillars varies by geographic location. There is, however, one study of how colonization of artificial constructs varies geographically and is modified by climate (Romero et al., in review).

The role of the host plant as the starting point for these interactions cannot be forgotten. Without the construct, there can be no colonization, and without the caterpillar, there can be no construct. The suitability of the host plant for the caterpillar is the crucial first step for the interaction. That suitability will depend on numerous plant traits (physical, chemical, nutritional, and architectural) and the nature of the natural enemies of the caterpillar that are associated with the host plant. All of these traits are influenced by predicted changing climate and soil factors (Pangga et al. 2013), as well as plant quality (e.g., DeLucia et al. 2012; Trebicki et al. 2017; Eisenach 2019; Pugnaire et al. 2019).

Regarding the availability of constructs, our casual observations of many biomes suggest that construct abundance, both in number and in type, varies geographically. For example, within the continental USA, there is a trend of decreasing abundance of leaf-tying caterpillars on deciduous oak species from the south (southern Missouri) to the north (the Great Lakes region of Michigan, Wisconsin, and

Minnesota). Very wet tropical forest (> 4 m rain annually) appears to harbor many fewer construct-building caterpillars than dry tropical forest. These wet forest habitats may be too wet if constructs are inundated in rainwater continuously, drowning caterpillars and colonists alike (see Blau 1980). In a Papua New Guinea forest (3500 mm annual rainfall), 84% of all caterpillars are leaf-tying, leaf-rolling, or web-spinning (Hrcek et al. 2013), while in Brazilian Cerrado (tropical savanna; 1400 mm annual rainfall), marked by short wet and very long dry seasons, 60% of the caterpillar species and 70% of the caterpillar individuals are construct-builders (not including leaf mines and plant galls) (Diniz et al. 2012). These Cerrado caterpillars are also more abundant during the dry than wet season, suggesting that desiccation is an important factor driving their abundance (consistent with our more local comparisons of slope aspect and canopy/understory as mentioned previously). More sites need to be sampled to clarify the relationship between annual rainfall and the frequency of construct building.

Construct builders at the highly seasonal Brazilian Cerrado study site are less parasitized than free feeders, suggesting a role of natural enemies in addition to climate (Diniz et al. 2012) (see also Braga and Diniz, Chapter "Trophic interactions of caterpillars in the seasonal environment of the Brazilian Cerrado and their importance in the face of climate change"). This savanna system may be unusual in this regard as construct-building caterpillars (excluding leaf miners and gallers) in wet tropical forests are generally more heavily parasitized than free-feeding species (Gentry and Dyer 2002; Connahs et al. 2011; Hrcek et al. 2013) and different construct-building species vary significantly in parasitism loads (Baer and Marquis 2020). However, we do not have sufficient sample size to know if there is a consistent effect of annual seasonality on parasitism levels. The seasons within a year might be also important. Spring leaf-rolling species on Missouri oaks had the highest level of parasitism compared to summer leaf-mining and leaf-tying species (Le Corff et al. 2000). Immune response to parasitism must also be considered and, in some cases, may supersede any differences in feeding guild (Smilanich et al. 2009) (see Smilanich and Muchoney, Chapter "Host Plant Effects on the Caterpillar Immune Response").

In the Midwest states of the USA, summer droughts are expected to become more frequent with climate change (Jin et al. 2018). Marquis et al. (2019a) document an almost complete decline of the leaf-tying caterpillar fauna in the east central portion of Missouri following a major drought. Although drought conditions were relieved the following year, the fauna did not recover to pre-drought levels for a full 5 years. Large-bodied caterpillar species recovered more quickly. Unfortunately, the direct and indirect effects (via loss of leaf ties) of the drought on potential colonizers of leaf ties were not followed. These results show, however, that construct builders are vulnerable to extreme weather events, and so potentially, are the arthropods that depend on those constructs.

Romero et al. (in review) conducted a global experiment in which equal numbers of artificial leaf rolls were made available for colonization by forest arthropods. The response by colonists was then correlated with local climate to predict how projections of future climate might affect the interactions. Five artificial leaf rolls (i.e., leaves rolled and clipped in place) on each of ten trees were paired with ten control trees, each with five unrolled, control leaves. The experiment was run for 10 days at each of 52 sites, spanning 11,790 km in latitude and 2895 m in elevation. Plant species were usually trees of the most common plant species at the site. At the end of the 10 days, artificial rolls and control leaves were collected, and the arthropods on those leaves were classified to order and trophic level, and their biomass determined by known relationships between body length and mass.

The results show that after only 10 days, arthropod biomass, abundance, diversity, and average body size were significantly greater in artificial leaf rolls than on unmanipulated, control leaves. Occupancy of leaf rolls by herbivores increased with increasing leaf size, decreased with predator abundance, and was higher in wetter climates. In contrast, predator diversity and overall biomass (reflecting individual body sizes) within shelters increased from sites that were less variable in temperature and precipitation to sites with increasing variability in local precipitation and temperature, regardless of latitude and elevation. These results occurred despite the fact that temperature and precipitation were constant during the 10-day time span of the experiment. These results suggest that predators are adapted to use shelters because of unpredictable climate within a season. Given the projected increase in instability of climate, the importance of shelters for predators is expected to increase, especially where temperature unpredictability is expected to increase. In turn, herbivores are predicted to increase their use of shelters in places on the globe where rainfall is expected to increase, and decrease their usage as rainfall declines in others.

In the global experiment by Romero et al. (in review), the researchers did not know the origins of the colonists, whether they were arriving from other parts of the same plant, from other plants, or via oviposition. If colonists are actively attracted to or selectively retained on plants with constructs, then we predict that the importance of these constructs as plant-level predictors of arthropod community structure will only increase in concert with increasing climate stressors.

Conclusions and Future Directions

We suggest that climate and natural enemies are the two most likely factors that drive (1) the local abundance of caterpillar constructs on plants and (2) the abundance and composition of colonists of those constructs. It will be difficult to disentangle these two drivers, as they are likely causally related. Climate, however, may have the larger, more encompassing effect on constructors and colonists. Romero et al. (in review) showed that herbivores as colonists primarily use leaf rolls as shelters against rainfall and predation, while predators use leaf rolls as shelters against variable temperature and rainfall. Thus climate has a direct impact on the use of constructs by predators, and an indirect effect on herbivores via its effect on predators.

What determines, however, the frequency of construct-building caterpillars at a given location in the first place? Only limited data exist at this time to answer this question. Both plant architecture (Marquis et al. 2002) and plant tissue quality (Marquis and Lill 2010) are important starting points. Studies of spatial variation in caterpillar abundance in four locations in Brazilian Cerrado show that both the local community composition of generalist predatory ants and the presence of extrafloral nectaries on plants determine the abundance and frequency of construct-building caterpillars on those plants (Sendoya and Oliveira 2015, 2017). Stireman et al. (2005) showed that parasitism of caterpillars by specialist braconid wasps was lowest in locations where rainfall is more variable from year to year. This result suggests that constructs, if they provide protection against parasitism, should be more common in environments with less inter-annual climate variability. But both predation and parasitism work in concert on caterpillars. Baer and Marquis (2020) found that when predation was high, parasitism was low, and vice versa, for a set of 24 leaf construct builders in a Costa Rican dry forest. Thus construct-building species may be most abundant in climatically predictable environments, where predation and parasitism together are high. This then sets the stage for strong impacts of constructs on colonists. Very high annual rainfall (> 4 m), however, may set an upper limit because caterpillars (and potentially other small colonizing arthropods) can literally drown in their constructs (e.g., Blau 1980).

There is much room for testing how strongly climate and natural enemies modulate caterpillar impacts on associated arthropod communities. As already mentioned, changes in leaf chemistry as a result of leaf manipulation, protection against UV radiation, amelioration of competition, sources of food and hosts, and location of mates can also be relevant. There is no reason to expect that only one factor is important for any given system. There are few quantifications of the internal versus external temperatures of caterpillar constructs, and all of them come from mesic temperate ecosystems. Humidity is higher in webs of *Hyphantria cunea* (Morris and Fulton 1970), and Joos et al. (1988), Alonso (1997), and Yule and Burns (2017) all showed temperatures were higher within than outside the constructs and that higher temperatures increased the growth rate of the caterpillars making the tents. The results of Romero et al. (in review) suggest that these measurements are needed across an entire season because conditions over the short term can be relatively even. Instead, temperature variability across the entire activity period is likely to be most important.

We presented a series of predictions about how construct architecture may influence colonization by colonists. We have tested the impact of construct architecture on the probability of parasitism and predation for the constructing caterpillar (Baer and Marquis 2020). However, there has been only one test of construct architectural impacts on colonization. Lo Presti and Morse (2013) show that the colonists, here parasitoids of the construct maker, were more likely to parasitize the leaf roller if that caterpillar had made rolls on smaller fern fronds. Tests can be made by correlating construct architecture and colonization of those constructs across caterpillar species, or within a caterpillar species if construct size and shape vary by instar (e.g., Moraes et al. 2012), or by caterpillar and host plant identity (e.g., Sendoya and Oliveira 2017; Lo Presti and Morse 2013). Testing can also be accomplished in appropriate cases by experimental manipulation (e.g., modification of caterpillar-made structures, addition of frass, addition or removal of interacting individuals (Green et al. 1998), and addition of artificial structures).

We know very little about interactions of species within constructs: What are the rules by which these mini-communities are assembled? Interactions may occur between construct builders, among construct builders and colonizers (as exemplified by parasitism and predation levels suffered by the builder), and among colonizers themselves. Studies of interactions between construct builders include those of the parsnip webworm on wild parsnip, *Pastinaca sativa* (Apiaceae) (Green et al. 1998), four species of leaf-tying caterpillars on *Quercus* (Sigmon 2015), and sound-producing Drepanidae on birch (*Betula papyrifera*) (Betulaceae) (Yack et al. 2001). Only 7.5% of *Drepana arcuata* intruders into conspecific webs successfully usurped the resident builder (Yack et al. 2001). Leaf-tying caterpillars colonizing leaf ties on *Quercus alba* made by another caterpillar displaced the original occupant in 24% of the cases (Sigmon 2015). In addition, there was a distinct ranking of competitive strengths among the four species of caterpillars tested on *Q. alba* (Sigmon 2015). Shelter traits may also influence the level of parasitism and predation, i.e., the interaction between the construct builder and the colonizer. Differences in shelter shape influence whether the caterpillars of two species of *Urbanus* (Hesperiidae) will be parasitized (Baer and Marquis 2021) (see also Lo Presti and Morse 2013). Interactions among other colonists have not been studied. Other outstanding questions include the following: What are the characteristics of the constructs that are built in a given location? What traits of constructs attract other arthropods to use them? How does the persistence of the construct influence interactions among constructors and colonists? To what degree are colonists arriving from other parts of the host plant versus other plants or via oviposition?

Many of these same considerations can be expanded to include cascading effects to the plant. Constructs represent "hotspots" of local damage, and they are not necessarily randomly distributed within plant canopies (e.g., Griffen et al. 2017; Seifert et al. 2020). How do interactions among species within constructs influence local damage levels to the plant? What are the effects on the amount and distribution of damage to plant parts caused by spatially nonrandom colonization of the host plant by construct builders? Are frass and silk significant sources of plant nutrients via uptake through the leaves? Finally, what are the impacts on ecosystem properties via changes in leaf damage, premature leaf fall, plant growth, and nutrient addition to the soil? Clever experimentation and even more clever monitoring techniques will be required to answer these questions because many of the interactions within the construct will not be readily visible to the observer.

Understanding the factors that structure arthropod communities on plants has been a major pursuit in community ecology, and yet there has been no synthesis since Strong et al. (1984). In that volume, construct-building Lepidoptera were not considered. Nor were they discussed in the most recent book on insect ecology (Price et al. 2011). Given their high relative abundance, high species richness, and their demonstrated impacts on arthropod composition on plants in systems studied to date, it seems important to know more about the ecology and natural history of construct-building caterpillars. Microlepidopteran species are probably the least known and, therefore, require the most baseline natural history data. Yet they are among the most abundant builders in most systems. The experiments to test the hypotheses presented in this chapter are straightforward, awaiting the ecologist who wishes to further understand factors that structure arthropod communities on plants.

Acknowledgments We thank Renan Moura, Eva Colberg, Suzanne Koptur, Paulo Oliveira, and Tatiana Cornelissen for the comments on earlier drafts and are grateful for financial and logistical support from the Missouri Department of Natural Resources, the Missouri Department of Conservation, the National Science Foundation, and the USDA Forest Service.

References

Alonso C (1997) Choosing a place to grow. Importance of within-plant abiotic microenvironment for *Yponomeuta mahalabella*. Entomol Exp Appl 83:171–180

Baer CS (2018) Shelter building and extrafloral nectar exploitation by a member of the *Aristotelia corallina* species complex (Gelechiidae) on Costa Rican acacias. J Lepid Soc 72:44–52

Baer CS, Marquis RJ (2014) Native leaf-tying caterpillars influence host plant use by the invasive Asiatic oak weevil through ecosystem engineering. Ecology 95:1472–1478

Baer CS, Marquis RJ (2020) Between predators and parasitoids: complex interactions among shelter traits, predation, and parasitism in a shelter-building caterpillar community. Funct Ecol 34:2186–2198

Baer CS, Marquis RJ (2021) Experimental shelter-switching shows caterpillar shelter type alters predation of skipper caterpillars (Hesperiidae). Beh Ecol (in press)

Balduf WV (1965) Observations on *Archips cerasivoranus* (Fitch) (Tortricidae: Lepidoptera) and certain parasites (Diptera: Hymenoptera). Ohio J Sci 65:60–70

Barber NA, Marquis RJ (2011) Light environment and the impacts of foliage quality on herbivorous insect attack and bird predation. Oecologia 166:401–409

Bird G, Kaczvinsky C, Wilson AE, Hardy NB (2019) When do herbivorous insects compete? A phylogenetic meta-analysis. Ecol Lett 22:875–883

Blau WS (1980) The effect of environmental disturbance on a tropical butterfly population. Ecology 61:1005–1012

Brack V Jr, Duffey JA (2006) Bats of Ravenna training and logistics site, Portage and Trumbull Counties, Ohio. Ohio J Sci 106:186–190

Bucheli S, Landry JF, Wenzel J (2002) Larval case architecture and implications of host-plant associations for North American *Coleophora* (Lepidoptera; Coleophoridae). Cladistics 18:71–93

Calderón-Cortés N (2020) Ecosystem Engineering by Insect Herbivores: Non-trophic Interactions in Terrestrial Ecosystems. In: Nuñez-Farfán J, Valverde PL (eds) *Evolutionary ecology of plant-herbivore interactions*. Springer, pp 147–172

Cardoso MZ (2008) Interação herbívoro-tricoma: o caso de *Heliconius charithonia* (L.) (Lepidoptera: Nymphalidae) e *Passiflora lobata* (Killip) Hutch. (Passifloraceae). Neotrop Ent 37:247–252

Connahs H, Aiello A, Van Bael S, Rodriguez-Castaneda G (2011) Caterpillar abundance and parasitism in a seasonally dry versus wet tropical forest of Panama. J Trop Ecol 27:51–58

Connor EF, Taverner MP (1997) The evolution and adaptive significance of the leaf-mining habit. Oikos 79:6–25

Cornelissen T, Cintra F, Santos JC (2016) Shelter-building insects and their role as ecosystem engineers. Neotrop Ent 45:1–12

Cory EN (1918) The peach-tree borer. The Maryland Agricultural Station, College Park, MD, Bulletin 176

Damman H (1987) Leaf quality and enemy avoidance by the larvae of a pyralid moth. Ecology 68:88–97
Davis DR (1964) Bagworm moths of the western hemisphere (Lepidoptera: Psychidae). Bulletin of the United States National Museum. 264 p.
DeLucia EH, Nabity PD, Zavala JA, Berenbaum MR (2012) Climate change: resetting plant-insect interactions. Plant Phys 160:1677–1685
DeVries PJ (1987) *The butterflies of Costa Rica and their natural history*. Princeton University Press, Princeton
Diniz IR, Hay JD, Rico-Gray V, Greeney HF, Morais HC (2012) Shelter-building caterpillars in the cerrado: seasonal variation in relative abundance, parasitism, and the influence of extra-floral nectaries. Arthopod-Plant Interact 6:583–589
Dyar HG (1900) Notes on the larval-cases of Lacosomidae (Perophoridae) and life-history of *Lacosoma chiridota* Grt. J NY Entomol Soc 8:177–180
Dyer LA, Singer MS, Lill JT, Stirman J III, Gentry GL, Marquis RJ, Ricklefs RE, Greeney HF, Wagner DL, Morais HC, Diniz IR, Kursar TA, Coley PD (2007) Host specificity of Lepidoptera in tropical and temperate forests. Nature 448:696–700
Eisenach C (2019) How plants respond to climate change: A new Virtual Special Issue of Plant, Cell & Environment. Cell & Environment:2537–2539
Faeth SH, Mopper S, Simberloff D (1981) Abundances and diversity of leaf-mining insects on three oak host species: effects of host-plant phenology and nitrogen content of leaves. Oikos 32:238–251
Fitzgerald TD (1995) *The tent caterpillars*. Cornell University Press, Ithaca
Freitas AV, Oliveira PS (1996) Ants as selective agents on herbivore biology: effects on the behaviour of a non-myrmecophilous butterfly. J Anim Ecol 65:205–210
Fukui A (2001) Indirect interactions mediated by leaf shelters in animal–plant communities. Popul Ecol 43:31–40
Gentry G, Dyer LA (2002) On the conditional nature of neotropical caterpillar defenses against their natural enemies. Ecology 83:3108–3119
Green ES, Zangerl AR, Berenbaum MR (1998) Reduced aggressive behavior: A benefit of silk-spinning in the parsnip webworm, *Depressaria pastinacella* (Lepidoptera: Oecophoridae). J Insect Behav 11:761–772
Greeney HF (2009) A revised classification scheme for larval hesperiid shelters, with comments on shelter diversity in the Pyrginae. J Res Lepid 41:53–59
Griffen BD, Riley ME, Cannizzo ZJ, Feller IC (2017) Indirect effects of ecosystem engineering combine with consumer behaviour to determine the spatial distribution of herbivory. J Anim Ecol 86:1425–1433
Henriques NR, Cintra F, Pereira CC, Cornelissen T (2019) Indirect effects of ecosystem engineering by insects in a tropical liana. Arthopod-Plant Interact 13:499–504
Hrcek J, Miller SE, Whitfield JB, Shima H, Novotny V (2013) Parasitism rate, parasitoid community composition and host specificity on exposed and semi-concealed caterpillars from a tropical rainforest. Oecologia 173:521–532
Ito F, Higashi S (1991) Variance of ant effects on the different life forms of moth caterpillars. J Anim Ecol 60:327–334
Ives WGH, Wong HR (1988) Tree and shrub insects of the prairie provinces. Canadian Forest Service, Northern Forest Center, Edmonton, Alberta, Canada. Information Report NOR-X-292, 327 p
Jin Z, Ainsworth EA, Leakey AD, Lobell DB (2018) Increasing drought and diminishing benefits of elevated carbon dioxide for soybean yields across the US Midwest. Glob Change Biol 24:e522–e533
Jones FM, Parks HB (1928) The bagworms of Texas. Bull Tex Agric Exp Stn 382:4–36
Jones CG, Lawton JH, Shachak M (1997) Positive and negative effects of organisms as physical ecosystem engineers. Ecology 78:1946–1957

Joos B, Casey TM, Fitzgerald TD, Buttemer WA (1988) Roles of the tent in behavioral thermoregulation of eastern tent caterpillars. Ecology 69:2004–2011

Kagata H, Ohgushi T (2004) Leaf miner as a physical ecosystem engineer: secondary use of vacant leaf mines by other arthropods. Ann Entomol Soc Am 97:923–927

Kendall RO (1965) Larval food plants and distribution notes for twenty-four Texas Hesperiidae. J Lepid Soc 19:1–33

Korb J (2003) Thermoregulation and ventilation of termite mounds. Naturwissenschaften 90:212–219

Le Corff J, Marquis RJ, Whitfield JB (2000) Temporal and spatial variation in a parasitoid community associated with the herbivores that feed on Missouri *Quercus*. Environ Entomol 29:181–194

Leiby RW (1922) Biology of the goldenrod gall-maker *Gnorimoschema gallaesolidaginis* Riley. J NY Entomol Soc 30:81–94

Lill JT (2004) Seasonal dynamics of leaf-tying caterpillars on white oak. J Lepid Soc 58:1–6

Lill JT, Marquis RJ (2003) Ecosystem engineering by caterpillars increases insect herbivore diversity on white oak. Ecology 84:682–690

Lill JT, Marquis RJ (2004) Leaf ties as colonization sites for forest arthropods. Ecol Ent 29:300–308

Lill JT, Marquis RJ (2007) Microhabitat manipulation: ecosystem engineering by shelter building insects. In: Cuddington KMD, Byers JE, Hastings A, Wilson WG (eds) Ecosystem engineers: concepts, theory, and applications in ecology. Elsevier, San Diego, pp 107–138

Lind EM, Jones MT, Long JD, Weiss MR (2001) Ontogenetic changes in leaf shelter construction by larvae of *Epargyreus clarus* (Hesperiidae), the silver-spotted skipper. J Lepid Soc 54:77–82

LoPresti EF, Morse DH (2013) Costly leaf shelters protect moth pupae from parasitoids. Arthropod Plant Interact 7:445–453

MacKay M (1959) Larvae of the North American Olethreutidae (Lepidoptera). Mem Entomol Soc Can 91(S10):5–338. https://doi.org/10.4039/entm9110fv

Marquis RJ, Lill JT (2006) Effects of arthropods as physical ecosystem engineers on plant based trophic interaction webs. In: Ohgushi T, Craig TP, Price PW (eds) Indirect interaction webs: non-trophic linkages through induced plant traits. Cambridge University Press, Cambridge, pp 246–274

Marquis RJ, Lill JT (2010) Impact of plant architecture versus leaf quality on attack by leaf-tying caterpillars on five oak species. Oecologia 163:203–213

Marquis RJ, Lill JT, Piccinni A (2002) Effect of plant architecture on colonization and damage by leaf-tying caterpillars of *Quercus alba*. Oikos 99:531–537

Marquis RJ, Lill JT, Forkner RE et al (2019a) Declines and resilience of communities of leaf chewing insects on Missouri oaks following spring frost and summer drought. Front Ecol Evol. https://doi.org/10.3389/fevo.2019.00396

Marquis RJ, Passoa SC, Lill JT, Whitfield JB, Le Corff J, Forkner RE, Passoa VA (2019b) An introduction to the immature Lepidoptera fauna of oaks in Missouri. USDA Forest Service, Forest Health Assessment and Applied Sciences Team, Morgantown, West Virginia. FHAAST-2018-05. 369 p

Martinsen GD, Floate KD, Waltz AM, Wimp GM, Whitham TG (2000) Positive interactions between leafrollers and other arthropods enhance biodiversity on hybrid cottonwoods. Oecologia 123:82–89

Minno MC (1981) The population biology of tropical and temperate butterflies of the genus *Pyrgus* (Hesperiidae). M.S. thesis, Department Entomology, University of California, Davis. 50 pp

Minno MC (1994) Immature stages of the skipper butterflies (Lepidoptera: Hesperiidae) of the United States: biology, morphology, and descriptions. Ph.D. dissertation. University of Florida, Gainesville, FL, 541 p.

Moore RG, Hanks LM (2004) Aerial dispersal and host plant selection by neonate *Thyridopteryx ephemeraeformis* (Lepidoptera: Psychidae). Ecol Ent 29:327–335.

Moraes AR, Greeney HF, Oliveira PS, Barbosa EP, Freitas AVL (2012) Morphology and behavior of the early stages of the skipper, *Urbanus esmeraldus*, on *Urera baccifera*, an ant–visited host plant. J Insect Sci 12:52

Morris RF (1972) Predation by wasps, birds, and mammals on *Hyphantria cunea*. Can Entomol 104:1581–1591

Morris RF, Fulton WC (1970) Models for the development and survival of *Hyphantria cunea* in relation to temperature and humidity: introduction. Mem Entomol Soc Can 102:1–60

Neunzig HH (1972) Taxonomy of *Acrobasis* larvae and pupae in eastern North America (Lepidoptera: Pyralidae) (No. 1457). US Department of Agriculture, Washington DC

Opler PA (1974) Oaks as evolutionary islands for leaf-mining insects: the evolution and extinction of phytophagous insects is determined by an ecological balance between species diversity and area of host occupation. Am Sci 62:67–73

Packard AS (1877) Insects as architects. In: *Half hours with insects*. Estes and Laurait, Boston, pp 295–320

Pangga IB, Hanan J, Chakraborty S (2013) Climate change impacts on plant canopy architecture: implications for pest and pathogen management. Eur J Plant Pathol 135:595–610

Pereira CC, Sperandei VDF, Henriques NR, Silva AAN, Fernandes GW, Cornelissen T (2020) Gallers as leaf rollers: ecosystem engineering in a tropical system and its effects on arthropod biodiversity. Ecol Entomol. https://doi.org/10.1111/een.12993

Powell JA (1975) Massive population levels of the cherry tree ugly-nest caterpillar. Panpac Entomol 51:94–95

Powell S, Costa AN, Lopes CT, Vasconcelos HL (2011) Canopy connectivity and the availability of diverse nesting resources affect species coexistence in arboreal ants. J Anim Ecol 80:352–360

Price P, Denno RF, Eubanks MD, Finke DL, Kaplan I (2011) Insect ecology: behavior, populations and communities. Cambridge University Press, Cambridge

Priest G, Camarota F, Powell S, Vasconcelos, Marquis RJ (2021) Ecosystem engineering in the arboreal realm: Heterogeneity of wood-boring beetle cavities and their usage by canopy ants. Oecologia 196:427-439

Pugnaire FI, Morillo JA, Peñuelas J, Reich PB, Bardgett RD, Gaxiola A, Wardle DA, Van Der Putten WH (2019) Climate change effects on plant-soil feedbacks and consequences for biodiversity and functioning of terrestrial ecosystems. Sci Adv 5:eaaz1834

Rathcke BJ, Poole RW (1975) Coevolutionary race continues: butterfly larval adaptation to plant trichomes. Science 187:175–176

Reinhart J, Marquis RJ. Ecosystem engineering and leaf quality affect arthropod community structure and diversity on white oak (*Quercus alba* L.) (In review)

Rhainds M, Davis DR, Price PW (2009) Bionomics of bagworms (Lepidoptera: Psychidae). Ann Rev Entomol 54:209–226

Riley CV (1869) Missouri report of the state entomologist on noxious insects, p 173

Romero GQ, Gonçalves-Souza T, Vieira C, Koricheva J (2015) Ecosystem engineering effects on species diversity across ecosystems: a meta-analysis. Biol Rev 90:877–890

Romero GQ, Thiago Gonçalves-Souza T, Roslin T, et al. Climate variability and drought modulate the role of refuges for arthropods: a global experiment (In review)

Rota J, Wagner DL (2008) Wormholes, sensory nets and hypertrophied tactile setae: the extraordinary defence strategies of *Brenthia* caterpillars. Anim Behav 76:1709–1713

Sagers CL (1992) Manipulation of host plant quality: herbivores keep leaves in the dark. Funct Ecol 6:741–743

Sane SP, Ramaswamy SS, Raja SV (2020) Insect architecture: structural diversity and behavioral principles. Curr Opin Insect Sci 42:39–46

Seifert CL, Lamarre GP, Volf M, Jorge LR, Miller SE, Wagner DL, Anderson-Teixerra KJ, Novotný V (2020) Vertical stratification of a temperate forest caterpillar community in eastern North America. Oecologia 192:501–514

Sendoya SF, Oliveira PS (2015) Ant-caterpillar antagonism at the community level: interhabitat variation of tritrophic interactions in a neotropical savanna. J Anim Ecol 84:442–452

Sendoya SF, Oliveira PS (2017) Behavioural ecology of defence in a risky environment: caterpillars versus ants in a Neotropical savanna. Ecol Entmol 42:553–564

Sherwood S, Webb MJ, Annan JD, Armour KC, Forster PM, Hargreaves JC, Hergerl G, Klein SA, Marvel KD, Rohling EJ, Watanabe M (2020) An assessment of Earth's climate sensitivity using multiple lines of evidence. Rev Geo, p e2019RG000678

Sidhu JK, Stout MJ, Blouin DC, Datnoff LE (2013) Effect of silicon soil amendment on performance of sugarcane borer, *Diatraea saccharalis* (Lepidoptera: Crambidae) on rice. Bull Entomol Res 103:656–664

Sigmon E (2015) Interspecific variation in aggressive fighting behavior of shelter-building caterpillars. J Insect Behav 28:403–416

Sigmon E, Lill JT (2013) Phenological variation in the composition of a temperate forest leaf tie community. Environ Entomol 42:29–37

Smilanich AM, Dyer LA, Gentry GL (2009) The insect immune response and other putative defenses as effective predictors of parasitism. Ecology 90:1434–1440

Steeves R, Nazari V, Landry JF, Lacroix CR (2008) Predispersal seed predation by a coleophorid on the threatened Gulf of St. Lawrence aster. Can Entomol 140:297–305

Stireman JO III, Dyer LA, Janzen DH et al (2005) Climate unpredictability and parasitism of caterpillars: implications of global warming. PNAS USA 102:17384–17387

Strong DR, Lawton JH, Southwood R (1984) Insects on plants, community patterns and mechanisms. Harvard University Press, 313 pp

Trębicki P, Dáder B, Vassiliadis S, Fereres A (2017) Insect–plant–pathogen interactions as shaped by future climate: effects on biology, distribution, and implications for agriculture. Insect Sci 24:975–989

Tuttle NM, Benson DP, Sparks DW (2006) Diet of the *Myotis sodalis* (Indiana bat) at an urban/rural interface. NE Nat 13:435–442

Tvardíková K, Novotný V (2012) Predation on exposed and leaf-rolling artificial caterpillars in tropical forests of Papua New Guinea. J Trop Ecol 28:331–341

USDA Forest Service (2019) Alien Forest Pest Explorer – species map. USDA Forest Service, Northern Research Station and Forest Health Protection https://www.nrs.fs.fed.us/tools/afpe/maps/

Vieira C, Romero GQ (2013) Ecosystem engineers on plants: indirect facilitation of communities by leaf-rollers at different scales. Ecology 94:1510–1518

Wagner DL (2005) Caterpillars in Eastern North America. Princeton University Press, Princeton

Wang HG, Marquis RJ, Baer CS (2012) Both host plant and ecosystem engineer identity influence leaf-tie impacts on the arthropod community of *Quercus*. Ecology 93:2186–2197

Wang HG, Wouk J, Anderson R, Marquis RJ. Feedback between leaf quality and leaf ties affects herbivory in eight species of *Quercus* (In review)

Weiss MR (2003) Good housekeeping: why do shelter-dwelling caterpillars fling their frass? Ecol Lett 6:361–370

Weiss MR, Lind EM, Jones MT, Long JD, Maupin JL (2003) Uniformity of leaf shelter construction by larvae of *Epargyreus clarus* (Hesperiidae), the silver-spotted skipper. J Insect Behav 16:465–480

Wright JP, Jones CG (2006) The concept of organisms as ecosystem engineers ten years on: progress, limitations, and challenges. BioScience 56:203–209

Yack JE, Smith ML, Weatherhead PJ (2001) Caterpillar talk: acoustically mediated territoriality in larval Lepidoptera. PNAS USA 20:11371–11375

Yoshioka T, Tsubota T, Tashiro K, Jouraku A (2019) A study of the extraordinarily strong and tough silk produced by bagworms. Nat Commun 10:1–11. https://doi.org/10.1038/s41467-019-09350-3

Yule K, Burns K (2017) Adaptive advantages of appearance: predation, thermoregulation, and color of webbing built by New Zealand's largest moth. Ecology 98:1324–1333

Part V
Caterpillar Foodwebs in a World of Rapidly Changing Climate

Caterpillar Patterns in Space and Time: Insights From and Contrasts Between Two Citizen Science Datasets

Grace J. Di Cecco and Allen H. Hurlbert

Introduction

Caterpillars are an important component of forest ecosystems as both herbivores on woody and herbaceous vegetation and as a food resource for higher trophic levels. Trends in the abundance, biomass, or phenology of caterpillars may thus have impacts that propagate up and down food chains. Region-specific combinations of climate and land use change may result in geographically variable responses by caterpillar populations, and understanding this geographic variation would help identify the most important global change drivers and their mechanisms. Geographically widespread data on occurrence has increased tremendously with the digitization of museum records (Nelson and Ellis 2018), the establishment of online repositories such as the Global Biodiversity Information Facility (GBIF; Telenius 2011), and the rise of popular citizen science platforms like iNaturalist (Seltzer 2019), resulting in millions of observations across the globe. Nevertheless, questions have been raised about the types of inferences that can be made from such data when information on the sampling effort underlying those records is biased or unknown (Isaac and Pocock 2015; Mair and Ruete 2016; Ries et al. 2019; Di Cecco et al. 2021).

Standardized monitoring schemes, on the other hand, are explicit about effort and are able to provide more accurate estimates of abundance and phenology, yet tend to yield fewer and more sparsely distributed data (Soroye et al. 2018). Given

G. J. Di Cecco
Department of Biology, University of North Carolina, Chapel Hill, NC, USA

A. H. Hurlbert (✉)
Department of Biology, University of North Carolina, Chapel Hill, NC, USA

Environment, Ecology, and Energy Program, University of North Carolina, Chapel Hill, NC, USA
e-mail: hurlbert@bio.unc.edu

this tradeoff between data collection approaches, it would be useful to know how well unstandardized occurrence records can recapitulate patterns of caterpillar abundance and phenology observed by more standardized monitoring schemes. If patterns are similar, this implies that the large volume of occurrence records can be used to infer patterns at greater geographic extents and with greater spatiotemporal resolution than would be possible from standardized monitoring data alone. If not, then more work will be needed to understand and deal with the various sources of bias that hamper opportunistic occurrence records.

Lepidoptera in general, and the caterpillar stage specifically, are well represented on iNaturalist. The "Caterpillars of Eastern North America" project on iNaturalist (https://www.inaturalist.org/projects/caterpillars-of-eastern-north-america) has over 400,000 observations from eastern North America alone at the time of publication. Unfortunately, while a number of geographically widespread monitoring programs exist for adult Lepidoptera (e.g. Fourth of July Butterfly Count, state- or nationwide butterfly atlases) or for single species (e.g., Monarch Larva Monitoring Project; Prysby and Oberhauser 2004), there have historically been no community-wide monitoring efforts for caterpillars as a group that have collected data across many regions. Community-wide monitoring is especially important when trying to understand variation in total resource availability for a higher trophic level like foliage-gleaning birds, rather than attempting to understand the dynamics of a particular taxonomic subset of caterpillars. To address this monitoring gap, the citizen science project *Caterpillars Count!* was created in 2015 to collect data on the phenology and abundance of foliage arthropods on woody vegetation (Hurlbert et al. 2019).

Here, we contrast patterns in the taxonomic representation, occurrence, and phenology of caterpillars derived from these two very different citizen science datasets, iNaturalist and Caterpillars Count!. We describe broadscale caterpillar patterns using each dataset and assess the strengths of each for inferring spatiotemporal variation in caterpillar occurrence across North America.

Datasets

iNaturalist

Participants of the iNaturalist citizen science project typically submit photo observations along with a date and georeferenced location of the observation. The observer can suggest a taxonomic identification, and then other iNaturalist users can agree or suggest alternative identifications. The Caterpillars of Eastern North America project is a collation of all larval Lepidoptera records in iNaturalist from Mexico, the United States, and Canada east of 100° W longitude. The number of records contributed to this iNaturalist project has grown steadily over time, with more than 59,000 unique observers contributing >206,000 photos as of June 2020.

We used all observations of larval lepidopterans including those not identified more finely than order. Taxonomic family was specified for 178,702 observations. iNaturalist records are the result of an unknown observation process that depends both on the number of users in space and time as well as their individually variable reporting behavior (Di Cecco et al. 2021).

Caterpillars Count!

Caterpillars Count! is a monitoring program in which participants conduct sets of standardized foliage arthropod surveys on woody vegetation by either carrying out visual inspection of an area of 50 leaves and their associated petioles and twigs or by striking a branch ten times over a beat sheet (Hurlbert et al. 2019). Each participating site has between 10 and 60 marked survey branches, all of which are ideally surveyed once every week or two from after leaf out at least through July. Survey branches are selected in a quasi-standardized fashion so as to capture representative vegetation within the area deemed of interest by each individual site coordinator (Hurlbert et al. 2019).

Participants identify all arthropods found to order and record their abundance and length in millimeters. They may optionally take photos of their observations, which are then submitted automatically to iNaturalist with the potential to receive finer-level taxonomic identifications. Through June 2020, 1,278 unique users have conducted 48,384 branch surveys at 116 sites from throughout the United States and Canada, mostly east of 100° W. Of the 9,981 total caterpillar observations from Caterpillars Count!, 605 were photographed, submitted to iNaturalist, and subsequently identified to the family level or below. All Caterpillars Count! observations shared with iNaturalist, including those that were not yet identified, were removed from the analyses representing iNaturalist patterns.

Family Composition

While Caterpillars Count! observations are specifically from trees and shrubs, iNaturalist observations are unconstrained by substrate and are more likely to capture caterpillars in gardens, on herbaceous vegetation, and even on the ground during a wandering phase. These differences in "sampling" process result in differential representation of Lepidoptera families within the two datasets.

Across both projects, Erebidae was one of the two most commonly represented families in eastern North America, but there were clear differences in family representation between projects as well (Fig. 1a; χ^2 = 320.6, df = 10, p < 2e-16). In the iNaturalist dataset, Nymphalidae, Papilionidae, Sphingidae, and Lasiocampidae were relatively overrepresented compared to Caterpillars Count!, and caterpillars in these families are frequently large, conspicuous, and found in gardens and yards.

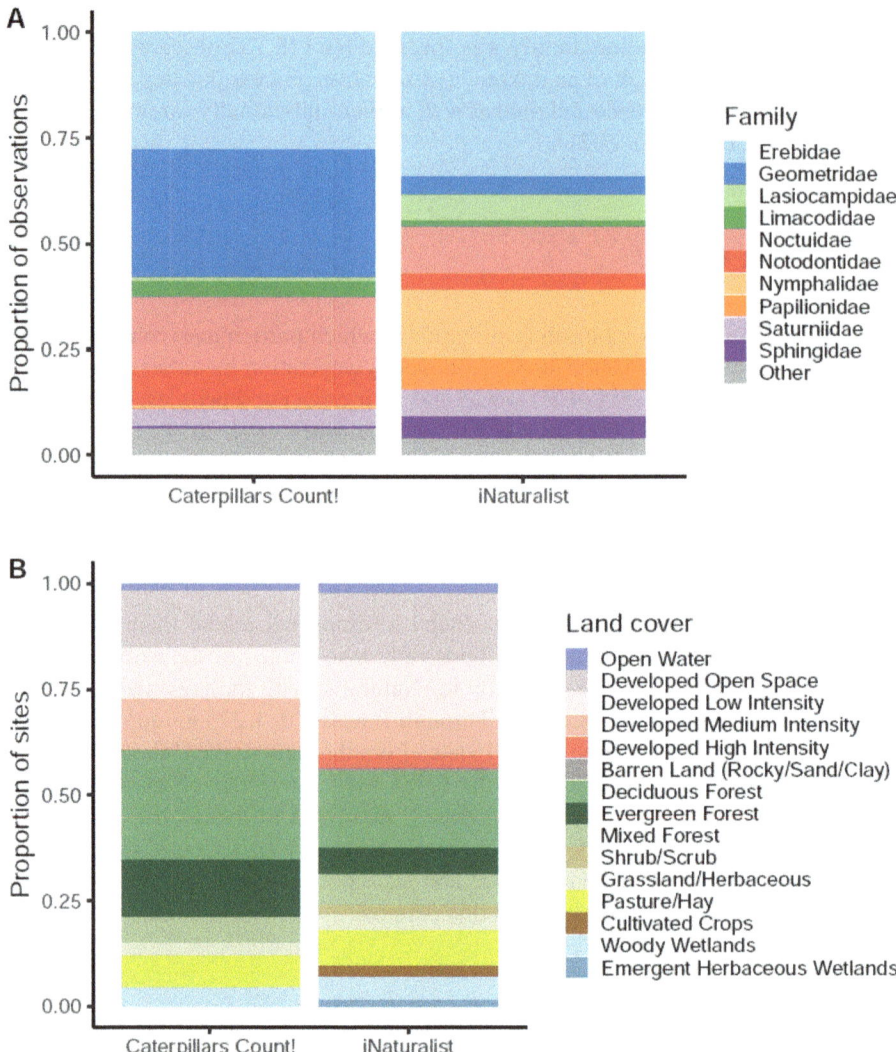

Fig. 1 Relative representation of the (**a**) most common caterpillar families and (**b**) land cover classes represented in the Caterpillars Count! (605 records identified to family, 66 sites) and iNaturalist datasets (178,702 records; 109,939 sites)

Conversely, Geometridae, and to a lesser extent, Notodontidae and Noctuidae were relatively overrepresented in the Caterpillars Count! dataset which was restricted to woody vegetation and which involved people actively searching for caterpillars, many of which are cryptic in form and color (Fig. 1a). To the extent that Caterpillars Count! participants are more likely to submit photos of showy or conspicuous caterpillars, the true proportion of these less conspicuous groups may be even greater.

A comparison of the land cover types within a 75 m radius of each Caterpillars Count! site and at the latitude and longitude of each iNaturalist observation based on data from the 30 m resolution National Land Cover Database from 2016 (Yang et al. 2018) highlights some of the differences in habitats represented between the two datasets (Fig. 1b). Forest habitats are more sampled by Caterpillars Count!, while high-intensity developed areas and open areas including cropland, pasture, grassland, and scrub are more sampled by iNaturalist users (although the small sample size of Caterpillars Count! sites precludes a chi-square test here).

Based on data from Caterpillars Count! as well as a study by Seifert et al. (2020) which examined caterpillar density across 15 common eastern North American tree species, the most commonly encountered families of caterpillars on woody vegetation were Geometridae, Erebidae, Noctuidae, Notodontidae, Depressariidae, and Tortricidae. Even restricting the comparison to these most commonly observed families, Geometridae is still strikingly underrepresented in iNaturalist (8.5% of observations compared to 36% in Caterpillars Count!), highlighting the impact of potential biases in the observation process in opportunistic versus survey-based citizen science datasets.

Geographic Patterns

We examined spatial patterns of caterpillar occurrence during June 2019 in both datasets using a uniform hexagonal grid (distance between cell centers of 285 km; per cell area of approximately 70,000 km^2). During this month-long window, there were 6,380 caterpillar observations submitted to iNaturalist, and we scaled the number of observations per hex cell by the total number of insect observations per hex cell in order to control for spatial variation in iNaturalist activity. During this period there were 3,882 Caterpillars Count! surveys conducted which reported 876 total caterpillars, and we calculated the number of caterpillars observed per survey within each hex cell. Importantly, this snapshot of caterpillar density or occurrence within the month of June may reflect slightly different points in time relative to forest leaf out depending on latitude.

Patterns of caterpillar occurrence were spatially heterogeneous. As a fraction of the total insect observations, iNaturalist caterpillar occurrence in June was highest in the Upper Midwest (Minnesota, Wisconsin, Michigan), as well as in Louisiana and Mississippi, typically making up fewer than 10% of all insect observations. Caterpillar occurrence was lowest across the Midwest (Missouri, Illinois, Indiana, Ohio), the southeastern states east of Mississippi, and easternmost Canada. Within the subset of hex cells that had sufficient Caterpillars Count! surveys, Michigan, Massachusetts, and North Carolina had the highest caterpillars observed per survey branch, and there was a positive correlation between iNaturalist caterpillar occurrence and Caterpillars Count! density estimates (Fig. 2c; $r = 0.729, p = 0.002, n = 15$ hex cells).

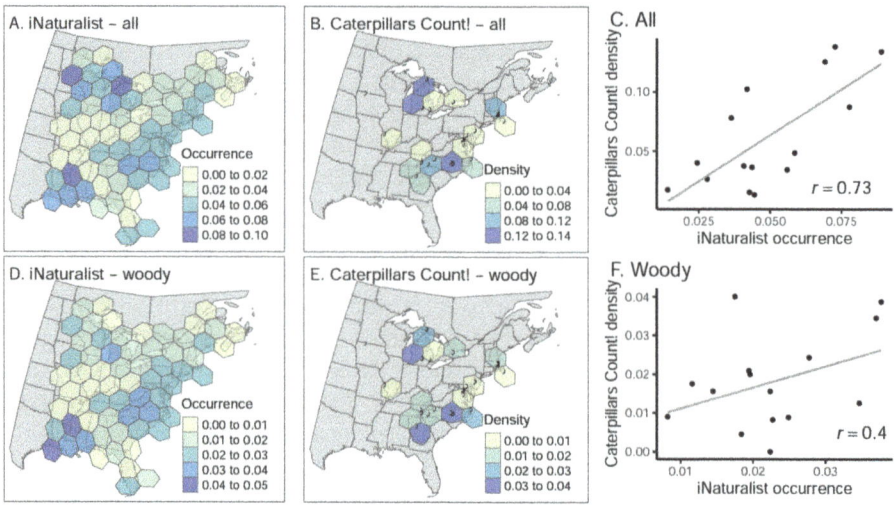

Fig. 2 Geographic variation in June 2019 in (a) iNaturalist caterpillar occurrence (proportion of all insect observations), (b) Caterpillars Count! caterpillar density (the number of caterpillars observed per branch survey), and (c) the relationship between the two at the scale of individual hex cells. (d–f) Same as (a–c) but for the subset of caterpillars from families well represented on woody vegetation (Erebidae, Geometridae, Notodontidae, Noctuidae, Depressariidae, and Tortricidae). The locations of Caterpillars Count! sites with a minimum of 20 surveys conducted in June 2019 are shown as open circles in panels (b) and (e)

Because many iNaturalist observations are presumably from gardens, yards, and nonwoody substrates, we expected that by filtering those observations to the most common families found on woody vegetation that the correlation between the two datasets would become stronger. Limiting analysis to the six most common families named above actually weakened the correlation between iNaturalist and Caterpillars Count! datasets (Fig. 2f; $r = 0.396$, $p = 0.144$, $n = 15$ hex cells). In particular, restricting the analysis to these families reduced iNaturalist caterpillar occurrence more severely than it did Caterpillar Count! density estimates in the Upper Midwest. While still positive, the weakness of this correlation between datasets may be due in part to the reduced number of iNaturalist observations within these focal families, and also highlights the remaining sources of uncertainty in the iNaturalist observation process, including geographic variation in the types of taxa, habitats, and substrates which observers sample.

Caterpillar density and diversity are well known to vary by host plant species (Futuyma and Gould 1979; Tallamy and Shropshire 2009; Singer et al. 2012; Shutt et al. 2019), and so the spatial patterns of caterpillar density in Fig. 2 are necessarily impacted by the plant species on which those observations were made. For the Caterpillars Count! dataset, unlike iNaturalist, we have information on host plant identity, and closer examination of three of the most commonly represented species reveals density patterns that appear to be host plant specific (Fig. 3). For example, on American beech (*Fagus grandifolia*), caterpillar density was highest at lower

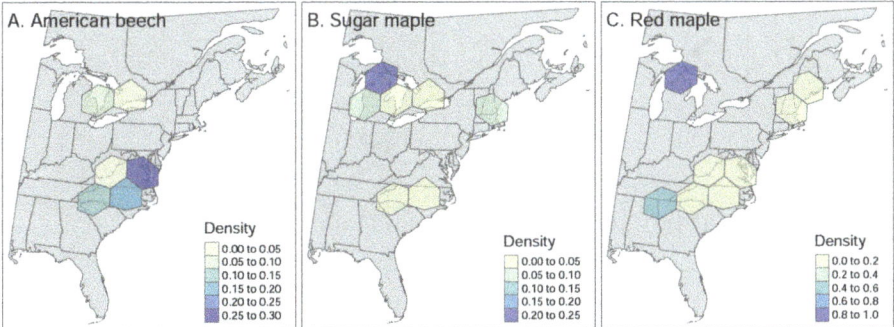

Fig. 3 Host-specific patterns of caterpillar density (number per survey branch) from Caterpillars Count! surveys on (**a**) American beech (*Fagus grandifolia*), (**b**) sugar maple (*Acer saccharum*), and (**c**) red maple (*Acer rubrum*)

latitudes, while on sugar maple (*Acer saccharum*) and red maple (*Acer rubrum*), density was highest at the northernmost site near Sault Ste. Marie, Ontario. Even within regions and on the same host plant, fine-scale gradients in temperature, humidity, and landscape context may lead to variation in caterpillar density (Kendeigh 1979; Jeffries et al. 2006; Reynolds et al. 2007; Smith et al. 2011; Seress et al. 2018), and thus these regional geographic patterns should be interpreted with caution pending more thorough characterization via the addition of more sampling sites within hex cells.

Phenology

Almost all of the work on the phenology of Lepidoptera has focused on modeling individual species and the variance between species (e.g., Diamond et al. 2014; Thorson et al. 2016; Belitz et al. 2020), but from the perspective of bird food, it is caterpillar phenology in aggregate that is of interest, and specifically the phenology during the avian nesting season (e.g., Visser et al. 2006; Lany et al. 2016; Shutt et al. 2019), which varies latitudinally but can broadly be considered to span May through July in North America. Thus, despite the often multimodal nature of aggregate caterpillar phenology from shortly after leaf out through the fall (Fig. 4), we estimate two phenometrics related to the period of peak caterpillar activity that should be most relevant to birds: (1) the peak caterpillar date between May 15 and July 30 and (2) the temporal centroid of caterpillar density or occurrence during this period. The former measure reflects when food resources for birds were ostensibly at their highest point, but may fail to capture shifts in the overall caterpillar distribution if there are multiple peaks of similar magnitude (Fig. 4). The latter measure captures variation in the center of mass of caterpillar observations even when the actual peak date does not vary.

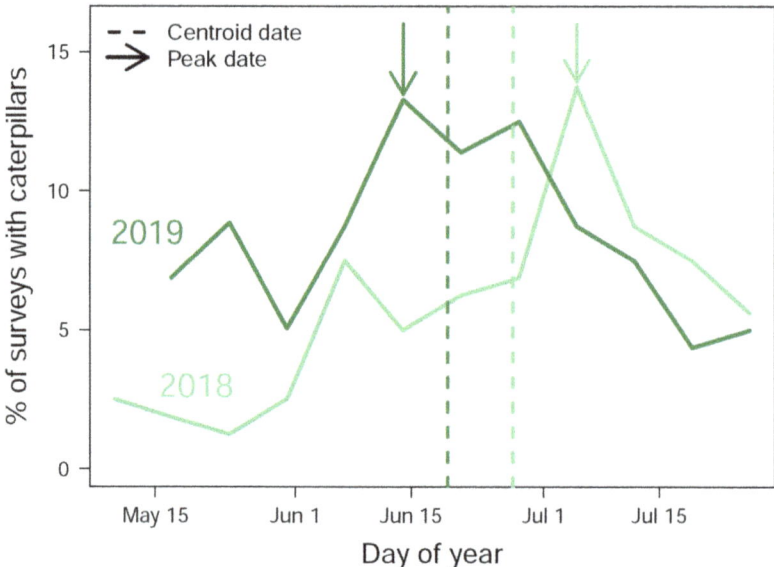

Fig. 4 Caterpillar phenology from Caterpillars Count! surveys at the North Carolina Botanical Garden in 2018 and 2019, illustrating the two phenometrics used, centroid date and peak date

For Caterpillars Count! data in 2019, we used data from all sites that had at least 20 branch surveys over at least 6 distinct sampling weeks spanning May 15 to July 30 and calculated the fraction of surveys detecting a caterpillar on any given date. When there were multiple Caterpillars Count! sites within one hex cell, we calculated the peak or centroid date for each site separately, and then averaged values across sites within a cell. For each hex cell with at least 100 iNaturalist insect observations per week, we calculated caterpillar phenology from iNaturalist data over the 6–10-week time window that matched the period of Caterpillars Count! data collection in that cell. Observations were binned by week to account for a "weekend effect," whereby users are more likely to contribute observations on Saturdays and Sundays (Courter et al. 2013; Di Cecco et al. 2021).

Because caterpillar phenology is tied to leaf out and both are closely related to spring temperatures (van Asch and Visser 2007; Uelmen et al. 2016), we also calculated average spring temperatures in each hex cell from March to June using Daymet climate data at 1 km resolution (Thornton et al. 2019) to investigate how standardized and opportunistic citizen science surveys capture relationships between caterpillar occurrence, density, and phenology and temperature.

Peak caterpillar date was poorly correlated between the datasets (Fig. 5a–c, $r = 0.330$, $p = 0.352$, $n = 10$ hex cells). The correlation was slightly stronger for centroid date (Fig. 5d–f, $r = 0.472$, $p = 0.169$, $n = 10$ hex cells), which also exhibited smoother variation with latitude (later dates at higher latitudes) than peak date. These results suggest that peak date in particular may be an overly sensitive phenometric, at least when estimated from an underlying distribution that is often

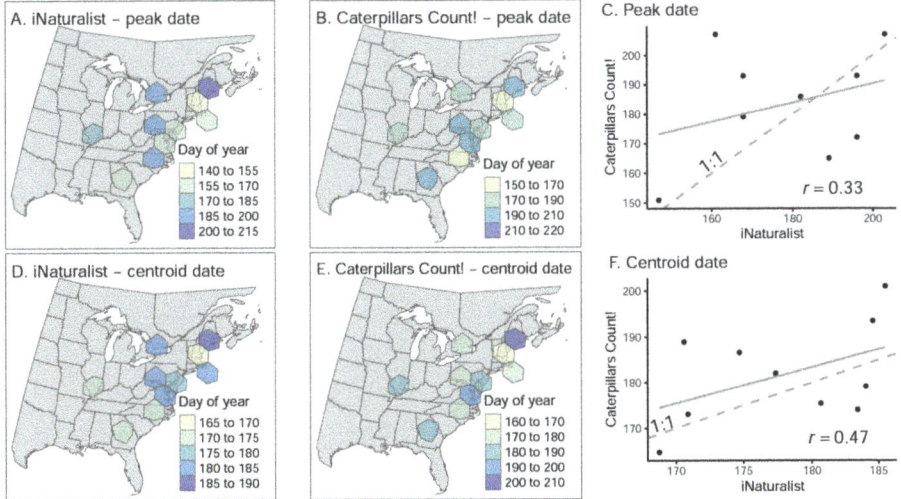

Fig. 5 Geographic variation in caterpillar peak date in 2019 based on data from (**a**) iNaturalist and (**b**) Caterpillars Count!. (**c**) The relationship between peak date estimates from the two datasets. Geographic variation in caterpillar centroid date in 2019 based on data from (**d**) iNaturalist and (**e**) Caterpillars Count!. (**f**) The relationship between centroid date estimates from the two datasets

multimodal and from a time series of only 6–10 weekly data points. Peak date is also expected to be more sensitive to single species outbreaks which may vary from year to year and region to region. Centroid date is a more promising phenometric, however, and one for which opportunistic and heterogeneous iNaturalist data may actually convey useful information about the geographic signal of forest caterpillar phenology.

Although the geographic variation in centroid date with latitude may be due, at least in part, to variation in spring temperature, phenometrics might also vary from cell to cell due to regional differences in the habitats, host plants, and Lepidoptera fauna sampled, and the nature of this cell-to-cell variation may differ in magnitude between iNaturalist and Caterpillars Count!. As such, it is perhaps not surprising that these two datasets exhibit different relationships between phenometrics, latitude, and spring temperature (e.g., centroid date-temperature relationship: iNaturalist, $R^2 = 0.223$, $p = 0.168$; Caterpillars Count!, $R^2 = 0.009$, $p = 0.791$, $n = 10$ hex cells). However, a comparison of phenometrics between years within the same hex cell is more likely to hold some of these sources of heterogeneity constant, and thus we might expect iNaturalist and Caterpillars Count! data to agree on the extent to which a year is early or late for a specific hex cell. As such, we identified all of the hex cells with sufficient Caterpillars Count! and iNaturalist data for 2 or more years from 2015 to 2019. For each dataset-cell-year and for each phenometric (peak date and centroid date), we calculated the difference between the phenometric in that year and the mean phenometric across years within that cell. All phenometrics within each dataset were thus represented as deviations from that cell's mean. We

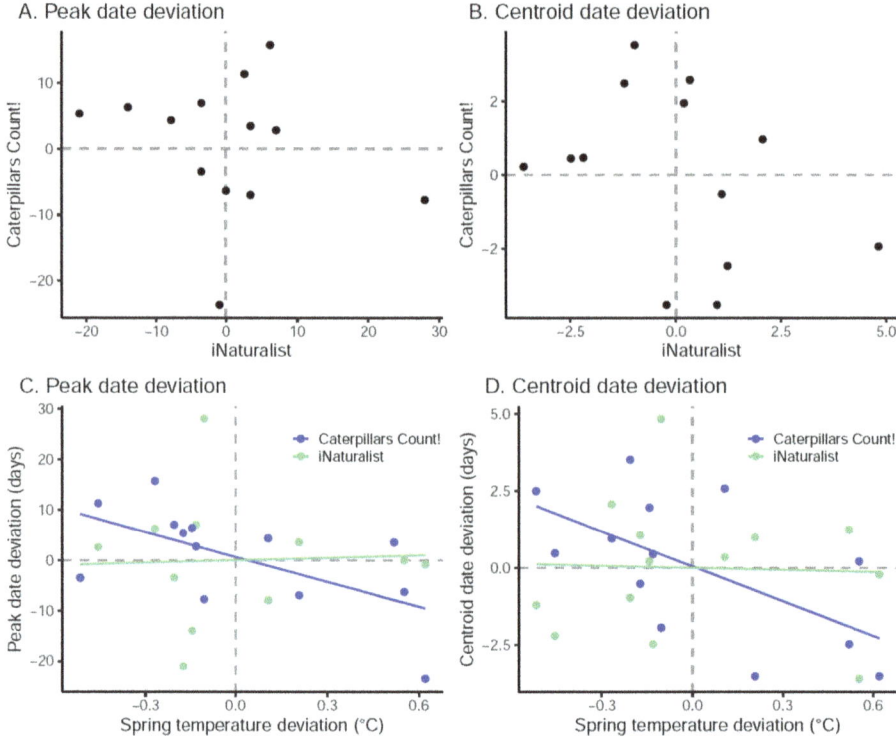

Fig. 6 The relationship between interannual deviations in phenology estimates between iNaturalist and Caterpillars Count! based on (**a**) peak date and (**b**) centroid date. (**c**) Peak date deviations and (**d**) centroid date deviations in caterpillar density as a function of interannual deviation in spring (March through June) temperature for iNaturalist (green) and Caterpillars Count! (blue)

were then able to ask whether there was agreement between iNaturalist and Caterpillars Count! in which years were early or late relative to that mean, as well as how interannual deviations in caterpillar phenology were related to interannual deviations in spring temperature.

Unfortunately, there was effectively no relationship between interannual deviations in either phenometric between the systematic monitoring of Caterpillars Count! and the opportunistic observations reported to iNaturalist (peak date, Fig. 6a, $r = -0.217$, $p = 0.477$; centroid date, Fig. 6b, $r = -0.354$, $p = 0.235$, $n = 13$ hex cell years). Within the Caterpillars Count! dataset, years with relatively early caterpillar phenology were also warmer, while years with later phenology were cooler (peak date, Fig. 6c, $R^2 = 0.367$, $p = 0.028$; centroid date, Fig. 6d, $R^2 = 0.364$, $p = 0.029$; blue lines, $n = 13$ hex cell years). While similar in variance explained, the parameter estimates differed substantially for the two phenometrics. A shift in centroid date by ~4 days per degree warming indicates the impact of temperature on the caterpillar fauna as a whole. In contrast, because peak date may be more sensitive to individual species outbreaks, a shift in peak date of ~17 days per degree suggests that a seasonally distinct caterpillar fauna is achieving higher abundances under warming than the fauna present during the peak period in cooler years (e.g., Fig. 4), either through direct effects of temperature or indirectly through phenological shifts in synchrony with host plants. Conversely, phenometric deviations within the iNaturalist dataset showed no relationship to interannual temperature deviations (peak date, Fig. 6c, $R^2 = 0.003$, $p = 0.868$; centroid date, Fig. 6d, $R^2 = 0.001$, $p = 0.902$; green lines, $n = 13$ hex cell years).

These results suggest that the observation process by which iNaturalist records are obtained has too many sources of variation to reliably characterize interannual variation in aggregate caterpillar phenology at the scale of these hex cells. Sources of variation include the numbers, spatial distribution, temporal distribution, and habitat representation of observations, phenological variation in the component caterpillar species among which there are early, late, and multivoltine species, as well as other aspects regarding the intent and behavior of individual iNaturalist users (e.g., taxonomic biases, activity of "superusers," etc., Di Cecco et al. 2021). Analyzing the phenology of individual Lepidoptera species with these types of data is more promising (Belitz et al. 2020), and as the iNaturalist dataset continues to grow exponentially, joint dynamic species distribution models and related approaches may improve the ability to detect signal from noise in multispecies assemblages (Thorson et al. 2016). At present, however, modeling caterpillar phenology in aggregate may require more standardized monitoring approaches like that used by the Caterpillars Count! project (Hurlbert et al. 2019).

Climate Change and Phenological Mismatch

Many organisms that depend directly or indirectly on temperature have been observed to shift phenology earlier in recent decades (Parmesan and Yohe 2003; Hurlbert and Liang 2012; Cook et al. 2012), including adult Lepidoptera (Kharouba et al. 2014; Diamond et al. 2014). The dependence of aggregate caterpillar phenology (as opposed to the phenology of any one species) on spring temperature (sometimes indirectly via the timing of green-up) has previously been shown at individual study sites with just one or two dominant caterpillar species (e.g., Visser et al. 2006; Burgess et al. 2018), but we show here that this applies across a much greater geographical extent spanning a much more diverse lepidopteran fauna. Current global climate change scenarios predict continued warming of anywhere from 1.7 to 4.8 °C globally by 2100 (Collins et al. 2013), suggesting that caterpillar phenology will continue to shift earlier. Such phenological shifts are expected both from the direct control of temperature on developmental rates (Knapp and Casey 1986; Gillooly et al. 2002) and from selection imposed by the earlier leaf out of vegetation and hence the earlier incorporation of secondary compounds that reduce leaf quality for herbivores later in the season (Feeny 1970; Martel and Kause 2002). The actual strength of selection for earlier phenology may depend on the relative degree of plasticity in responses by caterpillars and their host plants to changing temperature.

Continued shifts in caterpillar phenology, as well as potentially increased variability in phenology due to an increased frequency in extreme climatic events (Collins et al. 2013), may have negative consequences for organisms from higher trophic levels that depend heavily on caterpillars as a food resource. In particular, many species of foliage-gleaning birds have been found to rely heavily on caterpillars for raising their young (Holmes et al. 1979; Holmes and Schultz 1988; Sillett et al. 2000; Jones et al. 2003) and could face potentially reduced reproductive success if the reproductive period of high nestling resource demand does not shift in parallel with caterpillars. Negative fitness consequences have been demonstrated as a result of such phenological mismatch for resident great tits (*Parus major*) in the Netherlands (Visser et al. 2006; Reed et al. 2013), which have not advanced the timing of breeding as much as the shift in peak caterpillar date. Phenological mismatch is of even greater concern for long-distance migrants which have shown less sensitivity to interannual variation in breeding ground conditions compared to residents and short-distance migrants (Saino et al. 2011; Hurlbert and Liang 2012; Youngflesh et al. 2021) and which are presumably less able to accurately assess those conditions in distant breeding areas.

Several challenges confront research exploring the consequences of phenological mismatch between caterpillars and birds. First, the sensitivity of phenological responses appears to vary between species for both Lepidoptera (Kharouba et al. 2014; Diamond et al. 2014) and birds (Saino et al. 2011; Mayor et al. 2017), and also geographically within species (Hurlbert and Liang 2012; Youngflesh et al. 2021). This means that caterpillar phenology in aggregate may be difficult to predictively model without taking into account the individual responses of common species and species prone to outbreaks, which will vary regionally. Coupled with regional variation in avian sensitivity, an accurate understanding of phenological mismatch across a bird's geographic range will require the integration of avian and lepidopteran datasets spanning large extents and fine temporal resolution to fully unravel.

Examining the consequences of phenological mismatch also requires being able to quantify mismatch in a meaningful way. While the majority of studies that have compared caterpillar phenology with avian nesting phenology have focused on comparing shifts and differences between peak dates (e.g., Visser et al. 2006; Hinks et al. 2015), the height and width of caterpillar phenology curves may be equally or more important to birds than the timing alone (Shutt et al. 2019). Visser et al. (2015) found that an interaction between the height and timing of the caterpillar peak determined the number of nestlings fledged, with stronger seasonal selection during years with lower caterpillar peaks. Integrating the availability of high-value caterpillar prey over the entire nestling period may be the most relevant metric for predicting avian reproductive success, and comparing the integrated value of caterpillar availability during that period to the amount of nestlings would have experienced had the parents shifted reproduction earlier or later could provide a useful measure of phenological match. Both of these measures merit further research, especially with regard to whether and how the large volumes of existing opportunistic data might be integrated with information from structured surveys.

Conclusions

Caterpillars play a central role in forest ecosystems as both herbivores and as a food source for consumers, and citizen science datasets present an increasingly useful resource for understanding spatial and temporal patterns of caterpillar occurrence, abundance, and diversity. Projects like iNaturalist that consist of opportunistic photo observations have proven useful for mapping species distributions (Chardon et al. 2015; Fourcade 2016; Feldman et al. 2020) and for modeling the phenology of individual species (Taylor and Guralnick 2019; Barve et al. 2020). Estimates of caterpillar density, however, require knowledge of the total survey effort expended, and therefore are best obtained through standardized sampling protocols like those of the Caterpillars Count! project that report absences as well as presences. Nevertheless, we found that by scaling caterpillar observations by the total number of insect observations, iNaturalist was still able to recapitulate some of the geographic variation in density observed in the Caterpillars Count! dataset. As these citizen science data collection efforts continue into the future, they will provide a critical means of assessing abundance trends and the impacts of climate and land use change.

Finally, estimates of phenological timing of caterpillars in aggregate were less well correlated between the datasets across both space and time. Until methods are developed to better understand and deal with the sources of uncertainty and bias in the sampling process underlying opportunistic datasets like iNaturalist, geographically broadscale attempts to estimate phenological mismatch between caterpillars, and their avian predators will need to rely heavily on more systematic monitoring efforts.

Acknowledgments We thank the thousands of volunteers and amateur naturalists whose observations shared with citizen science projects like iNaturalist and Caterpillars Count! have enhanced scientific understanding of caterpillar diversity, abundance, and distribution. We also thank M. Singer, J. Forrest, and one anonymous reviewer for comments on an earlier draft of this chapter, and R. Marquis and S. Koptur for the invitation to contribute to this volume.

Literature Cited

Barve VV, Brenskelle L, Li D et al (2020) Methods for broad-scale plant phenology assessments using citizen scientists' photographs. Appl Plant Sci 8:e11315. https://doi.org/10.1002/aps3.11315

Belitz MW, Larsen EA, Ries L, Guralnick RP (2020) The accuracy of phenology estimators for use with sparsely sampled presence-only observations. Methods Ecol Evol 11:1273–1285. https://doi.org/10.1111/2041-210X.13448

Burgess MD, Smith KW, Evans KL et al (2018) Tritrophic phenological match–mismatch in space and time. Nat Ecol Evol 2:970–975. https://doi.org/10.1038/s41559-018-0543-1

Chardon NI, Cornwell WK, Flint LE et al (2015) Topographic, latitudinal and climatic distribution of *Pinus coulteri*: geographic range limits are not at the edge of the climate envelope. Ecography 38:590–601. https://doi.org/10.1111/ecog.00780

Collins M, Knutti R, Arblaster J, et al (2013) Long-term climate change: projections, commitments and irreversibility. In: Climate change 2013-the physical science basis: contribution of working group I to the Fifth Assessment report of the intergovernmental panel on climate change. Cambridge University Press, pp 1029–1136

Cook BI, Wolkovich EM, Davies TJ et al (2012) Sensitivity of spring phenology to warming across temporal and spatial climate gradients in two independent databases. Ecosystems 15:1283–1294. https://doi.org/10.1007/s10021-012-9584-5

Courter JR, Johnson RJ, Stuyck CM et al (2013) Weekend bias in citizen science data reporting: implications for phenology studies. Int J Biometeorol 57:715–720. https://doi.org/10.1007/s00484-012-0598-7

Diamond SE, Cayton H, Wepprich T et al (2014) Unexpected phenological responses of butterflies to the interaction of urbanization and geographic temperature. Ecology. https://doi.org/10.1890/13-1848.1

Di Cecco GJ, Barve V, Belitz MW, Stucky BJ, Guralnick RP, Hurlbert AH (2021) Observing the observers: how participants contribute data to iNaturalist and implications for biodiversity science. BioScience biab093. https://doi.org/10.1093/biosci/biab093

Feeny P (1970) Seasonal changes in oak leaf tannins and nutrients as a cause of spring feeding by winter moth caterpillars. Ecology 51:565–581. https://doi.org/10.2307/1934037

Feldman MJ, Imbeau L, Marchand P, et al (2020) Trends and gaps in the use of citizen science derived data as input for species distribution models: a quantitative review. bioRxiv 2020.06.01.127415. https://doi.org/10.1101/2020.06.01.127415

Fourcade Y (2016) Comparing species distributions modelled from occurrence data and from expert-based range maps. Implication for predicting range shifts with climate change. Ecol Inform 36:8–14. https://doi.org/10.1016/j.ecoinf.2016.09.002

Futuyma DJ, Gould F (1979) Associations of plants and insects in deciduous forest. Ecol Monog 49:33–50. https://doi.org/10.2307/1942571

Gillooly JF, Charnov EL, West GB et al (2002) Effects of size and temperature on developmental time. Nature 417:70–73

Hinks AE, Cole EF, Daniels KJ et al (2015) Scale-dependent phenological synchrony between songbirds and their caterpillar food source. Am Nat 186:84–97. https://doi.org/10.1086/681572

Holmes RT, Schultz JC (1988) Food availability for forest birds: effects of prey distribution and abundance on bird foraging. Can J Zool 66:720–728

Holmes RT, Schultz JC, Nothnagle P (1979) Bird predation on forest insects: an exclosure experiment. Science 206:462–463. https://doi.org/10.2307/1749326

Hurlbert AH, Liang Z (2012) Spatiotemporal variation in avian migration phenology: citizen science reveals effects of climate change. PLoS ONE 7:e31662. https://doi.org/10.1371/journal.pone.0031662

Hurlbert A, Hayes T, McKinnon T, Goforth C (2019) Caterpillars Count! A citizen science project for monitoring foliage arthropod abundance and phenology. Citizen Sci Theory Practice 4:1. https://doi.org/10.5334/cstp.148

Isaac NJB, Pocock MJO (2015) Bias and information in biological records. Biol J Linn Soc 115:522–531. https://doi.org/10.1111/bij.12532

Jeffries JM, Marquis RJ, Forkner RE (2006) Forest age influences oak insect herbivore community structure, richness, and density. Ecol Appl 16:901–912

Jones J, Doran PJ, Holmes RT (2003) Climate and food synchronize regional forest bird abundances. Ecology 84:3024–3032. https://doi.org/10.2307/3449971

Kendeigh SC (1979) Invertebrate populations of the deciduous forest: fluctuations and relations to weather. University of Illinois Press

Kharouba HM, Paquette SR, Kerr JT, Vellend M (2014) Predicting the sensitivity of butterfly phenology to temperature over the past century. Glob Change Biol 20:504–514. https://doi.org/10.1111/gcb.12429

Knapp R, Casey TM (1986) Thermal ecology, behavior, and growth of gypsy moth and eastern tent caterpillars. Ecology 67:598–608. https://doi.org/10.2307/1937683

Lany NK, Ayres MP, Stange EE et al (2016) Breeding timed to maximize reproductive success for a migratory songbird: the importance of phenological asynchrony. Oikos 125:656–666. https://doi.org/10.1111/oik.02412

Mair L, Ruete A (2016) Explaining spatial variation in the recording effort of citizen science data across multiple taxa. PLOS ONE 11:e0147796. https://doi.org/10.1371/journal.pone.0147796

Martel J, Kause A (2002) The phenological window of opportunity for early-season birch sawflies. Ecol Entomol 27:302–307. https://doi.org/10.1046/j.1365-2311.2002.00418.x

Mayor SJ, Guralnick RP, Tingley MW et al (2017) Increasing phenological asynchrony between spring green-up and arrival of migratory birds. Sci Rep 7:1–10. https://doi.org/10.1038/s41598-017-02045-z

Nelson G, Ellis S (2018) The history and impact of digitization and digital data mobilization on biodiversity research. Philos Trans R Soc Lond B Biol Sci 374. https://doi.org/10.1098/rstb.2017.0391

Parmesan C, Yohe G (2003) A globally coherent fingerprint of climate change impacts across natural systems. Nature 421:37–42. https://doi.org/10.1038/nature01286

Prysby M, Oberhauser KS (2004) Temporal and geographical variation in monarch densities: citizen scientists document monarch population patterns. In: Oberhauser KS, Solensky MJ (eds) Monarch butterfly biology and conservation. Cornell University Press, Ithaca, pp 9–20

Reed TE, Jenouvrier S, Visser ME (2013) Phenological mismatch strongly affects individual fitness but not population demography in a woodland passerine. J Anim Ecol 82:131–144. https://doi.org/10.1111/j.1365-2656.2012.02020.x

Reynolds LV, Ayres MP, Siccama TG, Holmes RT (2007) Climatic effects on caterpillar fluctuations in northern hardwood forests. Can J For Res 37:481–491. https://doi.org/10.1139/X06-211

Ries L, Zipkin EF, Guralnick RP (2019) Tracking trends in monarch abundance over the 20th century is currently impossible using museum records. PNAS 116:13745–13748. https://doi.org/10.1073/pnas.1904807116

Saino N, Ambrosini R, Rubolini D et al (2011) Climate warming, ecological mismatch at arrival and population decline in migratory birds. Proc R Soc B Biol Sci 278:835–842. https://doi.org/10.1098/rspb.2010.1778

Seifert CL, Lamarre GPA, Volf M et al (2020) Vertical stratification of a temperate forest caterpillar community in eastern North America. Oecologia 192:501–514. https://doi.org/10.1007/s00442-019-04584-w

Seltzer C (2019) Making biodiversity data social, shareable, and scalable: reflections on iNaturalist and citizen science. Biodivers Inf Sci Stand 3:e46670. https://doi.org/10.3897/biss.3.46670

Seress G, Hammer T, Bokony V et al (2018) Impact of urbanization on abundance and phenology of caterpillars and consequences for breeding in an insectivorous bird. Ecol Appl 28. https://doi.org/10.1002/eap.1730

Shutt JD, Burgess MD, Phillimore AB (2019) A spatial perspective on the phenological distribution of the spring woodland caterpillar peak. Am Nat 194:E109–E121. https://doi.org/10.1086/705241

Sillett TS, Holmes RT, Sherry TW (2000) Impacts of a global climate cycle on population dynamics of a migratory songbird. Science 288:2040–2042. https://doi.org/10.1126/science.288.5473.2040

Singer MS, Farkas TE, Skorik CM, Mooney KA (2012) Tritrophic interactions at a community level: effects of host plant species quality on bird predation of caterpillars. Am Nat 179:363–374. https://doi.org/10.1086/664080

Smith KW, Smith L, Charman E et al (2011) Large-scale variation in the temporal patterns of the frass fall of defoliating caterpillars in oak woodlands in Britain: implications for nesting woodland birds. Bird Study 58:506–511. https://doi.org/10.1080/00063657.2011.616186

Soroye P, Ahmed N, Kerr JT (2018) Opportunistic citizen science data transform understanding of species distributions, phenology, and diversity gradients for global change research. Glob Change Biol 24:5281–5291. https://doi.org/10.1111/gcb.14358

Tallamy DW, Shropshire KJ (2009) Ranking lepidopteran use of native versus introduced plants. Conserv Biol 23:941–947. https://doi.org/10.1111/j.1523-1739.2009.01202.x

Taylor SD, Guralnick RP (2019) Opportunistically collected photographs can be used to estimate large-scale phenological trends. bioRxiv 794396. https://doi.org/10.1101/794396

Telenius A (2011) Biodiversity information goes public: GBIF at your service. Nord J Bot 29:378–381. https://doi.org/10.1111/j.1756-1051.2011.01167.x

Thornton PE, Thornton MM, Mayer BW, et al (2019) Daymet: daily surface weather data on a 1-km grid for North America, Version 3. ORNL DAAC. https://doi.org/10.3334/ORNLDAAC/1328

Thorson JT, Ianelli JN, Larsen EA et al (2016) Joint dynamic species distribution models: a tool for community ordination and spatio-temporal monitoring. Global Ecology and Biogeography 25:1144–1158. https://doi.org/10.1111/geb.12464

Uelmen JA, Lindroth RL, Tobin PC et al (2016) Effects of winter temperatures, spring degree-day accumulation, and insect population source on phenological synchrony between forest tent caterpillar and host trees. For Ecol Manag 362:241–250. https://doi.org/10.1016/j.foreco.2015.11.045

van Asch M, Visser ME (2007) Phenology of forest caterpillars and their host trees: the importance of synchrony. Annu Rev Entomol 52:37–55. https://doi.org/10.1146/annurev.ento.52.110405.091418

Visser M, Holleman L, Gienapp P (2006) Shifts in caterpillar biomass phenology due to climate change and its impact on the breeding biology of an insectivorous bird. Oecologia 147:164–172. https://doi.org/10.1007/s00442-005-0299-6

Visser ME, Gienapp P, Husby A et al (2015) Effects of spring temperatures on the strength of selection on timing of reproduction in a long-distance migratory bird. PLOS Biol 13:e1002120. https://doi.org/10.1371/journal.pbio.1002120

Yang L, Jin S, Danielson P et al (2018) A new generation of the United States National Land Cover Database: requirements, research priorities, design, and implementation strategies. ISPRS J Photogramm Remote Sens 146:108–123. https://doi.org/10.1016/j.isprsjprs.2018.09.006

Youngflesh C, Socolar J, Amaral BR, Arab A, Guralnick RP, Hurlbert AH, LaFrance R, Mayor SJ, Miller DAW, Tingley MW (2021) Migratory strategy drives species-level variation in bird sensitivity to vegetation green-up. Nat Ecol Evol 5:987–994. https://doi.org/10.1038/s41559-021-01442-y

Impacts of Climatic Variability and Hurricanes on Caterpillar Diet Breadth and Plant-Herbivore Interaction Networks

Karina Boege, Ivonne P. Delgado, Jazmin Zetina, and Ek del-Val

Caterpillars of *Syssphinx* sp. (Saturniidae) (photo by Antonio López-Carretero, above); undetermined species of Aididae (photo by Juan Pablo Martinez, below)

Authors Karina Boege and Ek del-Val contributed equally in this chapter.

K. Boege · J. Zetina
Departamento de Ecología Evolutiva, Instituto de Ecología, Universidad Nacional Autónoma de México, Ciudad de México, Mexico

I. P. Delgado · E. del-Val (✉)
Instituto de Investigaciones en Ecosistemas y Sustentabilidad, Universidad Nacional Autónoma de México, Morelia, Michoacán, Mexico
e-mail: ekdelval@cieco.unam.mx

Introduction

Assessing the effects of climatic variability and meteorological extreme events on insect populations has been useful to understand the impacts that climate change can have on ecosystem processes and biodiversity conservation. Available evidence suggests that variations in temperature, precipitation, and extreme events such as hurricanes or acute droughts can affect nutrient cycling, species abundance and composition of biodiverse insect communities (Forkner et al. 2008; Lister and Garcia 2018; García-Robledo et al. 2016; Marquis et al. 2019), the strength of species interactions (Harrington et al. 1999), and the resilience of their networks (López-Carretero et al. 2014; Luviano et al. 2018). What has been less explored is how insects can respond and adapt to such environmental challenges, for example, by modifying their feeding habits and diet breadth, and thus their degree of specialization. In this chapter, we describe how the most abundant and oligotrophic caterpillars of a tropical dry forest show variability in their diet breadths, as a function of interannual variation in climatic variables and the incidence of hurricanes, influencing in turn the parameters of plant-herbivore interaction networks.

Given their evolutionary arms race with plants, Lepidopteran immature stages are specialized to feed on particular plant groups sharing the same chemical profiles (Bernays 2001), and this is particularly evident in the tropics due to the great number of plant species and phytochemical diversity (Forister et al. 2012; Forister et al. 2015). Trophic specialization can be advantageous in terms of efficiency of host plant choice and discrimination for oviposition decisions (Bernays 2001) and shorter times in host plant acceptance (Bernays 1998). Hence, specialized individuals can have greater fitness than generalist individuals under particular contexts (Egan and Funk 2006). However, broader diets or facultative switches to feed on different host plants can allow herbivores to cope with environmental variation in the availability and quality of their food plants (Rodrigues and Moreira 2004) and different sorts of human perturbation factors (Singer et al. 2008; Singer and Parmesan 2020). Diet breadth has also been related to antipredator mechanisms, where specialist caterpillars are better protected than generalists (Dyer 1995), although they are also more susceptible to parasitoids due to compromises in their immune systems (Smilanich et al. 2009). Moreover, mixing diets in generalist species has been found to increase herbivore fitness (Mody et al. 2007). This ecological flexibility can partially explain the intra- and interspecific variation in herbivore feeding degree of specialization observed in particular contexts. For example, feeding specialization of caterpillars can change through their development (Gaston et al. 1991, Hwang et al. 2007, but see Karowe 1989), across seasons (Rodrigues and Moreira 2004; López Carretero et al. 2018; Scherrer et al. 2016), among populations (Singer et al. 1989), or biogeographic regions (Dyer et al. 2007; Scriber et al. 2010; Forister et al. 2015).

Climatic Variability

Insect herbivores have been found to be particularly sensitive to climatic variability, as their abundance and species diversity can be affected by changes in temperature and precipitation. In particular, temperature can directly alter survival rates, development time, and reproductive success of caterpillars (Bale et al. 2002; Dewer and Watt 1992; Abarca and Lill 2015). Indirectly, both temperature and precipitation can also influence the availability and quality of plants as food for their consumers. For example, indirect effects of temperature occur through its influence in the synchronization between insect and plant phenologies (van Ash and Visser 2007), and hence the availability of food for recently emerged caterpillars (Marquis et al. 2019; Renner 2018). Depending on insect herbivore feeding and life cycle strategies, this can strongly influence the number of generations that multivoltine species can complete in a growing season (Hodkinson 1997; Bale et al. 2002) and thus alter population dynamics and population outbreaks.

Even when tissues of host plants are available, their quality can be strongly modified by variations in temperature and precipitation. It has been well documented, for example, that factors limiting plant growth, such as nutrient availability or water stress, can influence the production of plant defenses (Stamp 2003). Hence, variations in water availability can affect host plant quality and herbivore food preferences, indirectly influencing plant-herbivore networks (López-Carretero et al. 2014, 2018). Overall, climatic variability associated with global change can influence the evolutionary responses of herbivores and the persistence of their populations (Lister and García 2018; Marquis et al. 2019).

Diet Breadth of Caterpillars and Its Influence on Plant-Herbivore Networks

As a result of their evolutionary history, most insect herbivores are mainly specialized to feed on few host plant species, genera, or families with particular groups of secondary metabolites (Loxdale and Harvey 2016). However, there is a continuous gradient in the amount of host species they can feed on, and many herbivores can show some degree of plasticity in their feeding habits across environmental gradients (Forister et al. 2012). Hence, the level of herbivore specialization and ecological flexibility are likely to influence how populations are affected by climate variability. Herbivore species situated toward the generalist end of the gradient usually show high phenotypic plasticity and are more likely to adapt to fast climate change, although this also depends on the span of their life cycles and the speed of their evolutionary responses in a changing environment (Bale et al. 2002).

Whereas the effects of temperature on herbivore insects associated with climate change have been predicted to be stronger in polar and temperate regions, alterations of rainfall regimes may become of greater relevance in the tropics (Bale et al.

2002). Rainfall seasonality has already been reported as a climatic factor strongly influencing plant-herbivore interactions in tropical ecosystems. In particular, in tropical semi-deciduous and deciduous forests, herbivores are more specialized during the rainy seasons, with greater availability of resources than in the dry season, when food plants are either scarcer or of lower quality (López-Carretero et al. 2014, 2018; Scherrer et al. 2016). In addition, precipitation variability can alter the ability of parasitoids to track herbivore populations, and this is likely to influence herbivore outbreaks mediated by the absence of top-down forces controlling their populations (Stireman et al. 2005). Variation in the number of plant species used by herbivores as food can in turn affect plant-herbivore interaction networks. For example, in a semi-deciduous tropical forest, increased selectivity by herbivores during the rainy season resulted in increased modularity and reduced connectance of the plant-herbivore networks, compared to those parameters during the dry season (López-Carretero et al. 2014).

Extreme meteorological events can also have important impacts on insect communities and alter their interaction with plants. In particular, hurricanes are a fundamental part of tropical ecosystem dynamics and influence local and regional biodiversity (Manson and Jardel 2009; Pickett and White 1985; Showalter et al. 2017). These events can produce enough physical damage and change plant architecture, resource availability, and hence availability of niches for insect herbivores and predators (Walker and Willig 1999). Most studies, however, have focused on the assessment of hurricane effects on plants and vertebrates, and there is less evidence on the impacts of these meteorological events on insect herbivores (but see Hunter and Forkner 1999; Schowalter et al. 2017) and particularly, on the interactions they maintain with their food plants (Hunter and Forkner 1999; Luviano et al. 2018).

Overall, the ability of herbivores to switch hosts as a function of climate-driven changes in the availability and quality of their host plants can have a relevant adaptive value. A question that remains unexplored is how long-term interannual variation in climatic factors and extreme meteorological events may affect both herbivore's diet breadth and as a consequence, plant-herbivore networks. This information could provide some highlights to predict how this important group of insects may be affected by ongoing climate change. In this work, we assessed the influence of variation in precipitation and temperature within and across 11 years, and the incidence of two hurricanes on diet breadth of the most common generalist caterpillar species across three successional stages of the tropical dry forest of Chemela, Jalisco, Mexico.

Methods

Study Site

Our study was conducted in western Mexico in the surroundings of the Chamela-Cuixmala Biosphere Reserve (CCBR, 19°22′–19°39' N, 104°56′–105°10' W) between 2007 and 2018. The biosphere reserve consists of 13,142 ha of conserved land, and it is surrounded by a mosaic of secondary forests, induced cattle grazing pastures, and agricultural fields (Sánchez-Azofeifa et al. 2009). The vegetation within the reserve primarily consists of tropical dry forests (1149 plant species with an average canopy height of 6 m) and semi-deciduous forest established along larger streams (average canopy height of 10 m; Lott et al. 1987). The TDF found at the Chamela-Cuixmala region is considered one of the most diverse of its kind, with 1200 plant species and high levels of endemism (Lott et al. 1987; Trejo and Dirzo 2000). The invertebrate inventory is still very limited; 1877 arthropod species have been described in the reserve, 583 of which are lepidopterans (Pescador-Rubio et al. 2002). The average annual rainfall is 795.3 mm but varies greatly from year to year (from 366 to 1329 mm) and is mainly concentrated (87%) within the rainy season from June to October (Maass et al. 2018).

Hurricanes

On October 10, 2011, Hurricane Jova (*Category 2*, Saffir-Simpson scale) struck the coasts of Jalisco and Colima. During the hurricane, maximum wind intensities reached 205 km/h and lasted 168 h, providing 187.9 mm of precipitation in 2 days at the Chamela Biological Station (Luviano et al. 2018). Four years afterward, Hurricane Patricia (*Category 5*, Saffir-Simpson scale) also struck the coasts of Colima, Jalisco, Michoacán, and Nayarit in Mexico. It was categorized as the most intense tropical cyclone, reaching a sustained total wind of 325 km/h and the most intense observed in the western hemisphere (872 hPa) (Secretaría de Gobernación 2017).

Sampling Protocol

Between 2007 and 2018, we sampled lepidopteran larvae every year in nine plots corresponding to the experimental design of the CIECO-UNAM Tropical Forest Management project (MABOTRO). This project features a successional chronosequence of abandoned agricultural fields and cattle ranching pastures (Avila-Cabadilla et al. 2009). The plots we used correspond to a chronosequence of secondary succession, sampling plots of early secondary forest (6–9 years of abandonment at the beginning of the study), late secondary forest (13–16 years of abandonment at the beginning of the study), and mature forest, which have not been disturbed for more than 60 years. There were three replicates for each successional stage. Plot size was at least 1 ha (the mature forest plots are immersed in the biosphere reserve, hence are in fact larger), and the minimum distance between plots of the same successional stage was 3 km. Within each plot, in a defined area of 20 × 50 m, we established four parallel 2 × 20 m transects separated by 10 m. With the exception of lianas, all woody plants with stems ≥ 1 cm in diameter and ≥ 50 cm in height within these transects were labeled. No new tree recruitments of this size were observed. During the rainy season of each year, we surveyed caterpillars on all leaves and stems of all marked plants up to 3 m in height within transects. For adult trees, a subsample of three branches – consisting of approximately 100 leaves – was taken for each tree. Caterpillars were recorded and reared in the laboratory to confirm the trophic interaction with the host plant in which they were found and for subsequent taxonomic and molecular identification (see Villa-Galaviz et al. 2012, Boege et al. 2019 for details).

Caterpillar Diet Breadth

For those caterpillar species with more than 20 individuals recorded along the duration of the study, diet breadth was calculated as $DB = n/N$, where n is the number of plant species in which a caterpillar species was found and N is the number of plant species available in the transects of each plot.

Plant-Herbivore Interaction Networks

For each plot and year, we constructed a bipartite network and calculated the following network structure descriptors: network size (number of lepidopteran and plant species interacting), connectance (fraction of realized interactions from the possible total; Dunne et al. 2002), the number of compartments (network subgroups not connected with other subgroup, Tylianakis et al. 2007), and network specialization (H2, measures the deviance between the realized interactions for one species and the expected for each species with respect to the total possible interactions in the network; Blüthgen, 2006).

Statistical Analyses

To assess which climatic variables could be related to the variation observed in caterpillar diet breadth and plant-herbivore network parameters, we first performed a principal component analysis, considering the following variables for each year: the maximum and minimum precipitation during the rainy season (mm), the coefficient of variation in total precipitation during the rainy season (mm), the average maximum and minimum temperatures during the whole year (°C), the coefficients of variation for maximum and minimum temperatures, the duration of the dry season (days), and the number of days with more than 10 mm of precipitation during the rainy season. The onset of the rainy season was defined by the first rain > than 10 mm, and its end by the last rain >1 mm. Then, we used the scores of the first two principal components of the PCA to assess the influence of multivariate climatic factors on diet breadth and plant herbivore networks, as described next.

Changes in caterpillar diet breadth (log transformed) were evaluated using linear mixed effects models (library nlme) with the maximum likelihood method, considering sampling year as the random variable, successional stage as a covariate, and the scores of PC1 and PC2 and their interaction as the explanatory variables. We also evaluated changes in caterpillar diet breadth due to the effect of hurricanes in the region with a lme model using the maximum likelihood method, considering three periods of time as explanatory variables: (1) the period with no hurricane damage (pre-hurricanes), (2) the period after Jova and before Patricia (post-Jova), and (3) finally the period after the passage of hurricane Patricia (post-Patricia). The year of sampling was considered as the random variable. We simplified the models comparing the AIC to decide which model better explained our data.

To evaluate the changes in plant-caterpillar network parameters as a function of climatic variability, we also used linear mixed effect models considering the plot as a random variable and the successional stage as a covariate, with the maximum likelihood method. Explanatory variables were also the scores of PC1 and PC2 and their interaction. The parameters considered as response variables were network size, H2, the number of core lepidopteran species, the number of core plant species, the number of links, connectivity (log transformed), and the number of network compartments. We then performed another set of models evaluating the effect of hurricane on network parameters using three periods of time: pre-hurricanes, post-Jova, and post-Patricia as the explanatory variables. We compared AIC to decide which model better explain our data. All analyses were performed in R (R Development Core Team 2019).

Results

Climatic Variability

A high variability of climate in the Chamela region was registered in our 11-year study, as annual precipitation ranged between 655 mm in 2009 and 1289 mm in 2013. Also, the duration of the dry season lasted from 158 days in 2014 to 197 days in 2007. On top of that, two high-impact hurricanes struck the region Jova (Category 2) in 2011 and Patricia (Category 5) in 2015. When analyzing how environmental variables fluctuated and covaried during the span of our study using a principal component analysis, we found that the two first principal components explained 71% of the observed variation (PC1 41% and PC2 29%, respectively, Table 1). The first component was related to the average and variation coefficients in maximum and minimum daily temperatures, while PC2 was related to variations in rainfall and the span of dry periods (maximum annual precipitation, coefficient of variation in water precipitation during the rainy season, the duration of the dry season, and the

Table 1 Principal component analysis of the average values and coefficient of variation of temperature and precipitation variables in the region of Chamela, Mexico, registered between 2007 and 2018

	PC1	PC2	PC3	PC4	PC5	PC6
Eigenvalue	3.7202	2.6806	1.1049	0.9936	0.3641	0.0993
Proportion explained	0.4134	0.2978	0.1228	0.1104	0.0404	0.0110
Cumulative proportion	0.4134	0.7112	0.8340	0.9444	0.9848	0.9959
	Loadings					
Dry season duration	−0.3201	**0.4015**	−0.26035	0.1288	**−0.4682**	−0.3900
Max T	−0.3994	−0.2625	−0.0159	0.2757	**−0.6224**	0.2116
Min T	**0.4532**	−0.0013	−0.3785	0.2209	−0.1546	0.3514
CV T max	**0.4816**	0.1150	0.0820	0.1969	−0.1414	**−0.6960**
CV T min	**−0.4849**	0.0183	0.1344	−0.2562	0.2563	−0.2539
CV pp	−0.0321	**0.5990**	0.1172	0.0640	0.0397	0.2645
Min pp	−0.2487	−0.1149	**−0.4444**	**0.6478**	**0.5179**	−0.1261
Max pp	−0.0394	**0.4579**	**0.4600**	**0.4317**	0.1023	0.1935
# of days >10 mm	0.0325	**−0.4178**	**0.5840**	0.3796	−0.0595	−0.0857

PC loadings greater than 0.4 or smaller than −0.4 are highlighted in bold and were considered as the threshold to define the relative importance of traits for each PC

number of days with > pp. than 10 mm; Table 1, Fig. 1). As we describe next, this multivariate variation influenced the number of plants caterpillars fed on, their diet breadth, and network interactions (Table 2).

The Most Abundant Caterpillar Species at Chamela Tropical Dry Forest and Their Diet Breadth

During the span of the study (2007–2018), we found 13,716 caterpillars from 545 putative species (from which 216 were identified morphologically and/or with molecular markers; for details see Boege et al. 2019). The most abundant families were Erebidae, Staurniidae, Hesperiidae, Crambidae, and Nymphalidae (Fig. 2a) and only 32 species had more than 25 individuals (Fig. 2b). Diet breadth of these abundant species was rather wide, as they fed on 4 to 54 plant species belonging to 2 to 21 families (Fig. 3a), with a diet breadth ranging from 0.03 to 0.17 (Fig. 3b). However, when analyzed cautiously, the abundance of caterpillars feeding on those plant species was quite variable. For example, *Orgyia* sp. was found on 54 host species, but only 14 of them had more than 6 caterpillar individuals. The second most polyphagous species, *Misoria amra,* was found feeding on 41 plant species, but only 8 hosts had more than 6 individuals. Hence, caterpillar species are likely to be less generalist than it appears with most of the individuals within a population feeding on a reduced subset of the host species registered.

The number of plant species that caterpillars used as food varied through time and as a function of climatic variability. In particular, diet breadth was affected by

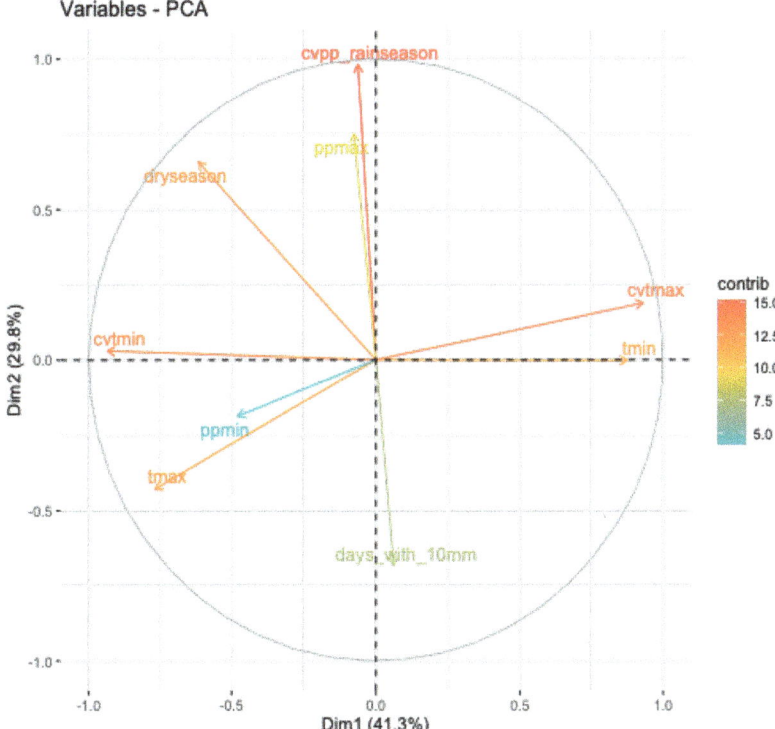

Fig. 1 Multivariate covariation in temperature and precipitation variables during 2007–2018 in the region of Chamela, Mexico. Tmax = average of maximum temperature, Tmin = average of minimum temperature, cvtmax = coefficient of variation in maximum temperature, cvtmin = coefficient of variation in minimum temperature, ppmax = maximum temperature registered during the rainy season, ppmin = minimum temperature registered during the rainy season, cvpp_rainyseason = coefficient of variation in precipitation registered during the rainy season, days with 10 mm = number of days in which precipitation was greater than 10 mm, dry season = length of the dry season (period since the last rains in the previous year until the onset of the rainy season with a precipitation event >10 mm)

PC1, related to temperature variables (Table 1). The greater the PC1, the greater was the diet breadth ($t_{6,568} = -3.65$, P = 0.01; Fig. 4a), which means that during the years with an average of cooler days, but increased variation in the maximum daily temperature, caterpillars expanded their diet breadth. In contrast, a negative relationship was found between the scores of PC2 and diet breadth ($t_{(6,568)} = -2.98$, P = 0.2; Fig. 4b), suggesting that during the years with high precipitation events but also greater variability in precipitation during the rainy season, caterpillars had smaller diet breadths. Interestingly, the duration of the preceeding dry season also reduced the number of plants used as food by caterpillars, as this variable had a positive and high loading in this PC (Table 1). The two hurricanes that hit the area in 2011 and 2015 significantly increased caterpillar diet breadth (Fig. 5).

Table 2 Most abundant caterpillar species of the tropical dry forest in Chamela, Jalisco, surveyed between 2007 and 2018

Morphospecies	Family	Species
O69	Crambidae	*Diaphania jairusalis*
O96	Crambidae	*Syllepis hortalis*
O161	Dalceridae	*Dalceridae* sp.
O287	Depressariidae	*Ethmia near similatella*
O10	Erebidae	*Hypercompe confusa*
O12	Erebidae	*Anomis editrix*
O18	Erebidae	*Hypercompe* sp.
O30	Erebidae	*Orgyia* sp.
O79	Erebidae	*Deinopa biligula*
O114	Erebidae	*Lophocampa citrina*
O347	Erebidae	*Eudesmia menea*
O57	Geometridae	Unidentified
O325	Geometridae	*Melanchroia vazquezae*
O44	Hesperiidae	*Misoria amra*
O177	Limacodidae	Unidentified
O2	Megalopygidae	Unidentified
O55	Notodontidae	*Dasylophia eminens*
O190	Notodontidae	*Cargida pyrrha*
O279	Nymphalidae	*Chlosyne gloriosa*
O27	Pyralidae	*Epipaschia superatalis*
O111	Riodinidae	*Emesis emesia*
O299	Saturniidae	*Hylesia* sp.
O358	Saturniidae	*Hylesia continua*
O24	Urodidae	*Wockia chewbacca*
O56	Unidentified	Unidentified
O130	Unidentified	Unidentified
O157	Unidentified	Unidentified
O246	Unidentified	Unidentified
O431	Unidentified	Unidentified

Plant-Caterpillar Interaction Networks

As a result of the variation in the number and identity of host plants used by caterpillars, plant-herbivore interaction networks changed through time, and some parameters were affected by the two hurricanes. When analyzing the importance of environmental variation explaining plant-herbivore network parameters, we found that PC1 negatively affected network size, H2, and the number of compartments, while it positively affected the number of links per species (Table 3, Fig. 6). Also, PC2 affected positively network size and the number of compartments (Fig. 7),

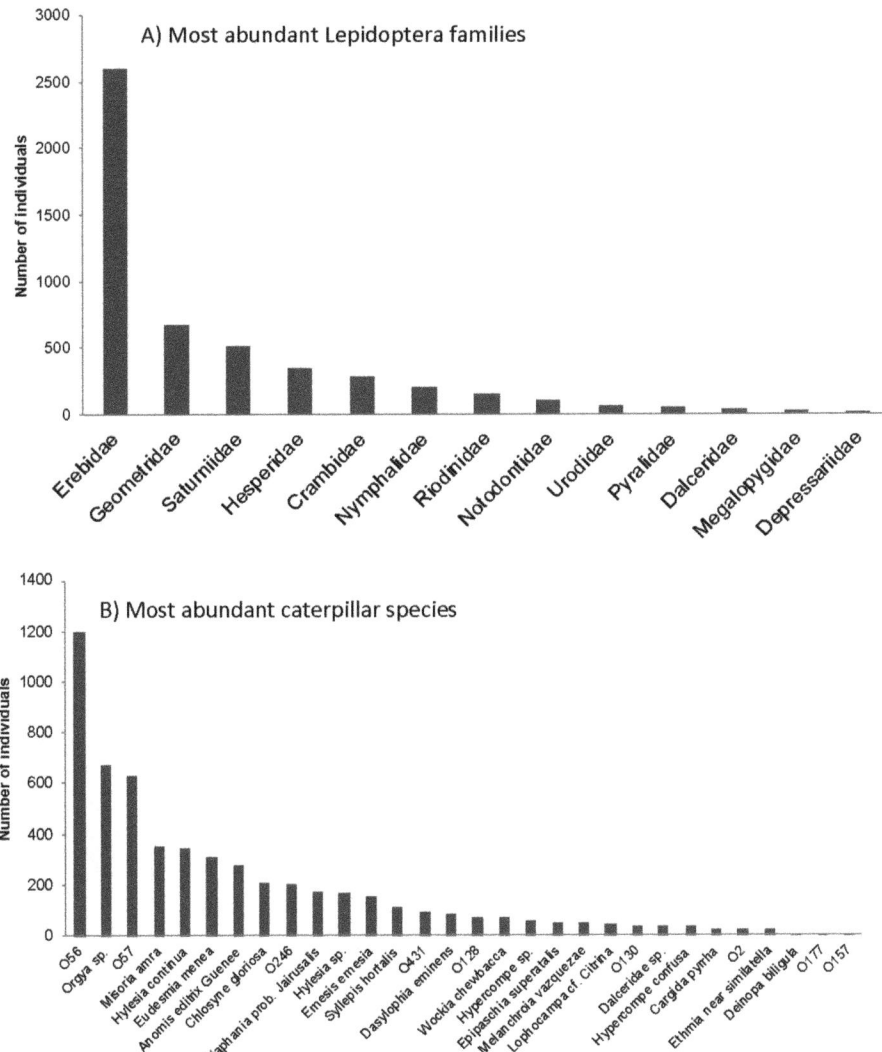

Fig. 2 Abundance of the most common caterpillar (**a**) families and (**b**) species found across an 11-year study in the tropical dry forest of Chamela, Mexico

suggesting years with high variation in precipitation during the rainy season with precipitation single events promoted larger and more specialized networks.

The impact of the two hurricanes also influenced plant-caterpillar network parameters. In particular, its size decreased after Jova and remained low after Patricia hurricanes, and the number of compartments and the network specificity (H2) was reduced after Jova and were even smaller after Patricia (Table 4, Fig. 8). Other parameters such as the number of links per species and network connectivity

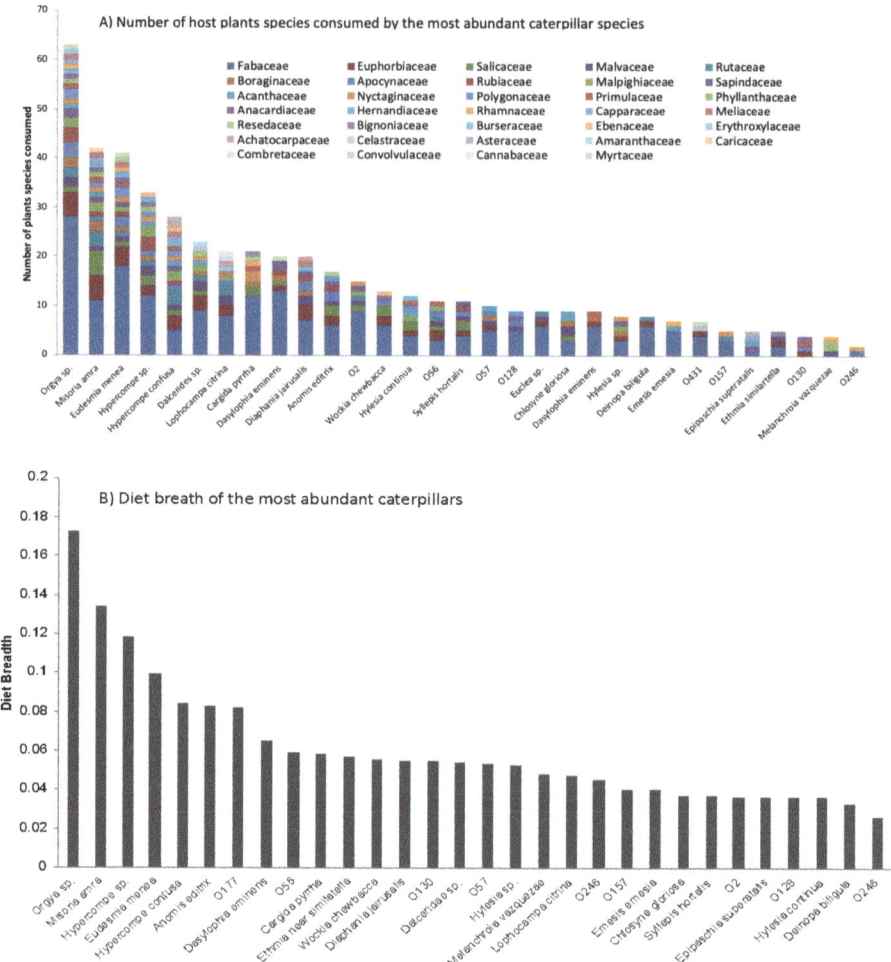

Fig. 3 (a) Number of host plant species used by the most abundant caterpillars and (b) their diet breadth in the tropical dry forest of Chamela, Mexico

increased after the impact of both hurricanes, while the number of core plants and core lepidopteran species remained unchanged (Plates 1 and 2).

Discussion

Tropical dry forests are characterized by a marked seasonality in rainfall, with dry periods that can extend from two up to six months (Sánchez-Azofeifa et al. 2014), depending on the geographical region. Hence, in this ecosystem plants are under a strong selection regime to either avoid, resist, or tolerate drought through different

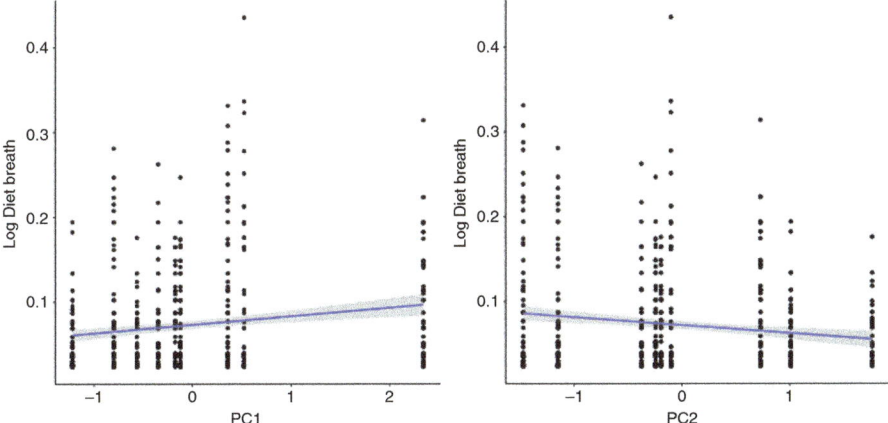

Fig. 4 Diet breadth (log transformed) related with the PC1(environmental variables related with temperature) and PC2 (environmental variables related with rainfall and dry periods). The line shown is the model prediction, and the gray areas correspond to the confidence interval

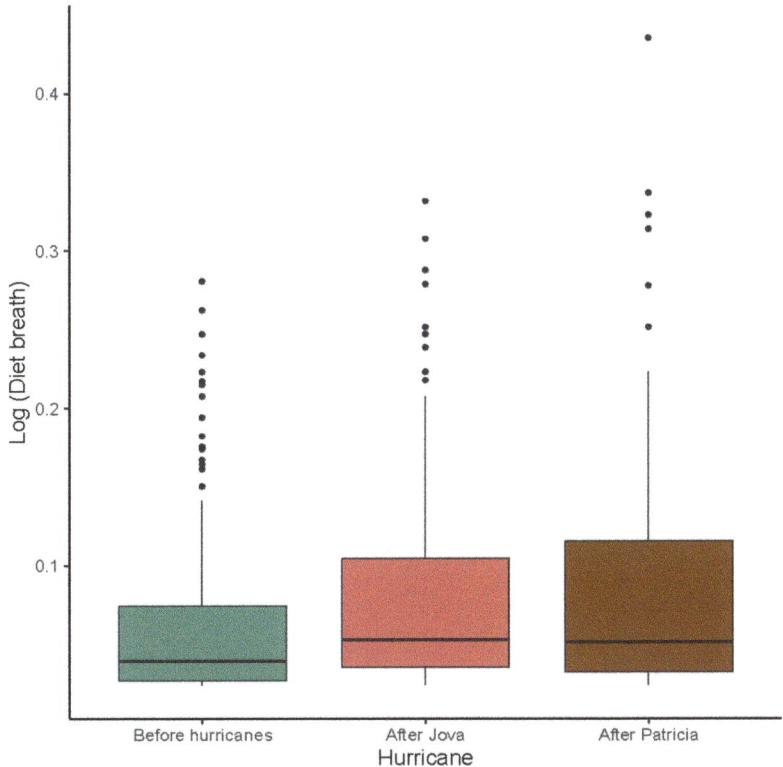

Fig. 5 Diet breadth amplitude (log transformed) before and after the hurricane passage

Table 3 Results from the mixed effects models evaluating different environmental variables upon network parameters

Network parameters	Df	PC1		PC2		PC1*PC2	
		F	p	F	p	F	p
Network size	1,65	5.12	0.03	8.7	0.004	8.04	**0.006**
H2	1,65	16.9	0.0001	2.78	0.1	6.16	**0.016**
Number of core lepidopterans	1,65	--	--	--	--	5.92	**0.02**
Number of core plants	1,65	--	--	--	--	5.92	**0.02**
Number of links (log)	1,65	5.11	**0.02**	--	--	--	--
Connectivity (log)	1,65	--	--	--	--	--	--
# compartments	1,66	25.5	**<0.0001**	7.52	**0.008**	--	--

We used the sampled plot as a random factor and successional stage as a covariate for all models with maximum likelihood method. F and p values for each model are shown

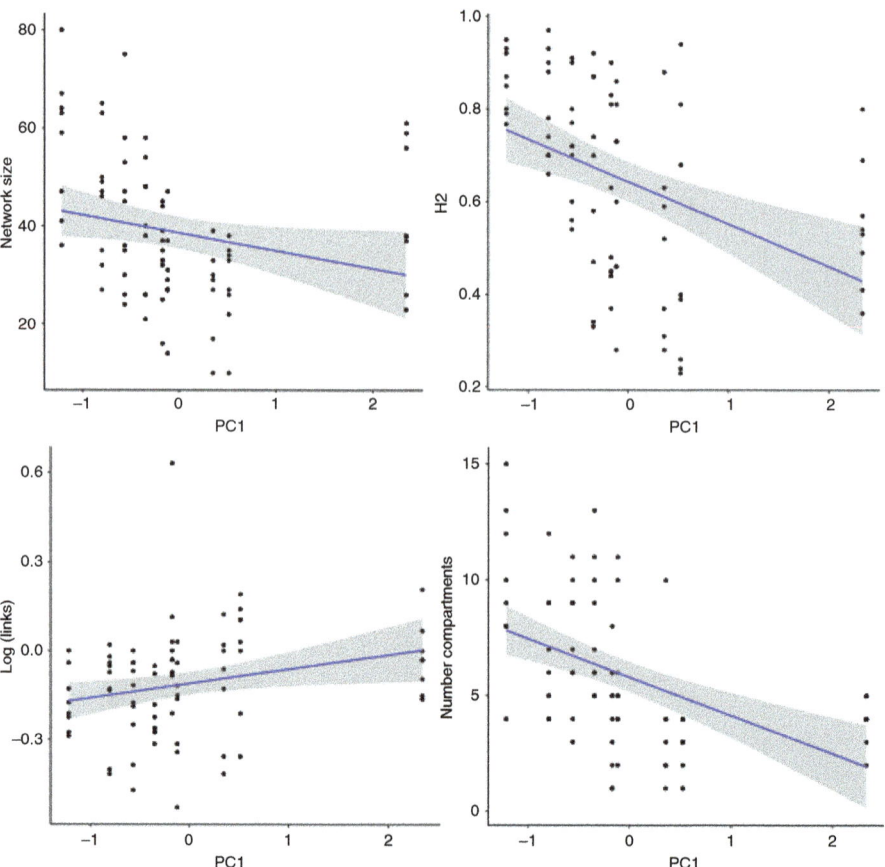

Fig. 6 Network parameters affected by PC1 (variables related with temperature). The lines shown are the model predictions, and the gray areas are the confidence intervals

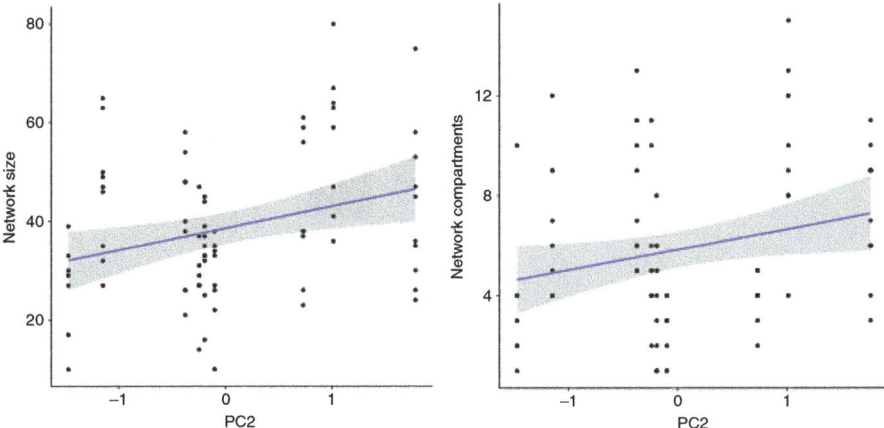

Fig. 7 Network parameters affected by PC2 (variables related with rainfall and dry periods). The lines shown are the model predictions, and the gray areas are the confidence intervals

Table 4 Results from the linear mixed effect models evaluating the effect of two hurricanes on plant-herbivore network parameters, considering the sampling year and the sampling plot as random factors and the successional stage as a covariate

Network parameters	F	Df	p
Network size	10.67	2,66	0.0001
H2	12.9	2,66	<0.0001
Number of core lepidopterans	2.45	2,66	NS
Number of core plants	2.41	2,66	NS
Number of links per species	4.75	2,66	0.01
Number of compartments	26.97	2,66	<0.0001
Connectivity (log)	15.1	2,66	<0.0001

water use strategies (Olivares and Medina 1992). In particular, plants face a trade-off between exploiting and using water to grow or store this resource to survive soil and tissue desiccation (Pineda García et al. 2015), which can influence the availability and quality of tissues as food for herbivores. For example, tolerance to drought is associated with greater leaf density and hence increased leaf toughness and also can promote longer periods of leaf availability before leaves are shed at the onset of the dry season (Markesteijn et al. 2011, Méndez-Alonso et al. 2012). In addition, resources stored during the previous growing season can influence resource allocation to growth, defense, and reproduction in the following growing season (Chapin III et al. 1990; Quiroz-Pacheco et al. 2020). However, the mechanistic links between plant water use strategies, the intensity of drought stress, and herbivore feeding choices in tropical dry forests are still poorly understood. In this study, we found that the multivariate covariation among different precipitation and temperature variables had a significant influence on the number of plant species used as food by the most abundant caterpillars. This, in turn, influenced their diet breadth, and as

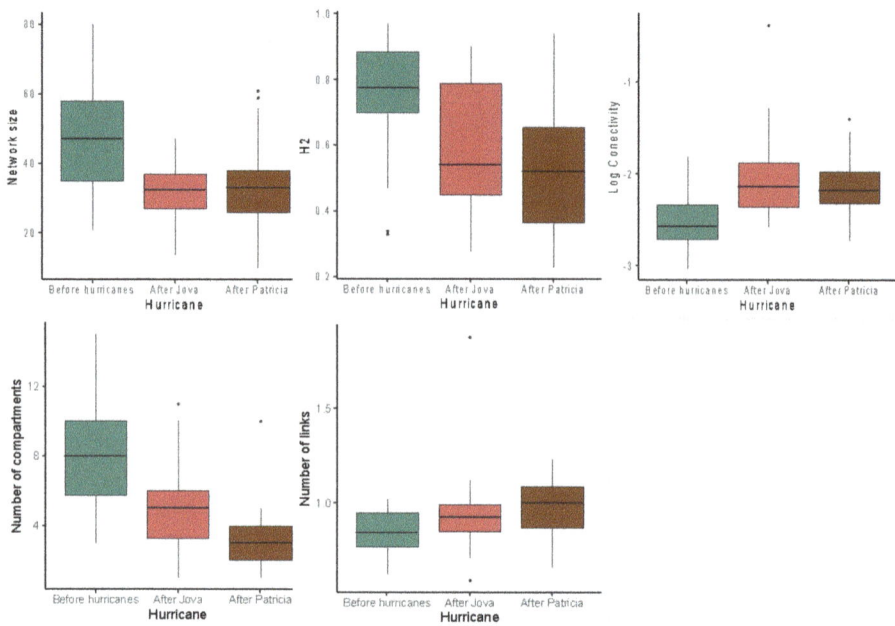

Fig. 8 Plant-herbivore network parameters significantly affected by hurricanes

a consequence, different parameters of the plant-herbivore interaction networks. Although these effects were rather modest, next we discuss some plausible explanations that need further experimental approaches to test the causality of the different climatic factors on the feeding choices, survival, and performance of caterpillars.

Influence of Temperature

Temperature has been identified as one of the main environmental factors influencing caterpillar survival and performance (Bale et al. 2002; Dewer and Watt 1992; Abarca and Lill 2015). However, it has been claimed that this climatic factor is rather stable in tropical regions and should affect more temperate and boreal herbivores (Bale et al. 2002). Nevertheless, available information of long-term data has revealed that an increase up to 2 °C in maximum temperatures has reduced arthropod abundances in the tropical wet forests in Puerto Rico (Lister and García 2018). Hence tropical insects also seem to be susceptible to variation in temperatures (García-Robledo et al. 2016). In this study we found that diet breadth was wider in slightly cooler years, with greater variation in the maximum daily temperature. Given the narrow thermal limits of tropical insects (Sunday et al. 2014), we propose that if some individuals indeed fail to survive during hot days, they were not registered in particular host plants, reducing their diet breadth in years with increased number of warmer days (and hence greater within-year variation in temperature). This also could explain the reduced network

Plates 1 and 2 Most abundant caterpillars in Chamela tropical dry forest (Jalisco, México). *Dasylophia eminens* and *Sylepis hortalis* (Crambidae); *Dalceridae* sp. (Dalceridae); *Ethmia* nr. *similatella* (Depresariidae); *Orgyia* sp. (Lymantriidae); *Eudesmia menea, Anomis editrix, Hypercompe* sp., *Lophocampa* cf. *Citrina*, and *Deinopa biligula* (Erebidae); *Melanchroia vazquezae* (Geometridae); *Misoria amra* (Hesperiidae); *Dasylophia eminens* and *Cargida pyrrha* (Notodontidae) and *Chlosyne gloriosa* (Nymphalidae); *Epipaschia superatalis* (Pyralidae); *Emesis emesia* (Riodinidae); *Hylesia continua* and *Hylesia* sp. (Saturniidae); *Wockia chewbacca* (Urodidae)

Plates 1 and 2 (continued)

size, the greater specialization of (H2), and an increased compartmentalization of plant-caterpillar interaction networks during those years. The influence of temperature on caterpillar survival in tropical dry forest, however, requires further experimental work. Detailed experiments controlling the maximum temperature while feeding caterpillars with different plant species could help link the effects of maximum temperatures on caterpillar survival as a function of the host plant species in which they feed. For example, in a series of controlled experiments on a common oligophagous butterfly from temperate forests, Abarca (2019) demonstrated that feeding on a high-quality host can mitigate detrimental effects of thermal stress such as mortality and reduced pupal mass.

Influence of Precipitation

Different components of the precipitation regime also affected the interaction between caterpillars and their food plants. In particular, the coefficient of variation in precipitation and the maximum precipitation during a single event registered in the rainy season, together with the length of the previous dry season, had a negative effect on the caterpillar's diet breadth. Previous works have reported that summer droughts can have significant effects on forest insect populations in temperate (Mattson and Haack 1987, Marquis et al. 2019) and tropical forests (Lister and García 2018). In addition, the influence of precipitation on diet breadth and specialization of plant-caterpillar networks has been reported for semi-deciduous tropical forests, showing that during the rainy season, plants have greater quality (López-Carretero et al. 2016) and caterpillars are more selective, reducing their diet breadth in contrast with what is observed during dry seasons (Scherrer et al. 2016; López-Carretero et al. 2018). Here, we provide further evidence that variation in precipitation within the rainy season can also influence these trophic interactions. The climatic variable with the greatest loading for PC2, the coefficient of variation of precipitation, is likely to produce prolonged availability of soil humidity for most plants across the rainy season. This, in turn, should promote increased plant growth, and, according to the carbon/nutrient balance hypothesis (Bryant et al. 1983), reduced concentration of carbon-based defenses in leaf tissues. Under this scenario, caterpillars should feed on their preferred host plants, reducing their diet breadth and increasing network specialization.

The other interesting pattern we found is the negative relationship between the length of the dry season and the caterpillar's diet breadth. As a consequence, a positive relationship was found between the span of the dry season and the specialization of plant-caterpillar networks and also with network size. These patterns suggest that leaf availability and/or quality for the caterpillar community in a given year was influenced by the length of the preceding dry season, affecting their diet breadth. We suggest different alternatives to explain this pattern. First, variation in dry season length can trigger host-specific mortality. Because the length of the dry season was mostly influenced by its onset during the previous year (between October and

December), variation in the timing of the last rain could promote a reduction in foliage availability at the end of the growing season for the last generation of caterpillars feeding on certain host plants. Resource acquisitive plant species shed their leaves earlier than drought tolerant species, resulting in differences in foliage availability windows. Thus, individuals feeding on plants with short availability windows would not be able to complete the development or would pupate at a small size. Pupal size is an important predictor of survival in temperate lepidopterans (Liu et al. 2007) and could also affect the probability of survival of tropical species. Due to the interaction between genetic and nongenetic factors in oviposition preferences of moths and butterflies for particular hosts promoting intraspecific variation in specialization (Karowe 1989; Karpinski et al. 2014), those genotypes failing to complete their development in host plant species shedding their leaves early could not be represented in the diet breadth registered during the following growing season. Hence, diet breadth in a given year preceded by a long dry season would be restricted to those plant species in which caterpillars were able to complete their development, promoting the observed negative relationship between the span of the dry season and the mean diet breadth at a community level. A second and not mutually exclusive alternative is that the timing of the onset of the dry season also affected the amount of resources that plants could store for the following reproductive and growing seasons. In this case, longer periods of drought could have reduced resources available for growth during the following growing season. Growth-defense trade-offs in turn would promote greater defenses in plants with shortage of stored resources for growth, influencing herbivore feeding choice for less defended plant species. This hypothesis, however, requires further experimental demonstration. Finally, the influence of climatic variability on the caterpillar's diet breadth and plant-herbivore networks can also be influenced by biotic factors, such as the presence of endosymbionts, which have been detected to strongly influence an insect's diet breadth (Wagner et al. 2015) or parasitoids. In particular, the latter have been found to reduce their capacity to track host populations with increased climatic variability and the associated unpredictable and amplified variance in host dynamics (Stireman et al. 2005).

Influence of Hurricanes

We found that the impact of the two hurricanes that hit the study area during the span of the study promoted an increase in the caterpillar's diet breadth. This could have been fostered by the massive loss of the entire host trees and/or foliage and disturbance in general, particularly during hurricane Patricia, probably forcing herbivores to feed not only on their preferred hosts but on what was available after these extreme perturbations. If caterpillar diet breadth increased after these extreme meteorological events, defoliation on particular plant species and herbivory at the community level should have decreased, as found in other tropical dry forest sites such as Florida after hurricane Andrew (Koptur et al. 2002) and in Puerto Rico after hurricane George (Angulo-Sandoval et al. 2004), although Hunter and Forkner (1999) found an increase of herbivory for maple and oak species in North Carolina after hurricane Opal. These investigations suggested that a decrease in herbivory after hurricanes can be due to a combination of changes in herbivore populations, predator populations, and/or leaf chemistry. Here we provide empirical evidence that changes in diet breadth could also help to understand those patterns. Hurricanes also affected plant-herbivore network parameters, decreasing specialization (H2), network size, and the number of network compartments. The remaining interacting species after the hurricanes also increased the number of links that caterpillars established with host plants, and the general connectivity of the network was also larger, providing a more robust network. Therefore, hurricane impacts were cumulative and lasting; however, due to landscape heterogeneity and to the presence of a large biosphere reserve in the area, lepidopterans could recolonize from the surroundings. Other studies looking into network structure after hurricane strikes have shown some resilience for hummingbird-plant interactions (Díaz-Infante et al. 2020), ant-plant interactions (Sánchez-Galván et al. 2012), and also for our system of caterpillar-plant interactions after the first hurricane (Luviano et al. 2018). Nevertheless, the recurrence of these extreme meteorological events seems to reduce network resilience. Because hurricanes are predicted to increase in the Pacific coast in terms of frequency and severity (Knutson et al. 2015), the plasticity of interacting species and caterpillar diet flexibility seem crucial to withstand future climatic scenarios.

Concluding Remarks

The relevance to establish a link between climatic variability and caterpillar diet breadth and their interaction networks with plants relies on the fact that herbivore's food choices can affect different components of their fitness, through increased risk of predation and disease, the likelihood of mating, thermal balance, and developmental time (Karban and Anurag 2002, Barber and Marquis 2011). This, in turn, can influence the evolution of herbivore foraging behavior and phenotypic plasticity in plant host use, as a function of climate and land use changes (Singer and Parmesan 2020).

Acknowledgments The authors thank I. Medina, C. Manrique, F. Gutierrez, E. Castro, L. Solis, A. Flores, N. Luviano, E. Cuevas, B. Mejía, I. Sosa, W. Mendoza, A. López-Carretero, and the students of the biotic interactions ecology laboratory at IIES-UNAM for fieldwork assistance. Special thanks to R. Pérez-Ishiwara for logistical support. We also thank the landowners, UNAM Chamela Biological Station, and the Cuixmala Foundation for kindly allowing access to their proprieties and M. Martínez-Ramos for allowing the incorporation of our study to the MABOTRO original design. This study was funded by PAPIIT-UNAM (IN208610, IN217507, IN211916, IN207016), by CONACyT, Mexico (Red Temática del Código de Barras de la Vida; Proyecto SEP-CB No. 220454), and by SEP CONACyT 2015 255544. Permission to collect larva and adult lepidopteran and plant specimens was given by the Secretaría de Medio Ambiente y Recursos Naturales (SGPA/DGVS/02005/08).

References

Abarca M (2019) Herbivore seasonality responds to conflicting cues: untangling the effects of host, temperature, and photoperiod. PLoSONE 14(9):e0222227

Abarca M, Lill JT (2015) Warming affects hatching time and early season survival of eastern tent caterpillars. Oecologia 179:901–912

Angulo-Sandoval P, Fernández-Marin H, Zimmerman JK, Aide TM (2004) Changes in patterns of understory leaf phenology and herbivory following hurricane damage. Biotropica 36:60–67

Avila-Cabadilla L, Stoner KE, Henry M, Alvarez-Añorve M (2009) Composition, structure and diversity of phyllostomid bat assemblages in different successional stages of a tropical dry forest. For Ecol Manage 258:986–996

Bale JS, Masters GJ, Hodkinson ID, Awmack C, Bezemer TM, Brown VK, Butterfield J, Buse A, Coulson JC, Farrar J, Good JEG, Harrington R, Hartley S, Jones TH, Lindroth RL, Press MC, Symrnioudis I, Watt AD, Whittaker JB (2002) Herbivory in global climate change research: direct effects of rising temperature on insects herbivores. Glob Chang Biol 8:1–16

Barber NA, Marquis RJ (2011) Leaf quality, predators, and stochastic processes in the assembly of a diverse herbivore community. Ecology 92:699–708

Bernays EA (1998) Evolution of feeding behavior in insect herbivores: success seen as different ways to eat without being eaten. Bio Science 48(1):35–44

Bernays EA (2001) Neural limitations in phytophagous insects: implications for dietbreadth and evolution of host affiliation. Annu Rev Entomol 46:703–727

Blüthgen N, Menzel F, Blüthgen N (2006) Measuring specialization in species interaction networks. BMC Ecol 6:9

Boege K, Villa-Galaviz E, López-Carretero A, Pérez-Ishiwara R, Zaldívar-Riverón A, Ibarra A, del-Val E (2019) Temporal variation in the influence of forest succession on caterpillar communities: a long-term study in a tropical dry forest. Biotropica 51:529–537

Bryant JP, Chapin FS III, Klein DR (1983) Carbon/nutrient balance of boreal plants in relation to vertebrate herbivory. Oikos 40:357–368

Chapin FS III, Schulze ED, Mooney HA (1990) The ecology and economics of storage in plants. Annu Rev Ecol Syst 21:423–447

Dewer RC, Watt AD (1992) Predicted changes in the synchrony of larval emergence and budburst under climatic warming. Oecologia 89:557–559

Díaz Infante S, Lara C, Arizmendi MC (2020) Temporal dynamics of the hummingbird-plant interaction network of a dry forest in Chamela, Mexico: a 30-year follow-up after two hurricanes. Peer J 8:e8338

Dunne JA, Williams RJ, Martinez ND (2002) Food-web structure and network theory, the role of connectance and size. Proc Natl Acad Sci U S A 99:12917–12922

Dyer LA (1995) Tasty generalists and nasty specialists? Antipredator mechanisms in tropical lepidopteran larvae. Ecology 76:1483–1496

Dyer LA et al (2007) Host specificity of Lepidoptera in tropical and temperate forests. Nature 448:696–699

Egan SP, Funk DJ (2006) Individual advantages to ecological specialization: insights on cognitive constraints from three conspecific taxa. Proc R Soc London Ser B 273:843–848

Forister ML, Dyer LA, Singer MS, Stireman JO, Lill JT (2012) Revisiting the evolution of ecological specialization, with emphasis on insect–plant interactions. Ecology 93:981–991

Forister ML et al (2015) The global distribution of diet breadth in insect herbivores. Proc Natl Acad Sci U S A 112:442–447

Forkner RE, Marquis RJ, Lill JT, Corff JL (2008) Timing is everything? Phenological synchrony and population variability in leaf-chewing herbivores of Quercus. Ecol Entomol 33:276–285

García-Robledo C, Kuprewicz EK, Staines CL, Erwin TL, Kress WJ (2016) Insect tolerance to global warming. Proc Natl Acad Sci U S A 113(3):680–685

Gaston KJ, Reavey D, Valladares G (1991) Changes in feeding habit as caterpillars grow. Ecol Entomol 16:339–344

Harrington R, Woiwod I, Sparks T (1999) Climate change and trophic interactions. Trends Ecol Evol 14:146–150. https://doi.org/10.1016/S0169-5347(99)01604-3

Hodkinson ID (1997) Progressive restriction of host plant exploitation along a climatic gradient: the willow psyllid Cacopsylla groenlandica in Greenland. Ecol Entomol 21:47–54

Hunter MD, Forkner RE (1999) Hurricane damage influences foliar polyphenolics and subsequent herbivory on surviving trees. Ecology 80:2676–2682

Hwang S, Hwang F, Shen T (2007) Shifts in developmental diet breadth of *Lymantria xylina* (Lepidoptera: Lymantriidae). J Econ Entomol 100:1166–1172

Karban R, Agrawal AA (2002) Herbivore offense. Annu Rev Ecol Syst 33:641–664

Karowe DN (1989) Facultative monophagy as a consequence of prior feeding experience: behavioral and physiological specialization in *Colias philodice* larvae. Oecologia 78:106–111

Karpinski A, Haenniger S, Schofl G, Heckel DG, Groot AT (2014) Host plant specialization in the generalist moth *Heliothis virescens* and the role of egg imprinting. Evol Ecol 28:1075–1093

Knutson TR, Sirutis JJ, Zhao M, Tuleya RE, Bender M, Vecchi CA, Villarini G, Chavas D (2015) Global projections of intense tropical cyclone activity for the late twenty-first century from dynamical downscaling of CMIP5/RCP4.5 scenarios. J Clm 28:7203–7224

Koptur S, Rodríguez MC, Oberbauer SF, Weekly C, Herndon A (2002) Herbivore-free time? Damage to new leaves of woody plants after hurricane Andrew. Biotropica 34:547–554

Lister BC, Garcia A (2018) Climate-driven declines in arthropod abundance restructure a rainforest food web. Proc Natl Acad Sci U S A 115:E10397–E10406

Liu Z, Gong P, Wu K, Wei W, Sun J, Li D (2007) Effects of larval host plants on over-wintering preparedness and survival of the cotton bollworm, *Helicoverpa armigera* (Hübner) (Lepidoptera: Noctuidae). J Insect Physiol 53:1016–1026

López-Carretero A, Díaz-Castelazo C, Boege K, Rico-Gray V (2014) Evaluating the spatiotemporal factors that structure network parameters of plant-herbivore interactions. PLoS One 9(10):e110430

López-Carretero A, Boege K, Díaz-Castelazo C et al (2016) Influence of plant resistance traits in selectiveness and species strength in a tropical plant-herbivore network. Am J Bot 103:1436–1448

López-Carretero A, Del-Val E, Boege K (2018) Plant-herbivore networks in the tropics. In: Dáttilo W, Rico-Gray V (eds) Ecological networks in the tropics. Springer Verlag, USA, pp 111–123

Lott EJ, Bullock SH, Solís-Magallanes JA (1987) Floristic diversity and structure of upland and arroyo forests in coastal Jalisco. Biotropica 19:228–235

Loxdale HD, Harvey JA (2016) The 'generalism' debate: misinterpreting the term in the empirical literature focusing on dietary breadth in insects. Biol J Linn Soc 119:265–282

Luviano N, Villa-Galaviz E, Boege K, Zaldivar-Riverón A, del-Val E (2018) Hurricane impacts on plant-herbivore networks along a successional chronosequence in a tropical dry forest. For Ecol Manage 426:158–163

Maass JM, Ahedo-Hernández R, Araiza S, Verduzco A, Martínez-Yrízar A, Jaramillo VJ, Parker G, Pascual F, García-Méndez G, Sarukhán J (2018) Long-term (33 years) rainfall and runoff dynamics in a tropical dry forest ecosystem in western Mexico: management implications under extreme hydrometeorological events. For Ecol Manage 426:7–17

Manson RH, Jardel E (2009) Perturbaciones y desastres naturales: impactos sobre las ecorregiones, la biodiversidad y el bienestar socioeconómico. In: Capital Natural de México, Vol. II: Estado de Conservación y Tendencias de Cambio (Ed: José Sarukhán., et al.). Comisión Nacional para el Conocimiento y Uso de la Biodiversidad. pp. 131–184

Markesteijn L, Poorter L, Paz H, Sack L, Bongers F (2011) Ecological differentiation in xylem cavitation resistance is associated with stem and leaf structural traits. Plant Cell Environ 34:137–148

Marquis RJ, Lill JT, Forkner RE, Le Corff J, Landosky JM, Whitfield JB (2019) Declines and resilience of communities of leaf chewing insects on Missouri oaks following spring frost and summer drought. Front Ecol Evol 7:396

Mattson WJ, Haack RA (1987) The role of drought in outbreaks of plant-eating insects. Bioscience 37:110–118

Méndez-Alonzo R, Paz H, Cruz Zuluaga R, Rosell JA, Olson ME (2012) Coordinated evolution of leaf and stem economics in tropical dry forest trees. Ecology 93:2397–2406

Mody K, Unsicker SB, Linsenmair KE (2007) Fitness related diet-mixing by intraspecific host-plant-switching of specialist insect herbivores. Ecology 88:1012–1020

Olivares E, Medina E (1992) Water and nutrient relations of woody perennials from tropical dry forests. J Veg Sci 3:383–392

Pescador-Rubio A, Rodríguez-Palafox A, Noguera FA (2002) Diversidad y estacionalidad de Arthropoda. In: Noguera-Alderte AN, Vega-Rivera JH, Quesada M (eds) Historia natural de Chamela. Instituto de Biología, UNAM, Mexico DF

Pickett STA, White PS (1985) The ecology of natural disturbance and patch dynamics. Academic Press- Science - 472 pages

Pineda-García F, Paz H, Meinzer FC, Angeles G (2015) Exploiting water versus tolerating drought: water-use strategies of trees in a secondary successional tropical dry forest. Tree Physiol 36:208–217

Quiroz-Pacheco E, Mora F, Boege K. Domínguez CA, Del-Val E (2020). Effects of herbivory and its timing on reproductive success of a tropical deciduous tree. Annals of Botany 126: 957–969

R Development Core Team (2019) R: a language and environment for statistical computing R foundation for statistical computing. Vienna, Austria

Renner SS, Zohner CM (2018) Climate change and phenological mismatch in trophic interactions among plants, insects, and vertebrates. Annu Rev Ecol Evol Syst 49:165–182

Rodrigues D, Moreira GRP (2004) Seasonal variation in larval host plants and consequences for *Heliconius erato* (Lepidoptera: Nymphalidae) adult body size. Austral Ecol 29:437–445

G. Arturo, Sánchez-Azofeifa Mauricio, Quesada Pablo, Cuevas-Reyes Alicia, Castillo Gumersindo, Sánchez-Montoya (2009) Land cover and conservation in the area of influence of the Chamela-Cuixmala Biosphere Reserve Mexico. Forest Ecology and Management 258(6):907–912. https://doi.org/10.1016/j.foreco.2008.10.030

Sánchez-Azofeifa A, Powers JS, Fernandes G, Quesada M (2014) Tropical dry forests in the Americas. Ecology, conservation, and management. CRC Press, Boca Raton, FL, 556 pp

Sánchez-Galván O, Díaz-Castelazo C, Rico-Gray V (2012) Effect of hurricane Karl on a plant-ant network occurring in coastal Veracruz, Mexico. J Trop Ecol 28:603–609

Scherrer S, Lepesqueur C, Vieira MC, Almeida-Neto M, Dyer L, Forister M, Diniz IR (2016) Seasonal variation in diet breadth of folivorous Lepidoptera in Brazilian cerrado. Biotropica 48(4):491–498

Schowalter TD, Willig MR, Presley SJ (2017) Post-hurricane successional dynamics in abundance and diversity of canopy arthropods in a tropical rainforest. Environ Entomol 46:11–20

Scriber JM (2010) Integrating ancient patterns and current dynamics of insect–plant interactions: taxonomic and geographic variation in herbivore specialization. Insect Sci 17:471–507

Secretaria de Gobernación (2017). https://www.gob.mx/segob/prensa/mensaje-de-los-titulares-de-cenapred-conagua-y-del-director-de-gestion-de-riesgos-de-la-coordinacion-nacional-de-proteccion-civil. Date consulted: 16-VIII-2017

Singer MC, Parmesan C (2020) Colonizations drive host shifts, diversification of preferences and expansion of herbivore diet breadth. bioRxiv:2020.03.31.017830. https://doi.org/10.1101/2020.03.31.017830

Singer MC, Thomas CD, Billington HL, Parmesan C (1989) Variation among conspecific insect populations in the mechanistic basis of diet breadth. Anim Behav 37:751–759

Singer MC, Wee B, Hawkins S, Butcher M (2008) Rapid natural and anthropogenic diet evolution: three examples from checkerspot butterflies. In: Tilmon KJ (ed) Specialization, speciation, and radiation: the evolutionary biology of herbivorous insects. University of California, Berkeley, pp 311–324

Smilanich AM, Dyer LA, Chambers JQ, Bowers MD (2009) Immunological cost of chemical defence and the evolution of herbivore diet breadth. Ecol Lett 12:612–621

Stamp N (2003) Out of the quagmire of plant defense hypotheses. Q Rev Biol 78(1):23–55

Stireman JO III, Dyer LA, Janzen DH, Singer MS, Lill JT, Marquis RJ, Ricklefs RE, Gentry GL, Hallwachs W, Coley PD, Barone JA, Greeney HF, Connahs H, Barbosa P, Morais HC, Diniz IR (2005) Climatic unpredictability and parasitism of caterpillars: implications of global warming. PNAS 102(48):17384–17387

Sunday JM, Bates AE, Kearney MR, Colwell RK, Dulvy NK, Longino JT, Huey RB (2014) Thermal-safety margins and the necessity of thermoregulatory behavior across latitude and elevation. Proc Natl Acad Sci U S A 111(15):5610–5615

Trejo I, Dirzo R (2000) Deforestation of seasonally dry tropical forest: a national and local analysis in Mexico. Biol Conserv 94:133–142

Tylianakis JM, Tscharntke T, Lewis OT (2007) Habitat modification alters the structure of tropical host-parasitoid food webs. Nature 445:202–205

van Asch M, Visser ME (2007) Phenology of forest caterpillars and their host trees: the importance of synchrony. Annu Rev Entomol 52(1):37–55

Villa-Galaviz E, Boege K, del-Val E (2012) Resilience in plant-herbivore networks during secondary succession. PLoS One 7:e53009

Wagner SM, Martinez AJ, Ruan YM, Kim KL, Lenhart PA, Dehnel AC et al (2015) Facultative endosymbionts mediate dietary breadth in a polyphagous herbivore. Funct Ecol 29:1402–1410

Walker LR, Willig MR (1999) An introduction to terrestrial disturbances. In: Walker LR (ed) Ecosystems of disturbed ground. Elsevier, Amsterdam, pp 1–15

Plant-Caterpillar-Parasitoid Natural History Studies Over Decades and Across Large Geographic Gradients Provide Insight Into Specialization, Interaction Diversity, and Global Change

Danielle M. Salcido, Chanchanok Sudta, and Lee A. Dyer

Introduction

In 1991, a first-year graduate student on a course in Costa Rica offered through the Organization for Tropical Studies (OTS) collected his first caterpillar in the wild – a geometrid caterpillar that was feeding on *Colubrina spinosa* (Rhamnaceae) (Fig. 1). It was surprising that none of the course staff nor any of the naturalists at the research station knew anything about this very abundant caterpillar. They also did not know if it specialized on that plant or what natural enemies might consume this common resource. It turns out that the caterpillar was *Cyclomia disparilis* (Geometridae: Ennominae), one of the most common caterpillars at La Selva Biological Station in Heredia Province, Costa Rica, with a distribution extending to forests in Guanacaste, where D.H. Janzen has been rearing caterpillars for over 40 years. Several insights from this experience in the tropics contributed to the start of a large rearing project at La Selva (Table 1). These insights included: (1) it seemed that immature stages of Lepidoptera in the tropics were mostly unknown, even for well-studied taxa such as the Papilionoidea; (2) very little was known about diet breadth of caterpillars of tropical moths, despite the importance of understanding specialization for much of ecological and evolutionary theory; and (3) considerable insight and inspiration can be gained from just discovering and describing caterpillars, their host plants, and their natural enemies. The plant-caterpillar-parasitoid rearing project at La Selva has continued since the early 1990s, with additional sites being added across the Americas in subsequent decades, and those datasets have been used successfully to address questions about specialization and interaction networks. Several decades later, it is a different experience to search for caterpillars at La Selva because of changes in caterpillar networks over time. Costa Rican caterpillars are simply harder

---

D. M. Salcido · C. Sudta · L. A. Dyer (✉)
Ecology, Evolution, Conservation Biology, University of Nevada, Reno, Reno, Nevada, USA
e-mail: dsalcido@nevada.unr.edu; csudta@nevada.unr.edu; ldyer@unr.edu

Fig. 1 *Cyclomia disparalis* (Geometridae, Ennominae) feeding on a leaf of *Colubrina spinosa* (Rhamnaceae). This is one of the most abundant caterpillar species at La Selva Biological Station Costa Rica and was the first caterpillar collected by the multi-site rearing project

to find in 2020 than they were in 1991 (Janzen and Hallwachs 2019; Salcido et al. 2020; Wagner 2020). The situation could be even worse for the wasps and flies that parasitize these caterpillars and appear to be declining rapidly at La Selva (Salcido et al. 2020).

Loss of caterpillars and parasitoids in Costa Rica is just one of many accumulating global change consequences for local ecosystems. Globally, the current decline of biological diversity is well documented (Biesmeijer 2006; Butchart et al. 2010; Keil et al. 2015; Scheele et al. 2019), and plants and insects are likely to be a significant component of that decline (Wagner 2020), but specifics of declines are far from clear (e.g., Wagner et al. 2021). For example, the recent uptick in the number of studies on insect "population" or abundance declines reveals no clear patterns for insects overall; rather there are geographic and taxonomic idiosyncrasies with some taxa or regions exhibiting increases, others showing dramatic declines, and others with no real trends (van Klink et al. 2020; Wagner 2020). For plant-caterpillar-parasitoid interactions, it is also true that we are losing plant and insect diversity, but it is difficult to say anything about abundances of very broad taxa, such as all insects or all Lepidoptera (e.g., saying "birds are declining" could be good for native species if common starlings (*Sturnus vulgaris*) are declining in North America, while the diversity of native species is increasing). Instead of studies focused on abundances of broad taxa or on other parameters related to population dynamics that were designed to examine species' population dynamics, it may be more fruitful to examine changes in local community parameters, such as diversity in a specific forest, and to uncover mechanistic details of diversity declines. Similarly, while it is very likely that we are losing genetic and functional diversity, there are many details that require more attention, such as which taxa, traits, or functions are declining most precipitously in a particular habitat. For example, there are few parameter estimates for the rates of decline of interaction diversity, which is a component of functional diversity within an ecosystem (Dyer et al. 2010). The loss of interaction diversity is a tragedy that extends beyond that of extant insect species losses (Stork 2018), since the functional roles they fill and the complex structure provided by diverse interactions maintain ecosystem function and opportunities for

Table 1 Study sites utilizing the rearing methods described in this chapter. While caterpillars.org shows a complete list of host genera from which plots can be centered on, at most sites, we sample more frequently from the following focal host genera: *Juniperus* (Cupressaceae), *Fouquieria* (Fouquieriaceae), *Quercus* (Fagaceae), *Ceanothus* (Rhamnaceae) and *Piper* (Piperaceae)

Country	State/province	Site description	Primary focal hosts	Site center	Sampling months[a]
United States	Arizona	Chiricahua Mountain Range, collection centered within 25 km of Southwest Research Station; Dataset initiated in 2001	*Juniperus* (Cupressaceae) *Fouquieria* (Fouquieriaceae) *Quercus* (Fagaceae)	31.88° N 109.21° W	July–October
		Santa Rita Mountain Range, collection centered within 25 km of Santa Rita Experimental Range; plot methods initiated in 2005	*Juniperus* (Cupressaceae) *Fouquieria* (Fouquieriaceae) *Quercus* (Fagaceae)	31.88° N 110.85°W	
	California	Eastern Sierra Nevada, centered within 15 km of Sagehen Creek Research Station; plot methods initiated in 2009	*Ceanothus* (Rhamnaceae)	39.52° N 119.81° W	June–October
	Nevada	Great Basin, centered within 100 km of Reno; plot methods initiated in 2009	*Juniperus* (Cupressaceae)	39.52° N 119.81° W	June–October
	Louisiana	New Orleans; initiated in 2000	*Quercus* (Fagaceae)	29.95° N 90.07° W	
	Florida	Florida Panhandle, centered within 30 km of Eglin Airforce Base; initiated in 2012	*Quercus* (Fagaceae)	30.62° N 86.55°W	June–November
	Alabama	Southern Alabama, centered within 30 km of Solon Dixon Forestry Center; initiated in 2015	*Quercus* (Fagaceae)	31.14° N 86.7° W	June–November

(continued)

Table 1 (continued)

Country	State/province	Site description	Primary focal hosts	Site center	Sampling months[a]
Mexico	Campeche, Yucatan, Quintana Roo	Yucatan Peninsula centered on Nuevo Becal; plot methods initiated in 2016	*Piper* (Piperaceae)	18.50° N 88.29° W	January–December
Costa Rica	Heredia	La Selva Biological Research Station; rearing initiated 1991, data for large-scale analyses initiated in 1998	*Piper* (Piperaceae); dozens of additional focal genera	10.42° N 88.29° W	January–December
Ecuador	Napo	North-Eastern Andes, centered on Yanayacu Biological Research Station; initiated in 2001	*Piper* (Piperaceae); dozens of additional focal genera	00.36°S 77.53°W	January–December
Peru	Chanchamayo	Fundo La Genova Research Station, La Merced; initiated in 2016	*Piper* (Piperaceae)	11.06° S 75.33°W	January–December
Brazil	Brasilia	Cerrado forests within 100 km around Brasilia; plot methods initiated in 2012	*Piper* (Piperaceae); dozens of additional focal genera	15.72° S 47.95°W	January–December

[a]Months listed represent a general window during which we collect and vary year to year depending on the onset of the growing season at each site, which depends on conditions like snowmelt or monsoons

evolution (Dyer 2018). In this chapter, we summarize an approach to natural history that allows for documenting trophic webs of plant-caterpillar-parasitoid interactions and, given long enough time series, provides data for testing hypotheses about the effects of specific global change parameters on species and interactions.

Diversity and Networks of Species Interactions

Metrics of diversity have been a staple of community ecology since its inception, and almost all metrics, including a variety of species diversity parameters and measures of functional and genetic diversity, have been examined in the context of the negative effects of global change on biotic communities. On the other hand, despite an explosion in ecological network studies, only a few studies have utilized time series to examine changes in network parameters over the past decades for any biotic communities (Salcido et al. 2020). Despite the fact that collecting and rearing caterpillars have been a staple of natural history for centuries (Wallace 1878; Hespenheide 2011; Greeney et al. 2012), the integration of network science to explain ecological phenomena is relatively new. Historically, network science within ecology was limited to studies of trophic interactions that estimated a small subset of current network parameters (Elton 1927; Pimm 1979; Paine 1980; Pimm et al. 1991). Relatively recent computational advances have made it feasible to calculate more sophisticated network metrics and have led to interesting observations, such as replicable scale-free patterns among empirically estimated networks (Ings et al. 2009). Examples of modern network topologies that have been quantified and contribute to ecological theory include mutualistic (Olesen and Jordano 2002), antagonistic (Tylianakis et al. 2007), and metabolomic networks (Albert et al. 2000; Kaling et al. 2018; Sedio et al. 2018; Sedio et al. 2020). Two issues that hinder inferences from most published networks are the lack of accurate natural history data and inadequate attention to the appropriate scale at which networks should be estimated (Dyer 2018). For example, many published pollination networks are actually visitation networks; thus, they include parasites and neutral visitors, which limit inferences about network parameters for actual mutualists (King et al. 2013); similar problems exist for caterpillar-enemy networks when plant-caterpillar affiliations are inferred based on simple field observations that are not accompanied by rearing. For the scaling issues, networks that are estimated based on published data for a region are essentially "metanetworks" that summarize smaller functional networks across the landscape (Dyer 2018). Comparisons of mutualistic and food web network topology (Thébault and Fontaine 2010) and bacteriophage networks (Flores et al. 2013) at different scales indicate that processes driving network structure are variable across scales and types of networks.

Why should caterpillar biologists utilize network approaches? In addition to providing a functional heuristic summary of caterpillar trophic interactions, the network approach allows for mathematical summaries of both species (nodes) and interactions (edges), estimates of interaction diversity, and tests of nonrandom patterns in network structure (Poisot et al. 2016). Interaction diversity quantifies the number of unique links connecting species within a network (Dyer et al. 2012) and has been associated with greater community stability, productivity, and ecosystem services (Mougi and Kondoh 2012; Poisot et al. 2016). Trophic interactions can provide important functional roles and ecosystem services, such as caterpillar population regulation by parasitoids; thus, it is likely that parameters from trophic interaction networks are related to ecosystem function. More generally, through mechanisms such as complementarity (Poisot et al. 2013) and redundancy (Peralta et al. 2014), interaction diversity increases the opportunity for interaction reorganization and the maintenance of ecosystem function during perturbations. Nevertheless, we still lack sufficient empirical data on interaction webs to guide conservation of complex ecosystems (Tylianakis et al. 2010).

Focus of This Chapter

We have answered macroecological and evolutionary questions (Fig. 2) with data collected from our rearing program and outline our approach below. We review the results from this approach that have contributed to understanding the evolution of specialization (Dyer 1995; Gentry and Dyer 2002), biological control (Dyer and Gentry 1999), patterns of insect defense (Gentry and Dyer 2002), latitudinal gradients in diet breadth and parasitism (Stireman et al. 2005; Dyer et al. 2007; Forister et al. 2015), and structure of food webs (Scherrer et al. 2016; Lepesqueur et al. 2018; Dell et al. 2019). Although the rearing methods were designed to address specific basic research questions in ecology and evolution, we also focus on how the long-term (albeit still limited) and large-scale nature of the rearing approach has allowed for hypothesis tests relevant to the effects of climate change on parasitoids (Stireman et al. 2005), declines in insect diversity (Salcido et al. 2020; Wagner 2020), changes in interaction diversity and other network parameters (Dell et al. 2019; Salcido et al. 2020), and erosion of ecosystem services (Stireman et al. 2005; Salcido et al. 2020). In addition to reviewing the results from this approach, we address questions about the changes in network parameters over time by examining two representative sites that have used this approach: La Selva Biological Station in Costa Rica and the Great Basin and Eastern Sierra Nevada site centered at Reno, Nevada (Table 1).

The Plant-Caterpillar-Parasitoid Network Approach

Collection of Immature Lepidoptera

Since 1991, we have used standardized methods for general collections of caterpillars and associated parasitoids from host plants, with some additions for standardizing caterpillar abundances and levels of parasitism for a site (i.e., plots and calculating person hours). The method consists of collecting externally feeding or shelter-building Lepidoptera species (excluding leaf miners and gallers), which includes over 40 families in the current database (Forister et al. 2016). Successful long-term datasets cataloguing ecological networks require a commitment to inclusive and diverse collaborative networks (Hampton and Parker 2011). For example, with respect to collection alone, we implement teams of local collaborators, principal investigators for individual sites, graduate and undergraduate students, Earthwatch volunteers, and other community scientists. Collecting is conducted throughout the year at tropical sites, while collections at temperate sites are focused on the growing season. For Arizona, the collecting season depends on the onsets of monsoons, whereas for the Great Basin and Sierra Nevada, sampling depends on snow melt (Table 1). To allow for standardized estimates of caterpillar-parasitoid abundances and interaction diversity, at all research sites, we collect in temporary

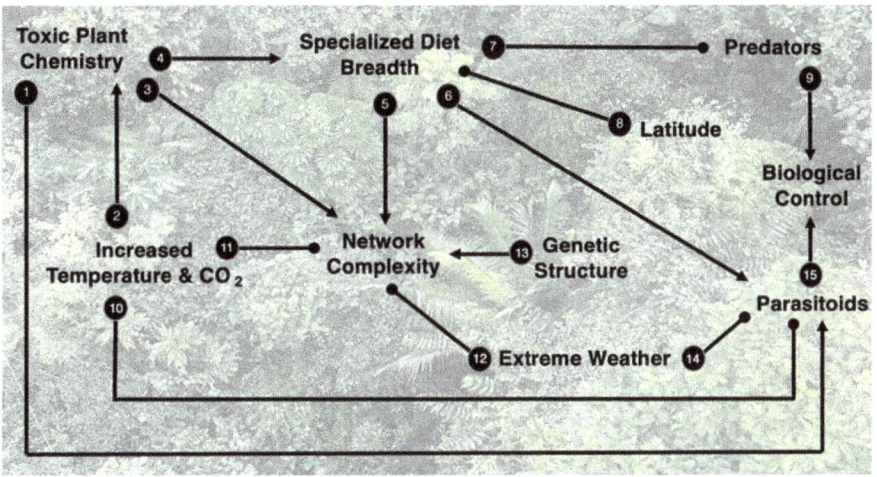

Edge	Publication	Edge	Publication
1	Slinn et al. 2018; Smilanich et al. 2009	8	Dyer et al. 2007; Forister et al. 2015
2	Dyer et al. 2013	9	Dyer & Gentry 1999; Dyer & Floyd 1993
3	Richards et al. 2016	10	Dyer et al. 2013; Salcido et al. 2020
4	Dyer 1995; Smilanich et al. 2009	11	Salcido et al. 2020
5	Dell et al. 2019; Lepesquer et al. 2018; Pardikes et al. 2018; Scherrer et al. 2016	12	Salcido et al. 2020
		13	Glassmire et al. 2016
6	Dyer 1995; Gentry & Dyer 2002	14	Stireman et al. 2005; Dyer et al. 2013; Salcido et al. 2020
7	Dyer 1995	15	Gentry & Dyer 2002

Fig. 2 Heuristic path diagram, summarizing relationships supported by data from the plant-caterpillar-parasitoid rearing research across the Americas. The diagram highlights some results uncovered using the rearing databases. Positive relationships are represented by pathways with arrowheads, negative relationships with bullet heads, and path coefficients correspond to publications listed in the table. *Paths 1 and 15*: Different classes (Smilanich et al. 2009a, b) and diverse mixtures (Slinn et al. 2018) of phytochemical compounds increase caterpillar mortality by parasitoids with consequences for biological control. *Paths 6, 7, 8, and 9*: The effect of plant chemistry on parasitism and network complexity is mediated by diet breadth, which is more constrained at lower latitudes; effects are greater for specialists versus generalists. *Paths 3, 10, 11, 12, and 14:* Changes in climate, such as increases in extreme weather events (Stireman et al. 2005; Salcido et al. 2020) or mean temperature and CO_2 (Dyer et al. 2013), negatively affect parasitism and network complexity, but increases in these parameters can also positively affect network complexity when mediated by variation in plant phytochemistry. *Paths 2, 4, 5, and 13*: Geographic variation in phytochemical defenses contribute to genetic structure of specialist herbivores (Glassmire et al. 2016), yielding geographic mosaics that modify network complexity through effects on plant-herbivore-parasitoid diversity. Many of these results also involved combining data with other rearing programs around the world

5-, 10-, 20-, or 25-m-diameter plots, depending on the density of plants and caterpillars at the site. The plots are chosen using the criteria as follows: each plot must be at least 10 m from a trail or road and are centered around 1 of over 100 focal plant genera (for the most well-sampled genera, see Table 1; also see Dyer et al. 2007, Forister et al. 2016, and www.caterpillars.org for complete lists of focal plant genera); these plant genera include herbs, shrubs, and trees that are common in the study area and include taxa for which we have the most data. Depending on the collection location, focal plants range from shrubs to trees that vary in maturity. Typically, the plot is divided into four equal wedges, and two to four investigators spend 60 min in each wedge looking for caterpillars on all the plants with foliage within reach (on average 2 meters above the ground). Different collection methods are used for caterpillar sampling in tropical and temperate sites. At tropical sites caterpillars are sampled using visual inspection, while canvas beat sheets are used at temperate sites. Once the plot has been searched, leaf abundance for all plants present in the plot are estimated via quick counts. Mean leaf sizes are available for all species in our focal genera or measured and calculated for species encountered without previous records in our dataset. This collecting strategy allows for intensive searches of specific hosts while providing basic data on caterpillar loads of individual plants (as opposed to biased searches for caterpillars already known to the collectors), parasitoid loads on caterpillar species, as well as quantitative data on the number of species per leaf area. These data are used to estimate caterpillar densities and to estimate a wide range of network parameters. Adjacent plots that are surveyed in the same period of time are joined to increase the scale of networks and to address scaling issues for networks and community interactions. All sites also include regular "general collecting," which involves collecting all externally free-feeding or shelter-building caterpillar species encountered in forest patches throughout the sites.

Rearing

All collected caterpillars are reared employing standardized methods at all sites, with slight variations depending on weather conditions, rearing facilities, and the number of available rearing days. We rear caterpillars individually in clear plastic cups or vials, clear plastic bags, or glass jars in our rearing barns or laboratories at ambient temperature and humidity. Every 1–3 days, we dispose of frass (or save it for analysis of chemistry or presence of viruses) and replace the foliage in each container and add vermiculite, soil, or rotten wood to the containers for caterpillar species that require a substrate for pupation. All pupae are checked twice daily to collect any adult Lepidoptera or parasitoids that have emerged from the pupae. These are allowed to fully harden before being stored in a freezer prior to pinning or prior to storage in alcohol. If a caterpillar dies or nothing emerges from a pupa, dissections are performed to search for parasitoids. Rearing success from larvae to adult varies across sites, taxa, and instar at which the caterpillar was collected. The mode for rearing success to adult across sites and families is 33%. To prevent disease transmission, all plastic bags are used only once, and jars, cups, and vials are cleaned with bleach between rearings. All species of caterpillars are photographed with a Canon EOS digital camera with macro lens or similar device and described

Fig. 3 A mix of caterpillars (some are still unidentified) from the Great Basin Sierra Nevada site. (**a**) *Juniperus osteosperma* (Cupressaceae) plot near Gardnerville in the Great Basin of Nevada. (**b**) *Ceanothus velutinus* (Rhamnaceae) plot in the Sierra Nevada, near Thomas Creek, Reno, Nevada. (**c**) *Chionodes* sp. (Gelechiidae) on manzanita – *Arctostaphylos patula* (Ericaceae). (**d**) *Recurvaria* sp. (Gelechiidae) with its silk shelter on *Amelanchier pallida* (Rosaceae). (**e**) *Sphinx sequoia* (Sphingidae) feeding on *J. osteosperma*. (**f**) *Mesogona* sp. (Noctuidae) on *A. patula*. (**g**) *Strymon melinus* (Lycaenidae). (**h**) *Apodrepanulatrix* sp. (Geometridae) on a flower of *C. velutinus*. (**i**) *Digrammia* sp. (Geometridae) camouflaged on a leaf of *J. osteosperma*. (**j**) *Cosmia* sp. (Noctuidae) feeding on *Amelanchier pallida* and parasitized by eulophid ectoparasitoids

in a thorough, standardized manner (e.g., Fig. 3c–j). Additional photographs and descriptions follow for subsequent instars and for pupae if time allows.

Curation and Taxonomic Identification

Adult lepidopteran specimens are pinned and curated using standard techniques (Forister et al. 2015 – supplemental material). Parasitoids are prepared according to the preferences of the taxonomists who identify them (pinned or in vials of alcohol), and eventually pinned specimens are properly curated and deposited in museums. After voucher specimens have been collected for identification and description purposes, a series of additional specimens are preserved in 95% ethanol for future molecular systematic work. All curated specimens receive a standard location label listing the country, province, collection site, elevation of the site, longitude and latitude, month and year of collection, and affiliation of collecting site. In addition, the unique voucher code connecting the specimen to its event-based record in the database is added below this, and another label is added when appropriate, identifying the specimen to family, genus, and species. Prior to taxonomic identification from specialists, species are morphotyped using morphological characters with reference to the image library or voucher specimens; for many individuals, when possible, morphospecies are confirmed using standard wing traits, dissections of genitalia, and in some cases DNA sequences. We typically do not use "barcoding" approaches (i.e., generating COI sequences), rather the focus is on full genomics data to address population genetics hypotheses or phylogenomics for generating phylogenetic hypotheses.

Specimens in liquid media include all the above standard information added to the vials. Every collection event (the act of finding a caterpillar) receives a unique voucher code. At 1–3 year intervals, specimens that are fully identified and curated are deposited to appropriate museums in Costa Rica, Ecuador, Brazil, and the United States, including the University of Nevada Museum of Natural History in Reno, Nevada. Voucher specimens examined by the various taxonomic authorities including newly established types and undescribed species are housed in collections of the taxonomists' preference to facilitate further description and systematic studies of the material, unless our permits specify otherwise. All Lepidoptera, Hymenoptera, and Diptera reared in this project are provisionally identified by the rearing and curation crews and then via a network of collaborating taxonomists (see www.caterpillars.org).

Climate Data

All research sites provide a base set of meteorological data consisting of precipitation, temperature, humidity, solar radiation, wind speed and direction, and evaporation dating back 20–100 years. The majority of sites include a central research station or nearby urban center that have established meteorological programs, including a combination of manually obtained data (e.g., maximum and minimum temperatures, precipitation, humidity, evaporation, and wind speed) and electronically obtained data (e.g., solar radiation, temperature, humidity, wind speed, and direction). These databases are managed by the research stations or urban centers and are freely available to the research community. To acquire climate data for sampling sites within the United States and outside the range of research station monitoring stations, we rely on PRISM data (Parameter-elevation Regressions on Independent Slopes Models; PRISM Climate Group, Oregon State University, http://prism.oregonstate.edu). PRISM data files are compiled and summarized in R using the prism package (v. 0.1.0). PRISM estimates daily, monthly, or annual climate data for point locations using a regression function that incorporates data from a network of monitors. For the purposes of analyses presented in this chapter, we have summarized the monthly means of precipitation, minimum temperature (T_{min}), and maximum temperature (T_{max}). Precipitation and temperature anomalies were defined as the frequency of monthly climate values greater than or equal to 2.5 standard deviations above the mean.

Data Management and Dissemination

Our data are compiled and organized in a manner compatible with other studies in other regions (e.g., see Stireman et al. 2005; Dyer et al. 2007: 200; Forister et al. 2015). This facilitates the comparison of plant-herbivore-parasitoid associations and interaction diversity across a wide range of ecosystems. These combined datasets constitute a powerful ecological tool. Our existing long-term database at http://www.caterpillars.org is updated every 3–5 years and expanded to include new approaches. Each caterpillar collected is registered. Collectors record several types of data directly into two different databases (on datasheets, field notebooks, or laptops) – rearing and plots databases. All of these data are entered into temporary databases at the field site and then are carefully checked and added to the main database and then later to a web database. Every individual parasitoid, and its rearing parameters, is linked to the record of its caterpillar host; each individual plant is linked to its associated herbivore; and all of these data are linked to the plot data or to specific GPS data.

Quantifying Diversity of Interactions

Using the 5–25-m-diameter circular plot data described above, we assess the richness of interactions, which will include all two-link (caterpillar-plant and parasitoid-caterpillar) and three-link (parasitoid-caterpillar-plant) chains found in a plot. Inferences made when analyzing interaction diversity as a response variable in models that compare effects between different habitats, elevations, and latitudes, depend on certain design-based and model-based assumptions. Generally, observations of interactions are assumed to be sampled without bias and have unknown probability distributions. Assumptions include the following: (1) interactions are sampled at random; (2) interactions are homogenously distributed within a plot; (3) any probability distribution of interaction abundance is possible (i.e., we do not assume log-normal distributions that characterize species abundance); and (4) interactions are equally likely to be encountered at all locations compared. We have used or are currently using this method at various Earthwatch-funded sites in Costa Rica, Ecuador, Arizona, Louisiana, and California/Nevada, as well as sites funded by other sources in Mexico, Peru, and Brazil. A large collaborative network participates in this approach (Table 1). In addition to local network parameters from plots, a regional network is defined as all species interactions recorded from plot and haphazard collection within a region, or a "metaweb."

Addressing Basic Questions in Ecology and Evolution

Caterpillar Defenses and the Evolution of Specialization

The observational approach to trophic interactions is a powerful method for generating rigorous hypothesis tests, especially when those tests are paired with experimental data (Fig. 2). The initial focus of the rearing dataset in Costa Rica was to pair caterpillar specialization data from collecting and rearing to antipredator defense experiments (Dyer and Floyd 1993; Dyer 1995, 1997) with broader observational data focused on parasitism levels for specialist versus generalist parasitoids and anti-parasitoid defenses in caterpillars. Our rearing data suggested that chemical defenses are among the most effective defenses of herbivores against natural enemies and that while predation exerts effective selective pressures for more specialized diets, high levels of parasitism may select for broader diets (Dyer 1995; Dyer and Gentry 1999; Gentry and Dyer 2002; Lampert et al. 2014). The higher parasitoid loads on specialists are interesting because most endoparasitoids of caterpillars spend all larval stages within host tissues; thus, some have hypothesized that the negative effects of sequestered secondary metabolites may be more pronounced for parasitoids than for predators (Petschenka and Agrawal 2016). In contrast to this prediction, the vulnerable host and safe haven hypotheses, developed from a decade of rearing results in Costa Rica (Gentry and Dyer 2002), posit that (1) predators avoid specialists that have sequestered toxic secondary metabolites, (2) phytochemically defended plants may host specialist herbivores that are immunocompromised due to sequestration, and (3) as a consequence, specialists on toxic hosts are better hosts for parasitoids, resulting in higher levels of parasitism (Smilanich et al. 2009a, b; Lampert et al. 2010; Slinn et al. 2018).

The large rearing projects have allowed for expanded tests related to insect defense theory (reviewed by Greeney et al. 2012), including predictive models for biological control based on larval insect defensive traits (Dyer and Gentry 1999). Caterpillar feeding is made dangerous by natural enemies, but insects protect themselves with a diverse suite of defenses that are employed before, during, or after encounters with enemies. Testing hypotheses about how caterpillars repel, escape, or avoid detection by predators (and even applying results to guide biological control) requires experimental data, but rearing projects help put these defenses into context and provide data relevant to how these defenses may prevent parasitism. For example, observation of anti-parasitoid behaviors such as postattack strategies through the removal or destruction of parasitoid eggs or larvae has come from natural history observational data that can be collected in large rearing projects. Of course, focused studies on individual species will provide much more detail on specific defensive mechanisms, but such studies are often one of the beneficial products of the multi-species rearing approach (e.g., Hansen et al. 2017). One conclusion from inferences about caterpillars provided by the rearing studies has not changed: the basic natural history of most caterpillars is still unknown, especially in the tropics (Hespenheide 2011; Greeney et al. 2012). With global change intensifying, especially habitat loss and climate change, the ever-increasing devaluations of descriptive taxonomic and natural history (Futuyma 1986; Futuyma and Moreno 1988; Greene 1994; Noss 1996; Greeney et al. 2012) have ensured that many aspects of caterpillar defenses and countless interesting details of plant-caterpillar-enemy interactions will not be discovered. Continuation of large-scale rearing projects facilitates the documentation of life history traits relevant to defense against enemies, in addition to answering other questions in evolutionary ecology.

Interaction Diversity Across Environmental Gradients

At all of the rearing sites, we have documented interaction diversity and tested prominent diversity hypotheses as well as hypotheses about specialization (Dyer et al. 2007; Martén-Rodríguez et al. 2010; Forister et al. 2012). Most notably, this approach allows accurate measures of consumer specialization via ensuring consumption with the rearing experiments and sampling at relatively large spatial and temporal scales (Forister et al. 2015). A major challenge in community ecology and evolutionary biogeography is to explain the latitudinal gradient in diversity, and for herbivorous insects, specialization may be one mechanism that allows for greater species richness at a given locale. We tested this hypothesis by combining our data with collaborators' rearing datasets and comparing host specialization by caterpillars for eight different New World Forest sites ranging in latitude from 15°S to 45°N (Dyer et al. 2007; four of the sites in this comparison are described in this chapter; Table 1). We found that larval diets of tropical Lepidoptera are more specialized than their temperate forest counterparts: tropical species on average feed on fewer plant taxa than temperate species (Fig. 2). As a result of this latitudinal gradient in specialization, there is greater turnover in caterpillar species composition (greater β diversity) between tree species in tropical faunas than in temperate faunas (Fig. 2). Hypothesized associations between diet breadth and other variables correlated with latitude, such as temperature and plant diversity, were confirmed with studies of specific taxa across elevational (e.g., geometrid caterpillars, Rodriguez-Saona et al. 2010) or disturbance gradients (macrolepidoptera in Louisiana; G.L. Gentry, unpublished manuscript; Dell et al. 2019). Greater effect sizes of consumers on resources in the tropics can contribute to higher specialization in tropical faunas in a variety of ways. For example, there may be higher beta-diversity of plant defensive compounds in tropical versus temperate forests, and there are more diverse and chronic pressures from natural enemy communities – both conditions make feeding on tropical plants more difficult (Coley and Barone 1996; Connahs et al. 2009; Salazar and Marquis 2012; War et al. 2012; Martin et al. 2013). If this is true, then similar patterns could be apparent for higher trophic levels (e.g., parasitoids; see Hawkins 1994 for a literature-based test), due to well-defended caterpillars and high levels of predation on most insects, which will influence patterns of interaction diversity. Overall, tropical plant-herbivore-parasitoid webs should be characterized by higher degrees of specialization, higher node diversity, and greater evenness among edges.

Global Change and Trophic Interaction Networks

Climate Change and Parasitism

When examining interaction webs across gradients, as discussed in the previous section, some important gradients to examine in the Anthropocene are temporal and spatial gradients related to climate change, such as variation in temperature and precipitation. Climate change studies that examine effects of climate on insects typically focus on direct effects of climate change parameters, especially temperature, on insect population dynamics (Stange and Ayres 2010). One conclusion from much of this research is that insect outbreaks are expected to increase in frequency and intensity (Dyer et al. 2012). Additional studies have examined indirect effects, such as disruption of community interactions (e.g., Nooten et al. 2014), changes in insect host plant chemistry (Dyer et al. 2013), and phenological asynchrony (Donoso et al. 2016; Mayor et al. 2017; Renner and Zohner 2018). Many of these studies have also focused on mean changes in climate parameters, but it is quite clear that the impacts of climatic variability are equally large. For example, we used parasitism data from our rearing sites and combined these data with ten other sites to compare caterpillar-parasitoid interactions across a latitudinal and climatic gradient that included a clear range of variation in year-to-year precipitation (Stireman et al. 2005). We found that a moderate increase in this rainfall variation was associated with a biologically significant decrease in levels of parasitism – from approximately 27% to less than 7% over 15 geographically dispersed databases across the Americas. When dipteran versus hymenopteran parasitoids were considered separately, we found that the overall pattern was driven by the more specialized parasitic Hymenoptera. Phenological asynchrony was likely a mechanism causing this pattern because specialist consumers are vulnerable to slight changes in host populations, while asynchrony among generalist parasitoids is less consequential due to greater dispersal abilities and ease of switching to alternative hosts. While caterpillar populations are able to persist after floods and droughts via diapause, dispersal, or other adaptations, small parasitic wasps are possibly unlinked from their host phenology during such weather events. We experimentally confirmed this type of asynchrony in laboratory experiments, where increases in CO_2 and temperature quickly delinked the development of caterpillars and their parasitoids, such that the caterpillars developed too fast and the parasitoid populations crashed (Dyer et al. 2015). Given the important role of parasitoids in regulating insect herbivore populations in natural and managed systems, we predicted an increase in the frequency and intensity of caterpillar outbreaks through phenological asynchrony as climates become more variable (Stireman et al. 2005; Dyer et al. 2015). As we outline below, these predictions are being corroborated by our rearing data at one of our sites.

Global Change, Caterpillar Declines, and Network Erosion

Recently, Humberto Garcia Lopez, a long-term caterpillar collector and researcher at La Selva, noted that it is fairly obvious that caterpillars and their parasitoids have steeply declined since he first started working with us in the 1990s. Careful data analysis has confirmed this (Salcido et al. 2020; Figs. 4–7), and the rate of loss of associated parasitoid richness is particularly noteworthy (Figs. 4d and 7b), causing reduction in parasitism frequency (Fig. 4f). This decrease in parasitism levels was predicted by previous papers using the rearing datasets (Stireman et al., 2005) and ancillary experiments (Dyer et al. 2013) to demonstrate that high variation in climate parameters and increases in extreme weather events, such as floods and

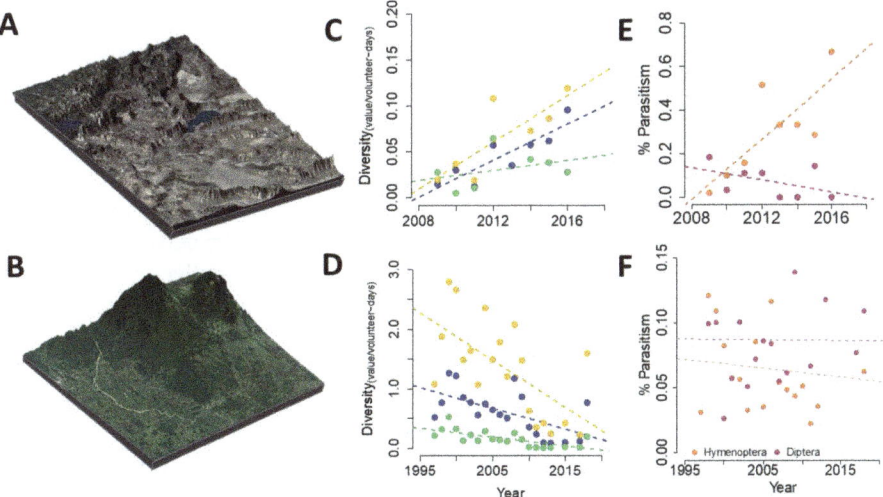

Fig. 4 Time series data at two collection sites from the rearing projects across the Americas. (**a**) The shorter time series are data from the Great Basin and Sierra Nevada (9 years, 2009–2017). (**b**) A longer time series uses data from La Selva Biological Research Station Braulio Carrillo National Forest (22 years, 1997–2018; Salcido et al. 2020). Diversity (Simpson's species or interaction equivalents) trends for caterpillars (blue points), associated parasitoids (green), and interaction diversity (yellow) for the Great Basin (**c**) versus La Selva (**d**). Dotted lines depicting declines or increases are based on a Bayesian regression, with 95% credible intervals. Note the scale of the y-axis is different for the two sites (**e, f**). Percent parasitism for hymenopteran (orange) and dipteran (pink) parasitoids at the Great Basin (**e**) versus La Selva (**f**) sites

droughts, are associated with substantial declines in parasitism – or even extinction of parasitoids within experimental microcosms (Dyer et al. 2013). Phenological asynchrony is one mechanism linking extreme weather events to declines in parasitism – caterpillars change their development rates or go into diapause in response to changes in weather conditions or due to floods, and parasitoids are unable to locate their hosts efficiently. Such asynchrony is more likely to affect parasitic Hymenoptera, which are more specialized and dependent on specific phenologies of one host versus the more generalist parasitic flies. Data from across the rearing sites corroborate this prediction (Stireman et al. 2005), and data from La Selva indicate that parasitic wasp diversity is declining more than that for tachinid flies (Fig. 4f).

The global change parameters most commonly examined with respect to the loss of biodiversity are climate change, habitat loss, fragmentation, invasive species, pesticides, and pollution. For La Selva, all of these are likely to affect both caterpillars and their parasitoids and have likely contributed to their declines. Lowland tropical forests are facing rapid conversion to high-intensity, high-output agriculture from former homesteads and pastureland. La Selva is no exception. Similarly, all of our sites will eventually house very different tri-trophic webs due to these negative impacts. The very clear effect of climate change variables on caterpillar and parasitoid webs at La Selva was uncovered using structural equation models (Fig. 7b). While large shifts in climate variables are clearly responsible for these shifts and for the loss of parasitism as an ecosystem service, the path coefficient from time to caterpillar and parasitoid richness (which corrects for the effects of climate) was still quite large, suggesting that other unmeasured factors have contributed to these declines. These other factors likely include increased pesticide use (Fagan et al. 2002; de la Cruz et al. 2014) in Sarapiquí and the fragmented nature of La Selva.

In contrast, when we compare these patterns of diversity loss to our temperate site in the Great Basin, *diversity* does not appear to be declining (Fig. 4c), but there are steep declines in caterpillar *abundances*, as measured by the encounter frequency for the most common caterpillar families (Fig. 5a). The networks have also changed, but the changes are not as severe – there is a much smaller drop in interaction diversity from the beginning of the rearing effort to recent times (Fig. 6). There appear to be some effects of climate change parameters, particularly the large negative effects of increases in precipitation anomalies and minimum temperatures on caterpillar and interaction richness, which were similar for both sites (Fig.7). However, the limited time series (and associated lack of model fit due to limited data) make it difficult to be confident about those inferences. Furthermore, the negative effect on ecosystem services provided via biocontrol by parasitoids found at La Selva was not observed at the Great Basin site (Fig. 4e). It is clear that the length of the time series is very important for making appropriate inferences (Janzen and Hallwachs 2019) – if the La Selva dataset is cut to the past 9 years, the signal also diminishes to weak or nonexistent. This lack of appropriate time series could also be the case for other decade-long studies that do not detect signals of insect declines (Crossley et al. 2020; Outhwaite et al. 2020).

The limited temporal extent for the Great Basin database is likely to be one of the more important explanations for inconsistencies among changes over time in

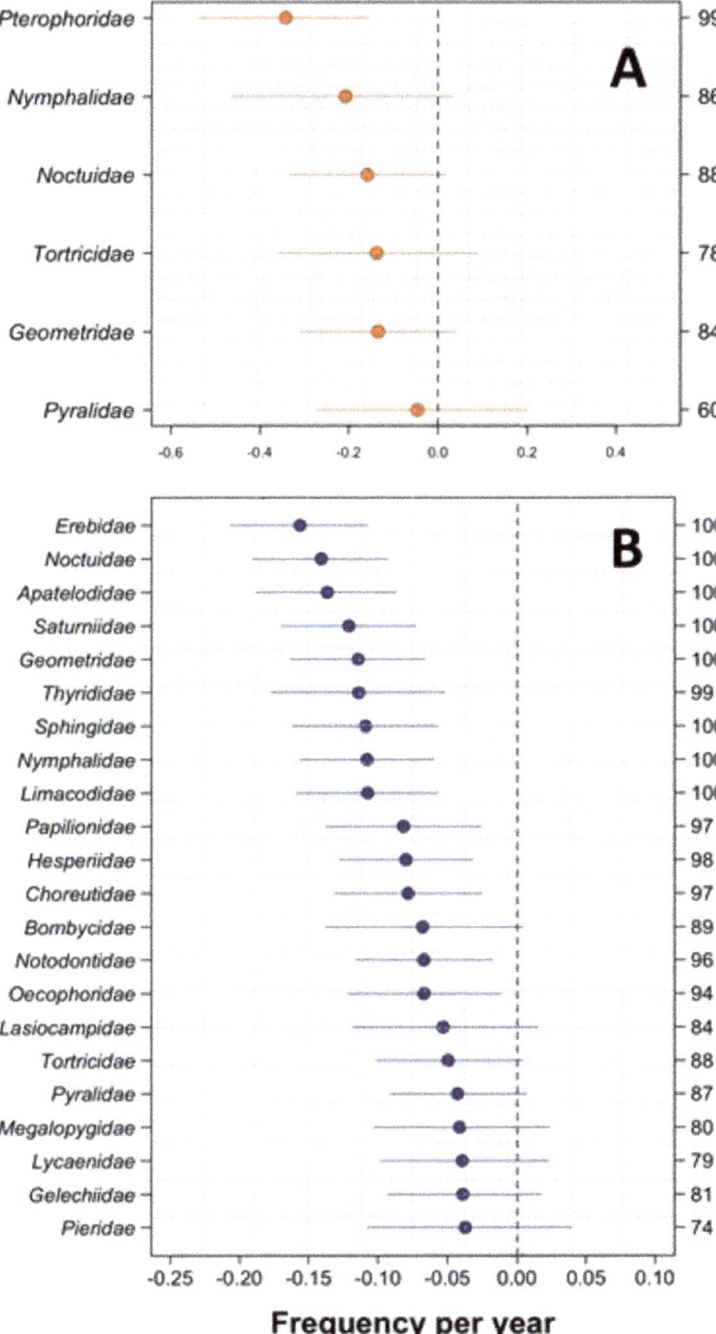

Fig. 5 Family-level patterns in caterpillar densities (encounter frequencies) across the years for (a) the Great Basin and Sierra Nevada site and (b) the La Selva, Costa Rica site (Salcido et al. 2020) displayed as point estimates for beta coefficients and associated 80% credible intervals. Family names are listed on the left margin, and probabilities of a decline are on the right margin. Units of the year coefficient are on a log-odds scale

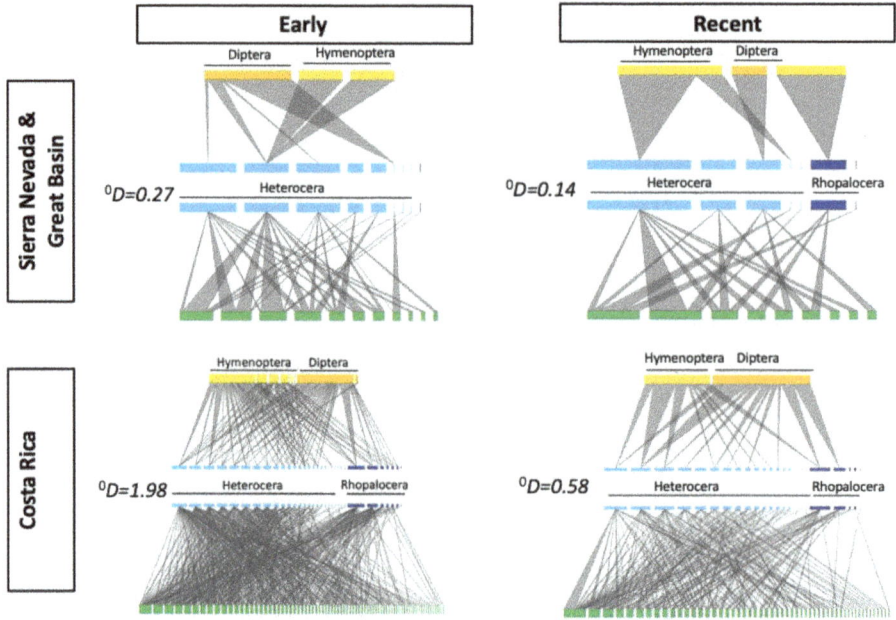

Fig. 6 Networks summarizing interactions for La Selva, Costa Rica, and the Great Basin datasets. For La Selva, the networks summarize the first (early) and last (recent) quartiles of the 20-year time series. The Sierra Nevada and Great Basin networks summarize the first and last quartiles of the 9-year dataset. Although the time series are different and the particulars of the network changes are distinct, it is clear that both sites are characterized by less reticulate webs in recent years. Mean interaction richness (^0D) is included to the left of each network and was calculated by summarizing all unique plant-caterpillar and caterpillar-parasitoid interactions

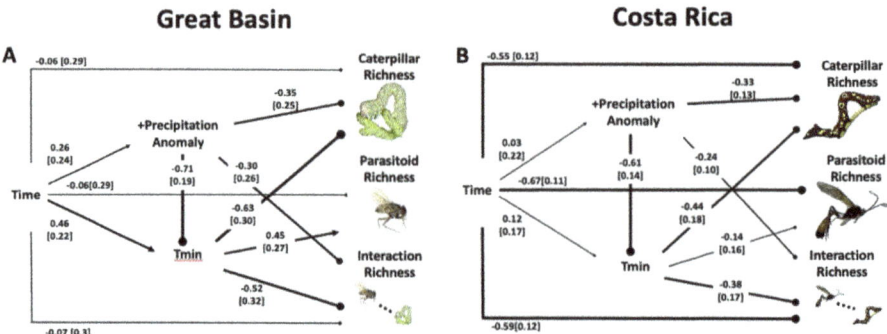

Fig. 7 Structural equation models (SEM) estimating the effects of climate variables on caterpillar, parasitoid, and interaction richness for the Great Basin and Costa Rica sites. The same models (reported in Salcido et al., 2020, for Costa Rica) were examined for each site and focus on associations between time (year), richness, daily minimum temperatures, and precipitation anomalies; Costa Rica model fit: χ^2: 0.02, p = 0.89, df = 1; Great Basin model fit: χ^2: 3.2, p = 0.07, df = 1. Path coefficients are standardized, and the width of the arrows is scaled based on the magnitude of path coefficients. Arrows represent positive associations and lines with circle represent negative associations.

diversity (Fig. 4c), abundance (Fig. 5a), and richness (Figs. 6, 7a) observed from data collected in the Great Basin – data from La Selva exhibited declines across all measures. However, it is also likely that site-level differences may explain the different patterns at the two sites. The 16 km^2 of available sampling area at La Selva is quite small compared to the spatial extent of sampling sites in Nevada and California (~32,000 km^2 centered on Reno), and it is clear that trophic interactions are likely to exhibit substantial site-to-site variation in their response to climate variables (Pardikes et al. 2015; Nice et al. 2019), so in addition to longer time series, future analyses using the Great Basin rearing data would benefit from greater sampling within elevational bands or within multiple specific focal sites across the Great Basin and Sierra Nevada. While models are likely to be context dependent as shown comparing these two sites, it is interesting to consider the consistently large effect of rising minimum temperature on caterpillar richness for both sites, which is consistent with patterns observed elsewhere (Nice et al. 2019).

Conclusion

Rearing projects fall squarely into the realm of natural history, and all ecological studies benefit from improved natural history. The methods described here are straightforward and transferable, and fortunately the number of similar caterpillar rearing projects continues to grow globally (Forister et al. 2015; Staude et al. 2016; Seifert et al. 2020). Whether the goal is to describe all of the species of caterpillars and their associated host plants and parasitoids in a region or to document the unwillingness of interaction accumulation curves to asymptote (Price et al. 1995), rearing caterpillars to adult moths or parasitoids is a worthwhile and important endeavor. Beyond the potential impacts to ecology and related fields, our rearing projects across the Americas have brought the worlds of metamorphosis, parasitism, species discovery, plant chemistry, and caterpillar frass to well over 1000 community volunteers (Dyer et al. 2012). It has not escaped our notice that none of our time series are long enough to provide confident estimates of caterpillar declines, parasitoid reductions, or network erosion. Yet, sites like Costa Rica for which we have the most data provide strong evidence that the parasitoids are disappearing from our tropical food webs due to climate change and other factors, and the networks of interactions are dramatically changing. With continued focus and commitment, all of our sites across the Americas are inching closer to more thorough time series, allowing for careful estimation of how network parameters are changing across latitude, along elevational gradients, within disturbance gradients, and for different subtaxa of Lepidoptera, Hymenoptera, and Diptera. The datasets have come a long way since Dyer, the first-year graduate student who collected that striking geometrid caterpillar back in the early 1990s (Fig. 1), first started this rearing project with his collaborator, Grant Gentry. The authors of this chapter will all continue to search for rare and common little caterpillars alike, and hopefully others will have the chance to be surprised by a mix of common little green *Eois* (Geometridae)

Fig. 8 A third instar *Hemeroplanes triptolemus* (Sphingidae) observed on its hostplant, *Fischeria panamensis* (Asclepiadaceae) in Fundo La Genova, La Merced, Chanchamayo, Peru. A *Piper* (Piperaceae) shrub is in the background. This caterpillar was not collected

caterpillars and rare crazy snake mimics like *Hemeroplanes triptolemus* (Sphingidae) (Fig. 8). Rearing everything from the big, showy sphingids to smaller, more secretive geometrids has a lot to tell us about science, and we are quickly losing a large chunk of caterpillar and parasitoid diversity, so now is the time to ramp up those rearing efforts.

Acknowledgments We thank a myriad of collaborators for their substantive contributions to this work, including G. Gentry, J.O. Stireman, S. Shaw, J. Whitfield, J. Miller, J. Brown, L. Richards, A. Smilanich, A. Glassmire, N. Pardikes, H. Slinn, J. Dell, C. Lepesqueur, S. Scherrer. T. Walla, H. Greeney, G. Rodriguez-Castañeda, Humberto Garcia-Lopez, Wilmer Simbaña, the Hitchcock Center for Chemical Ecology, and the UNR Plant-Insect Group. It takes networks of people to build networks of caterpillar interactions, and many others were important contributors to the approaches and results described here, and a huge part of that network was made possible by Earthwatch Institute and thousands of volunteers. Funding was also provided by the National Science Foundation, the Strategic Environmental Research and Development Program, and the University of Nevada.

References

Albert R, Jeong H, Barabási A-L (2000) Error and attack tolerance of complex networks. Nature 406(6794):378–382

Biesmeijer JC (2006) Parallel declines in pollinators and insect-pollinated plants in Britain and the Netherlands. Science 313(5785):351–354

Butchart SHM, Walpole M, Collen B, van Strien A, Scharlemann JPW, Almond REA, Baillie JEM, Bomhard B, Brown C, Bruno J et al (2010) Global biodiversity: indicators of recent declines. Science 328(5982):1164–1168

Coley PD, Barone JA (1996) Herbivory and plant defenses in tropical forests. Annu Rev Ecol Syst 27:305–335

Connahs H, Rodríguez-Castañeda G, Walters T, Walla T, Dyer L (2009) Geographic variation in host-specificity and parasitoid pressure of an herbivore (Geometridae) associated with the tropical genus *Piper* (Piperaceae). J Insect Sci 9(28)

Crossley MS, Meier AR, Baldwin EM, Berry LL, Crenshaw LC, Hartman GL, Lagos-Kutz D, Nichols DH, Patel K, Varriano S et al (2020) No net insect abundance and diversity declines across US Long Term Ecological Research sites. Nat Ecol & Evol 4(10):1368–1376

de la Cruz E, Bravo-Durán V, Ramírez F, Castillo LE (2014) Environmental hazards associated with pesticide import into Costa Rica, 1977–2009. J Environ Biol 35(1):43–55

Dell JE, Salcido DM, Lumpkin W, Richards LA, Pokswinski SM, Loudermilk EL, O'Brien JJ, Dyer LA (2019) Interaction diversity maintains resiliency in a frequently disturbed ecosystem. Front Ecol Evol 7:145

Donoso I, Stefanescu C, Martínez-Abraín A, Traveset A (2016) Phenological asynchrony in plant–butterfly interactions associated with climate: a community-wide perspective. Oikos 125(10):1434–1444

Dyer LA (1995) Tasty generalists and nasty specialists? Antipredator mechanisms in tropical lepidopteran larvae. Ecology 76(5):1483–1496

Dyer LA (1997) Effectiveness of caterpillar defense against three species of invertebrate predators. J Res Lepidoptera 34:48–68

Dyer LA (2018) Multidimensional diversity associated with plants: a view from a plant–insect interaction ecologist. Am J Bot 105(9):1439–1442

Dyer LA, Floyd T (1993) Determinants of predation on phytophagous insects: the importance of diet breadth. Oecologia 96(4):575–582

Dyer LA, Gentry G (1999) Predicting natural-enemy responses to herbivores in natural and managed systems. Ecol Appl 9(2):402–408

Dyer LA, Singer MS, Lill JT, Stireman JO, Gentry GL, Marquis RJ, Ricklefs RE, Greeney HF, Wagner DL, Morais HC et al (2007) Host specificity of Lepidoptera in tropical and temperate forests. Nature 448(7154):696–699

Dyer LA, Walla TR, Greeney HF, Stireman JO, Hazen RF (2010) Diversity of interactions: a metric for studies of biodiversity. Biotropica 42(3):281–289

Dyer LA, Wagner DL, Greeney HF, Smilanich AM, Massad TJ, Robinson ML, Fox MS, Hazen RF, Glassmire AE, Pardikes NA et al (2012) Novel insights into tritrophic interaction diversity and chemical ecology using 16 years of volunteer-supported research. Am Entomol 58(1):15–19

Dyer LA, Richards LA, Short SA, Dodson CD (2013) Effects of CO_2 and temperature on tritrophic interactions. PLoS One 8(4):e62528

Dyer L, Massad T, Forister M (2015) The question of scale in trophic ecology. In: Ecology T (ed) Bottom-Up and Top-Down Interactions across Aquatic and Terrestrial Systems. Cambridge University Press, Cambridge, MA, pp 288–317

Elton CS (1927) Animal ecology. Sidgwick & Jackson, London

Fagan WF, Lewis MA, Neubert MG, Driessche PVD (2002) Invasion theory and biological control. Ecol Lett 5(1):148–157

Flores CO, Valverde S, Weitz JS (2013) Multi-scale structure and geographic drivers of crossinfection within marine bacteria and phages. ISME J 7(3):520–532

Forister ML, Dyer LA, Singer MS, Stireman JO, Lill JT (2012) Revisiting the evolution of ecological specialization, with emphasis on insect–plant interactions. Ecology 93(5):981–991

Forister ML, Novotny V, Panorska AK, Baje L, Basset Y, Butterill PT, Cizek L, Coley PD, Dem F, Diniz IR et al (2015) The global distribution of diet breadth in insect herbivores. Proc Natl Acad Sci 112(2):442–447

Forister ML, Cousens B, Harrison JG, Anderson K, Thorne JH, Waetjen D, Nice CC, De Parsia M, Hladik ML, Meese R et al (2016) Increasing neonicotinoid use and the declining butterfly fauna of lowland California. Biol Lett 12(8):20160475

Futuyma DJ (1986) Evolutionary biology, 2nd edn. Sinauer Associates, Sunderland

Futuyma DJ, Moreno G (1988) The evolution of ecological specialization. Annu Rev Ecol Syst 19(1):207–233

Gentry GL, Dyer LA (2002) On the conditional nature of neotropical caterpillar defenses against their natural enemies. Ecology 83(11):3108–3119

Glassmire AE, Jeffrey CS, Forister ML, Parchman TL, Nice CC, Jahner JP, Wilson JS, Walla TR, Richards LA, Smilanich AM et al (2016) Intraspecific phytochemical variation shapes community and population structure for specialist caterpillars. New Phytol 212(1):208–219

Greene HW (1994) Systematics and natural history, foundations for understanding and conserving biodiversity. Am Zool 34(1):48–56

Greeney HF, Dyer LA, Smilanich AM (2012) Feeding by lepidopteran larvae is dangerous: a review of caterpillars' chemical, physiological, morphological, and behavioral defenses against natural enemies. Invertebr Surviv J 9(1):7–34

Hampton SE, Parker JN (2011) Collaboration and productivity in scientific synthesis. Bioscience 61(11):900–910

Hansen AC, Glassmire AE, Dyer LA, Smilanich AM (2017) Patterns in parasitism frequency explained by diet and immunity. Ecography 40(7):803–805

Hawkins BA (1994) Pattern and process in host-parasitoid interactions. Cambridge University Press, Cambridge

Hespenheide HA (2011) Introduction to the natural history and interaction diversity of neotropical plant-caterpillar-parasitoid systems. Ann Entomol Soc Am 104(6):1029–1032

Ings TC, Montoya JM, Bascompte J, Blüthgen N, Brown L, Dormann CF, Edwards F, Figueroa D, Jacob U, Jones JI et al (2009) Review: ecological networks – beyond food webs. J Anim Ecol 78(1):253–269

Janzen DH, Hallwachs W (2019) Perspective: where might be many tropical insects? Biol Conserv 233:102–108

Kaling M, Schmidt A, Moritz F, Rosenkranz M, Witting M, Kasper K, Janz D, Schmitt-Kopplin P, Schnitzler J-P, Polle A (2018) Mycorrhiza-triggered transcriptomic and metabolomic networks impinge on herbivore fitness. Plant Physiol 176(4):2639–2656

Keil P, Storch D, Jetz W (2015) On the decline of biodiversity due to area loss. Nat Commun 6(1):8837

King C, Ballantyne G, Willmer PG (2013) Why flower visitation is a poor proxy for pollination: measuring single-visit pollen deposition, with implications for pollination networks and conservation. Methods Ecol Evol 4(9):811–818

Lampert EC, Dyer LA, Bowers MD (2010) Caterpillar chemical defense and parasitoid success: *Cotesia congregata* parasitism of *Ceratomia catalpae*. J Chem Ecol 36(9):992–998

Lampert EC, Dyer LA, Bowers MD (2014) Dietary specialization and the effects of plant species on potential multitrophic interactions of three species of nymphaline caterpillars. Entomol Exp Appl 153(3):207–216

Lepesqueur C, Scherrer S, Vieira MC, Almeida-Neto M, Salcido DM, Dyer LA, Diniz IR (2018) Changing interactions among persistent species as the major driver of seasonal turnover in plant-caterpillar interactions. Schädler M, editor. PLoS One 13(9):e0203164

Martén-Rodríguez S, Fenster CB, Agnarsson I, Skog LE, Zimmer EA (2010) Evolutionary breakdown of pollination specialization in a Caribbean plant radiation. New Phytol 188(2):403–417

Martin EA, Reineking B, Seo B, Steffan-Dewenter I (2013) Natural enemy interactions constrain pest control in complex agricultural landscapes. Proc Natl Acad Sci 110(14):5534–5539

Mayor SJ, Guralnick RP, Tingley MW, Otegui J, Withey JC, Elmendorf SC, Andrew ME, Leyk S, Pearse IS, Schneider DC (2017) Increasing phenological asynchrony between spring green-up and arrival of migratory birds. Sci Rep 7(1):1902

Mougi A, Kondoh M (2012) Diversity of interaction types and ecological community stability. Science 337(6092):349–351

Nice CC, Forister ML, Harrison JG, Gompert Z, Fordyce JA, Thorne JH, Waetjen DP, Shapiro AM (2019) Extreme heterogeneity of population response to climatic variation and the limits of prediction. Glob Chang Biol 25(6):2127–2136

Nooten SS, Andrew NR, Hughes L (2014) Potential impacts of climate change on insect communities: a transplant experiment. PLoS One 9(1):e85987

Noss RF (1996) The naturalists are dying off. Conserv Biol 10(1):1–3

Olesen JM, Jordano P (2002) Geographic patterns in plant–pollinator mutualistic networks. Ecology 83(9):2416–2424

Outhwaite CL, Gregory RD, Chandler RE, Collen B, Isaac NJB (2020) Complex long-term biodiversity change among invertebrates, bryophytes and lichens. Nat Ecol & Evol 4(3):384–392

Paine RT (1980) Food webs: linkage, interaction strength and community infrastructure. J Anim Ecol 49(3):667–685

Pardikes NA, Shapiro AM, Dyer LA, Forister ML (2015) Global weather and local butterflies: variable responses to a large-scale climate pattern along an elevational gradient. Ecology 96(11):2891–2901

Peralta G, Frost CM, Rand TA, Didham RK, Tylianakis JM (2014) Complementarity and redundancy of interactions enhance attack rates and spatial stability in host–parasitoid food webs. Ecology 95(7):1888–1896

Petschenka G, Agrawal AA (2016) How herbivores coopt plant defenses: natural selection, specialization, and sequestration. Curr Opin Insect Sci 14:17–24

Pimm SL (1979) The structure of food webs. Theor Popul Biol 16(2):144–158

Pimm SL, Lawton JH, Cohen JE (1991) Food web patterns and their consequences. Nature 350(6320):669

Poisot T, Mouquet N, Gravel D (2013) Trophic complementarity drives the biodiversity–ecosystem functioning relationship in food webs. Ecol Lett 16(7):853–861

Poisot T, Stouffer DB, Kéfi S (2016) Describe, understand and predict: why do we need networks in ecology? Funct Ecol 30(12):1878–1882

Price P, Diniz I, Morais H, Evelyn S, Marques A (1995) The abundance of insect herbivore species in the tropics: the high local richness of rare species. Biotropica 27(4):468–478

Renner SS, Zohner CM (2018) Climate change and phenological mismatch in trophic interactions among plants, insects, and vertebrates. Annu Rev Ecol Evol Syst 49(1):165–182

Richards LA, Glassmire AE, Ochsenrider KM, Smilanich AM, Dodson CD, Jeffrey CS, Dyer LA (2016) Phytochemical diversity and synergistic effects on herbivores. Phytochem Rev 15(6):1153–1166

Rodriguez-Saona CR, Musser RO, Vogel H, Hum-Musser SM, Thaler JS (2010) Molecular, biochemical, and organismal analyses of tomato plants simultaneously attacked by herbivores from two feeding guilds. J Chem Ecol 36(10):1043–1057

Salazar D, Marquis RJ (2012) Herbivore pressure increases toward the equator. Proc Natl Acad Sci 109(31):12616–12620

Salcido DM, Forister ML, Garcia Lopez H, Dyer LA (2020) Loss of dominant caterpillar genera in a protected tropical forest. Sci Rep 10(1):1–10

Scheele BC, Pasmans F, Skerratt LF, Berger L, Martel A, Beukema W, Acevedo AA, Burrowes PA, Carvalho T, Catenazzi A et al (2019) Amphibian fungal panzootic causes catastrophic and ongoing loss of biodiversity. Science 363(6434):1459–1463

Scherrer S, Lepesqueur C, Vieira MC, Almeida-Neto M, Dyer L, Forister M, Diniz IR (2016) Seasonal variation in diet breadth of folivorous Lepidoptera in the Brazilian cerrado. Biotropica 48(4):491–498

Sedio BE, Parker JD, McMahon SM, Wright SJ (2018) Comparative foliar metabolomics of a tropical and a temperate forest community. Ecology 99(12):2647–2653

Sedio BE, Devaney JL, Pullen J, Parker GG, Wright SJ, Parker JD (2020) Chemical novelty facilitates herbivore resistance and biological invasions in some introduced plant species. Ecol Evol 10(16):8770–8792

Seifert CL, Volf M, Jorge LR, Abe T, Carscallen G, Drozd P, Kumar R, Lamarre GPA, Libra M, Losada ME et al (2020) Plant phylogeny drives arboreal caterpillar assemblages across the Holarctic. Ecol Evol 10(24):14137–14151

Slinn HL, Richards LA, Dyer LA, Hurtado PJ, Smilanich AM (2018) Across multiple species, phytochemical diversity and herbivore diet breadth have cascading effects on herbivore immunity and parasitism in a tropical model system. Front Plant Sci 9:656

Smilanich AM, Dyer LA, Chambers JQ, Bowers MD (2009a) Immunological cost of chemical defence and the evolution of herbivore diet breadth. Ecol Lett 12(7):612–621

Smilanich AM, Dyer LA, Gentry GL (2009b) The insect immune response and other putative defenses as effective predictors of parasitism. Ecology 90(6):1434–1440

Stange EE, Ayres MP (2010) Climate change impacts: insects. eLS

Staude HS, Mecenero S, Oberprieler RG, Sharp A, Sharp IC, Williams MC, Maclean M (2016) An illustrated report on the larvae and adults of 962 African Lepidoptera species. Results of the Caterpillar Rearing Group: a novel, collaborative method of rearing and recording lepidopteran life-histories. Metamorphosis 27:46–59

Stireman JO, Dyer LA, Janzen DH, Singer MS, Lill JT, Marquis RJ, Ricklefs RE, Gentry GL, Hallwachs W, Coley PD et al (2005) Climatic unpredictability and parasitism of caterpillars: Implications of global warming. Proc Natl Acad Sci 102(48):17384–17387

Stork NE (2018) How many species of insects and other terrestrial arthropods are there on Earth? Annu Rev Entomol 63(1):31–45

Thébault E, Fontaine C (2010) Stability of ecological communities and the architecture of mutualistic and trophic networks. Science 329(5993):853–856

Tylianakis JM, Tscharntke T, Lewis OT (2007) Habitat modification alters the structure of tropical host–parasitoid food webs. Nature 445(7124):202–205

Tylianakis JM, Laliberté E, Nielsen A, Bascompte J (2010) Conservation of species interaction networks. Biol Conserv 143(10):2270–2279

van Klink R, Bowler DE, Gongalsky KB, Swengel AB, Gentile A, Chase JM (2020) Meta-analysis reveals declines in terrestrial but increases in freshwater insect abundances. Science 368(6489):417–420

Wagner DL (2020) Insect Declines in the Anthropocene. Annu Rev Entomol 65(1):457–480

Wagner DL, Fox R, Salcido DM, Dyer LA (2021) A window to the world of global insect declines: Moth biodiversity trends are complex and heterogeneous. Proc Natl Acad Sci 118(2):e2002549117

Wallace AR (1878) Tropical nature, and other essays. Macmillan & Company, London & New York

War AR, Paulraj MG, Ahmad T, Buhroo AA, Hussain B, Ignacimuthu S, Sharma HC (2012) Mechanisms of plant defense against insect herbivores. Plant Signal Behav 7(10):1306–1320

Part VI
Synthesis

Synopsis and the Future of Caterpillar Research

Robert J. Marquis and Suzanne Koptur

The most extraordinary instance of imitation I ever met with was that of a very large caterpillar, which...startled me by its resemblance to a small snake. The first three segments behind the head were dilatable at the will of the insect, and had on each side a large black pupillated spot, which resembled the eye of the reptile: it a poisonous or viperine species mimicked, and not an innocuous or colubrine snake, this was proved by the imitation of keeled scales on the crown, which was produced by the recumbent feet, as the caterpillar threw itself backwards. (HW Bates 1863, The Naturalist on the River Amazons, p. 509)

Caterpillar Biology and Ecology in a Tritrophic World

Beginnings Along the Amazon

Henry Bates was one of the early naturalists who first documented the unusual appearance and behavior of caterpillars, as well as interactions between caterpillars, their host plants, and their natural enemies. This early research was foundational to all that appears in this volume. It involved observations of adult oviposition, host plant associations, the behavior of caterpillars on their host plants and with their natural enemies, rearing caterpillars to adulthood, recording changes in morphology and behavior from one instar to the next, and discovering the environmental conditions necessary for successful pupation and eclosion to adulthood.

R. J. Marquis (✉)
Department of Biology, Whitney R. Harris World Ecology Center,
University of Missouri–St. Louis, St. Louis, MO, USA
e-mail: robert_marquis@umsl.edu

S. Koptur
Department of Biological Sciences, International Center for Tropical Botany,
Florida International University, Miami, FL, USA
e-mail: kopturs@fiu.edu

As a result of Bates' work, a number of major ideas in ecology and evolution were first born. These include tritrophic interaction theory (Hairston et al. 1960; Price et al. 1980; Abdala-Roberts et al. 2019), coevolution between plants and angiosperms (Brues 1924; Ehrlich and Raven 1964), mimicry theory (Bates 1862), chemical ecology (e.g., Brower et al. 1968; Reichstein et al. 1968), and some of the first evidence for adaptations as the outcome of natural selection (Darwin 1862).

Advances in Technology

In the absence of analog or digital photography, early caterpillar biologists documented their subjects by drawing and/or painting them on paper. We have come a long way since then: technology is making the study of caterpillars and their role in an anthropocentric world much more feasible. Since Stamp and Casey's (1993) landmark volume, numerous advances allow us to understand more completely the biology of caterpillars. To start, we now have the ability to identify caterpillars to species level, using DNA barcoding, without resorting to rearing to adulthood. This technology can be an enormous time-saving, particularly for projects studying the community ecology of caterpillars in temperate regions where faunas are more well known. It is also most convenient when the exact conditions have not been discovered that result in successful pupation and eclosion. Still, nothing can take the place of identification based on updated taxonomy. Family and subfamily taxonomic treatments at the global level (e.g., Marquis et al. 2019b) are sorely needed at this time.

We can sequence caterpillar genes and modify those genes in traditional model organisms (*Drosophila melanogaster*), to assemble the evolutionary history of adaptations allowing caterpillars to feed on toxic host plants (Karageorgi et al. 2019; Groen and Whiteman, Chapter "Ecology and Evolution of Secondary Compound Detoxification Systems in Caterpillars", this volume). Gene sequencing and directed biochemistry allow us to understand how plant material is metabolized once ingested. The network analysis of plants, their herbivores, and associated natural enemies, not widely available as tool to ecologists in 1993, when the Stamp and Casey volume was published (Ings and Hawes 2018), allows us to picture the role of caterpillars in ecosystems (Salcido et al., Chapter "Plant-Caterpillar-Parasitoid Natural History Studies Over Decades and Across Large Geographic Gradients Provide Insight into Specialization, Interaction Diversity, and Global Change", this volume). The techniques and associated data analysis methods are now available to quantify the metabolome of entire host plants with the goal of discovering just which plant traits influence host plant use by caterpillars (e.g., Endara et al., Chapter "Impacts of Plant Defenses on Host Choice by Lepidoptera in Neotropical Rainforests", this volume). Even more recent than network analysis of species, the development of network analysis of secondary metabolites promises to reveal the evolutionary history of secondary metabolite pathways (Sedio 2017). This should bring us one step closer to understanding the coevolutionary history between vascular plants and Lepidoptera.

We now have the tools to characterize the microbiome, the rich assemblage of microorganisms found within and on the surface of caterpillars, and how they influence the ability of the insect to successfully complete development (Hammer and Bowers 2015; Hammer et al. 2017). Finally, many of the questions that caterpillar biologists are puzzled by require large teams of researchers at multiple, protected locations to collect sufficiently large data sets to answer those questions. These teams have been assembling over the years and are making rapid advances. Large data sets are being gathered, requiring advances in statistical analysis (Braga and Diniz, Chapter "Trophic Interactions of Caterpillars in the Seasonal Environment of the Brazilian Cerrado and Their Importance in the Face of Climate Change", this volume; Boege et al., Chapter "Impacts of Climatic Variability and Hurricanes on Caterpillar Diet Breadth and Plant-Herbivore Interaction Networks", this volume; Salcido et al., Chapter "Plant-Caterpillar-Parasitoid Natural History Studies Over Decades and Across Large Geographic Gradients Provide Insight into Specialization, Interaction Diversity, and Global Change", this volume). Citizen science is just coming into its own, allowing geographically widespread collection of original data and observations (e.g., iNaturalist: Pierce and Dankowicz, Chapter "The Natural History of Caterpillar-Ant Associations", this volume) and making important contributions in the area of bird and insect conservation. Certainly, the research outlined in Chapter "Caterpillar Patterns in Space and Time: Insights From and Contrasts Between Two Citizen Science Datasets" (Di Cecco and Hurlbert) and Chapter "Plant-Caterpillar-Parasitoid Natural History Studies Over Decades and Across Large Geographic Gradients Provide Insight into Specialization, Interaction Diversity, and Global Change" (Salicido et al.) reveals the major contributions that the general public can make to caterpillar science. At a smaller scale, but no less important, many important questions require teams of geneticists, organic chemists, biochemists, microbiologists, and neurobiologists to answer questions that intrigue the caterpillar biologist (Groen and Whiteman, Chapter "Ecology and Evolution of Secondary Compound Detoxification Systems in Caterpillars", this volume).

Forces Driving Coevolution

Major advances in ecology and evolution are represented on the pages of this book, with a focus on how caterpillars relate to neighboring trophic levels. A tritrophic view of the caterpillar world is one that drives much of current research in caterpillar ecology. Caterpillars are "sandwiched" between their neighboring trophic levels (Abdala and Mooney 2015). There is mounting evidence that their biology thus represents a compromise between dealing with nasty food plants and nasty natural enemies (Lill and Weiss, Chapter "Host Plants as Mediators of Caterpillar-Natural Enemy Interactions", this volume). The pages of this book are replete with examples of these trade-offs.

Contrary to the more traditional view of reciprocal adaptation between plants and insects (Ehrlich and Raven 1964), caterpillars and other herbivorous insects may readily switch host plants over evolutionary time, at least ones that share similar defensive profiles (Endara et al., Chapter "Impacts of Plant Defenses on Host Choice by Lepidoptera in Neotropical Rainforests", this volume). In the *Inga* system (Fabaceous tropical tree), while herbivores collectively select for rapid divergence in host defenses, the lack of congruence between phylogenies of hosts and herbivores supports frequent host shifts (see also Dobler et al. 2012). These results call into question how often reciprocal coevolution comes into play between plants and their insect herbivores. In contrast, results from the *Inga* system suggest that insects are jumping to new hosts repeatedly, imposing little diversifying selection on their hosts. Shaking the foundation even more, Singer et al. (Chapter "Predators and Caterpillar Diet Breadth: Appraising the Enemy-Free Space Hypothesis", this volume) present evidence that predators must be considered as possible instigators of host plant shifts (see also Murphy 2004). Thus, secondary plant chemistry may no longer be the single linchpin upon which co-diversification depends. Importantly, it is likely that there is some combinatorial effect of the first and third trophic levels that influence host plant shifts (Lill and Weiss, Chapter "Host Plants as Mediators of Caterpillar-Natural Enemy Interactions", this volume).

Defenses Against Natural Enemies

A traditional view of caterpillars as prey is that they employ one of two general strategies to escape their natural enemies. One strategy is to forgo chemical and physical defense, escaping natural enemies by visual camouflage, either by crypsis (background matching) or by masquerade, resembling an inedible object (Skelhorn et al. 2010; Higginson et al. 2012; see examples in Wagner and Hoyt, Chapter "On Being a Caterpillar: Structure, Function, Ecology, and Behavior", this volume). Alternatively, they may sequester or manufacture defensive compounds de novo to make themselves distasteful, often accompanied by aposematic coloration. Bowers (Chapter "Sequestered Caterpillar Chemical Defenses: From "Disgusting Morsels" to Model Systems", this volume) has spent a career studying this latter strategy, building on the early theoretical writings of Darwin, Bates, and Wallace, and on experiments by Rothschild and the Browers, to understand the details of the phenomenon: how do host plant chemistry and caterpillar species identity together affect sequestration level, how much is sequestered, and how do host plant age and caterpillar instar affect the process? Despite the fact that sequestration occurs throughout the Lepidoptera phylogeny (and in many other insect orders), Bowers' chapter reveals that we know much less about the actual adaptive advantages of chemical sequestration in terms of escape from natural enemies.

In some cases, sequestration may actually reduce the immunocompetence of caterpillars against parasitoids (Ode, Chapter "Caterpillars, Plant Chemistry, and Parasitoids in Natural vs. Agroecosystems", this volume; Smilanich and Muchoney,

Chapter "Host Plant Effects on the Caterpillar Immune Response", this volume). This begs the question as to the adaptive targets of sequestration, whether the enemies are vertebrates, predatory arthropods, parasitoids, or some combination thereof. Singer et al.'s analysis (Chapter "Predators and Caterpillar Diet Breadth: Appraising the Enemy-Free Space Hypothesis", this volume) suggests that sequestration is most effective against vertebrate predators. This question is not only important to ecologists and caterpillar biologists, but is immensely important in agriculture. We rely upon natural enemies, either introduced or native, to provide at least partial control of caterpillar pests in crop systems (Garfinkel et al. 2020). At the same time, we select for plant phenotypes that are sometimes elevated and sometimes diminished in secondary compounds. What is the corresponding effect of altered host chemistry on caterpillar phenotypes, and in turn, the vulnerability of the caterpillars to their natural enemies? Is vulnerability reduced compared to less managed ecosystems? Detailed studies of these interactions in agricultural systems are limited to a relatively few crop species, perhaps no more than 15 in total (Ode, Chapter "Caterpillars, Plant Chemistry, and Parasitoids in Natural vs. Agroecosystems", this volume).

Kauer et al. (Chapter "Surface Warfare: Plant Structural Defenses Challenge Caterpillar Feeding", this volume), Groen and Whiteman (Chapter "Ecology and Evolution of Secondary Compound Detoxification Systems in Caterpillars", this volume), Pierce and Dankowicz (Chapter "The Natural History of Caterpillar-Ant Associations", this volume), and Koptur et al. (Chapter "Caterpillar Responses to Ant Protectors of Plants", this volume) suggest a third adaptive option for caterpillar lineages in the face of natural enemy attack: chemicals may be synthesized or sequestered and used not as deterrents but as camouflage such that natural enemies do not recognize the larvae as prey. We do not know whether chemistry itself can provide sufficient camouflage or whether chemistry, morphology, color, and behavior may work in concert to provide protection from enemies. This hypothesis has been the subject of little study outside of caterpillar interactions with ants, where chemistry is known to be important in the evolution of caterpillars that co-op ant behavior for their own protection. Here ants, normally predators, become protectors, providing caterpillars with enemy-free space (Pierce and Dankowicz, Chapter "The Natural History of Caterpillar-Ant Associations", this volume). Although most diverse and widespread in the Riodinidae and Lycaenidae, caterpillar-ant symbioses have evolved at least 25 times in disparate lines scattered across the Lepidoptera phylogeny.

There is yet another strategy evolved by caterpillars to provide protection against natural enemies, and in some cases to mitigate low host plant quality and ameliorate a stressful abiotic environment, all simultaneously. Numerous clades of caterpillars have evolved the ability to use plant parts plus silk, and sometimes frass, to build "shelters" on their host plant (e.g., Braga and Diniz, Chapter "Trophic Interactions of Caterpillars in the Seasonal Environment of the Brazilian Cerrado and Their Importance in the Face of Climate Change", this volume). These shelters can provide protection against predators (e.g., Baer and Marquis 2020, 2021), but they may increase susceptibility to attack by parasitoids (e.g., Gentry and Dyer 2002).

The factors that determine the balance between reduced versus increased susceptibility depending on the natural enemy type are not known. The presence of these caterpillar-built shelters can have important implications for arthropod community structure on their host plants. The shelters are often subsequently occupied by other arthropods, often resulting in novel communities of host-associated arthropods not seen on shelter-free host plants. Marquis et al. (Chapter "The Impact of Construct Building by Caterpillars on Arthropod Colonists in a World of Climate Change", this volume) present a predictive model of the impacts of such "shelters" on community structure, depending on the structure of the shelter and the behavior of the caterpillar shelter-builder. Shelter-building caterpillars comprise a large portion of the diversity of Lepidoptera, but their effects on other arthropods through sharing of host plants have been studied only recently.

Development of Caterpillars in a Tritrophic World

A true frontier for future caterpillar research is the study of how developmental stage (instar) influences the myriad interactions between the caterpillar and its adjoining trophic levels. Each chapter of this book touches on this topic either directly or indirectly. Most current ecological data come from the study of late instars, as they are the easiest to work with. Early instars are often overlooked in ecological sampling, difficult to identify when found, and just as easily lost when collected. And yet profound effects are revealed when instar is incorporated into experimental and sampling designs.

Caterpillars frequently vary dramatically from one instar to the next not only in size but in color, shape, setal covering, and internal morphology. Often, changes are so dramatic that the uninitiated could understandably classify early and late instars as different species (Plates 1 and 2). Lepidopteran larvae show more diversity of form among instars and among species than all other herbivorous orders. It is this diversity of form that draws us to them. Part of this diversity must arise from the fact that hemimetaboly constrains development in other phytophagous orders: the Orthoptera, Phasmatodea, and Hemiptera. The Coleoptera, which are holometabolous, are speciose, but their larvae are generally not as diverse in form as those of Lepidoptera, and they show relatively less change with ontogeny. Most herbivorous species of Coleoptera are internal plant feeders. These observations lead to the reasonable hypothesis, yet to be tested, that interactions between host plants, caterpillars, and the third trophic level have given rise to much of the diversity of form and function seen in caterpillars, both during development and across species (see also Wagner and Hoyt, Chapter "On Being a Caterpillar: Structure, Function, Ecology, and Behavior", this volume). It also suggests that the greatest diversity in form will be found in externally feeding caterpillars.

These changes in external and internal caterpillar morphology influence how the caterpillar interacts with its host plants, natural enemies, and other non-predator arthropods. Caterpillars change in behavior with instar (e.g., parts of leaf consumed,

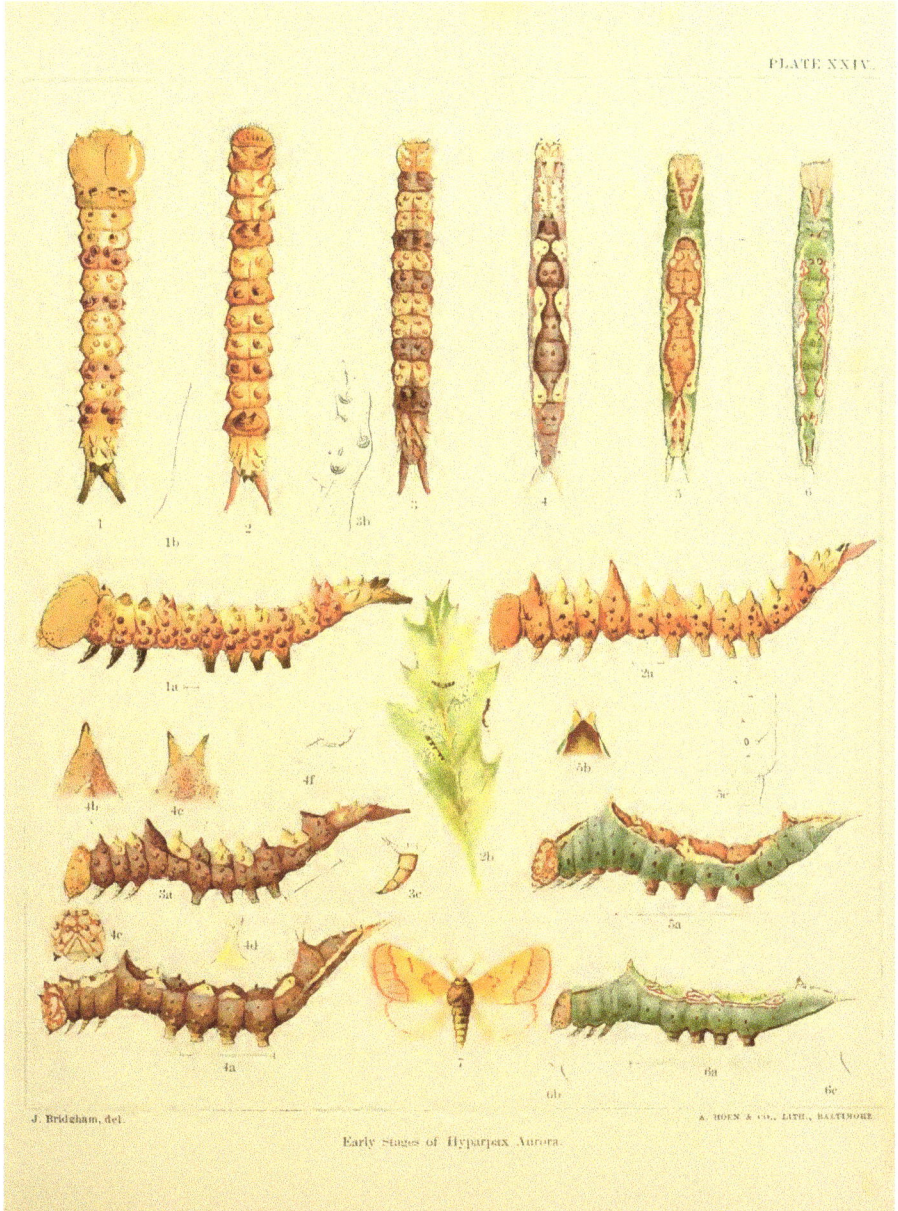

Plate 1 Variation in size, color, and morphology from one instar to the next in *Hyparpax aurora* (Smith) (Notodontidae) (Plate XXIV of Packard (1895))
Fig. 1 Stage I (first instar), dorsal view; (1a) side view; (1b) dorsal piliferons tubercule
Fig. 2 End of Stage I; (2a) side view; (2b) freshly hatched larvae, natural size
Fig. 3 Stage II (second instar); (3a) side view; (3b) third abdominal segment, side view; (3c) a thoracic leg
Fig. 4 Stage III (third instar); (4a) side view; (4b) dorsal tubercle; (4c) front view of the same; (4d) subdorsal tubercle; (4e) face; (4f) natural size
Fig. 5 Stage IV (fourth instar); (5a) side view; (5b) dorsal tubercle of the eighth abdominal segment; (5c) third abdominal segment, side view
Fig. 6 Last stage (fifth instar); (6a) side view; (6b) dorsal tubercle of the first abdominal and (6c) eighth abdominal segment
Fig. 7 Adult male, natural size

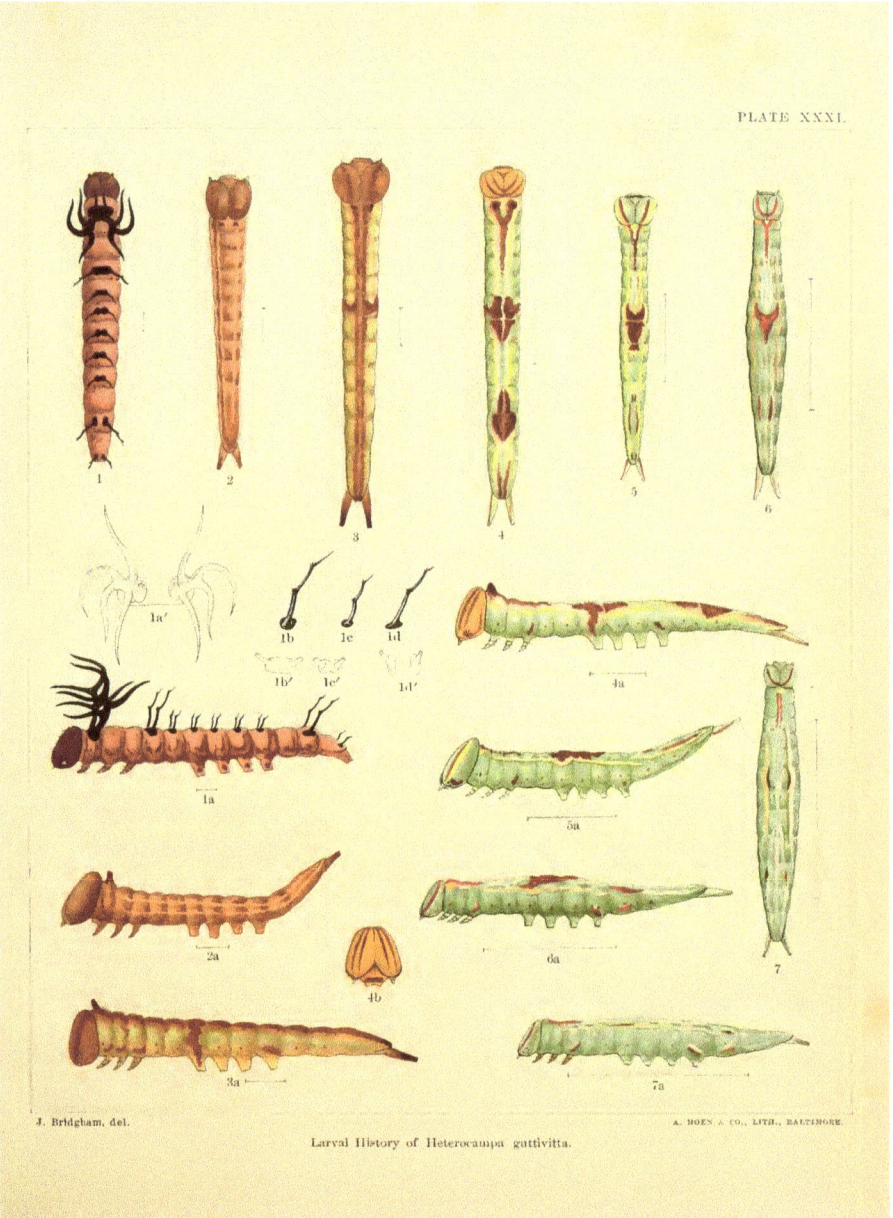

Plate 2 Variation in size, color, and morphology with instar in *Heterocampa guttivitta* (Smith) (now *Cecrita guttivitta* Walker) (Notodontidae) (Plate XXXI of Packard (1895))

Fig. 1 Stage I (first instar); (1a) side view; (1a') prothoracic antlers; (1b, 1b') antlers on the first abdominal segment; (1c, 1c') antlers on the second to the seventh abdominal segments; (1d, 1d') antlers on the eighth abdominal segment

Fig. 2 Stage II (second instar); (2a) side view

Fig. 3 End of Stage II; (3a) side view

Fig. 4 (4a) Stage III (third instar)

Fig. 5 (5a) Stage IV (fourth instar)

Fig. 6 (6a) End of Stage IV

Fig. 7 (7a) Stage V (fifth instar)

level of aggregation, ballooning, internal vs. external feeding, shelter structure) (Zalucki et al. 2002). Food quality experienced by early instars can influence the subsequent phenotype of later instars, even so far as affecting caterpillar coloring and morphology (Akino et al. 2004; Koptur et al. 2015) and larval diapause (Hunter and McNeil 1997).

The tritrophic world in which caterpillars exist looks profoundly differently from the view of a 1-mm-long first instar than that of an 8-cm-long (or longer) fifth instar caterpillar (e.g., Packard 1895). To early instar caterpillars, leaf hairs can be an impenetrable, dangerous chaparral, in some cases full of barbs that can puncture the cuticle (Gilbert 1971), preventing access to the actual leaf surface (Zalucki et al. 2002). Even on glabrous leaves, the surface texture takes on a vastly different landscape for the early versus late instar caterpillar (Kauer et al., Chapter "Surface Warfare: Plant Structural Defenses Challenge Caterpillar Feeding", this volume). We see, as a result, different strategies for dealing with plant morphological defenses as caterpillars mature (e.g., Keathley and Potter 2011; Kariyat et al. 2018; Boege et al. 2019; Kauer et al., Chapter "Surface Warfare: Plant Structural Defenses Challenge Caterpillar Feeding", this volume).

The landscape of natural enemies and abiotic threats also changes with instar (Boege et al. 2019). Early instar caterpillars may be too small to be vulnerable to bird predation, but they are susceptible to predation by arthropods (Singer et al. 2017). As they mature, they are subject to a shifting community of parasitoids from early to late instars (Stireman and Shaw, Chapter "Natural History, Ecology, and Human Impacts on Caterpillar Parasitoids", this volume). Instar affects the likelihood of predation by ants, as demonstrated frequently for caterpillars found on plants with extrafloral nectaries (Koptur et al., Chapter "Caterpillar Responses to Ant Protectors of Plants", this volume and references therein). Instar also affects sequestration (Quintero and Bowers 2018; Jones et al. 2019), de novo chemical synthesis (Frankfater et al. 2009), susceptibility to predation by non-ant predators (e.g., Schwenk et al. 2010; Singer et al. 2017; Baer and Marquis 2020), and parasitism (Lill 1999; Stireman and Shaw, Chapter "Natural History, Ecology, and Human Impacts on Caterpillar Parasitoids", this volume). How acoustical strategies (Yack, Chapter "Acoustic Defence Strategies in Caterpillars", this volume; Pierce and Dankowicz, Chapter "The Natural History of Caterpillar-Ant Associations", this volume) and chemical signaling to attract ants (Pierce and Dankowicz, Chapter "The Natural History of Caterpillar-Ant Associations", this volume) (and perhaps repel other predator types) change with instar is an area ripe for future research.

Consideration of the role of developmental stage is important because success prior to the pupal stage is a consequence of the cumulative demographic impacts of various ecological factors at all instars. Thus, a study that delimits the importance of an ecological factor for a particular instar may under- or overestimate the strength or totally overlook the impact of other factors at earlier and later instars. Our understanding of the factors that shape the ecology and evolution of caterpillars, their interactions with other trophic levels (Boege et al. 2019), and those factors that impinge on their ability to survive in a changing world would be incomplete if our studies are limited to a few developmental stages. A key first step in an important research program would be to document changes with instar by sampling species in individual clades or across the Lepidoptera tree of life. The next step would be to link such changes with changing selective pressures from one instar to the next.

Caterpillars in a Changing World

There is mounting evidence that entire communities of insects are under threat (Van Klink et al. 2020; Wagner 2020; Wagner et al. 2021). Studies of individual species of Lepidoptera across the years demonstrate that butterflies in particular can be critically threatened by habitat destruction (e.g., adonis blue butterfly: Thomas 1983; Palos Verdes blue butterfly: Arnold 1987; monarch butterflies: Brower et al. 2012). If this were not sufficiently alarming news, there is mounting evidence from across the globe that entire regional faunas are declining in some places (e.g., California: Forister et al. 2011; Great Britain: Warren et al. 2001). Salcido et al. (Chapter "Plant-Caterpillar-Parasitoid Natural History Studies Over Decades and Across Large Geographic Gradients Provide Insight into Specialization, Interaction Diversity, and Global Change", this volume and associated references) demonstrate declines in diversity of caterpillar genera at their Costa Rican wet forest site over the last 20 years. Accumulating sufficient data to show a decline at a regional level is difficult because natural fluctuations that are likely to occur in the absence of human intervention (Marquis et al. 2019a; Schowalter et al. 2021; Boege et al., Chapter "Impacts of Climatic Variability and Hurricanes on Caterpillar Diet Breadth and Plant-Herbivore Interaction Networks", this volume). Long-term data sets, like those described here (Braga and Diniz, Chapter "Trophic Interactions of Caterpillars in the Seasonal Environment of the Brazilian Cerrado and Their Importance in the Face of Climate Change", this volume; Boege et al., Chapter "Impacts of Climatic Variability and Hurricanes on Caterpillar Diet Breadth and Plant-Herbivore Interaction Networks", this volume; Salcido et al., Chapter "Plant-Caterpillar-Parasitoid Natural History Studies Over Decades and Across Large Geographic Gradients Provide Insight into Specialization, Interaction Diversity, and Global Change", this volume), are necessary to demonstrate statistically significant declines and discern their root causes.

All of the main interactions described in this book are potentially influenced by one or more factors driving global change. Uncovering the root causes for such declines, however, is exceedingly difficult because there are so many candidates (Wagner et al. 2021). The problem becomes even more knotty when one considers that climatic events (droughts, severe winter and dry seasons, late spring freezes) that cause local, temporary declines may actually be increasing in frequency because of human-driven climate change. One promising approach for better understanding the patterns and drivers of caterpillar abundance and phenology over large geographic scales is the analysis of data from citizen science projects such as iNaturalist and Caterpillars Count! (Hurlbert et al. 2019; Di Cecco and Hurlbert, Chapter "Caterpillar Patterns in Space and Time: Insights From and Contrasts Between Two Citizen Science Datasets", this volume). If the rate of accumulation of these data continues according to current trends, they will be a tremendous source of information regarding where and in what environmental contexts declines or phenological shifts are of greatest concern.

The consequences for such losses are potentially enormous. Just the loss of Lepidoptera alone would have a major impact on insect and overall biodiversity (Wagner and Hoyt, Chapter "On Being a Caterpillar: Structure, Function, Ecology, and Behavior", this volume). However, caterpillars, their pupae, and resulting adults (especially moths) provide major sustenance for insectivorous birds (Hurlbert et al. 2021), multitudes of rodent species, and the majority of the world's 11,000 species of bats. In addition, the vast diversity of hymenopteran parasitoids, likely the most species rich clade of the insect family tree, as well as the largest clades of dipteran parasitoids (Stireman and Shaw, Chapter "Natural History, Ecology, and Human Impacts on Caterpillar Parasitoids", this volume), predominantly rely on Lepidoptera larvae as a food source.

We end with a plea for the continued use of caterpillars as the subjects of both research and education, as well as for the necessary infrastructure for conducting such research and education. On the research side, continued documentation of the basic natural history of caterpillars, their host plants, and their natural enemies provides the information needed for understanding broad patterns of ecology and evolution, and our understanding of the growing human impacts on natural and managed ecosystems (Salcido et al., Chapter "Plant-Caterpillar-Parasitoid Natural History Studies Over Decades and Across Large Geographic Gradients Provide Insight into Specialization, Interaction Diversity, and Global Change", this volume). On the education side, teaching the life cycle of caterpillars in elementary school introduces children to the biology of insects, metamorphosis, comparative anatomy, biodiversity, and natural history (Clayborn et al. 2020). Teaching the same to adults reinforces messages learned earlier and provides an introduction to the diversity of the most speciose clade of macroscopic life on Earth. Both can lead to an increased appreciation of nature, which is so critical for support for conservation in an increasingly damaged world. Finally, it is important to recognize the critical role of museums and field stations for both research and education. Museums house voucher collections and are foci for education of the public. Field stations provide access to adjacent natural areas and basic laboratory facilities for caterpillar field research. Field stations also provide educational opportunities for students of all ages and all education levels. Together, these two forms of institutions facilitate the advances necessary for understanding the role of caterpillars in current and future ecosystems.

References

Abdala-Roberts L, Mooney KA (2015) Plant and herbivore evolution within the trophic sandwich. In: Trophic interactions: bottom-up and top-down interactions in aquatic and terrestrial ecosystems. Cambridge University Press, pp 340–364

Abdala-Roberts L, Puentes A, Finke DL, Marquis RJ, Montserrat M, Poelman EH, Rasmann S, Sentis A, van Dam NM, Wimp G, Mooney K (2019) Tri-trophic interactions: bridging species, communities and ecosystems. Ecol Lett 22(12):2151–2167

Akino T, Nakamura KI, Wakamura S (2004) Diet-induced chemical phytomimesis by twig-like caterpillars of *Biston robustum* Butler (Lepidoptera: Geometridae). Chemoecology 14:65–174

Arnold RA (1987) Decline of the endangered Palos Verdes blue butterfly in California. Biol Conserv 40(3):203–217

Baer CS, Marquis RJ (2020) Between predators and parasitoids: complex interactions among shelter traits, predation, and parasitism in a shelter-building caterpillar community. Funct Ecol 34:2186–2198

Baer CS, Marquis RJ (2021) Experimental shelter-switching shows caterpillar shelter type alters predation of skipper caterpillars (Hesperiidae). Behav Ecol. in press

Bates HW (1862) XXXII. Contributions to an insect fauna of the Amazon Valley. Lepidoptera: Heliconidæ. Trans Linnean Soc Lond 23(3):495–566

Bates HW (1863) The naturalist on the River Amazons. Murray, London

Boege K, Agrawal AA, Thaler JS (2019) Ontogenetic strategies in insect herbivores and their impact on tri-trophic interactions. Curr Opin Insect Sci 32:61–67

Browe LP, Taylor OR, Williams EH, Slayback DA, Zubieta RR, Ramirez ML (2012) Decline of monarch butterflies overwintering in Mexico: is the migratory phenomenon at risk? Insect Cons Divers 5(2):95–100

Brower LP, Ryerson WN, Coppinger LL, Glazier SC (1968) Ecological chemistry and the palatability spectrum. Science 161(3848):1349–1350

Brues CT (1924) The specificity of food-plants in the evolution of phytophagous insects. Am Nat 58(655):127–144

Carper AL, Enger M, Bowers MD (2019) Host plant effects on immune response across development of a specialist caterpillar. Front Ecol Evol 7:208

Clayborn J, Koptur S, O'Brien G (2020) Plugging students into nature through butterfly gardening: a reconciled ecological approach to insect conservation. Child Youth Environ 30(2):30–71

Darwin C (1862) In: Darwin FI (ed) Life and letters of Charles Darwin. J. Murray, London, pp 391–392

Dobler S, Dalla S, Waschal V, Agrawal AA (2012) Community-wide convergent evolution in insect adaptation to toxic cardenolides by substitutions in the Na, K-ATPase. PNAS USA 109:13040–13045

Ehrlich ER, Raven PH (1964) Butterflies and plants: a study in coevolution. Evolution 18:586–608

Forister ML, Jahner JP, Casner KL, Wilson JS, Shapiro AM (2011) The race is not to the swift: long-term data reveal pervasive declines in California's low-elevation butterfly fauna. Ecology 92(12):2222–2235

Frankfater C, Tellez MR, Slattery M (2009) The scent of alarm: ontogenetic and genetic variation in the osmeterial gland chemistry of *Papilio glaucus* (Papilionidae) caterpillars. Chemoecology 19:81–96

Garfinkel MB, Minor ES, Whelan CJ (2020) Birds suppress pests in corn but release them in soybean corps within a mixed prairie/agriculture system. The Condor 122:1–12

Gentry G, Dyer LA (2002) On the conditional nature of neotropical caterpillar defenses against their natural enemies. Ecology 83:3108–3119

Gilbert LE (1971) Butterfly-plant coevolution: has *Passiflora adenopoda* won the selectional race with Heliconiine butterflies? Science 172(3983):585–586

Hairston NG, Smith FE, Slobodkin LB (1960) Community structure, population control, and competition. Am Nat 94(879):421–425

Hammer TJ, Bowers MD (2015) Gut microbes may facilitate insect herbivory of chemically defended plants. Oecologia 179:1–14. https://doi.org/10.1007/s00442-015-3327-1

Hammer T, Janzen D, Hallwachs W, Jaffe S, Fierer N (2017) Caterpillars lack a resident gut microbiome. Proc Natl Acad Sci U S A 114:9641–9646. https://doi.org/10.2307/26487619

Higginson AD, De Wert L, Rowland HM, Speed MP, Ruxton GD (2012) Masquerade is associated with polyphagy and larval overwintering in Lepidoptera. Biol J Linn Soc 106:90–103

Hunter MD, McNeil JN (1997) Host-plant quality influences diapause and voltinism in a polyphagous insect herbivore. Ecology 78(4):977–986

Hurlbert A, Hayes T, McKinnon T, Goforth C (2019) Caterpillar count! A citizen science project for monitoring foliage arthropod abundance and phenology. Citizen Sci: Theor Prac 4:1–12

Hurlbert AH, Olsen AM, Sawyer MM, Winner PM (2021) Avian diet database. https://doi.org/10.5281/zenodo.4960688

Ings TC, Hawes JE (2018) The history of ecological networks. In: Dáttilo W, Rico-Gray V (eds) Ecological networks in the tropics. Springer, Cham. https://doi.org/10.1007/978-3-319-68228-0_2

Jones PL, Petschenka G, Flecht L, Agrawal AA (2019) Cardenolide intake, sequestration, and excretion by the monarch butterfly along gradients of plant toxicity and larval ontogeny. J Chem Ecol 45:264–277

Karageorgi M, Groen SC, Sumbul F, Pelaez JN, Verster KI, Aguilar JM, Hastings AP, Bernstein SL, Matsunaga T, Astourian M, Guerra G (2019) Genome editing retraces the evolution of toxin resistance in the monarch butterfly. Nature 574(7778):409–412

Kariyat RR, Hardison SB, Ryan AB, Stephenson AG, De Moraes CM, Mescher MC (2018) Leaf trichomes affect caterpillar feeding in an instar-specific manner. Comm Integrat Biol 11(3):1–6

Keathley CP, Potter DA (2011) Behavioral plasticity of a grass-feeding caterpillar in response to spiny-or smooth-edged leaf blades. Arthro-Plant Inter 54:339–349

Koptur S, Jones IM, Peña JE (2015) The influence of host plant extrafloral nectaries on multitrophic interactions: an experimental investigation. PLoS One 10(9):e0138157. https://doi.org/10.1371/journal.pone.0138157

Lill JT (1999) Structure and dynamics of a parasitoid community attacking larvae of *Psilocorsis quercicella* (Lepidoptera: Oecophoridae). Environ Entomol 28(6):1114–1123

Marquis RJ, Lill JT, Forkner RE, Le Corff J, Landosky J, Whitfield JB (2019a) Impacts of late season frost and summer drought on communities of leaf chewing insects on Missouri oaks. Front Ecol Evol 7:396

Marquis RJ, Passoa SC, Lill JT, Whitfield JB, Le Corff J, Forkner RE, Passoa VA (2019b) An introduction to the immature Lepidoptera fauna of oaks in Missouri. USDA Forest Service, Forest Health Assessment and Applied Sciences Team, Morgantown, West Virginia. FHAAST-2018-05. 369 p

Murphy SM (2004) Enemy-free space maintains swallowtail butterfly host shift. PNAS USA 52:18048–18052

Packard AS (1895) Monograph of the bombycine moths of America north of Mexico, including their transformations and origin of the larval markings and armature. Part I. Family I.-Notodontidae. Mem Natl Acad Sci 7:1–291

Price P, Bouton C, Gross P, McPheron B, Thompson J, Weiss A (1980) Interactions among three trophic levels: influence of plants on interactions between insect herbivores and natural enemies. ARES 11:41–65. https://doi.org/10.1146/annurev.es.11.110180.000353

Quintero C, Bowers MD (2018) Plant and herbivore ontogeny interact to shape the preference, performance and chemical defense of a specialist herbivore. Oecologia 187(2):401–412

Reichstein TV, Von Euw J, Parsons JA, Rothschild M (1968) Heart poisons in the monarch butterfly. Science 161(3844):861–866

Schowalter TD, Pandey M, Presley SJ, Willig MR, Zimmerman JK (2021) Arthropods are not declining but are responsive to disturbance in the Luquillo Experimental Forest, Puerto Rico. Proc Natl Acad Sci 118(2):e2002556117

Schwenk WS, Strong AM, Sillett TS (2010) Effects of bird predation on arthropod abundance and tree growth across an elevational gradient. J Avian Biol 41(4):367–377

Sedio BE (2017) Recent breakthroughs in metabolomics promise to reveal the cryptic chemical traits that mediate plant community composition, character evolution and lineage diversification. New Phytol 214:952–958

Singer MS, Clark RE, Lichter-Marck LH, Johnson ER, Mooney KA (2017) Predatory birds and ants partition caterpillar prey by body size and diet breadth. J Anim Ecol 86(6):1363–1371

Skelhorn J, Rowland HM, Speed MP, Ruxton GD (2010) Masquerade: camouflage without crypsis. Science 327:51

Stamp NE, Casey TM (eds) (1993) Caterpillars – ecological and evolutionary constraints on foraging. Chapman and Hall, New York

Thomas JA (1983) The ecology and conservation of *Lysandra bellargus* (Lepidoptera: Lycaenidae) in Britain. J Appl Ecol 20:59–83

Van Klink R, Bowler DE, Gongalsky KB, Swengel AB, Gentile A, Chase JM (2020) Meta-analysis reveals declines in terrestrial but increases in freshwater insect abundances. Science 368(6489):417–420

Wagner DL (2020) Insect declines in the Anthropocene. Annu Rev Entomol 65:457–480

Wagner DL, Grames EM, Forister ML, Berenbaum MR, Stopak D (2021) Insect decline in the Anthropocene: death by a thousand cuts. PNAS 118(2):e2023989118

Warren MS, Hill JK, Thomas JA, Asher J, Fox R, Huntley B, Roy DB, Telfer MG, Jeffcoate S, Harding P, Jeffcoate G (2001) Rapid responses of British butterflies to opposing forces of climate and habitat change. Nature 414(6859):65–69

Zalucki MP, Clarke AR, Malcolm SB (2002) Ecology and behavior of first instar larval Lepidoptera. Annu Rev Entomol 47(1):361–393

Correction to: The Natural History of Caterpillar-Ant Associations

Naomi E. Pierce and Even Dankowicz

Correction to:
Chapter 11 in: R. J. Marquis, S. Koptur (eds.), *Caterpillars in the Middle*, Fascinating Life Sciences,
https://doi.org/10.1007/978-3-030-86688-4_11

Chapter 11, "The Natural History of Caterpillar-Ant Associations" was previously published non-open access. It has now been changed to open access under a CC BY 4.0 license and the copyright holder updated to 'The Author(s)'. The book has also been updated with this change.

The updated version of the chapter can be found at
https://doi.org/10.1007/978-3-030-86688-4_11

© The Author(s) 2022
R. J. Marquis, S. Koptur (eds.), *Caterpillars in the Middle*, Fascinating Life Sciences, https://doi.org/10.1007/978-3-030-86688-4_21

Subject Index

A
ABC transporters, 135
Abiotic pressure, 361
Abiotic stressors, 463
Abundance/scarcity (boom/bust), 32
Accumulation/sequestration, of plant toxins, 406
Acoustic crypsis, 197, 200
Acoustic cues, 197, 205, 209
Acoustic stimuli, 196, 197
Adaptation, 34–46, 84–85
Adenosoma, 41
Adult Lepidoptera, 542
Agricultural intensification, 254, 255
Agroecosystems, 396, 404
AhR nuclear translocator (ARNT), 128
Airborne vibrations, 196
Alarm, 197, 199, 206, 208–210, 216
Alaskan swallowtail butterfly, 439
Alkaline caterpillar midgut, 18
Alkaloids, 400
Allelochemicals, 280, 288–290
American grasshopper/American bird grasshopper, 166
Amino acid, 19, 21, 47, 68, 83, 97, 127, 142, 285
Ant associates, 327
Ant attraction, 101
Ant plants, 435
Ant protectors, caterpillar responses to, 298, 299
Anterior tentacle organs (ATOs), 355
Antimicrobial peptides (AMPs), 451, 452, 454, 474
Antioxidant response element (ARE), 128
Antipredator strategies, in insects, 197, 199
Ant-plant relationships, 435
Ant-tended plants, cryptic caterpillars on, 300–302
Aphid mimicry, 358
Aphid species, 280, 281, 284
Aphnaeine lycaenids, 21
Aphytophagy, 360
Aposematic signals, 199, 215
Apple leaf miner, 200, 208
Appraisal, 286
Argentine ant, 281
Arms race paradigm, 104
Arms races, 139
Arthropod community structure, 517, 518, 522, 523, 526, 528, 530
Arthropods, 436, 617
Artificial constructs, 521, 528
Artificial defoliation, 301
Asiatic oak weevil, 524, 525
ATPa, 142

B
Background matching, 299
Ballooning, 49, 520, 617
Balloon setae, 355
Bates, Henry, 181, 609
Beet armyworm, 474
Beetles, 242, 511, 514, 515, 524, 614
Behavioral adaptations, 35
Behavioral efficiency, 284, 285, 288
Belay line, 46, 49
Benzoxazinoid glycosides, 133
Benzoxazinoids, 133

Bernays' hypothesis, 276
Biomass, 12, 32, 50, 236, 430, 434, 517, 520, 523, 525, 530
Biotic associations, 439
Biotic defense, 300
Birds, 299
Bottom-up agents of selection, 139–141, 143, 144
Bottom-up effects of plant defense, 396
Bottom-up pressures, 34–36
Brazilian savanna, 489
Brown ambrosia aphid, 284
Brussels lace, 283
Bullet ant, 280

C

Cabbage looper larva, 73
Cabbage moth, 205
Cabbage moth caterpillar, 132, 205, 399, 402
Camouflage, 275, 279, 282, 283, 288, 290, 299
Cannibalism, 30, 31
Carboxyl/cholinesterases (CCEs), 131
Cardiac glycosides (CGs), 116, 117, 400
Carnivory, 30
CAS/CYS enzymes, 144
Case making/case bearer, 40, 48, 49
Cellular immune system, 21
Cerrado, 488
Cerrado biome, 489
Cerrado dry season, 486
Cervical gland (adenosoma), 41
CG sequestration, 147
Chaetotaxy, 15
Chemical camouflage, 311
Chemical defense, 97–99, 148, 464, 465, 468, 472
Chemical sensing, 119
Chemical sequestration, 279
Chemical signals, 339
Chemoreception, 15, 18, 119
Chemosensory proteins (CSPs), 121
Chewing herbivores, 137
Chi-squared tests, 305
Chitin binding domains (CBDs), 124
Chlorogenic acid (CA), 470
Chlorophyll content, 102, 103
Citizen science, viii
Climate change, 256, 257, 411–413, 517, 529, 559, 560
Climate conditions, 489
Climatic data, 490
Climatic events, 618
Climatic variability, 558, 559, 563, 564

Climatic variables, 562
Cocoons, 47
Coevolution theory, 274
Colonial Audubon, 6
Colonists of leaf constructs on Missouri *Quercus*, 522, 523
Commensalism, 516
Community-wide monitoring, 542
Conifer-feeding caterpillars, 35
Conjugation, 131–134
Constitutive Androstane Receptor (CAR), 128
Construct building by caterpillars
Construct traits, 526, 528
Coprophagy, 29
Cowpea, 122, 285
CRISPR, 120
Crop (foregut), 18, 19, 41
Crop (pests), 15, 33
Crypsis, vii, 197, 200
Cue, 196, 197, 200, 205, 207, 209, 217
Cuticle, 66
Cuticular hydrocarbon (CHC), 341, 357
Cyanogenic glycosides, 118
CYP450s, 129, 131
CYP6AE, 129
CYP6AE19, 130
CYP6B8, 130

D

Darwin, 6, 165
Data collection approaches, 542
Defense posture, 299
Defense response induction, prevention of, 122, 123
Defense strategies, 102, 106, 500
Defense trait, 438
Defensive behaviors, 250
Defensive secretions, 42
Defensive sound and vibration detection, 209
Deimatic displays, 195, 199
Detecting sounds and vibrations made by non-predators, 208
Detection of near-field sounds, 206
Detection of solid-borne vibrations, 207, 208
Detoxification, 116, 126–128, 410
Detritivore, 30, 333
Detritivory, 29
Developmental defenses, 102
Development (larval), 26, 27, 35, 500
D-gluconic acid, 122
Diapause, 28
Dietary specialization, 115, 116, 274, 276, 277, 280, 282–291

Subject Index

Diet breadth, 497, 569
Diet breadth amplitude, 569
Digestive tract, 18, 22, 238
Dipteran parasitoids, 242–246
Direct defenses, 400
Dish organs, 21, 352
Dispersal, 33, 49, 51, 254, 596
Disruptive coloration, 299
DMNT-induced damage, 124
DNA barcoding, 251, 610
Dorsal nectary organ, 326
Dose-dependent compounds, 430
Drought, 529
Dyar's rule, 22, 23

E
Eastern lubber grasshopper, 173, 281
Ecoimmunology, 449
Ecological immunology, 449
Ecological specialization, 115
Ecosystem engineering, 514, 516, 527
Ecosystem function, 12
Ectoparasitism, 227
Ecuador, 94, 95, 98
Egg parasitoids, 236
Encapsulation, 396, 451, 452, 454, 458–460, 464–466, 468, 471
Enemy-free space, 249, 407, 408
Enemy-free space hypothesis, 276–279, 282–287, 289–291
Environmental variables, 570
Epicuticular layer, 68
Epicuticular plant waxes, 69–71, 433
Esophagus, 18
Eulophid parasitoids, 231
Excretion, 134, 136, 137
Exocrine glands, 20, 21, 41, 357
Exocrine secretions, 36, 42, 300
Extrafloral nectar (EFN), 101, 301, 435, 436
Extreme climatic events, 552
Extreme meteorological events, 560

F
Facultative ant associates, 327
Facultative carnivory, 30
Fall army worm, 74
Far-field sounds, 196, 205
Feculae, 20, 41, 44, 45, 49, 50
Fecundity, 283, 436, 473, 474, 503
Feeding, 65, 68, 83, 85
Feeding decisions, 279, 284
Field stations, 619

Filament encapsulation assay, 452
Fire ant encounters, 306
Fire ants, 308, 310
Flies, 170, 227, 241, 279
Foliage-dwelling ants, 299
Food webs, 248, 258
Foregut, 18, 19, 41
Forest bitterberry, 67
Forest habitats, 545
Fragmentation, 253, 254
Frass, 510, 516, 527, 532
French Guiana, 97, 98, 105
Fruit (borer/predator), 41
Furanocoumarins, 400

G
Galls, 510, 514, 521, 525, 527, 529
Gene regulation, 23, 476
Gene sequencing, 610
Genista broom moth, 281
Geometric Framework for Nutrition, 460
Geometric growth, 23
Glandular trichomes, 72, 75–77, 83
Global Biodiversity Information Facility, 541
Global climate change, 106–108
Glucosinolates (GSLs), 116, 117, 400, 401, 409
Glucosinolate-specific transporter (GTR) genes, 146
Glycoalkaloids, 142
Google Scholar, 277
$Gr66$, 120
Gram-negative bacterium, 470
Granulocytes, 451
Greenhouse gases, 412
Green peach aphid, 281, 286
Ground cherry, 284, 470
Gustatory receptors, 120
Gut, viii, 21
Guttation droplets, 436

H
Habitat fragmentation, 254
Habitat loss, 253, 254
Haplodiploidy, 228
Hearing, 196, 199, 205, 206, 209, 217
Heliconiine species, 175
Hemocyte concentrations, 466
Herbaceous vegetation, 543
Herbivore-induced plant volatiles (HIPVs), 279, 285, 288, 402, 426, 437
Herbivory, 72, 80, 86, 400

Herbivorous insects, 612
Hindgut, 18–20
Holometaboly, vii
Hormone receptor-like in 96 (HR96), 128
Host feeding ecology, 251
Host manipulation, 285, 286
Host plant, 96, 426, 429
Host plant selection, 96
Host specificity, 285, 288
Hurricanes, 561, 572, 576, 577
Hydrolysis, 129–131
Hydrophobicity, 68
Hymenopteran parasitoids, 228–232, 242, 436
Hypermetamorphosis, 241
Hyperparasitism, 237, 238
Hyperparasitoids, 402

I

Idiobiont parasitoid species, 257
Idiobionts, 227
Immune assay, 452
Immune responses, 405, 408, 410–412, 470, 474–475
Immunity, 408, 410
Immunocompromised host hypothesis, 408
Immunosuppressive effects, 465
Implant encapsulation, 466
iNaturalist, 549, 550, 553
iNaturalist caterpillar occurrence, 546
iNaturalist dataset, 543
iNaturalist project, 542
Indirect defense, 398, 400
Indol-3-carbinol (I3C), 128
Induced defenses, 74, 467
Information processing hypothesis, 284
Information resources, 437, 438
Inga system, 612
Innate immune response, 451
Insect abundance, 489
Insect herbivore, 274–279, 282, 283, 286–288, 290, 291, 435
Insect immunity, 451, 452, 454, 458
Insectivorous, 39, 167, 198, 210, 299, 434, 437
Insectivory, 50
Instar, 617
Instar caterpillars, 617
Integumental spines/chalazae/scoli, 40
Intensive agricultural practices, 255
Internal (feeder/feeding), 40
Invasive red imported fire ants, 303
Invasive species, 255–257, 524
Invertebrate predators, 13, 21, 33, 46, 148, 195
Ionotropic receptors, 120

Ir8a, 120
Iridoid glycosides (IGs), 177, 400, 470–472
Iridoid glycosides, as model system, 171–173
ISI Web of Science, 277
Isothiocyanates (ITCs), 117, 118

J

Jasmonic acid (JA), 122
JcDV, 470–472

K

Koinobiont strategy, 227, 438

L

Ladybug beetles, 281
Larval phenotype, 39, 40
Leaf bags, 510
Leaf folds, 510, 518, 519, 527
Leaf mines, 510, 527, 529
Leafminers, 13, 29
Leaf quality, 498, 522, 524, 525
Leaf rolls, 510, 515, 518, 520, 523–525, 528–530
Leaf shelters, 524
Leaf ties, 510, 518, 520–525, 527, 529, 532
Leguminous host plant, 349
Lethal compounds, 428–430
Lichens, 283
Life cycle, 243, 511
Linear mixed effect models, 571
Local mate competition, 230
Long-tailed skipper, 306
Looping (loss of anterior prolegs), 35

M

MABOTRO, 561
Macrocentrines, 235
Macrodecomposition, 13
Macroevolutionary patterns, 105
Macronutrient composition, 460, 461
Macronutrients, 453–454, 458, 461–463
Major allergen (MA) protein, 143
Major facilitator superfamily (MFS), 146
Malpighian tubules, 240
Mandibles, 14–16, 23, 31, 46, 66, 212, 213, 245
Mandibular gland, 20, 137
Manipulation, 357–361
Masquerade, vii, 282, 289
Mdr50, 147

Melanization, 451, 459, 464–466, 471, 472, 475
Merian, 5, 6
Metabarcoding, 247
Metagenomics, 247
Metamorphosis, 6, 23
Meteorines, 234
Meteorological extreme events, 558
Microbial interactions, 137, 138
Microbiome, 21, 138, 139, 474, 475
Microgastrine braconid cocoons, 229
Middorsal gland, 21
Midgut, 40, 124
Migratory, 26, 33, 185
Mimic/mimicry, 23, 24, 26, 27, 36–39, 44, 45, 199, 210, 216, 237, 238, 298, 299, 333, 335, 336, 346, 347, 353, 357, 358, 521, 602, 610
Mitogen-activated protein kinase (MAPK), 128
Monarchs, 180, 471
Mow, 75, 81, 85, 427
Multi-CBD chitin binding proteins, 124
Multidrug transporters (Mdrs), 134
Multiple immune parameters, 452
Mummy wasps, 235
Mustard aphid/turnip aphid, 286
Mutualism, 328, 357–361

N
Na+/K+ATPase alpha, 142
Nasty host hypothesis, 408, 416–417, 464
Nasty hosts, 407
National Land Cover Database, 545
Natural enemies, vii, 279, 282, 612
Natural enemy complex, 23, 396, 430
Natural history, 517, 518, 521, 532
Natural selection, 36, 280, 286
Near-field sounds, 196, 200, 205–207
Nests, 47
Network parameters, 570, 571
Neural constraints hypothesis, 284
Newcomber's gland, 21
Nicotine, 407
Nitrile-specifier proteins (NSPs), 123
Non-feeding caterpillars, 50
Non-glandular trichomes, 72, 74, 79–82
Non-receptor chemosensory gene families, 121, 122
Non-sequestering milkweed-feeding caterpillars, 429
Non-trophobiotic, 326, 328
NSD, 307
Nutrient cycling, 13, 516

O
Oaks, 518, 522, 524, 527–529
Obligate ant associates, 327
Odorant-binding proteins (OBPs), 121
Odorless-2 (od-2), 75
Olfactory receptors, 119, 120
Oligophagous butterfly, 575
Ontogeny, 427, 432
Organic anion transporting polypeptides (Oatps), 136
Organization for Tropical Studies (OTS), 583
Oriental silk moth, 20
Outbreak species, 549, 550
Oviposition, 71
Oxidation, 129–131
Oxidoreductases, 131
Ozone, 413

P
Palatability, 167, 168, 170
Panama, 95, 98
Papilionoidea, 583
Parasites, 33, 34, 450–452, 458, 469–473
Parasitism, vii, 328, 360, 438, 518, 529, 531, 532
Parasitoid loss, consequences of, 258
Parasitoidism, 228
Parasitoids, 33, 34, 195, 197, 200, 205–208, 215–217, 226, 275, 277–279, 282, 285, 429, 433, 436, 450–452, 463, 468, 469, 471, 475, 499
Parasitoids community patterns, 247–249
Passifloraceae, 84, 178
Passion vine, 280
Pathogens, 33, 34, 463, 469, 470
Pattern recognition receptors (PRRs), 451
Perforated cupola organs (PCOs), 341
Peritrophic membrane, 80, 124, 125
Phenological changes, 23, 503
Peritrophin A, 124
Peru, 98
Peruvian Amazon, 104
Phase I, 118, 129–131
Phase II, 131–134
Phase III, 134, 136, 137
Phenolics, 105, 466, 467
Phenological defenses, 103, 104
Phenological match/mismatch, 411–413, 552
Phenological tracking, 103
Phenoloxidase (PO), 451, 475
Phenotypic plasticity, 25, 51, 559
Physical barriers, 124, 125
Physical barriers to secondary compound, 124, 125

Physical defenses, 66, 250
Physical manipulation of environment, 514
Physiological efficiency hypothesis, 274, 276
Physiological responses, 40–43
Phytomimesis, 72
Plant allelochemicals, 7
Plant architecture, 517, 522, 525, 531
Plant breeding, 400–402
Plant-caterpillar interaction networks, 563, 566, 567
Plant community structure, 396
Plant compounds, 22, 35, 50, 120, 121, 173
Plant defense, 65, 75–77, 517
Plant domestication, 405
Plant domestication-reduced defense, 400
Plant-herbivore interaction networks, 559, 560, 562, 572
Plant-herbivore network, 559, 560
Plant leafing patterns, 486
Plant nutritional and defensive traits, 106
Plant parasite, 29
Plant phenology, 498
Plant phenotypes, 439
Plant secondary metabolites, 274, 411–413
Plant toxin-mediated interactions, 405–411
Plant toxins, 400–402
Plant volatile-mediated interactions, 402–405
Plant waxes, 66, 68
Plasmatocytes, 451
Plasmocytes, 21
Plasticity, 559, 577
Polyembryony, 235
Polyphagous caterpillars, 426
Polyphagy, 283, 288, 289
Polyphenism, 26, 27, 288
Polyphenol oxidases (PPOs), 400
Pore cupola organs (PCOs), 341
Precipitation, 564
Precursor toxins, diversion strategies for, 123, 124
Predaceous caterpillars, 30, 31, 33, 49
Predation, 464, 516–518, 525, 527, 530–532
Predators, 33, 34, 396, 429, 433, 436
Predatory, 200, 298, 356
Prepupal larvae, 44
Primary defense, 50
Prognathous mouthparts, 35
Protozoans, 451
Proventriculus (proventricular valve), 18, 19
Pugnacious bodyguards, 300
Pyrrolizidine alkaloids (PAs), 471

Q
Quantitative defenses, 430

R
Rainfall, 529–531
Reactive oxygen species (ROS), 128
Recruitment signals, 199, 209, 216
Reduction, 129–131
Regurgitant, 41, 137, 146, 210, 402–404
Regurgitation, 438
Resting posture, 44, 45
Riodinid butterflies, 21
Riodinidae larvae, 103
Risk assessment, 207, 208
Rodents, 34, 195, 209, 619

S
Sack bearer, 41
Salicylic acid (SA), 122, 413
Salivary gland, 20, 137
Saponins, 97, 99
Saprophagy, 29
Seasonality, 487, 491–498
Secondary defense, 517
Secondary metabolite pathways, 610
Secondary setae, 40, 42
Seed (borer/predator), 29
Self-medication, 462, 463, 472
Semicrystalline fibers, 47
Sending signals to non-predators, 216
Sending signals to predators, 210, 215
Sensilla, 17
Sensory receptors, 196, 197, 200, 205, 206, 209
Sequestered caterpillar chemical defenses, 165–167, 180–181
Sequestration, viii, 174, 280–282, 406–409, 429, 430, 464–466, 468, 475, 476, 612
SG-HM hypothesis, 430, 431
Shelter-building behavior, 494
Shelter-building caterpillars, 614
Shelter-building species, 492
Shelters, 432, 434
Short hindgut, 19
Silk, 47, 49, 50
Silken pads, 49
Silver-spotted skipper, 432
Single domain major allergen (SDMA) proteins, 143
Skipper caterpillars, 516, 526
SlCXE10, 131
SlCXE7, 131
Slow-growth high-mortality hypothesis, 430
Soil nitrogen, 182
Solid-borne vibrations, 196, 197, 200, 205, 207, 209, 215, 217

Songbirds, 33
Sound and vibration reception, 200, 205–208, 210, 215, 216
Sound and vibration reception in caterpillars, 200
Sound production, 196, 209, 210, 216–217
Specialist and generalist immunity, 475, 476
Species richness, 28, 100, 232, 492
Spin communal nests, 49
Standardized monitoring schemes, 541
Startle sounds, 210
Stellate trichomes, 79
Steroid and Xenobiotic Receptor (SXR), 128
Steroidal glycosides, 142
Straw moth, 71, 97, 98, 127, 128, 130, 131, 284, 285, 470, 476
Stridulation, 341
Structural defenses, 65
Sublethal compounds, 430, 431
Substrate-borne acoustic signals, 342
Subterranean caterpillars, 227
Sulphur butterfly, 303
Sulphur butterfly caterpillar, 301, 312, 313
Sulphur caterpillars, 306
Sundews, 35
Survivorship curve, 23, 32
Synchronization, 103
Synergistic and combinatorial phytochemistry, 467–469

T
Tachinid fly life cycles, 243
Tachinids, 246
Tangible resources, 435–437
Target site insensitivity (TSI), 116, 125, 126
Tarweeds, 35
Temperature, 564, 572
Tent, 50
Tentacle nectary organs (TNOs), 21, 354
Tentacle organs (TOs), 342
Terpenoids, 124
Theclinae-Polyommatinae Assemblage, 345–349
Tobacco budworm, 71, 97, 98, 127, 128, 130, 131, 285, 470, 476
Tobacco hornworm, 78
Top-down agents of selection, 145–150
Top-down pressures, 36, 37, 39, 40
Toxin resistance, 141, 143
Trail pheromones, 20
Transient receptor potential channels, 120, 121

Trench warfare, 143
Trichoid sensilla, 196
Trichomes, 72–74, 78, 80, 83, 84, 100, 101, 431–433
Trigonalid wasps, 238
Tritrophic interaction theory, 610
Tritrophic interactions, 426, 428–438
Trophic cascades, 426
Trophobiotic, 326, 328
Tropical dry forests, 568
True bugs, 170, 173, 279, 515, 614
Tupelo leaf miners, 211, 215
Turtle-shelled wasps, 236
Twig-mimicking geometrids, 44

U
Unpalatability, 170–171, 178–179
Urticaceae, 82

V
Vascular plants, 28, 611
Velvetbean moth, 123, 132, 138, 285
Vibrations/solid-borne vibrations, 197
Vibratory signaling, 325, 326, 342, 343
Vibroacoustic environments, 196
Volatiles (from frass), 44
Volatile organic compounds (VOCs), 429
Vulnerable host hypothesis, 408, 409, 464

W
Wallace, Alfred R., 169, 182–183, 587, 612
Walnut sphinx, 210, 212, 215
Waxes, 66, 67
Western paper wasp, 281
White mustard, 286, 401
Wolf spider, 281
Wondrous Transformation of Caterpillars, 6
Wood boring, 227
Woody plants, 518, 519
Woolly bears, 471

X
Xenobiotic metabolism, 128
Xenobiotic response elements (XRE), 128

Y
Yellow sulphur caterpillar, 311
Young leaves, 95

Taxonomic Index

A
Abaeis nicippe, 300
Abies balsamea, 70
Abraxas grossulariata, 168
Acacia, 323, 324
Acer rubrum, 547
Acer saccharum, 547
Acherontia atropos, 170
Achroia grisella, 22
Acinetobacter, 138
Acleris, 512–513
Acontiinae/arctiine, 29, 48
Acrobasis indigninella, 511
Acrodipsas, 347
Acrolepiopsis assectella, 32
Acromyrmex, 321
Acronicta, 23, 37, 46
Acronicta americana, 37, 43
Acronicta funeralis, 49
Acronicta impleta, 37
Acronicta radcliffei, 37
Acronictinae/acronictine, 32
Acrospila gastralis, 330
Actia interrupta, 243
Adejeania sp., 243
Adeliini, 233, 234
Adelpha iphiclus, 48
Adelpha serpa, 25
Adelpha tracta, 231
Adenostoma, 283
Aegeria acerni, 511
Agathidinae, 233
Agathis malvacearum, 230
Agathis pumila, 412
Aglais urticae, 257
Agrilus planipennis, 256

Alchornea cordifolia, 330, 332
Aleiodes, 235
Aleiodes dentatis, 345
Aleiodes pallida, 352
Allium ampeloprasum, 70
Alomya, 236
Amelanchier pallida, 591
Amorpha juglandis, 210, 212, 215
Amydria, 331
Amydria anceps, 321
Amyops ingens, 336
Anagrus nilaparvatae, 404
Anartia jatrophae, 175, 176, 181, 455, 466, 471
Ancylis divisana, 510
Anicia checkerspot, 172
Anomaloninae, 236, 237
Anomis editrix, 566, 573
Anoplolepis, 331
Anoplolepis gracilipes, 337
Anteos clorinde, 310
Anthene, 347
Antheraea polyphemus, 201, 210, 211, 215
Anthocharis, 28
Anthrocera filipindulae, 168
Anthropocene, 253–258
Anticarsia gemmatilis, 123, 132, 138, 285
Antispila nysaefoliella, 211, 215
Apantales, 251
Aphnaeinae, 325, 345, 351, 352, 361
Aphidae, 280, 281, 284, 286
Apiaceae, 42
Apocynaceae, 178, 180
Apocynum cannabinum, 147
Apodrepanulatrix sp., 591
Apotolype blanchardi, 45

Aquilegia pyrenaica, 76
Aquilegia vulgaris, 76
Arabidopsis, 413
Arabidopsis lyrata, 82
Arabidopsis thaliana, 127, 437
Arachnidomyia aldrichi, 244
Arachnis, 28
Araneae, 511
Archips cerasivorana, 49, 512–513, 527
Arctia caja, 43
Arctiinae, 173, 234
Arctostaphyllus patula, 591
Arctostaphylos pungens, 42
Ardiosteres, 321
Argidae, 98, 102, 105
Arhopala, 347
Arhopala madytus, 343
Arhopala japonica, 358
Aricoris arenarum, 338, 356
Aristolochiaceae, 179
Aristolochia erecta, 179
Aristotelia, 533
Arsenura, 24
Arsenura batesii, 24
Ascia monuste, 300
Asclepias curassavica, 128, 131, 134–136, 145, 147, 149, 180, 182, 472
Asclepias fruticosa, 147
Asclepias incarnata, 128, 472
Asclepias syriaca, 82, 145
Asimina, 520
Aslauga spp., 342, 350, 356
Aspalathus, 352
Asobara tabida, 410
Asteraceae, 526
Asthenidia, 42
Astragalus canadensis, 475
Athalia rosae, 146
Atropa belladonna, 170
Atta, 321
Azteca, 300

B

Bacillus subtilis, 453, 462
Bacillus thuringiensis, 457, 470, 474
Bacopa monnieri, 466
Baltimore checkerspot, 172, 176, 181
Banchinae, 236
Baratrachedra, 328
Barbarea vulgaris, 71
Batrachedra, 332
Batrachedra myrmecophila, 322, 332
Battus philenor, 179

Bemisia tabaci, 137
Besseya alpina, 178
Betula pubescens ssp. *czerepanovii*, 459, 466, 467
Betula papyrifera, 532
Betula populifolia, 38
Biblidine, 37, 43
Bignoniaceae, 171, 175, 177
Biston, 26
Biston betularia, 26
Biston robustum, 72
Blastobasidae, 328
Blondelia, 245
Blondeliini, 244, 256
Boerhavia erecta, 45
Boletobiinae, 334, 335
Bombycidae, 37, 43, 119
Bombyliidae, 242, 244
Bombyx mori, 20, 70, 119, 120, 135, 137, 146, 454
Brachymeria intermedia, 239
Braconidae, 232–237, 239
Braconid parasitoids, 229
Braconids, 233, 237
Bradysia pauper, 80
Brassicaceae, 69, 117, 401
Brassica juncea, 401
Brassica napus, 68, 70, 401
Brassica nigra, 81, 149
Brassica oleracea, 71, 149, 409, 467
Brassica oleracea var. *viridis*, 68
Brassica spp., 401, 437
Brenthia sp., 15, 25, 43, 48
Brevicoryne brassicae, 146
Byrsonima pachyphylla, 495
Byrsonima spp., 494
Bryolymnia viridata, 38
Bucculatricidae/bucculatricid, 49, 233
Buckeye, 172, 176, 179–181
Buprestidae, 256
Bursera, 98

C

Cabbage, 433
Cactoblastis cactorum, 256
Caligo, 241
Callophrys, 347
Calophasia lunula, 170, 175–177, 182
Calycopidina, 348
Calycopidini/calycopidine, 37
Calydnini, 325
Cameraria, 25
Camponotus brutus, 330

Camponotus floridanus, 301, 302, 307, 312, 313
Camponotus japonicus, 358
Camponotus obscuripes, 358
Camponotus spp., 309, 311, 346, 347, 357, 361
Campopleginae, 236
Campoplegine ichneumonids, 237
Camptocosa parallela, 281
Capsicum annuum, 74, 78
Capsicum spp., 133
Carboniclava alpicoides, 39
Cardiochilinae, 233
Cargida pyrrha, 566, 573
Caria ino, 356
Caripeta hilumaria, 45, 51
Caryocaraceae, 492
Cassia, 310
Cassia fasciculata, 300
Cassia fruticosa, 310
Castilleja hispida, 181
Castilleja integra, 178
Castilleja levisecta, 181
Catalpa bignonioides, 177, 465
Catalpa sphinx, 171, 173, 184, 475
Catalpa spp., 171
Catocala aholiba, 45
Catocala ilia, 38
Catocala illecta, 21
Catopsilia, 324, 337
Catopsilia pyranthe, 337
Ceanothus, 283
Ceanothus velutinus, 591
Cecrita guttivitta, 616
Cecropia, 300, 323, 329
Cecropia obtusifolia, 330
Centrosema virginiana, 304
Ceratomia catalpae, 171, 173, 175, 177, 184, 425, 428–430, 455, 465, 475
Ceratomia undulosa, 184, 410, 455, 466
Cerconota achatina, 494, 498
Chalcididae, 239
Chalcidoidea, 228, 231, 232, 236, 238–242
Chalcopasta howardi, 45, 51
Chamaecrista (syn. *Cassia*) *fasciculata*, 300
Chamaecleini, 48
Chamise, 283
Charaxinae, 25
Chelone glabra, 165, 181
Cheloninae, 234
Chelonus, 230
Chetogena spp., 463
Chilades kedonga, 329
Chilades lajus, 347

Chilo partellus, 403
Chilopsis linearis, 175, 177
Chilo suppressalis, 404
Chioides catillus catillus, 486
Chionodes sp., 591
Chlamydastis platyspora, 486, 498
Chloridea, 16
Chloridea virescens, 14, 16
Chloropidae, 432, 437
Chlosyne, 28
Chlosyne gloriosa, 566, 573
Choreutidae, 28
Choristoneura fumiferana, 70
Choristoneura spp., 70, 243
Chrysauginae, 332
Chrysocharis laricinellae, 240, 412
Chrysochus auratus, 147
Chrysodeixis, 411
Chrysodeixis includens, 77
Chrysomela populi, 146
Chrysomelidae, 404
Chrysoperla carnea, 147
Chrysopidae, 328
Chrysotachina, 245
Cicadidae, 332
Cicinnus, 41
Cicinnus melsheimeri, 41
Cigaritis takanonis, 352
Cirrhochrista saltusalis, 333
Cleorodes lichenaria, 283
Cleridae, 242
Cnidoscolus urens, 85
Cobubatha, 42
Cobubatha dividua, 43
Coccinellid, 281
Colastes, 234
Colchicaceae, 170
Coleophora laricella, 240, 412
Coleophora xyridella, 48
Coleophoridae, 45, 233, 322, 332
Coleoptera, 614, 242, 515, 524, 511, 514
Colubrina spinosa, 583, 584
Comadia redtenbacheri, 25
Compsilura concinnata, 246, 250, 256
Conchylodes nolckenialis, 330
Copidosoma, 240
Copidosoma floridanum, 145, 407
Copidosoma sosares, 407
Cordia, 362
Cordia alliodora, 330
Corrachia, 364
Corylus, 168
Cosmia sp., 591
Cosmopterigidae/cosmopterigid, 43, 496

Cossidae, 237
Cossus cossus, 20
Cotesia congregata, 407, 410, 463
Cotesia glomerata, 256, 409, 467
Cotesia marginiventris, 403, 404
Cotesia rubecula, 403
Cotesia sesamiae, 403
Cotesia sp., 229
Cotton boll weevil, 310
Coxina spp., 323, 329
Crambidae, 236, 322, 333
Crambidia, 333
Crambidia casta, 333
Crataegus, 168
Cremastinae, 236
Crematogaster, 322, 330–332, 352, 353, 363
Crematogaster mimosae, 329
Crematogaster nigriceps, 329
Crematogaster striatula, 330
Cristina's timema, 283
Cryptinae, 238
Ctenopelmatinae, 236
Cucullia verbasci, 168
Cucumis sativus, 78
Curetinae, 325, 344, 345
Curetis, 344
Curetis thetis, 344
Cyanocitta cristata, 172
Cyclomia disparalis, 584
Cyclotorna monocentra, 331
Cyclotorna spp., 322, 331
Cyclotornidae, 322, 327, 328, 331, 332
Cydia pomonella, 71, 511
Cyrtepistomus castaneus, 527, 524

D

Dalceridae, 566, 573
Dalcerina tijucana, 486
Dalea, 36
Dalmatian toadflax, 170, 182
Danainae, 281
Danaus gilippus, 173, 175
Danaus plexippus, 19, 27, 33, 82, 121, 135, 137, 173, 175, 205–207, 243, 454, 455, 458, 471, 472
Daphnia magna, 136, 473
Daphnis nerii, 136
Dasylophia eminens, 566, 573
Datana drexelli, 37
Datana eileena, 42
Datana eileena-perspicua, 42
Datura, 407
Datura stramonium, 81

Datura wrightii, 77, 120
Daucus carota, 70
Death's head hawkmoth, 170
Deinopa biligula, 566, 573
Delphacidae, 404, 405
Deloneura, 325, 353
Depressaria pastinacella, 130, 140, 143, 406, 407, 512–513
Depressaria radiella, 130, 131, 144, 512–513
Depressariidae, 545, 546
Desmodium, 304
Desmodium incanum, 304
Desmodium tortuosum, 305
Dexiinae, 242
Diabrotica virgifera virgifera, 404
Diadegma semiclausum, 145, 407
Dianesiini, 325
Diadegma semiclausum, 455
Diaphania jairusalis, 566
Dibrachys cavus, 241
Dibrachys microgastri, 250
Dichordophora, 40
Dichordophora phoenix, 39, 51
Digrammia continuata, 47
Digrammia sp., 591
Diloba caeruleocephala, 168
Dinomyrmex, 322
Diptera, 170, 227, 241, 279
Diurnea fagella, 213, 215
Dolichoderus, 322, 331, 337, 350
Dolichoderus gibbosoanalis, 332
Dolichogenidea homoeosomae, 397
Doryctinae, 233
Dorylus, 363
Drepana arcuata, 200, 201, 207, 208, 213, 215, 532
Droseraceae, 35
Drosophila, 121, 124–127, 129, 134–136, 141, 147, 148
Drosophila grimshawi, 127
Drosophila melanogaster, 119, 121, 124–127, 129, 133–136, 141, 142, 147, 148, 610
Drosophila mojavensis, 127
Drosophila subobscura, 142
Drosophila virilis, 127
Drypetes, 118
Dyops spp., 300, 323, 329

E

Eacles sp., 486
Eciton, 363
Ectatomma tuberculatum, 102

Eggerthella lenta, 139
Egira, 23
Elachistidae, 244, 487, 492–494, 496, 499
Elaphria, 38
Eldana saccharina, 70
Emarginea, 42
Emarginea percara, 43, 51
Emesidini, 325
Emesis aurimna, 354
Emesis emesia, 566, 573
Empyreuma pugione, 136
Encyrtidae, 239, 240
Ennominae/ennomine, 25, 43
Ennomos subsignarius, 27
Enterobacter spp., 137, 138
Enterococcus faecalis, 456, 470
Enterococcus mundtii, 474
Enterococcus sp., 138
Eois, 408, 601
Eois apyraria, 408, 468
Eois nympha, 408, 468
Epargyreus clarus, 426, 427, 431–433, 527
Epipaschia superatalis, 566, 573
Epiplemidae, 39
Epipyropidae, 30, 242
Epirrita autumnata, 456, 458, 459, 466, 467
Eptesicus fuscus, 524
Erebidae, 93, 96, 104, 173, 175–177, 179, 182, 234, 235, 244, 323, 333–335, 401, 543, 545, 546
Ericaceae, 401, 591
Ericameria ericoides, 36
Erinnyis ello, 85
Eriocraniidae, 233, 234
Eriocranioidea, 233
Erora, 347
Erysimum, 117
Erythroxylum tortuosum, 495
Estigmene acrea, 17, 70, 179
Ethmia nr. *similatella*, 566, 573
Eublemma, 328
Eublemma albifascia, 323, 334
Eublemma spp., 334
Eucaterva variaria, 175–177
Eucelatoria, 245
Euceros, 238
Eucerotinae, 238
Euchaetes egle, 122
Euchrysops cnejus, 361
Eudesmia menea, 566, 573
Euliphyra, 350, 351
Eulophidae, 232, 240
Eulophus, 240
Eumaeus, 348

Euonymus europaeus, 168
Eupackardia calleta, 42
Eupelmidae, 242
Eupelmus vesicularis, 242
Euphorbiaceae, 85
Euphorinae, 233–235
Euphydryas, 28, 171, 172
Euphydryas anicia, 178
Euphydryas colon, 183
Euphydryas editha, 181
Euphydryas editha taylori, 181
Euphydryas phaeton, 172, 174–176, 181, 452, 455, 465, 471, 475
Eupithecia, 30, 31, 40
Euplectrus, 240
Euploea core, 136
Euproctis chrysorrhoea, 42
Eurata semiluna, 494, 495
Eurema, 300
Eustema opaca, 494, 495
Eurybiina, 324, 354
Eurybiini, 325
Exoristinae, 242, 245
Exothecinae, 233, 234

F
Fabaceae, 312, 349, 397
Fagaceae, 412
Fagus grandifolia, 518, 546, 547
Feniseca, 364
Feniseca tarquinius, 350
Ficus, 98
Fischeria panamensis, 602
Flies, 227, 228, 241, 242, 244–247, 254
Formica, 333, 336
Formica podzolica, 439
Formicidae, 511
Fouquieria, 585–586

G
Galleria mellonella, 22, 25, 452, 454
Gelechiidae, 230, 235, 328, 492, 496
Gelechioidea, 96, 103
Genea, 245
Gentianaceae, 339
Geocoris sp., 437
Geometridae, 235, 244, 328, 544–546
Glaucopsyche lygdamus, 349
Gloriosa, 170
Gloriosa superba, 170
Gloveria, 42
Glycine max, 70, 409

Glyptapantales, 251
Gnamptodontinae, 233, 234
Glycine max, 70
Glyptapantales, 251
Gnamptodontinae, 233, 234
Goniini, 245
Gracillariidae, 233, 234, 244
Graminicides, 183
Grammia incorrupta, 454–456, 460, 463, 466, 471, 475
Graphogastrini, 244
Grapholitini, 43
Grotella tricolor, 45, 51
Guaiacum angustifolium, 39
Gymnelia salvini, 42
Gynaephora groenlandica, 27

H
Halia vauaria, 168
Halysidota tessellaris, 37
Haploa, 28
Harrisimemna, 46
Harrisimemna trisignata, 46
Helianthus, 413
Helianthus annuus, 78, 397
Helicobia, 244
Helicobia sp., 244
Heliconiidae, 167
Heliconiinae, 80
Heliconius charithonia, 84
Heliconius erato, 179
Heliconius melpomene, 135, 137
Heliconius pachinus, 82, 84
Heliconius sara-sapho, 280
Heliconius spp., 82, 178, 280
Helicopini, 325
Helicoverpa armigera, 70, 80, 83, 120, 121, 129, 132, 133, 140, 141, 146, 404
Helicoverpa assulta, 133
Helicoverpa spp., 30, 133
Helicoverpa zea, 14, 69, 75, 122, 124, 130, 133, 436, 474
Heliothis, 97, 98
Heliothis subflexa, 284, 457, 470, 476
Heliothis virescens, 71, 97, 98, 127, 128, 130, 131, 284, 285, 470, 476
Heliothinae, 27, 30
Heliozelidae, 234
Hemeroplanes triptolemus, 602
Hemileucinae/hemileucine, 20
Hemiptera, 170, 173, 279, 515, 614
Hemocytes, 21, 451, 452, 458, 465, 466, 470, 471, 475

Hepialidae/hepialid, 37
Hepialoidea, 233
Herminiinae/herminiine, 42
Herpetogramma theseusalis, 527
Hesperiidae, 304, 324, 336, 337, 496, 518, 520, 526, 527, 532
Heterocampa guttivitta, 230, 616
Heterophleps, 45
Heterorhabditis bacteriophora, 25
Homodes spp., 323, 335
Homoeopteryx, 42
Homoeosoma electellum, 397
Homolobinae, 233, 235
Homoptera, 23
Hormiinae, 233, 234
Hyblaea, 16
Hylesia continua, 566, 573
Hylesia nigricans, 201, 208, 211
Hylesia sp., 566, 573
Hyles lineata, 12
Hymenoptera, 170, 227, 232, 241, 242, 247, 248, 253, 495
Hymenothrix wrightii, 45
Hyparpax aurora, 615
Hypena madefactalis, 231
Hypercompe confusa, 566
Hypercompe scribonia, 16, 31
Hypercompe sp., 566, 573
Hyphantria cunea, 512, 526, 531
Hypocala andremona, 27
Hypochrysops, 360
Hypochrysops byzos, 360
Hypochrysops polycletus, 360
Hypochrysops pythias, 360
Hypolycaena, 346
Hypolycaena othona, 213, 216
Hypolycaenini, 346
Hyposmocoma, 49
Hyposmocoma molluscivora, 30
Hyposoter, 237
Hyposoter ebeninus, 145
Hystrichophora spp., 321, 329

I
Ichneumonidae, 231–233, 236–239, 410
Ichneumoninae, 236, 237
Ichneumonoidea, 228, 230, 232, 233, 243
Ichneutinae, 233, 234
Idalus lineosus, 494, 495
Inga, 95–98, 100, 103, 104, 108, 109
Inga capitata, 102
Inga jenmanii, 105
Inga thibaudiana, 102

Iphierga, 321
Ipomoea aquatica, 409
Ippa, 321
Iridomyrmex, 321, 328, 357
Iridomyrmex humilis, 281
Iridomyrmex mayri, 358
Iridomyrmex purpureus, 331
Iridomyrmex rufoniger, 347

J
Jalmenus daemeli, 347
Jalmenus eichhorni, 328
Jalmenus evagoras, 328, 343, 359, 360
Juglans nigra, 512–513
Juniperus, 511, 585–586
Juniperus osteosperma, 591
Junonia coenia, 172, 175–177, 179–181, 429, 455, 456, 464, 465, 468, 471, 475
Junonia coenia densovirus (JcDV), 455, 456, 470–472

K
Keiferia lycopersicella, 75, 76, 83
Kipepeo kedonga, 329
Klebsiella, 138
Kolana ergina, 486
Krameria ramosissima, 39

L
Lacanobia oleracea, 410
Lachnocnema, 340
Lachnocnema laches, 350
Lagenaria siceraria, 79
Laportea canadensis, 82
Larentiinae/larentiine, 42
Lasiocampidae, 27, 543
Leptinotarsa decemlineata, 137, 142
Lasius, 331
Lasius emarginatus, 342
Lepidochrysops, 346, 347
Lepisiota capensis, 352
Leptinotarsa decemlineata, 137, 142
Leskia, 245
Lespesia archippivora, 243
Libytheana carinenta, 33
Limacodidae, 235, 496
Limenitidinae, 11, 42
Linaria dalmatica, 170, 175, 177, 182
Lipaphis erysimi, 286
Liphyra, 350, 351
Liphyra brassolis, 350, 351

Liphyra grandis, 350
Liptenini, 330
Lissonotus, 237
Lithophane, 23, 30
Lithophane lepida, 38
Lithosiine, 28
Lobesia botrana, 476
Lonomia, 47
Lophocampa caryae, 43
Lophocampa citrina, 566, 573
Lophocampa ingens, 456, 467, 468
Lotongus calathus, 324, 336
Lotus corniculatus, 168
Lotus pentaphyllum, 168
Lycaeides argyrognomon, 357–358
Lycaeides melissa, 453, 460, 468, 474
Lycaenidae, 99, 235, 320, 324, 325, 327, 333, 338, 339, 343
Lycaeninae, 325, 349
Lycopersicon esculentum, 75, 81, 83, 84
Lycopersicon hirsutum, 76, 83
Lycopersicon hirsutum f. *glabratum*, 83
Lycopersicon spp., 80, 83, 413, 432
Lycopersicum pennellii, 83
Lycosa carolinensis, 180
Lydella, 245
Lygropia cernalis, 330
Lymantria dispar, 42, 239, 256, 401
Lymantriidae, 499
Lyonetiidae, 233
Lypha, 245

M
Macalla sp., 330
Macaranga, 362
Macaria aemulataria, 439
Macaria wauaria, 168
Machimia tentoriferella, 512–513
Macrocentrinae, 233, 235
Macrocentrus, 235
Maculinea, 338, 346
Maieta guianensis, 330
Malacosoma americana, 512–513
Malacosoma americanum, 28, 32, 41, 48
Malacosoma californicum pluviale, 458
Malacosoma disstria, 33, 244
Mamestra brassicae, 132, 205, 399, 402
Mamestra configurata, 401
Manduca sexta, 22, 23, 74, 75, 78–81, 83–85, 120, 138, 281, 407, 432, 433, 451, 453, 454, 456, 459, 461, 463, 466, 467, 470
Mantis religiosa, 181

Mantispidae, 242
Mantis religiosa, 181
Margaroniini, 333
Marmara, 25, 29
Mechanitis isthmia, 84, 337
Mechanitis menapis, 85
Mechanitis polymnia, 142, 143
Medicago sativa, 454, 460, 468, 475
Megalopyge lanata, 485
Megalopygidae, 234, 496
Melanchroia vazquezae, 566, 573
Melitaea cinxia, 135, 456, 465, 470, 474
Menzie's goldenbush, 428
Mesochorinae, 238
Mesogona sp., 591
Meris alticola, 43
Mermithidae, 245
Mesochorinae, 238
Meteorini, 233–235
Meteorus sp., 234
Metopiinae, 236
Metzneria lappella, 230
Microgastri, 250
Microgastrinae, 233, 251
Microplitis demolitor, 410
Micropterigoidea, 233
Miletinae, 349–351, 361
Miracinae, 325
Miletus, 350
Miletus biggsii, 340
Mimallonidae, 42, 496
Minute pirate bug, 436
Miracinae, 233, 234
Mischocyttarus flavitarsis, 281
Misogada unicolor, 231
Misoria amra, 564, 566
Momphidae, 234
Monarch, 173, 175, 178–180, 182, 183, 185
Moraceae, 36
Morpho, 42
Morus alba, 70, 120
Muesebeckiini, 233, 234
Munona robpuschendorfi, 330
Mymaridae, 240, 241
Mymariform, 241
Myotis sodalis, 524
Myrascia, 41
Myrmecophiles, 327, 329
Myrmecophytes, 435
Myrmecoxenous, 328
Myrmecozela, 331
Myrmecozela ochraceella, 321
Myrmecozelinae, 321
Myrmica, 339, 346

Myrmica rubra, 146
Myrmicaria opaciventris, 330
Myrmicinae, 331
Myrtaceae, 401
Mythimna unipuncta, 33
Myzus persicae, 281, 286

N

Nanohymenoptera, 236
Nematoda, 245
Nemeobiinae, 325
Nemoria, 40
Nemoria arizonaria, 26, 27
Nemoria bifilata planuscula, 38
Neoponera villosa, 356
Neostauropus, 336
Nephelodes minians, 243
Nepticulidae, 233, 234
Nepticuloidea, 233
Neuroptera, 242
Neuropteroidea, 242
Nicotiana attenuata, 120, 433
Nicotiana spp., 407
Nicotiana tabacum, 74, 83, 437, 459
Nilaparvata lugens, 404
Niphanda fusca, 345, 347
Niphopyralis, 322
Niphopyralis aurivillii, 333
Niphopyralis chionesis, 333
Niphopyralis myrmecophila, 333
Noctuidae, 25, 96, 120, 123, 234, 235, 244, 323, 329, 496, 544–546
Notodontidae, 323, 336, 544–546
Nudina artaxidia, 323
Nyctaginaceae, 39
Nymphalidae, 19, 82, 84, 85, 181, 231, 234, 244, 336, 337, 429, 543
Nymphidiini, 324, 325, 354–356

O

Oaks, 515, 517, 519–522, 524–526, 528
Oboronia punctatus, 333
Oecophoridae, 235, 322, 332, 492, 496
Oedemopsini, 237
Oecophylla, 322, 323, 330, 332, 334, 335, 338, 346, 350, 351, 357, 363
Oecophylla longinoda, 330, 332–334, 336
Oecophylla smaragdina, 333, 335
Ogyris, 360
Ogyris amaryllis, 361
Ogyris barnardi, 361
Ogyris olane, 361

Ogyris oroetes, 361
Omphalea, 33
Omphalocera, 520
Omphalocera munroei, 434
Operophtera brumata, 412
Ophioninae, 236
Ophryocystis elektroscirrha, 145, 185, 458, 472, 476
Orgilinae, 233, 234
Orgyia manto, 37
Orgyia sp., 73, 564, 566
Origanum vulgare, 339
Orius tristicolor, 437
Orobanchaceae, 178, 181
Orthoptera, 170, 614
Orthosia alurina, 26
Orussoidea, 227
Oryza rufipogon, 404
Oryza sativa, 86, 397, 404
Osmanthedon domaticola, 322
Ostrinia nubilalis, 69, 70
Oxteninae, 42
Oxytenis, 42

P

Pachypodistes goeldii, 322, 332
Pantoea, 138
Pantographa limita, 515
Papilio canadensis, 130
Papilio cresphontes, 23
Papilio glaucus, 130
Papilio homerus, 240
Papilio machaon aliaska, 439
Papilio major, 178, 180
Papilio multicaudatus, 130
Papilionidae, 240, 543
Papilio polyxenes, 70, 130, 143
Papilio spp., 130
Papilio troilus, 23
Papilionoidea, 336, 337, 583
Paracles rudis, 401
Paralobesia viteana, 16
Paraponera, 310
Paraponera clavata, 280
Parasemia plantaginis, 466
Pardasena diversipennis, 85
Passiflora adenopoda, 82
Passiflora biflora, 179
Passifloraceae, 84, 178
Passiflora incarnata, 80, 82, 84
Passiflora lobata, 82, 84
Passiflora spp., 84, 280
Passionflower, 179

Pasteuria ramosa, 473
Pastinaca sativa, 512–513, 532
Pelargonium× *hortorum*, 77
Peridroma saucia, 33
Perilampidae, 241
Perisceptis carnivora, 30, 49
Perisoreus canadensis, 130, 172
Petunia hybrida, 77
Phaseolus, 413
Phasmatodea, 614
Phaseolus lunatus, 397
Phaseolus vulgaris, 80, 305
Pheidole, 330, 333
Phengaris, 338, 346, 347, 357
Phengaris arion, 339, 346
Phengaris rebeli, 346
Phobetron hipparchia, 486
Phoebis philea, 97, 300, 303, 310, 312, 337
Phoebis sennae, 300, 310, 312
Phoebis spp., 300, 307, 311, 313
Phoradendron juniperinum, 38
Phoridae, 242, 244
Phorocera, 245
Phrynosoma platyrhinos, 84
Phrynosomatidae, 84
Phthorimaea operculella, 69, 76, 77, 83
Phycita nebulella, 511
Phyllocnistis, 29
Phyllonorycter malella, 200, 208
Phyllotreta armoraciae, 146
Physalis spp., 284, 470
Phytodietini, 237
Picea glauca, 70
Pieridae, 324, 336, 399, 402, 409, 410
Pieris brassicae, 69, 71, 72, 128, 146, 402, 410, 467
Pieris oleracea, 256
Pieris rapae, 69, 81, 120, 121, 123, 132, 138, 140, 145, 146, 256, 401–403, 409, 467
Pieris spp., 118, 132, 143, 146
Pimpla hypochondriaca, 410
Pimplinae, 237, 238
Pimplini, 237
Pinus, 28
Pinus edulis, 467
Pinus rigida, 38
Piper, 98, 100, 408, 585–586, 602
Piperaceae, 408
Piper cenocladum, 408
Piper imperiale, 408, 468
Pipevine swallowtail, 429
Plagodis alcoolaria, 38
Plantago, 407

Plantago lanceolata, 173, 175, 177–183, 464–466, 471
Plantago major, 172, 464, 465, 471
Plantaginaceae, 165, 170, 173, 175, 177, 178, 181, 182
Platygastroidea, 232, 236
Platynota rostrana, 486
Plebejus idas, 342
Plodia interpunctella, 459
Plutella spp., 118, 131, 146
Plutella xylostella, 68–71, 82, 120, 121, 123, 124, 144, 147, 401, 407, 433, 455
Plutellidae, 69, 82, 120, 236, 399, 401, 407
Poaceae, 397, 403, 404
Podalia annulipes, 486
Pogonomyrmex rugosus, 83, 433
Polideini, 245
Polistes comanchus, 275
Polygrammate, 25
Polygrammate hebraeicum, 25, 26
Polyommatini, 346
Polyommatus icarus, 358
Polyrhachis, 331
Polyrhachis bicolor, 333
Polyrhachis dives, 332
Polytela gloriosae, 170
Populus, 28
Poritiinae, 325, 351–353, 361
Porizontinae, 236
Pristomyrmex punctatus, 358
Proboscidea louisianica, 459
Prochoerodes lineola, 229
Prodoxus, 28
Prolimacodes, 39, 42
Prolimacodes badia, 39
Proteaceae, 492
Protium, 98
Prunus, 28, 512–513
Prunus spinosa, 168
Pseudocedrela, 309
Pseudomonas, 137, 138
Pseudomonas aeruginosa, 456, 470
Pseudomyrmex, 329
Pseudoplusia, 411
Pseudoplusia includens, 411
Pseudotelphusa quercinigracella, 523
Psychidae, 321, 331
Psylliodes chrysocephala, 138
Pteromalidae, 241, 242
Pueraria montana, 432
Putranjiva, 118
Pyralidae, 234–236, 242, 322, 330, 332, 333, 492, 496

Pyralinae, 333
Pyrisitia, 300
Pyrrharctia, 28

Q

Quercus alba, 512, 518, 522, 523, 532
Quadrus cerealis, 468
Quercus imbricaria, 512–513
Quercus macrocarpa, 512–513
Quercus robur, 412
Quercus rubra, 38
Quercus sp., 28, 38, 509, 517, 520–524, 532, 585–586
Quercus stellata, 512–513

R

Ramalina, 283
Recurvaria sp., 591
Ribes nigrum, 168
Ribes rubrum, 168
Riodinidae, 96, 235, 324, 327, 333, 338, 339, 353, 354, 356, 496
Riodinini, 325
Rogadinae, 233–235, 237
Romalea guttata, 173, 281
Rosaceae, 339
Rosema dentifera, 323, 329
Roupala montana, 493, 495
Rubiaceae, 492

S

Saccharum, 70
Saccharum alopecuroides, 70
Salix, 28, 168
Saltmarsh caterpillar, 179
Sarcophagidae, 244, 245
Saturniidae, 24, 244, 324, 329, 496
Satyrium, 347
Saucrobotys futilalis, 49
Scaptomyza, 127
Scaptomyza flava, 121, 126, 127
Scaptomyza nigrita, 128
Scelionidae, 236
Schinia, 27, 29, 30
Schistocerca americana, 284
Scolitantides orion, 215–217
Scrophulariaceae, 168
Semiothisa aemulataria, 201, 208
Semutophila saccharopa, 321, 331, 357
Senna chapmanii, 301–303, 307, 310, 311

Senna mexicana, 337
Senna mexicana chapmanii, 301
Senna occidentalis, 300
Senna spp., 300, 310, 311
Serratia marcescens, 456, 470
Sesiidae, 235, 322, 328
Setomorpha, 331
Setomorpha melichrosta, 321
Shirozua jonasi, 345, 347
Sicya morriscaria, 38
Sinapis alba, 286, 401
Siphonini, 244
Skipper caterpillars, 304, 306, 307, 515, 526
Solanaceae, 83–85, 142, 270
Solanum acerifolium, 85
Solanum aethiopicum, 78
Solanum anguivi, 67
Solanum berthaultii, 76
Solanum carolinense, 78, 81
Solanum elaegnifolium, 81
Solanum glaucescens, 67
Solanum grandiflorum, 78
Solanum grandifolium, 79
Solanum lycopersicum, 79, 137, 437
Solanum melongena, 79
Solanum spp., 84, 142, 432
Solanum tarijense, 77
Solanum tuberosum, 74
Solenopsis invicta, 301, 302, 307, 310, 312, 313
Somatolophia ectrapelaria, 43, 51
Sorbus, 168
Sorghum bicolor, 67
Spalgis, 350, 364
Sparganothis sulfureana, 401
Speyeria, 28
Sphinctini, 237
Sphingidae, 75, 78, 85, 244, 432, 496, 543
Sphinx sequoia, 591
Spilomelinae, 333
Spilosoma congrua, 175, 176, 179, 184
Spilosoma obliqua, 70
Spodoptera, 69
Spodoptera androgea, 15
Spodoptera exempta, 452, 460, 462
Spodoptera exempta nucleopolyhedrovirus, 453
Spodoptera eridania, 140
Spodoptera exigua, 19, 33, 76, 83, 145, 281, 397, 403, 404, 409, 474
Spodoptera frugiperda, 22, 69, 74, 123, 132, 133, 137, 138, 285, 404, 408, 409, 474

Spodoptera littoralis, 77, 119, 120, 124, 131, 171, 404, 453, 454, 460–463
Spodoptera littoralis nucleopolyhedrovirus, 453
Spodoptera litura, 132
Spodoptera spp., 132
Spragueia, 29
Squamata, 84
Stalachtis, 325, 355
Stathmopoda sp., 322, 328, 332
Stauropus, 323, 336
Stauropus fagi, 336
Stenachroia myrmecophila, 322
Stenoma cathosiota, 494
Sternus vulgaris, 584
Stiria, 23, 29
Stiria intermixta, 24
Stiriinae/stiriine, 29
Stiropiini, 234
Sturnidae, 584
Strymon melinus, 39, 591
Sturmia bella, 257
Styx, 364
Sugar cane borer, 70
Syllepis hortalis, 566, 573
Symmachiini, 325
Symphyta, 227, 234
Sympiesis sericeicornis, 200
Sympistis, 40
Sympistis badistriga, 231
Synanthedon acerni, 511
Synanthedon exitiosa, 511
Synargis calyce, 355
Synchlora, 28, 48
Synchlora faseolaria, 51
Synchlorine Geometridae, 48, 49
Synchlorini/synchlorine, 38, 40
Syssphinx mexicana, 324, 329

T
Tachinidae, 228, 232, 238, 239, 241, 242, 248, 252
Tachininae, 242, 245
Tachinini, 245
Taraka, 364
Taraka spp., 350
Tarchon felderi, 449
Telea polyphemus, 201
Telenomus, 236
Tenthredinoidea, 234
Tetramorium aculeatum, 330
Tetraponera penzigi, 329

Thaumetopoea pityocampa, 43, 244
Thaumetopoeinae, 26, 27
Theclinae-Polyommatinae, 345
Theroa, 36
Thisbe irenea, 356
Thurberiphaga, 40, 41
Thurberiphaga diffusa, 41, 51
Thyrididae, 37, 496
Thysanoptera, 515
Tilia americana, 515
Timena cristinae, 283
Timema podura, 283
Timema sp., 283
Tineidae, 321, 328, 333
Tischeriidae, 234
Toadflax defoliator, 170, 176, 182
Tobacco, 433
Tolype prop. *innocens*, 486
Tomato, 432
Tortricidae, 235, 236, 321, 328, 331, 357, 496, 545, 546
Torymidae, 239
Trachypetidae, 232
Trichogramma bournieri, 403
Trichogramma brassicae, 71
Trichogramma minutum, 433
Trichogrammatidae/trichogrammatid, 236, 240, 241
Trichoplusia, 137
Trichoplusia ni, 40, 73, 81, 124, 132, 135, 141, 406, 454
Triclema lamias, 347
Trictena atripalpis, 31
Trigonalidae/Trigonalyidae, 227, 238
Trigonospila brevifacies, 256
Triphassa, 333
Triplaris melaenodendron, 330
Tryphoninae, 237
Tryphonine ichneumonids, 237
Turtlehead, 165, 181

U
Ugni molinae, 401
Urania, 33
Urbanus esmeraldus, 520, 527
Urbanus proteus, 304, 306, 307
Urera baccifera, 520
Uresiphita reversalis, 281
Urodidae, 566, 573
Uroleucon ambrosiae, 284
Urticaceae, 82

V
Vaccinium corymbosum, 401
Vachellia, 321, 362
Vachellia cornigera, 329
Vachellia drepanolobium, 328, 329
Valerian, 171
Valerianaceae, 171
Valeriana officinalis, 171
Vanessa atalanta, 82
Vanessa cardui, 27, 33, 173, 175, 181, 456, 473
Vanessa gonerilla, 20
Verbascum, 168
Vespidae, 273, 275
Vigna, 361
Vigna radiata, 70
Vigna unguiculata, 122, 285
Vireo griseus, 12
Vitus vinifera, 476
Vochysiaceae, 492
Voriini, 242

W
White peacock, 176, 181
Wild cotton, 436
Wolf spider, 180
Winthemia sp., 243
Wockia chewbacca, 566, 573
Wurthiini, 333

X
Xanthium strumarium, 230
Xenorhabdus nematophila, 453, 462

Y
Yponomeutidae, 23

Z
Zabuella paucipuncta, 355
Zanthoxylum coriaceum, 302
Zale, 35
Zea mays, 69, 437
Zizeeria knysna, 358
Zygaena, 168
Zygaena filipendulae, 168
Zygaenidae, 235
Zygaenoidea/zygaenoid, 23, 45

GPSR Compliance

The European Union's (EU) General Product Safety Regulation (GPSR) is a set of rules that requires consumer products to be safe and our obligations to ensure this.

If you have any concerns about our products, you can contact us on

ProductSafety@springernature.com

In case Publisher is established outside the EU, the EU authorized representative is:

Springer Nature Customer Service Center GmbH
Europaplatz 3
69115 Heidelberg, Germany

www.ingramcontent.com/pod-product-compliance
Ingram Content Group UK Ltd.
Pitfield, Milton Keynes, MK11 3LW, UK
UKHW021446190426
11946UKWH00022B/47